P9-DUI-657

Phosphate Metabolism and Cellular Regulation in Microorganisms

Phosphate Metabolism and Cellular Regulation in Microorganisms

Edited by

Annamaria Torriani-Gorini
Department of Biology
Massachusetts Institute of Technology
Cambridge, Massachusetts

Frank G. Rothman
Division of Biology and Medicine
Brown University
Providence, Rhode Island

Simon Silver
Department of Microbiology and Immunology
The University of Illinois College of Medicine
Chicago, Illinois

Andrew Wright
Department of Molecular Biology and Microbiology
Tufts University
Boston, Massachusetts

Ezra Yagil
Department of Biochemistry
The George S. Wise Faculty of Life Sciences
Tel Aviv University
Tel Aviv, Israel

AMERICAN SOCIETY FOR MICROBIOLOGY
Washington, D.C.

1913 I St., N.W.
Washington, DC 20006

Library of Congress Cataloging-in-Publication Data

Phosphate metabolism and cellular regulation in microorganisms.

Sponsored by the American Society for Microbiology and the International Union for Biochemistry.
Includes indexes.
1. Phosphates—Metabolism—Congresses. 2. Microbial metabolism—Congresses. 3. Phosphates—Metabolism—Regulation—Congresses. I. Torriani-Gorini, Annamaria. II. American Society for Microbiology. III. International Union of Biochemistry. [DNLM: 1. Escherichia Coli—metabolism—congresses. 2. Microbiology—congresses. 3. Phosphates—metabolism—congresses. QW 52 P575 1986]
QR92.P45P47 1987 576′.1133 87–12607

ISBN 0–914826–94–8

Printed in the United States of America

Cover: "Old Harbor, Concarneau," original drawing for the Pho-86 Symposium by Jacotte Mauffret

Hiroshi Inouye

Hiroshi Inouye, a productive contributor to the studies of the Pho regulon over the last decade, died on 24 July 1986.

Inouye, who was to be a convener of one of the sessions at the Pho-86 Symposium, was sorely missed. The presentations at the symposium contained numerous references to his fundamental contributions, and we looked forward to his participation in the discussions.

After receiving his education through a Masters Degree in Japan, Inouye obtained his Ph.D. at the Weizmann Institute in Israel and then joined Jon Beckwith's laboratory at Harvard in 1973. He took a position as Assistant Professor of Biology at Temple University at the beginning of 1984.

Inouye is best known for his discovery that alkaline phosphatase is made initially in precursor form, the first case of such a finding for a periplasmic protein, and for the development of *phoA* plasmid vectors which have become widely used in recent years. What is less well known about Inouye is that from his undergraduate major on, he maintained a deep interest in philosophy in general, and philosophy of science in particular. As recently as 1981, he published an article in a philosophical journal. His many friends will miss both their interactions with a man of such broad interests and the continuing scientific contributions he would have made.

CONTENTS

Preface

The importance of phosphates and phosphate metabolism in organisms is of course well recognized, but the subject is so wide and so all-encompassing that a general meeting would be forbidding. However, a limited and well-defined get-together of researchers interested in microbial phosphate metabolism seemed reasonable. I proposed the idea of a symposium on inorganic phosphate (P_i) metabolism in microorganisms to Harry Rosenberg at Perth in 1982. He suggested that we wait 2 years. We waited 4 years, and that was wise: a lot of good work was in progress and has in fact appeared during that time. The chapters which follow testify to the fact that an overview of the field, restricted to microorganisms, was timely and novel and should be of wide interest.

An enthusiastic and efficient organizing committee was formed: P. L. Boquet, H. Halvorson, N. Rao, F. Rothman, S. Silver, A. Torriani, A. Wright, and E. Yagil. We met frequently at the Massachusetts Institute of Technology with a beautiful view of the Charles River (but not as beautiful as the rock-studded, seagull-dotted, windswept bay we overlooked in Concarneau!). Fortunately, Yagil was working at the National Institutes of Health for the year, and Boquet was at Harvard.

The location of the symposium at Concarneau in Bretagne (France) was a stroke of genius which we owe to F. Gros.

Back in France, at the Commissariat à l'Énergie Atomique (C.E.A.), Boquet obtained the invaluable help of Yves Le Gal, director of the Marine Biological Laboratory in Concarneau, the oldest and most beautiful of its kind in France. The laboratory belongs to the Collège de France which, together with the Municipality of Concarneau, contributed the use of the conference building and the libraries to our meeting.

Boquet had the hardest and most time-consuming task of managing the local funding and organization. The Centre National de la Recherche Scientifique and the C.E.A. contributed more than 30% of the needed funds. The American Society for Microbiology and the International Union for Biochemistry were the cosponsors and major contributors in the United States. All of the contributors are gratefully acknowledged.

And the sun did shine on the picturesque coves of Bretagne.

The subject was restricted to microorganisms, but a topic on PP_i in plants slipped in as well. The first section dealt mainly with genetic aspects of regulation of phosphatases in *Escherichia coli* and yeasts. Some participants were getting nervous about so much genetics, but the sections following were more diversified—more biochemistry, nuclear magnetic resonance studies, protein structures, P_i transport, and P_i reserves, as well as protein excretion. Finally an attempt was made at integrating phosphate metabolism into general cellular global metabolism: SOS, DNA repair, carbon circuits, nitrogen assimilation, and stable RNA synthesis.

The editors were surprised to receive 80% of the papers at Concarneau. This allowed a rapid review and release of the book by the efficient staff at the American Society for Microbiology.

The last meeting on phosphate metabolism took place in 1958 in Baltimore. Much has changed in 28 years. A Pho '91 symposium will probably prove as interesting and fruitful as Pho '86. The genetic elements of the Pho regulon of *Escherichia coli* should be completely understood by then, with new and interesting surprises. Polyphosphate and PP_i metabolism will be clarified, and their roles in various microorganisms should appear very diversified.

Global metabolic networks and P_i metabolism will become yet more intricate and inseparably meshed with all of microbial metabolism as we learn more!

For the Editorial Board:
ANNAMARIA TORRIANI-GORINI

Acknowledgments

The Symposium that led to this monograph was cosponsored by the American Society for Microbiology (USA), the Centre National de la Recherche Scientifique (France), the Commissariat a l'Energie Atomique (France), the International Union of Biochemists (USA), and the International Union of Microbiological Societies (UK). We gratefully acknowledge the support of the National Science Foundation (grant DMB 85-124-37) and the National Institutes of Health (grant AI22022).

Financial support was also derived from a number of industrial contributors (cited here alphabetically):

Ajinomoto Co., Inc. (Japan)
Brucker, Inc. (USA)
Boehringer-Mannheim (France)
Bristol-Myers Co. (USA)
Cetus (USA)
Elf-Biorecherches (France)
Gist Brocades (Delft, Netherlands)
Hoffmann-La Roche, Inc. (USA)
Mallinckrodt (USA)
Monsanto (USA)
New England BioLabs, Inc. (USA)
Takara Shuro Co., Ltd. (Central Research Laboratories, Japan)

We are particularly indebted to the College de France and to the Marinarium (Marine Biological Laboratory) of Concarneau and its director, Yves Le Gal, for the local organization of the symposium. We also thank the City of Concarneau for generously lending the seminar rooms at the Conference Center.

I. PHOSPHATE REGULATION IN *ESCHERICHIA COLI*

The *pho* Regulon in *Escherichia coli*

BEN LUGTENBERG

Botanical Laboratory, Department of Plant Molecular Biology, Leiden University, 2311 VJ Leiden, The Netherlands

Phosphorus is an essential element for the cell. It is present in nucleic acids, phospholipids, lipopolysaccharides, and various cytoplasmic solutes.

Under conditions of sufficient P_i, this molecule diffuses through OmpC and OmpF pores in the outer membrane and is then recognized by the phosphate inorganic transport (Pit) carrier, which transports it through the cytoplasmic membrane (Fig. 1A). Under these conditions cells might even store phosphorus in the form of polyphosphate granules.

To cope with conditions of low phosphate concentrations, *Escherichia coli* and other microorganisms have developed an emergency system. This system is now known as the *pho* regulon. The first indication of this system was the discovery by Annamaria Torriani in 1958 that the enzyme alkaline phosphatase is induced when cells are starved for a phosphorus source.

We now know that the *pho* regulon consists of a large number of genes. Apparently this is the price that *E. coli* is willing to pay for being able to scavenge traces of usable phosphate sources from the surrounding medium.

The genes of the *pho* regulon are scattered around the chromosome, and the products of these genes can be found in all cellular compartments. When we follow the fate of low concentrations of a number of phosphorus sources, we see that many of the P_i starvation-induced proteins play a role in their uptake (Fig. 1B).

P_i and other P-containing solutes are recognized by the outer membrane pore protein PhoE which facilitates diffusion of P_i through the outer membrane six to eight times more efficiently than the constitutive pore proteins OmpC and OmpF do. P_i is then bound by the periplasmic P_i-binding protein and subsequently transported through the cytoplasmic membrane by the phosphate-specific transport (Pst) system.

Uptake of low concentrations of glycerol 3-phosphate is mediated by the periplasmic glycerol 3-phosphate-binding protein and a high-affinity cytoplasmic membrane carrier.

The presence of PhoE protein facilitates the passage of linear polyphosphates through the outer membrane. This phosphorus source cannot pass the cytoplasmic membrane unless it is first depolymerized by alkaline phosphatase.

Also, the phosphorus groups of 5′-AMP and 3′-AMP need to be removed by 5′ nucleotidase and 2′,3′-cyclic phosphodiesterase, respectively, before they can pass the cytoplasmic membrane. All the P_i starvation-induced uptake systems mentioned so far have in common a need for the *phoB* gene for expression.

The effects of P_i limitation can more or less be mimicked by using mutants that express the *pho* regulon constitutively. However, various uptake systems might act differently under conditions of real P_i limitation. It seems likely, for example, that P_i limitation affects the synthesis of the P-rich lipopolysaccharide, causing it to be shorter and less negatively charged. Under those conditions P sources undergo less steric hindrance by lipopolysaccharide when they try to reach the neighboring PhoE protein pore. Even more important, the negatively charged P sources undergo less repulsion than by wild-type lipopolysaccharide, the lipid A core region of which is rich in phosphorus groups (for a discussion see reference 1).

The regulation of the *pho* operon is extremely complex. At least four genes, *phoB, phoR, phoU*, and *phoM*, are involved. In recent years this study has largely benefited from the use of genetic fusions between the promoters under study and the *lacZ* structural gene, a system elegantly developed by M. Casadaban and J. Beckwith. More recently, a variation on the original system was used by B. Wanner to link so-far-unknown P_i-regulated promoters to *lacZ*.

It should be noted also that P_i concentration-regulated systems exist which are not dependent on *phoB*. More specifically, it appears that aminopeptidase N and three polyphosphate de-

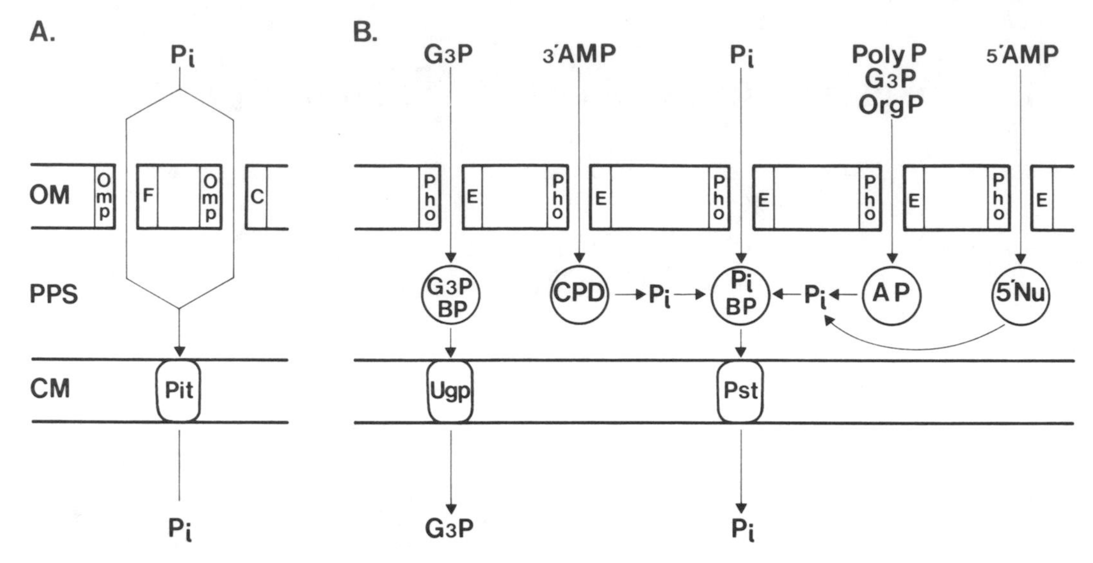

FIG. 1. Uptake pathway of various phosphorus sources through the cell envelope of *E. coli* K-12 under conditions of sufficient phosphate (A) and phosphate starvation (B). Abbreviations: AP, alkaline phosphatase; BP, binding protein; CM, cytoplasmic membrane; CPD, 2′,3′-cyclic phosphodiesterase; G3P, glycerol 3-phosphate; 5′Nu, 5′ nucleotidase; OM, outer membrane; OmpC and OmpF, outer membrane pore proteins; OrgP, organic phosphate; PhoE, PhoE pore protein; Pit, low-affinity system for P_i transport; Poly P, linear polyphosphate; PPS, periplasmic space; Pst, phosphate-specific transport system; Ugp, high-affinity uptake system for G3P.

polymerases at the outside of the cytoplasmic membrane, as well as a polyphosphate depolymerase with a pH optimum of 2.5, are P_i starvation-inducible proteins which do not need *phoB* for expression.

LITERATURE CITED

1. **Korteland, J., and B. Lugtenberg.** 1984. Increased efficiency of the outer membrane PhoE protein pore in *Escherichia coli* K-12 mutants with heptose-deficient lipopolysaccharide. Biochim. Biophys. Acta **774:**119–126.

The Birth and Growth of the Pho Regulon

A. TORRIANI-GORINI

Department of Biology, Massachusetts Institute of Technology, Cambridge, Massachusetts 02139

On 11 April 1949, Jacques Monod and I were rushing in and out of the 37°C room every 30 min to dilute by one-half with fresh medium a culture of *Escherichia coli* growing in glycerol and lactose. The purpose was to keep the culture in the exponential phase of growth so that we could understand the kinetics of adaptation to lactose. From these experiments the bactogene (32), a device for maintaining a culture in continuous exponential phase, was born. Such a device (a rotating flask with an inlet of fresh medium and an outlet of exponentially growing culture) turned out to be useful for many other purposes, one of which was to challenge the following statement by Virtanen and Winkler (62): "The adaptive enzymes which are necessary only in definite nutritional conditions seem, as a rule, to decrease or disappear with the lowering of the nitrogen content of the cells."

The challenge entailed a variety of conditions in which the growth rate (established by input of fresh medium) and the growth-limiting factors (nitrogen, carbon, sulfur, phosphate) were modified. The idea was to obtain various steady-state levels of utilization of one of the growth factors and to measure the level of various inducible enzymes (β-galactosidase and amylomaltase) and constitutive enzymes (phosphatase and protease).

Was Virtanen and Winkler's statement correct? The experiment was never completed because it produced a new and unexpected result. Using conditions in which P_i was the growth-limiting factor, we studied the synthesis of an acid phosphatase known to exist in *E. coli* (41). If acid phosphatase was constitutive, it should not disappear according to Virtanen's predictions. Unexpectedly, the activity suddenly increased when a slow doubling time was established. The phosphatase activity was measured with *p*-nitrophenyl phosphate at pH 4.5 (the expected pH optimum for this acid phosphatase). To fully develop the yellow color of the *p*-nitrophenol liberated, the reaction was stopped by the addition of KOH to bring the pH above 9. At that pH value, the yellow *p*-nitrophenol concentration kept on rapidly increasing. Was this due to spontaneous hydrolysis of the *p*-nitrophenyl phosphate at high pH, or was another enzyme activated? Since the substrate alone in buffer was stable, the hydrolysis was probably due to an enzyme. Treating the extract at 90°C for 10 min before adding the substrate should have eliminated the activity, but it did not. A new enzyme was born—an amazing one (at least in 1953). It was stable at 90°C, it was synthesized when the cells had stopped growing, and it was possibly regulated by a ubiquitous molecule like P_i (55). A major question arose: Was its formation due to a negative regulation? The word "repression" had not yet been coined by Vogel (63), and Monod did not really like such an idea. Enzyme synthesis, he thought, was regulated positively by induction: ". . . the synthesis of many, and perhaps of all enzymes, is sensitive to specific effects, . . . exerted by compounds belonging to the family of their substrates, justifying the assumption that induction mechanisms control the synthesis of all enzymes and insure the balanced, efficient equilibrium of cellular syntheses. . . ." (Monod, 1953, Rome, unpublished lecture).

Furthermore, since the synthesis of the new enzyme started in nongrowing cultures, the differential rate of synthesis (the Monod plot, Δ-enzyme/Δ-bacterial growth) was not applicable. The finding was tabled and I was dismissed.

Luigi Gorini and I left for the United States in 1955, and for 2 years at New York University my attention was diverted to studying the inducibility of polysaccharide depolymerases (58, 59). In 1957, we moved to Boston and I joined A. M. Pappenheimer at Harvard. The positions offered were by no means stable, nor were they particularly good compared with our stable positions in France at the Centre National de la Recherche Scientifique where I was Chargée de Recherche at the Institut Pasteur. But we had the adventurous spirit and the social and family needs of the Italian emigrants. So we went back to Paris, sold the apartment, and sailed for the United States. We finally reached Brookline, Massachusetts. Jeana and Cy Levinthal offered the mansard (top floor) of their house as living quarters. It was large and charming. Brookline was neither Milan nor Paris nor New York, but the Levinthals' house was always filled with interesting people and so were our laboratories. Leo Szilard was a frequent guest. We thought that he was a Hungarian musician. Instead, he was a most exciting person with whom to discuss one's scientific projects.

The inspiring association of the Levinthals and their friends in the Boston area revived the phosphatase story. I was at Harvard writing a paper (as advised by Pappenheimer) on the properties of the alkaline phosphatase repressed by P_i, following an abstract I had published in 1958 (A. Torriani, Fed. Proc. **18:**339, 1958), when Martin Pollock wrote to me from London about a short paper by T. Horiuchi to be published in *Nature* on the same topic—the existence of a new enzyme regulated by P_i (19). So I rushed my paper to *Biochimica Biophysica Acta,* and it appeared in 1960 (56). Horiuchi developed his interest in other fields. I persisted with alkaline phosphatase. At that time (1959), Levinthal was interested in finding an ideal system to show that there was collinearity between the genetic map of the gene and the corresponding amino acid sequence of a protein. He thought that alkaline phosphatase must be the best choice since it was easy to measure! He invited me to join his new group at the Massachusetts Institute of Technology.

To analyze collinearity, it was essential to isolate mutants, to map the structural gene, and to isolate, purify, and analyze the protein. With Al and Sue Garen we isolated more than 2,000 alkaline phosphatase-negative mutants; they all mapped in a single cistron (*phoA*) at 9 min on the *Escherichia coli* genetic map. Garen and Levinthal studied the chemical properties of the enzyme (16).

The peptide fingerprint analysis of the purified enzyme (43) gave an interesting result: the number of peptides obtained was only half the number expected from the molecular weight. Thus, alkaline phosphatase was probably a dimer of identical subunits; the subunits had intrachain disulfide bonds and were held together by noncovalent bonds (44, 45). Monomerization of alkaline phosphatase could be obtained by the reduction of the inrachain disulfide bonds or by low pH (45). The monomers were inactive and temperature labile (44), but could refold and dimerize in the presence of Zn^{2+} and Mg^{2+} to regain the enzymatic activity (45).

The dimeric property of the protein invited studies on the in vivo and in vitro intracistronic complementation among *phoA* mutants (46, 47, 61). The dimerization reaction was remarkably specific and was not influenced by the presence of high concentrations of other proteins (56, 61).

A hybrid phosphatase was first created in vivo in *Serratia marcescens* carrying an *E. coli* $phoA^+$ plasmid. The two parental phosphatases had different electrophoretic mobilities, and the hybrid protein had an intermediary one (24, 50). This evidence of hybrid formation in vivo and in vitro showed that alkaline phosphatase protein was eminently useful for proving the current hypothesis of complementation by Brenner (6) and by Fincham (13). Alkaline phosphatase activity in heterogenotes in vivo was paralleled by that obtained by hybridization in vitro (47). A rather elaborate (and laborious) study of the level of complementation with various mutant strains showed a lack of correlation between the capacity to complement and the position of the mutation on the *phoA* genetic map.

The activity recovered by formation of the hybrid protein depended on two factors: (i) the affinity of the two mutant subunits for each other to form a dimer and (ii) the specific activity of the hybrid dimer formed (12). The results stressed the importance of protein folding and the superstructure of the molecule at a time (1963–1965) when the linear order of DNA was dominating the scene.

As for the collinearity between the genetic map and the amino acid sequence, the preliminary results obtained with alkaline phosphatase in 1961 were only suggestive (see Rothman, this volume); the proof of it was firmly established in 1964 by Yanofsky et al. (70), using the tryptophan synthetase system. Our attention was shifted to the synthesis of the enzyme itself and to the mechanism which controlled it.

The enzymatic activity was found to be inhibited by metal-binding agents (36). By analogy with mammalian phosphatases, this led to investigations on the nature and the role of the metal. Zinc was firmly bound to the enzyme molecule (51). Alkaline phosphatase was crystallized in 1964 (29), and detailed physiological, chemical, and structural properties of the enzyme (34) were furnished by Vallee and his group at Harvard Medical School (51) and by Schlesinger and his associates at St. Louis (40, 44–46). However, the amino acid sequence was slow in coming: a few amino acids were missing in the middle of the sequence, which was completed only in 1981 (5). Recent results on the crystallography, quaternary structure, and metal-binding sites of alkaline phosphatase are in this volume (see papers by Coleman and by Wyckoff).

Alkaline phosphatase had in store other provocative features. Malamy and Horecker (28) were interested in understanding how phosphate esters were utilized by *E. coli* grown to P_i starvation. Their reasoning was that, since phosphate esters were considered not to enter the cell as such, it followed that alkaline phosphatase must be located at the surface. They found that, in the absence of P_i, the uptake of P_i from ^{32}P-labeled glucose 6-phosphate was rapid and quantitative. However, in the presence of excess P_i (5 mM) the uptake of ^{32}P from glucose 6-phosphate in the cell was only partially reduced. Two alternatives were suggested: either glucose 6-phosphate was taken up directly, or the $^{32}P_i$ produced was not in equilibrium with the external medium (28), i.e., the hexose phosphate was hydrolyzed in the periplasm and P_i was utilized via a transport system in the cellular membrane. Experiments with lysozyme-produced spheroplasts settled the question: alkaline phosphatase was in the periplasmic space and available for glucose 6-phosphate hydrolysis (30, 35). This discovery had two important impacts on our investigations: (i) it modified our approach and concept of the mechanisms of transport of molecules across membranes, and (ii) it opened the floodgate to the studies of exported proteins in bacteria.

Bennett and Malamy, pursuing the studies of the transport of phosphorylated compounds into *E. coli,* isolated mutants (*pit*) resistant to arsenate (2), with an altered pattern of P_i transport (68). These results led to the identification of two transport systems for P_i in *E. coli* (42, 69): Pit (phosphate inorganic transport) and Pst (phosphate-specific transport). Furthermore, one of the Pst gene products (the phosphate-binding protein determined by *phoS*), like alkaline phosphatase, was proved to be periplasmic and controlled by P_i (31, 33).

This brings us back (1958) to the original question raised when P_i was first observed to affect the synthesis of alkaline phosphatase: How? The first obvious approach was to look for mutants in which the alkaline phosphatase synthesis was indifferent to the P_i level in the growth medium. The constitutive mutants were very easy to detect as yellow colonies (the yellow color was from *p*-nitrophenol formed when *p*-nitrophenyl phosphate was added to the plate and hydrolyzed by the cells, making alkaline phosphatase in high phosphate) on a sea of white colonies. Constitutive mutants could be selected on a solid medium containing β-glycerol phosphate as carbon source and a high level of P_i (60). Such a method gave a large crop of mutants. When analyzed, they were grouped into two classes, mapping at opposite sides of the circular *E. coli* chromosome. One class mapped close to *phoA* at 9 min (R1a and R1b, or C2 and C3, or *phoR68* and *phoR69*) and produced a low level of enzyme. The other class of mutants mapped at 83 min and produced a high level of constitutivity (R2a and R2b, or *phoS* and *phoT,* now *pstS* and *pstA*) (11, 17, 33).

Among the mutants isolated, a few had a Pho^- phenotype and could not be complemented by *phoA*: that defined a gene with a positive function which mapped close to *phoA* (4, 14, 15, 23, 28, 53).

The system was thus not very different from the classical inducible β-galactosidase, since it was harboring a positive control (*phoB*). Monod's statement was correct after all: "enzyme synthesis is regulated by induction. The inducer may be an anti-repressor."

If the product of *phoB* was required to activate *phoA,* at which level of the synthetic cycle was it acting? Using cell-free, mRNA-programmed synthesis of alkaline phosphatase monomers (from induced and repressed cells), we proved (9, 10) that *phoB* product, as expected from the current Jacob-Monod hypothesis (22), was not an activator of the messenger, nor was it required at the translational level. The amount of ^{14}C monomers synthesized in vitro was about 4% of the expected level in vivo. They were immunologically recognized as dimers, after a rescue with ^{12}C monomers (34, 37, 44, 56). The need for the *phoB* gene product to transcribe *phoA* was later proved in vitro using phage ϕ80 *phoA* DNA (21, 37) and in vivo with a deletion of *phoB* (7).

The in vitro synthesis study produced a novel result, expected perhaps of an exported protein, but not yet observed. The subunit of the enzyme had a larger molecular weight than predicted from the purified molecule (20). The protein was an immunologically reactive precursor with molecular weight higher than the expected 94,000. This made a lot of sense when it was proved that, to process this precursor to the mature form, it was necessary to clip off the signal sequence, a process which was performed by a proteolytic enzyme found in the outer membrane (8) and absent in the in vitro system.

With additional research, more genes were found to be regulated by P_i; besides *phoA* and *phoS* (31), there were *phoE* (52), the gene of porin E, and *ugpA* and *ugpB* (48), the genes of *sn*-glycerol 3-phosphate uptake and binding. By analogy with the Arg regulon, we called these genes the Pho regulon (66).

How many genes in *E. coli* are induced by phosphate starvation (*psi*)? With a clever survey method, Wanner and Letterel (66) found some 20 P_i-regulated promoters.

Of the phosphate-regulated genes, some are induced by the product of *phoB* and others are not (64). Only the *phoB*-regulated genes are considered members of the Pho regulon.

How is the *phoB* gene regulated? It is autoregulated (49) and phosphate regulated (18, 49, 53), but it also required a functional product of the *phoR* gene (27). In this cascade of events (Fig. 1) the product of *phoR* could have either a positive effect or a negative one. This double function was first suggested (14, 15) by the observation of two classes of mutants of *phoR,* one with a high constitutive level of alkaline phosphatase (*phoR69*) and one with a low level (50% of wild-type derepressed, like *phoR68*). It was subsequently found that another regulatory

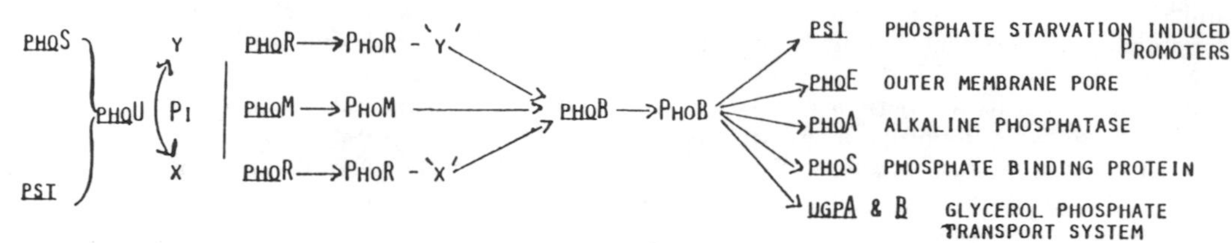

FIG. 1. Scheme of regulation of the phosphate (*pho*) regulon. The genes belonging to the *pho* regulon, as represented by *phoA* (structural gene for alkaline phosphatase), are derepressed by P_i limitation. Their expression is positively regulated by *phoB*, *phoM*, and *phoR* and is negatively regulated by *phoR*, *phoS*, *pst*, and *phoU*. The *phoS* and *pst* genes are involved in the transport of P_i. They may regulate the *pho* regulon by influencing the level of a hypothetical effector(s) through the mediation of *phoU*. PhoR (the product of *phoR*) presumably binds to a nucleotide cofactor X to form the inducer PhoR-X or to a nucleotide cofactor Y to form the repressor PhoR-Y, both of which influence the expression of *phoB*. In the absence of PhoR, the *phoM* gene product activates the expression of *phoB*, independent of P_i levels. High levels of PhoB turn on the expression of genes like *phoE*, *phoA*, and *phoS*. The product of the *phoU* gene may change X to Y directly, or it may regulate the gene encoding this function. (From reference 39.)

gene, *phoM* (8, 65), which is not phosphate regulated (25, 26, 54) is supplying the positive function lost in the *phoR68* mutants (25, 66); however, the *phoM* promoter is weak. These results are summarized in Table 1, which shows that double mutants *phoR68* and *phoM* cannot activate the regulon (no alkaline phosphatase induction) in spite of being $phoB^+$. A double mutant *phoR69 phoM* is constitutive, having lost the negative function of PhoR protein. Thus *phoB* requires a functional *phoR* product to become an inducer.

Wild-type cells fully induced by P_i starvation for alkaline phosphatase synthesis must contain a full complement of *phoR* and *phoB* products. However, the addition of repressing levels of P_i, "phosphate shock," results in a rapid (15 min) shutoff of *phoA* expression and an instant revival of growth (Fig. 2). This shutoff takes less than one-fifth of a generation time and depends on *phoR* function since it does not occur in *phoR68* mutants or in any other of the constitutive mutants tested (*phoR69* or *phoU35*).

This rapid effect of P_i could be explained by an enzymatic phosphorylation. Either the inducer proteins (PhoB or PhoR, or both) or a cofactor

TABLE 1. Regulation of alkaline phosphatase (AP)[a]

Relevant genotype	AP sp act	
	High P_i	Low P_i
Wild type	0.1	38.0
phoB	0.1	0.1
phoM	0.1	36.0
phoR68 (C_2)	15.0	12.0
phoR68 phoM	0.1	0.1
phoR69 (C_3)	36.0	37.0
phoR69 phoM	35.0	39.0
phoU35	82.0	70.0

[a] Strains are isogenic except for mutations in the *pho* genes. The wild type is W3110 F^-. The cultures were grown overnight in MOPS medium with 4 mg of glucose per ml and 2 mM (high) or 0.1 mM (low) P_i. Levels of activity are expressed as nanomoles of *p*-nitrophenol produced by 1 ml of 2×10^8 toluenized cells in 1 min from *p*-nitrophenyl phosphate. (From Torriani and Ludtke [57].)

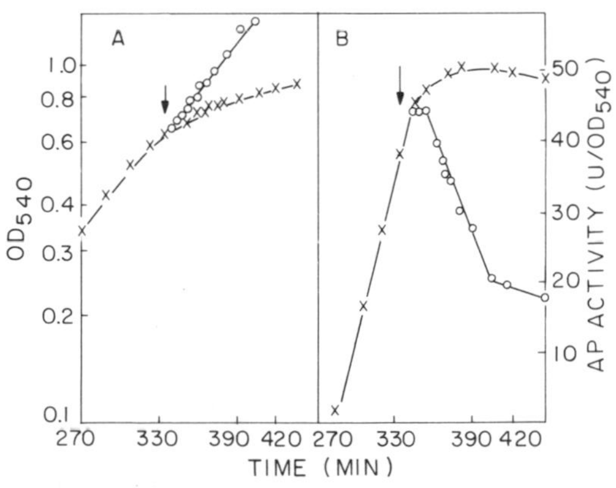

FIG. 2. Phosphate "shock" experiment. (A) Growth (optical density at 540 nm, OD_{540}) of *E. coli* 6300 F^- wild type in morpholinepropanesulfonic acid (MOPS)-buffered synthetic medium (39) containing glucose (4 mg/ml) and K_2HPO_4 (0.2 mM). When growth slowed down as a result of P_i depletion, K_2HPO_4 to 2 mM was added to the culture (at the arrow); growth resumed immediately. (B) Activity (units/OD_{540}) of alkaline phosphatase (AP) from samples of culture A. Symbols: X, control in limiting phosphate; O, growth and activity after addition of K_2HPO_4, 2 mM.

TABLE 2. Pattern of some specific ^{32}P nucleotide spots from TLC autoradiograms[a]

Spot[b]	W3110		Δ*phoA8*		*phoU35*		*phoR68*		Predicted function on *phoA* regulation
	Ex[c] (0.1)	St (38)	Ex (0.1)	St (0.1)	Ex (82)	St (70)	Ex (15)	St (12)	
S1a	+	+	+	–	–	–	+	+	None
S1b	+	–	+	–	–	–	+	–	Negative
S2	–	+	–	+	+	+	–	+	Positive
S3	+	–	+	–	–	–	+	–	Negative
S8	+	–	+	–	–	–	+	–	Negative
S15	–	+	–	–	+	+	–	+	Positive
S13	+	+	+	+	+	+	+	+	None

[a] The strains used were W3110 F^-. Cell growth, nucleotide extraction, and analysis were reported previously (39).

[b] Spots S1a and S1b were resolved from S1 (39) by a different solvent system (Ga/Sb) (3).

[c] Ex, Sample withdrawn during exponential phase of growth in excess phosphate = repressibility conditions. St, Sample withdrawn during stationary phase of growth after phosphate depletion = inducibility conditions. Numbers in parentheses show alkaline phosphatase (units per optical density at 540 nm) as in Table 1.

could be phosphorylated to produce the repressor, as in the case of the *gln* ALG system (35a). Results obtained earlier suggested that the cofactor could be a nucleotide. In 1970, Wilkins (67), in J. Gallant's laboratory, observed that mutants of *E. coli* with an altered nucleotide biosynthetic pathway could produce alkaline phosphatase in a medium with excess phosphate when starved either for pyrimidine or for guanine. Wilkins concluded that this escape synthesis was due to an alteration of the cellular nucleotide pool, rather than a change of the internal P_i pool. The mutants employed by Wilkins were capable of only a very low level of alkaline phosphatase escape synthesis: about 1 to 2% of the amount produced by the derepressed cells. Possibly, the low level of enzyme synthesis was due to a low amount of a positive nucleotide cofactor present in these inducible strains.

We reasoned that a constitutive mutant should constantly produce a high level of positive cofactor(s), particularly the *phoU* mutant, which produces a high level of alkaline phosphatase (Table 1). Recently, it became possible to analyze the *E. coli* nucleotide pool, by a sensitive method (3) which resolved 30 to 40 nucleotides on two-dimensional thin-layer chromatography (TLC) plates. We compared the nucleotide pool of *E. coli* wild-type W3110, growing in conditions of alkaline phosphatase repression (exponential growth in P_i excess) or induction (P_i depleted, stationary phase), with the pool of various mutants, either alkaline phosphatase negative (Δ*pho*A8) or alkaline phosphatase constitutive (*phoR68* and *phoU35*), under similar growth conditions (39).

Formic acid extracts of ^{32}P-labeled cells were analyzed by TLC (39). Some ^{32}P spots were present only when the wild-type cells had the potential to synthesize alkaline phosphatase (during P_i depletion) and were absent in conditions adverse to the synthesis (repression in excess P_i); these spots were classified as positive cofactors (Table 2). During the exponential growth phase, the wild type and mutant *phoR68* showed specific spots on TLC (S3 and S8, Table 2) which were absent in the stationary phase and were thus classified as negative cofactors.

The presence of active alkaline phosphatase in the cells did not alter the spot pattern. In fact, a *phoA* mutant (Δ*phoA8*) devoid of alkaline phosphatase activity presents a pattern similar to that of the wild-type strain. Furthermore, the level of P_i in the medium and the growth conditions (exponential or stationary growth phase) had no effect on the spot pattern as seen in the constitutive mutant *phoU35*. The observed modifications were under genetic control. In all growth conditions, mutant *phoU35*, constitutive for alkaline phosphatase, produced two specific spots (S2 and S15 in Table 2) which have the properties of positive cofactors. However, another alkaline phosphatase constitutive mutant, *phoR68*, behaved like the wild type and produced the positive cofactors (S2 and S15) only in conditions of P_i starvation (Table 2).

TLC analysis of formic acid extracts treated with sodium periodate demonstrated that most of the spots (25 to 30) were sensitive to oxidation, as expected for most of the nucleotides. Some of these nucleotides, extracted individually from the TLC spots, were treated with snake venom phosphodiesterase. The results revealed that all the nucleotides of interest (classified as positive or negative effectors in Table 2) were substrates of phosphodiesterase and were either di- or polyphosphorylated, except for S2. Figure 3 diagrammatically indicates the position of a few selected nucleotides on TLC plates

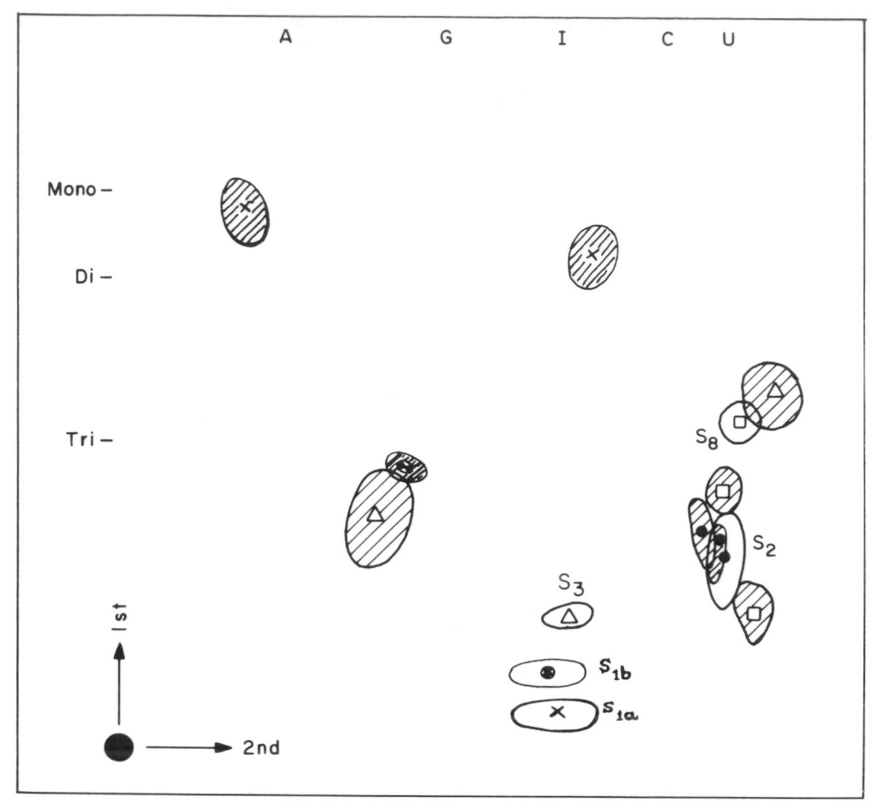

FIG. 3. Diagram of a two-dimensional TLC autoradiogram of ^{32}P nucleotides before and after treatment with phosphodiesterase (39). The solvent system was Tb/Sb (3). The spots were isolated, treated with phosphodiesterase, and analyzed by TLC again. Plain spots are the original untreated nucleotides. Shaded spots are the products of the phosphodiesterase treatment. Symbols: ×, S1a; *, S1b; ●, S2; △, S3; □, S8.

(plain spots) and the positions assumed by their products after phosphodiesterase treatment (shaded spots). From these observations, the following scheme can be derived for the formation of positive or negative cofactors by phosphodiesterase action:

$$\begin{array}{l} X \leftarrow S3^{-} \\ \qquad \downarrow \\ \qquad S8^{-} \rightrightarrows S2^{+} \leftarrow S1b^{+} \rightarrow X \end{array}$$

Each nucleotide (−, negative cofactor; +, positive cofactor) upon phosphodiesterase treatment gives two products (X = unknown nucleotide intermediate). Only S2 is not hydrolyzed.

If these specific nucleotides are cofactors in the regulation of *phoA* expression in vivo, they should produce a similar effect in vitro; i.e., individual nucleotides with the pattern of a positive cofactor (e.g., S2 and S15 isolated from the constitutive mutant *phoU35* during exponential growth) should overcome repression by P_i of alkaline phosphatase synthesis. To check this hypothesis, cells of K10 wild type grown in a P_i-rich medium (LB) were permeabilized and used as a source of DNA, mRNA, ribosomes, and enzymes in an in vitro Zubay system (71) which supplied all other factors including an energy-generating system producing a high level of P_i. The synthesis of alkaline phosphatase in the control plasmolyzed cells was poor (Table 3) (about 1% of derepressed wild type), but was similar to the amount observed in previous experiments in vitro (21, 37) and to the level of escape synthesis of repressed mutants in vivo (67). When the nucleotide S2 or S15 was added, the synthesis increased five- to eightfold (Table 3). The addition of any one of the other 28 nucleotides from the TLC plates, ppGpp, ppGppp, or PP_i, had no effect. The nucleotides classified as negative cofactors were not tested in vitro since *phoA* was already highly repressed.

In these experiments, the function of the positive cofactor(s) is presumably to bypass the P_i repression. The permeabilized cells should contain only a low level of *phoB* product (18, 49), since the gene was repressed by P_i during growth of the cells in LB broth. Guan et al. (18) observed that expressed cells contained ca. 3%

TABLE 3. In vitro synthesis by permeabilized (1) cells of *E. coli* K10

Prepn[a]	Activity (U/4 × 10^8 cells)	
	Phosphatase	β-Galactosidase
Viable cells + IPTG	43.00	7,672
Permeabilized cells + IPTG	0.35	506
Permeabilized cells (no IPTG)		
+ S13	0.35	10
+ S2	2.79	10
+ S15	1.75	10

[a] S13 is 1 of 28 nucleotides from TLC plates. S2 and S15 are nucleotides classified as positive factors (Table 2). IPTG, Isopropyl-β-D-thiogalactopyranoside. From Rao et al. (39).

of the derepressed level of *phoB* product. Since *phoB* and *phoR* constitute an operon, as observed by Shinagawa et al. (this volume), *phoR* gene product should also be present at a low level in these repressed cells. All or most of the low-molecular-weight factors should be lost from the permeabilized cells. This could explain the low level of *phoA* gene expression in vitro. The increased synthesis of alkaline phosphatase (up to 7% of the derepressed level) upon addition of nucleotides (S2 or S15) could be explained by activation of *phoA* zeither directly or via the product of the regulatory genes of positive control, PhoR and PhoB.

The mechanism by which the nucleotides affect alkaline phosphatase induction is a matter of speculation. Possibly, they bind to the regulatory protein PhoR and switch the negative function to the positive one (Fig. 1). Since PhoR is probably a dimer (38), it would dimerize only if bound to a nucleotide. We can hypothesize that the repressor form is the more phosphorylated dimer and the activator form is the less phosphorylated monomer.

We observed in vivo that in low P_i or in *phoU* mutants there were no negative cofactors (S3 or S8) and only the less phosphorylated forms (S2 and S15) were present. According to the above hypothesis, the PhoR monomers should be activated to $PhoR^A$ (e.g., PhoRS2 or PhoRS15). $PhoR^A$ will induce *phoB* or will activate PhoB to $PhoB^A$ by donating the cofactor. In high phosphate, the PhoR product is possibly bound to the more phosphorylated nucleotide(s) and locked into an inactive dimeric complex and PhoB is not activated:

Low P_i: PhoRS2 + PhoB → PhoR + PhoBS2

High P_i: PhoRS2 + P_i → PhoR$\overset{S8}{\overline{S2\text{-}S2}}$PhoR + PhoB

The results of the "P_i shock" experiment favor the hypothesis of a rapid dephosphorylation of the existing PhoB, possibly by exchanging the positive cofactor with PhoR. Alternatively, the P_i shock could be interpreted as a rapid loss of PhoB product to a level too low to induce alkaline phosphatase. The loss should still be the consequence of P_i metabolism.

Genes *phoB* and *phoR* are negatively regulated by the P_i transport system Pst (*pstS, pstC, pstA, pstB*). In our hypothesis (Fig. 1) the Pst transport feeds P_i into a private pool utilized by the *phoU* gene to provoke the phosphorylation of the nucleotide cofactors.

In *phoU* mutants the phosphorylating enzyme may be missing, and thus the positive cofactor (S2 or S15) can never be phosphorylated to a negative one (S8) and form the $PhoR^R$ complex. On the other hand, in mutant *phoR68* only the $PhoR^R$ form can exist, and PhoM, which is not P_i regulated, supplies the positive function. Mutant *phoR69* has only $PhoR^A$ (the activator form) and cannot make $PhoR^R$ even in high P_i. This hypothesis complements the one presented by Shinagawa et al. (this volume), in which a signal molecule (possibly the product of *phoU*) interacts with PhoB.

The fundamental question of the function of P_i will probably be answered soon.

This work was supported by National Science Foundation grant DMB 85-19214. We thank the Undergraduate Research Opportunity Program at the Massachusetts Institute of Technology for financial support of D. Kaffire and G. Huang, who performed the phosphate shock experiments, and Angeline Li, who performed some of the TLC experiments.

LITERATURE CITED

1. **Ben-Hamida, F., and F. Gros.** 1971. Transcription and translation mechanism in a "permeabilized" *E. coli* system. Biochimie **53:**71–80.
2. **Bennett, R. L., and M. H. Malamy.** 1970. Arsenate resistant mutants of *Escherichia coli* and phosphate transport. Biochem. Biophys. Res. Commun. **10:**496–503.
3. **Bochner, B. R., and B. N. Ames.** 1982. Complete analysis of cellular nucleotides by two-dimensional thin layer chromatography. J. Biol. Chem. **257:**9759–9769.
4. **Bracha, M., and E. Yagil.** 1973. A new type of alkaline phosphatase negative mutant in *Escherichia coli* K12. Mol. Gen. Genet. **122:**53–60.
5. **Bradshaw, R. A., F. Cancedda, L. H. Ericsson, P. A. Neumann, S. P. Piccoli, M. J. Schlesinger, K. Shrieffer, and K. A. Walsh.** 1981. Amino acid sequence of *Escherichia coli* alkaline phosphatase. Proc. Natl. Acad. Sci. USA **78:**3473–3477.
6. **Brenner, S.** 1959. The mechanism of gene action, p. 304. *In* G. E. W. Walstenholme and C. M. O'Connor (ed.), Symposium on Biochemistry of Human Genetics, CIBA Foundation and International Union of Biological Sciences. Churchill, London.
7. **Brickman, E., and J. Beckwith.** 1975. Analysis of the regulation of *Escherichia coli* alkaline phosphatase synthesis using deletions and φ80 transducing phage. J. Mol. Biol. **96:**307–316.
8. **Chang, C. N., H. Inouye, P. Model, and J. Beckwith.** 1980. Processing of alkaline phosphatase precursor to the ma-

ture enzyme by an *Escherichia coli* membrane preparation. J. Bacteriol. **142:**726–728.
9. **Dohan, F. C., Jr., R. H. Rubman, and A. Torriani.** 1969. *In vitro* synthesis of alkaline phosphatase monomers directed by *Escherichia coli* messenger. Cold Spring Harbor Symp. Quant. Biol. **34:**768.
10. **Dohan, F. C., Jr., R. H. Rubman, and A. Torriani.** 1971. *In vitro* synthesis of *Escherichia coli* alkaline phosphatase monomers. J. Mol. Biol. **58:**469–479.
11. **Echols, H., A. Garen, S. Garen, and A. Torriani.** 1961. Genetic control of repression of alkaline phosphatase in *Escherichia coli*. J. Mol. Biol. **3:**425–438.
12. **Fan, D. P., M. J. Schlesinger, A. Torriani, K. J. Barrett, and C. Levinthal.** 1966. Isolation and characterization of complementation products of *Escherichia coli* alkaline phosphatase. J. Mol. Biol. **15:**32–48.
13. **Fincham, J. R. S.** 1960. Genetically controlled differences in enzyme activity. Adv. Enzymol. **22:**1.
14. **Garen, A., and H. Echols.** 1962. Genetic control of induction of alkaline phosphatase synthesis in *Escherichia coli*. Proc. Natl. Acad. Sci. USA **48:**1398–1402.
15. **Garen, A., and H. Echols.** 1962. Properties of two regulating genes for alkaline phosphatase. J. Bacteriol. **83:**297–300.
16. **Garen, A., and C. Levinthal.** 1960. A fine structure genetic and chemical study of the enzyme alkaline phosphatase of *Escherichia coli*. I. Purification and characterization of alkaline phosphatase. Biochim. Biophys. Acta **38:**470–483.
17. **Garen, A., and N. Otsuji.** 1964. Isolation of a protein specified by a regulator gene. J. Mol. Biol. **8:**841-852.
18. **Guan, C. D., B. Wanner, and H. Inouye.** 1984. Analysis of regulation of *phoB* expression using a *phoB-cat* fusion. J. Bacteriol. **156:**710–717.
19. **Horiuchi, T., S. Horiuchi, and D. Mizuno.** 1959. A possible negative feedback phenomenon controlling formation of alkaline phosphorous esterase in *Escherichia coli*. Nature (London) **183:**1529–1530.
20. **Inouye, H., and J. Beckwith.** 1977. Synthesis and processing of an *Escherichia coli* alkaline phosphatase precursor *in vitro*. Proc. Natl. Acad. Sci. USA **74:**1440–1444.
21. **Inouye, H., C. Pratt, J. Beckwith, and A. Torriani.** 1977. Alkaline phosphatase synthesis in a cell-free system using DNA and RNA templates. J. Mol. Biol. **110:**75–87.
22. **Jacob, F., and J. Monod.** 1961. Genetic regulatory mechanisms in the synthesis of proteins. J. Mol. Biol. **3:**318–356.
23. **Kreuzer, K., C. Pratt, and A. Torriani.** 1975. Genetic analysis of regulatory mutants of alkaline phosphatase of *E. coli*. Genetics **81:**459–468.
24. **Levinthal, C., E. R. Signer, and K. Fetherolf.** 1962. Reactivation and hybridization of reduced alkaline phosphatase of *Escherichia coli*. Proc. Natl. Acad. Sci. USA **48:**1230–1237.
25. **Ludtke, D., J. Bernstein, C. Hamilton, and A. Torriani.** 1984. Identification of the *phoM* gene product and its regulation in *Escherichia coli* K-12. J. Bacteriol. **159:**19–25.
26. **Makino, D., H. Shinagawa, and A. Nakata.** 1984. Cloning and characterization of the alkaline phosphatase positive regulatory gene (*pho*M) of *Escherichia coli*. Mol. Gen. Genet. **195:**381–390.
27. **Makino, D., H. Shinagawa, and A. Nakata.** 1985. Regulation of the phosphate regulon of *Escherichia coli* K12: regulation and role of the regulatory gene *pho*R. J. Mol. Biol. **184:**231–240.
28. **Malamy, M. H., and B. L. Horecker.** 1961. The localization of alkaline phosphatase in *Escherichia coli* K-12. Biochem. Biophys. Res. Commun. **5:**104–108.
29. **Malamy, M. H., and B. L. Horecker.** 1964. Purification and crystallization of the alkaline phosphatase of *Escherichia coli*. Biochemistry **3:**1893–1897.
30. **Malamy, M. H., and B. L. Horecker.** 1964. Release of alkaline phosphatase of *Escherichia coli* upon lysozyme spheroplast formation. Biochemistry **3:**1889–1893.
31. **Medveczky, N., and H. Rosenberg.** 1970. The phosphate-binding protein of *Escherichia coli*. Biochim. Biophys. Acta **211:**158–168.
32. **Monod, J.** 1950. La technique de culture continue. Theorie et applications. Ann. Inst. Pasteur (Paris) **79:**390–410.
33. **Morris, H., M. J. Schlesinger, M. Bracha, and E. Yagil.** 1974. Pleiotropic effects of mutations involved in the regulation of *Escherichia coli* K-12 alkaline phosphatase. J. Bacteriol. **119:**583–592.
34. **Nesmeyanova, M. A., A. V. P. Vitvitsky, and A. A. Bogdanov.** 1966. A study of the formation of alkaline phosphatase macrostructure in the course of its biosynthesis. Biokhimiya **31:**902–909.
35. **Neu, H. C., and L. A. Heppel.** 1965. The release of enzymes from *Escherichia coli* by osmotic shock and during the formation of spheroplasts. J. Biol. Chem. **240:** 3685–3692.
35a. **Ninfa, A. J., and B. Magasanik.** 1986. Covalent modification of the *gln* G product, NRI, but the *gln* L product, NRII, regulates the transcription of the *gln* ALG operon in *E. coli*. Proc. Natl. Acad. Sci. USA **83:**5909–5913.
36. **Plocke, D. J., C. Levinthal, and B. Vallee.** 1962. Alkaline phosphatase of *Escherichia coli*: a zinc metalloenzyme. Biochemistry **1:**373–378.
37. **Pratt, C.** 1980. Kinetics and regulation of cell-free alkaline phosphatase synthesis. J. Bacteriol. **143:**1265–1274.
38. **Pratt, C., and J. Gallant.** 1972. A dominant constitutive *pho*R mutation in *E. coli*. Genetics **72:**217–226.
39. **Rao, N. N., E. Wang, J. Yashphe, and A. Torriani.** 1986. Nucleotide pool in *pho* regulon mutants and alkaline phosphatase synthesis in *Escherichia coli*. J. Bacteriol. **165:**205–211.
40. **Reynolds, J. A., and M. J. Schlesinger.** 1968. Hydrogen ion equilibria of conformational state of *Escherichia coli* alkaline phosphatase. Biochemistry **7:**2080–2085.
41. **Roche, J., and Nguyen Van Thoai.** 1950. Phosphatase alcaline. Adv. Enzymol. **10:**83–122.
42. **Rosenberg, H., R. G. Gerdes, and K. Chegwidden.** 1977. Two systems for the uptake of phosphate in *Escherichia coli*. J. Bacteriol. **131:**505–511.
43. **Rothman, F., and R. Byrne.** 1963. Fingerprint analysis of alkaline phosphatase of *Escherichia coli* K12. J. Mol. Biol. **6:**330–340.
44. **Schlesinger, M. J.** 1965. The reversible dissociation of alkaline phosphatase of *Escherichia coli*. II. Properties of the subunits. J. Biol. Chem. **240:**4293–4298.
45. **Schlesinger, M. J., and K. Barrett.** 1965. The reversible dissociation of the alkaline phosphatase of *Escherichia coli*. J. Biol. Chem. **240:**4284–4292.
46. **Schlesinger, M. J., and C. Levinthal.** 1963. Hybrid protein formation of *Eshcerichia coli* alkaline phosphatase leading to *in vivo* complementation. J. Mol. Biol. **7:**1–12.
47. **Schlesinger, M. J., A. Torriani, and C. Levinthal.** 1963. *In vitro* formation of enzymatically active hybrid proteins from *Escherichia coli* alkaline phosphatase CRM's. Cold Spring Harbor Symp. Quant. Biol. **28:**539–542.
48. **Schweizer, H., T. Grussenmeyer, and W. Boos.** 1982. Mapping of two *ugp* genes coding for the *pho* regulon-dependent *sn*-glycerol-3-phosphate transport system of *Escherichia coli*. J. Bacteriol. **150:**1164–1171.
49. **Shinagawa, H., K. Makino, and A. Nakata.** 1983. Regulation of the pho regulon in *Escherichia coli* K12. Genetic and physiological regulation of the positive regulatory gene *pho*B. J. Mol. Biol. **168:**477–488.
50. **Signer, E. R., A. Torriani, and C. Levinthal.** 1961. Gene expression in intergeneric merozygotes. Cold Spring Harbor Symp. Quant. Biol. **26:**31–34.
51. **Simpson, R. T., and B. L. Vallee.** 1968. Two differential classes of metal atoms in alkaline phosphatase of *Escherichia coli*. Biochemistry **7:**4343–4350.
52. **Tomassen, J., and B. Lugtenberg.** 1980. Outer membrane protein e of *Escherichia coli* K-12 is co-regulated with alkaline phosphatase. J. Bacteriol. **143:**151–157.
53. **Tommassen, J., P. De Geus, and B. Lugtenberg.** 1982.

Regulation of the pho regulon of *E. coli* K-12. Cloning of the regulatory genes *pho*B and *pho*R and identification of their gene products. J. Mol. Biol. **157**:265–274.
54. **Tommassen, J., P. Hiemstra, P. Overduin, and B. Lugtenberg.** 1984. Cloning of *pho*M, a gene involved in regulation of the synthesis of phosphate limitation inducible proteins in *Escherichia coli* K-12. Mol. Gen. Genet. **195**:190–194.
55. **Torriani, A.** 1960. Influence of inorganic phosphate in the formation of phosphatases by *Escherichia coli*. Biochim. Biophys. Acta **38**:460–479.
56. **Torriani, A.** 1968. Alkaline phosphatase subunits and their dimerization in vitro. J. Bacteriol. **96**:1200–1207.
57. **Torriani, A., and D. N. Ludtke.** 1985. The Pho regulon of *E. coli* K12, p. 224–242. *In* M. Schaechter, F. C. Neidhardt, J. Ingraham, and N. O. Kjeldgaard (ed.), The molecular biology of bacterial growth. Jones and Bartlett Publishers, Boston.
58. **Torriani, A., and A. M. Pappenheimer, Jr.** 1960. Degradation of pneumococcal polysaccharides S_3 and S_8 by inducible depolymerases from *Bacillus palustris*. Bull. Soc. Chim. Biol. **42**:1619–1626.
59. **Torriani, A., and A. M. Pappenheimer, Jr.** 1962. Inducible polysaccharide depolymerases of *Bacillus palustris*. J. Biol. Chem. **273**:3–13.
60. **Torriani, A., and F. Rothman.** 1961. Mutants of *Escherichia coli* constitutive for alkaline phosphatase. J. Bacteriol. **81**:835–836.
61. **Torriani, A., M. J. Schlesinger, and C. Levinthal.** 1963. Intracistronic complementation of alkaline phosphatase from *Escherichia coli*. Colloq. Int. CNRS **124**:297–302.
62. **Virtanen, A. I., and U. Winkler.** 1949. Effect of decrease in the protein content of cells on the proteolytic enzyme system. Acta Chim. Scand. **3**:272–278.
63. **Vogel, H. J.** 1958. Comments on the possible role of repressors and inducers of enzyme formation in development, p. 479–484. *In* L. McElroy and B. Glass (ed.), Symposium on the chemical basis of development. Johns Hopkins Press, Baltimore.
64. **Wanner, B. L.** 1983. Overlapping and separate controls on the phosphate regulon in *E. coli* K12. J. Mol. Biol. **166**: 283–308.
65. **Wanner, B. L., and J. Bernstein.** 1982. Determining the *phoM* map location in *Escherichia coli* K-12 by using a nearby transposon Tn*10* insertion. J. Bacteriol. **150**:429–432.
66. **Wanner, B., and P. Letterel.** 1980. Mutants affecting alkaline phosphatase expression: evidence for multiple positive effectors of the phosphate regulon in *Escherichia coli*. Genetics **96**:353–366.
67. **Wilkins, A.** 1972. Physiological factors in the regulation of alkaline phosphatase synthesis in *Escherichia coli*. J. Bacteriol. **110**:616–623.
68. **Willsky, G. R., R. L. Bennett, and M. H. Malamy.** 1973. Inorganic phosphate transport in *Escherichia coli*: involvement of two genes which play a role in alkaline phosphatase regulation. J. Bacteriol. **113**:529–539.
69. **Willsky, G. R., and M. H. Malamy.** 1980. Characterization of two genetically separable inorganic phosphate transport systems in *Escherichia coli*. J. Bacteriol. **144**: 356–365.
70. **Yanofsky, C., B. C. Carlton, J. R. Guest, D. R. Helinski, and U. Henning.** 1964. On the colinearity of gene structure and protein structure. Proc. Natl. Acad. Sci. USA **51**:266–272.
71. **Zubay, G., D. A., Chambers, and L. C. Cheong.** 1970. Cell-free studies on the regulation of the *lac* operon, p. 375–392. *In* J. R. Beckwith and D. Zipser (ed.), The *lac* operon. Cold Spring Harbor Laboratory, Cold Spring Harbor, N.Y.

Bacterial Alkaline Phosphatase Gene Regulation and the Phosphate Response in *Escherichia coli*

BARRY L. WANNER

Department of Biological Sciences, Purdue University, West Lafayette, Indiana 47906

A large amount is now known concerning the molecular control of individual genes. Yet, comparatively little is known about how whole sets of genes are coregulated. It seems likely that the molecular interactions which regulate such gene sets, or regulons, are also important in regulating cell growth itself. The best understood general (or global) control systems concern catabolite repression, nitrogen control, oxygen regulation, the stringent response, the SOS and adaptive responses, and the heat-shock response in bacteria. Here, I will briefly discuss some aspects of bacterial alkaline phosphatase regulation and the "phosphate response" in *Escherichia coli*.

The Phosphate Response

Cell growth requires several elemental nutrients from the environment. Somehow these assimilations are balanced with their growth need, especially with respect to synthesis of macromolecules (Fig. 1). The complexity of cell growth and the ability of bacteria to adapt so well imply that cells have evolved regulatory mechanisms to cope with a limitation of essential macronutrients. I will refer here to the manner in which *E. coli* responds to phosphate limitation as the phosphate response. Similar adaptive responses occur as a result of imbalances in catabolism, nitrogen and oxygen availability, protein synthesis, DNA metabolism, and temperature.

Like nitrogen, phosphorus compounds are major essential constituents in biological systems. The phosphorus cycle, however, is relatively simpler because phosphorus is assimilated only in the +5 valence state found naturally in phosphates and phosphonates. The three principal chemical forms of phosphate are orthophosphates (P_is), pyrophosphates (PP_is), and metaphosphates. The last occurs only in highly polymeric polyphosphate structures, which may act as energy reserves in bacteria (18). In spite of its relative abundance in nature, phosphate is a limiting growth factor for many organisms because much of its natural supply occurs as insoluble salts. To cope with phosphate limitation, bacteria such as *E. coli* have evolved complex regulatory systems to assimilate phosphorus compounds very efficiently. Whereas *E. coli* can utilize P_i or organic phosphates for growth, some other bacteria, such as *Bacillus cereus* (20) and *Pseudomonas aeruginosa* (12), can also use phosphonates such as ciliatine, a natural C-P bond compound, as a phosphorus source. (See also reference 39a.)

The regulatory interactions within the phosphate gene system of *E. coli* suggest that it may also be important for cells to maintain a "phosphate balance," similar to the carbon, nitrogen, and energy balances. Such a balance would most likely be closely related to energy metabolism because of the biological role of high-energy phosphate bonds. The concept of a phosphate balance is also supported by the manner of regulation of genes within the phosphate regulon. The transcription of several phosphate-regulated promoters in *E. coli* is also specifically induced by other environmental stresses, including nitrogen or carbon starvation, anoxia, or UV irradiation (Table 1); and induction by these other stresses, at least in some cases, involves the same regulatory elements (*phoB* and *phoR*) which control phosphate regulation of bacterial alkaline phosphatase synthesis (40, 46). While the chemical nature of the phosphate balance is only speculative, it probably involves metabolic intermediates and high-energy phosphate compounds.

Multiple Controls in the Phosphate Regulon

Bacterial alkaline phosphatase is a classic example of an inducible protein. Its differential rate of synthesis increases nearly 1,500-fold (Table 2) when *E. coli* is starved of phosphate. (Actually, induction by phosphate limitation does not occur in all *E. coli* strains [19]). Although bacterial alkaline phosphatase regulation has been studied since the late 1950s, a detailed understanding of its molecular control is still lacking. Early genetic studies showed that two loci, R1 (*phoB-phoR*) and R2 (Pst-*phoU*), negatively regulate its synthesis (Fig. 2). R1 mutants were of three types: (i) R1a mutants were constitutive but made only one-third the fully induced amount of enzyme, even when starved; (ii) R1b, such as *phoR69*, mutants were fully constitutive; and (iii) R1c (*phoB*) mutants were uninducible. R2 mutants behaved like classic repressor mutants and either made the higher

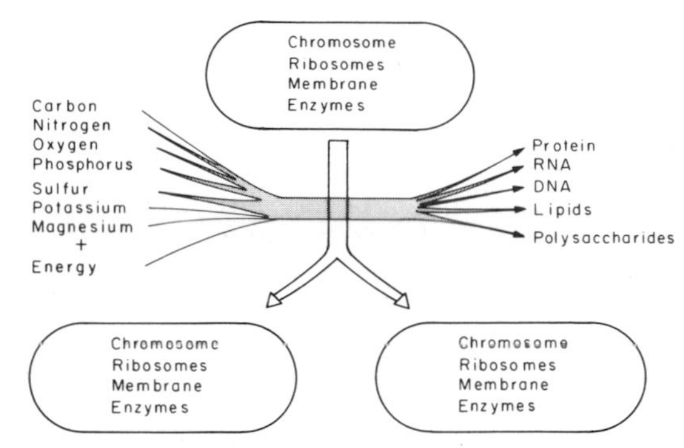

FIG. 1. Balanced growth of a cell. The stippled arrows depict the various metabolic pathways leading to synthesis of the major cellular constituents. The crossing arrows show that their syntheses are coupled to the availability of the macronutrients and their requirement for cell growth.

amount of enzyme when starved or were already fully constitutive. However, R1a and R1c alleles were epistatic to R2 mutations; i.e., the appropriate double mutants make the lower amounts of enzyme characteristic of the R1 allele, and most R1 mutations were of the R1a type, including R1 amber alleles. Later, R2 mutants were shown to be defective in the phosphate-specific transport (Pst) system (49, 50); they probably regulate bacterial alkaline phosphatase synthesis indirectly, perhaps via the synthesis of an effector molecule (Fig. 3). (Pst mutants transport phosphate [P_i] via the Pit [phosphate inorganic transport; 33] system and can, therefore, still grow on low concentrations of P_i.) Wilkins (48) showed that bacterial alkaline phosphatase synthesis was induced by guanine starvation in a manner dependent upon the *phoR* allele, and he suggested that its expression was regulated by some guanine compound. The biochemical nature of such an effector molecule(s) and its metabolism have yet to be elucidated.

In 1974, Morris et al. (30) proposed a model to explain the molecular control of bacterial alkaline phosphatase synthesis. The PhoB product acted as a transcription activator which also interacted with the PhoR product via protein-protein interaction, and the PhoR-PhoB com-

TABLE 1. Phosphate-regulated genes and promoters in *E. coli* K-12

Promoter	Inducing conditions[a]	Function[b]
phoA	$-P_i$	Bacterial alkaline phosphatase structural gene
phoBR[c]	$-P_i$	Activators and repressor
phoE	$-P_i$	Outer membrane porin
phoM	None	Alternative activator
phoS (Pst-*phoU*)	$-P_i$	Phosphate-binding protein
aphA	$-P_i$, anaerobiosis	Acid phosphatase structural gene
himA (psiP)[d]	UV, $-P_i$	Integration host factor
pepN	$-P_i$	Aminopeptidase N
ugpA,B (psiB, C)[e]	$-P_i$, −C, TYE	*sn*-Glycerol 3-phosphate transport
psiD	$-P_i$	Carbon-phosphorus lyase (39a)
psiE	$-P_i$, −C, −N	Unknown
psiF,K	$-P_i$, −C, TYE	Unknown
psiG,I,J,N,R	$-P_i$, −C, −N, TYE	Unknown
psiH,L	$-P_i$, UV, −C, −N, TYE	Unknown
psiO	$-P_i$, −C, −N, anaerobiosis	Unknown

[a] The symbols $-P_i$, −C, −N, TYE, and UV indicate induction by phosphorus-, carbon-, or nitrogen-limited growth, a tryptone-yeast extract medium, or UV irradiation. Anaerobiosis means induction occurs during the logarithmic growth phase when grown anaerobically (in the absence of oxygen and in the presence of carbon dioxide). None refers to a constitutive promoter. The *psiM* and *psiO* promoters are also induced by phosphate starvation or TYE; they are not listed here because they were not tested for carbon or nitrogen control (see references 40, 42, 45 and 47).

[b] See reference 4 for bacterial alkaline phosphatase protein sequence and references 3, 15, and 17 for partial *phoA* nucleotide sequence. The complete DNA sequence for *phoA* and a downstream open reading frame (possibly *psiF*) was determined by C. N. Chang (personal communication). See reference 32 for *phoE* nucleotide sequence and references 1, 22, 35, 37, and 38 for Pst-*phoU* DNA sequence. The *aphA* product was defined in reference 11, and its regulation was defined by P. Boquet (personal communication). See reference 29 for UV induction of *himA* and references 40 and 42 for unpublished results cited; *himA* is associated as a distal gene with the phenylalanine tRNA synthetase *pheST* operon for which the complete nucleotide sequence is known (27, 28). See references 26 and 31 for *pepN* control and reference 34 for *ugpA, B*.

[c] Makino et al. (25) showed that a *phoR-lacZ* fusion is regulated by the *phoB* promoter. Also, Mu d1 insertions in *phoB* are polar and totally abolish PhoR function by complementation (unpublished data). See references 13 and 36 for induction of *phoB*.

[d] Reference 42.

[e] See legend to Fig. 2.

TABLE 2. Regulation of *phoA*, *psiE*, and *psiO* transcription

Strain	Phosphate concn[a]	Promoter[b]		
		phoA	*psiE*	*psiO*
Wild type	Excess	0.2	5.2	0.3
Wild type	Limited	172	169	13.8
phoB	Excess	0.1	4.6	0.3
	Limited	0.1	47.2	17.0
phoR[c]	Excess	56.2	5.0	0.4
	Limited	47.0	45.0	14.8
phoR phoM[d]	Excess	1.6	4.1	0.2
	Limited	1.7	39.2	15.0

[a] Bacteria were grown for 10 doublings or longer in 0.4% glucose morpholinepropanesulfonic acid medium with 2 mM K_2HPO_4 (excess) or 0.1 mM K_2HPO_4 (limited). Samples were taken during the logarithmic phase or 3 to 5 h after onset of starvation for enzyme assays.

[b] Units are nanomoles of nitrophenol made per minute at 28°C in standard assays, given per unit of optical density at 420 nm. Data show the amounts of β-galactosidase made in *phoA-lacZ*, *psiE-lacZ*, or *psiD-lacZ* fusion strains, respectively. (Quantitatively similar data were found when bacterial alkaline phosphatase was measured in *phoA*$^+$ strains [40].)

[c] Data are shown for an R1a type mutant.

[d] An amber mutant was used.

plex was not an activator. In PhoR1a mutants, this interaction did not occur, and hence PhoB activated transcription, albeit somewhat inefficiently. However, phosphate starvation produced an effector which bound to the PhoB-PhoR complex, converting it to an efficient transcription activator. The PhoR1b mutants were altered such that the PhoB-PhoR complex activated transcription even in the absence of the effector.

The Morris et al. (30) model was accepted in principle when I joined Jon Beckwith's laboratory as a postdoctoral fellow in the fall of 1975. My interests at the time, however, were not in bacterial alkaline phosphatase as a regulatory system but rather in its characteristics as a secreted protein. About this time, Inouye and Beckwith (16) showed that bacterial alkaline phosphatase was made in vitro as a larger precursor protein, and others were studying the role of signal sequences in protein secretion. I searched for unlinked mutants in the hope of uncovering nonregulatory mutants which might be altered in protein localization.

Since bacterial alkaline phosphatase mutants could be readily detected, I searched for phosphatase-negative mutants, using a method which avoided mutations within its structural gene (46). An R1a-type *(phoR68)* parental strain was used for the convenience of assaying the mutants. By testing over 50,000 nitrosoguanidine- or UV-mutagenized colonies, about 50 independent mutants were identified and characterized for regulatory versus nonregulatory defects (46; B. L. Wanner and J. Beckwith, unpublished data). The regulatory mutants showed decreased expression of *phoA-lacZ* transcriptional fusions, and regulatory defects were always overcome by the *phoA*(Bin) *pho-1003* mutation (46), in which *phoA* expression is *phoB* independent. (Although the notation Bin was adopted to indicate PhoB-independent transcription by the *pho-1003* promoter mutation, transcription was

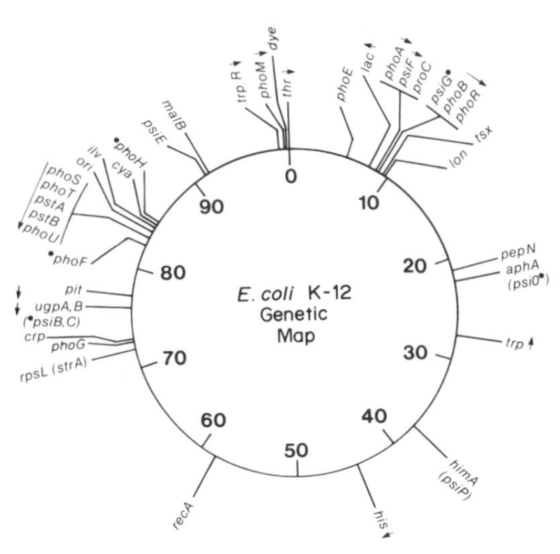

FIG. 2. Genetic map of the phosphate regulon. The circular map, nomenclature, and map positions are in general according to references 2 and 42. The *thr*, *lac*, *proC*, *tsx*, *trp*, *his*, *recA*, *rpsL*, *ori*, *ilv*, *malB*, and *trpR* loci are illustrated for reference. The arrows show the directions of transcription. All other genes shown are somehow related to phosphate metabolism. The acronym *pho* refers to a phosphate gene, which may have either a transport or a regulatory function; Pst-*phoU* indicates the cluster of Pst genes and the regulatory gene *phoU*. The *phoM* locus encodes four open reading frames which are transcribed as shown (H. Shinagawa, personal communication). The *psi* promoters show phosphate-starvation-inducible transcription (47). Twelve other *psi* promoters are not shown here. The *psiB* and *psiC* promoters probably correspond to *ugpA* and *ugpB* (B. L. Wanner and F. C. Neidhardt, unpublished data); *psiP* is probably identical to *himA* (40, 42). The physiological function of several *psi* genes is unknown. The *psiO* promoter was mapped only by Hfr crosses; other *pho* or *psi* genes were mapped by P1 crosses (40, 42, 45; unpublished data). Genes less well mapped are marked with an asterisk. See reference 47 for mapping of *psiF* and *psiG*, reference 5 for *pepN*; reference 35 for *ugpA ugpB* region; references 1 and 37 for the orientation of the Pst-*phoU* gene cluster *phoS phoT(phoW, pstC) pstA(phoT) pstB phoU* (alternative nomenclature is in parentheses); and reference 42 for *psiE*. The sources for other data are given in the footnotes to Table 1.

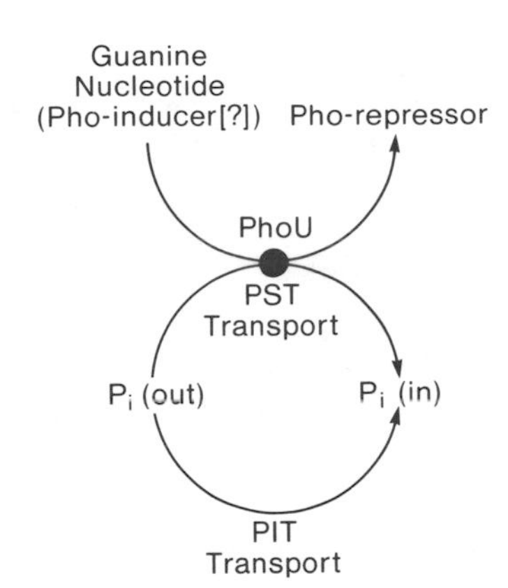

FIG. 3. Phosphate transport and regulation of the phosphate gene system. Phosphate transport via the Pst system is somehow coupled to the molecular repression of several *pho* genes. Inorganic phosphate enters *E. coli* via either the high-affinity Pst system or the low-affinity phosphate inorganic transport (Pit) system (33). A functional Pst system and PhoU are required for repression, while Pit mutants are without effect on phosphate control. It is postulated that a low level of transport via a repressed Pst system (which itself is induced during P_i limitation [50]) is sufficient to maintain repression via an interaction with the phoU product. This may involve direct protein-protein interactions at or near the membrane but, in any case, it probably affects the metabolism of a small guanine effector molecule (since guanine limitation also abolishes repression [48; Wanner, unpublished data]). The PhoU product may be an enzyme which synthesizes the hypothetical Pho repressor or degrades the Pho inducer, P_i (out) and P_i (in) refer to P_i outside or inside the cells.

subsequently shown to be PhoM and PhoR independent, as well [Wanner, unpublished data]. Both the *phoA-lacZ* fusions [34] and the *phoA*(Bin) mutation [46] were isolated by Aparna Sarthy.) Of interest at the time were several nonregulatory mutants (*perA*) which did not affect *phoA-lacZ* transcription and affected *phoA*$^+$ and *phoA*(Bin) expression similarly. The mutational defect in the nonregulatory mutants mapped extremely close to *ompB* at 73 min (46) and was probably in a locus now called *envZ* (14). Eleven of the 30 regulatory mutants were tested further as indicated below.

Since *phoB* mutants were anticipated, the presence of a *phoB* mutation was tested. All the regulatory mutants showed decreased expression of a *phoA-lacZ* fusion, were complemented by a φ80 *phoB*$^+$ transducing phage (6), made normal amounts of bacterial alkaline phosphatase under the control of the *phoA*(Bin) promoter mutation (15, 46), and could be transduced to a phosphatase-positive phenotype with a donor *phoB*$^+$ *phoR*$^+$ lysate. In spite of these results, four mutants were different because they (unlike the others) could also yield *phoB*$^+$ transductants when used as donors. Since these mutants were clearly regulatory in nature, I resumed studying them later after joining Annamaria Torrini's laboratory at the end of 1978.

I called the new mutants *phoM* mutants to indicate that their phosphatase phenotype was masked in PhoR$^+$ strains. One (the *phoM451* mutant) was suppressed by an amber suppressor (44), and I later showed that *phoM* deletion mutants have the same phenotype as the original point mutants (43). In 1980, I proposed a model in which multiple positive activators regulate bacterial alkaline phosphatase synthesis by two pathways, essentially as shown in Fig. 4. The *phoM* locus has since been cloned and studied in five laboratories (7, 21, 24, 39), including my own (B. L. Wanner, J. Bacteriol., in press; unpublished data), and the complete nucleotide sequence is now known (K. Makino, H. Shinagawa, and A. Nakata, personal communication). The DNA sequence shows four open reading frames, and ORF-3 complements *phoM* mutations. Also, five distinct alkaline phosphatase phenotypes (including "wild type") are associated with mutations in the *phoM* locus (unpublished data). Presumably, some novel phenotypes correspond to mutations in the other open reading frames, as reported elsewhere (42a, 42b).

We have made some recent progress in understanding the hypothetical trigger for the PhoM pathway. A *cis*-dominant *phoA* mutation (which is probably in the "phosphate box" [23] for the *phoA* promoter) was discovered in which bacterial alkaline phosphatase is no longer made during phosphate starvation. Although this

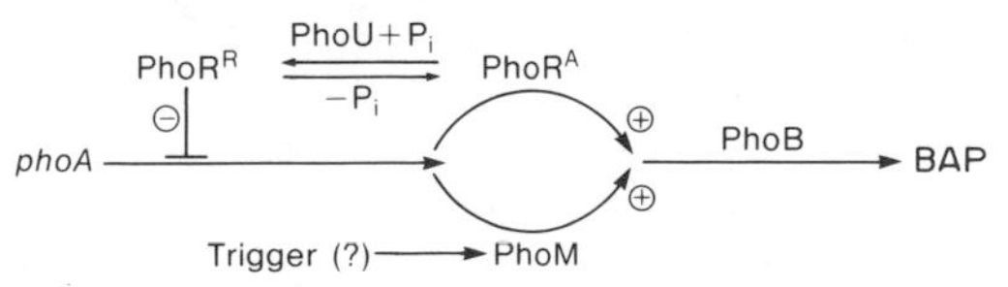

FIG. 4. Regulation of bacterial alkaline phosphate synthesis in *E. coli*. Two pathways control bacterial alkaline phosphatase synthesis: PhoR regulates *phoA* transcription in response to the phosphate supply, and the alternative PhoM pathway is active in *phoR* mutants. The PhoR protein is a bifunctional protein which acts as an activator (in the PhoR^A form) and as a repressor (in the PhoR^R form). The PhoU product converts PhoR between its activator and repressor forms, perhaps via synthesis of effector molecules by the Pst system as proposed in Fig. 3 (44). An alternative trigger induces PhoM-dependent synthesis (see text). Both PhoM and PhoR regulate the amount of the PhoB product made, and this induction may account for part of the control over *phoA* transcription (13, 36). BAP refers to bacterial alkaline phosphatase. A plus sign signifies gene or gene product activation, and a minus sign means inhibition or repression.

phoA mutant is induced over 30-fold by other growth conditions and the induction is both PhoB and PhoM dependent, the induction requires that the cells still be PhoR negative. Interestingly, Pst mutations bypass the PhoR-mediated repression (but do not lead to induction), even though phosphate limitation abolishes bacterial alkaline phosphatase synthesis (unpublished data). That is, in this mutant repression and induction are uncoupled, thus ruling out the possibility that it is solely the amount of the PhoB product which regulates bacterial alkaline phosphatase synthesis (13, 36, 44). Further studies of this mutant are now in progress.

Approach with *lacZ* Fusions

The dual PhoR and PhoM controls over bacterial alkaline phosphatase synthesis were particularly intriguing. Since the synthesis of three periplasmic proteins, including *phoA* and *phoS*, was known to be phosphate regulated (30, 50), I considered that other genes may be similarly controlled and that PhoR may regulate one subset of *pho* genes and PhoM regulates a different subset. To test this and other possibilities, I constructed strains in which *lacZ* was fused to any of several *psi* promoters (47). Previously, Sarthy et al. (34) used a two-step Mu-*lacZ* method (8) for making *phoA-lacZ* fusions. One criterion they used to show that *lacZ* was fused to the *phoA* promoter was induction of β-galactosidase synthesis by phosphate starvation (34). Subsequently, Casadaban and Cohen (10) constructed the Mu d1(Ap^r *lacZ*) transposon (Fig. 5), which made it possible to construct *lacZ* fusions in a single step, and Malcolm Casadaban gave me a sample during a return visit to Beckwith's laboratory in 1978. This phage made it possible to identify *lacZ* fusions directly on the basis of their mode of regulation. By screening about 20,000 Mu d1 lysogens for phosphate-regulated *lacZ* expression, I identified 54 independent phosphate-starvation-inducible (Psi) *lacZ* transcriptional fusions, including 8 *phoA-lacZ* fusions. These Psi promoter-*lacZ* fusions correspond to 22 or more phosphate-regulated promoters in *E. coli* (40, 42, 45, 47; unpublished data).

The Phosphate Gene System

It is now known that the synthesis rates of more than 80 proteins are markedly enhanced when cells are stressed by phosphate starvation. Also, these protein "spots" show different regulatory patterns with respect to their induction and repression by various *pho* regulatory elements (Wanner and Neidhardt, unpublished data), in a manner analogous to the various regulatory classes of *psi* promoters (Table 3). As shown by using the *psi-lacZ* fusions, these

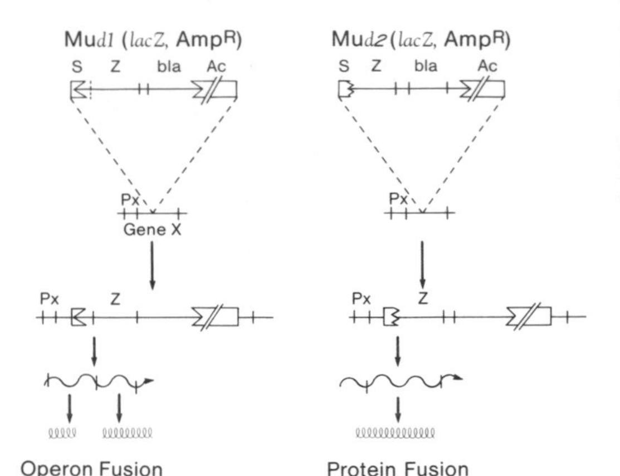

FIG. 5. Method used to construct *psi-* and *pho-lacZ* fusions. The bacteriophage transposons Mu d1 (10) and Mu d2 (9) were used to generate fusions in the phosphate gene system. Transcriptional fusions of *lacZ* to the *psi* promoters were identified by their phosphate-regulated synthesis of β-galactosidase (47). Transpositions into *pho* genes were identified by their mutant phenotype (47; Wanner, unpublished data). The open bars represent bacteriophage Mu DNA sequences. The Mu d1 transposon lacks a promoter for *lacZ* but contains a suitable ribosome-binding site. Fusions constructed with Mu d1 show the amount of transcription into *lacZ* from an exogenous promoter, P_x. The Mu d2 transposon lacks both a promoter and ribosome-binding site for *lacZ*. As a consequence, protein-*lacZ* fusions are generated with the Mu d2 transposon and the amount of *lacZ* expressed reflects the amount of translation (and transcription) initiated at P_x. Nomenclature is according to reference 10.

"stress-induced" proteins are probably controlled by more than 20 separate transcription promoters. While all the *psi* promoters are highly regulated by the medium phosphate, their induction ratios vary from 8- to 1,000-fold (45; see Table 2 for the *phoA, psiE*, and *psiO* promoters).

The *psi* genes belong to a variety of regulatory classes based upon both their physiological and genetic controls (see Tables 1 and 3). Presumably, the controls reflect the normal gene function. However, except for *phoA, psiB, psiC* (*ugpA, ugpB*), *psiD*, and *psiP* (*himA*), the *psi* genes identified with Mu d1 are unknown. Nevertheless, several regulatory features are quite striking. (i) Most *psi* promoters are expressed constitutively in *phoU* mutants, and except for *psiE*, the same promoters also show constitutive expression in Pst mutants. (ii) Some show *phoB*-dependent induction and others do not. (iii) Some show *phoB*-dependent constitutive synthesis in *phoU* mutants even though their induction is *phoB* independent. (iv) Induction by phosphate, nitrogen, or carbon starvation is *phoB* dependent for some *psi* promoters. (v)

TABLE 3. Controls in the phosphate regulon[a]

Class	Promoter	Activation by	Repression by
1	*phoA, phoE, phoS, psiB, psiC (ugpA, ugpB), psiD*	PhoB, PhoM, PhoR	PhoR, PhoU,[b] Pst
2	*psiE*	PhoB, PhoR	PhoU, Cya, Crp, Crr
3	*psiH, psiJ, psiK*		PhoR, PhoU[c]
4	*psiO*		Cya, Crp, Lon

[a] Only mutations whose gene assignments were made are indicated. See references 40, 42, and 45 for complete descriptions.

[b] Formerly called *phoT35; phoF* also (42) causes *phoA* to be expressed constitutively, but it was not tested with the other promoters.

[c] Constitutive expression of the *psiH, psiJ,* and *psiK-lacZ* fusions in *phoR* and *phoU* mutants requires the PhoB and PhoM products, in a manner which is phenotypically similar to *phoA* regulation. However, unlike *phoA* transcription, phosphate starvation-inducible expression of the *psiH, psiJ,* and *psiK* promoters does not require the PhoB or PhoR products (45).

Both *psiE* and *psiO* show elevated expression in either *cya* or *crp* mutants, but only *psiO* is induced by anaerobiosis; and *cya* and *crp* mutants were selected as Lac$^+$ mutants of these fusion strains. (vi) Induction of *psiO* shows a dramatic temporal lag of about one-half generation under several inducing conditions (Fig. 6). (vii) Expression of *psiO* is enhanced in *lon* mutants, which are frequently isolated as Lac$^+$ mutants on repressing media. (viii) Expression of *psiE* is *phoB* and *phoR* dependent but *phoM* independent (Fig. 7). Indeed, the *psiE* promoter corresponds to one class which was sought after. The failure to find a *phoM*-dependent, *phoR*-independent promoter probably occurred because only phosphate-regulated *lacZ* fusions were tested. (See references 40, 42, 45, and 47 for further information. Only general characteristics are given here. Although not every promoter was examined in the same detail, of those studied closely, no two *psi* promoters shared all the same features [40, 42, 45, 47; unpublished data].)

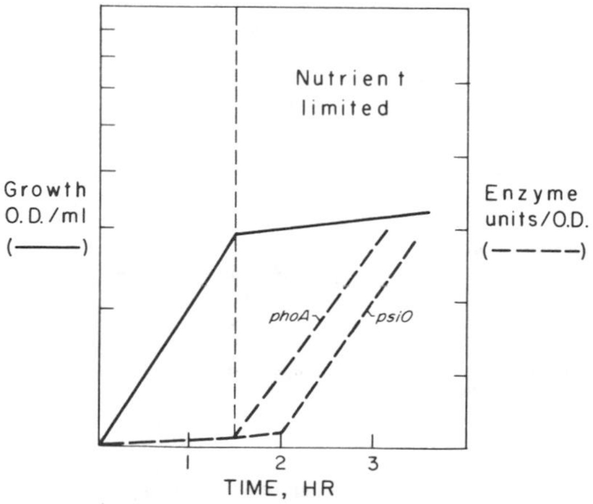

FIG. 6. Temporal control in the phosphate regulon. Bacterial alkaline phosphatase synthesis is induced abruptly upon phosphate starvation while induction of *psiO* occurs about one-half generation later, under a variety of growth conditions (40; unpublished data).

Conclusions

The genes of the phosphate regulon can be arbitrarily characterized with regard to three features. (i) The expression of several *pho* or *psi* genes is highly inducible by phosphate limitation, although many are also induced by a variety of other physiological conditions. (ii) Some gene products are involved in transport of P_i or phosphorylated compounds, including the Pst and Pit systems, PhoE, PhoA (which has a higher activity as a phosphotransferase than a hydrolase), and UgpA UgpB; but among these only Pst has a regulatory function. (iii) Other genes including *phoB, phoF, phoG, phoH, phoM, phoR, phoU, cya, crp, himA*, and *lon* have distinct regulatory roles, which may differ with individual promoters. Future studies of the molecular interactions in this complex control system are likely to reveal not only how phosphate regulates gene expression and, perhaps,

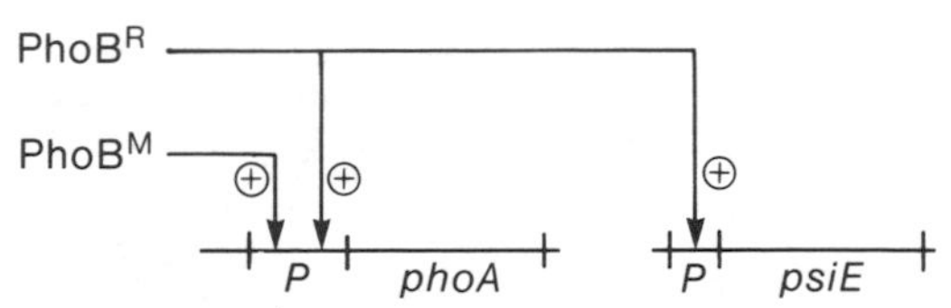

FIG. 7. Promoter-specific PhoB-dependent control of *phoA* and *psiE*. PhoB activates *phoA* transcription via either the PhoM or PhoR pathway, denoted PhoB^M and PhoB^R, respectively. The PhoB^R pathway makes about fourfold more *phoA* transcript than the phoB^M pathway, which suggests that different promoters are activated. Since only the PhoB^R pathway activates *psiE* expression (40, 42, 45), there must also exist a promoter specificity in PhoB^M and PhoB^R. The different activator forms of PhoB may involve alternative effector molecules or PhoB may act together with another protein factor, such as an activated PhoR product, to stimulate transcription only at certain promoters. *P* denotes the promoter regions.

new information about how cells grow, but also novel molecular control mechanisms (41).

I greatly appreciate the communication of unpublished results by P. Boquet, C. N. Chang, A. Nakata, and H. Shinagawa. B. Chang, C. Schmellik-Sandage, D. Stark, M. Wilmes, and D. Young contributed to recent unpublished results from my laboratory.

Support was provided by Public Health Service Award GM35392.

ADDENDUM IN PROOF

The *psiD* locus probably encodes a carbon-phosphorus lyase activity which allows use of phosphonates as a phosphorus source (39a). Unexpectedly, the *E. coli* K-12 strains which were in general used to study, clone, and sequence the *phoM* locus have since been shown to contain a mutant *phoM* region (42b). *phoR* mutants with a "wild-type" *phoM* region are not constitutive for bacterial alkaline phosphatase and instead show a novel "clonal variation" phenotype (42a). The variation itself is somehow controlled by cyclic AMP and its receptor protein (Wanner et al., submitted for publication). It is now unlikely that the *phoM* product(s) ever substitutes directly for the *phoR* product.

LITERATURE CITED

1. **Amemura, M., K. Makino, H. Shinagawa, A. Kobayashi, and A. Nakata.** 1985. Nucleotide sequence of the genes involved in phosphate transport and regulation of the phosphate regulon in *Escherichia coli*. J. Mol. Biol. **184:** 241–250.
2. **Bachmann, B. J.** 1983. Linkage map of *Escherichia coli* K-12, edition 7. Microbiol. Rev. **47:**180–230.
3. **Boidol, W., M. Simonis, M. Töpert, and G. Siewert.** 1982. Recombinant plasmids with genes for the biosynthesis of alkaline phosphatase of *Escherichia coli*. Mol. Gen. Genet. **185:**510–512.
4. **Bradshaw, R. A., F. Cancedda, L. H. Ericsson, P. A. Neumann, S. P. Piccoli, M. J. Schlesinger, K. Shriefer, and K. A. Walsh.** 1981. Amino acid sequence of *Escherichia coli* alkaline phosphatase. Proc. Natl. Acad. Sci. USA **78:** 3473–3477.
5. **Braun, V., K. Gunther, K. Hankte, and L. Zimmerman.** 1983. Intracellular activation of albomycin in *Escherichia coli* and *Salmonella typhimurium*. J. Bacteriol. **156:**308–315.
6. **Brickman, E., and J. Beckwith.** 1975. Analysis of the regulation of *E. coli* alkaline phosphatase subunits using deletions and φ80 transducing phages. J. Mol. Biol. **96:**307–316.
7. **Buxton, R. S., and L. S. Drury.** 1984. Identification of the *dye* gene product, mutational loss of which alters envelope protein composition and also affects sex factor expression in *Escherichia coli* K-12. Mol. Gen. Genet. **194:**241–247.
8. **Casadaban, M. J.** 1976. Transposition and fusion of the *lac* genes to selected promoters in *E. coli* using bacteriophage lambda and Mu. J. Mol. Biol. **194:**541–555.
9. **Casadaban, M. J., and J. Chou.** 1984. *In vivo* formation of gene fusions encoding hybrid β-galactosidase proteins in one step with a transposable Mu-*lac* transducing phage. Proc. Natl. Acad. Sci. USA **81:**535–539.
10. **Casadaban, M. J., and S. N. Cohen.** 1979. Lactose genes fused to exogenous promoters in one step using a Mu-*lac* bacteriophage: *in vivo* probe for transcriptional control sequences. Proc. Natl. Acad. Sci. USA **76:**4503–4533.
11. **Dassa, E., M. Cahu, B. Desjoyaus-Cherel, and P. Boquet.** 1982. The acid phosphatase with optimum pH of 2.5 of *E. coli*. J. Biol. Chem. **257:**6669–6676.
12. **Dumora, C., A. M. Lacoste, and A. Cassaigne.** 1983. Purification and properties of 2-aminoethylphosphonate: pyruvate aminotransferase from *Pseudomonas aeruginosa*. Eur. J. Biochem. **133:**119–125.
13. **Guan, C.-d., B. L. Wanner, and H. Inouye.** 1983. Analysis of regulation of *phoB* using a *phoB-cat* fusion. J. Bacteriol. **156:**710–717.
14. **Hall, M. H., and T. J. Silhavy.** 1981. Genetic analysis of the *ompB* locus in *Escherichia coli* K-12. J. Mol. Biol. **151:** 1–15.
15. **Inouye, H., W. Barnes, and J. Beckwith.** 1982. Signal sequence of alkaline phosphatase of *Escherichia coli*. J. Bacteriol. **149:**434–439.
16. **Inouye, H., and J. Beckwith.** 1977. Synthesis and processing of alkaline phosphatase precursor *in vitro*. Proc. Natl. Acad. Sci. USA **74:**1440–1444.
17. **Kikuchi, Y., K. Yoda, M. Yamasaki, and G. Tamura.** 1981. The nucleotide sequence of the promoter and the amino-terminal region of alkaline phosphatase structural gene (*phoA*) of *Escherichia coli*. Nucleic Acids Res. **9:** 5671–5678.
18. **Kulaev, I. S., and V. M. Vagabov.** 1983. Polyphosphate metabolism in microorganisms. Adv. Microb. Physiol. **24:** 83–171.
19. **Kuo, M.-H., and H. J. Blumenthal.** 1961. Absence of phosphate repression by inorganic phosphate in some micro-organisms. Nature (London) **190:**29–31.
20. **LaNauze, J. M., J. R. Coggins, and H. B. F. Dixon.** 1977. Aldolase-like imine formation in the mechanism of action of phosphonoacetaldehyde hydrolase. Biochem. J. **165:** 409–411.
21. **Ludtke, D., J. Bernstein, C. Hamilton, and A. Torriani.** 1984. Identification of the *phoM* gene product and its regulation in *Escherichia coli* K-12. J. Bacteriol. **159:**19–25.
22. **Magota, K., N. Otsuji, T. Miki, T. Horiuchi, S. Tsunasawa, J. Kondo, F. Sakiyama, M. Amemura, T. Morita, H. Shinagawa, and A. Nakata.** 1984. Nucleotide sequence of the *phoS* gene, the structural gene for the phosphate-binding protein of *Escherichia coli*. J. Bacteriol. **157:**909–917.
23. **Makino, K., H. Shinagawa, M. Amemura, and A. Nakata.** 1986. Nucleotide sequence of the *phoB* gene, the positive regulatory gene for the phosphate regulon of *Escherichia coli* K-12. J. Mol. Biol. **300:**900–909.
24. **Makino, K., H. Shinagawa, and A. Nakata.** 1984. Cloning and characterization of the alkaline phosphatase positive regulatory gene (*phoM*) of *Escherichia coli*. Mol. Gen. Genet. **195:**381–390.
25. **Makino, K., H. Shinagawa, and A. Nakata.** 1985. Regulation of the phosphate regulon of *Escherichia coli* K-12: regulation and role of the regulatory gene *phoR*. J. Mol. Biol. **184:**231–240.
26. **McCaman, M., A. McPartland, and M. Villarejo.** 1982. Genetics and regulation of peptides N in *Escherichia coli* K-12. J. Bacteriol. **152:**848–854.
27. **Mechulam, Y., G. Fayat, and S. Blanquet.** 1985. Sequence of the *Escherichia coli pheST* operon and identification of the *himA* gene. J. Bacteriol. **163:**787–791.
28. **Miller, H. I.** 1985. Primary structure of the *himA* gene of *Escherichia coli*: homology with DNA-binding protein HU and association with the phenylalanyl-tRNA synthetase operon. Cold Spring Harbor Symp. Quant. Biol. **44:** 691–698.
29. **Miller, H. I., M. Kirk, and H. Echols.** 1981. SOS induction and autoregulation of the *himA* gene for site-specific recombination in *Escherichia coli*. Proc. Natl. Acad. Sci. USA **78:**6754–6758.
30. **Morris, H., M J. Schlesinger, M. Bracha, and E. Yagil.** 1974. Pleiotropic effects of mutations involved in the regulation of *Escherichia coli* K-12 alkaline phosphatase. J. Bacteriol. **119:**583–592.
31. **Murgier, M., and S. Gharbi.** 1982. Fusion of the *lac* genes to the promoter for the aminopeptidase N gene of *Escherichia coli*. Mol. Gen. Genet. **187:**316–319.
32. **Overbeeke, N., H. Bergmans, F. Van Mansfeld, and B. Lugtenberg.** 1983. Complete nucleotide sequence of *phoE*, the structural gene for the phosphate limitation inducible outer membrane pore protein of *Escherichia coli* K12. J. Mol. Biol. **163:**513–532.

33. **Rosenberg, H., R. G. Gerdes, and K. Chegwidden.** 1977. Two systems for the uptake of phosphate in *Escherichia coli*. J. Bacteriol. **131:**505–511.
34. **Sarthy, A., A. Fowler, I. Zabin, and J. Beckwith.** 1979. Use of gene fusions to determine a partial signal sequence of alkaline phosphatase. J. Bacteriol. **139:**932–939.
35. **Schweizer, H., and W. Boos.** 1983. Cloning of the *ugp* region containing the structural genes for the *pho* regulon-dependent *sn*-glycerol-3-phosphate transport system of *Escherichia coli*. Mol. Gen. Genet. **192:**177–186.
36. **Shinagawa, H., K. Makino, and A. Nakata.** 1983. Regulation of the *pho* regulon in *Escherichia coli* K-12: genetic and physiological regulation of the positive regulatory gene *phoB*. J. Mol. Biol. **168:**477–488.
37. **Surin B., D. Jans, A. Fimmel, D. Shaw, G. Cox, and H. Rosenberg.** 1984. Structural gene for the phosphate-repressible phosphate-binding protein of *Escherichia coli* has its own promoter: complete nucleotide sequence of the *phoS* gene. J. Bacteriol. **157:**772–778.
38. **Surin, B. P., H. Rosenberg, and G. B. Cox.** 1985. Phosphate-specific transport system of *Escherichia coli*: nucleotide sequence and gene-polypeptide relationships. J. Bacteriol. **161:**189–198.
39. **Tommassen, J., P. Hiemstra, P. Overduin, and B. Lugtenberg.** 1984. Cloning of *phoM*, a gene involved in regulation of the synthesis of phosphate limitation inducible proteins in *Escherichia coli* K12. Mol. Gen. Genet. **195:**190–194.
39a. **Wackett, L. P., B. L. Wanner, C. P. Venditti, and C. T. Walsh.** 1987. Involvement of the phosphate regulon and the *psiD* locus in carbon-phosphorus lyase activity of *Escherichia coli* K-12. J. Bacteriol. **169:**1753–1756.
40. **Wanner, B. L.** 1983. Overlapping and separate controls on the phosphate regulon in *Escherichia coli* K12. J. Mol. Biol. **166:**283–308.
41. **Wanner, B. L.** 1985. Phase mutants: evidence of a physiologically regulated "change-in-state" gene system in *Escherichia coli*, p. 103–122. *In* I. Herskowitz and M. Simon (ed.), Genome rearrangement. Alan R. Liss, Inc., New York.
42. **Wanner, B. L.** 1986. Novel regulatory mutants of the phosphate regulon in *Escherichia coli* K-12. J. Mol. Biol. **300:**1000–1009.
42a. **Wanner, B. L.** 1986. Bacterial alkaline phosphatase clonal variation in some *Escherichia coli* K-12 *phoR* mutant strains. J. Bacteriol. **168:**1366–1371.
42b. **Wanner, B. L.** 1987. Control of *phoR*-dependent bacterial alkaline phosphatase clonal variation by the *phoM* region. J. Bacteriol. **169:**900–903.
43. **Wanner, B. L., and J. Bernstein.** 1982. Determining the *phoM* map location in *Escherichia coli* K-12 by using a nearby transposon Tn*10* insertion. J. Bacteriol. **150:**429–432.
44. **Wanner, B. L., and P. Latterell.** 1980. Mutants affected in alkaline phosphatase expression: evidence for multiple positive regulators of the phosphate regulon in *Escherichia coli*. Genetics **96:**353–366.
45. **Wanner, B. L., and R. McSharry.** 1982. Phosphate-controlled gene expression in *Escherichia coli* K12 using Mu*dl*-directed *lacZ* fusions. J. Mol. Biol. **158:**347–363.
46. **Wanner, B. L., A. Sarthy, and J. Beckwith.** 1979. *Escherichia coli* pleiotropic mutant that reduces amounts of several periplasmic and outer membrane proteins. J. Bacteriol. **140:**229–239.
47. **Wanner, B. L., S. Wieder, and R. McSharry.** 1981. Use of bacteriophage transposon Mu*dl* to determine the orientation for three *proC*-linked phosphate-starvation-inducible (*psi*) genes in *Escherichia coli* K-12. J. Bacteriol. **146:**93–101.
48. **Wilkins, A. D.** 1972. Physiological factors in the regulation of alkaline phosphatase synthesis in *Escherichia coli*. J. Bacteriol. **110:**616–623.
49. **Willsky, G. R., R. L. Bennett, and M. H. Malamy.** 1973. Inorganic phosphate transport: involvement of two genes which play a role in alkaline phosphatase regulation. J. Bacteriol. **113:**529–539.
50. **Willsky, G. R., and M. H. Malamy.** 1976. Control of the synthesis of alkaline phosphatase and the phosphate-binding protein in *Escherichia coli*. J. Bacteriol. **127:**595–609.

Structure and Function of the Regulatory Genes for the Phosphate Regulon in *Escherichia coli*

HIDEO SHINAGAWA, KOZO MAKINO, MITSUKO AMEMURA, AND ATSUO NAKATA

Department of Experimental Chemotherapy, The Research Institute for Microbial Diseases, Osaka University, 3-1, Yamadaoka, Suita, Osaka, Japan 565

The existence of at least 25 genes which are inducible by phosphate starvation has been shown in *Escherichia coli*. These genes were designated *psi* (phosphate starvation inducible) genes. Some were identified to be under the control of the same set of regulatory genes as *phoA*, but others were shown to be under different genetic control (29, 31). Therefore, the system should be called the Psi stimulon, since the genes, including the *psi* genes, are controlled by the same physiological signal but by different genetic systems. The phosphate (*pho*) regulon may be regarded as a subset of the Psi stimulon.

A number of genes, including *phoA, pstS (phoS), phoE*, and *ugpB*, which code for alkaline phosphatase, phosphate-binding protein, outer membrane porin e, and *sn*-glycerol 3-phosphate-binding protein, respectively, are related to transport and assimilation of phosphate and are under the same physiological and genetic control. They constitute a single *pho* regulon (26, 27). With limited phosphate, they are positively regulated by the products of *phoB* and *phoR*, and with excess phosphate, they are negatively regulated by the product of *phoR*. The product of *phoM* can substitute for a positive regulatory function of PhoR in a *phoR*-defective strain (30). In addition, these genes are negatively regulated by the genes constituting the Pst operon such as *pstS(phoS), pstC(phoW), pstA(phoT), pstB*, and *phoU* (2, 23).

The regulatory gene *phoB* is genetically and physiologically regulated in very much the same way as the genes in the *pho* regulon (9, 19). Since *phoB* and *phoR* constitute an operon, *phoR* itself is regulated coordinately with *phoB* (12, 13, 16; B. L. Wanner, personal communication). Therefore, these regulatory genes are regulated both positively and negatively by their gene products.

In this article, we describe our recent studies on the regulatory mechanism of the *pho* regulon and present our model for regulation based on current understanding of the regulon.

Structure of the *phoB-phoR* Operon

Structure of the *phoB* gene. The nucleotide sequence of the 3-kilobase chromosomal DNA containing the *phoB* and *phoR* genes has been determined (12, 13). In the regulatory region of the operon, we found a consensus nucleotide sequence shared by the regulatory regions of the *phoA, pstS*, and *phoE* genes, which we proposed to name the phosphate (*pho*) box (Fig. 1). Since all of these genes are positively regulated by the *phoB* gene product, this suggested that transcription of the *phoB-phoR* operon is also regulated positively by its own products, in agreement with the finding of Guan et al. (9).

The open reading frame corresponding to the *phoB* gene can code for a protein with 229 amino acid residues. The amino-terminal sequence of the four amino acids of the purified PhoB'-'LacZ hybrid protein (22) was compared with that deduced from the DNA sequence, and it agreed with the sequence following the Met residue. Therefore, we concluded that the PhoB protein consists of 228 amino acid residues with an M_r of 26,161. This value is slightly smaller than those determined by the minicell (30,000) and maxicell methods (31,000) (14, 24).

Immediately distal to the coding region of *phoB*, a sequence whose transcript can form a stem-and-loop structure was found. This sequence may serve as a transcriptional attenuator since expression of *phoB* is more efficient than that of *phoR* (16, 24). No promoterlike structure has been found in the intercistronic region between *phoB* and *phoR*, which is in agreement with the previous finding that the transcription of the *phoR* gene is initiated at the promoter of the *phoB* gene (16).

The PhoB protein is homologous to many regulatory proteins of various bacteria. Extensive homology was found in the amino acid sequences of the PhoB protein and the gene products of several bacterial species which are involved in the regulation of gene expression or signal transduction. Among them, the PhoB protein shows very high homology (ca. 40%) with the products of the *ompR* and *dye* genes of *E. coli* (4, 6) and the *virG* gene of *Agrobacterium tumefaciens* (S. C. Winans, personal communication), all of which are considered to be transcriptional activators of the groups of functionally related genes. An open reading frame (ORF2) found upstream of the *phoM* gene also belongs to this group (3). The products of *spo0A*

```
phoA : CAGTAAAAAGTTAATCTTTTCAACAGCTGTCATAAAGTTGTCACGGCCGAGACTTATAGTCGCTTTGTTTTTATTTTTTAA-
                                                               (-10)          -mRNA
       TGTATTTGTACATGGAGAAAATAAAGTG
                    (RB)        Met

phoB : ATAACCTGAAGATATGTGCGACGAGCTTTTCATAAATCTGTCATAAATCTGACGCATAATGACGTCGCATTAATGATCGCA-
                                                               (-10)          -mRNA
       ACCTATTTATTACAACAGGGCAAATCATG
                       (RB)      Met

phoE : TACCACATTTTAAGAATATTATTAATCTGTAATATATCTTTAACAATCTCAGGTTAAAAACTTTCCTGTTTTCAACGGGAC-
                                                               (-10)
       TCTCCCGCTGAATATTCGCGCGTTAATTAAAATCAGGAATGAAAATG
                                         (RB)     Met

pstS : CTCTCTGTCATAAAACTGTCATATTCCTTACATATAACTGTCACCTGTTTGTCCTATTTTGCTTCTCGTAGCCAACAAACA-
                                                               (-10)           -mRNA
       ATGCTTTATGAATCCTCCCAGGAGACATTATG
                           (RB)     Met

Consensus sequence       G    A A      C
(the pho box)      : CT TCATA A CTGTCA    (TGACANATATNTGCA, CTGTCATAAANCTGTCATAAA)
                         T    T T      T
```

FIG. 1. Comparison of the regulatory regions of *phoA*, *phoB*, *phoE*, and *pstS*. The transcriptional initiation sites of these genes which are known are shown by boldface letters. Sequences characteristic for ribosome binding sites (RB) and for Pribnow box (−10) are also shown. A consensus sequence for the promoter regions of the *pho* regulon (the *pho* box) is shown at the bottom. The alternative sequences for the *pho* box are shown in parentheses. (Adapted from reference 12.)

(8) and *spo0F* (28), both of which are required for initiation of sporulation in *Bacillus subtilis*, of *ntrC*, which is a transcriptional regulator for the genes involved in nitrogen metabolism of *Klebsiella pneumoniae* (5), and of *cheB* and *cheY* of *Salmonella typhimurium*, which are involved in signal transduction in chemotaxis (21), are homologous only to the amino-terminal half of the PhoB protein.

The PhoB-like protein family may consist of two functional domains, since the total size of the Spo0F and CheY proteins corresponds to the amino-terminal halves of the PhoB-like proteins and the NtrC, Spo0A, and CheB proteins are homologous to only the amino-terminal halves. It has been shown that removal of the amino-terminal half of the CheB protein results in a much increased methylesterase activity of the protein, indicating that the amino-terminal half of CheB is a regulatory domain that affects the catalytic activity in the carboxy-terminal domain responding to the chemotactic signal (20). A putative DNA-binding sequence has been suggested in the carboxy-terminal portion of the NtrC protein (5). Since all of the PhoB-like proteins are likely to be transcriptional activators which modulate gene expression responding to physiological changes, the amino-terminal halves of them may contain a domain for signal reception, and the carboxy-terminal halves may contain a domain which directly interacts with DNA for activation of transcription.

Structure of the *phoR* gene. A translational open reading frame was identified about 60 base pairs downstream of the termination codon of the *phoB* gene, and it can code for a protein consisting of 431 amino acid residues with an M_r of 49,666 (13). This value agrees well with the M_r of the PhoR protein determined by the minicell (25) and maxicell methods (unpublished data). Since the PhoR protein deduced from the nucelotide sequence of the gene revealed a long stretch of hydrophobic residues (ca. 60 residues) in the amino-terminal region, it may be a membrane protein.

A possible stem-and-loop structure of the transcript, followed by several Us characteristic of Rho-independent terminators, was found distal to the *phoR* coding region. Therefore, the *phoB* and *phoR* genes are likely to constitute an operon.

Homology of the PhoR protein with other regulatory proteins of bacteria. Amino acid sequences of regulatory proteins of several bacteria were found to share substantial homology with that of the PhoR protein (13). They include the products of *envZ* of *E. coli* involved in osmoregulation of outer membrane proteins (4), of *virA* of *A. tumefaciens* Ti plasmid involved in the regulation of genes required for the transfer of the plasmid (Winans, personal communication), and of *ntrB* of *K. pneumoniae* involved in the regulation of the genes for nitrogen assimilation (11). Except for NtrB, they contain a long stretch of hydrophobic amino acids in the amino-terminal region. Highest homology was detected near the carboxy-terminal regions of these proteins. The PhoR and EnvZ proteins were detected in the membrane fraction of the cell (unpublished data; D. E. Comeau and M. Inouye, personal communication). Therefore, PhoR, EnvZ, and VirA are likely to be membrane proteins that may receive physiological signals and transfer the information to transcriptional activators such as PhoB, OmpR, and VirG by modifying their function, rather than transcriptional regulators that directly interact with DNA. Homology to the carboxy terminus of PhoR was also found in the Cpx (1), NifL (M. J. Merrick et al., this volume), and CheA proteins (17).

The *pho* box as a consensus sequence for the promoters of the *pho* regulon. The genes in the *pho* regulon, including *phoA, pstS, phoE*, and *phoB*, have been shown to be regulated physiologically by phosphate in the medium and genetically by *phoB* and several regulatory genes such as *phoR, phoM*, and the genes in the Pst operon. Therefore, they are likely to share a common regulatory element in the promoter region. In agreement with this idea, a consensus sequence, which we named the *pho* box, was found in the promoter regions of all of them as shown in Fig. 1 (12). A well-conserved 18-nucleotide sequence, CTG/TTCATAA/TAA/TCTGTCAC/T, was found 11 nucleotides upstream of the putative Pribnow boxes in all cases. The *pho* box is duplicated in tandem in the regulatory region of *pstS* as shown in Fig. 1. Since active forms of many regulatory proteins are dimeric, and their target sequence often has a twofold symmetry (18), the consensus sequence of the *pho* box may also have twofold symmetry or may be resolved into a direct repeat depending on the choice of bases at a few nonconserved sites as shown in Fig. 1.

Function of the PhoB Protein

To study the function of the PhoB protein, we purified the protein and examined its biochemical properties in vitro (details to be published elsewhere). A PhoB′-′LacZ fusion protein was purified from the cells carrying the fusion gene by monitoring β-galactosidase activity of the fusion protein, and an antiserum was prepared by immunizing rabbits with the fusion protein. The serum thus prepared showed specificity for both the PhoB protein and β-galactosidase (22). The PhoB protein was purified from overproducing cells by immunologically monitoring the protein fractions during the purification steps.

We obtained the PhoB protein with approximately 90% purity. The M_r as determined by sodium dodecyl sulfate-polyacrylamide gel electrophoresis was about 29,000, which agrees well with the values previously determined by the maxicell (14) and minicell (24) methods. It binds specifically to the DNA fragments containing the promoter regions of the *phoA, pstS*, and *phoB* genes. Footprinting experiments with the PhoB protein suggest that the protected regions cover the *pho* box. It activates the in vitro transcription of these genes; the initiation of transcription occurred at positions that reflect the transcription in vivo. Therefore, it is highly likely that the PhoB protein recognizes and binds to the *pho* box of the genes in the *pho* regulon and activates the transcription of these genes in vivo.

Function of the PhoR Protein

The *phoR* gene has been considered to serve dual regulatory functions for the *pho* regulon, as a repressor with excessive phosphate and as an activator with limited phosphate (7, 16, 30). This hypothesis was based on the isolation of two kinds of *phoR* mutants (7). In one class of *phoR* mutants, represented by *phoR68* and *phoR17*, alkaline phosphatase is expressed constitutively but at a low level. These *phoR* genes were considered to be defective both as an activator and as a repressor. Another class of *phoR* mutants, exemplified by *phoR69*, synthesizes alkaline phosphatase constitutively but at a high level. This mutant gene product was considered to function as an activator but to be defective as a repressor. The *phoM-phoR* double mutants with the *phoR68* mutation did not synthesize alkaline phosphatase while those with the *phoR69* mutation synthesized the enzyme (15, 27, 30). Therefore, these results suggest that the positive regulatory function in the former class of *phoR* mutants is provided by the *phoM* gene, while the latter class of the *phoR* mutants retain the positive regulatory function of the gene.

How does the gene product of *phoR* act as an activator with limited phosphate, and as a repressor with excess phosphate, for the *pho* regulon? Two models can be proposed for the function of the PhoR protein. First, a complex of PhoB-PhoR with an active form of PhoR may activate transcription of the *phoB-phoR* operon and the *pho* regulon. The repressor form of PhoR or its complex with PhoB may repress the transcription. Second, PhoR during phosphate limitation or PhoM may modify the function of the PhoB protein so that the PhoB protein functions as a transcriptional activator. In the first model the PhoB, PhoR, and PhoM proteins are all soluble proteins that directly interact with the *pho* box to facilitate specific transcription. In the second model, PhoR and PhoM can be informational mediators in the membrane that modify the function of the PhoB protein by responding to physiological signals. The activated PhoB protein itself should activate transcription of the regulon without participation of either the PhoR or PhoM protein. Our recent results favor the second model. The structures of the PhoR and PhoM proteins deduced from the nucleotide sequences indicate that these proteins are membrane proteins (3, 13). The purified PhoB protein in the absence of the PhoR protein activates transcription of the *pho* genes in vitro (manuscript in preparation).

Structure and Function of the *phoM* Gene

The *phoM* gene was identified by isolating mutants which could not synthesize alkaline phosphatase in a *phoR* mutant (30). The function of this gene can substitute for the positive regulatory function of the *phoR* gene for the *pho* regulon. The dual function of the *phoR* gene dominates over the positive function of the *phoM* gene because *phoM* mutants do not affect the levels of alkaline phosphatase in *phoR*$^+$ strains and the phenotype of *phoM* becomes apparent only in *phoR*-defective strains.

The *phoM* gene has been cloned and the gene product has been identified as a protein with an M_r of 55,000 to 60,000 (10, 15, 25). To study the regulation of *phoM, phoM'-'lacZ* fusion genes were constructed (10, 15). The fusion gene we constructed was recently found to be lacking the promoter by sequencing the *phoM* region (3). The results obtained by Ludtke et al. (10) and by us, using a reconstructed *phoM'-'lacz* fusion gene, suggest that the *phoM* expression is not regulated by phosphate.

The experiments with the multicopy plasmid carrying the *phoM* gene suggest that not only the intact PhoR protein dominates over the PhoM function but also the mutant PhoR proteins interfere with the PhoM function to various degrees, depending on the kinds of *phoR* mutations (15). Introduction of the multicopy plasmid carrying the *phoM* gene into *phoR* mutants elevated *phoA* expression to different degrees characteristic of the kinds of *phoR* mutants, but it did not affect *phoA* expression in *phoR*$^+$ strains or in any other *pho* regulatory mutants.

The results obtained by using the fusion gene and multicopy *phoM*$^+$ gene are consistent with the model that both positive and negative regulatory functions of the PhoR protein dominate over the positive function of the PhoM protein, and the PhoM protein forms an oligomer in the cell (15).

We recently completed sequencing the *phoM* gene and its flanking regions (3). The 4.7-kilobase chromosomal segment we sequenced contains four open reading frames (ORFs) in the order of (*trp*)-ORF1-ORF2-ORF3(*phoM*)-ORF4-*dye*-(*thr*) clockwise on the standard *E. coli* genetic map. Since these ORFs are preceded by a putative promoter sequence upstream of ORF1 and are followed by a putative terminator distal to ORF4, they seem to constitute an operon.

The M_r and the primary structure of the putative ORF2 protein are highly homologous (ca. 40%) to those of several regulatory proteins such as the PhoB, Dye, and OmpR proteins. Therefore, it may also be a transcriptional activator. The ORF3 corresponding to the *phoM* gene can code for a protein with an M_r of 52,116. The PhoM protein contains two stretches of highly hydrophobic residues in the amino-terminal and central regions and, therefore, is likely to be a membrane protein. Ludtke et al. (10) identified the proteins encoded by plasmid pM142, which carries ORF1 to ORF4, by the maxicell method. The four proteins with M_rs of 17,000, 28,000, 55,000, and 49,000 identified by them may correspond to the products of ORF1, ORF2, *phoM*, and ORF4, respectively.

Since the *phoM* gene can substitute the positive regulatory function of the *phoR* gene, homology between the PhoM and PhoR proteins was examined. Highly homologous sequences were detected only in the carboxy-terminal halves of the proteins (ca. 30% identity). Although no significant homology was found in their amino-terminal halves, they contain a long stretch of hydrophobic residues in the regions, indicating that they may be membrane proteins (C. Ronson, personal communication).

We examined whether any of the ORFs other than *phoM* is involved in the regulation of *phoA* expression. A plasmid carrying only the *phoM* gene was introduced into a *phoR* strain that has a deletion of the entire ORF1-ORF4 region (kindly provided by B. L. Wanner), and the resulting strain became PhoA$^+$. Therefore, none of the ORFs other than *phoM* seems to be required for the positive regulation of the *pho* regulon. Although the *phoM* gene carried on a multicopy plasmid did not affect the regulation of *phoA* (15), a PhoM hyperproducing cell was constitutive for *phoA* expression (unpublished data). Therefore, it is possible that this operon is inducible by some unidentified physiological condition, and when this operon is induced the product(s) of this operon may activate the *pho* regulon in the wild-type strain.

Concluding Remarks

The *pho* regulon is regulated by multiple regulatory genes in a very complex manner. These genes seem to constitute a cascade regulatory network (9, 16, 19). The work presented here may be able to eliminate or modify the previous models for the regulon (9, 16, 24, 27, 30). Incorporating these new findings, we propose the following model, illustrated in Fig. 2.

Since the purified PhoB protein binds to the *pho* box and activates the transcription of the genes in the *pho* regulon in vitro (in preparation), it would be a transcriptional activator for the *pho* regulon as well as for the *phoB-phoR* operon. Since the PhoR and PhoM proteins are likely to be membrane proteins (3, 15), it is difficult to think that they act as transcriptional regulator for *phoB* or that they form a complex with PhoB to function as transcriptional regulator for the *pho* regulon. Since the *phoB-phoR*

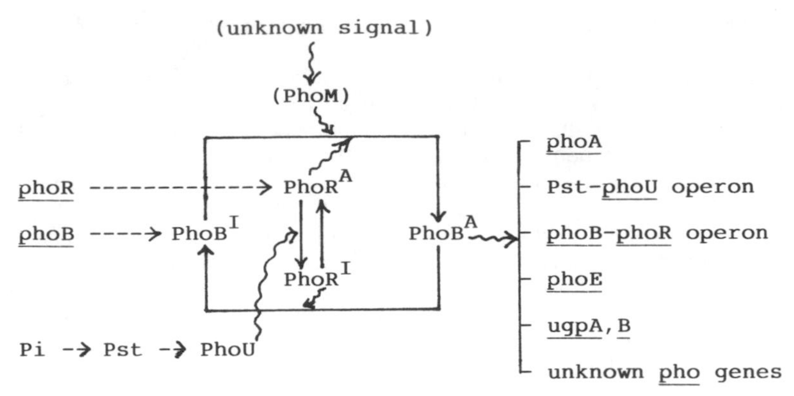

FIG. 2. Model for the regulatory network of the *pho* regulon. Transcription of the *pho* genes is activated by the active form of the *phoB* gene product ($PhoB^A$). The PhoB protein is synthesized in the inactive form ($PhoB^I$) and is activated by the function of the activator form of the *phoR* gene product ($PhoR^A$). The *phoR* gene product is synthesized in the activator form and is converted into the inactive form ($PhoR^I$) by the function of PhoU that has received the phosphate signal via the Pst pathway. $PhoR^I$ converts $PhoB^A$ into $PhoB^I$. The signal transmission may be achieved by synthesizing an effector molecule(s) or by sequential covalent modification of the proteins involved, or by both mechanisms. The levels of external phosphate are monitored by the phosphate-binding protein (PstS) in the periplasm and the signal is transmitted through the membrane by the products of *pstC, pstA*, and *pstB* (2; Nakata et al., this volume). In the absence of $PhoR^A$ and $PhoR^I$ PhoM converts $PhoB^I$ into $PhoB^A$ irrespective of external phosphate levels. It is possible that the *phoM* operon is inducible by some unknown physiological condition and the induced level of the product induces the *pho* regulon and some unknown regulon. The genes in the *pst-phoU* region seem to constitute an operon (Nakata et al., this volume). The existence of unknown *pho* genes has been shown by Wanner and McSharry (31).

operon is repressed with excess phosphate (9, 16, 19), the induction by phosphate limitation should involve activation of a small amount of the inactive PhoB protein by the PhoR protein. This leads to the accumulation of a large amount of the active PhoB protein in the cell, which may be required for full induction of the *pho* regulon (9, 19). Thc PhoB protein may shuttle between the PhoR protein in the membrane where it is modified and the *pho* boxes where it activates transcription.

The PhoR protein may function as an informational mediator that receives a phosphate signal transmitted by the function of the products of the Pst operon, including *phoU* (2, 23; Nakata, Amemura, Makino, and Shinagawa, this volume), and transmit the signal to the PhoB protein either by modifying the protein or by producing a signal molecule that interacts with the PhoB protein. Since the signal is produced by excess phosphate in the medium, a defect in any genes in the Pst operon or in *phoR* will fix the PhoB protein in the active state ($PhoB^A$), and as a consequence, the cell will become constitutive for *phoA* expression. The activation of PhoB can be achieved either by the activator form of the PhoR protein ($PhoR^A$), which is caused by the absence of a phosphate signal, or by the PhoM protein in the absence of the functional PhoR protein, as in *phoR68* and *phoR17* mutants (15, 16, 19, 27, 30). The PhoB protein as it is synthesized may be in the inactive form ($PhoB^I$) since *phoA* is not expressed in *phoR68-phoM* double mutants (15, 19, 27, 30). The active form of the PhoB protein is converted to the inactive form by the function of the inactivator form of the PhoR protein ($PhoR^I$) that has received the phosphate signal transmitted through the function of the Pst operon (2; Nakata et al., this volume). The nature of the phosphate signal(s) is not known at present.

Since we have recently constructed strains which greatly overproduce the PhoB, PhoR, and PhoM proteins and the products of the Pst operon, studies on their biochemical functions as well as their biological functions will be much facilitated. The accumulated data about the regulon enabled us to refine the models for the regulation. The discovery of many pairwise proteins homologous to PhoB/PhoR, such as OmpR/EnvZ, VirG/VirA, NtrC/NtrB, and CheB, CheY/CheA indicates the existence of a common mechanism for the global regulatory systems, and the *pho* system will contribute much to the understanding of the mechanisms by which bacterial cells adapt to environmental changes.

We thank B. L. Wanner for supplying bacterial strains and communicating unpublished results, and A. Ishihama and H. Aiba for technical advice.

This work was supported by a Grant-in-Aid for Scientific Research from the Ministry of Education, Science, and Culture of Japan, by a grant from the Mochida Memorial Foundation for Medical and Pharmaceutical Research, and by a travel grant to H.S. from the Yamada Foundation for the Promotion of Science.

LITERATURE CITED

1. **Albin, R., R. Weber, and P. M. Silverman.** 1986. The Cpx protein of *Escherichia coli* K12. Immunological detection of the chromosomal *cpxA* gene product. J. Biol. Chem. **261:**4698–4705.
2. **Amemura, M., K. Makino, H. Shinagawa, A. Kobayashi, and A. Nakata.** 1985. Nucleotide sequence of the genes involved in phosphate transport and regulation of the phosphate regulon in *Escherichia coli*. J. Mol. Biol. **184:** 241–250.
3. **Amemura, M., K. Makino, H. Shinagawa, and A. Nakata.** 1986. Nucleotide sequence of the *phoM* region of *Escherichia coli*: four open reading frames that may constitute an operon. J. Bacteriol. **168:**294–302.
4. **Comeau, D. E., K. Ikenaka, K. Tsung, and M. Inouye.** 1985. Primary characterization of the protein products of the *Escherichia coli ompB* locus: structure and regulation of synthesis of the OmpR and EnvZ proteins. J. Bacteriol. **164:**578–584.
5. **Drummond, M., P. Whitty, and J. Wootton.** 1986. Sequence and domain relationships of *ntrC* and *nifA* from *Klebsiella pneumoniae*: homologies to other regulatory proteins. EMBO J. **5:**441–447.
6. **Drury, L. S., and R. S. Buxton.** 1985. DNA sequence analysis of the *dye* gene of *Escherichia coli* reveals amino acid homology between the Dye and OmpR proteins. J. Biol. Chem. **260:**4236–4242.
7. **Echols, H., A. Garen, S. Garen, and A. Torriani.** 1961. Genetic control of repression of alkaline phosphatase in *E. coli*. J. Mol. Biol. **3:**425–438.
8. **Ferrari, F. A., K. Trach, D. LeCoq, J. Spence, E. Ferrari, and J. A. Hoch.** 1985. Characterization of the *spo0A* locus and its deduced product. Proc. Natl. Acad. Sci. USA **82:** 2647–2651.
9. **Guan, C.-D., B. Wanner, and H. Inouye.** 1983. Analysis of regulation of *phoB* expression using a *phoB-cat* fusion. J. Bacteriol. **156:**710–717.
10. **Ludtke, D., J. Bernstein, C. Hamilton, and A. Torriani.** 1984. Identification of the *phoM* gene product and its regulation in *Escherichia coli* K-12. J. Bacteriol. **159:**19–25.
11. **McFarlane, S. A., and M. Merrick.** 1985. The nucleotide sequence of the nitrogen regulation gene *ntrB* and the *glnA-ntrBC* intergenic region of *Klebsiella pneumoniae*. Nucleic Acids Res. **13:**7591–7606.
12. **Makino, K., H. Shinagawa, M. Amemura, and A. Nakata.** 1986. Nucleotide sequence of the *phoB* gene, the positive regulatory gene for the phosphate regulon of *Escherichia coli* K-12. J. Mol. Biol. **190:**37–44.
13. **Makino, K., H. Shinagawa, M. Amemura, and A. Nakata.** 1986. Nucleotide sequence of the *phoR* gene, a regulatory gene for the phosphate regulon of *Escherichia coli*. J. Mol. Biol. **192:**549–556.
14. **Makino, K., H. Shinagawa, and A. Nakata.** 1982. Cloning and characterization of the alkaline phosphatase positive regulator gene (*phoB*) of *Escherichia coli*. Mol. Gen. Genet. **187:**181–186.
15. **Makino, K., H. Shinagawa, and A. Nakata.** 1984. Cloning and characterization of the alkaline phosphatase positive regulatory gene (*phoM*) of *Escherichia coli*. Mol. Gen. Genet. **195:**381–390.
16. **Makino, K., H. Shinagawa, and A. Nakata.** 1985. Regulation of the phosphate regulon of *Escherichia coli* K-12. Regulation and role of the regulatory gene *phoR*. J. Mol. Biol. **184:**231–240.
17. **Mutoh, N., and M. I. Simon.** 1986. Nucleotide sequence corresponding to five chemotaxis genes in *Escherichia coli*. J. Bacteriol. **165:**161–166.
18. **Pabo, C. O., and R. T. Sauer.** 1984. Protein-DNA recognition. Annu. Rev. Biochem. **53:**293–321.
19. **Shinagawa, H., K. Makino, and A. Nakata.** 1983. Regulation of the *pho* regulon in *Escherichia coli* K 12. Genetic and physiological regulation of the positive regulatory gene *phoB*. J. Mol. Biol. **168:**477–488.
20. **Simms, S. A., M. G. Keane, and J. Stock.** 1985. Multiple forms of the CheB methylesterase in bacterial chemosensing. J. Biol. Chem. **260:**10161–10168.
21. **Stock, A., D. E. Koshland, Jr., and J. Stock.** 1985. Homologies between the *Salmonella typhimurium* CheY protein and proteins involved in the regulation of chemotaxis, membrane protein synthesis and sporulation. Proc. Natl. Acad. Sci. USA **82:**7989–7993.
22. **Sugita, T., H. Shinagawa, K. Makino, and A. Nakata.** 1985. Use of a *phoB'-'lacZ* fusion gene to determine the N-terminal amino acid sequence of the PhoB protein and to prepare antiserum against the protein. J. Biochem. **97:** 1247–1250.
23. **Surin, B. P., H. Rosenberg, and G. B. Cox.** 1985. Phosphate-specific transport system of *Escherichia coli*: nucleotide sequence and gene-polypeptide relationships. J. Bacteriol. **161:**189–198.
24. **Tommassen, J., P. De Geus, B. Lugtenberg, J. Hackett, and P. Reeves.** 1982. Regulation of the *pho* regulon of *Escherichia coli* K-12. Cloning of the regulatory genes *phoB* and *phoR* and identification of their gene products. J. Mol. Biol. **157:**265–274.
25. **Tommassen, J., P. Hiemstra, P. Overduin, and B. Lugtenberg.** 1984. Cloning of *phoM*, a gene involved in regulation of the synthesis of phosphate limitation inducible proteins in *Escherichia coli* K12. Mol. Gen. Genet. **195:**190–194.
26. **Tommassen, J., and B. Lugtenberg.** 1982. *pho*-regulon of *Escherichia coli* K12: a minireview. Ann. Microbiol. (Paris) **133A:**243–249.
27. **Torriani, A., and D. N. Ludtke.** 1985. The *pho* regulon of *Escherichia coli*, p. 224–242. *In* M. Schaechter, F. C. Neidhart, J. Ingraham, and N. O. Kjeldgaard (ed.), The molecular biology of bacterial growth. Jones and Bartlett Publishers, Boston.
28. **Trach, K. A., J. W. Chapman, P. J. Piggot, and J. A. Hoch.** 1985. Deduced product of the stage 0 sporulation gene *spo0F* shares homology with the Sop0A, OmpR, and SfrA proteins. Proc. Natl. Acad. Sci. USA **82:**7260–7264.
29. **Wanner, B. L.** 1983. Overlapping and separate controls on the phosphate regulon in *Escherichia coli* K12. J. Mol. Biol. **166:**283–308.
30. **Wanner, B. L., and P. Latterell.** 1980. Mutants affected in alkaline phosphatase expression: evidence for multiple positive regulators of the phosphate regulon in *Escherichia coli*. Genetics **96:**353–366.
31. **Wanner, B. L., and R. McSharry.** 1982. Phosphate-controlled gene expression in *Escherichia coli* K12 using Mu*d1*-directed *lacZ* fusions. J. Mol. Biol. **158:**347–363.

Expression of Outer Membrane PhoE Protein in *Escherichia coli* K-12

JAN TOMMASSEN

Department of Molecular Cell Biology and Institute for Molecular Biology, State University of Utrecht, 3584 CH Utrecht, The Netherlands

The outer membrane of gram-negative bacteria functions as a permeation barrier for harmful compounds. The membrane contains a number of proteins which form general diffusion pores to allow the passage of small hydrophilic solutes with molecular weights of up to about 600 (14). Under standard laboratory conditions, *Escherichia coli* K-12 produces two distinct pore proteins, specified by the *ompF* and *ompC* genes. Another pore protein, designated PhoE protein (29), was first discovered in pseudorevertants of mutants lacking both the OmpF and OmpC proteins (4, 7, 32). It was subsequently demonstrated that the synthesis of this protein can be induced in wild-type strains by growth under P_i starvation (16), suggesting that the protein is somehow particularly involved in the uptake of P_i or phosphate-containing compounds. Indeed, by measuring the rate of nutrient uptake in cells producing only one pore protein species, it has been shown that PhoE protein forms more efficient pores than OmpF and OmpC for these compounds (9) and for negatively charged compounds in general (8, 13, 17). The preference of the PhoE protein pores for anions, in contrast to the cation-preferring OmpF and OmpC pores, has also been demonstrated in vitro after reconstitution of the protein pores in liposomes (12) and lipid bilayer membranes (1, 3).

Studies on the structure-function relationship of PhoE protein are described by van der Ley and Tommassen (this volume). In this paper, I describe studies on the expression of PhoE protein, including regulation of the synthesis of the protein and the transport to its correct cellular compartment, the outer membrane.

Regulation of the Synthesis of PhoE Protein

The synthesis of PhoE protein is coregulated with the synthesis of several other P_i starvation inducible proteins in a single regulon, designated the *pho* regulon (28). The *phoE* structural gene was localized at min 6 of the chromosomal map (29), well separated from other structural genes of the regulon, including *phoA* (alkaline phosphatase), *phoS* (P_i-binding protein), and *ugpB* (glycerol 3-phosphate-binding protein). The products of at least three regulatory genes, *phoB, phoR*, and *phoM* (33), and a functional phosphate-specific transport system for the uptake of P_i are involved in this regulation. On the basis of the phenotypes of regulatory gene mutants, we have proposed a cascade model for this regulation system (30). According to this model (Fig. 1), the *phoB* gene product is an activator required for transcription of the structural genes, whereas transcription of *phoB* is regulated by the *phoR* and *phoM* gene products. Evidence for this model was obtained after cloning of the regulatory genes and identification of their products. It was shown that the *phoR* gene product represses the synthesis of the *phoB* gene product in minicells (26). Furthermore, by using *phoB-lacZ* (22) or *phoB-cat* (5) gene fusions, it was shown that the synthesis of the *phoB* product is induced under P_i limitation as predicted by this model (Fig. 1). It should be noted that P_i is probably not the effector molecule that influences the state of the *phoR* gene product. In mutants with a defective phosphate-specific transport system, a high intracellular P_i level can be maintained by another uptake system, the phosphate inorganic transport system (20). Still, these mutants are derepressed for the *pho* regulon. It was recently shown that the intracellular nucleotide pools are changed upon P_i limitation and that the *phoU* gene is involved in these alterations (18). It seems likely that these nucleotides are the actual effectors of the mode of action of the *phoR* gene product.

Analysis of the *phoE* Promoter Region

According to the model presented in Fig. 1, the *phoB* gene product will interact with the promoter regions of all structural genes of the *pho* regulon. Therefore, these genes may have a common feature in their promoter regions with which the *phoB* gene product interacts. Comparison of the nucleotide sequences of *phoS, phoA*, and *phoE* led to the identification of a 17-base-pair (bp) fragment in the corresponding promoter regions with a high degree of homology (10, 24) (Fig. 2). This fragment, designated hereafter as a "pho box", was followed in all these genes by a potential Pribnow box at a distance of 11 bp and was considered to consist of an anomalous −35 region (involved in the binding of the σ factor of RNA polymerase holoen-

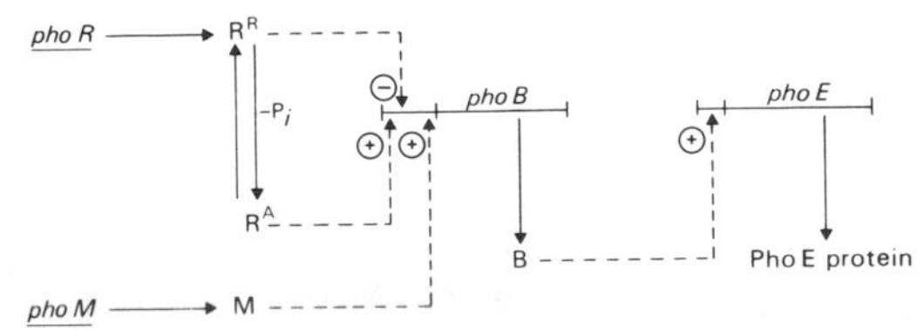

FIG. 1. Model for *pho* regulation as proposed previously (30). B, R, and M are the products of the genes *phoB*, *phoR*, and *phoM*, respectively. B is an activator which is essential for transcription of *phoE* and the other *pho* structural genes. In the presence of sufficient P_i, R is in a repressor form, R^R, which prevents transcription of *phoB*. Under P_i limitation, R is in an activator form, R^A. Both R^A and M stimulate transcription of *phoB*.

zyme), flanked on either side by a palindromic pentanucleotide, CTGTC (24). Potential Pribnow boxes were also identified 33 to 35 bp downstream of the pho boxes (10).

We have sequenced the promoter regions of two other *pho*-regulated genes. The *ugp* locus consists of at least four phosphate limitation-inducible genes and has been cloned by Schweizer and Boos (21). The products of these genes are involved in the uptake of glycerol 3-phosphate. The promoter region preceding the *ugpD* gene contained several fragments with some homology to the pho box (manuscript in preparation). The promoter, identified by S1 nuclease mapping, is shown in Fig. 2. At distances of 11 and 34 bp upstream of this promoter, regions

```
           CTGTC -35 CTGTC            PB1      PB2
phoS       CTGTCATAAACTGTCA-11bp-TATAAC  TATTTT
phoA       CTGTCATAAAGTTGTCA-11bp-TATAGT  TAATGT
phoE       CTGTAATATATCTTTAA-11bp-TAAAAA  GACTCT
                          |__________33-35bp__|
ugpD       AAGTTATTTTTCTGTAA-11bp-CATGTT
           CTATCTTACAAATGTAA-33-35bp-CATGTT
phoE
K. pn      ATGTCATAAATATTTAA-11bp-TAAAAA
```

FIG. 2. Homology in the promoter regions of *phoS*, *phoA*, and *phoE*, according to Surin et al. (24) and Magota et al. (10). The homologous regions are followed by putative Pribnow boxes at distances of 11 bp (PB1) and 33 to 35 bp (PB2). The pentanucleotides on either side of the putative −35 region (24) are boxed. The promoter CATGTT of the *ugp* gene was determined by S1 nuclease mapping (manuscript in preparation), and the regions located 11 and 34 bp upstream of this promoter are aligned with the pho box. A potential pho box in the *K. pneumoniae* (K. pn) *phoE* gene (P. van der Ley et al., in preparation) is also shown.

with some, although weak, homology to the pho box are found.

The *phoE* gene of *Klebsiella pneumoniae* was cloned by an in vivo cloning procedure and appears to be normally expressed in *E. coli* K-12 as a part of the *pho* regulon (C. Verhoef et al., manuscript in preparation). In Fig. 3, the promoter region of this gene is compared with the promoter region of the *E. coli* K-12 *phoE* gene. The pho box appears to be rather well conserved although it should be noted that the pentanucleotide ATTTA downstream of the putative anomalous −35 region has only two bases homologous to the consensus CTGTC sequence. Also the Pribnow box 11 bp downstream of the pho box (indicated as PB1 in Fig. 3) is conserved in the *K. pneumoniae phoE* gene, in contrast to the Pribnow box 35 bp downstream (PB2).

To determine whether the sequence TAAAAA (PB1 in Fig. 3), located 11 bp downstream of the pho box in the *phoE* gene of *E. coli* K-12, actually functions as the −10 region of the *phoE* promoter, S1 nuclease mapping was performed. Two strong signals, flanked by several weak signals, were observed, as indicated in Fig. 3. It is not clear whether all the indicated bases actually serve as start points of transcription or whether there is only one initiation point and the other signals arise by degradation of the mRNA. In this respect, it is interesting to note that only the first of the indicated bases is a purine, which usually serves as start point of transcription (19). In any case, since the transcription start point is usually located 5 to 9 bp downstream of the Pribnow box (19), it appears unlikely that the sequence indicated as PB1 in Fig. 3 is the actual −10 region of the *phoE* promoter. Rather, the sequence CAGGTT, indicated as PB3 in Fig. 3, probably functions as the −10 region.

To identify additional sequences within the *phoE* promoter region that are important for the expression of the gene, a deletion analysis was performed. Starting from the *Eco*RV site, indicated in Fig. 3, deletions were created by limited digestion with exonuclease BAL 31. Eighteen deletion mutants were studied in more or less detail. In six mutants, the expression of PhoE protein was not influenced. This phenotype could be correlated by restriction enzyme analysis to the presence of the *Sau*3A site indicated in Fig. 3. In the other mutants, the expression of PhoE protein under P_i limitation was either drastically reduced (<5% of the wild-type level) or, in four cases, totally abolished. In all these mutants, the indicated *Sau*3A site was removed, suggesting that a region in the DNA close to this *Sau*3A site is important for the expression of PhoE protein. To define this region more accurately, the nucleotide sequences of eight mutant alleles were established. The endpoints of these

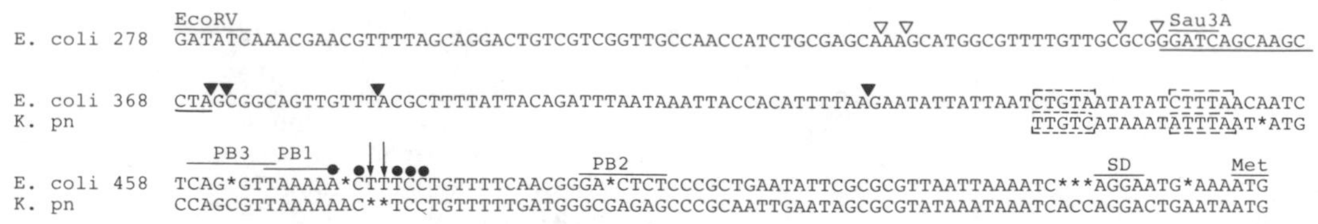

FIG. 3. Nucleotide sequences of the promoter regions of the *E. coli* K-12 *phoE* gene (18) and the *K. pneumoniae* (K. pn.) *phoE* gene (van der Ley et al., in preparation). The numbering of the nucleotides is as described by Overbeeke et al. (15). The ATG initiation codons (Met) and the ribosome binding sites (SD) are indicated. PB1 and PB2 indicate the putative Pribnow boxes as in Fig. 2. The palindromic pentanucleotides on either side of the −35 region in the pho box are boxed. The two arrows indicate the major start points of transcription in the *E. coli phoE* gene as determined by S1 nuclease mapping (J. Tommassen et al., manuscript in preparation). The bases indicated by points (●) gave additional weak signals in these experiments. PB3 indicates the most likely −10 region of the *phoE* gene as deduced from these S1 mapping data. The triangles indicate the positions of the end points of deletion mutations which were created with BAL 31 exonuclease, starting from the *Eco*RV site: (▽) mutations which did not affect expression of PhoE protein; (▼) mutations in which the expression of PhoE protein was drastically reduced.

deletions are indicated in Fig. 3. These results show that at least part of the 15-bp sequence GGATCAGCAAGCCTA (underlined in Fig. 3), located approximately 100 bp upstream of the transcription start point, is important for the expression of the PhoE protein. Whereas the pho box appears to be actually involved in the binding of the *phoB* product (H. Shinagawa, personal communication), the function of this upstream region remains to be elucidated.

Intragenic Export Information in PhoE Protein

Expression of PhoE protein requires export of the protein to the outer membrane. Obviously, a signal sequence which is present in PhoE protein and in other outer membrane and periplasmic proteins plays an important role in this process. However, a signal sequence does not determine the ultimate localization of an exported protein since replacement of the signal sequence of PhoE protein by the signal sequence of the periplasmic enzyme β-lactamase did not affect the localization of the protein (31). In addition, a signal sequence does not seem to contain all the information required for transport through the cytoplasmic membrane, since the cytoplasmic enzyme β-galactosidase was not exported from the cytoplasm when fused to the signal sequence of outer membrane protein LamB (11). To identify additional export information in LamB protein, gene fusions have been constructed containing larger promoter-proximal portions of the *lamB* gene (6, 23). This approach appeared to be very fruitful, since cell fractionation experiments suggested that such hybrid proteins were exported to the outer membrane provided that they contained a sufficiently large N-terminal LamB portion. However, recent results obtained with a *phoE-lacZ* fusion strongly suggest that standard cell fractionation experiments are not reliable when applied to localize such hybrid proteins at the subcellular level (27). To localize intragenic export information in PhoE protein, we constructed a *phoE-lacZ* fusion encompassing 85% of the *phoE* gene. Induction of the synthesis of the hybrid protein was lethal to the cells and led to the accumulation of precursors of other exported proteins (27). This result suggested that attempts of the cells to export the hybrid protein failed, resulting in a lethal blocking of the export machinery. However, after cell disintegration, the hybrid protein cofractionated with the outer membranes. On the other hand, immunocytochemical labeling on ultrathin cryosections of the cells (27) as well as protease-accessibility experiments (J. Tommassen and T. de Kroon, manuscript in preparation) convincingly demonstrated that the hybrid protein was not exported from the cytoplasm. Apparently, cell fractionations may lead to artifacts when applied to localize hybrid proteins of this kind, probably because these proteins form aggregates which cofractionate with the outer membranes, or because they become associated with the membranes during the fractionation procedure. Although the results described above can be interpreted to mean that the C-terminal 15% of PhoE protein contains export information, I prefer the explanation that β-galactosidase cannot be exported from the cytoplasm when fused to any outer membrane protein fragment, because it contains sequences or conformations which prevent export.

If the latter interpretation is correct, the gene fusion technology cannot be employed to identify additional export information. We have therefore created a set of mutants, with deletions internal to the *phoE* gene, with the aid of exonuclease BAL 31 (2). Ten deletion mutants were analyzed in detail. The extent of the deletions was determined with the aid of restriction enzymes (Fig. 4). Together, these deletions

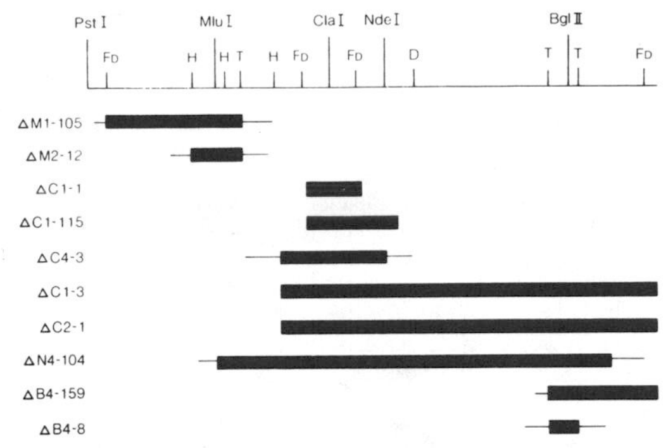

FIG. 4. Analysis of deletions internal to the *phoE* gene (2). The upper line represents the part of the *phoE* gene that corresponds to the mature PhoE protein. The deletions were created with exonuclease BAL 31, starting from the indicated *Mlu*I, *Cla*I, *Nde*I, and *Bgl*II sites. The extent of the deletions was determined with the aid of the restriction enzymes *Fok*I (F) *Hpa*II (H), *Taq*I (T), *Fnu*DII (FD), and *Dde*I (D). The DNA deleted in the mutant alleles is indicated by a thick line. When the exact boundaries of the deletions are unknown, the thin lines indicate the maximal extent of the deletions.

cover the complete *phoE* gene, except for the signal sequence and the N-terminal 11 amino acids of the mature protein. Pulse-label and pulse-chase experiments revealed that the polypeptides encoded by the mutant alleles were all processed and, therefore, probably exported from the cytoplasm. Two of these polypeptides could not be localized further because they were rapidly degraded in the cells. Immunocytochemical labeling on ultra-thin cryosections showed that the other mutant proteins all accumulated in the periplasm (2). Since the deletions cover almost the complete *phoE* gene, these results suggest that the PhoE protein contains no information required for export through the cytoplasmic membrane except for the signal sequence. Since none of the mutant proteins is inserted into the outer membrane, these results also suggest that PhoE protein does not contain a discrete insertion sequence for assembly into the outer membrane. Rather, the total conformation of the protein contributes to this process (Fig. 5).

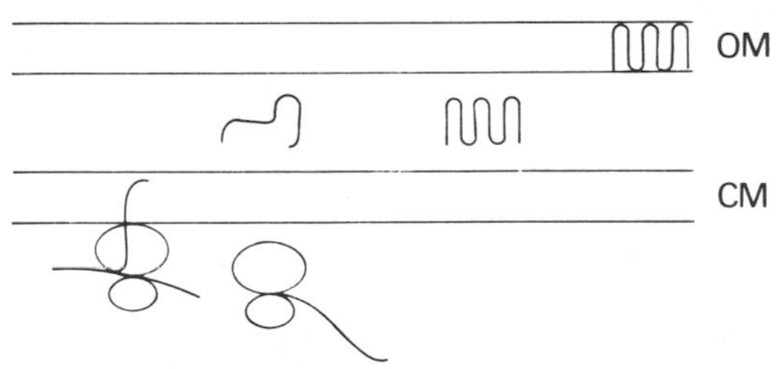

FIG. 5. Model for the export of PhoE protein to the outer membrane (OM) as proposed in reference 25. Proteins are cotranslationally or posttranslationally (T. de Vrije, J. Tommassen, and B. de Kruyff, manuscript submitted for publication) transported through the cytoplasmic membrane (CM). The signal sequence, which is cleaved off during or shortly after transport, contains the only information required for export. β-Galactosidase when fused to a signal sequence cannot pass the CM, because it contains sequences or conformations which prevent export. In the periplasm, outer membrane proteins fold into a configuration, allowing insertion into the OM via hydrophobic interactions. No specific insertion sequence is required for this step, but the total (secondary/tertiary) structure triggers insertion.

I thank D. Bosch, P. de Geus, T. de Kroon, M. Koster, J. Leunissen, B. Lugtenberg, P. Overduin, and P. van der Ley for their contributions in different aspects of the work described here.

LITERATURE CITED

1. **Benz, R., R. P. Darveau, and R. E. W. Hancock.** 1984. Outer membrane protein PhoE from *Escherichia coli* forms anion-selective pores in lipid bilayer membranes. Eur. J. Biochem. **140**:319–324.
2. **Bosch, D., J. Leunissen, J. Verbakel, M. de Jong, H. van Erp, and J. Tommassen.** 1986. Periplasmic accumulation of truncated forms of outer membrane PhoE protein of *Escherichia coli* K-12. J. Mol. Biol. **189**:449–455.
3. **Dargent, B., W. Hofmann, F. Pattus, and J. P. Rosenbusch.** 1986. The selectivity filter of voltage-dependent channels formed by phosphoporin (PhoE protein) from *E. coli*. EMBO J. **4**:773–778.
4. **Foulds, J., and T. Chai.** 1978. New major outer membrane protein found in an *Escherichia coli tolF* mutant resistant to bacteriophage Tulb. J. Bacteriol. **133**:1478–1483.
5. **Guan, C.-D., B. Wanner, and H. Inouye.** 1983. Analysis of regulation of *phoB* expression using a *phoB-cat* fusion. J. Bacteriol. **156**:710–717.
6. **Hall, M. N., M. Schwartz, and T. J. Silhavy.** 1982. Sequence information within the *lamB* gene is required for proper routing of the bacteriophage λ receptor protein to the outer membrane of *Escherichia coli* K-12. J. Mol. Biol. **156**:93–112.
7. **Henning, U., W. Schmidmayr, and I. Hindennach.** 1977. Major proteins of the outer cell envelope membrane of *Escherichia coli* K-12: multiple species of protein I. Mol. Gen. Genet. **154**:293–298.
8. **Korteland, J., P. de Graaff, and B. Lugtenberg.** 1984. PhoE protein pores in the outer membrane of *Escherichia coli* K-12 not only have a preference for P_i and P_i-containing solutes but are general anion-preferring channels. Biochim. Biophys. Acta **778**:311–316.
9. **Korteland, J., J. Tommassen, and B. Lugtenberg.** 1982. PhoE protein pore of the outer membrane of *Escherichia coli* K-12 is a particularly efficient channel for organic and inorganic phosphate. Biochim. Biophys. Acta **690**:282–289.
10. **Magota, K., N. Otsuji, T. Miki, T. Horiuchi, S. Tsunasawa, J. Kondo, F. Sakiyama, M. Amemura, T. Morita, H. Shinagawa, and A. Nakata.** 1984. Nucleotide sequence of the *phoS* gene, the structural gene for the phosphate-binding protein of *Escherichia coli*. J. Bacteriol. **157**:909–917.
11. **Moreno, F., A. V. Fowler, M. Hall, T. J. Silhavy, I. Zabin, and M. Schwartz.** 1980. A signal sequence is not sufficient to lead β-galactosidase out of the cytoplasm. Nature (London) **286**:356–359.
12. **Nikaido, H., and E. Y. Rosenberg.** 1983. Porin channels in *Escherichia coli*: studies with liposomes reconstituted

from purified proteins. J. Bacteriol. **153:**241–252.
13. **Nikaido, H., E. Y. Rosenberg, and J. Foulds.** 1983. Porin channels in *Escherichia coli*: studies with β-lactams in intact cells. J. Bacteriol. **153:**232–240.
14. **Nikaido, H., and M. Vaara.** 1985. Molecular basis of bacterial outer membrane permeability. Microbiol. Rev. **49:**1–32.
15. **Overbeeke, N., H. Bergmans, F. van Mansfeld, and B. Lugtenberg.** 1983. Complete nucleotide sequence of *phoE*, the structural gene for the phosphate limitation inducible outer membrane pore protein of *Escherichia coli* K-12. J. Mol. Biol. **163:**513–532.
16. **Overbeeke, N., and B. Lugtenberg.** 1980. Expression of outer membrane protein e of *Escherichia coli* K-12 by phosphate limitation. FEBS Lett. **112:**229–232.
17. **Overbeeke, N., and B. Lugtenberg.** 1982. Recognition site for phosphorus-containing compounds and other negatively charged solutes on the PhoE protein pore of the outer membrane of *Escherichia coli* K-12. Eur. J. Biochem. **126:**113–118.
18. **Rao, N. N., E. Wang, J. Yashphe, and A. Torriani.** 1986. Nucleotide pool in *pho* regulon mutants and alkaline phosphatase synthesis in *Escherichia coli.* J. Bacteriol. **166:**205–211.
19. **Rosenberg, M., and D. Court.** 1979. Regulatory sequences involved in the promotion and termination of RNA transcription. Annu. Rev. Genet. **13:**319–353.
20. **Rosenberg, H., R. G. Gerdes, and K. Chegwidden.** 1977. Two systems for the uptake of phosphate in *Escherichia coli.* J. Bacteriol. **131:**505–511.
21. **Schweizer, H., and W. Boos.** 1984. Characterization of the *ugp* region containing the genes for the *phoB* dependent *sn*-glycerol-3-phosphate transport system of Escherichia coli. Mol. Gen. Genet. **197:**161–168.
22. **Shinagawa, H., K. Makino, and A. Nakata.** 1983. Regulation of the *pho* regulon of *Escherichia coli* K-12. Genetic and physiological regulation of the positive regulatory gene *phoB*. J. Mol. Biol. **168:**477–488.
23. **Silhavy, T. J., H. A. Shuman, J. Beckwith, and M. Schwartz.** 1977. The use of gene fusions to study outer membrane protein localization in *Escherichia coli.* Proc. Natl. Acad. Sci. USA **74:**814–817.
24. **Surin, B. P., D. A. Jans, A. L. Fimmel, D. C. Shaw, G. B. Cox, and H. Rosenberg.** 1984. Structural gene for the phosphate-repressible phosphate-binding protein of *Escherichia coli* has its own promoter: complete nucleotide sequence of the *phoS* gene. J. Bacteriol. **157:**772–778.
25. **Tommassen, J.** 1986. Fallacies of *E. coli* cell fractionations and consequences thereof for protein export models. Microb. Pathogen. **1:**225–228.
26. **Tommassen, J., P. de Geus, B. Lugtenberg, J. Hackett, and P. Reeves.** 1982. Regulation of the *pho* regulon of *Escherichia coli* K-12. Cloning of the regulatory genes *phoB* and *phoR* and identification of their gene products. J. Mol. Biol. **157:**265–274.
27. **Tommassen, J., J. Leunissen, M. van Damme-Jongsten, and P. Overduin.** 1985. Failure of *E. coli* K-12 to transport PhoE-LacZ hybrid proteins out of the cytoplasm. EMBO J. **4:**1041–1047.
28. **Tommassen, J., and B. Lugtenberg.** 1980. Outer membrane protein e of *Escherichia coli* K-12 is co-regulated with alkaline phosphatase. J. Bacteriol. **143:**151–157.
29. **Tommassen, J., and B. Lugtenberg.** 1981. Localization of *phoE*, the structural gene for major outer membrane protein e of *Escherichia coli* K-12. J. Bacteriol. **147:**118–123.
30. **Tommassen, J., and B. Lugtenberg.** 1982. *Pho*-regulon of *Escherichia coli* K-12: a minireview. Ann. Microbiol. (Paris) **133A:**243–249.
31. **Tommassen, J., H. van Tol, and B. Lugtenberg.** 1983. The ultimate localization of an outer membrane protein of *Escherichia coli* K-12 is not determined by the signal sequence. EMBO J. **2:**1275–1279.
32. **van Alphen, W., N. van Selm, and B. Lugtenberg.** 1978. Pores in the outer membrane of *Escherichia coli* K-12. Involvement of proteins b and e in the functioning of pores for nucleotides. Mol. Gen. Genet. **159:**75–83.
33. **Wanner, B., and P. Latterell.** 1980. Mutants affecting alkaline phosphatase expression: evidence for multiple positive effectors for the phosphate regulon in *E. coli.* Genetics **96:**353–366.

Acid Phosphatase (pH 2.5) of *Escherichia coli*: Regulatory Characteristics

ELIETTE TOUATI,† ELIE DASSA,† JANIE DASSA, AND PAUL L. BOQUET

Service de Biochimie, Département de Biologie, C.E.N. Saclay, 91191 Gif-sur-Yvette, France

In one of the first publications describing the properties of alkaline phosphatase in *Escherichia coli*, Torriani (22) also reported the presence in this bacterium of an acid phosphatase activity with more restricted specificity. Five different acid phosphatases have been actually found in *E. coli* and have been shown to be localized in the periplasmic space (14), but compared with alkaline phosphatase, such systems have received much less attention. One of these was identified as a 2′,3′ cyclic phosphodiesterase showing a 3′ nucleotidase activity and was purified by Anraku (1–3). The gene for this enzyme is *cpdB* (4, 5). A 5′ nucleotidase (14) is encoded by gene *ush* (6, 7). The periplasm also contains one or several related hexose phosphatases and an "unspecific" acid phosphatase (14, 20, 25), but the literature relative to the properties of these enzymes is somehow confusing since the corresponding proteins have not been purified to homogeneity and their genes have never been identified. All such enzymes show an optimum for activity around pH 5; however, phosphatases working under strongly acidic conditions have also been reported.

In 1952, Courtois and Manet (9) first reported the existence in *E. coli* of a peculiar activity producing P_i from inositol hexaphosphate under very acidic conditions. Later, Hafckensheid (16) showed that homogenates of freeze-dried *E. coli* cells were able to hydrolyze *p*-nitrophenyl phosphate (PNPP) between pH 2 and pH 3. More recently, our laboratory has shown the presence of an acid phosphoanhydride phosphohydrolase activity (EC 3.6.1.11) in supernatant fluids of crude cellular extracts of *E. coli* K-12 precipitated with formic acid (13, 21). When purified to homogeneity, this enzyme does not hydrolyze phosphodiesters and phosphomonoesters, with the exception of 2,3-bis-phosphoglycerate and PNPP. It shows a sharp optimum for activity at pH 2.5 and is unusually soluble and stable in strongly acidic solutions (12). This protein, which is released from cells by osmotic shock (13), is very likely responsible for the activities described in references 9 and 16.

†Present address: Département de Biotechnologies, U.P.M.T.G., Institut Pasteur, 75015 Paris, France.

The structural gene for this phosphatase, *appA*, was identified by a mutation causing the synthesis of an enzyme with a thermolabile activity (11). This gene was cloned in vivo on a mini-Mu replicon "phasmid" (8a, 15), and a small DNA fragment containing *appA* was subcloned into the multicopy plasmid pBR322 (8). By random insertions of the particular transposon Tn*phoA* (17) into one such recombinant plasmid, several *appA-phoA* protein fusions were generated in vivo. All such hybrid proteins were exported to the periplasm, where they displayed alkaline phosphatase activity, but no more acid phosphatase activity. These fusions were used to determine the precise location of *appA* on the cloned DNA fragment and its direction of transcription; *appA* was shown to possess its own promoter located immediately upstream from its N-terminal end (8a). In addition, *appA-phoA* fusions allowed a visualization of the activity of the *appA* promoter in colonies growing in the presence of the chromogenic substrate XP (5-bromo-4-chloro-3-indolyl phosphate), which is not possible with the original acid phosphatase (XP is not a substrate for this enzyme, and plate assays with PNPP at pH 2.5 kill the cells).

Several growth conditions that promote the synthesis of the pH 2.5 acid phosphatase have already been described (12). This paper presents new understanding of the complex pattern of regulation of this enzyme, obtained by experiments with different recombinant plasmids and with fusions to alkaline phosphatase. It also raises some questions about the actual physiological role of the enzyme.

When Is *appA* Expressed?

Alkaline phosphatase shows a broad range of substrates and belongs to the Pho regulon, which comprises several proteins whose synthesis is coordinately induced under only one specific condition, i.e., the limitation of P_i in the medium (for review, see reference 23). By contrast, the expression of the pH 2.5 acid phosphatase is promoted by several conditions (12). In nonlimiting rich or synthetic media the enzyme is expressed only when the culture enters the stationary phase (Fig. 1), and it accumulates for 6 to 10 h in the absence of apparent net protein

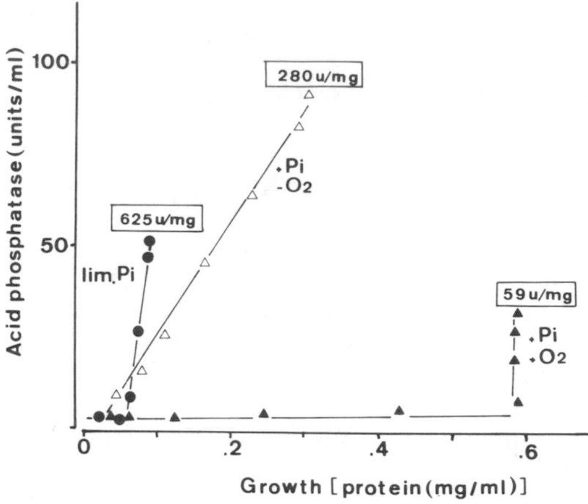

FIG. 1. Differential rate of acid phosphatase synthesis according to growth conditions. Cells of *E. coli* K10 were grown aerobically in the TEB phosphate-limited medium (●) already described (12) or in the same medium supplemented with P_i (10 mM) and referred to as TEBP (▲). Anaerobic growth in TEBP (△) was obtained with a layer of paraffin oil at the top of the culture tube and slow stirring (24). Activity and protein determinations were as described (12, 24). The boxed numbers refer to the specific activities (in units per milligram of protein) measured after 24 h of growth at 37°C (plateau value). One unit of enzyme catalyzes the hydrolysis of 1 nmol of PNPP per min at 37°C.

synthesis. This accumulation, however, corresponds to de novo synthesis. The activity per cellular mass reaches a reproducible plateau value which is maintained over a period of 2 to 3 days and is characteristic of the strain and of the growth medium used. Another condition causing enzyme synthesis was oxygen deprivation (see Fig. 1). Bacteria maintained under anaerobic conditions produced acid phosphatase constitutively. A comparison of the specific activities at their plateau values (Fig. 1) shows that P_i limitation remains the most efficient stimulus for acid phosphatase synthesis, and it may be asked whether the corresponding regulatory mechanism does not share some components with that of alkaline phosphatase.

Does *appA* Belong to the Pho Regulon?

Bacterial strains with mutations in the *phoB* gene, which encodes the general positive effector of the Pho regulon (23), were grown in a high- or a low-phosphate medium and compared with related *phoB*$^+$ strains for expression of both acid and alkaline phosphatases (Table 1). Point mutations or deletions in *phoB* or *phoA* as well as a large deletion of both *phoB* and *phoA* did not change the expression of *appA*. The possibility that other mutations of the Pho regulon might control *appA* independently from *phoB* was considered. Mutations in *phoS, phoR, phoM*, or a combination of *phoR* and *phoM* had no significant effect on acid phosphatase levels measured either in a low- or in a high-phosphate medium. Consequently, this enzyme is regulated by a mechanism sensitive to the condition of phosphate starvation which probably differs

TABLE 1. Expression of acid phosphatase (pH 2.5) and alkaline phosphatase in the presence of different mutations affecting the Pho regulon[a]

Strain	Relevant genotype or phenotype	Sp act (U/mg of protein)			
		Acid phosphatase		Alkaline phosphatase	
		Low P_i	High P_i	Low P_i	High P_i
K10	Wild type	412	76	168	3.6
E15	*phoA8*	493	104	2.6	2.0
H2	*phoB52*	466	66	3.1	3.6
MC4100	Wild type	1,082	36	906	2.3
MpH2	Δ*(phoA,B)*	932	42	3.4	2.4
SBS1355	Δ*phoA20 phoB*$^+$	894	45	2.8	1.0
SBS877 (K10)	Wild type	432	68	183	3.4
SBS876	*phoR68*	571	92	147	94
C5	*phoR17*	458	87	111	109
C86	*phoS21*	752	68	155	58
C90	*phoT9*	170	22	404	137
BW576	Wild type	752	89	203	3.3
SBS873	*phoR68*	655	113	125	102
BW486	*phoM451*	988	140	411	2.8
BW494	*phoM451* *phoR68*	705	141	1.0	0.7

[a] Bacteria were harvested in the stationary phase of growth in TEB (low-phosphate) or TEBP (high-phosphate) medium (see legend of Fig. 1). The specific activity of each phosphatase was determined on whole cells with PNPP as previously reported (9).

TABLE 2. Effect of *appR* on the expression of acid and alkaline phosphatases[a]

Strain	Relevant genotype or phenotype	Sp act (U/mg of protein)			
		Acid phosphatase		Alkaline phosphatase	
		Low P_i	High P_i	Low P_i	High P_i
SBS816 (K10)	*appR*+ *srl*::Tn*10*	1,007	50	107	2.0
SBS814	*appR1699 srl*::Tn*10*	291	11	1,332	2.2
Gy3442	*appR*+ *cysC*	533	73	532	1.0
SBS817	*appR90 cysC*+	108	20	2,035	8.1
SBS1367	*appR*+ *phoT9 srl*::Tn*10*	1,138	323	405	464
SBS1366	*appR90 phoT9 srl*::Tn*10*	341	95	1,475	312
ANS50	*appR*+ *iap*+	1,707	92	1,110	0.2
ANS51	*appR90 iap*	675	20	3,977	0.2

[a] The conditions for growth, enzyme determinations, and expression of results are the same as in Table 1.

greatly from that of the alkaline phosphatase. However strain C90 (*phoT*) showed a threefold reduction in acid phosphatase and an enhanced alkaline phosphatase level in phosphate-limited medium, but this phenotype was not linked to the *phoT* mutation, as shown below.

Allelic Differences in a New Gene, *appR*, Change the Expression of Both Acid and Alkaline Phosphatases but in Opposite Ways

We recently reported that a large variation in acid phosphatase expression among strains of laboratory collections was accounted for by allelic differences in a new gene called *appR* which is located at min 59 on the *E. coli* linkage map (24). Such genetic modifications are pleiotropic, and their mechanism of action on the expression of *appA* is not well understood. In a set of isogenic *appR* and *appR*+ strains grown in a high- or a low-phosphate medium, we have measured the levels of acid and alkaline phosphatases in the stationary phase. When compared with *appR*+ strains, all *appR* strains tested showed a 3- to 4-fold reduction in acid phosphatase level in both media, but alkaline phosphatase, by contrast, was elevated 5- to 10-fold in the low-phosphate medium (Table 2). Several usual strains in fact were shown to harbor mutations in *appR*, for example, strain C90, the *phoT9* derivative of strain K10 (see Table 1). When the original 59-min region of the parent strain K10 (*appR*+) was introduced back into C90, using the nearby transposon *srl*::Tn*10*, acid and alkaline phosphatase levels were both modified. The effect of this *appR90* allele on acid and alkaline phosphatase expression in isogenic pairs was not influenced by the presence in the membrane of the *tet* protein coded for by Tn*10*. The "isogenic" strains ANS50 and ANS51 are reported to differ by a mutation in gene *iap*, which determines the proportion of alkaline phosphatase isozymes and is located near min 59 (18). In fact, they also differ by an *appR* mutation which is very likely the *appR90* mutation originally evidenced in C90 from which the 59-min region of the chromosome has been transferred (24). This effect of the allelic state of *appR* on the expression of *phoA* and *appA* shows that particular attention must be paid when comparing strains, even from the same lineage.

Acid Phosphatase Regulation by P_i Is Promoter Dependent

Three different recombinant plasmids containing the same 4.6-kilobase fragment of chromosomal DNA encompassing *appA* were constructed (Fig. 2A). In plasmid pPB1132, a *Bam*HI-*Bgl*II fragment originating from a Mu d II4042 recombinant phasmid was inserted into the unique *Bam*HI site of pBR322. The construction and the restriction map of this plasmid were recently reported (8a). In such a plasmid which overexpresses acid phosphatase by more than 100-fold, the *appA* gene shows the same orientation as the *tet* gene into which it is inserted. Plasmid pJD1301 contains the same insert, but in the opposite orientation, and overexpresses acid phosphatase by only 10 times. This large difference in expression according to orientation may be explained by an important contribution of the *tet* promoter in plasmid pPB1132. To test this possibility, we constructed plasmid pJD1235, which is a derivative of pPB1132 carrying a deletion of the *tet* promoter. After mutagenesis of pPB1132 in vivo with transposon Tn*phoA* (17), one such transposon was found inserted into the N-terminal region of the *tet* gene with the *'phoA* end opposite to the *tet* promoter. This intermediate plasmid, pPB1164, was digested with *Eco*RI to remove most of the Tn*phoA* and the *tetp* regions and was religated to give pPB1235. We also recently reported that the cloned DNA region located upstream from *appAp* does not contain any other promoter able to transcribe *appA* (8a).

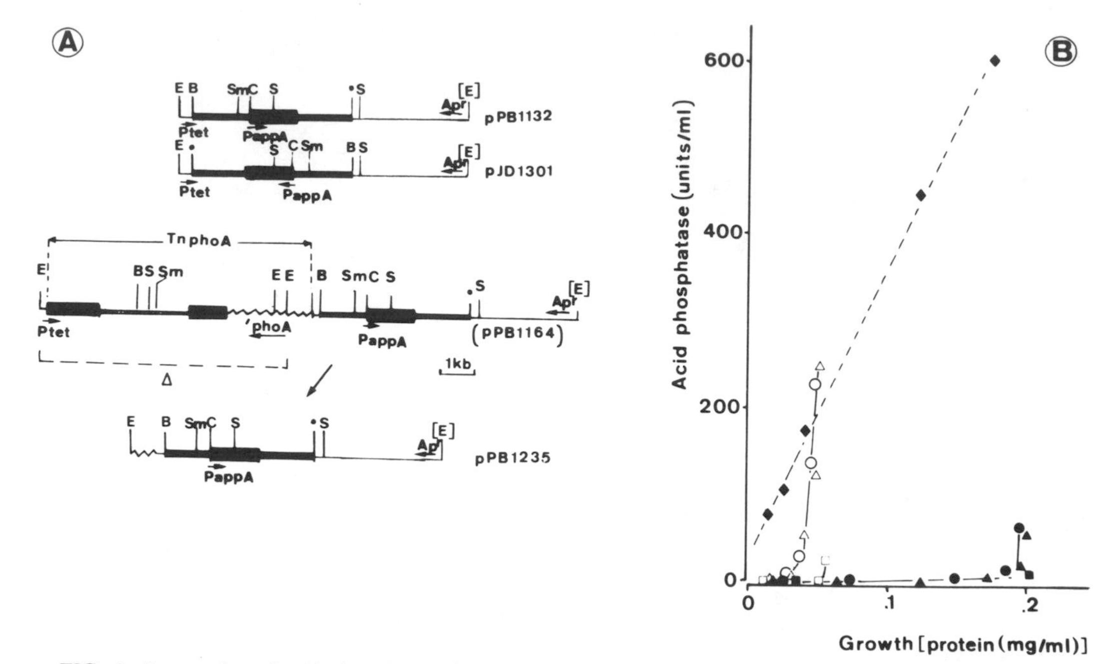

FIG. 2. Expression of acid phosphatase from the structural gene *appA* on multicopy recombinant plasmids. (A) Structure of the three recombinant plasmids used in the experiment: the same *Bam*HI-*Bgl*II fragment (thick bar) of chromosomal DNA harboring the *appA* region originating from a Mu d II4042 recombinant phasmid was inserted into the *Bam*HI site of pBR322 (—) in the two opposite orientations, yielding plasmids pPB1132 and pJD1301. The direction of transcription of the interrupted *tet* gene and that of *appA* (represented by the thick portion of the inserted DNA segment), starting from their own promoters, are indicated by arrows. Plasmid pJD1235, which harbors a deletion of the *tet* promoter, was derived from plasmid pPB1164, which is the same as pPB1132 but contains a Tn*phoA* insert in the N-terminal *tet* region at the position and in the orientation shown. The entire *tet* promoter was removed by digestion with *Eco*RI and religation. The remaining portion of the *phoA* gene (squiggle line) does not contain any promoter. Symbols for restriction sites are: E, *Eco*RI; B, *Bam*HI; Sm, *Sma*I; C, *Cla*I; and S, *Sal*I. (B) Cells of strain CC118 *(recA)* harboring plasmid pJD1301 (○, ●), pJD1235 (△, ▲), pPB1132 (◆), or pBR322 (□, ■) were grown under aerobic conditions either in low-phosphate medium (open symbols) or in high-phosphate medium (closed symbols) in the presence of ampicillin (200 μg/ml). The differential rate of acid phosphatase synthesis was measured as in the experiment of Fig. 1.

Consequently, it can be assumed that, in pPB1235, *appA* is transcribed only from its own promoter. Plasmids pPB1132, pJD1301, and pJD1235 were introduced into strain CC118 (*recA appA*$^+$ *appR*), and transformants grown in a high- or a low-phosphate medium were assayed for their differential rate of acid phosphatase synthesis (Fig. 2B). In a control experiment with strain CC118 harboring only plasmid pBR322 (and *appA*$^+$ on its chromosome), phosphate starvation elicited a normal synthesis of enzyme from chromosomal origin. In strains harboring plasmids pJD1301 and pJD1235, P_i prevented enzyme synthesis in the exponential phase, and limited synthesis was seen in the stationary phase. In the phosphate-limited medium, phosphate starvation immediately elicited with the two plasmids an important and equal synthesis of enzyme. By contrast, the expression from plasmid pPB1132 was not inhibited by P_i and proceeded constitutively, a result in agreement with a predominant expression of *appA* from the constitutive *tet* promoter in this plasmid. This promoter dependence for the regulation of *appA* expression by phosphate availability strongly suggests a control at the level of transcription.

Fusions to Alkaline Phosphatase for the Analysis of Regulation

The genetic study of acid phosphatase regulation is presently hampered by the inability to detect this activity directly on growing colonies, and the conditions for the assay with PNPP are lethal. This disadvantage can be overcome if the original activity is replaced by another one easily detectable in vivo with a chromogenic substrate. Preliminary attempts to construct on a recombinant plasmid an *appA-lacZ* operon fusion with the *lacZ* DNA region originally present in phage Mu d1 have shown that the portion of the *trp* region located upstream of *lacZ* in such fusions contains a weak promoter which by itself is able to transcribe substantially *lacZ* on the

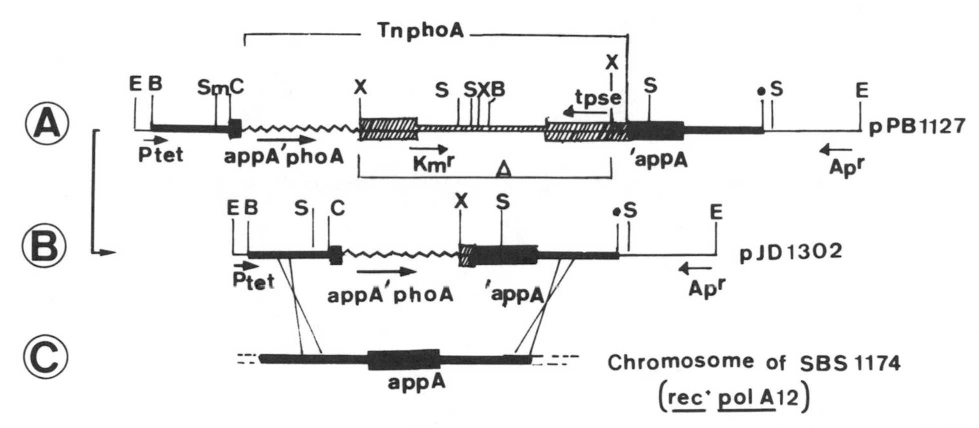

FIG. 3. Construction of a stable *appA-phoA* fusion and its introduction into the chromosome. (A) Plasmid pPB1127 containing an *appA'-phoA* fusion was obtained by insertion of Tn*phoA* into the *appA* gene in plasmid pPB1132 (8a). This plasmid expresses the fusion (alkaline phosphatase) constitutively since it is also transcribed from the *tet* promoter. The symbols for restriction sites are as in Fig. 2 plus X for *Xho*I. (B) Elimination of most of the Tn*5* region of the transposon was obtained by digestion with *Xho*I and self-religation, yielding plasmid pJD1302, which itself was introduced into the conditional *polA* strain SBS1174 (Δ*phoA polA12 appA*$^+$ *recA*$^+$). (C) Recombination between plasmid pJD1302 and the chromosome of SBS1174 by homologous *appA* flanking regions was performed at 42°C (low plasmid copy number state). The fusion, once inserted into the chromosome, was further transduced with P1 into other strains, with the neighboring *pyrD* marker.

plasmid (E. Touati and A. Danchin, unpublished data). We recently reported the construction on the recombinant plasmid pPB1132 (see Fig. 2) of *appA-phoA* protein fusions obtained by insertions of transposon Tn*phoA* (17) into the *appA* gene (8a). Such fusions led to the synthesis of hybrid proteins with N-terminal fragments of acid phosphatase of variable sizes bound to a large functional C-terminal fragment of alkaline phosphatase. Strains containing these plasmids showed a high alkaline phosphatase activity as a consequence of the export of the hybrid proteins promoted by a signal contained within the N-terminal part of the *appA* gene. Such a fusion was transferred from plasmid pPB1127 into the chromosome by homologous recombination, but to avoid further transpositions of the neighboring Tn*5*, most of the latter was previously removed by digestion with *Xho*I and religation (Fig. 3). The deleted plasmid, pJD1302, was introduced into strain SBS1174 (Δ*phoA polA12 appA*$^+$), in which the *polA* function necessary to maintain a high plasmid copy number is thermosensitive at 42°C, and was allowed to recombine by growth at this temperature in the presence of ampicillin at low concentration (20 μg/ml). Plating this culture on high-phosphate medium plus XP and ampicillin at a high concentration (200 μg/ml) and incubating at 30°C resulted in the appearance of XP-negative clones which expressed constitutively a high level of acid phosphatase at 30°C as a result of the presence of *appA* on the recombined plasmid in the multicopy state. Such clones appeared to originate from endogenous homologous recombination of the *appA* flanking regions. The *appA-phoA* fusion now located on the chromosome in place of *appA* was transduced with phage P1 from one of the recombined clones into other strains with a linked marker (Fig. 3). The results in Table 3 show that alkaline phosphatase originating from this fusion on plasmid pJD1302 was expressed

TABLE 3. Expression of alkaline and acid phosphatase from the hybrid protein encoded by an *appA'-phoA* fusion[a]

Strain	Relevant genotype or phenotype	Sp act (U/mg of protein)			
		Acid phosphatase		Alkaline phosphatase	
		Low P_i	High P_i	Low P_i	High P_i
Mph2	Δ*pho(A,B) appA*$^+$	432	10.2	<0.1	<0.1
Mph2 pJD1302	*appA'-phoA* fusion on plasmid	192	2.0	694	360
SBS1182	Same as Mph2 but *appA'-phoA* fusion on the chromosome	<0.1	<0.1	16.2	<0.1

[a] The conditions for growth and enzyme determinations are as in Tables 1 and 2 except that pJD1302 was maintained with 200 μg of ampicillin per ml. Results are expressed as in Tables 1 and 2.

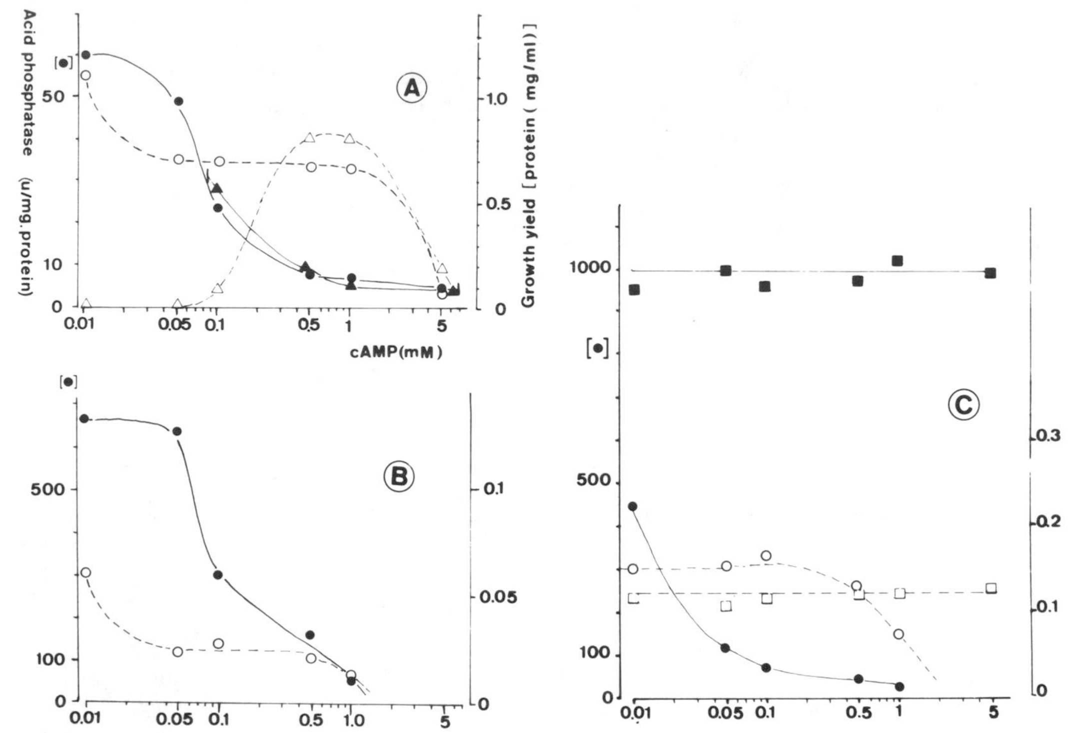

FIG. 4. Inhibition of acid phosphatase expression by cAMP. Experiments were performed with strains CA234 (Δ*cya-854 appA*$^+$) and POP3Δcrp (Δ*crp-39*). The enzymatic specific activity was measured at the plateau value for different concentrations of cAMP (▲, ●, ■). The growth yield (as protein) is indicated for each measure (○, △, □). (A) Aerobic growth of CA234 in high-phosphate medium with glucose (○, ●) or succinate (△, ▲) as carbon source; (B) anaerobic growth of CA234 in high-phosphate medium with glucose; (C) aerobic growth of CA234 (○, ●) and of POP3Δcrp (□, ■) in low-phosphate medium with glucose. The filler (●) outside the log scale for cAMP concentration corresponds to acid phosphatase level without cAMP addition.

constitutively. According to its orientation, this fusion was probably mainly transcribed from the exogenous *tet* promoter. When reinserted into the chromosome, however, the expression of the fusion was actually regulated by phosphate availability in the same way as acid phosphatase itself. It allowed dark-blue staining of colonies growing on phosphate-limited plates in the presence of XP and only very low staining with added phosphate, giving the possibility for a selection of regulatory mutants.

Negative Regulation by cAMP

Acid phosphatase expression in a given strain depends on the nature of the carbon source used for growth. Glucose (known to cause "catabolite repression") allowed the highest yields while succinate was inefficient. That cyclic AMP (cAMP) and its receptor the CAP protein, rather than the carbon source itself, are directly involved in this regulation is shown by the experiments reported in Fig. 4. In the cyclase-deficient strain CA8306 (Δ*cya-854 appA*$^+$) growing in a nonlimiting minimal medium with glucose as carbon source, the expression of the acid phosphatase measured in the stationary phase under aerobic conditions was progressively inhibited with increasing concentrations of cAMP in the growth medium. The 50% inhibitory concentration of cAMP was around 0.1 mM (Fig. 4A). When glucose was replaced by succinate, growth was possible above a threshold concentration of 0.1 mM cAMP in the medium, and the specific activity of the enzyme in this case was entirely superimposable with that observed with glucose, showing that succinate has no specific effect. When the cells were grown under anaerobic conditions in the same nonlimiting medium with glucose as carbon source to obtain a constitutive synthesis of enzyme (see Fig. 1), the inhibitory effect of cAMP was very similar (Fig. 4B). In a phosphate-limited medium, cAMP showed the same type of inhibition, but the 50% inhibition was shifted to 0.02 mM (Fig. 4C). This increased sensitivity of phosphate-starved Cya$^-$ cells to exogenously supplied cAMP, also detectable with other systems such as *lac* expression (Touati and Boquet, unpublished data),

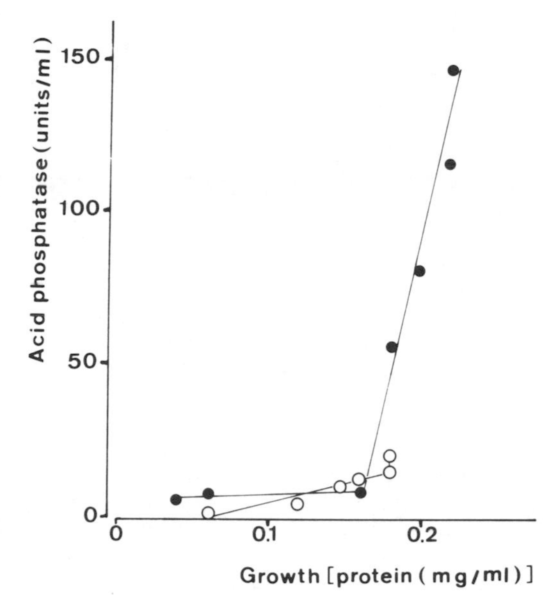

FIG. 5. Differential rate of acid phosphatase expression in a cyclase-deficient strain. Strain CA8306 (Δ*cya-854*) was grown in high-phosphate medium in the presence of 0.5 mM cAMP (○) or no addition (●). Activity and protein determinations were as in Fig. 1.

may be due to a better penetration of cAMP or to a reduced degradation under such conditions. The level of acid phosphatase in the CAP-deficient strain POP3Δcrp(Δ*crp*-39) was independent of added cAMP and as high as in the Cya$^-$ strain without added cAMP (Fig. 4C). Consequently, the inhibition of *appA* expression by CAP and cAMP was independent of the growth conditions used to promote enzyme synthesis.

The hypothesis that cAMP and CAP might be the only regulators of *appA* expression was considered. If this hypothesis is correct, then acid phosphatase must be expressed constitutively in a Cya$^-$ strain grown in the absence of cAMP. However, in such a strain grown aerobically in a nonlimiting medium, the enzyme was made only in the stationary phase of growth as in Cya$^+$ strains (Fig. 5), showing the presence of another control independent from cAMP inhibiting its expression in the exponential phase.

To learn whether the effect of cAMP on *appA* expression was also promoter dependent, plasmids pPB1132 and pJD1301 (see Fig. 2) were introduced into the cyclase-deficient strain SBS1254, and acid phosphatase levels were measured in transformants with increasing concentrations of added cAMP (Fig. 6). When the *appA* promoter only was used for transcription as in pJD1301, cAMP inhibited enzyme expression exactly as in the case of a chromosomal gene (see Fig. 4). In contrast, when the additional *tet* promoter was predominantly used, cAMP stimulated acid phosphatase expression up to threefold. This last result can be explained by a previously observed effect of cAMP which increased the number of copies of the plasmid (Ullman, personal communication; E. Touati and A. Danchin, submitted for publication). Consequently, the negative effect of cAMP on acid phosphatase appears to depend on the promoter used and is very likely exerted at the level of transcription.

When the *appA-phoA* fusion obtained as described in Fig. 3 was introduced in place of *appA* into the chromosome of a Cya$^-$ strain, the expression of the alkaline phosphatase originating from the hybrid protein was itself negatively regulated by cAMP and showed the same dependence as did the original acid phosphatase (Fig. 7).

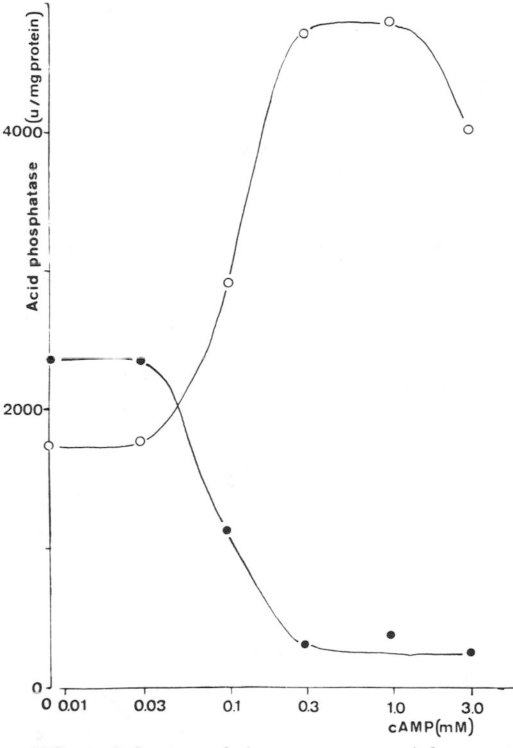

FIG. 6. Influence of the promoter used for *appA* transcription on the effect of cAMP. Cells of strain SBS1254 (Δ*phoA,B* Δ*cya appA*$^+$) were transformed with plasmid pPB1132, in which *appA* is transcribed from both the *appA* and the *tet* promoters (○), and plasmid pJD1301, in which *appA* is transcribed only by its own promoter (●), and grown in TEB (phosphate-limited) medium in the presence of variable amounts of cAMP. Acid phosphatase specific activity was measured in the stationary phase according to the concentration of added cAMP.

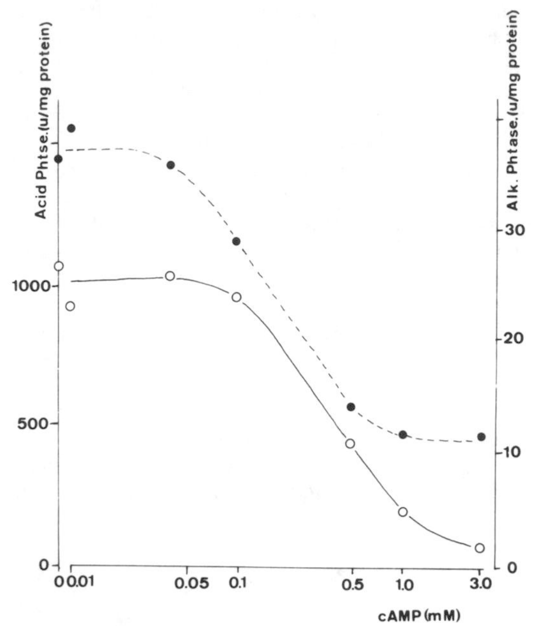

FIG. 7. Regulation of the expression of a chromosomal *appA-phoA* fusion by cAMP. Cells of strain SBS1254 (Δ*phoA,B* Δ*cya appA*$^{+}$) grown in TEB (low-phosphate) medium with cAMP at different concentrations were assayed for acid phosphatase (○) as in the experiment of Fig 4. Cells from the isogenic strain SBS1255 (Δ*phoA,B* Δ*cya appA'-phoA*) harboring the fusion on the chromosome in place of *appA*, and grown in the same conditions, were assayed for alkaline phosphatase activity (●) as described (8a).

When Is Acid Phosphatase Used?

This complex mode of regulation does not shed much light on the role of the enzyme in the periplasmic space. Its efficient induction (or derepression) by phosphate starvation raises the question of a possible role in providing the cell with P_i hydrolyzed from molecules with phosphoanhydride bonds such as, for instance, polyphosphates, the best substrate for the enzyme so far tested in vitro (10). An easily transducible "labeled" deletion of *appA* was constructed on a recombinant plasmid from a particular insertion of Tn*5* into the C-terminal portion of the gene (Fig. 8). This deletion, which removes the gene coding for the Tn*5* transposase and brings the *neo* gene (Km^{r}) in place of *appA*, was introduced from the plasmid into the chromosome by the method described in the Fig. 3 legend. It was further transduced into a set of isogenic strains differing only by the *phoA-phoB* region which were compared for growth on synthetic solid medium with limited amounts of different phosphorylated compounds as the sole source of phosphate (Table 4). In a *pho(A,B)*$^{+}$ background (SBS926 and SBS1370) a deletion of *appA* had no effect, and all substrates tested were degraded by the alkaline phosphatase. Strains deleted for both *phoA* and *phoB* (SBS1186 and SBS1371) did not grow on any of the phosphorylated substrates even in the presence of a functional *appA* gene. If, in the same genetic background, *appA* was replaced on the

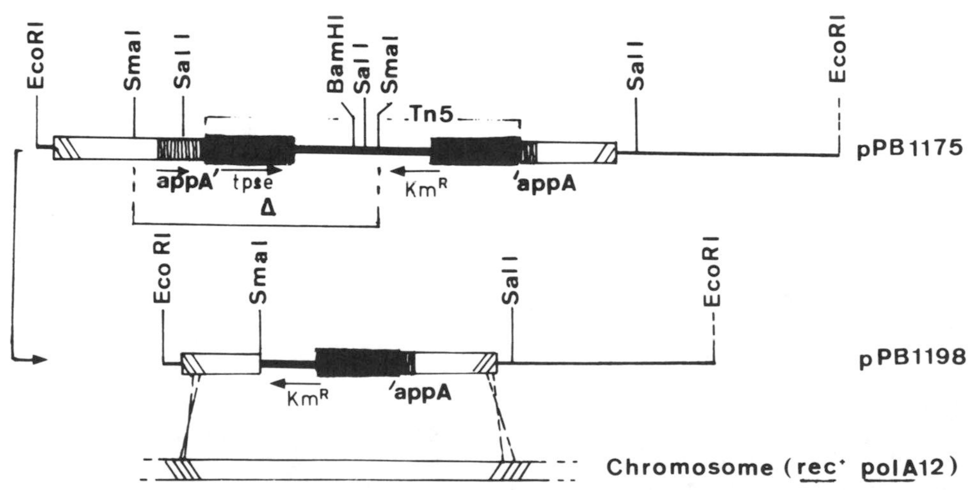

FIG. 8. Construction of a "labeled" deletion of *appA*. Transposon Tn*5* was introduced into strain CC118 (F^{-} *recA*) harboring plasmid pPB1132 (see Fig. 2) by mating with strain SBS922 (F-106::Tn*5*), and Tn*5* hops from the episome into the plasmid were selected as resistant to 300 μg of kanamycin per ml. Plasmid pPB1175 in which Tn*5* was inserted into the *appA* gene was used for the construction of the deletion between the two *Sma*I sites, yielding plasmid pPB1198. This labeled deletion was introduced from pPB1198 into the chromosome of strain SBS1174 *(polA12)* in place of *appA* by homologous recombination by the same method as described in the Fig. 3 legend and further transduced with P1 into other strains, selecting for Km^{r} colonies.

TABLE 4. Comparison of the roles of *appA* and *phoA* gene products for growth in the presence of different phosphorylated compounds as sole source of P_i

Strain	Relevant genotype or phenotype	Phosphatase activity in P_i-limited medium[a]		Plate growth assays with indicated phosphorylated compounds as sole source of P_i[b]				
		pH 2.5	pH 8.5	Poly-3P	GTP	PEA	PNPP	Glyc-3P
SBS1371	Δ*appA* Δ*pho(A,B)*	1.0	0.6	−	−	−	−	−
SBS1186	*appA*$^+$ Δ*pho(A,B)*	932	0.4	−	−	−	−	−
SBS1370	Δ*appA pho(A,B)*$^+$	0.9	891	+	+	+	+	+
SBS926	*appA*$^+$ *pho(A,B)*$^+$	1,082	906	+	+	+	+	+
SBS1182	*appA'-phoA* Δ*pho(A,B)*	1.8	16	±	±	±	±	±
SBS1355	*appA*$^+$ Δ*phoA20* (*phoB*$^+$)	875	0.6	±	−	−	−	−
SBS1362	Δ*appA* Δ*phoA20* (*phoB*$^+$)	1.0	0.5	±	−	−	−	−
SBS1356	*appA'-phoA* Δ*phoA20* (*phoB*$^+$)	1.2	23	+	+	+	+	+

[a] Expressed as nanomoles of paranitrophenol formed per minute per milligram of protein.

[b] Poly-3P, Linear tripolyphosphate; GTP, guanosine triphosphate; PEA, phosphorylethanolamine; PNPP, *p*-nitrophenyl phosphate; Glyc-3P, glycerol 3-phosphate. The relative growth of patches was evaluated after 48 h at 37°C.

chromosome by the *appA-phoA* fusion described above, slow growth was observed on all compounds, as a consequence of the alkaline phosphatase activity expressed from this fusion. In a *phoB*$^+$ *phoA*$^-$ background, no difference was seen between an *appA*$^+$ (SBS1355) and an *appA* strain (SBS1362). Replacing *appA* by an *appA-phoA* fusion, by contrast, allowed a wild-type growth rate. Consequently, acid phosphatase in vivo cannot replace alkaline phosphatase for hydrolysis of the phosphorylated substrates and in particular is not responsible for the degradation of exogenously supplied short-chain polyphosphates. A difference in growth yield between isogenic *phoB* and *phoB*$^+$ strains may be due to the absence or presence of the *phoE* pore protein, of the *phoS* protein, or of both.

Discussion

Anoxia and prolonged incubation in the stationary phase are conditions able to promote the synthesis of the pH 2.5 acid phosphatase of *E. coli*, but limiting the availability of P_i in the growth medium is more efficient in terms of enzyme yields. This synthesis, however, is not under the control of the positive regulator needed for alkaline phosphatase transcription, the *phoB* gene product which is the general effector of the Pho regulon (23). Although mutations in genes *phoU* and *phoV* were not tested, the lack of influence of mutations in *phoR, phoM, phoT*, and *phoS* is a strong indication that acid phosphatase regulation by P_i is triggered by a mechanism entirely different from that for alkaline phosphatase. Acid phosphatase shares with aminopeptidase and some *psi* (phosphate starvation inducible) genes (26, 27) the property of being regulated by exogenous P_i levels in a *phoB*-independent pathway. Among such *psi-lac* fusions, some are also "induced" by oxygen deprivation and could be submitted to the same regulatory mechanism as *appA*.

We have shown that acid phosphatase expressed from a multicopy recombinant plasmid containing *appA* is controlled by phosphate availability as the chromosomal gene. This is only true, however, if in this plasmid *appA* is transcribed only from its own original promoter, *appAp. Its constitutive expression when transcribed from the exogenous tet* promoter indicates that the last step in phosphate control probably affects the initiation of *appA* transcription at P appA. In phosphate-limited medium, the amount of enzyme made from plasmid-borne *appA* genes transcribed from *appAp* only, when compared with that of the chromosomal gene, roughly corresponds to 20 to 30 gene copies per cell (see Fig. 3). Consequently, if a positive regulator coded for by a single chromosomal gene is involved, it is not represented in limiting amounts. Alternatively, the gene for such a regulator could be contained in the DNA regions flanking *appA* on the recombinant plasmids.

Alkaline phosphatase activity from a hybrid protein encoded by a chromosomal *appA-phoA* fusion is shown to be regulated by P_i as the original acid phosphatase. Strains harboring such a fusion can be excellent tools for the direct selection of *appA* regulatory mutants on plates containing the chromogenic substrate XP.

The negative control afforded by cAMP and the CAP protein is independent of the conditions used to promote *appA* transcription. It is also operative on recombinant plasmids (provided the *appA* promoter alone is used), indicating that cAMP does not affect a posttranscriptional event. That the CAP-cAMP complex directly interacts with the *appA* promoter is suggested by the existence of a sequence close to the CAP-binding consensus sequence, which overlaps the −10 region of this promoter (Touati and Danchin, in preparation). It is, however, surprising that, when present in the cell, CAP and cAMP bound to this site do not inhibit the progression of polymerases initiating at the upstream exogenous *tet* promoter (see Fig. 6).

Strains deleted for *appA* are not affected in their ability to use phosphoanhydride-containing molecules as sole source of phosphate, even in the absence of alkaline phosphatase. This raises the question of the real physiological role of this enzyme and of the existence in the periplasm of local acidic domains allowing its activity. On one hand, the enzyme might possess another unraveled function under other pH conditions, the acid phosphatase activity being due, for instance, to a structural change in the protein by acid. It might, on the other hand, behave as a "safeguard enzyme" able to prevent the cell from temporarily acidic conditions, by buffering the periplasm with phosphate released from exogenous or membrane-bound phosphoanhydride-containing compounds such as polyphosphates. A structural and functional analysis of the protein is now necessary to elucidate this question.

LITERATURE CITED

1. **Anraku, Y.** 1964. A new cyclic phosphodiesterase having a 3′ nucleotidase activity from *Escherichia coli* B. I. Purification and properties of the enzyme. J. Biol. Chem. **239:**3412–3418.
2. **Anraku, Y.** 1964. A new cyclic phosphodiesterase having a 3′ nucleotidase activity from *Escherichia coli* B. II. Further studies on substrate specificity and mode of action of the enzyme. J. Biol. Chem. **239:**3420–3424.
3. **Anraku, Y.** 1966. Cyclic phosphodiesterase of *Escherichia coli*, p. 130–149. *In* D. Cantoni and B. Davis (ed.), Procedures in nucleic acid research. Harper & Row, Publishers, Inc., New York.
4. **Beacham, I. R.** 1979. Periplasmic enzymes in gram-negative bacteria. Int. J. Biochem. **10:**877–883.
5. **Beacham, I. R., and S. Garrett.** 1980. Isolation of *Escherichia coli* mutants (*cpdB*) deficient in periplasmic 2′:3′-cyclic phosphodiesterase and genetic mapping of the *cpdB* locus. J. Gen. Microbiol. **119:**31–34.
6. **Beacham, I. R., R. Kahana, L. Levy, and E. Yagil.** 1973. Mutants of *Escherichia coli* K-12 "cryptic" or deficient in 5′-nucleotidase (uridine diphosphate-sugar hydrolase) and 3′-nucleotidase (cyclic phosphodiesterase) activity. J. Bacteriol. **116:**957–964.
7. **Beacham, I. R., and E. Yagil.** 1976. Genetic location of the gene (*ush*) specifying periplasmid uridine 5′-diphosphate glucose hydrolase (5′-nucleotidase) in *Escherichia coli* K-12. J. Bacteriol. **128:**487–489.
8. **Bolivar, F., R. L. Rodriguez, P. J. Green, H. L. Betlach, H. L. Heynecker, H. W. Boyer, J. H. Crosa, and S. Falkow.** 1977. Construction and characterization of new cloning vehicles. II. A multipurpose cloning system. Gene **2:**95–113.

8a. **Boquet, P. L., C. Manoil, and J. Beckwith.** 1987. Use of Tn*phoA* to detect genes for exported proteins in *Escherichia coli*: identification of the plasmid-encoded gene for a periplasmic acid phosphatase. J. Bacteriol. **169:**1663–1669.

9. **Courtois, J. E., and L. Manet.** 1952. Recherches sur la phytase. XVII. Les phytases du colibacille. Bull. Soc. Chim. Biol. **34:**265–278.
10. **Dassa, E., and P. L. Boquet.** 1981. Is the acid phosphatase of *Escherichia coli* with optimum pH of 2.5 a polyphosphate depolymerase? FEBS Lett. **135:**148–150.
11. **Dassa, E., and P. L. Boquet.** 1985. Identification of the gene *appA* for the acid phosphatase (pH optimum 2.5) of *Escherichia coli*. Mol. Gen. Genet. **200:**68–73.
12. **Dassa, E., M. Cahu, B. Desjoyaux-Cherel, and P. L. Boquet.** 1982. The acid phosphatase with optimum pH of 2.5 of *Escherichia coli*: physiological and biochemical study. J. Biol. Chem. **257:**6669–6676.
13. **Dassa, E., C. Tétu, and P. L. Boquet.** 1980. Identification of the acid phosphatase (optimum pH 2.5) of *Escherichia coli*. FEBS Lett. **113:**275–278.
14. **Dvorak, H. F., R. W. Brockman, and L. A. Heppel.** 1967. Purification and properties of two acid phosphatase fractions isolated from osmotic shock fluid of *Escherichia coli*. Biochemistry **6:**1743–1751.
15. **Groisman, E. A., B. A. Castilho, and M. J. Casadaban.** 1984. *In vivo* DNA cloning and adjacent gene fusing with a mini-Mu-lac bacteriophage containing a plasmid replicon. Proc. Natl. Acad. Sci. USA **81:**1480–1483.
16. **Hafckensheid, J. C. M.** 1968. Properties of an acid phosphatase in *Escherichia coli*. Biochim. Biophys. Acta **167:** 582–589.
17. **Manoil, C., and J. Beckwith.** 1985. Tn*phoA*: a transposon probe for protein export signals. Proc. Natl. Acad. Sci. USA **82:**8129–8133.
18. **Nakata, A., M. Yamaguchi, K. Izutani, and M. Amemura.** 1978. *Escherichia coli* mutants deficient in the production of alkaline phosphatase isozymes. J. Bacteriol. **134:**287–294.
19. **Neu, H. C.** 1967. The 5′ nucleotidase of *Escherichia coli*. I. Purification and properties. J. Biol. Chem. **242:**3896–3904.
20. **Rogers, D., and F. J. Reithel.** 1960. Acid phosphatases of *Escherichia coli*. Arch. Biochem. Biophys. **89:**97–104.
21. **Tétu, C., E. Dassa, and P. L. Boquet.** 1979. Unusual pattern of nucleoside polyphosphate hydrolysis by the acid phosphatase (optimum pH 2.5) of *Escherichia coli*. Biochem. Biophys. Res. Commun. **87:**314–322.
22. **Torriani, A.** 1960. Influence of inorganic phosphate in the formation of phosphatases by *Escherichia coli*. Biochim. Biophys. Acta **38:**460–479.
23. **Torriani, A., and D. N. Ludtke.** 1985. The Pho regulon of *Escherichia coli*, p. 224–242. *In* M. Schaechter, F. C. Neidhardt, J. Ingraham, and N. O. Kjeldgaard (ed.), The molecular biology of bacterial growth. Jones and Bartlett Publishers, Boston.
24. **Touati, E., E. Dassa, and P. L. Boquet.** 1986. Pleiotropic mutations in *appR* reuce pH 2.5 acid phosphatase expression and restore succinate utilisation in CRP-deficient strains of *Escherichia coli*. Mol. Gen. Genet. **202:**257–264.
25. **van Hofsten, B., and J. Porath.** 1962. Purification and some properties of an acid phosphatase from *Escherichia coli*. Biochim. Biophys. Acta **64:**1–12.
26. **Wanner, B. L., and P. Latterell.** 1980. Mutants affected in alkaline phosphatase expression: evidence for multiple positive regulators of the phosphate regulon in *Escherichia coli*. Genetics **96:**353–366.
27. **Wanner, B. L.** 1983. Overlapping and separate controls on the phosphate regulon in *Escherichia coli* K12. J. Mol. Biol. **166:**283–308.

II. PHOSPHATE REGULATION IN DIVERSE MICROORGANISMS

Introduction

H. O. HALVORSON[1] AND A. NAKATA[2]

Brandeis University, Waltham, Massachusetts 02254,[1] and Osaka University, Osaka, Japan[2]

During the latter part of the last century, it was recognized that the enzymatic properties of cells could be manipulated. In 1882, Wortmann (23) observed that microorganisms would produce amylase only when grown on starch, and 17 years later Ducleaux (4) found that *Aspergillus* would produce protease only when grown on a medium containing protein, such as milk. It was not clear, however, from these early experiments whether changes in enzymes occurred because of a Darwinian selection or an inherent capacity in each cell in the population. In 1900, Dienart (3) showed that adaptation in yeast to galactose fermentation could occur in the absence of cell division. Many years later, Benzer (1), in Jacques Monod's laboratory, showed that adaptation occurred in each and every cell in the population of *Escherichia coli*, a conclusion directly demonstrated later by Ganesan and Rotman (5). From subsequent investigations, Jacob and Monod (8) went on to clarify the mechanism of regulation of β-galactosidase synthesis as negative, to describe the function of a bacterial operon (in which a number of contiguous genes could be subjected to the control of a common regulatory gene), and to set the framework for modern manipulation of enzyme expression through studies of regulation at the genetic level.

Coordinate control of gene expression was extended by Maas and Clark (13) to metabolic pathways in which relevant genes are scattered over the bacterial chromosome. They proposed the term regulon to describe "a system in which the production of enzymes can be controlled by a single repressor substance." The complex *pho* regulon in *E. coli*, which is controlled by the repressor substance P_i, is described in this volume. In response to P_i limitation, the synthesis of approximately 85 cellular proteins in *E. coli* becomes affected, as measured by two-dimensional gels (7).

A similar complex regulatory mechanism for the synthesis of enzymes involved in phosphate metabolism by P_i has been described in diverse bacteria (6), in *Saccharomyces cerevisiae*, and in *Neurospora* (for review, see reference 15). Of these latter two, the regulation of P_i metabolism is best understood in *S. cerevisiae*.

In *S. cerevisiae*, phosphate regulation involves five principal enzymes for the acquisition and metabolic integration of P_i: an exocellular acid phosphatase (20), a phosphate permease(s) (19), polyphosphate kinase (6, 22), and alkaline phosphatase and polyphosphatase located in the yeast vacuole (9, 12). These enzymes regulate intracellular concentrations of P_i by a cyclic pathway of polyphosphate synthesis and degradation (12), and this cycle is thought to be an important factor in the regulation of cellular homeostasis (2). All of these enzymes are regulated by external growth conditions of P_i (2, 6). When the cells are starved for P_i, all of these enzymes are highly synthesized under derepressed conditions (21).

The genetic system for phosphate metabolism in *S. cerevisiae* is a dispersed gene system consisting of numerous structural and regulatory genes. This complex system involves both positive (*PHO4*) and negative (*PHO80*) gene product effectors (for current review, see Oshima, this volume). At the molecular level, understanding of the regulation of phosphate metabolism in *S. cerevisiae* is beginning to emerge. The most detailed information available is on acid phosphatase (Hinnen et al., this volume; Parent et al., this volume).

The extent to which regulation of phosphate metabolism has been conserved during evolution has not been determined, since we have only limited information concerning genetic regulation of this system in other microorganisms. The only bacterium other than *E. coli* for which we have extensive genetic information, *Bacillus subtilis*, contains many elements of the *pho* regulon of *E. coli*. These include alkaline phosphatase (*phoR*) (10, 14), repressible alkaline phosphatase and alkaline phosphodiesterase (*phoP*) (11), constitutive alkaline phosphatase (*phoS, phoT*) (17, 18), and two sporulation-associated phosphatases (*sapA, sapB*) (16). It is of particular interest that *Bacillus licheniformis* contains tandem duplicate genes for alkaline phosphatase, one produced during vegetative growth and the other produced only during sporulation (Hulett, this volume).

LITERATURE CITED

1. **Benzer, S.** 1953. Induced synthesis of enzymes in bacteria analyzed at the cellular level. Biochim. Biophys. Acta **11:** 383–395.
2. **Dawes, E. A., and P. J. Senior.** 1973. The role and regulation of energy reserve polymers in microorganisms. Adv. Microb. Physiol. **10:**135–266.
3. **Dienart, F.** 1900. Sur la fermentation du galactose et sur l'accontumance des levures a ce sucre. Ann. Inst. Pasteur (Paris) **14:**139–189.
4. **Ducleaux, E.** 1899. Sur le lait congelé. Ann. Inst. Pasteur (Paris) **10:**393–402.
5. **Ganesan, A. K., and B. Rotman.** 1964. Measurement of activity of single molecules of β-D-galactosidase. J. Mol. Biol. **10:**337–340.
6. **Harold, F. M.** 1966. Inorganic polyphosphates in biology: structure, metabolism, and function. Bacteriol. Rev. **30:** 772–794.
7. **Ingram, J. L., O. Maaløe, and F. C. Neidhardt.** 1983. Growth of the bacterial cell. Sinauer Associates, Inc., Sunderland, Mass.
8. **Jacob, F., and J. Monod.** 1962. On the regulation of gene activity. Cold Spring Harbor Symp. Quant. Biol. **26:**193–211.
9. **Katchman, B. J., and W. O. Felty.** 1955. Phosphorus metabolism in growing cultures of *Saccharomyces cerevisiae*. J. Bacteriol. **69:**607–615.
10. **LeHégarat, J.-C., and C. Anagnastopoulos.** 1969. Localisation chromosomique d'un gene gouvernant la synthese d'une phosphatase alcaline chez *Bacillus subtilis*. C.R. Acad. Sci. **269:**2048–2050.
11. **LeHégarat, J.-C., and C. Anagnastopoulos.** 1973. Purification, subunit structure and properties of two repressible phosphohydrolases of *Bacillus subtilis*. Eur. J. Biochem. **39:**525–539.
12. **Liss, E., and P. Langen.** 1960. Uber ein hochmolekulares Polyphosphät der Hefe. Biochem. Z. **333:**193–201.
13. **Maas, W., and A. J. Clark.** 1964. Studies on the mechanism of repression of arginine biosynthesis in *Escherichia coli*. J. Mol. Biol. **8:**365–370.
14. **Miki, T., Z. Minimi, and Y. Ikeda.** 1965. The genetics of alkaline phosphatase formation in *Bacillus subtilis*. Genetics **52:**1093–1100.
15. **Oshima, Y.** 1982. Regulatory circuits for gene expression: the metabolisms of galactose and phosphate, p. 159–180. *In* J. N. Strathern, E. W. Jones, and J. R. Broach (ed.), The molecular biology of the yeast *Saccharomyces*: metabolism and gene expression. Cold Spring Harbor Laboratory, Cold Spring Harbor, N.Y.
16. **Piggot, P. J., and R. S. Buxton.** 1982. Bacteriophage PBSX-induced deletion mutants of *Bacillus subtilis* 168 constitutive for alkaline phosphatase. J. Gen. Microbiol. **128:**663–669.
17. **Piggot, P. J., and J. A. Hoch.** 1985. Revised genetic linkage map of *Bacillus subtilis*. Microbiol. Rev. **49:**158–179.
18. **Piggot, P. J., and S. Y. Taylor.** 1977. New types of mutation affecting formation of alkaline phosphatase by *Bacillus subtilis* in sporulation conditions. J. Gen. Microbiol. **102:**69–80.
19. **Roomans, G. M., and G. W. F. H. Borst-Pauwels.** 1979. Interaction of cations with phosphate uptake by *Saccharomyces cerevisiae*. Effects of surface potential. Biochem. J. **178:**521–527.
20. **Schmidt, G., G. Bartsch, M. C. Laumont, J. Herman, and M. Liss.** 1963. Acid phosphatase of bakers yeast: an enzyme of the external cell surface. Biochemistry **2:**126–131.
21. **Toh-e, A., Y. Ueda, S. Kakimota, Y. Oshimo, A. A. Toh-e, and Y. Oshima.** 1973. Isolation and characterization of acid phosphatase mutants in *Saccharomyces cerevisiae*. J. Bacteriol. **113:**727–738.
22. **Urech, K., M. Duerr, T. H. Boller, A. Wiemken, and J. Schwencki.** 1978. Localization of polyphosphate in vacuoles of *Saccharomyces cerevisiae*. Arch. Microbiol. **116:**275–278.
23. **Wortmann, J.** 1882. Ulersuchunger über das diastatische Ferment der Bakterium. Z. Tschr. Physiol. Chem. **6:**287–296.

Alkaline Phosphatase from *Bacillus licheniformis*: Proteins and Genes

F. MARION HULETT

Department of Biological Sciences, University of Illinois at Chicago, Chicago, Illinois 60680

Alkaline phosphatase (orthophosphoric-monoester phosphohydrolase [alkaline optimum], EC 3.1.31) from procaryotes has been most extensively studied in *Escherichia coli*. Secretion of alkaline phosphatase in *E. coli* into the periplasmic space has been well documented (23, 24, 27). A single gene is responsible for *E. coli* alkaline phosphatase (1, 4), and both the gene and the protein have been well characterized (2, 20, 21). The alkaline phosphatase system is more complex in *Bacillus* species. We have shown that there are at least two structural genes for alkaline phosphatase in *B. licheniformis* MC14 (12, 13, 18). Expression of these genes in *E. coli* showed that they both code for 60,000-molecular-weight proteins which cross-react with anti-alkaline phosphatase. In vitro transcription studies suggested that different RNA polymerase holoenzymes were required to transcribe these genes.

Localization studies in *B. licheniformis* MC14 have shown that there are membrane-bound and secreted forms of alkaline phosphatase (see Fig. 1). A peripherally bound alkaline phosphatase (salt extractable) is present on the inner leaflet of the cytoplasmic membrane (23a, 31), and an integrally bound form (requires detergent extraction) is present on the outer leaflet (30). Secreted forms include a soluble alkaline phosphatase secreted through the membrane which can be released by removal of the cell wall with lysozyme (labeled "periplasmic" on Fig. 1) (11) and a truly extracellular form (29). The alkaline phosphatase from each of these locations has been purified to homogeneity and characterized (11, 14–16, 28, 30). There are no discernible chemical-physical differences in the enzymes from these different locations. The amount of alkaline phosphatase synthesized and its location depend on the phase of growth and the growth conditions (9, 29). *B. licheniformis* MC14 synthesizes 15 times more alkaline phosphatase activity than is reported for other *Bacillus* species (19) and under specific culture conditions secretes 99% of this activity into the medium.

More recently, similar localization studies have been carried out in another strain of *B. licheniformis*, 749C. Those results (10) and immuno-electron microscopy data (5) showed that the distribution in strain 749C was similar to that in MC14.

In *B. subtilis* the localization studies have been less rigorous; however, membrane-bound and secreted forms of alkaline phosphatase have been reported (3, 6). Genetic analysis of alkaline phosphatase in *B. subtilis* is fairly extensive. These analyses are complicated by two facts. (i) There is a vegetative alkaline phosphatase and a sporulation alkaline phosphatase. (ii) There are phosphodiesterases which hydrolyze *p*-nitrophenyl phosphate and 5-bromo-4-chloro-3-indolyl phosphate, the substrates used to screen for alkaline phosphatase. The vegetative form is produced in greater quantities than the sporulation form but is repressed by P_i at concentrations which do not affect synthesis of the sporulation enzyme. It is still not clear whether *B. subtilis* has one gene under complex regulation or two genes which differ in regulation but differ very little in the sequences which affect the mature protein. Comparison of the sporulation and vegetative alkaline phosphatases showed no discernible chemical-physical differences (7, 8), but the cloning of two structural genes for alkaline phosphatase from *B. licheniformis* suggests the possibility that multiple structural genes with similar protein gene products also exist in *B. subtilis*.

We are in the process of constructing strains of *B. licheniformis* MC14 which retain only one alkaline phosphatase gene. This will enable us to unambiguously correlate the time of expression and the final location of the protein gene product with each alkaline phosphatase gene. We comment briefly here on the alkaline phosphatase proteins and on the genes responsible for their production.

Results

Alkaline phosphatase is the major secreted protein. When cobalt (0.1 mM) is added to a minimal salts defined medium at the onset of alkaline phosphatase production (when the phosphate concentration in the medium reaches 0.075 mM), over 99% of the enzyme activity is present in the culture medium 2.75 h after cobalt addition (17). This activity remains in the supernatant after centrifugation at 100,000 × *g*. A 2.24-fold purification yields a homogeneous enzyme with approximately 10% recovery of the initial activity. Figure 2 shows a Coomassie blue-stained gel of concentrated medium (lanes 1 and 2) prior to any purification and of purified medium alkaline phosphatase (lanes 3 and 4). The major secreted protein is alkaline phosphatase.

Comparison of the Mg^{2+}-extracted membrane-associated alkaline phosphatase with the soluble secreted alkaline phosphatase by limited proteolysis. No immunological or structural differences were observed between the cell-bound alkaline phosphatase(s) and the secreted alkaline phosphatase. We expected that the enzyme species most likely to differ significantly would be that peripherally associated with the inner leaflet of the cytoplasmic membrane, the salt-extractable alkaline phosphatase (31).

However, fluorograms of inner leaflet membrane or secreted enzyme subjected to *Staphylococcus aureus* V-8 limited proteolysis show no significant difference in the digestion pattern. Figure 3 shows such a fluorogram. Lanes 1 and 2 contain a limited protease digest of Mg^{2+}-extractable membrane alkaline phosphatase, and lanes 3 and 4 show secreted enzyme treated in the identical manner. Labeled peptides generated by limited proteolysis of either enzyme (peripherally membrane associated or secreted) are similar as judged by migration on sodium dodecyl sulfate gels.

Pulse-chase experiments revealed an immunologically related cytosol protein whose disappearance from the cytosol was correlated with the appearance of the secreted alkaline phosphatase in the medium. Cells were grown as described for Fig. 2. [^{35}S]Methionine was added 1 min after the cobalt addition. After 1 min, the chase was initiated by the addition of unlabeled methionine, and samples were taken at 2-min intervals. The samples were fractionated as described in the legend to Fig. 4 and subjected to sodium dodecyl sulfate-gel electrophoresis. Gels were analyzed by fluorography. The bottom frame of Fig. 4 shows the labeled protein pattern of whole-cell soluble proteins while the top frame shows the secreted proteins of samples taken at 2-min intervals during the chase period. Arrows indicate the 63,000- and 60,000-molecular-weight proteins in each fraction that are immunoprecipitatable by anti-alkaline phosphatase. Note that the disappearance of the 63,000-molecular-weight protein in the whole-cell

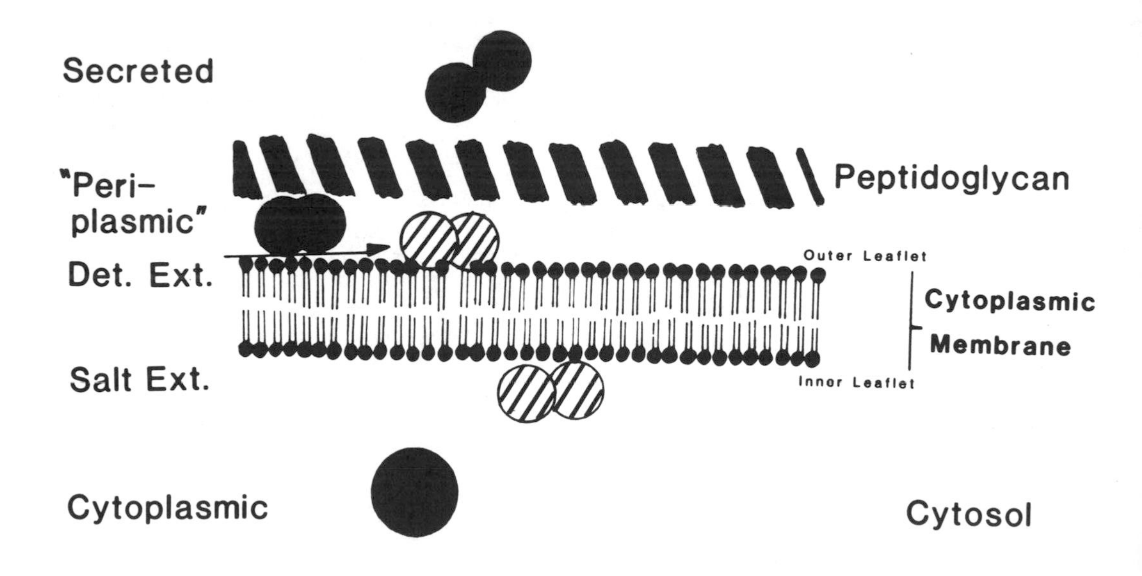

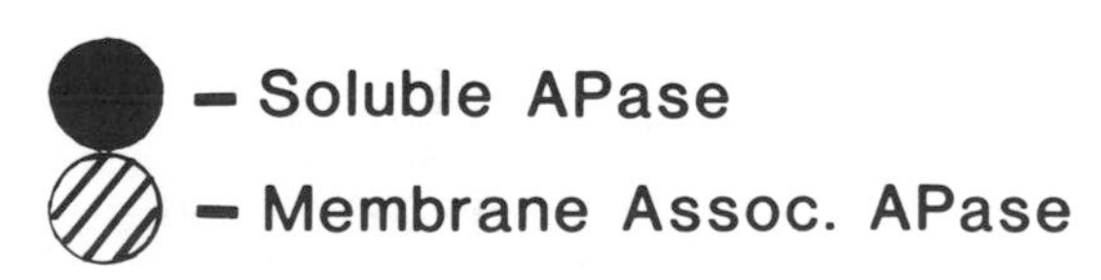

FIG. 1. Distribution of alkaline phosphatase in *B. licheniformis* MC14. Double circles represent active dimers. Single circles represent inactive monomers. Dark circles represent proteins which are soluble. Light circles represent membrane-associated alkaline phosphatase. The predominant species at any one time depends on the phase of growth and the growth conditions.

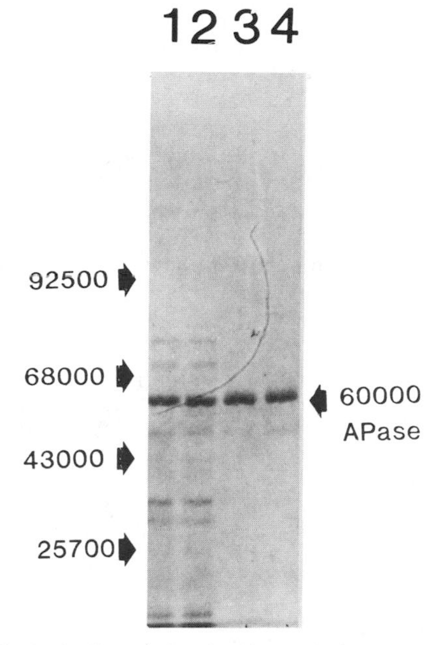

FIG. 2. Sodium dodecyl sulfate gel of concentrated medium fraction and purified secreted alkaline phosphatase. Lanes 1 and 2 contain concentrated medium (25 μl per lane). Lanes 3 and 4 contain purified secreted alkaline phosphatase (24 μg per lane).

soluble fraction correlates with the appearance of the 60,000-molecular-weight alkaline phosphatase subunit in the growth medium (top frame). The 60,000-molecular-weight protein in the whole-cell soluble fraction (bottom frame) which cross-reacts with the anti-alkaline phosphatase was released from whole cells as a soluble active alkaline phosphatase dimer when protoplasts were made. This enzyme has been described previously (11). Further, the 63,000-molecular-weight protein was localized in the cytosol of the lysed protoplasts. After protoplast formation, washed protoplasts were lysed and the cytosol fraction was separated from the membrane fraction by centrifugation at 100,000 × *g* (1 h). The cytosol fraction contained no alkaline phosphatase activity. The 63,000-molecular-weight inactive anti-alkaline phosphatase cross-reaction protein was isolated from the cytosol fraction by immunoadsorption. Thus, the 63,000-molecular-weight protein which is chased from the whole-cell soluble fraction as the secreted alkaline phosphatase appears in the medium as an inactive cytosol protein 3,000 molecular weight larger than, but antigenically similar to, the secreted alkaline phosphatase.

Genes responsible for alkaline phosphatase production in *B. licheniformis*. To examine the possibility that the multiple forms of alkaline phosphatase in *B. licheniformis* and their subcellular membrane association or secretion might be the result of selective expression of multiple structural genes under different regulatory control, we cloned the structural gene(s) for alkaline phosphatase.

The alkaline phosphatase gene of *B. licheniformis* MC14 was inserted into the *Pst*I site of pMK2004 (pMH8) and cloned as described previously (12). It was further subcloned in a 4.2-kilobase *Eco*RI-*Xho*I fragment inserted into the *Eco*RI-*Xho*I sites on pMK2004, to yield plasmid pMH81 (Fig. 5). The 0.97-kilobase *Pvu*II$_3$-*Hin*dIII$_3$ fragment within the coding region of this gene in pMH81 (Fig. 1) was used as a probe in the Southern transfer analysis. DNA flanking the *Bgl*II$_1$ site on pMH8 was implicated as a second region of hybridization by the Southern analysis (data not shown). A *Sal*I deletion of pMH8 (pMH87) was constructed (Fig. 5) (18). This deletion plasmid lacks more than half the coding region of the alkaline phosphatase gene subcloned in pMH81 (APase I) and contains 1.1

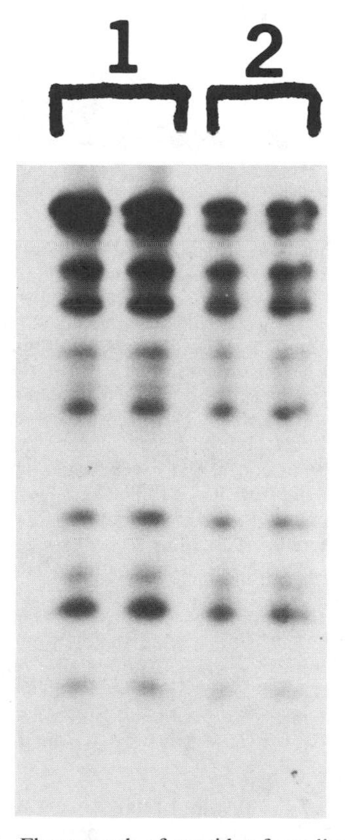

FIG. 3. Fluorograph of peptides from limited proteolysis of Mg^{2+}-extracted membrane and secreted alkaline phosphatase electrophoretically resolved on sodium dodecyl sulfate gels. (1) Two lanes containing identical digests of Mg^{2+}-extracted membrane alkaline phosphatase. (2) Two lanes containing identical digests of secreted alkaline phosphatase.

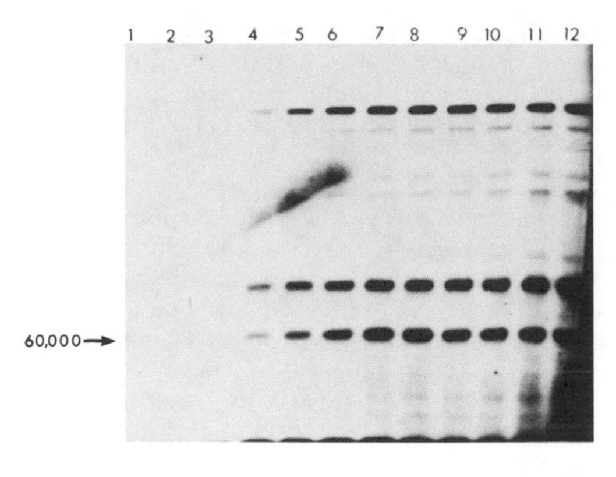

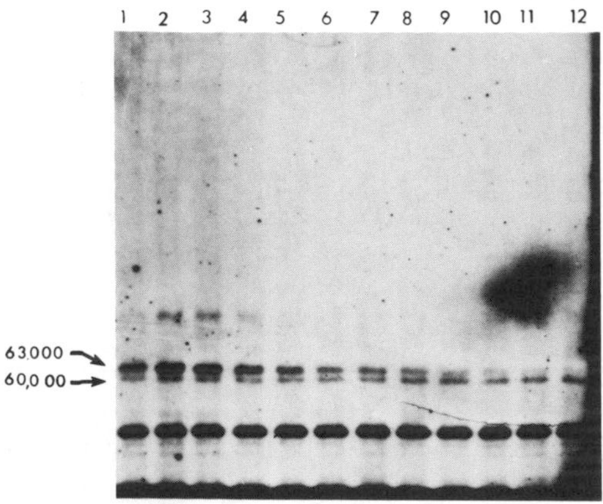

FIG. 4. Fluorogram of sodium dodecyl sulfate gels showing that the kinetics of disappearance of the 63,000-molecular-weight cell-bound soluble protein identified by anti-alkaline phosphatase correlates with the kinetics of appearance of the 60,000-molecular-weight secreted alkaline phosphatase subunit. Cells were grown as described above. [^{35}S]Methionine was added at the same time cobalt was added (at 6.5 h of growth). After a 1-min pulse, excess unlabeled methionine was added. Samples were taken at 2-min intervals during a 22-min chase. Cells were separated from the medium by centrifugation (100,000 × *g*, 1 h). Cells were lysed, and the soluble fraction was separated from the membrane fraction (100,000 × *g*, 1 h). The bottom picture shows a sodium dodecyl sulfate gel of the whole-cell soluble fractions; 1 through 12 correspond to 0 to 22 min after the chase was initiated. The top picture shows the labeled secreted proteins from the corresponding sample. Anti-alkaline phosphatase immunoprecipitates the 63,000- and 60,000-molecular-weight proteins of the whole-cell soluble fraction and the 60,000-molecular-weight protein from the medium.

kilobases of the original DNA insert that is also in pMH81. When this deletion plasmid was used to transform *E. coli* Xph90a *(phoA)* and plated on Neopeptone–5-bromo-4-chloro-3-indolyl phosphate–kanamycin plates, the colonies turned blue in 1 day. The possibility that the alkaline phosphatase phenotype of *E. coli* Xph90a transformed with pMH87 was dependent on DNA previously assigned to the coding region of APase I was eliminated by constructing and analyzing deletion plasmids of pMH87 that retained none of the APase I gene. Such deletion plasmids of pMH87 complement *phoA E. coli*.

Deletions of pMH87 were constructed by digestion of pMH87 with *Sal*I, followed by exonuclease III and nuclease S1 digestion (18). Deletion plasmids were used to transform *E. coli* Xph90a. Plasmids from blue and from white colonies were isolated, and the extent of the deletion in each was mapped. Colonies containing plasmids in which all of the DNA from the coding region of the alkaline phosphatase gene in pMH81 (APase I) was deleted remained blue (Fig. 5). Further deletion (from *Sal*I$_1$) up to a point approximately 400 bases before *Xho*I$_2$ also showed alkaline phosphatase (APase II) production when used to transform Xph90a. Deletion plasmids in which all DNA from *Sal*I$_2$ to a point approximately 260 bases before *Xho*I$_2$ or further was removed showed no alkaline phosphatase (APase II) production when used to transform *E. coli* Xph90a (Fig. 5). Therefore, one terminus of the APase II gene is mapped at 6.1 to 6.3 on pMH87.

When the proteins from *E. coli* cultures containing either pMH81 or pMH87 were subjected to immunoblot analysis with anti-alkaline phosphatase followed by ^{125}I-labeled goat anti-rabbit immunoglobulin, an autoradiogram of each showed an M_r 60,000 protein (18). This suggests that APase II and APase I code for an M_r 60,000 protein that cross-reacts with anti-alkaline phosphatase (18).

Preliminary characterization of the promoter for APase I and APase II. Deletion mapping and in vitro transcription studies suggest that the promoter for APase I lies between *Pvu*II$_3$ and *Pvu*II$_4$ on pMH81 and that a σ^{55}-containing RNA polymerase holoenzyme is required for transcription (18).

Deletion mapping and in vitro transcription studies showed that the promoter for APase II lies between *Hin*dIII$_3$ and *Xho*I$_2$ approximately 300 base pairs from *Xho*I$_2$ and that it requires a σ^{37}-containing RNA polymerase holoenzyme for transcription (18).

Analysis of these promoters should allow us to determine whether one of these genes is responsible for the alkaline phosphatase that is made in small amounts in rich medium during logarithmic growth and localized on the inside of the cytoplasmic membrane, whereas the other is responsible for the production of the alkaline phosphatase that constitutes 50% of the total protein secreted by this organism when the P_i concentration in the medium decreases below 0.075 mM and the culture is entering the stationary phase of growth (29).

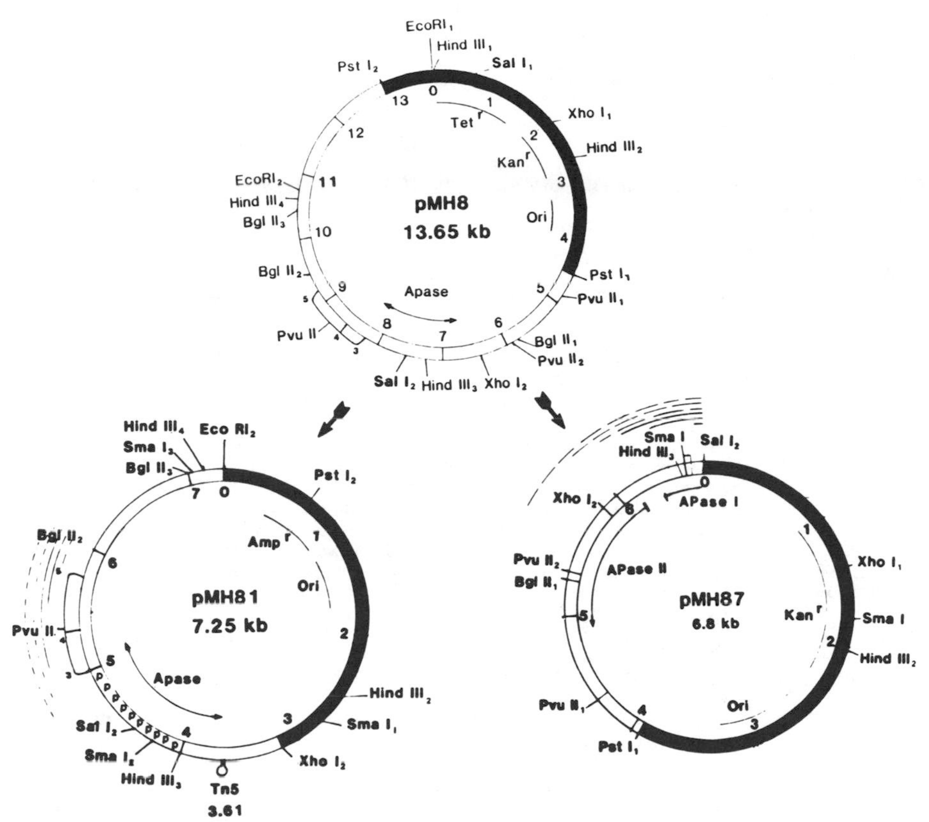

FIG. 5. Subcloning of two fragments from pMH8. The 4.2-kilobase *Xho*I$_2$-*Eco*RI$_2$ fragment from pMH8 was inserted into the *Eco*RI-*Xho*I sites of pMK2004 to yield pMH81 (12). The coding region for APase I on pMH81 is indicated. A *Sal*I$_2$ to *Sal*I$_1$ deletion of pMH8 yielded pMH87. Xph90a containing pMH87 can also hydrolyze 5-bromo-4-chloro-3-indolyl phosphate. The pMH81 fragment used as the probe for Southern hybridization (*Pvu*II$_3$ to *Hin*dIII$_3$) is indicated by ppp. Exonuclease III/nuclease S1 deletion mapping: for pMH81, solid lines indicate deletion distance from *Bgl*II$_2$ in pMH81 that resulted in retention of the APase phenotype when the deletion plasmid was used to transform Xph90a, and dashed lines indicate lengths of DNA deleted that resulted in APase$^-$ phenotype; for pMH87, solid lines indicate the deletion distance from *Sal*I$_2$ that allowed retention of the APase phenotype when the constructed plasmid was used to transform Xph90a, and dashed lines indicate lengths of DNA deleted that resulted in APase$^-$ phenotype.

Wong et al. (32) suggested that the role of the σ^{37} promoter may include, in addition to expression of sporulation-specific genes (22, 25, 26), expression of genes encoding extracellular enzymes as well as of genes regulated by growth phase. We now can determine whether the alkaline phosphatase gene requiring σ^{37}, APase II, is responsible for the secreted alkaline phosphatase species of *B. licheniformis*, whereas the σ^{55}-requiring APase I encodes the internal alkaline phosphatase.

Having cloned the alkaline phosphatase genes, we now can generate structural-gene mutants in APase I or APase II in *B. licheniformis* to determine the final destination of each gene product. It will be of interest to determine whether divergence of duplicated genes in *Bacillus* species has resulted in altered regulation of the gene and location of the protein product.

Our working hypothesis envisions that the unique distribution of alkaline phosphatase in *B. licheniformis* (membrane associated on the inner and outer leaflet of the cytoplasmic membrane and secreted) is the result of multiple structural genes, which differ significantly in regulation and in the sequences important for localization of the protein gene product, but to a lesser extent in sequences responsible for the primary structure of the mature protein gene product.

I thank Karen Stuckmann, Donald B. Spencer, Thomai Sanopoulou, Fereshteh Abedinpour, Pei-Zhi Wang, Michael Sussman, and Jung-Wan K. Lee for their contributions to this work.

This investigation was supported by Public Health Service grant GM-21909 from the National Institute of General Medical Sciences.

LITERATURE CITED

1. **Berg, P. E.** 1981. Cloning and characterization of the *Escherichia coli* gene coding for alkaline phosphatase. J.

Bacteriol. **146:**660–767.

2. **Bradshaw, R. A., F. Cancedda, L. H. Ericsson, P. A. Neumann, S. P. Piccoli, M. J. Schlesinger, K. Shriefer, and K. A. Walsh.** 1981. Amino acid sequence of *Escherichia coli* alkaline phosphatase. Proc. Natl. Acad. Sci. USA **78:** 3473–3477.
3. **Cashel, M., and E. Freese.** 1964. Excretion of alkaline phosphatase by *Bacillus subtilis*. Biochem. Biophys. Res. Commun. **16:**541–544.
4. **Echols, H., A. Garen, S. Garen, and A. Torriani.** 1961. Genetic control of repression of alkaline phosphatase in *E. coli*. J. Mol. Biol. **3:**425–438.
5. **Ghosh, A., S. Vallespir, and B. K. Ghosh.** 1984. Specificity of subcellular distribution of alkaline phosphatase in *Bacillus licheniformis* 744/C. Can. J. Microbiol. **30:**113–125.
6. **Ghosh, B. K., J. T. M. Wouters, and J. O. Lampen.** 1971. Distribution of the sites of alkaline phosphatase(s) activity in vegetative cells of *Bacillus subtilis*. J. Bacteriol. **108:** 928–937.
7. **Glenn, A. R.** 1975. Alkaline phosphatase mutants of *Bacillus subtilis*. Aust. J. Biol. Sci. **28:**323–330.
8. **Glenn, A. R., and J. Mandelstam.** 1971. Sporulation in *Bacillus subtilis* 168. Comparison of alkaline phosphatase from sporulating and vegetative cells. Biochem. J. **123:** 129–138.
9. **Glynn, J. A., S. D. Schaffel, J. M. McNicholas, and F. M. Hulett.** 1977. Biochemical localization of alkaline phosphatase of *Bacillus licheniformis* as a function of culture age. J. Bacteriol. **129:**1010–1019.
10. **Guan, T., A. Ghosh, and B. K. Ghosh.** 1984. Subcellular localization of alkaline phosphatase in *Bacillus licheniformis* 749/C by immunoelectron microscopy with colloidal gold. J. Bacteriol. **159:**668–777.
11. **Hansa, J. G., M. Laporta, M. A. Kuna, R. Reimschuessel, and F. M. Hulett.** 1981. A soluble alkaline phosphatase from *Bacillus licheniformis* MC14: histochemical localization, purification and characterization and comparison with the membrane-associated alkaline phosphatase. Biochim. Biophys. Acta **675:**340–401.
12. **Hulett, F. M.** 1984. Cloning and characterization of the *Bacillus licheniformis* gene coding for alkaline phosphatase. J. Bacteriol. **158:**978–982.
13. **Hulett, F. M.** 1985. Two structural genes for alkaline phosphatase in *Bacillus licheniformis* which require different RNA polymerase holoenzymes for transcription: a possible explanation for complex synthesis and localization data, p. 124–134. *In* J. Hoch and P. Setlow (ed.), Molecular biology of microbial differentiation. American Society for Microbiology, Washington, D.C.
14. **Hulett, F. M., and L. L. Campbell.** 1971. Purification and properties of an alkaline phosphatase of *Bacillus licheniformis*. Biochemistry **10:**1364–1370.
15. **Hulett, F. M., and L. L. Campbell.** 1971. Molecular weight and subunits of the alkaline phosphatase of *Bacillus licheniformis*. Biochemistry **10:**1371–1376.
16. **Hulett, F. M., S. D. Schaffel, and L. L. Campbell.** 1976. Subunits of alkaline phosphatase of *Bacillus licheniformis*: chemical physiochemical, and dissociation studies. J. Bacteriol. **128:**651–757.
17. **Hulett, F. M., K. Stuckmann, D. B. Spencer, and T. Sanopoulou.** 1986. Purification and characterization of the secreted alkaline phosphatase of *Bacillus licheniformis* MC14: identification of a possible precursor. J. Gen. Microbiol. **132:**2387–2395.
18. **Hulett, F. M., P.-Z. Wang, M. Sussman, and J.-W. Lee.** 1985. Two alkaline phosphatase genes positioned in tandem in *B. licheniformis* MC14 which require different RNA polymerase holoenzymes for transcription. Proc. Natl. Acad. Sci. USA **82:**1035–1039.
19. **Hydrean, C., A. Ghosh, M. Nallin, and B. K. Ghosh.** 1977. Interrelationship of carbohydrate metabolism and APase synthesis in *Bacillus licheniformis* 759/C. J. Biol. Chem. **252:**6806–7812.
20. **Inouye, H., W. Barnes, and J. Beckwith.** 1982. Signal sequence of alkaline phosphatase of *Escherichia coli*. J. Bacteriol. **149:**434–439.
21. **Inouye, H., S. Michaelis, A. Wright, and J. Beckwith.** 1981. Cloning and restriction mapping of the alkaline phosphatase structural gene (*phoA*) of *Escherichia coli* and generation of deletion mutants in vitro. J. Bacteriol. **146:**668–675.
22. **Losick, R., and J. Pero.** 1981. Cascades of sigma factors. Cell **25:**582–584.
23. **Malamy, M., and B. L. Horecker.** 1961. The localization of alkaline phosphatase in *E. coli* K_{12}. Biochem. Biophys. Res. Commun. **5:**104–108.

23a. **McNicholas, J. M., and F. M. Hulett.** 1977. Electron microscope histochemical localization of alkaline phosphatase(s) in *Bacillus licheniformis*. J. Bacteriol. **129:** 501–515.

24. **Michaelis, S., and J. Beckwith.** 1982. Mechanism of incorporation of cell envelope proteins in *Escherichia coli*. Annu. Rev. Microbiol. **36:**435–465.
25. **Moran, C. P., Jr., N. Lang, C. D. B. Banner, W. G. Haldenwang, and R. Losick.** 1981. Promoter for a developmentally regulated gene in *Bacillus subtilis*. Cell **25:** 783–791.
26. **Moran, C. P., Jr., N. Lang, and R. Losick.** 1981. Nucleotide sequence of a *Bacillus subtilis* promoter recognized by *Bacillus subtilis* RNA polymerase containing σ^{37}. Nucleic Acids Res. **9:**5979–5990.
27. **Neu, H. C., and L. A. Heppel.** 1965. The release of enzymes from *Escherichia coli* by osmotic shock and during the formation of spheroplasts. J. Biol. Chem. **240:** 3685–3692.
28. **Schaffel, S., and F. M. Hulett.** 1978. Alkaline phosphatase from *Bacillus licheniformis* solubility dependent on magnesium, purification and characterization. Biochim. Biophys. Acta **526:**457–467.
29. **Spencer, D. B., C.-P. Chen, and F. M. Hulett.** 1981. Effect of cobalt on synthesis and activation of *Bacillus licheniformis* phosphatase. J. Bacteriol. **145:**926–933.
30. **Spencer, D., J. Hansa, K. Stuckmann, and F. M. Hulett.** 1982. A membrane associated alkaline phosphatase from *Bacillus licheniformis* MC14 requiring detergent for solubilization: lactoperoxides ^{125}I and molecular weight determination. J. Bacteriol. **150:**826–834.
31. **Spencer, D. B., and F. M. Hulett.** 1981. Lactoperoxides-^{125}I localization of salt-extractable alkaline phosphatase on the cytoplasmic membrane of *Bacillus licheniformis*. J. Bacteriol. **145:**934–945.
32. **Wong, S. L., C. W. Price, D. S. Goldfarb, and R. A. Doi.** 1984. The subtilisin F gene of *Bacillus subtilis* is transcribed from a σ^{37} promoter *in vivo*. Proc. Natl. Acad. Sci. USA **81:**1184–1188.

Regulatory Circuit for Phosphatase Synthesis in *Saccharomyces cerevisiae*

KAZUYA YOSHIDA, ZYUNROU KUROMITSU, NOBUO OGAWA, KOUHEI OGAWA, AND YASUJI OSHIMA

Department of Fermentation Technology, Faculty of Engineering, Osaka University, 2-1 Yamadaoka, Suita-shi, Osaka 565, Japan

Saccharomyces cerevisiae has two species of acid phosphatase (EC 3.1.3.2) on the cell wall (11): one is coded for by the *PHO3* gene and is not regulated by P_i (15), and the other is repressible acid phosphatase (rAPase), of which the major fraction is protein p60 encoded by the *PHO5* gene and the minor fraction is composed of proteins p58 and p56 encoded by, respectively, *PHO10* and *PHO11* (10, 13). rAPase is derepressed upon P_i starvation. *S. cerevisiae* also has two species of an enzyme that hydrolyzes *p*-nitrophenylphosphate at alkaline pH (18): specific *p*-nitrophenylphosphatase and nonspecific repressible alkaline phosphatase (rALPase; EC 3.1.3.1) encoded by the *PHO8* gene (7) and localized mostly in the vacuole (21). A kinetic study of P_i transport has revealed that *S. cerevisiae* has two systems for P_i transport: one with a low K_m (8.2 μM) for external P_i and the other with a high K_m (770 μM) (16). The low-K_m system is repressible by P_i in the medium and is thought to be specified by the *PHO84* gene (16, 20).

Although the phosphatase system is complicated, being composed of numerous structural and regulatory genes (Table 1; for review, see reference 11), it offers considerable advantages over other systems in the study of genetic regulatory mechanisms in *S. cerevisiae*. These advantages are as follows: (i) activities of rAPase, of rALPase, and of the repressible P_i transport system are controlled coordinately in re-sponse to a simple effector, P_i; (ii) differences between the repressed and derepressed levels of rAPase activities are significant, i.e., 500-fold (19) and 60-fold for the repressible P_i transport system (16), while 2- to 3-fold for rALPase (18); and (iii) rAPase and rALPase activities of colonies are both easily determined by specific staining methods (18, 19). With these advantages, the regulatory system for phosphatase synthesis was subjected to genetic analysis and a model was proposed (11). This communication describes the genetic model briefly and presents an evaluation of it obtained by measuring transcripts originating from the regulatory genes.

Genetic Regulatory Model

Phosphatase genes. By effective staining procedures for detecting activities of acid and alkaline phosphatases of colonies on a test plate, various mutants were isolated. Some had the constitutive phenotype for the production of the repressible enzymes while others lacked these activities. These mutations were subjected to allelism or complementation tests, dominance-recessiveness tests with the corresponding wild-type allele or with another allelic mutation of a different phenotype, coarse and fine mapping of the loci, and epistasis-hypostasis tests between constitutive and blocked mutations occurring in two different loci. In addition, hybridization selection of mRNAs and DNA fragments by using a cloned *PHO5* DNA revealed two other loci, *PHO10* and *PHO11*, encoding isozymes of rAPase (10, 13), and a DNA fragment possibly encoding the gene for specific *p*-nitrophenylphosphatase (*PHO13*) was identified during the course of cloning the *PHO8* gene (Y. Kaneko, Y. Tamai, A. Toh-e, and Y. Oshima, unpublished data). The genes and their possible functions in the phosphatase system deduced from the above studies are summarized in Table 1.

Genetic model. A previous publication (11) described a genetic hierarchy of the regulatory genes for transmittance to the structural genes of the signals due to the presence or absence of exogenous P_i. This model is illustrated in Fig. 1. Expression of the *PHO3* gene to produce the constitutive acid phosphatase requires the function of two additional complementary genes, *PHO6* and *PHO7* (not shown in Fig. 1). In the case of rAPase (structural genes, *PHO5*, *PHO10*, and *PHO11*) and the low-K_m P_i transport system (*PHO84*), a genetic system consisting of at least five genes, *PHO2*, *PHO4*, *PHO80*, *PHO81*, and *PHO85*, participates in the regulation of gene expression in response to the P_i signals. Recessive mutations at any of the *PHO2*, *PHO4*, or *PHO81* loci block the derepression of the four structural genes, whereas a recessive mutation in *PHO80* or *PHO85* results in their constitutive expression. Dominant mutations also occur in the *PHO4* and *PHO81* loci

TABLE 1. Genes and their possible functions in the phosphatase system

Gene[a]	Chromosome location	Function of gene or its product
Structural genes		
PHO3	II	Constitutive APase
PHO5	II	rAPase; p60
PHO10		rAPase; p58
PHO11		rAPase; p56
PHO8	IV	rALPase
PHO13		Specific *p*-nitrophenylphosphatase
PHO84		P_i transport (permease, low-K_m system)
Genes affecting the expression of the structural genes		
Regulatory genes		
PHO2	IV	Specific factor for expression of *PHO5* and *PHO84*
PHO6		Required for *PHO3* expression
PHO7		Required for *PHO3* expression
PHO4	VI	Positive factor for the repressible enzymes
PHO80[b]	XV	Negative factor for the repressible enzymes
PHO81	VII	Mediator for the repressible enzymes
PHO85	XVI	Factor required for *PHO80* expression (thought to be a subunit of negative factor)[c]
Gene for processing enzyme: *PHO9*[d]	XVI	Maturation of rALPase (thought to be a specific factor for *PHO8* expression)[c]

[a] Gene symbols are described in wild-type form.
[b] Allelic with *TUP7*.
[c] Comments proposed in the original genetic model.
[d] Allelic with *PEP4*.

which give the constitutive phenotype for the repressible enzymes. The transcription of *PHO8* to produce rALPase is also regulated in coordination with the four structural genes through the same regulatory system, except for the *PHO2* product, while the function of another gene, *PHO9*, is indispensable for the *PHO8* expression.

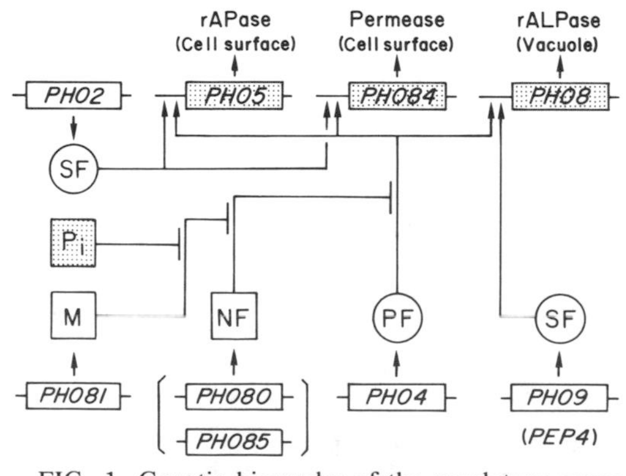

FIG. 1. Genetic hierarchy of the regulatory genes in the phosphatase system. The genes in the dotted boxes represent the structural genes for the repressible enzymes. Regulatory factors are PF, positive factor; NF, negative factor; M, mediator; and SF, specific factor. The systems for producing acid phosphatase specified by the *PHO3* gene and for P_i transport with the high K_m value were omitted from the figure.

The genetic model (11) proposes that a regulatory molecule (positive factor) specified by *PHO4* is indispensable for the transcription of the structural genes for the enzymes. With a repressively high amount of P_i in the medium, a complex of the *PHO80* and *PHO85* gene products (negative factor) aggregates with the positive factor, preventing it from activating the transcription of the structural genes. Under derepressive conditions, i.e., a low P_i concentration in the medium, the *PHO81* product (mediator) reacts with the negative factor, thereby releasing the positive factor, which can then combine with the *PHO2* protein (specific factor) to turn on the genes for the rAPase production and the repressible P_i transport system, or with the *PHO9* protein (another specific factor) to turn on the gene for the rALPase production.

Evaluation of the Model by Molecular Analysis

Posttranscriptional control of *PHO8* expression by *PHO9*. Considerable biochemical and molecular evidence clearly indicates that the on-off control of the structural genes in the phosphatase system by these regulatory genes, except for the *PHO9* gene, is at the level of mRNA synthesis (8, 10, 13). Since it has been shown that *PHO9* is identical to *PEP4* (7) and the function of *PEP4* is supposed to be required for the conversion of the inactive precursor to the active mature enzyme of at least four vacuolar hydrolases (proteinases A and B, carboxypeptidase Y, and RNase) along with rALPase (5, 22), the *PHO9* product may not be involved in the transcriptional control of *PHO8*. This was confirmed by Northern hybridization of the *PHO8* transcript with the cloned *PHO8* DNA as probe, in which the RNA samples prepared from the cells of the *pho9* mutant showed the same pattern of hybridization as those from the wild-type cells (6). Only a basal amount of transcripts was observed in the *pho4* mutant, while the derepressed amount was observed in the *PHO80* mutant, irrespective of whether these mutants were grown in high-P_i or low-P_i medium.

Expression of *PHO2* and *PHO4* is low-level constitutive. Since the *pho2* and *pho4* mutations were epistatic over the other regulatory genes, except for a constitutive *PHO83* mutation by insertion of a 6-kilobase (kb) Ty element at the promoter region of *PHO5* (17), the genetic model posits that the *PHO4* product must interact directly with the structural genes. The *PHO2* product is also indispensable for the expression of the structural genes but not for *PHO8*. These positive regulatory factors should be titrated out by the product(s) of *PHO80* or *PHO85* or both, under the repressible condition. These arguments suggest that the *PHO4* and *PHO2* genes must be transcribed in a limited amount and stoichiometrically, irrespective of the presence or absence of P_i in the growth medium.

To test these possibilities, the *PHO2* and *PHO4* genes were cloned with an *Escherichia coli* host. Partially digested *Sau*3A1 fragments of chromosomal DNA of *S. cerevisiae* P-28-24C (*pho3-1* but wild type for the regulatory system) were connected to the *Bam*HI site of YEp13 (10.7 kb; a chimeric plasmid consisting of pBR322, the *LEU2* gene of *S. cerevisiae*, and the replication origin of 2μm DNA [3]). Plasmid DNAs were prepared from the gene bank and used to transform *S. cerevisiae* NA93-3A (*pho2-8 leu2-3,112*; only the relevant genotype is described) and *S. cerevisiae* W754-3A (*pho4-1 leu2-3,112*) to the leucine prototrophic phenotype (Leu$^+$). The Leu$^+$ yeast transformants were screened for acid phosphatase activity (rAPase in this case) by staining colonies on a test plate. Some phosphatase-positive (Pho$^+$) clones were isolated at random, and the plasmids were purified by successive transformation of the *E. coli* and *S. cerevisiae* hosts.

One of the plasmids prepared from the Leu$^+$ Pho$^+$ transformants of NA93-3A for the *PHO2* cloning contained a 9.6-kb *Sau*3A1 fragment. The complementing fragment was delimited to a 2.3-kb *Alu*I-*Hpa*I fragment (Fig. 2A). That the

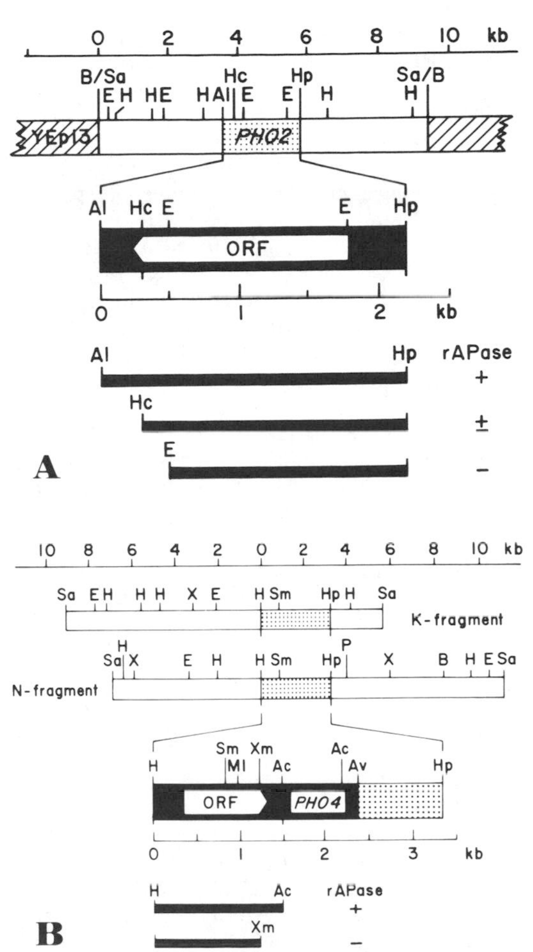

FIG. 2. Restriction maps of the *S. cerevisiae* DNA fragments cloned on YEp13 and their complementation of the *pho2* (panel A) and *pho4* (panel B) mutations. Open boxes (with dotted region) are the cloned *S. cerevisiae* DNAs, and the shaded regions in panel A indicate the YEp13 DNA. The open arrow with ORF in the closed box indicates the open reading frame found in the region. The thick lines are the fragments used in the complementation tests. Restriction sites shown are Ac, *Acc*I; Al, *Alu*I; Av, *Ava*I; B, *Bam*HI; E, *Eco*RI; H, *Hin*dIII; Hc, *Hin*cII; Hp, *Hpa*I; Ml, *Mlu*I; P, *Pst*I; Sa, *Sau*3A1; Sm, *Sma*I; X, *Xho*I; and Xm, *Xma*I. B/Sa or Sa/B indicates the junction of *Bam*HI and *Sau*3AI cohesive ends.

fragment bears the *PHO2* gene was confirmed by the integration of a plasmid DNA containing the 3.6-kb *Hin*dIII-*Hin*dIII *PHO2* fragment on YIp32 (6.7 kb; a yeast integrative vector composed of pBR322 and a yeast DNA fragment bearing the *LEU2* gene [2]) at or close to the *PHO2* locus (data not shown). The nucleotide sequence of the 2.3-kb *Alu*I-*Hpa*I fragment was determined, and only one large open reading frame was found, which was 1,557 base pairs (bp) long, extending from just 9 bp upstream of the *Eco*RI site to 74 bp beyond the *Hin*cII site.

In the case of *PHO4* cloning, a few yeast transformants showing the Leu$^+$ Pho$^+$ phenotype were detected by staining, because of their slow growth. When we isolated one of the slow-growing Leu$^+$ Pho$^+$ colonies, its phenotype was found to be due to a plasmid harbored in the cells. The plasmid carried a large DNA fragment of ca. 14.5 kb (K-fragment [Fig. 2B]). The effective region for complementation of the *pho4* mutation was delimited to the 1.5-kb *Hin*dIII-*Acc*I region by subcloning the fragment on YCp19 (a low-copy vector having the *TRP1* and *URA3* markers for the *S. cerevisiae* host and the Ap^r marker for the *E. coli* host [4]). The nucleotide sequence of the 2.4-kb *Hin*dIII-*Ava*I fragment containing the 1.5-kb *Hin*dIII-*Acc*I region was determined (data not shown). The 1.5-kb *Hin*dIII-*Acc*I region contains one open reading frame, consisting of 936 bp. That the open reading frame encodes the *PHO4* protein was confirmed by the observations that the 1.2-kb *Hin*dIII-*Xma*I fragment could not complement the *pho4* mutation (Fig. 2B) and that a chimeric plasmid constructed by recloning of the 4.3-kb *Hin*dIII fragment containing the open reading frame on YIp32 was integrated into chromosome VI of *S. cerevisiae* at or close to the *PHO4* locus (data not shown).

Since the original 14.5-kb fragment conferred the slow-growing phenotype to the host, while no growth inhibition was observed with plasmids bearing the 4.3-kb *Hin*dIII *PHO4* fragment, and the K-fragment was isolated by forced selection, we were concerned about the possibility of structural modification of the *PHO4* gene. Therefore, we tried the *PHO4* cloning again from the same yeast chromosomal DNA with the *E. coli*-Charon 28 system (12) and plaque hybridization, with ^{32}P-labeled 4.3-kb *Hin*dIII *PHO4* DNA of the K-fragment as the hybridization probe. Five independent recombinant phage clones were isolated at random, and their yeast DNA was confirmed to contain the *PHO4* gene by complementation of the *pho4* mutation after recloning them on YEp13. The restriction map of the yeast DNA fragment (N-fragment) compiled from the five recombinant phage clones differed from that of the K-fragment (Fig. 2B). However, the 3.3-kb *Hin*dIII-*Hpa*I region (dotted region) of the N-fragment showed exactly the same restriction map as the K-fragment, and this region was shared by all five yeast DNA fragments cloned on the recombinant phages. It thus appeared that the 3.3-kb *Hin*dIII-*Hpa*I region of the K-fragment contains no significant modifications.

To test the transcription of the *PHO2* and *PHO4* genes, Northern hybridization of poly(A)$^+$ RNAs prepared from the wild-type cells for the regulatory system, P-28-24C, cultivated in nutrient low-P$_i$ or nutrient high-P$_i$ medium was performed, with ^{32}P-labeled DNAs of locus-specific chimeric plasmids as probes. For detection of the *PHO2* transcript, a chimeric plasmid constructed by cloning the 1.9-kb *Hin*cII-*Hpa*I fragment (Fig. 2A) at the *Bam*HI site of YIp5 (5.54 kb, Ap^r Tc^r *URA3* [2]) was used as the probe, and a similar plasmid constructed by replacing the shorter *Hin*dIII-*Bam*HI fragment of YIp5 with the 1.5-kb *Hin*dIII-*Acc*I *PHO4* fragment (Fig. 2B) was used as the *PHO4* probe. The results indicated that the *PHO2* transcript of 1.85 kb in molecular size and the *PHO4* transcript of 1.7 kb were produced at low levels but constitutively, irrespective of the cultivation conditions of the cells (Fig. 3). The amounts of mRNA were, respectively, 20 to 30% of that of the *URA3* transcript in the nutrient medium, which was suggested to be the average for the yeast genes (1).

Transcription of *PHO81* is under the regulation of the *PHO* regulatory system. To investigate the mode of action of the *PHO81* gene, the transcription of *PHO81* was determined. Plasmid DNAs were extracted from the gene bank constructed for the *PHO8* and *PHO2* clonings and used for transformation of *S. cerevisiae* strain NA95-4B (*pho81 leu2-3,112*) to the Leu$^+$ phenotype. Pho$^+$ transformants were then detected from among the Leu$^+$ transformants by replicating them on a nutrient low-P$_i$ plate and by staining the colonies. Several transformants complementing the *pho81* mutation were isolated. Plasmid DNA bearing a 12.4-kb DNA fragment (Fig. 4) was extracted from one of these transformants. The 12.4-kb fragment was delimited to the 2.8-kb *Bam*HI-*Sau*3A1 fragment

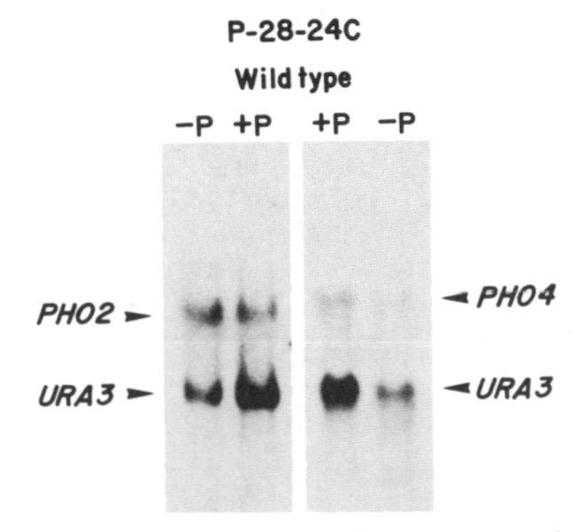

FIG. 3. Northern hybridization of poly(A)$^+$ RNA prepared from the wild-type cells for the regulatory system, grown in nutrient high-P$_i$ (+P) or nutrient low-P$_i$ (−P) medium with ^{32}P-labeled fragments of chimeric plasmids bearing the *PHO2* or *PHO4* gene. Detailed procedures are described in the text.

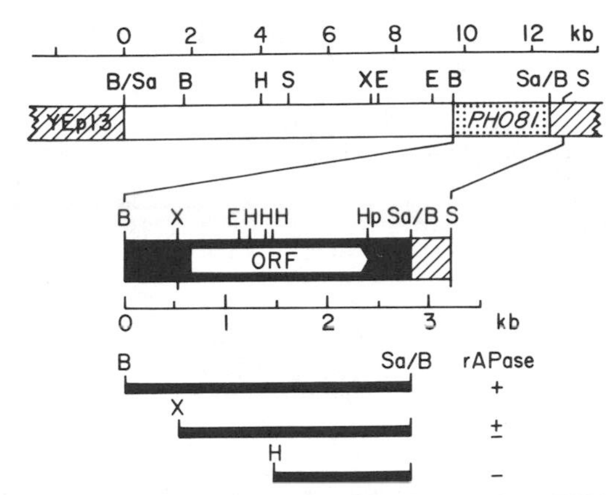

FIG. 4. Restriction maps of the *S. cerevisiae* DNA fragments cloned on YEp13 and their complementation of the *pho81* mutation. The open boxes, shaded regions, closed box with ORF arrow, and symbols for restriction sites, with the addition of S for the *Sal*I site, are the same as in Fig. 2.

by complementation of the *pho81* mutation, and the nucleotide sequence of the 2.8-kb fragment was determined (data not shown). The 2.8-kb fragment (actual length, 2,841 bp) contains only one large open reading frame, consisting of 1,809 bp. Complementation tests with the *Xho*I-*Sau*3A1 and *Hin*dIII-*Sau*3A1 fragments in the 2.8-kb region (Fig. 4) indicate that the open reading frame encodes the *PHO81* protein.

The 2.8-kb *Xho*I-*Sal*I fragment bearing the open reading frame of *PHO81* was cloned at the *Sal*I site of YIp5. The chimeric plasmid was used as probe for Northern analysis of the *PHO81* transcript after nick translation with ^{32}P. Cells of P-28-24C and a *PHO81*^{c} mutant, O106-M30, showing constitutive production of the repressible enzymes were grown on nutrient high-P_i or nutrient low-P_i medium to the late log phase. Poly(A)$^{+}$ RNAs prepared from the cells were electrophoresed, blotted, and hybridized with the probe. Results clearly indicated that the level of *PHO81* mRNA produced was higher than that of *URA3* in the cells cultivated in low-P_i medium and was almost comparable to that of the structural genes of phosphatases, while that of the cells cultivated in high P_i was severely repressed (Fig. 5). The *PHO81*^{c} mutant showed an increased amount of the *PHO81* transcript, but was still sensitive to the P_i signals. Results of Northern hybridizations with poly(A)$^{+}$ RNAs prepared from the other regulatory mutants showed that the *PHO81* gene could not be derepressed in the *pho4* mutant, but could in low level by the *pho2* mutant, and was constitutive in the *pho80* and *pho85* mutants (data not shown). These observations were entirely different from those obtained from the *PHO2* and *PHO4* genes.

Discussion

The results of Northern hybridization indicate that the *PHO2* and *PHO4* genes, the most epistatic regulatory genes in the functional hierarchy of the regulatory genes, are transcribed at low levels but constitutively, as suggested by the genetic model. Similar measurements of mRNAs of the *PHO80* and *PHO85* genes by Northern hybridization performed by A. Toh-e of Hiroshima University (personal communication) have also shown that *PHO80* is transcribed constitutively at low level. *PHO80* transcription, however, does not occur in the *pho85* mutant cells. This fact indicates that *PHO85* does not encode a subunit of the negative factor, but provides a factor necessary for *PHO80* expression (85F; Fig. 6). The most striking observation was, however, that the transcription of *PHO81* is under the control of P_i through the function of *PHO4, PHO80*, and, therefore, the *PHO81* gene itself. This indicates that the regulatory system is a kind of closed circuit.

With these observations, the original genetic model for phosphatase regulation should be revised as illustrated in Fig. 6. In the original genetic model, the *PHO81* gene was placed at the most hypostatic position in the functional hierarchy of the regulatory genes. It was, therefore, proposed that P_i must interact with the *PHO81* product (mediator). In the closed circuit model, however, it is possible to speculate that P_i inhibits the function of the positive factor or the mediator, or both, or that it stimulates the function of the negative factor. If only the

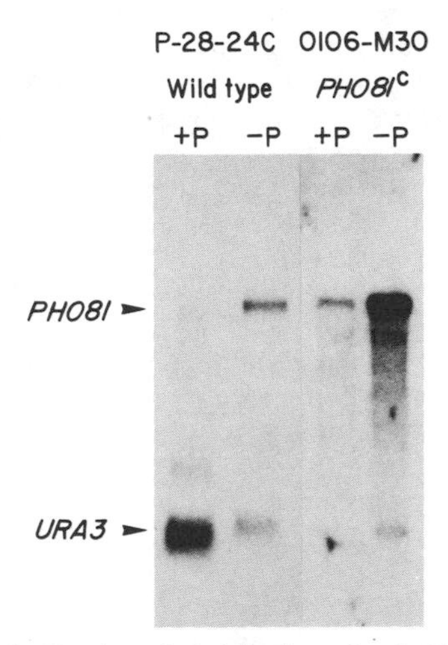

FIG. 5. Northern hybridization of poly(A)$^{+}$ RNAs prepared from the cells of wild-type strain P-28-24C, or the *PHO81*^{c} mutant O106-M30, grown in nutrient high-P_i (+P) or nutrient low-P_i (−P) medium with ^{32}P-labeled DNA of the chimeric plasmid constructed by connecting the 2.8-kb *Xho*I-*Sal*I fragment (Fig. 4) with YIp5 as probe.

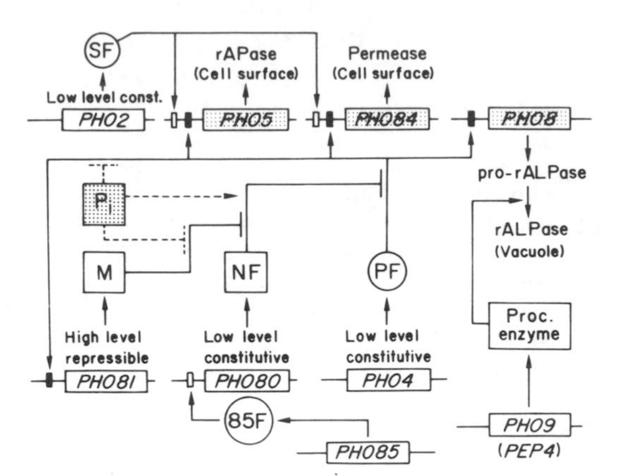

FIG. 6. Revised model for the function of regulatory factors in the phosphatase system. 85F is a factor encoded by *PHO85* and required for the *PHO80* expression. The other symbols are the same as those in Fig. 1.

PHO81 product senses P_i signals, the closed circuit model suggests that the P_i level for turn-on of the structural genes should be significantly lower than the turn-off level, because the amount of the *PHO81* product is expected to be lower in the repressed cells than in the derepressed cells.

A closed circuit in the regulatory system has also been described in the *GAL* system in *S. cerevisiae* (14). Transcription of the tightly linked genes, *GAL7-GAL10-GAL1*, encoding three galactose-metabolizing enzymes is regulated by the interplay of the positive factor encoded and produced by *GAL4* and the negative factor produced by *GAL80* with the signals due to the presence and absence of galactose. In this case, transcription of *GAL80* is the opposite of what might be expected: in the presence of substrate (galactose), a greater amount of transcript is produced (14), while the mRNA level of *GAL4* is unaffected by galactose (9). The mode of transcription of the *GAL* regulatory genes suggests that the regulatory system provides a mechanism which guarantees the expression of the *GAL* structural genes only in response to a sufficient quantity of the effector (galactose).

The 5′-upstream regions of the open reading frames for the *PHO5*, *PHO8*, *PHO84*, and also *PHO81* genes should have the specific sequence for binding the positive factor encoded by *PHO4* and for that of *PHO2* in the regions of *PHO5* and *PHO84*. The possible consensus sequence is now being sought. Since the amounts of the transcript of the *PHO2*, *PHO4*, and *PHO80* genes were low but also constant, these genes might be autoregulated. However, no homologous sequences were detected in the 5′-upstream region of their open reading frames by a graphic matrix analysis.

We thank A. Toh-e of the Department of Fermentation Technology, Hiroshima University, for his communication of unpublished results. Appreciation is also extended to H. Araki and S. Okada of our laboratory for their discussions and technical assistance.

This work was supported by grants from the Ministry of Education, Science, and Culture of Japan.

LITERATURE CITED

1. **Bach, M.-L., F. Lacroute, and D. Botstein.** 1979. Evidence for transcriptional regulation of orotidine-5′-phosphate decarboxylase in yeast by hybridization of mRNA to the yeast structural gene cloned in *Escherichia coli*. Proc. Natl. Acad. Sci. USA **76:**386–390.
2. **Botstein, D., S. C. Falco, S. E. Stewart, M. Brennan, S. Scherer, D. T. Stinchcomb, K. Struhl, and R. W. Davis.** 1979. Sterile host yeasts (SHY): a eukaryotic system of biological containment for recombinant DNA experiments. Gene **8:**17–24.
3. **Broach, J. R., J. N. Strathern, and J. B. Hicks.** 1979. Transformation in yeast: development of a hybrid cloning vector and isolation of the *CAN1* gene. Gene **8:**121–133.
4. **Harashima, S., A. Takagi, and Y. Oshima.** 1984. Transformation of protoplasted yeast cells is directly associated with cell fusion. Mol. Cell. Biol. **4:**771–778.
5. **Jones, E. W., G. S. Zubenko, and R. R. Parker.** 1982. *PEP4* gene function is required for expression of several vacuolar hydrolases in *Saccharomyces cerevisiae*. Genetics **102:**665–677.
6. **Kaneko, Y., Y. Tamai, A. Toh-e, and Y. Oshima.** 1985. Transcriptional and posttranscriptional control of *PHO8* expression by *PHO* regulatory genes in *Saccharomyces cerevisiae*. Mol. Cell. Biol. **5:**248–252.
7. **Kaneko, Y., A. Toh-e, and Y. Oshima.** 1982. Identification of the genetic locus for the structural gene and a new regulatory gene for the synthesis of repressible alkaline phosphatase in *Saccharomyces cerevisiae*. Mol. Cell. Biol. **2:**127–137.
8. **Kramer, R., and N. Andersen.** 1980. Isolation of yeast genes with mRNA levels controlled by phosphate concentration. Proc. Natl. Acad. Sci. USA **77:**6541–6545.
9. **Laughon, A., and R. F. Gesteland.** 1982. Isolation and preliminary characterization of the *GAL4* gene, a positive regulator of transcription in yeast. Proc. Natl. Acad. Sci. USA **79:**6827–6831.
10. **Lemire, J. M., T. Willcocks, H. O. Halvorson, and K. A. Bostian.** 1985. Regulation of repressible acid phosphatase gene transcription in *Saccharomyces cerevisiae*. Mol. Cell. Biol. **5:**2131–2141.
11. **Oshima, Y.** 1982. Regulatory circuits for gene expression: the metabolism of galactose and of phosphate, p. 159–180. *In* J. N. Strathern, E. W. Jones, and J. R. Broach (ed.), The molecular biology of the yeast Saccharomyces: metabolism and gene expression. Cold Spring Harbor Laboratory, Cold Spring Harbor, N.Y.
12. **Rimm, D. L., D. Horness, J. Kucera, and F. R. Blattner.** 1980. Construction of coliphage lambda Charon vectors with *Bam*HI cloning sites. Gene **12:**301–309.
13. **Rogers, D. T., J. M. Lemire, and K. A. Bostian.** 1982. Acid phosphatase polypeptides in *Saccharomyces cerevisiae* are encoded by a differentially regulated multigene family. Proc. Natl. Acad. Sci. USA **79:**2157–2161.
14. **Shimada, H., and T. Fukasawa.** 1985. Controlled transcription of the yeast regulatory gene *GAL80*. Gene **39:**1–9.
15. **Tait-Kamradt, A. G., K. J. Turner, R. A. Kramer, Q. D. Elliott, S. J. Bostian, G. P. Thill, D. T. Rogers, and K. A. Bostian.** 1986. Reciprocal regulation of the tandemly duplicated *PHO5/PHO3* gene cluster within the acid phosphatase multigene family of *Saccharomyces cerevisiae*.

Mol. Cell. Biol. **6:**1855–1865.
16. **Tamai, Y., A. Toh-e, and Y. Oshima.** 1985. Regulation of inorganic phosphate transport systems in *Saccharomyces cerevisiae*. J. Bacteriol. **164:**964–968.
17. **Toh-e, A., Y. Kaneko, J. Akimaru, and Y. Oshima.** 1983. An insertion mutation associated with constitutive expression of repressible acid phosphatase in *Saccharomyces cerevisiae*. Mol. Gen. Genet. **191:**339–346.
18. **Toh-e, A., H. Nakamura, and Y. Oshima.** 1976. A gene controlling the synthesis of non specific alkaline phosphatase in *Saccharomyces cerevisiae*. Biochim. Biophys. Acta **428:**182–192.
19. **Toh-e, A., and Y. Oshima.** 1974. Characterization of a dominant, constitutive mutation, *PHOO*, for the repressible acid phosphatase synthesis in *Saccharomyces cerevisiae*. J. Bacteriol. **120:**608–617.
20. **Ueda, Y., and Y. Oshima.** 1975. A constitutive mutation, *phoT*, of the repressible acid phosphatase synthesis with inability to transport inorganic phosphate in *Saccharomyces cerevisiae*. Mol. Gen. Genet. **136:**255–259.
21. **Wiemken, A., M. Schellenberg, and K. Urech.** 1979. Vacuoles: the sole compartments of digestive enzymes in yeast (*Saccharomyces cerevisiae*)? Arch. Microbiol. **123:** 23–35.
22. **Zubenko, G. S., F. J. Park, and E. W. Jones.** 1983. Mutations in *PEP4* locus of *Saccharomyces cerevisiae* block final step in maturation of two vacuolar hydrolases. Proc. Natl. Acad. Sci. USA **80:**510–514.

Molecular Aspects of Acid Phosphatase Synthesis in *Saccharomyces cerevisiae*

ALBERT HINNEN, WAJEEH BAJWA, BERND MEYHACK, AND HANS RUDOLPH

Biotechnology Department, CIBA-GEIGY Ltd., CH-4002 Basel, Switzerland

At least three structural genes per haploid genome code for acid phosphatase activity in *Saccharomyces cerevisiae*: *PHO3, PHO5*, and *PHO11* (for review, see reference 6). Each of these genes shows different levels of expression at both the transcript and the protein level (2).

The *PHO5* and *PHO3* genes are closely linked on chromosome II (5, 11). These two genes show an evolutionary relationship as they share 82% homology in their DNA coding sequences (1). In contrast, only short blocks of homology are observed in their upstream flanking sequences. This observation is in good agreement with distinct physiological and biochemical properties: under derepressed conditions the *PHO5* gene product is increased whereas the *PHO3* gene product remains essentially unaffected (2, 7) or even decreases slightly (9).

Several mutants affecting *PHO5* expression have been isolated and characterized. They show that the appearance of the *PHO5* gene product depends on several transacting unlinked regulatory genes: *PHO2, PHO4, PHO80, PHO81*, and *PHO85* (6).

Our effort was directed toward a search for specific promoter sequences which are involved in expression and regulation of the different members of the acid phosphatase gene family. Three main lines have been pursued: (i) DNA sequence comparisons of the different promoters, (ii) fusion of promoter upstream elements to promoters of constitutively expressed genes, and (iii) deletion mapping of the regulated *PHO5* promoter.

Promoter Sequence Comparisons

To obtain a better understanding of expression of the regulated acid phosphatase genes, we decided to sequence the unlinked *PHO11* gene including its flanking promoter and transcription termination area. By comparing the *PHO11* promoter sequence with the available sequence information of *PHO3* and *PHO5*, we hoped to be able to distinguish between functional sequence elements and important regulatory components. Using published information (4), we isolated the gene by screening a gene library prepared from *S. cerevisiae* wild-type strain S288C, using a 2-kilobase *Bam*HI-*Pst*I fragment containing *PHO5* as a radioactive probe (1). Subcloning in M13mp8 and M13mp9 resulted in single-stranded templates which were sequenced by the dideoxy chain termination method using synthetic oligonucleotides (manuscript in preparation). Figure 1 shows the DNA sequence of the protein coding part of *PHO3, PHO5*, and *PHO11* and the deduced amino acid sequence. From a preliminary analysis of the three sequences the following conclusions can be drawn. (i) The similarity between the three genes is very high (>80% at the amino acid or nucleotide level). *PHO3* and *PHO5* show the highest degree of homology, *PHO5* and *PHO11* are intermediate, and *PHO3* and *PHO11* exhibit the lowest degree of similarity. (ii) Base pair changes which give rise to amino acid substitutions usually replace amino acids of the same charge, chain length, or hydrophobicity. (iii) The longest stretch of amino acids with absolute identity among all the three genes is 74 residues long (between positions 372 and 445). This area might be of absolute importance for maintaining the functional or structural integrity, or both.

A comparison of the promoter regions of *PHO3, PHO5*, and *PHO11* led to the identification of several homology boxes, as was first pointed out by others (10). Our analysis was performed with the three acid phosphatase genes from *S. cerevisiae* wild-type strain S288C with extended DNA sequence data of the *PHO11* promoter (up to position −458) (see Fig. 2). The additional information uncovers an important block of sequence conservation between the *PHO5* and the *PHO11* promoter which coincides with a functional regulatory element as determined by deletion mapping. The conclusions from this promoter comparison will be discussed together with the data from the *PHO5* promoter analysis (see below).

PHO5-PHO3 Promoter Fusions

In contrast to *PHO5* and *PHO11*, the *PHO3* gene is not inducible by low P_i concentrations. We therefore used the *PHO3* protein coding region and its adjacent downstream promoter part (up to position −128 from the ATG) as an assay gene. This includes promoter sequences of *PHO3* containing a TATA motif (position −96) and the transcription initiation region (around

```
                                                          15
PHO 3 MET PHE LYS SER VAL VAL TYR SER VAL LEU ALA ALA ALA LEU VAL
      ATG TTT AAG TCT GTT GTT TAT TCG GTT CTA GCC GCT GCT TTA GTT
PHO 5  *  PHE  *   *  VAL  *   *   *  ILE  *   *   *  SER  *  ALA
      ATG TTT AAA TCT GTT GTT TAT TCA ATT TTA GCC GCT TCT TTG GCC
PHO11  *  LEU  *   *  ALA  *   *   *  ILE  *   *   *  SER  *  VAL
      ATG TTG AAG TCA GCC GTT TAT TCA ATT TTA GCC GCT TCT TTG GTT

                                                          30
PHO 3 ASN ALA GLY THR ILE PRO LEU GLY GLU LEU ALA ASP VAL ALA LYS
      AAT GCA GGT ACA ATT CCC CTC GGA GAG TTA GCC GAT GTT GCC AAA
PHO 5  *   *   *   *   *   *   *   *  LYS  *  ALA  *  VAL ASP  *
      AAT GCA GGT ACC ATT CCC TTA GGC AAA CTA GCC GAT GTC GAC AAG
PHO11  *   *   *   *   *   *   *   *  LYS  *  SER  *  ILE ASP  *
      AAT GCA GGT ACC ATA CCC CTC GGA AAG TTA TCT GAC ATT GAC AAA

                                                          45
PHO 3 ILE GLY THR GLN GLU ASP ILE PHE PRO PHE LEU GLY GLY ALA GLY
      ATT GGC ACT CAG GAA GAC ATA TTC CCA TTC CTG GGT GGT GCC GGG
PHO 5  *   *   *   *  LYS ASP  *   *   *   *   *   *   *  ALA  *
      ATT GGT ACC CAA AAA GAT ATC TTC CCA TTT TTG GGT GGT GCC GGA
PHO11  *   *   *   *  THR GLU  *   *   *   *   *   *   *  SER  *
      ATC GGA ACT CAA ACG GAA ATT TTC CCA TTT TTG GGT GGT TCT GGG

                                                          60
PHO 3 PRO TYR PHE SER PHE PRO GLY ASP TYR GLY ILE SER ARG ASP LEU
      CCA TAC TTC TCT TTC CCT GGC GAC TAT GGT ATT TCT CGT GAC TTG
PHO 5  *   *  TYR  *   *   *   *   *   *   *   *   *   *   *   *
      CCA TAC TAC TCT TTC CCT GGC GAC TAT GGT ATT TCT CGT GAT TTG
PHO11  *   *  TYR  *   *   *   *   *   *   *   *   *   *   *   *
      CCA TAC TAC TCT TTC CCT GGT GAC TAT GGT ATT TCT CGT GAT TTG

                                                          75
PHO 3 PRO GLU GLY CYS GLU MET LYS GLN LEU GLN MET LEU ALA AGA HIS
      CCT GAA GGT TGT GAA ATG AAA CAA TTG CAA ATG CTT GCC AGA CAT
PHO 5  *   *  GLY  *   *   *   *   *  LEU  *   *  VAL GLY  *   *
      CCT GAA GGT TGT GAA ATG AAG CAA CTG CAA ATG GTT GGT AGA CAT
PHO11  *   *  SER  *   *   *   *   *  VAL  *   *  VAL GLY  *   *
      CCG GAA AGT TGT GAA ATG AAG CAA GTG CAA ATG GTT GGT AGA CAC

                                                          90
PHO 3 GLY GLU ARG TYR PRO THR TYR SER LYS GLY ALA THR ILE MET LYS
      GGT GAA AGA TAC CCA ACT TAC AGT AAA GGT GCT ACC ATC ATG AAA
PHO 5  *   *   *   *   *   *  VAL  *  LEU ALA LYS THR  *  LYS SER
      GGT GAA AGA TAC CCT ACT GTC AGT CTG GCT AAG ACT ATC AAG AGT
PHO11  *   *   *   *   *   *  VAL  *  LYS ALA LYS SER  *  MET THR
      GGT GAA AGA TAC CCC ACT GTC AGC AAA GCC AAA AGT ATC ATG ACA

                                                         105
PHO 3 THR TRP TYR LYS LEU SER ASN TYR THR ARG GLN PHE ASN GLY SER
      ACA TGG TAT AAG TTG AGC AAT TAC ACA CGT CAA TTC AAC GGC TCA
PHO 5  *   *   *   *   *   *   *   *   *  ARG  *   *  ASN  *  SER
      ACA TGG TAT AAG TTG AGC AAT TAC ACT CGT CAA TTC AAC GGC TCA
PHO11  *   *   *   *   *   *   *   *   *  GLY  *   *  SER  *  GLN
      ACA TGG TAC AAA TTG AGT AAC TAT ACC GGT CAA TTC AGC GGA CAA

                                                         120
PHO 3 LEU SER PHE LEU ASN ASP ASP TYR GLU PHE PHE ILE ARG ASP ASP
      TTG TCA TTC TTG AAC GAT GAT TAC GAG TTT TTC ATC CGT GAT GAC
PHO 5  *   *   *   *   *   *   *   *   *   *   *   *   *   *  ASP
      TTG TCA TTC TTG AAC GAT GAT TAC GAG TTT TTC ATC CGT GAT GAC
PHO11  *   *   *   *   *   *   *   *   *   *   *   *   *   *  THR
      TTG TCT TTC TTG AAC GAT GAC TAC GAA TTT TTC ATT CGT GAC ACC

                                                         135
PHO 3 ASP ASP LEU GLU MET GLU THR THR PHE ALA ASN SER ASP ASN VAL
      GAT GAT TTG GAA ATG GAA ACC ACT TTT GCC AAC TCA GAC AAT GTT
PHO 5 ASP ASP  *   *   *   *   *   *  PHE  *   *   *  ASP ASP  *
      GAT GAT TTG GAA ATG GAA ACC ACT TTT GCC AAC TCG GAC GAT GTT
PHO11 LYS ASN  *   *   *   *   *   *  LEU  *   *   *  VAL ASN  *
      AAA AAC CTA GAA ATG GAA ACC ACA CTT GCC AAT TCG GTC AAT GTT

                                                         150
PHO 3 LEU ASN PRO TYR THR GLY GLU MET ASP ALA LYS ARG HIS ALA ARG
      TTG AAT CCA TAC ACT GGT GAG ATG GAT GCT AAG AGA CAT GCT CGT
PHO 5  *   *   *   *   *   *   *   *  ASN  *   *   *   *   *   *
      TTG AAC CCA TAC ACT GGT GAA ATG AAC GCC AAG AGA CAT GCT CGT
PHO11  *   *   *   *   *   *   *   *  ASN  *   *   *   *   *   *
      TTG AAC CCA TAT ACC GGT GAG ATG AAT GCT AAG AGA CAC GCT CGT

                                                         165
PHO 3 GLU PHE LEU ALA GLN TYR GLY TYR MET PHE GLU ASN GLN THR SER
      GAG TTT TTA GCG CAA TAT GGC TAC ATG TTC GAA AAT CAA ACC AGT
PHO 5 ASP  *   *   *   *   *   *   *   *  VAL  *   *   *   *   *
      GAC TTC TTG GCT CAA TAC GGT TAC ATG GTC GAA AAC CAA ACC AGT
PHO11 ASP  *   *   *   *   *   *   *   *  VAL  *   *   *   *   *
      GAT TTC TTG GCG CAA TAT GGC TAC ATG GTC GAA AAC CAA ACC AGT

                                                         180
PHO 3 PHE PRO ILE PHE ALA ALA SER SER GLU ARG VAL HIS ASP THR ALA
      TTC CCA ATT TTC GCC GCT AGT TCT GAA AGG GTT CAT GAC ACT GCT
PHO 5  *  ALA VAL  *  THR SER ASN  *  LYS  *  CYS  *   *   *   *
      TTC GCC GTT TTT ACC TCT AAT TCT AAG AGA TGT CAT GAC ACT GCT
PHO11  *  ALA VAL  *  THR SER ASN  *  ASN  *  CYS  *   *   *   *
      TTT GCC GTT TTT ACG TCT AAC TCG AAC AGA TGT CAT GAT ACT GCC

                                                         195
PHO 3 GLN TYR PHE ILE ASP GLY LEU GLY ASP GLN PHE ASN ILE SER LEU
      CAA TAT TTC ATT GAT GGT TTA GGT GAC CAA TTC AAC ATC TCC TTG
PHO 5  *   *   *   *   *   *   *   *   *  GLN  *   *   *  THR  *
      CAA TAT TTC ATT GAT GGT TTA GGT GAC CAA TTC AAC ATC ACC TTG
PHO11  *   *   *   *   *   *   *   *   *  LYS  *   *   *  SER  *
      CAG TAT TTC ATT GAC GGT TTG GGT GAT AAA TTC AAC ATA TCC TTG

                                                         210
PHO 3 GLN THR VAL SER GLU ALA MET SER ALA GLY ALA ASN THR LEU SER
      CAG ACT GTC AGT GAA GCC ATG TCC GCC GGC GCA AAC ACT TTG AGT
PHO 5  *   *  VAL  *   *   *  GLU  *   *   *   *   *   *   *   *
      CAG ACT GTC AGT GAA GCT GAA TCC GCT GGT GCC AAC ACT TTG AGT
PHO11  *   *  ILE  *   *   *  GLU  *   *   *   *   *   *   *   *
      CAA ACC ATC AGT GAA GCC GAG TCT GCT GGT GCC AAT ACT CTG AGT

                                                         225
PHO 3 ALA GLY ASN ALA CYS PRO GLY TRP MET LYS THR ALA ASN ASP ASP
      GCT GGT AAT GCG TGC CCA GGA TGG ATG AAG ACT GCT AAC GAT GAC
PHO 5  *  CYS ASN SER  *   *  ALA  *  ASP TYR ASP ALA  *   *   *
      GCT TGT AAC TCA TGT CCT GCT TGG GAC TAC GAT GCC AAT GAT GAC
PHO11  *  HIS HIS SER  *   *  ALA  *  GLN TYR ASP VAL  *   *   *
      GCC CAC CAT TCG TGT CCT GCT TGG CAG TAT GAT GTC AAC GAT GAC

                                                         240
PHO 3 ILE LEU ASP LYS TYR ASP THR THR TYR LEU ASP ASP ILE ALA LYS
      ATT TTG GAC AAA TAC GAT ACC ACA TAC TTG GAT GAC ATT GCC AAG
PHO 5  *  VAL ASN GLU  *   *   *  THR  *   *  ASP ASP  *   *   *
      ATT GTA AAT GAA TAC GAC ACA ACC TAC TTG GAT GAC ATT GCC AAG
PHO11  *  LEU LYS LYS  *   *   *  LYS  *   *  SER GLY  *   *   *
      ATT TTG AAA AAA TAT GAT ACC AAA TAT TTG AGT GGT ATT GCC AAG

                                                         255
PHO 3 ARG LEU ASN LYS GLU ASN LYS GLY LEU ASN LEU THR SER LYS ASP
      AGA TTA AAC AAA GAA AAC AAG GGT TTG AAT TTG ACC TCA AAG GAC
PHO 5  *   *   *   *   *   *   *   *   *   *   *   *   *  THR  *
      AGA TTG AAC AAG GAA AAC AAG GGT TTG AAC TTG ACC TCA ACT GAC
PHO11  *   *   *   *   *   *   *   *   *   *   *   *   *  SER  *
      AGA TTA AAC AAG GAA AAC AAG GGT TTG AAT CTG ACT TCA AGT GAT

                                                         270
PHO 3 ALA ASN THR LEU PHE ALA TRP CYS ALA TYR GLU LEU ASN ALA ARG
      GCC AAC ACT TTG TTT GCA TGG TGT GGA TAC GAA TTG AAC GCT AGA
PHO 5  *  SER  *   *   *  SER  *   *   *  PHE  *  VAL  *   *  LYS
      GCT AGT ACT TTA TTC TCG TGG TGT GCA TTT GAA GTG AAC GCT AAA
PHO11  *  ASN  *   *   *  ALA  *   *   *  TYR  *  ILE  *   *  ARG
      GCA AAC ACT TTT TTT GCA TGG TGT GCA TAT GAA ATA AAC GCT AGA

                                                         285
PHO 3 GLY TYR SER ASP VAL CYS ASP ILE PHE THR GLU ASP GLU LEU VAL
      GGC TAC AGT GAT GTT TGT GAT ATC TTC ACC GAA GAT GAA TTG GTA
PHO 5  *   *   *   *  VAL  *   *   *   *   *  LYS  *   *   *   *
      GGT TAC AGT GAT GTC TGT GAT ATT TTC ACC AAG GAT GAA TTA GTC
PHO11  *   *   *   *  ILE  *   *   *   *   *  LYS  *   *   *   *
      GGT TAC AGT GAC ATC TGT AAC ATC TTC ACC AAA GAT GAA TTG GTC

                                                         300
PHO 3 ARG TYR SER TYR GLY GLN ASP LEU VAL SER PHE TYR GLN ASP GLY
      CGT TAC TCA TAC GGC CAG GAC CTG GTA TCG TTT TAC CAG GAT GGA
PHO 5 HIS TYR  *   *  TYR  *   *   *  HIS THR TYR  *  HIS GLU  *
      CAT TAC TCC TAC TAC CAA GAC TTG CAC ACT TAT TAC CAT GAG GGT
PHO11 ARG PHE  *   *  GLY  *   *   *  GLU THR TYR  *  GLN THR  *
      CGT TTC TCC TAC GGC CAA GAC TTG GAA ACT TAT TAT CAA ACG GGA

                                                         315
PHO 3 PRO GLY TYR ASP MET ILE ARG SER VAL GLY ALA ASN LEU PHE ASN
      CCA GGT TAT GAT ATG ATC AGA TCC GTC GGT GCC AAC TTG TTT AAC
PHO 5  *   *   *   *  ILE ILE LYS  *   *   *  SER  *   *   *   *
      CCA GGT TAC GAC ATT ATC AAG TCT GTC GGT TCC AAC TTG TTC AAT
PHO11  *   *   *   *  VAL VAL ARG  *   *   *  ALA  *   *   *   *
      CCA GGC TAT GAC GTC GTC AGA TCC GTC GGT GCC AAC TTG TTC AAC

                                                         330
PHO 3 ALA THR LEU LYS LEU LEU LYS GLN SER GLU THR GLN ASP LEU LYS
      GCT ACT TTG AAG TTG TTA AAG CAA AGT GAA ACT CAA GAC TTA AAA
PHO 5  *  SER VAL  *   *   *   *   *   *   *  ILE  *   *  GLN  *
      GCC TCA GTC AAA TTA TTA AAG CAA AGT GAG ATT CAA GAC CAA AAG
PHO11  *  SER VAL  *   *   *   *   *   *   *  VAL  *   *  GLN  *
      GCT TCA GTG AAA CTA CTA AAG GAA AGT GAG GTC CAG GAC CAA AAG

                                                         345
PHO 3 VAL TRP LEU SER PHE THR HIS ASP THR ASP ILE LEU ASN TYR LEU
      GTC TGG TTG AGT TTT ACC CAC GAT ACC GAT ATC CTA AAC TAT TTG
PHO 5  *   *   *   *   *   *   *   *   *   *   *   *   *  PHE  *
      GTT TGG TTG AGT TTT ACC CAC GAT ACC GAT ATC CTA AAC TTT TTG
PHO11  *   *   *   *   *   *   *   *   *   *   *   *   *  TYR  *
      GTT TGG TTG AGT TTC ACC CAC GAT ACC GAT ATT CTG AAC TAT TTG

                                                         360
PHO 3 THR THR ALA GLY ILE ILE ASP ASP LYS ASN ASN LEU THR ALA GLU
      ACC ACC GCT GGT ATA ATT GAC GAC AAA AAC AAC TTA ACT GCC GAA
PHO 5  *   *  ALA  *   *   *   *   *   *   *   *   *   *   *   *
      ACC ACC GCT GGT ATA ATT GAC GAC AAA AAC AAC TTA ACT GCC GAA
PHO11  *   *  ILE  *   *   *   *   *   *   *   *   *   *   *   *
      ACC ACT ATC GGC ATA ATC GAT GAC AAA AAT AAC TTG ACC GCC GAA

                                                         375
PHO 3 TYR VAL PRO PHE MET GLY ASN THR PHE HIS LYS SER TRP TYR VAL
      TAC GTT CCA TTC ATG GGC AAC ACC TTC CAT AAG TCC TGG TAC GTT
PHO 5 TYR  *   *   *   *  GLY  *   *   *   *  ARG  *   *   *   *
      TAC GTT CCA TTC ATG GGC AAC ACT TTC CAC AGA TCC TGG TAC GTT
PHO11 HIS  *   *   *   *  GLU  *   *   *   *  ARG  *   *   *   *
      CAT GTT CCA TTC ATG GAA AAC ACT TTC CAC AGA TCC TGG TAC GTT

                                                         390
PHO 3 PRO GLN GLY ALA ARG VAL TYR THR GLU LYS PHE GLN CYS SER ASN
      CCT CAA GGT GCT CGT GTC TAC ACC GAA AAA TTC CAA TGT TCT AAC
PHO 5  *   *   *   *   *   *   *   *   *   *   *   *   *   *   *
      CCT CAA GGT GCT CGT GTC TAC ACC GAA AAA TTC CAA TGT TCT AAC
PHO11  *   *   *   *   *   *   *   *   *   *   *   *   *   *   *
      CCA CAA GGT GCT CGT GTT TAC ACT GAA AAG TTC CAG TGT TCC AAT

                                                         405
PHO 3 ASP THR TYR VAL ARG TYR VAL ILE ASN ASP ALA VAL VAL PRO ILE
      GAC ACC TAC GTC AGA TAC GTC ATT AAC GAT GCT GTC GTT CCA ATT
PHO 5  *   *   *   *   *   *   *   *   *   *   *   *   *   *   *
      GAC ACC TAC GTC AGA TAC GTC ATT ACC GAT GCT GTT GTT CCA ATT
PHO11  *   *   *   *   *   *   *   *   *   *   *   *   *   *   *
      GAC ACC TAT GTT AGA TAC GTC ATC AAC GAT GCT GTC GTT CCA ATT

                                                         420
PHO 3 GLU THR CYS SER THR GLY PRO GLY PHE SER CYS GLU ILE ASN ASP
      GAA ACC TGT TCC ACC GGC CCA GGG TTC TCT TGT GAA ATC AAT GAT
PHO 5  *   *   *   *   *   *   *   *   *   *   *   *   *   *   *
      GAA ACC TGT TCC ACT GGT CCA GGG TTC TCT TGT GAA ATC AAT GAC
PHO11  *   *   *   *   *   *   *   *   *   *   *   *   *   *   *
      GAA ACC TGT TCT ACT GGT CCA GGG TTC TCC TGT GAA ATA AAT GAC

                                                         435
PHO 3 PHE TYR ASP TYR ALA GLU LYS ARG VAL ALA GLY THR ASP PHE LEU
      TTC TAC GAC TAT GCT GAA AAG AGA GTA GCT GGT ACT GAC TTC CTA
PHO 5  *   *   *   *   *   *   *   *   *   *   *   *   *   *   *
      TTC TAC GAC TAT GCT GAA AAG AGA GTA GCC GGT ACT GAC TTC CTA
PHO11  *   *   *   *   *   *   *   *   *   *   *   *   *   *   *
      TTC TAC GAC TAT GCT GAA AAG AGA GTA GCC GGT ACT GAC TTC CTA

                                                         450
PHO 3 LYS VAL CYS ASN VAL SER SER VAL SER ASN VAL THR GLU LEU THR
      AAG GTC TGT AAC GTC AGC AGT GTC AGT AAC GTC ACC GAA TTG ACC
PHO 5  *   *   *   *   *   *   *   *   *   *  SER  *   *   *   *
      AAG GTC TGT AAC GTC AGC AGC GTC AGT AAC TCT ACT GAA TTG ACC
PHO11  *   *   *   *   *   *   *   *   *   *  SER  *   *   *   *
      AAG GTC TGT AAC GTC AGC AGC GTC AGT AAC TCT ACT GAA TTG ACC

                                                         465
PHO 3 PHE TYR TRP ASP TRP ASN THR THR HIS TYR ASN ASP THR LEU LEU
      TTC TAC TGG GAC TGG AAT ACT ACT CAC TAC AAC GAT ACC CTA TTA
PHO 5  *  TYR  *   *   *   *   *  THR  *   *   *  ALA SER  *   *
      TTC TAC TGG GAC TGG AAC ACT ACT CAT TAC AAC GCC AGT CTA TTG
PHO11  *  PHE  *   *   *   *   *  LYS  *   *   *  ASP THR  *   *
      TTT TTC TGG GAC TGG AAT ACC AAG CAC TAC AAC GAC ACT TTA TTA

          467
PHO 3 LYS GLN
      AAA CAA TAA
PHO 5 ARG  *
      AGA CAA TAG
PHO11 LYS  *
      AAA CAG TAA

*) indicates same amino acid as for PHO 3 in all three genes
```

FIG. 1. Comparison of the protein coding sequences of PHO3, PHO5, and PHO11.

position −32). To this downstream *PHO3* promoter element were fused the various upstream segments of the *PHO5* promoter as indicated in Fig. 3A. The whole construction was integrated at the *PHO3/PHO5* locus on chromosome II as described (8). This integration leads to a replacement of the *PHO5/PHO3* gene cluster by the *PHO3* gene controlled by various *PHO3/PHO5* hybrid promoters. These promoter/gene arrangements are called *PHO3HP*.

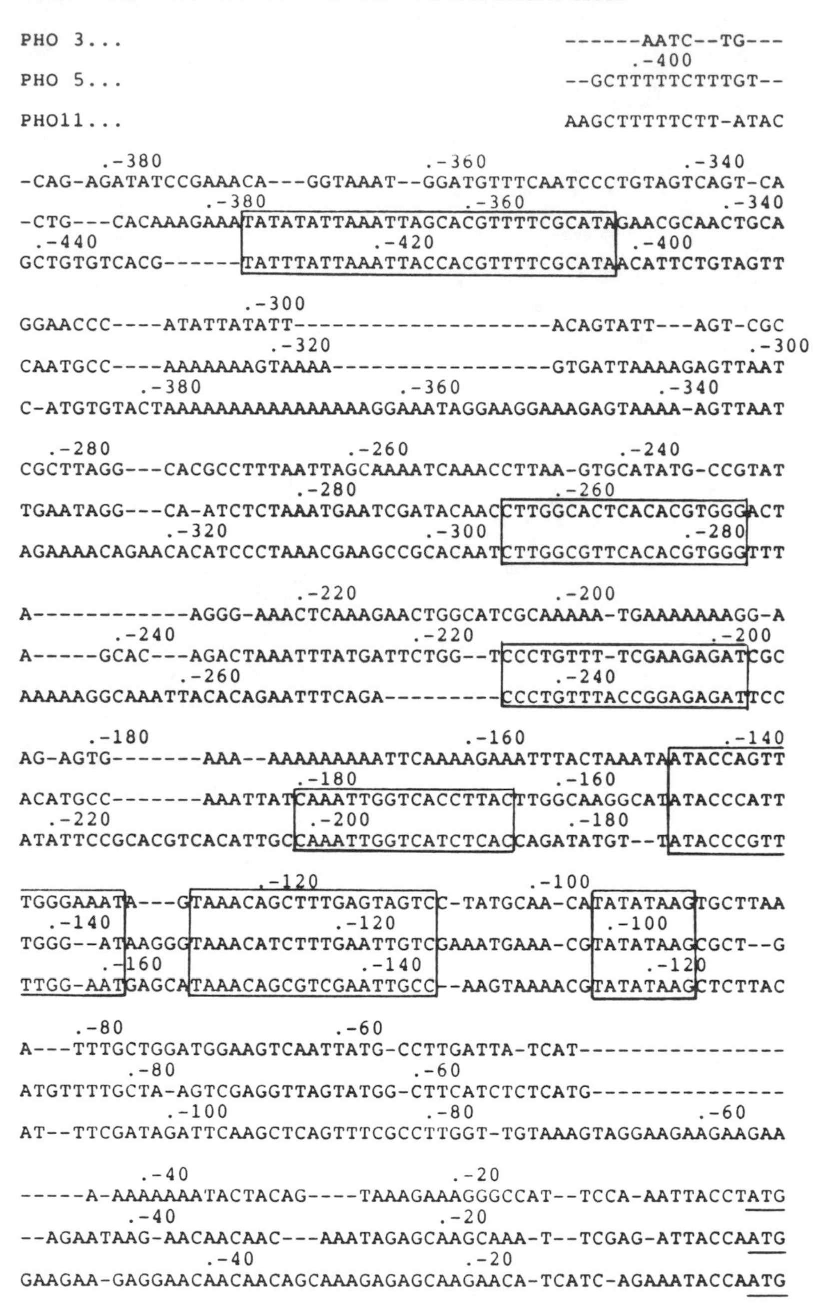

FIG. 2. Promoter elements of PHO3, PHO5, and PHO11. Sequences were aligned to maximize homology among the three genes.

Transcripts from *PHO3HP* were analyzed by S1 nuclease protection experiments and Northern blots. S1 protection experiments indicated that the *PHO5* sequences brought to various distances from the TATA box of the *PHO3* gene did not affect the transcription initiation site(s) of *PHO3* (results not shown).

An S1 mapping analysis was done to quantitate the amount of mRNA produced from *PHO3HP* under repressed and derepressed conditions. Poly(A)$^+$ RNA was isolated from strains IH2, IH530, WB10, WB13, WB15, WB24, and WB24(FP).

Figure 3B shows the results obtained from all of the above-mentioned strains, using a *PHO3*-specific probe. The *URA3* transcripts shown in the same gel served as an internal reference. It can be seen that *PHO3* transcripts are not

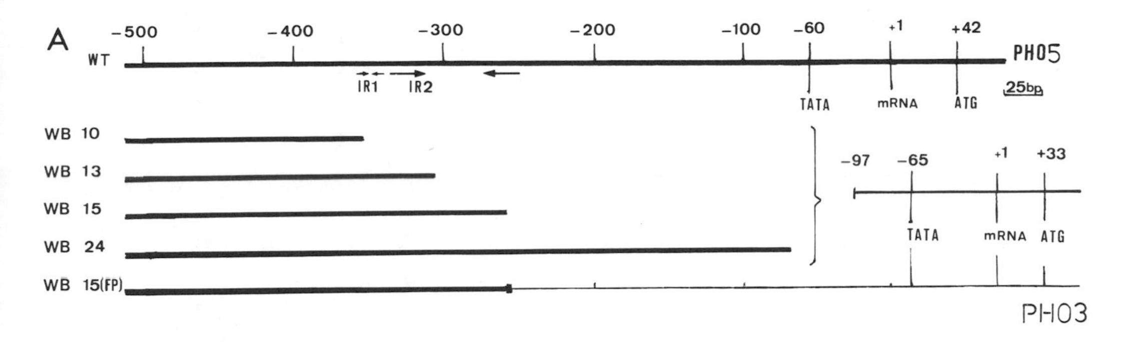

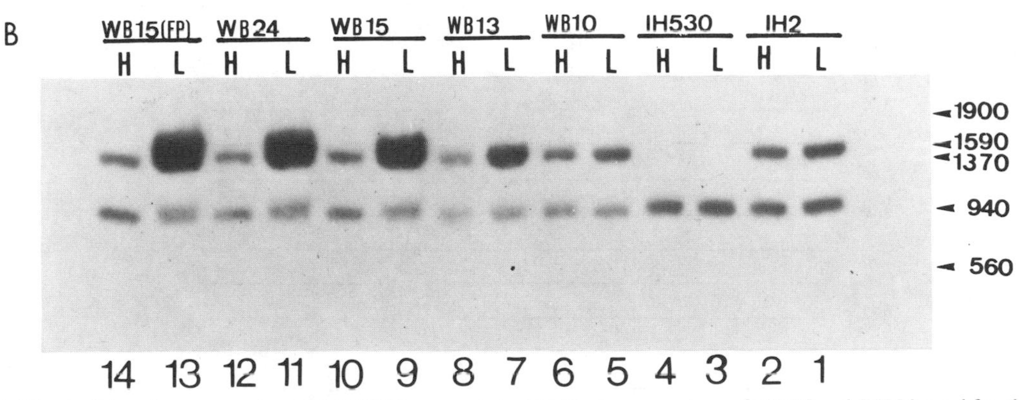

FIG. 3. S1 nuclease mapping of *PHO3*HP transcripts. (A) Upstream regions of *PHO5* and *PHO3* used for the hybrid promoter constructions. Thick line on the top represents *PHO5* (wild type) promoter sequences. Numbered thick lines represent the sequences of *PHO5* used for the fusion to *PHO3* (represented by thin lines). Numbering begins from the transcription initiation sites. (B) Total RNA from the *S. cerevisiae* strains, grown in high P_i (H) and low P_i (L), was hybridized to a *PHO3*-specific DNA fragment. The amount of RNA used for hybridization was as follows: 200 μg for lanes 1 and 3 to 14; 500 μg for lane 2. Source of RNA is shown above each lane. The upper band represents *PHO3* specific transcripts; the band at about 900 bp shows the *URA3* transcripts used as standards.

affected by P_i (Fig. 3B, lanes 1 and 2), as is expected from a constitutively expressed gene. Strain IH530, in which *PHO5* and *PHO3* are disrupted with *URA3*, did not show any *PHO3* transcripts. However, strains carrying *PHO3HP* were differentially affected by P_i as well as by the length of the *PHO5* 5′ flanking sequences placed in front of the *PHO3* TATA box. *PHO3HP* in strain WB10 has *PHO5* flanking sequences beyond position −359 (numbered from the major transcription initiation site[s] of *PHO5*). Transcripts in this strain were produced constitutively at a level similar to that of normal *PHO3* (Fig. 3B; compare lanes 1 and 2 with 5 and 6). WB13 showed a fourfold derepression in low P_i (Fig. 3B, lanes 7 and 8). *PHO3HP* in this strain has *PHO5* sequences beyond −310. However, *PHO3HP* from WB15 and WB24, which have *PHO5* sequences beyond −260 and −70, respectively, showed a 40-fold derepression which is comparable to the *PHO5* wild-type gene (results not shown). In WB24(FP) the length of the *PHO5* sequences in *PHO3HP* is the same as in WB24; however, the length of the *PHO3* promoter part is increased to 270 base pairs (bp) upstream from the TATA box. This shows that the distance between the TATA box of *PHO3* and the regulatory element of *PHO5* is not critical.

Functional Analysis of the *PHO5* Promoter

The *PHO5* promoter was analyzed by constructing a series of small deletions which cover the functionally relevant part of the *PHO5* 5′ flanking region. Using the cotransformation technique, all deletions were transplaced at the *pho5/URA3* locus of strain IH20 described previously (8). A summary of all the deletion mutations is given in Fig. 4. Activity was substantially reduced by deletions in each of three regions, one approximately 370 bp upstream of the ATG, the second between −255 and −174, and the third between −111 and −89.

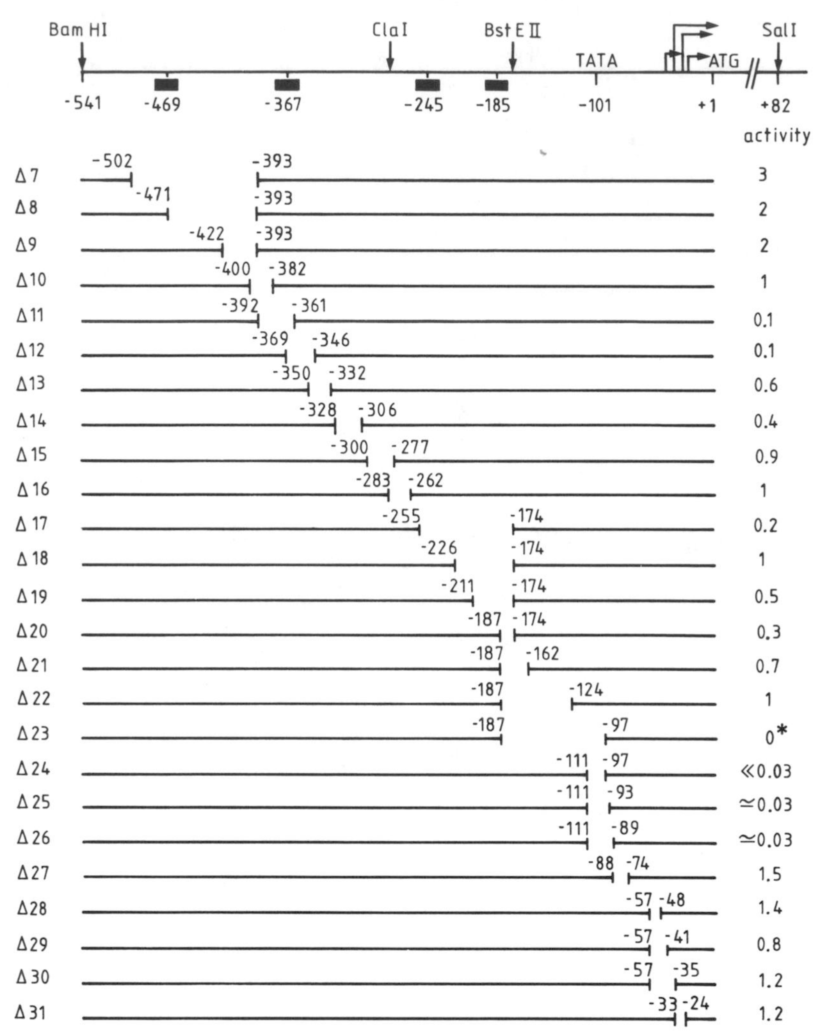

FIG. 4. *PHO5* promoter and deletion derivatives. The restriction map (top) shows the 623-bp *Bam*HI-*Sal*I fragment bearing the wild-type promoter. The four mRNA start sites are marked by arrows. The positions of the TATA box and the first ATG are given. The black boxes represent the four 19-bp dyads. The lower part presents all promoter mutants employed in this study. The number indicates the nucleotides deleted in a particular construct. Each deletion carries an 8-bp synthetic *Eco*RI site (GGAATTCC) and has been transplaced to the *PHO5* locus of strain IH20. The column to the right displays the relative amount of total *PHO5* mRNA (wild type = 1).

A first region, located approximately 370 bp upstream of the ATG, is characterized by the partially overlapping deletions Δ11 and Δ12. Both reduce transcriptional activation 10-fold. The flanking deletions Δ10 and Δ13, expressing essentially wild-type RNA levels, reduce the size of this element further to 31 bp (between positions −382 and −350).

The boundaries of a second region are more difficult to assign. Removal of 82 bp, between −255 and −174 (Δ17), results in a fivefold reduction of total *PHO5* mRNA. Two deletions, Δ19 and Δ20, which have the same 3′ border as deletion Δ17 but delete only 38 and 14 bp, respectively, lead to a less pronounced reduction in transcriptional activity. The alleles Δ21 and Δ22 retain essentially wild-type function, but do not unambiguously define the right border of the critical region. On the 5′ side, deletions Δ15 and Δ16 both allow full expression, suggesting that nucleotides around −260 mark the left border of the important sequences.

Further analysis of the deletion mutants reveals a complex structure for the *PHO5* promoter in the Δ17 region. This region apparently contains two critical regulatory elements: one located close to the 5′ end and the other near the 3′ border of Δ17. The approximate location of the 3′ element is evident from mutant Δ20, in which only 14 bp are removed from the pro-

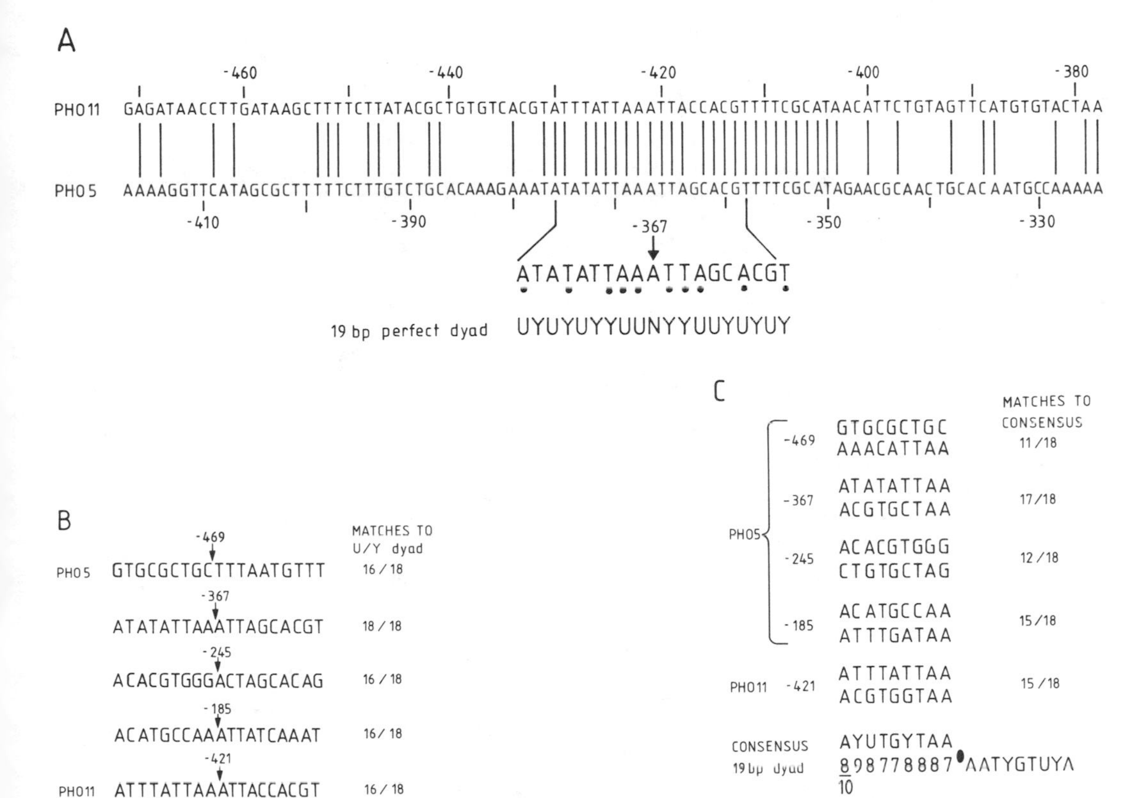

FIG. 5. Short dyad symmetric sequences that are present four times in the *PHO5* and once in the *PHO11* promoter. Panel A compares *PHO5* and *PHO11* promoter sequences. The numbering indicates the distance from the corresponding ATG. A specific 19-bp dyad around position −367 in *PHO5* is displayed in a purine (U)-pyrimidine (Y) notation. Dots mark the matching nucleotides in the specific dyad, N represents the core nucleotide in the general, perfect dyad. Panel B shows all motifs found in *PHO5* and *PHO11* looking by computer for homology to the U/Y dyad. The core positions are marked by arrows and the matches to the U/Y-dyads are indicated. In panel C the five dyads of panel B are examined for a consensus sequence by comparing the individual arms of the dyads. The resulting consensus dyad is shown at the bottom and the fitting of each motif to the consensus is indicated.

moter, reducing transcriptional activation about threefold. However, this defect caused by Δ20 can be restored progressively by increasing the size of the deletion in either direction (see Δ20, Δ19, Δ18, and Δ20, Δ21, Δ22). Only the relatively large deletion Δ17 is again less inducible and exhibits more functional damage than Δ20. This indicates the presence of a second regulatory element within Δ17 just upstream of Δ18. We conclude from these data that the *PHO5* promoter contains, in addition to the 31-bp element mapped between deletions Δ10 and Δ13, two other regulatory elements: one in the region from −262 to −226 and the other around −187 to −174.

PHO5 and *PHO11* Upstream Regions Contain Short, Related Sequences with Dyad Symmetry

A comparison of the promoter sequences of *PHO5* and *PHO11*, including about 350 bp upstream of the ATG, has been published (10). We cloned the *PHO11* gene and extended the sequence analysis to about −458 bp (see Fig. 5A).

In the upstream regions of both promoters we found a strikingly conserved region (Fig. 5A). Specifically, 94 bases of the *PHO5* promoter (−324 to −417) are compared with *PHO11* sequences between −378 and −471 upstream of the ATG. Both DNAs contain a cluster of 29 nucleotides with nearly perfect homology (27 identical bases) flanked by sequences matching only at random. This conserved region almost exactly coincides with the critical regulatory element identified between Δ10 and Δ13 by the deletion approach.

Looking at this sequence in more detail, we found that 19 bases of the cluster exhibit approximate twofold rotational symmetry, as shown in Fig. 5A. Two arms, each 9 bp long, form a dyad centered at position −367 with five matching nucleotides. The symmetry becomes perfect when the actual DNA sequence is converted into the purine-pyrimidine notation, as is done in Fig. 5A.

Variants of this totally symmetric purine-pyrimidine motif were found at three other positions in the *PHO5* promoter by computer analysis. Two variants, at −245 and −185, lie within the Δ17 region, each close to one boundary. The third one is located at −469. All motifs found including the one in *PHO11* are listed in Fig. 5B.

The strong relationship of these five sequences becomes evident when all 10 individual dyad arms are examined for an underlying consensus sequence, as illustrated in Fig. 5C. Six nucleotides, three forming the core of the dyad, two in the middle of an arm, and the last one at the outside, are highly conserved. The other three positions show a strong bias to either a purine or a pyrimidine base. Two features of these dyads—their occurrence in *PHO5* and *PHO11* and their presence in all three regulatory regions of the *PHO5* promoter—suggest that they represent binding sites for a regulatory protein (complex).

In addition to these presumptive regulatory elements, two additional regions which affect promoter function were observed. (i) Four deletions (Δ23 to Δ26) show a very dramatic effect upon transcription, while neighboring deletions (Δ22, Δ27) exhibit wild-type activity; this 35-bp region includes a TATATAA motif at −101 which is in perfect agreement with the canonical sequence (3). (ii) Five mutants were obtained which showed altered initiation of *PHO5* mRNA; the significance of these latter findings will be discussed elsewhere (H. Rudolph and A. Hinnen, manuscript in preparation).

The functional analysis of the *PHO5* promoter by promoter fusions to *PHO3* and by analyzing the effect of small internal deletions within *PHO5* compares favorably with the data obtained by DNA sequence comparison between the promoters of *PHO3*, *PHO5*, and *PHO11*. The three clusters of approximate twofold rotational symmetry either are located within the *PHO5*/*PHO11* homology boxes or overlap with them. Furthermore, deletions within the promoter which show severe effects on transcriptional activity of *PHO5* either partially or completely removed sequences which are conserved between the *PHO5* and *PHO11* promoter. These sequence elements are interesting target areas for possible protein/DNA interactions. More direct in vitro experiments (retardation assays, DNA footprinting) combined with genetic approaches will help us to understand the action of the many different regulatory proteins (*PHO2*, *PHO4*, *PHO6*, *PHO7*, *PHO80*, *PHO81*, *PHO85*) on the promoter function of the differentially expressed members of the acid phosphatase gene family in *S. cerevisiae*.

LITERATURE CITED

1. **Bajwa, W., B. Meyhack, H. Rudolph, A.-M. Schweingruber, and A. Hinnen.** 1984. Structural analysis of two tandemly repeated acid phosphatase genes in yeast. Nucleic Acids Res. **12:**7721–7739.
2. **Bostian, K. A., J. J. Lemire, L. E. Cannon, and H. O. Halvorson.** 1980. In vitro synthesis of repressible yeast acid phosphatase: identification of multiple mRNAs and products. Proc. Natl. Acad. Sci. USA **77:**4504–4508.
3. **Breathnach, R., and P. Chambon.** 1981. Organization and expression of eukaryotic split genes coding for proteins. Annu. Rev. Biochem. **50:**349–383.
4. **Kramer, R. A., and N. Anderson.** 1980. Isolation of yeast genes with mRNA levels controlled by phosphate concentration. Proc. Natl. Acad. Sci. USA **77:**6541–6545.
5. **Meyhack, B., W. Bajwa, H. Rudolph, and A. Hinnen.** 1982. Two yeast acid phosphatase genes are the result of a tandem duplication and show different degrees of homology in their promoter and coding sequences. EMBO J. **1:**675–680.
6. **Oshima, Y.** 1982. Regulatory circuits for gene expression: the metabolism of galactose and phosphate, p. 159–180. *In* J. N. Strathern, E. W. Jones, and J. R. Broach (ed.), The molecular biology of the yeast *Saccharomyces*: metabolism and gene expression. Cold Spring Harbor Laboratory, Cold Spring Harbor, N.Y.
7. **Rogers, D. T., J. M. Lemire, and K. A. Bostian.** 1982. Acid phosphatase polypeptides in *Saccharomyces cerevisiae* are encoded by a differentially regulated multigene family. Proc. Natl. Acad. Sci. USA **79:**2157–2161.
8. **Rudolph, H., I. Rauseo-Koenig, and A. Hinnen.** 1985. One-step gene replacement in yeast by cotransformation. Gene **36:**87–95.
9. **Tait-Kamradt, A. G., K. J. Turner, R. A. Kramer, Q. D. Elliott, S. J. Bostian, G. P. Thill, D. T. Rogers, and K. A. Bostian.** 1986. Reciprocal regulation of the tandemly duplicated *PHO5*/*PHO3* gene cluster within the acid phosphatase multigene family of *Saccharomyces cerevisiae*. Mol. Cell. Biol. **6:**1855–1865.
10. **Thill, G. P., R. A. Kramer, K. J. Turner, and K. A. Bostian.** 1983. Comparative analysis of the 5′-end region of two repressible acid phosphatase genes in *Saccharomyces cerevisiae*. Mol. Cell. Biol. **3:**570–579.
11. **Toh-e, A., S. Kakimoto, and Y. Oshima.** 1975. Genes coding for the structure of the acid phosphatases in *Saccharomyces cerevisiae*. Mol. Gen. Genet. **141:**81–83.

Regulation of the Phosphatase Multigene Family of *Saccharomyces cerevisiae*

S. A. PARENT, A. G. TAIT-KAMRADT, J. LEVITRE, O. LIFANOVA, AND K. A. BOSTIAN

Division of Biology and Medicine, Brown University, Providence, Rhode Island 02912

Our work is directed toward elucidating the molecular basis of phosphate control and its assimilation in the yeast *Saccharomyces cerevisiae*. This system involves five principal enzymes: an acid phosphatase, a phosphate permease, a polyphosphate kinase, an alkaline phosphatase, and a polyphosphatase. These enzymes regulate intracellular concentrations of P_i by a cyclic pathway of polyphosphate synthesis and degradation (6, 7) which is thought to be an important factor in the regulation of cellular homeostasis.

The five enzymes are regulated by the external concentration of P_i (6). Under nonlimiting conditions they are present in basal amounts. However, when cells are starved for P_i, the enzymes are derepressed, and internal reserves of polyphosphate are depleted. Both acid phosphatase and an inducible phosphate permease participate in phosphorus acquisition (16, 21). When P_i is replenished, polyphosphate is rapidly synthesized and the enzymes are repressed in a coordinated manner (6).

Little is known about the mechanism of this regulation or its metabolic effectors, although the phosphatases are regulated by a complex but genetically well-defined system. The acid and alkaline phosphatases of *S. cerevisiae* include several isozymes which are encoded by separate and distinct structural genes. Three acid phosphatase genes (*PHO5*, *PHO3*, and *PHO11*) (1, 2) and one alkaline phosphatase gene (*PHO8*) (9) have been isolated, their flanking control regions have been sequenced and characterized (2, 17), and the level of their regulation has been demonstrated to be predominantly transcriptional (1, 4, 9, 11, 13). Transcription of *PHO5* and *PHO11* is induced during growth in low P_i, while *PHO3* transcription occurs only during high-P_i growth (1, 4). Several other loci are involved in the genetic control of this complex regulatory system. These include *PHO2, PHO4, PHO6, PHO7, PHO80, PHO81*, and *PHO85*, as described below (Fig. 1).

The acid phosphatase structural genes *PHO5* and *PHO11* are regulated along with an unlinked phosphatase locus, *PHO10*, by several loci. *PHO4* and *PHO81* positively regulate the phosphate-repressible acid phosphatases (*PHO5, PHO10,* and *PHO11*) and alkaline phosphatase (*PHO8*) (21); recessive *PHO4* and *PHO81* mutations are uninducible for these enzymes. However, *PHO2* recessive mutants are uninducible for the acid phosphatases (*PHO5, PHO10,* and *PHO11*) and the permease (*PHO84*) (16, 21), while *PHO80* or *PHO85* recessive mutants express all of these enzymes (*PHO5, PHO10, PHO11, PHO8*, and *PHO84*) constitutively (16, 22). Both *PHO6* and *PHO7* are required for expression of the constitutive acid phosphatase, *PHO3* (19).

On the basis of genetic and biochemical data, a model has been proposed for the control of the yeast phosphatase gene family (11, 18). This model involves the simultaneous posttranslational functioning of regulatory proteins by direct molecular interactions. Simply stated, the negative factors (*PHO80* and *PHO85*) compete successfully, via *PHO81* in the presence of corepressor, with the *PHO5* controlling region, at an upstream activation site, for binding of the positive transcriptional factors (*PHO4* and *PHO2*). In the absence of corepressor, the positive factors are freed to bind the promoter and activate transcription (18). We have begun to test this model by cloning and characterizing the regulatory genes and their products. The positive-acting *PHO4* and *PHO2* genes have been cloned and analyzed, providing further insight into the regulatory mechanism. RNA blot hybridizations and in vivo studies of the genes overexpressed in a variety of mutant strains suggest that excess *PHO4* is sufficient for positive control of acid phosphatase.

We are also examining the mechanism of *PHO5* control of *PHO3*. The *PHO3* gene, previously thought to be constitutive, has been shown to be expressed under conditions of high-P_i growth, a reversed regulation from that of *PHO5* (14, 15). Rearrangement of the relative gene positions, and mutation and deletion analysis of *PHO5*, indicate a role for the *PHO5* protein in regulating expression of *PHO3* (15). Potentially, the genes regulating *PHO5* may also be involved in the expression of *PHO3*.

Results

Regulation of *PHO5*: properties of the cloned *PHO4* and *PHO2* genes. In our initial studies of this system, we developed in vitro assays to cha-

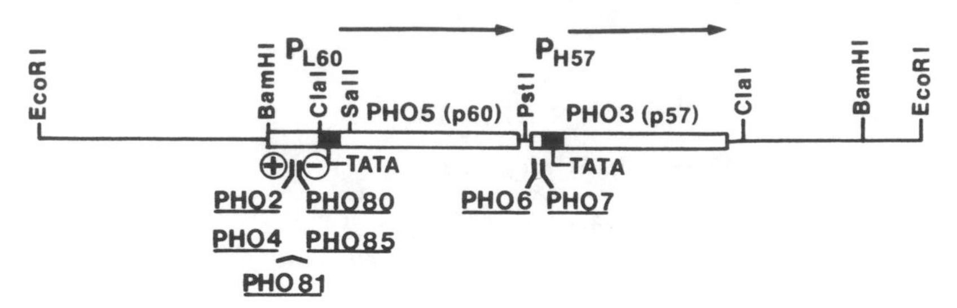

FIG. 1. Regulation of the *PHO5/PHO3* gene cluster by various *PHO* regulatory genes. +, Positive regulator; −, negative regulator.

racterize repressible acid phosphatase mRNA. Using antibodies prepared against purified repressible acid phosphatase, we identified three mRNAs (p60, p58, and p56) in RNA from low-P_i-grown cells, by immunoprecipitation of in vitro translation products (Fig. 2A). Three acid phosphatase genes were cloned from a genomic yeast DNA bank in bacteriophage λ. RNA blot hybridizations with the cloned acid phosphatase genes indicated that they are regulated transcriptionally (Fig. 2B). Using the above assays, we

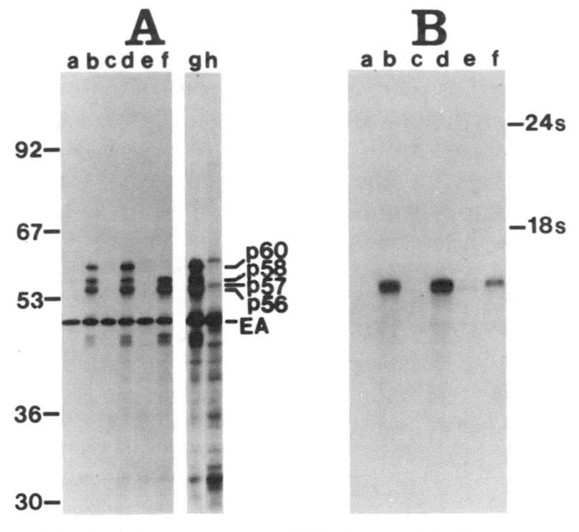

FIG. 2. Measurement of distinct acid phosphatase mRNAs and transcripts in wild-type and *pho3 pho5* strains. Total cellular RNAs were prepared, transferred to nitrocellulose, and hybridized as described previously (4, 15). RNAs were translated in a wheat germ cell-free system using L-[^{35}S]methionine, and specific cell-free translation products were analyzed by immunoprecipitation followed by polyacrylamide gel electrophoresis (4, 5). (A) Immunoprecipitations of in vitro translations of total RNAs from a variety of strains grown in low and high P_i, respectively, resolved by sodium dodecyl sulfate-polyacrylamide gel electrophoresis. Lanes: a and b, H42 (*PHO5 PHO3*); c and d, P28-24C (*PHO5 pho3-1*); e and f, P142-4A (*pho5-1 pho3-1*); h and g, the same as a and b, but a darker exposure. (B) RNA blot hybridizations of RNAs corresponding to A, hybridized with ^{32}P-labeled *PHO5/PHO3* DNA.

established that regulation of the repressible acid phosphatase genes occurs through posttranslational interactions of the regulatory gene factors (11). A 57-kilodalton cross-reactive translation product of RNA from high-P_i-grown cells was identified as the in vitro product of the *PHO3* gene (Fig. 2A, lane h).

More recently, we have begun to characterize the positive-acting regulatory genes (10). The *PHO4* and *PHO2* genes were cloned from plasmid libraries of wild-type genomic yeast DNA fragments by genetic complementation of mutant alleles. Southern hybridization of total genomic DNA prepared from wild-type strains and integrative transformants has shown that *PHO4* and *PHO2* consist of unique single-copy yeast DNA sequences (10; unpublished data).

The *PHO4* transcript was identified as a 1.7-kilobase polyadenylated RNA by RNA blot hybridization using double-stranded and strand-specific hybridization probes derived from various portions of the cloned region. Examination of transcript levels in derepressed and repressed cells indicated that the *PHO4* transcript is constitutively expressed at very low abundance. No appreciable differences in *PHO4* RNA levels were observed in mutants defective in acid phosphatase expression (*pho2, pho81, pho4*). Both constitutive mutants, *pho80* and *pho85*, showed a 1.5- to 2-fold elevation in *PHO4* RNA in low-P_i-grown cells, probably reflecting a general stabilization of RNA under phosphate starvation conditions (10). Preliminary data indicate that *PHO2* is also constitutively transcribed at a low level (data not shown).

The effect of *PHO4* gene dosage on acid phosphatase expression was examined by transforming wild-type and mutant strains with a variety of plasmids containing the cloned gene. Plasmid YESH4 is a derivative of plasmid YEp24 (pBR322 derivative containing the yeast *URA3* gene and 2μm plasmid origin of replication) carrying a wild-type *PHO4* gene on a *Hin*dIII-*Sal*I DNA fragment (Fig. 3). This plasmid overexpresses *PHO4* mRNA by 10- to 30-fold in a wild-type cell. In addition, *PHO4*

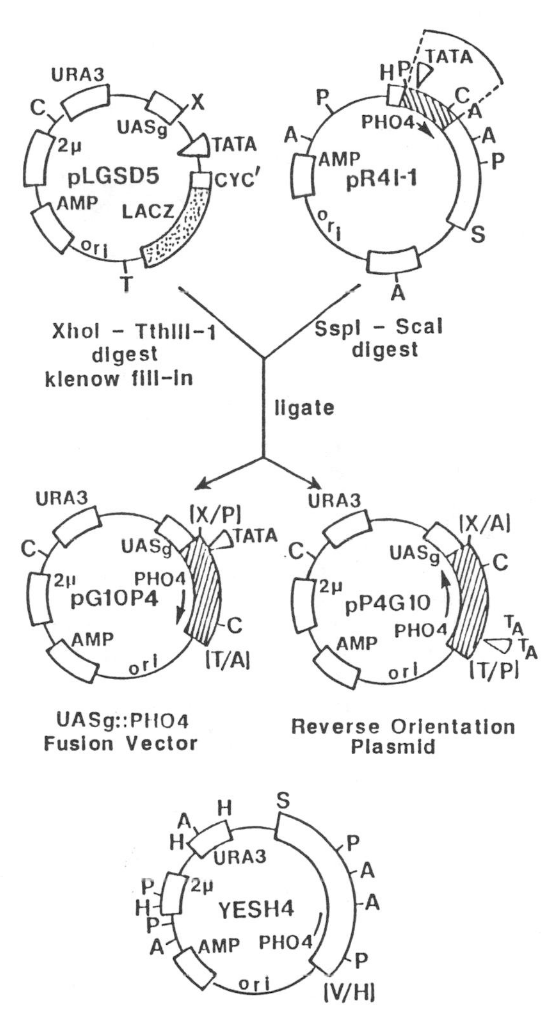

FIG. 3. *PHO4* plasmids used for in vivo studies. *PHO4* gene fusions and other plasmid constructs were made by standard recombinant DNA techniques (10, 14, 15). Restriction enzymes are A, *Sca*I; H, *Hin*dIII; P, *Ssp*I; S, *Sal*I; T, *Tth*III-1; U, *Pvu*II.

coding sequences and the putative TATA box were fused to the *GAL10* upstream activation sequence (UAS) of plasmid pLGSD5, yielding plasmid pG10P4 (Fig. 3). This plasmid overexpressed *PHO4* RNA several hundredfold in galactose-grown cells. Plasmid pP4G10 is a pLGSD5 derivative containing the *GAL10* UAS at the 3′ end of the *PHO4* gene (Fig. 3). pP4G10 transformants constitutively overexpress *PHO4* RNA at levels similar to those of galactose-grown pG10P4 transformants. Immunoblots using antibodies prepared against a *PHO4-lacZ* fusion protein showed that the *PHO4* protein is overexpressed in these transformants (data not shown).

Wild-type transformants containing the various multicopy *PHO4* plasmids were assayed for acid phosphatase expression by liquid assays after growth under high- and low-P_i conditions (Table 1). The level of derepressed acid phosphatase activity in YESH4 transformants is approximately two- to threefold higher than that seen in wild-type cells, indicating that multiple copies of the *PHO4* gene enhance the level of derepressed acid phosphatase expression. In high-P_i-grown transformants, the level of acid phosphatase expression is threefold greater than that seen in wild-type cells, indicating that multiple copies of the *PHO4* gene also confer partial constitutive expression in a wild-type strain. The level of derepressed acid phosphatase expression in wild-type pP4G10 transformants is approximately 3-fold higher than that seen in wild-type cells, while the constitutive level of expression in wild-type pP4G10 transformants is about 1.7-fold less than its derepressed level (Table 1). Constitutive acid phosphatase expression also occurs when *PHO4* transcription is regulated by the *GAL10* UAS. While wild-type pG10P4 transformants express acid phosphatase activity that approximates that of wild-type cells when grown in high- and low-P_i medium containing dextrose, galactose-grown pG10P4 transformants express acid phosphatase constitutively (Table 1).

These data suggest that, in a wild-type strain, *PHO4* may be a limiting factor determining the derepressed level of acid phosphatase expression. The varying degrees of constitutive expression seen in the transformants implies that the level of *PHO4* protein available may be crucial for *PHO5* repression.

When acid phosphatase expression is examined in a *pho4* mutant bearing the same multicopy *PHO4* plasmids, an intriguing picture emerges. Unlike the wild-type transformant, *pho4* mutants bearing plasmid YESH4 completely repress acid phosphatase in high-P_i medium (Table 1). Derepressed *pho4* pP4G10 transformants express eightfold higher acid phosphatase activity than the derepressed wild-type cells. The level of high-P_i expression in *pho4* pP4G10 transformants is reduced to twice that of the wild type, compared with fivefold higher constitutive levels for the same plasmid in a *PHO4* strain. The *pho4* pG10P4 transformants express very low levels of acid phosphatase on dextrose-low-P_i medium and high levels on galactose-low-P_i medium (Table 2). Acid phosphatase activity in the *pho4* pG10P4 transformants grown on galactose-high-P_i medium was reduced to the wild-type repressed level. The reduced levels of constitutive expression of *pho4* transformants bearing multiple copies of the *PHO4* or the UASg::*PHO4* genes relative to their wild-type counterparts suggest that the mutant *pho4*

TABLE 1. Acid phosphatase expression in strains carrying multicopy *PHO4* plasmids[a]

Strain	Plasmid	Gene on plasmid	Acid phosphatase activity[b]			
			Dextrose		Galactose	
			Low P_i	High P_i	Low P_i	High P_i
Wild type	YEp24		1.0	0.2	2.8	0.2
Wild type	YESH4	*PHO4*	2.8	0.6		
Wild type	pG10P4	UAS-GAL::*PHO4*	1.3	0.2	6.3	5.3
Wild type	pP4G10	pBR::*PHO4*	3.6	2.2	7.2	6.1
pho4	YEp24		0.2	0.1		
pho4	YESH4	*PHO4*	1.4	0.1		
pho4	pP4G10	pBR::*PHO4*	8.4	0.5		

[a] Acid phosphatase regulatory mutants were crossed to GG100-14D (α *pho5-1 pho3 ura3 his3 trp1*) to obtain acid phosphatase mutant strains bearing the *ura3* marker. All other strains and media were as described previously (4, 5, 10, 14). *S. cerevisiae* strains were transformed by published procedures (8, 10). Liquid acid phosphatase assays were performed as described previously (11). Results are normalized to the derepressed level of dextrose-grown wild-type cells. UAS-GAL, Upstream activating site of the *GAL10* gene.

[b] Expressed as relative activity in enzyme units per optical density at 660 nm.

allele studied interferes with multicopy constitutive acid phosphatase expression.

Previous genetic data established that *PHO4*^{c} constitutive mutants were epistatic to *pho81* mutants while *pho2* nonderepressible mutants were epistatic to *PHO4*^{c} (20, 22), suggesting that *PHO4* functions upstream of *PHO2* in regulating *PHO5* expression. We therefore examined the effect of elevated or hyperexpression of a wild-type copy of the *PHO4* gene in nonderepressible *pho2* and *pho81* mutants and in *pho80* and *pho85* constitutive mutants. Plasmids pG10P4 and pP4G10 were transformed into *pho2, pho81, pho80*, and *pho85* strains, and the level of acid phosphatase activity in the transformants grown on high- and low-P_i medium containing galactose or dextrose was determined. Interestingly, hyperexpression of the *PHO4* gene suppresses the *pho81* mutation (Table 2) and the transformants constitutively express acid phosphatase. More

TABLE 2. Effects of *PHO4* overexpression on acid phosphatase activity in various strains[a]

Strain	Plasmid	Gene on plasmid	Relative acid phosphatase activity			
			Dextrose		Galactose	
			Low P_i	High P_i	Low P_i	High P_i
Wild type	YEp24		1.0	0.1	1.0	0.1
Wild type	YESH4	*PHO4*	1.0	0.3	1.0	0.2
Wild type	pG10P4	UAS-GAL::*PHO4*	1.0	0.1	1.0	0.7
Wild type	pP4G10	pBR::*PHO4*	1.0	0.7	1.0	0.7
Wild type	pJR2	*PHO2*	1.0	0.3	1.0	0.2
pho4	YEp24		<0.1	<0.1	<0.1	<0.1
pho4	YESH4	*PHO4*	1.0	<0.1	—	—
pho4	pG10P4	UAS-GAL::*PHO4*	0.1	<0.1	1.0	0.1
pho4	pP4G10	pBR::*PHO4*	1.0	0.2	1.0	0.1
pho4	pJR2	*PHO2*	<0.1	<0.1	—	—
pho81	YEp24		0.1	<0.1	<0.1	<0.1
pho81	pG10P4	UAS-GAL::*PHO4*	0.1	0.1	0.8	0.2
pho81	pP4G10	pBR::*PHO4*	1.0	0.4	0.8	0.2
pho81	pJR2	*PHO2*	0.3	0.1	0.1	<0.1
pho2	YEp24		<0.1	<0.1	<0.1	<0.1
pho2	pG10P4	UAS-GAL::*PHO4*	<0.1	<0.1	1.0	0.5
pho2	pP4G10	pBR::*PHO4*	1.0	0.3	1.0	0.5
pho2	pJR2	*PHO2*	1.0	<0.1	1.0	<0.1

[a] Acid phosphatase assays on solid growth medium were performed as described previously (10). Results are normalized to the derepressed level of dextrose-grown wild-type cells. The minimum amount of measurable acid phosphatase activity is one-tenth the derepressed level of dextrose-grown wild-type cells. Plate assays are less sensitive than the liquid assays described in Table 1, and very high derepressed or constitutive levels appear like the wild-type derepressed level. UAS-GAL, Upstream activating site of the *GAL10* gene.

intriguing, *PHO4* overexpression also suppresses a *pho2* mutation. Galactose-grown *pho2* pG10P4 transformants constitutively express acid phosphatase, as do dextrose- or galactose-grown *pho2* pP4G10 transformants. Overexpression of *PHO4* does not suppress *pho80* or *pho85* constitutive mutations, nor does it substantially alter the level of activity in the high-P_i-grown mutant cells (data not shown).

Preliminary results with the *PHO2* plasmid pJR2 (a pBR322 derivative containing the 2μm origin of replication and the yeast *URA3* and *PHO2* genes) show no suppression of *pho4* or *pho81* mutations (Table 2). However, the levels of overexpression may be too low to show an observable effect. Further studies analogous to those described above for *PHO4* are under way.

Regulation of *PHO3*: involvement of *PHO5* and its regulatory circuit. We are also studying the organization and regulation of the *PHO3* gene. Data with the cloned genes indicate that *PHO3* is expressed only when cells are grown in high-P_i medium, a reverse regulation to that of *PHO5*. We verified that the reciprocal expression of *PHO3* and *PHO5* observed on a high-copy-number plasmid (13) also applies to the genomic copies of the two genes (14, 15).

We have also demonstrated that altering *PHO5* expression through mutations in *trans*-acting regulatory genes is sufficient to either block or make constitutive transcription of *PHO3*. *PHO3* expression occurs only when *PHO5* is transcriptionally inactive, indicating the lack of dependence of *PHO3* repression on the concentration of P_i in the growth medium and the direct relationship between expression of the two genes (15). This does not, however, make clear whether *PHO5* is directly involved in the expression of *PHO3* or whether the two genes are controlled by an overlapping regulatory circuit.

The regions needed for *PHO5* and *PHO3* expression were determined by subcloning experiments. These showed that *PHO3* is transcribed from its own promoter and not that of *PHO5*, and that all sequences necessary for expression of *PHO3* are contained within a *Pst*-*Bam*HI restriction fragment (Fig. 4, pDR530; 15). The 3′ end of *PHO5* is within 350 base pairs of the transcription start site for *PHO3*. The subcloning also showed that *PHO3* expression is constitutive whenever sequences necessary for *PHO5* transcription are eliminated (Fig. 4, pSB15-1; 15). Sequence data on the three cloned acid phosphatase genes, *PHO3, PHO5*, and *PHO11*, show 66% homology in their promoter regions (2, 4, 15, 17). Five short regions of the *PHO5* and *PHO11* promoters, 15 to 20 nucleotides in length, show greater than 80% homology to one another (*HOM1* to *HOM5*) (Fig. 5). These

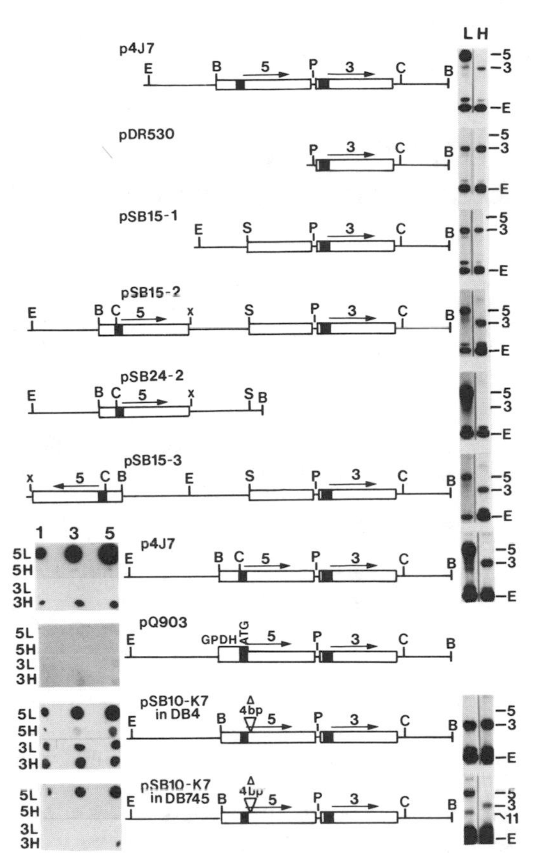

FIG. 4. *PHO3* and *PHO5* expression in strains transformed with various plasmids. *PHO5* gene fusions and other plasmid constructs were made by standard recombinant DNA techniques (10, 14, 15). Plasmid maps show the acid phosphatase composition of the plasmids used. Restriction enzymes are B, *Bam*HI; C, *Cla*I; E, *Eco*RI; P, *Pst*I; S, *Sal*I; x, destroyed *Pst*I and *Pvu*II sites. The right-hand panels show in vitro translations of RNA isolated from low-P_i-grown (L) or high-P_i-grown (H) DB4 transformants (*pho5 pho3*) or, in the case of plasmid pSB10-K7, in DB4 or DB745 (*PHO5 PHO3*) transformants. *PHO5*, *PHO3*, and enolase translation products are marked 5, 3, and E, respectively. Left-hand panels show RNA blot hybridizations of 1, 3, and 5 μg of RNA (dots 1, 3, and 5, respectively) from high-P_i-grown transformants (H) or low-P_i-grown transformants (L), using *PHO5* (5H or 5L) or *PHO3* (3H or 3L) specific oligonucleotide hybridization probes.

separate the UAS (beyond *Cla*I in *PHO5*) from the TATA box. The two most proximal regions (*HOM1* and *HOM2*) are also found in *PHO3* (Fig. 5; 15).

The tight linkage of the *PHO5* and *PHO3* genes and their orientations suggested that *PHO5* transcription might repress *PHO3* transcription via promoter occlusion or transcription readthrough. To test this, *PHO5* was placed on a *PHO3* plasmid at a site 1.4 kilobases away. In

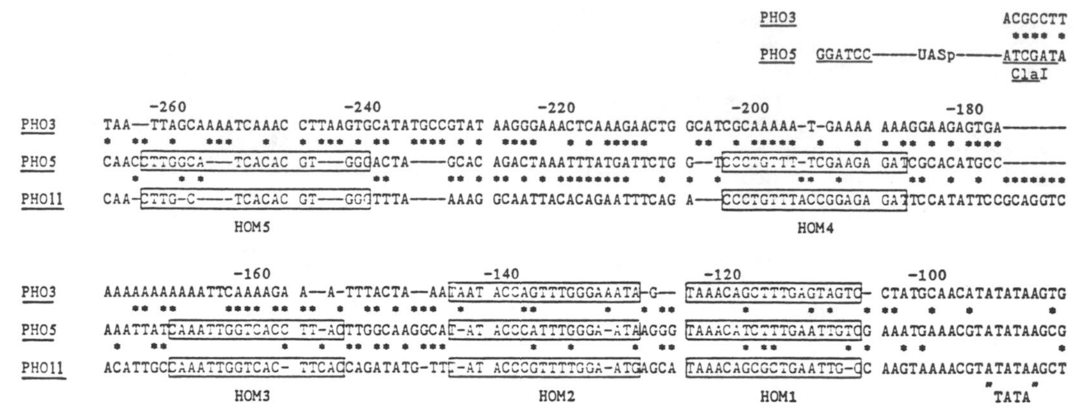

FIG. 5. Nucleotide sequence of the *PHO5*, *PHO11*, and *PHO3* proximal promoter regions. Sequences are aligned to maximize homology between the three genes. Nonmatched bases are indicated by an asterisk.

this position, *PHO5* was shown to regulate expression of *PHO3* in either orientation (Fig. 4, pSB15-2 or pSB15-3). Thus, *PHO5* need not be immediately adjacent to *PHO3* to regulate *PHO3* transcription.

To clarify the involvment of *PHO5* in repressing *PHO3* expression, the constitutive yeast glyceraldehyde-3-phosphate dehydrogenase promoter was fused to the *PHO5* gene at its ATG on a plasmid which also contained *PHO3* (Fig. 4, pQ903). Two oligonucleotides specific for either *PHO3* or *PHO5* were synthesized and were used in dot-blot experiments to measure the levels of *PHO5* and *PHO3* RNA in the transformants. As a result of the extensive homology of the acid phosphatase genes, probes of Northern blots with entire genes show high cross-hybridization, causing problems in distinguishing between mRNAs because of similarities in size. Reciprocal regulation of *PHO5* and *PHO3* can be clearly seen in control transformants with the intact *PHO5*/*PHO3* locus (Fig. 4, p4J7). When the pQ903 plasmid was transformed into a *pho5 pho3* strain, *PHO5* was expressed constitutively. The *PHO5* oligonucleotide is complementary to sequences upstream of the ATG initiator codon and therefore does not hybridize to the hybrid transcript. *PHO3* expression was not observed in high-P_i- or low-P_i-grown transformants (Fig. 4, pQ903). Thus in a wild-type regulatory background, constitutive expression of *PHO5* is sufficient to repress *PHO3* expression.

To test whether the *PHO5* protein itself is involved in the regulation of *PHO3* transcription, a frameshift mutation in *PHO5* was created by making a 4-base-pair deletion within the N-terminal coding sequence of *PHO5* in the plasmid pSB10-K7 (Fig. 4). RNA prepared from *pho5 pho3* transformants was analyzed by dot-blot hybridizations and in vitro translations. RNA from the mutant *PHO5* gene is present in low-P_i-grown cells, but the *PHO5* protein is not synthesized, as expected. *PHO3* expression was constitutive (Fig. 4, pSB10-K7 in DB4). The plasmid was also transformed into a strain with a wild-type *PHO5* allele, and expression of *PHO3* was analyzed. *PHO5* protein and mRNA were both detected in low-P_i-grown cells. Under these conditions, *PHO3* transcription did not occur (Fig. 4, pSB10-K7 in DB745; 15), indicating that production of *PHO5* protein from a single genomic copy is sufficient to block transcription of *PHO3*.

Discussion

Any model for the regulation of repressible acid phosphatase must take into account recently obtained molecular information as well as the genetic data. To summarize, it is clear that the level of *PHO5* expression is directly proportional to structural gene dosage up to 80 copies per cell, although the regulation of its expression is not altered (3, 13). The regulatory genes, *PHO4* and *PHO2*, however, are present at single copy and in wild-type cells are transcribed constitutively at very low levels (10; unpublished data). Increases in *PHO4* expression in wild-type cells elevate derepressed levels of acid phosphatase activity and confer constitutive expression to various degrees. The presence of a *pho4* allele lowers the constitutive level of acid phosphatase activity, and hyperexpression of the *PHO4* gene also suppresses *pho81* and *pho2* mutations, leading to constitutive expression.

The simplest interpretation of these results is that *PHO4* interacts with the *PHO5* UAS or some undefined UAS binding protein independent of the other phosphatase regulatory gene products. The remaining regulatory factors simply function, via a P_i-mediated signal, to control the balance between free and bound *PHO4* protein. Accordingly, *PHO81* mediates the response of the complex to corepressor, and the *PHO2* product facilitates the release of *PHO4*. A more complicated interpretation is that overproduction of *PHO4* protein compensates for the defect in the products of either *pho2* or *pho81* mutant apparent when normal levels of *PHO4* are present. Alternatively, *PHO2* may facilitate the binding of *PHO4* to the UAS site by direct interaction, but when *PHO4* is overexpressed, this cooperative assistance is not required. Thus, *PHO4* may act prior to or cooperatively with one of these gene products at the *PHO5* UAS.

The data concerning overexpression of *PHO4* in a *pho4* strain are more difficult to interpret. Similar interallelic effects in diploids are usually taken as evidence for mixed subunit interactions. This has been found for *pho4* and *PHO82*^{c} alleles (18). Since the wild-type *PHO4* protein is substantially overproduced in the *pho4* mutant, competitive interactions of this sort would require overproduction of the mutant *pho4* protein as well. This implies that *PHO4* autogenously regulates its own production. If true, this might also explain the lack of dosage limitation in expressing multiple copies of *PHO5*. It is unclear, however, conforming to the model presented above, why the suppression observed for high-P_i-grown cells is not also observed in cells grown in low-P_i medium. There are many facets of this regulatory system that are thus as yet unexplained by existing models.

Inferences can also be drawn from comparison of the nucleotide sequences of the promoter regions of the acid phosphatase genes. In particular, the high degree of conservation of the *HOM1* and *HOM2* regions in these genes suggests a role for the two regions in regulating acid phosphatase expression. Sequences considerably upstream of these sites (UAS) which are lacking in *PHO3* mediate the gross derepression of *PHO5* (15). The more proximal region, on the other hand, separating the UAS from the TATA region, may be involved in fine tuning the levels of *PHO5* expression and also in regulating *PHO3*, via repressor control. Examples of this form of promoter control have been well documented in *S. cerevisiae*, most notably, $\alpha 2$ negative regulation of **a**-specific genes. During the course of our studies on the regulation of *PHO5*, we observed that a derepression of acid phosphatase mRNA is biphasic and is transiently repressed in response to physiological signals (5). This autoregulatory control may operate through the proximal promoter region, eliciting a similar response in the *PHO3* gene, which lacks the upstream UAS and, therefore, remains unexpressed under low-P_i growth conditions. When cells are starved for P_i, *PHO5* mRNA and enzyme levels show a transient increase and decrease which is followed by full derepression (5, 11). The transient repression of *PHO5* transcription occurs as internal polyphosphate reserves are consumed and external organophosphates are metabolized. This autoregulation requires de novo protein synthesis, presumably the *PHO5* protein or other proteins undergoing depression, or both.

We have begun to identify *trans*-acting mutations affecting expression of *PHO3*, in an attempt to further define *PHO3* control and, possibly, better understand the autoregulation of *PHO5*. A *pho3 pho5* strain was transformed with pSB15-1, which contains the *PHO3* gene, and transformants were mutagenized with ethyl methanesulfonate. Several potential mutants have been identified. Their properties, and the allelism of the respective genes to *PHO6* and *PHO7*, are being determined. It seems unlikely that a totally independent control mechanism has evolved for *PHO3*. Rather, adaptations of the extant circuit controlling *PHO5* are probably used. These mutations may define genes functioning in the control of both *PHO3* and *PHO5*. The mechanisms of this regulation and precisely which factors are involved remain to be determined.

We thank Jennifer Rideout and Tikva Amran for technical assistance, Ronit Koren for her involvement in the *PHO4* expression studies, Steven Hanes for photographic assistance, and Stephen Sturley for helpful discussions and comments on the manuscript.

This work was supported in part by Public Health Service grant GM32496 to K.A.B. from the National Institute of General Medical Sciences.

LITERATURE CITED

1. **Andersen, N., G. Thill, and R. Kramer.** 1983. RNA and homology mapping of two DNA fragments with repressible acid phosphatase genes from *Saccharomyces cerevisiae*. Mol. Cell. Biol. **3**:562–569.
2. **Bajwa, W., B. Meyhack, H. Rudolph, A.-M. Schweingruber, and A. Hinnen.** 1984. Structural analysis of the two tandemly repeated acid phosphatase genes in yeast. Nucleic Acids Res. **12**:7721–7739.
3. **Bergman, L. W., M. C. Stranathan, and L. H. Preis.** 1986. Structure of the transcriptionally repressed phosphate-repressible acid phosphatase gene (*PHO5*) of *Saccharomyces cerevisiae*. Mol. Cell. Biol. **6**:38–46.
4. **Bostian, K. A., J. M. Lemire, L. E. Cannon, and H. O. Halvorson.** 1980. *In vitro* synthesis of repressible yeast acid phosphatase. Identification of multiple mRNAs and products. Proc. Natl. Acad. Sci. USA **77**:4504–4508.
5. **Bostian, K. A., J. M. Lemire, and H. O. Halvorson.** 1983. Physiological control of repressible acid phosphatase gene transcripts in *Saccharomyces cerevisiae*. Mol. Cell. Biol. **3**:839–853.
6. **Dawes, E. A., and P. J. Senior.** 1973. The role and regulation of energy reserve polymers in micro-organisms. Adv. Microb. Physiol. **10**:135–166.
7. **Erecenska, M., M. Stubbs, Y. Miyata, C. M. Ditrey, and D. F. Wilson.** 1977. Regulation of cellular metabolism by intracellular phosphate. Biochim. Biophys. Acta **462**:20–35.

8. **Hinnen, A., J. B. Hicks, and G. R. Fink.** 1978. Transformation of yeast. Proc. Natl. Acad. Sci. USA **75:**1929–1933.
9. **Kaneko, Y., Y. Tamai, A. Toh-e, and Y. Oshima.** 1985. Transcriptional and posttranscriptional control of *PHO8* expression by *PHO* regulatory genes in *Saccharomyces cerevisiae*. Mol. Cell. Biol. **5:**248–252.
10. **Koren, R., J. LeVitre, and K. A. Bostian.** 1986. Isolation of the positive-acting regulatory gene *PHO4* from *Saccharomyces cerevisiae*. Gene **41:**271–280.
11. **Lemire, J. M., T. Willcocks, H. O. Halvorson, and K. A. Bostian.** 1985. Regulation of repressible acid phosphatase gene transcription in *Saccharomyces cerevisiae*. Mol. Cell. Biol. **5:**2131–2141.
12. **Matsumoto, K., T. Uno, and T. Ishikawa.** 1984. Regulation of repressible acid phosphatase by cyclic AMP in *Saccharomyces cerevisiae*. Genetics **108:**53–66.
13. **Rogers, D. T., J. M. Lemire, and K. A. Bostian.** 1982. Acid phosphatase polypeptides in *Saccharomyces cerevisiae* are encoded by a differentially regulated multigene family. Proc. Natl. Acad. Sci. USA **79:**2157–2161.
14. **Tait-Kamradt, A. G., J. Lemire, R. Koren, and K. A. Bostian.** 1985. Regulation of the phosphatase multigene family in *S. cerevisiae*, p. 595–610. *In* R. Calendar and L. Gold (ed.), Sequence specificity in transcription and translation. Alan R. Liss, New York.
15. **Tait-Kamradt, A., K. Turner, R. Kramer, Q. Elliott, S. Bostian, G. Thill, D. Rogers, and K. Bostian.** 1986. Reciprocal regulation of the tandemly duplicated *PHO5/PHO3* gene cluster within the acid phosphatase multigene family of *Saccharomyces cerevisiae*. Mol. Cell. Biol. **6:**1855–1865.
16. **Tamai, Y., A. Toh-e, and Y. Oshima.** 1985. Regulation of inorganic phosphate transport systems in *Saccharomyces cerevisiae*. J. Bacteriol. **164:**964–968.
17. **Thill, G., R. Kramer, K. Turner, and K. A. Bostian.** 1983. Comparative analysis of the 5′ end regions of two repressible acid phosphatase genes in *Saccharomyces cerevisiae*. Mol. Cell. Biol. **3:**570–579.
18. **Toh-e, A., S. Inouye, and Y. Oshima.** 1981. Structure and function of the *PHO82-pho4* locus controlling the synthesis of repressible acid phosphatase of *Saccharomyces cerevisiae*. J. Bacteriol. **145:**221–232.
19. **Toh-e, A., S. Kakimoto, and Y. Oshima.** 1975. Two new genes controlling the constitutive acid phosphatase synthesis in *Saccharomyces cerevisiae*. Mol. Gen. Genet. **141:**81–83.
20. **Toh-e, A., S. Kobayashi, and Y. Oshima.** 1978. Disturbance of the machinery for the gene expression by acid pH in the repressible acid phosphatase system in *Saccharomyces cerevisiae*. Mol. Gen. Genet. **162:**139–149.
21. **Toh-e, A., Y. Ueda, S. Kakimoto, and Y. Oshima.** 1973. Isolation and characterization of acid phosphatase mutants in *Saccharomyces cerevisiae*. J. Bacteriol. **113:**727–738.
22. **Ueda, Y., A. Toh-e, and Y. Oshima.** 1975. Isolation and characterization of recessive, constitutive mutations for repressible acid phosphatase synthesis in *Saccharomyces cerevisiae*. J. Bacteriol. **122:**911–922.

III. PROTEIN SECRETION AND USE OF ALKALINE PHOSPHATASE

Introduction: Secretion of Phosphate-Related Proteins

JON BECKWITH

Department of Microbiology and Molecular Genetics, Harvard Medical School, Boston, Massachusetts 02115

Proteins involved in phosphate metabolism have played a major role in the evolution of the study of protein secretion in bacteria. The discovery of the periplasmic space 25 years ago came about through the finding that alkaline phosphatase was located in a compartment between the bacterial inner and outer membranes (6). Some of the earliest attempts to understand how proteins cross the bacterial cytoplasmic membrane utilized alkaline phosphatase as a model system. The possibility that the amino acid composition of a protein was important in determining its ability to be secreted was tested by incorporating amino acid analogs into the enzyme and looking for effects on its export. However, no effects were seen (9). In addition, preliminary experiments suggested that polysomes involved in the synthesis of alkaline phosphatase might be preferentially localized to the cytoplasmic membrane (2), a fact confirmed many years later (11).

The first periplasmic enzyme which was demonstrated to be made as a precursor with a hydrophobic peptide extension was alkaline phosphatase (4). The simultaneous discovery of a precursor for the outer membrane lipoprotein (5) provided significant stimulus to the field of protein secretion, since the results suggested a similarity in mechanism of secretion in procaryotes and eucaryotes. Alkaline phosphatase has continued to provide an important system for studying protein secretion. A major genetic approach to this problem has involved the use of gene fusions in which the amino-terminal regions of cell envelope proteins are fused to β-galactosidase. The properties of such fusion strains have allowed the isolation and characterization of mutants which alter secretion. Such fusions of alkaline phosphatase to β-galactosidase have yielded a collection of signal sequence mutants (8) and a mutant which defines a new gene involved in secretion, *secD* (2a).

Two other proteins involved in phosphate metabolism, PhoE, an outer membrane protein, and PhoS, a periplasmic protein, have also provided fruitful material for the study of protein export. A series of gene fusions have been constructed with *phoE*, in which the signal sequence for a protein destined for one cell envelope compartment is fused to the mature portion of a protein destined for another. In all cases, the ultimate location of the protein is determined by the mature portion of the protein, not the signal sequence (12, 13). These results strengthen the conclusion that there are additional internal signals which direct certain proteins to the outer membrane. Detailed analysis of the secretion of PhoS protein in strains carrying multicopy *phoS* plasmids has contributed to the evidence that there are distinct pools of precursors of exported proteins (1). These correspond perhaps to export-competent and export-incompetent fractions (10).

Finally, the *phoA* gene has provided an important source of vectors and systems for new kinds of gene fusion studies. A plasmid has been constructed which carries the promoter and signal sequence coding region of *phoA* and has restriction sites at the end of the sequence which allow the construction of gene fusions (H. Inouye, personal communication; Yoda et al., this volume). Since the *phoA* promoter is a highly active one, this vector allows potential high-level expression and secretion of foreign proteins. Another class of vectors carries a *phoA* gene missing its promoter and the DNA coding for the signal sequence, plus some amino acids from the mature portion of the protein (3). Substitution of DNA which codes for an export signal when fused in the proper frame to this *phoA* vector allows expression and secretion of alkaline phosphatase. Since the enzyme is active only when it is exported to the periplasm, the vector provides a means of detecting export signals in other genes. This approach has been generalized with the development of an in vivo system, Tn*phoA*, for generating such fusions (7). Tn*phoA*, a derivative of Tn*5*, allows the ready isolation of alkaline phosphatase fusions to bacterial genes or to genes which have been cloned into bacteria. Not only does it permit the detection of export signals, but studies to be described here indicate that Tn*phoA* also provides a means of analyzing membrane protein topology.

LITERATURE CITED

1. **Anba, J., J.-M. Pages, and C. Lazdunski.** 1986. Mode of transfer of the phosphate-binding protein through the cytoplasmic membrane in *Escherichia coli*. FEMS Microbiol. Lett. **34:**215–219.
2. **Cancedda, R., and M. J. Schlesinger.** 1974. Localization of polyribosomes containing alkaline phosphatase nascent polypeptides on membranes of *Escherichia coli*. J. Bacteriol. **117:**290–301.

2a. **Gardel, C., S. Benson, J. Hunt, S. Michaelis, and J. Beckwith.** 1987. *secD*, a new gene involved in protein export in *Escherichia coli*. J. Bacteriol. **169:**1286–1290.

3. **Hoffman, C. S., and A. Wright.** 1985. Fusions of secreted proteins to alkaline phosphatase: an approach for studying protein secretion. Proc. Natl. Acad. Sci. USA **82:**5107–5111.
4. **Inouye, H., and J. Beckwith.** 1977. Synthesis and processing of alkaline phosphatase precursor *in vitro*. Proc. Natl. Acad. Sci. USA **74:**1440–1444.
5. **Inouye, S., S. Wang, J. Sekizawa, S. Halegoua, and M. Inouye.** 1977. Amino acid sequence for the peptide extension on the prolipoprotein of the *Escherichia coli* outer membrane. Proc. Natl. Acad. Sci. USA **74:**1004–1008.
6. **Malamy, M., and B. Horecker.** 1961. The localization of alkaline phosphatase in *E. coli* K12. Biochem. Biophys. Res. Commun. **5:**104–108.
7. **Manoil, C., and J. Beckwith.** 1985. Tn*phoA*: a transposon probe for protein export signals. Proc. Natl. Acad. Sci. USA **82:**8129–8133.
8. **Michaelis, S., J. F. Hunt, and J. Beckwith.** 1986. Effects of signal sequence mutations on the kinetics of alkaline phosphatase export to the periplasm in *Escherichia coli*. J. Bacteriol. **167:**160–167.
9. **Morris, H., and M. J. Schlesinger.** 1972. Effects of proline analogs on the formation of alkaline phosphatase in *Escherichia coli*. J. Bacteriol. **111:**203–210.
10. **Ryan, J. P., and P. J. Bassford, Jr.** 1985. Post-translation export of maltose-binding protein in *Escherichia coli* strains harboring *malE* signal sequence mutations and either prl^+ or *prl* suppressor alleles. J. Biol. Chem. **260:**14832–14837.
11. **Smith, W. P., P.-C. Tai, R. C. Thompson, and B. D. Davis.** 1977. Extracellular labeling of nascent polypeptides traversing the membrane of *Escherichia coli*. Proc. Natl. Acad. Sci. USA **74:**2830–2834.
12. **Tommassen, J., A. P. Pugsley, J. Korteland, J. Verbakel, and B. Lugtenberg.** 1984. Gene encoding a hybrid OmpF-PhoE pore protein in the outer membrane of *Escherichia coli* K12. Mol. Gen. Genet. **197:**503–508.
13. **Tommassen, J., H. van Tol, and B. Lugtenberg.** 1983. The ultimate localization of an OM protein of *Escherichia coli* K12 is not determined by the signal sequence. EMBO J. **2:**1275–1279.

New Insights into the Export Machinery through Studies on the Synthesis of Phosphate-Binding Protein in *Escherichia coli*

JAMILA ANBA, JEAN-MARIE PAGES, ALAIN BERNADAC, AND CLAUDE LAZDUNSKI

Centre de Biochimie et de Biologie Moléculaire du Centre National de la Recherche Scientifique, 13402 Marseille Cedex 9, France

In recent years, the mechanism of protein secretion has been extensively studied in both eucaryotic and procaryotic organisms. The availability of in vitro systems proved helpful in the delineation of the early steps of synthesis and translocation of nascent chains through the rough endoplasmic reticulum membrane (7, 10, 24). A coupling between synthesis and export mediated by a nucleoprotein complex named SRP (signal recognition particle) (24) and a so-called "docking protein" (10) has thus been demonstrated. It has been proposed that the purpose of translational pausing induced by these factors is to prevent the accumulation of presecretory proteins in the cytoplasm (24).

Studies on the early steps in protein export, despite the power of the genetic approach (11, 21), have lagged behind for procaryotic cells. To try to obtain new insights with regard to these steps, studies were started in our laboratory several years ago on the synthesis and export of PhoS, the phosphate-binding protein which participates in the high-affinity phosphate transport system (see Rosenberg et al., this volume). This system appeared to be particularly well suited for the study of protein export for the following reasons: (i) PhoS synthesis is induced under conditions of P_i starvation where the cells are growing slowly and the actively transcribed genes are mostly those which belong to the Pho regulon (see Shinagawa et al., this volume), and furthermore, the synthesis of PhoS can be rapidly repressed simply by adding P_i to the medium; (ii) the *phoS* promoter is very strong; (iii) by using strains carrying multicopy plasmids containing *phoS*, hyperproduction can be obtained, and the level of hyperproduction can be modulated depending upon the copy number, which itself depends on the plasmid size; (iv) pre-PhoS contains only three methionine residues and they are all located within the signal sequence, thus allowing its cleavage from the protein to be readily monitored.

Studying PhoS synthesis and export in the hyperproducing cells under phosphate limitation was used as an alternative to a cell-free protein export system (which was until recently unavailable).

Earlier studies from our laboratory contributed to the demonstration that exported proteins are produced on membrane-bound polysomes (5, 19, 23), that alterations in membrane fluidity often result in inhibition of processing of precursor forms (9, 15, 18), and that the electrochemical potential across the cytoplasmic membrane is required for this processing (16, 17). This report presents our recent results obtained with the PhoS system mentioned above.

Results

Existence of two PhoS precursor pools under conditions of hyperproduction. Plasmids bearing *phoS* either alone (pSN5182) or with other genes from the *pho* regulon (pSN507) have been constructed by Morita et al. (13) (Table 1). The extent of PhoS hyperproduction and overproduction in strains bearing these plasmids appears to be related to the copy number and thus to *phoS* gene dosage (13). We have now determined that pSN5182 allows a fourfold higher production of PhoS than pSN507, thereby causing more saturation of export sites and more accumulation of pre-PhoS (Table 2).

Escherichia coli ANCC75 harboring the plasmid pSN5182 (referred to as hyperproducer strain) hyperproduces PhoS when grown under phosphate limitation. The localization of PhoS has been investigated on subsets of the same cultures at various times after induction, by electron microscopy (1, 6) and by cell fractionation (14). Ultrathin sections from frozen cells showing PhoS localized in the periplasm are shown in Fig. 1. At early times after induction, PhoS was detected only in the periplasmic space. By 1 h after transfer to phosphate-limiting medium, when many exported proteins in addition to proteins belonging to the "*pho* regulon" are still expressed, pre-PhoS was found to be mostly accumulated in the cytoplasm and in the cytoplasmic membrane (14). Protein analysis by sodium dodecyl sulfate-polyacrylamide gel electrophoresis indicated that, at late times of induction, pre-PhoS was indeed accumulated both in the cytoplasm and in the inner membrane (14). The trypsin accessibility of pre-PhoS was investigated by using spheroplasts. The results demonstrate that the membrane-associated pre-PhoS

TABLE 1. Relevant characteristics of bacterial strains and plasmids

Strain or plasmid	Relevant characteristics	Source
E. coli K-12		
C600	F^- *leu thr thi lacY*	2
ANCC75	*leu purE trp his argG rpsL met thi phoS*	12
Plasmids		
pSN5182	Tc^r, *phoS* gene	12
pSN507	Ap^r, Tc^r, *phoS, pstA(phoT), pstB,* and *phoU* genes	12
pTD101	Ap^r, *lep* gene	8
pAJ202	Tc^r, *phoS* gene, ori^- pColA	2
pColA	*caa cai cal*	13a

is not accessible to trypsin and thus has not been translocated (4).

To probe the conformation of the various forms, the sensitivity to trypsin of mature PhoS, membrane-associated pre-PhoS, and cytoplasmic pre-PhoS was also compared. The results suggest a difference in conformation between membrane-associated pre-PhoS and cytoplasmic pre-PhoS since the former is more trypsin sensitive than the latter. Mature PhoS is resistant to trypsin (4).

We then asked whether both the cytoplasmic

TABLE 2. Effect of *phoS* gene dosage on the production of PhoS protein

Plasmid carrying the *phoS* gene[a]	M_r ($\times 10^3$)	PhoS/EF-Tu ratio
pSN5182	4.95	13: hyperproducer
pSN507	12.6	4: overproducer

[a] These plasmids have been described by Morita et al. (13).

and membrane-associated pre-PhoS could be processed and exported posttranslationally, and we compared the kinetics of processing under conditions of normal production and hyperproduction (Fig. 2). Under conditions of PhoS overproduction, a biphasic exponential decay was observed. This is consistent with our previous report of a slow two-step cleavage of cytoplasmic pre-PhoS and rapid maturation of membrane-associated pre-PhoS (2, 14). The slow two-step cleavage for the cytoplasmic pre-PhoS is not accompanied by export even when proper growth conditions are restored (14). In contrast, the membrane-associated pre-PhoS can be exported. It thus seems that only the latter (although it has not been processed cotranslationally) has a conformation compatible with export. There was a marked retardation of processing in the PhoS overproducer as compared with the

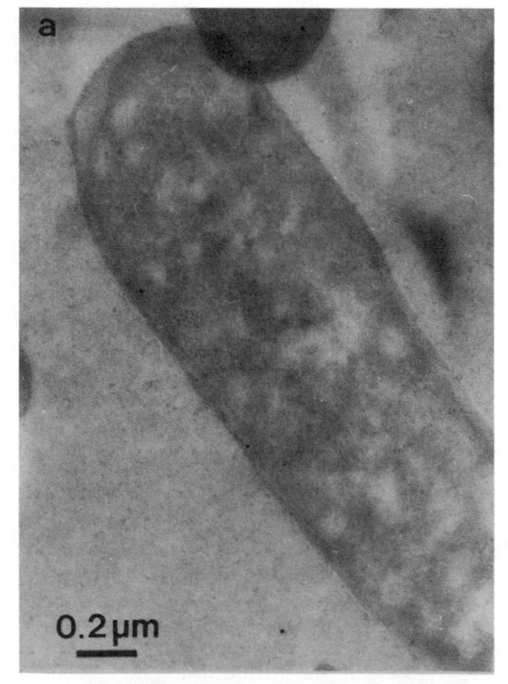

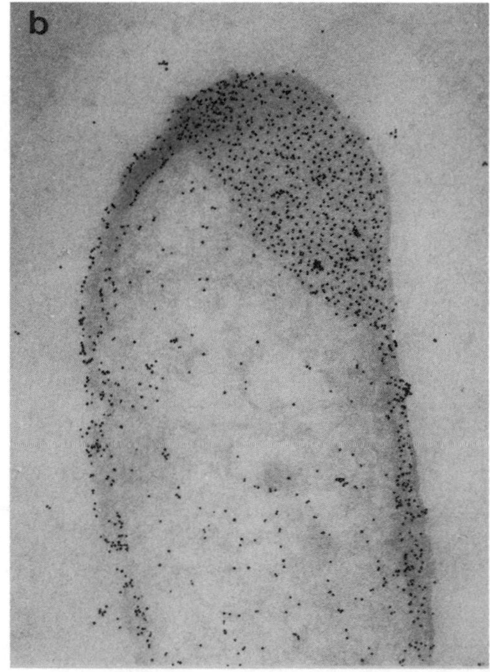

FIG. 1. Ultrathin sections of *E. coli* ANCC75(pSN5182). Cell samples were removed at various times after transfer to low-phosphate medium: (a) control noninduced cells, (b) 3 h after transfer. Ultrathin sections were prepared and incubated for 60 min at room temperature in the presence of anti-PhoS serum (1:100 dilution) and then with gold-coated protein A (1:20 dilution) for 30 min as previously described (1, 6).

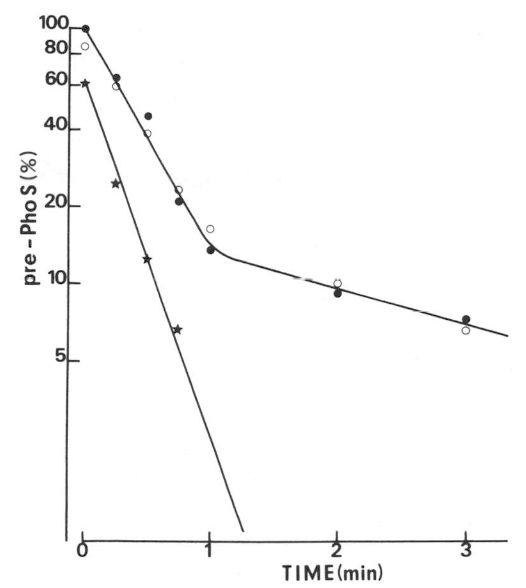

FIG. 2. Kinetics of pre-PhoS processing. At 2 h after transfer in low-phosphate medium, cells of strains C600 (★), C600(pAJ202) (○), and C600(pAJ202, pTD101) (●) (for description of plasmids see references 2 and 8) were pulse-labeled for 15 s with ^{14}C-labeled amino acids, and an excess of nonradioactive amino acids was then added with 1 mM phosphate. Samples were subsequently removed during the chase. Immunoprecipitation with antiserum directed against PhoS was carried out, and the immunoprecipitates were analyzed by sodium dodecyl sulfate-polyacrylamide gel electrophoresis. The relative amount of pre-PhoS was determined by densitometer scanning of fluorograms from three independent experiments.

wild type: the half-life of membrane-associated pre-PhoS was extended from 11 to 22 s. It was demonstrated that the delay in maturation of precursor forms of both PhoS and β-lactamase was not due to signal peptidase activity being overwhelmed but to a delay in completion of synthesis of precursor polypeptide chains (3). Overproduction of leader peptidase did not alter the half-life of the membrane-bound pre-PhoS (2) (Fig. 2).

Direct evidence for a coupling between synthesis and export of PhoS. We recently reported that, during the induction of PhoS synthesis, a delay in the completion of polypeptide chain elongation can be detected. This delay is related to the extent of jamming of export sites by pre-PhoS or by other exported proteins. These results suggest that a component required for completion of pre-PhoS polypeptide becomes limiting (3). This component probably acts at an early step in the export pathway.

It has been previously shown in our laboratory that translation is a nonuniform process in *E. coli*; pause sites in polypeptide elongation, in relation to tRNA availability, have been demonstrated (22). By using pulse-labeling with [^{35}S]methionine and chase as previously described, (i) the migration of intermediates in sodium dodecyl sulfate-polyacrylamide gel electrophoresis can be observed and (ii) the time required to complete the elongation of nascent polypeptide chains can be evaluated. The patterns of pause sites have been compared for normal PhoS producer strains and for PhoS overproducer strains bearing pAJ202 (Fig. 3). There are two main differences. First, there is a 10-s delay in completion of nascent chains in the overproducer strain (40 s as compared with 30 s) and second, with this strain, the relative importance of a pause site located at about 8 kilodaltons is increased as compared with that observed with the normal strain. The level of this pause is increased by the production of another exported protein, β-lactamase, in strains carrying both pAJ202 and pTD101; moreover, there is a 20-s delay for completion of nascent chains in this case. These results provide additional experimental support for the hypothesis that the saturation of export sites induces a delay in polypeptide elongation at a very early step. Although a translation arrest does not appear to exist, there is a marked pause in translation when nascent chains containing about 70 to 80 amino acid residues have been assembled.

Conclusion

The use of PhoS hyperproducing strains under conditions of growth with phosphate limitation as an alternative to an in vitro protein export system has been fruitful. It has been demonstrated that (i) precursor PhoS which is released into the cell cytoplasm under conditions of overproduction cannot be exported posttranslationally even when export sites become available; (ii) the membrane-associated precursor that is accumulated when export sites become saturated is probably located at the inner side of the cytoplasmic membrane and can be exported when the export sites become available; (iii) there is a coupling between synthesis and export since saturation of export sites causes a delay in completion of nascent polypeptide chains; and (iv) overproduction of leader peptidase can neither relieve the accumulation of membrane-associated pre-PhoS nor speed up processing of the signal peptide.

All of these results are compatible with the idea that an early interaction of nascent chains of exported proteins with some component of the export machinery must occur to prevent cleavage of the signal peptide by cytoplasmic proteases. Indeed, the half-life of pre-PhoS is only 400 s in the cytoplasm.

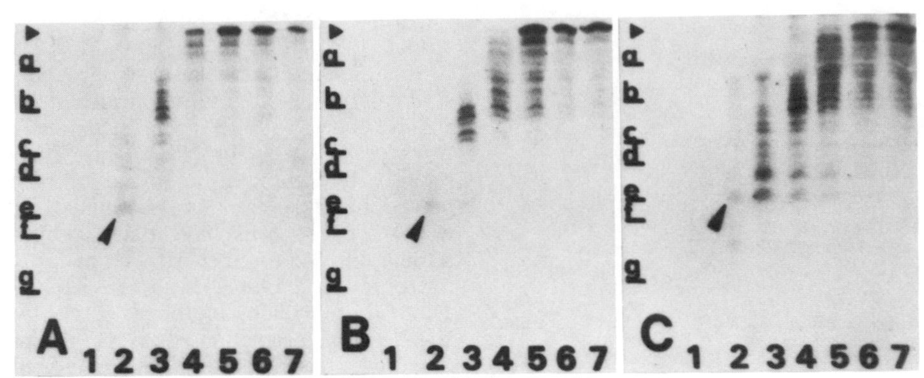

FIG. 3. Patterns of pause sites for normal PhoS producer and overproducer strains. At 3 h after transfer in low-phosphate medium, cells of strains C600 (A), C600(pAJ202) (B), or C600(pAJ202, pTD101) (C) (for description of plasmids see references 2 and 8) were pulse-labeled for 5 s with [^{35}S]methionine (lane 1), and an excess of unlabeled methionine was then added. Samples were removed during the chase at 10 s (lane 2), 20 s (lane 3), 30 s (lane 4), 40 s (lane 5), 50 s (lane 6), and 60 s (lane 7). Immunoprecipitation by antiserum against PhoS was carried out, and the immunoprecipitates were analyzed by urea-sodium dodecyl sulfate-polyacrylamide gel electrophoresis and fluorography. Triangles and arrowheads indicate, respectively, the migration of pre-PhoS and of the intermediate of about 8 kilodaltons. Relative molecular mass standards (in kilodaltons): a, 30; b, 20.1; c, 17; d, 14.4; e, 8.15; f, 6.2; g, 2.5. Panel A corresponds to an overexposed film (obtained with the normal producer strain).

It has recently been reported that PhoA and OmpA (20, 25) can be exported posttranslationally. We have demonstrated that posttranslational export of cytoplasmic pre-PhoS cannot occur in vivo because of the rapid proteolytic degradation of the signal peptide. It thus seems likely that an SRP-like system is present in *E. coli* to ensure that the signal peptide will be protected from cytoplasmic proteases very early after synthesis.

An early interaction between nascent chains and the membrane is also probably required to allow the formation of an export-compatible conformation of the precursor proteins. There are two lines of evidence for this: cytoplasmic pre-PhoS cannot be exported even when export sites become available, and the conformation of membrane-bound pre-PhoS appears to be different from that of periplasmic PhoS, as reflected in their different levels of trypsin sensitivity (the latter being fully resistant).

To conclude, the export of periplasmic proteins appears to require a number of distinct stages: synthesis, translocation, maturation, conformational change, and release in the periplasmic space. The latter step has recently been demonstrated by Minsky et al. (12). In this complex pathway, our results demonstrated that the cleavage of the signal peptide is not the rate-limiting step, that an early shielding of this signal is probably required, that some sort of coupling between synthesis and export probably exists, and that newly synthesized membrane-bound pre-PhoS has a conformation different from that of periplasmic mature PhoS.

We are grateful to Peter Howard for a critical reading and to M. Payan for carefully preparing the manuscript.

This work was supported by grants from the Centre National de la Recherche Scientifique, INSERM (CRL no. 82.1022), and the Fondation pour la Recherche Médicale.

LITERATURE CITED

1. **Anba, J., A. Bernadac, J. M. Pagès, and C. Lazdunski.** 1984. The periseptal annulus in *Escherichia coli*. Biol. Cell **50:**273–278.
2. **Anba, J., C. Lazdunski, and J. M. Pagès.** 1986. Lack of effect of leader peptidase overproduction on the processing *in vivo* of exported proteins in *Escherichia coli*. J. Gen. Microbiol. **132:**689–696.
3. **Anba, J., C. Lazdunski, and J. M. Pagès.** 1986. Direct evidence for a coupling between synthesis and export in *Escherichia coli*. FEBS Lett. **196:**9–12.
4. **Anba, J., J. M. Pagès, and C. Lazdunski.** 1986. Mode of transfer of the phosphate binding protein through the cytoplasmic membrane in *Escherichia coli*. FEMS Microbiol. Lett. **34:**215–219.
5. **Baty, D., A. Bernadac, Y. Berthois, and C. Lazdunski.** 1981. Synthesis and export of tem-β-lactamase in *Escherichia coli*. FEBS Lett. **127:**161–165.
6. **Bernadac, A., and C. Lazdunski.** 1981. Immunoferritin labelling of ultrathin frozen sections of gram-negative bacterial cells. Biol. Cell. **41:**211–216.
7. **Blobel, G., and B. Dobberstein.** 1975. Presence of proteolytically processed and nonprocessed nascent im munoglobulin light chains on membrane-bound ribosomes of murine myeloma. J. Cell Biol. **67:**835–851.
8. **Date, T., and W. Wickner.** 1981. Isolation of the *Escherichia coli* leader peptidase gene and effects of leader peptidase overproduction *in vivo*. Proc. Natl. Acad. Sci. USA **78:**6106–6110.
9. **Lazdunski, C., D. Baty, and J. M. Pagès.** 1979. Procaine, a local anesthetic interacting with the cell membrane, inhibits the processing of precursor forms of periplasmic proteins in *Escherichia coli*. Eur. J. Biochem. **96:**49–57.
10. **Meyer, D., E. Krause, and B. Dobberstein.** 1982. Secretory protein translocation across the membranes. The role of the docking protein. Nature (London) **297:**647–650.
11. **Michaelis, S., and J. Beckwith.** 1982. Mechanism of incorporation of cell envelope proteins in *Escherichia coli*.

Annu. Rev. Microbiol. **36**:435–465.
12. **Minsky, A., R. G. Summers, and J. R. Knowles.** 1986. Secretion of β-lactamase into the periplasm of *Escherichia coli*: evidence for a distinct release step associated with a conformational change. Proc. Natl. Acad. Sci. USA **83**: 4180–4184.
13. **Morita, F., M. Amemura, K. Makina, H. Shinagawa, K. Magota, N. Otsuji, and A. Nakata.** 1983. Hyperproduction of phosphate-binding protein, PhoS, and pre-PhoS in *Escherichia coli* carrying a cloned *phoS* gene. Eur. J. Biochem. **130**:427–435.
13a. **Morlon, J., D. Caraud, and C. Lazdunski.** 1982. Physical map of pColA-CA31, an amplifiable plasmid, and location of colicin A structural gene. Gene **17**:317–321.
14. **Pagès, J. M., J. Anba, A. Bernadac, H. Shinagawa, A. Nakata, and C. Lazdunski.** 1984. Normal precursors of periplasmic proteins accumulated in the cytoplasm are not exported post-translationally in *Escherichia coli*. Eur. J. Biochem. **143**:499–505.
15. **Pagès, J. M., and C. Lazdunski.** 1981. Action of phenethyl alcohol on the processing of precursor forms of periplasmic proteins in *Escherichia coli*. FEMS Microbiol. Lett. **12**:65–69.
16. **Pagès, J. M., and C. Lazdunski.** 1982. Maturation of exported proteins in *Escherichia coli*. Requirement for energy, site and kinetics of processing. Eur. J. Biochem. **124**:561–566.
17. **Pagès, J. M., and C. Lazdunski.** 1982. Membrane potential ($\Delta\Psi$) depolarizing agents inhibit maturation. FEBS Lett. **149**:51–54.
18. **Pagès, J. M., M. Piovant, S. Varenne, and C. Lazdunski.** 1978. Mechanistic aspects of the transfer of nascent periplasmic proteins across the cytoplasmic membrane in *Escherichia coli*. Eur. J. Biochem. **86**:589–602.
19. **Piovant, M., S. Varenne, J. M. Pagès, and C. Lazdunski.** 1978. Preferential sensitivity of syntheses of exported proteins to translation inhibitors of low polarity in *Escherichia coli*. Mol. Gen. Genet. **164**:265–274.
20. **Rhoads, D. B., P. C. Tai, and B. D. Davis.** 1984. Energy requiring translocation of the OmpA protein and alkaline phosphatase of *Escherichia coli* into inner membrane vesicles. J. Bacteriol. **159**:63–70.
21. **Silhavy, T. J., S. A. Benson, and S. D. Emr.** 1983. Mechanisms of protein localization. Microbiol. Rev. **47**: 313–344.
22. **Varenne, S., J. Buc, R. Lloubès, and C. Lazdunski.** 1984. Translation is a non-uniform process. Effect of tRNA availability on the rate of elongation of nascent polypeptide chains. J. Mol. Biol. **180**:549–576.
23. **Varenne, S., M. Piovant, J. M. Pagès, and C. Lazdunski.** 1978. Evidence for synthesis of alkaline phosphatase on membrane-bound polysomes in *Escherichia coli*. Eur. J. Biochem. **86**:603–606.
24. **Walter, P., and G. Blobel.** 1981. Translocation of proteins across the endoplasmic reticulum. III. Signal recognition protein (SRP) causes signal sequence-dependent and site-specific arrest of chain elongation that is released by microsomal membranes. J. Cell Biol. **91**:557–561.
25. **Zimmermann, R., and W. Wickner.** 1983. Energetics and intermediates of the assembly of protein OmpA into the outer membrane of *Escherichia coli*. J. Biol. Chem. **258**: 3920–3925.

Alkaline Phosphatase as a Tool for Analysis of Protein Secretion

CHARLES S. HOFFMAN,† YOLANTA FISHMAN, AND ANDREW WRIGHT

Department of Molecular Biology and Microbiology, Tufts University, Health Sciences Campus, Boston, Massachusetts 02111

This paper describes our studies on the alkaline phosphatase of *Escherichia coli* K-12, leading to the development of the *phoA* gene as a tool for studying protein secretion (5). The *phoA* gene codes for a polypeptide containing 471 amino acids (3), the first 21 of which form a signal peptide that is removed during secretion of the polypeptide to the periplasm (9). All of the alkaline phosphatase activity of the cell appears to reside in this location. Each molecule of active enzyme contains two polypeptide monomers complexed with Zn^{2+} and Mg^{2+} (11). Each monomer contains four cysteine residues, forming two intrachain disulfide bonds. The amino-terminal arginine residue of the secreted polypeptide is cleaved from some of the chains, exposing an amino-terminal threonine residue (7). The two different monomers combine to give an active dimer that gives rise to a series of three isozymes (AA, BB, and AB) which can be separated by electrophoretic methods (12).

The variability at the amino terminus of the mature protein suggests that it might be possible to modify the amino terminus further without affecting enzymatic activity. We found, in fact, that removal of up to the first 12 or 13 amino-terminal amino acids of alkaline phosphatase (PhoA) had no measurable effect on enzyme activity (5). Replacement of these amino acids with large segments of other secreted proteins, by gene fusion, gave rise to active enzyme. In all of these cases alkaline phosphatase activity was absolutely dependent on the presence of an amino-terminal signal sequence, indicating that secretion of PhoA is essential for formation of active enzyme. This paper describes the properties of a series of *phoA* fusions which indicate that fusion of genes specifying secreted products to the *phoA* gene (deleted for its signal sequence coding region) give rise to alkaline phosphatase activity. Our results, along with those of others (8), indicate that *phoA* is a powerful tool for the analysis of determinants of protein secretion through the construction of gene fusions.

†Present address: Department of Genetics, Harvard Medical School, Boston, MA 02115.

Deletion of the *phoA* Amino-Terminal Coding Sequence

The source of the DNA used for alteration of the *phoA* coding sequence was plasmid pHI1 (6). A series of deletions extending from an *Hpa*I site 400 nucleotides upstream of the *phoA* promoter was generated by exonuclease BAL 31 treatment of *Hpa*I-cut plasmid pHI1 DNA. The resulting DNA was ligated with *Pst*I linkers, treated with *Pst*I, and then inserted into a series of vectors constructed by Talmadge et al. (14). Each vector possesses a unique *Pst*I site immediately downstream of the *bla* (β-lactamase) promoter and signal sequence coding region. The ligated DNA was used to transform a *phoA* deletion strain (AW1061) to tetracycline resistance. $PhoA^+$ transformants were readily recognized by their blue color on XP-containing indicator medium (XP = 5-bromo-4-chloro-3-indolyl phosphate). Analysis of plasmid DNA indicated, as expected, that they contained the *bla* promoter and signal sequence coding region fused to *phoA*. The fusion sites within *phoA* varied as indicated in Table 1 from that in pCH4 (which contained the last three codons of the signal sequence) to that in pCH9 (which lacked all of the signal sequence coding region and 39 codons beyond).

Structure, Function, and Localization of *bla-phoA* Fusion Proteins

Cells carrying the plasmids indicated in Table 1, with the exception of those carrying pCH9, all produced about 1,300 to 1,400 U of alkaline phosphatase activity. This indicates that loss of the first 13 amino acids has little effect on enzyme function and agrees with the observation that proteolytic removal of the amino terminus of alkaline phosphatase has only a minor effect on activity (11).

Cells carrying pCH9 produced only 80 U of phosphatase activity, indicating that deletion of the first 39 amino acids has a significant effect on enzyme activity. The crystal structure of alkaline phosphatase, determined by Wyckoff et al. (16; Wyckoff, this volume), indicates that the amino-terminal regions of the two polypeptide chains in the enzyme are in contact with each

TABLE 1. Structure of β-lactamase-PhoA fusion proteins[a]

Plasmid	Vector	Amino acid and nucleotide sequences of fusion junctions
pCH2	pKT287	14 His Pro Glu Thr Ala Ala Ala Gln CAC CCA GAA ACG GCT GCA GCT CAG
pCH4	pKT287	−3 His Pro Glu Thr Ala Ala Ala Thr CAC CCA GAA ACG GCT GCA GCG ACA
pCH9	pKT280	40 His Pro Leu Gln Arg Asp CAC CCG CTG CAG CGC GAT
pCH39	pKT279	6 His Arg Cys Ser Pro CAC CGC TGC AGC CCT
pCH40	pKT280	14 His Pro Leu Gln Pro Gln CAC CCG CTG CAG CCT CAG
pCH58	pKT218	14 Met Ser Ile Gln Ala Ala Ala Gln ATG AGT ATT CAA GCT GCA GCT CAG

[a] The first five DNA sequences begin with the *bla* codons immediately following the leader peptidase cleavage site. Since the pCH58-encoded protein possesses an incomplete signal sequence and is therefore unprocessed, the displayed sequence begins at the translational start site. The number above each final residue presented indicates its position in the wild-type PhoA polypeptide. *Pst*I linker DNA is underlined.

other and are important for dimerization. The absence of the first 39 amino acids of mature PhoA from the fusion protein produced by plasmid pCH9 may therefore affect dimer formation or may lead to instability of the fusion protein.

The fusion protein of plasmid pCH39 behaved anomalously in that some of it ran as a dimer and some ran as a monomer when subjected to sodium dodecyl sulfate-polyacrylamide gel electrophoresis after boiling in the absence of reducing agent (Fig. 1, lane 4). The pCH39 product is unique in containing a cysteine residue near its amino terminus, encoded by the *Pst*I linker joining the *bla* signal sequence coding region to *phoA*. It is likely that this cysteine forms an interchain disulfide bond linking the two polypeptide monomers together. The properties of this derivative thus support the conclusions from the crystal structure which indicate juxtaposition of the amino termini of the two chains (16).

The amino terminus of alkaline phosphatase can be altered with little effect on its activity. Such flexibility is essential for a protein if it is to be used as an enzymatically active component in hybrid proteins. In the fusions discussed above, the phosphatase amino terminus was replaced by the β-lactamase signal sequence plus a few additional amino acids. Another *bla-phoA* fusion which contains two-thirds of the *bla* gene also produces a hybrid protein which is enzymatically active (Fig. 1, lane 6).

Greater than 90% of the PhoA-containing polypeptides specified by the *bla-phoA* fusions

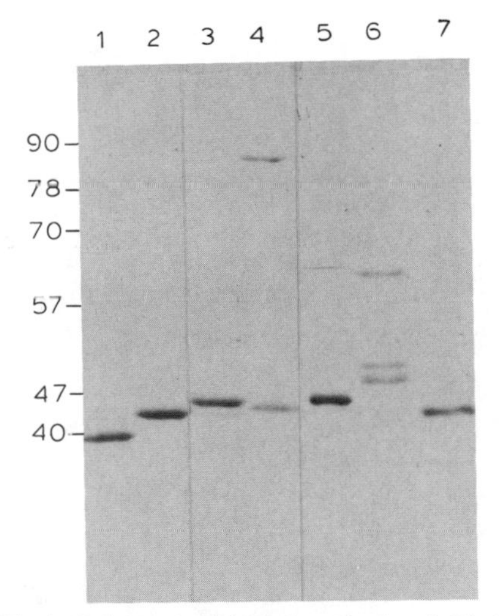

FIG. 1. Polyacrylamide gel analysis of Bla-PhoA fusion proteins. Cells containing *bla-phoA* fusion plasmids were grown in minimal medium, labeled with [^{35}S]methionine, and subjected to cold osmotic shock (5). The PhoA-specific polypeptides were immunoprecipitated from the shock fluids. The reducing agent β-mercaptoethanol was omitted from the sodium dodecyl sulfate-boiling buffer used to dissolve precipitated proteins. The bacterial host, AW1061, is an *E. coli* K-12 derivative with a chromosomal deletion eliminating the *phoA*, *phoB*, and *phoR* genes. Host strain AW1046 is deleted for the *phoA* gene only. Plasmids are described in Table 1. Lanes: 1, AW1061(pCH9); 2, AW1061 (pCH40); 3, AW1061(pCH2); 4, AW1061(pCH39); 5, AW1061(pCH4); 6, AW1061(pCH38); 7, AW1046 (pHI1).

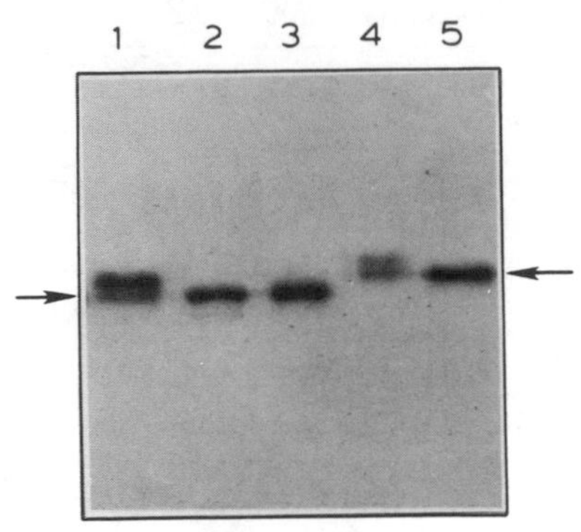

FIG. 2. Identification of Bla-PhoA preproteins expressed by plasmids pCH2 and pCH4. Cells were labeled and subjected to cold osmotic shock as described in the legend to Fig. 1. PhoA-specific polypeptides were immunoprecipitated both from the shock fluid (periplasmic fraction) and from the shock pellet (nonperiplasmic fraction). Lanes: 1, AW1061(pCH2), shock pellet; 2, AW1061(pCH2), shock fluid; 3, AW1046(pHI1), shock fluid (wild-type protein as a size standard); 4, AW1061(pCH4), shock fluid; 5, AW1061(pCH4), shock pellet. The arrow on the left indicates the position of the mature form of the pCH2-encoded hybrid protein. The arrow on the right indicates the position of the mature form of the pCH4-encoded hybrid protein.

described above was released from cells by osmotic shock. Therefore, it appears that these proteins are localized in the periplasm. Immune precipitation of the proteins with anti-alkaline phosphatase serum followed by analysis on sodium dodecyl sulfate-polyacrylamide gels showed that these polypeptides were of the expected size (Fig. 1) and further that the signal sequence had been removed in each case (Fig. 2).

Secretion Is Essential for Phosphatase Activity

All of the fusions giving rise to active phosphatase, described above, contain a signal sequence coding region. We inserted the *'phoA* sequence from plasmid pCH2 into plasmid pKT218 (14), which encodes only the first four amino acids of the β-lactamase signal sequence. Cells carrying the resultant plasmid, pCH58, produced no active phosphatase, but they did produce a PhoA-related polypeptide identified by immune precipitation. A significant but variable fraction (about 50%) of this product was released from cells by cold osmotic shock. However, the observed release was considerably less than with fusions possessing intact signal sequences (more than 90%). Manoil and Beckwith (8; Manoil et al., this volume) independently demonstrated that lack of a signal sequence prevents formation of active phosphatase as a result of lack of secretion. Our results agree with those of Michaelis et al. (10), who demonstrated that point mutations in the *phoA* signal sequence coding region which block PhoA secretion give a cytoplasmic product without activity. These authors suggested that the reducing environment of the cytoplasm might prevent formation of the intrachain disulfide bonds essential for enzyme activity.

Although a signal sequence is essential for phosphatase secretion, its source seems relatively unimportant. Signal sequences coded by the *bla*, *ompF*, and *lamB* genes are as efficient as the PhoA signal sequence itself for phosphatase secretion (5).

Sequences other than normal signal sequences can also promote phosphatase secretion. Fusions of *phoA* to the *tet* gene of pBR322 gave rise to phosphatase activity in vivo (8). The *tet* gene encodes the tetracycline resistance protein (Tet) which is associated with the inner membrane. To try to define sequences within the Tet protein that specify secretion, we fused various amounts of the amino-terminal coding region of *tet* to *phoA*. Fusions containing the first 34 amino acids of the *tet* gene product were enzymatically active. Localization of the fusion protein by immunoprecipitation indicated that it was largely present in the cytoplasm and in the cytoplasmic membrane (Fig. 3). The presence of

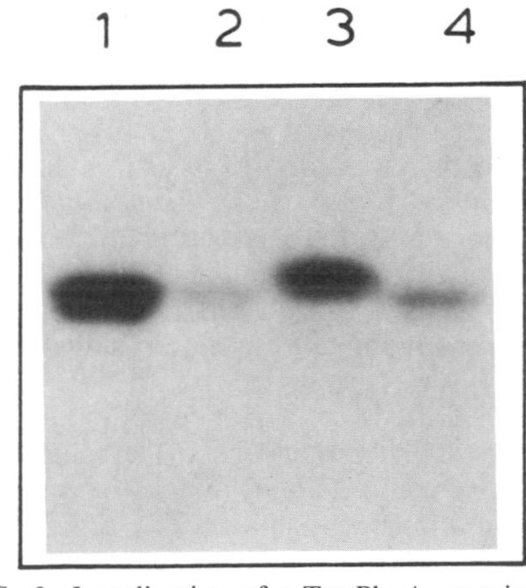

FIG. 3. Localization of a Tet-PhoA protein. Cells containing plasmid pCH90, which contains a *tet-phoA* fusion with the first 34 codons of the pBR322 *tet* gene fused to the *phoA* sequences from plasmid pCH2, were grown and labeled as described in the Fig. 1 legend. The labeled cells were subjected to osmotic shock, followed by freeze-thaw lysis to yield the periplasmic and cytoplasmic fractions. The remaining membrane-associated proteins were fractionated by the Triton extraction method of Schnaitman (13), giving inner and outer membrane fractions. The PhoA-specific material from each fraction was immunoprecipitated and subjected to sodium dodecyl sulfate-polyacrylamide gel electrophoresis. Lanes: 1, freeze-thaw-released fraction (cytoplasmic proteins); 2, shock fluid (periplasmic proteins); 3, Triton X-100 soluble fraction (inner membrane proteins); 4, Triton X-100 insoluble fraction (outer membrane proteins).

```
        (-21)  +
PhoA     Met Lys Gln Ser Thr Ile Ala Leu Ala Leu Leu Pro Leu Leu Phe Trp Pro

         (+1)  +                                                          -
Tet      Met Lys Ser Asn Asn Ala Leu Ile Val Ile Leu Gly Thr Val Thr Leu Asp


                          (+1)
                  + (-1)   +          -                 -       +   (+13)
PhoA     Val Thr Lys Ala Arg Thr Pro Glu Met Pro Val Leu Glu Asn Arg Ala Ala

                                                                    (+34)
                                                                    +   -
Tet      Ala Val Gly Ile Gly Leu Val Met Pro Val Leu Pro Gly Leu Leu Arg Asp
```

FIG. 4. Structural comparison of the alkaline phosphatase signal sequence and the amino terminus of the tetracycline resistance protein. Hydrophobic residues are underlined. Positively charged residues are denoted with a plus sign; negatively charged residues are indicated with a minus sign. Numbers in parentheses indicate the position of the residue in the PhoA amino terminus (from −21 to +13) or in the Tet amino terminus (from +1 to +34).

active enzyme suggests that, for some fraction of the hybrid proteins, the phosphatase moiety of the proteins associated with the cytoplasmic membrane is oriented into the periplasm. Although the amino terminus of the Tet protein is not a standard signal sequence, it resembles a signal sequence in that it possesses a positively charged residue followed by a core of hydrophobic amino acids (Fig. 4). Thus, the Tet amino terminus seems sufficient to allow secretion of at least a fraction of the Tet-PhoA fusion protein.

lamB-phoA Fusions

Secreted proteins of *E. coli* fall into two broad classes, those localized in the periplasm and those localized in the outer membrane. Both types of proteins are produced with amino-terminal signal sequences which are interchangeable and which contain no specific information for localization (5, 15). Gene fusions between genes encoding outer membrane proteins and *lacZ* have been used in attempts to define regions of outer membrane proteins that determine protein localization. The *lamB* gene specifies a well-characterized outer membrane protein of *E. coli* which acts as a porin for maltodextrans and as a receptor for bacteriophage lambda. Studies carried out using *lamB-lacZ* fusions have indicated that one such determinant may lie within the first 49 amino acids of the mature LamB protein (1, 2).

We were interested in knowing whether the properties of *lamB-phoA* fusion products would be useful for analysis of protein localization in the outer membrane. A series of *lamB-phoA* fusions were constructed containing different amounts of the *lamB* gene, and these fusions gave rise to phosphatase activity in vivo, indicating that the LamB sequences were providing secretion information.

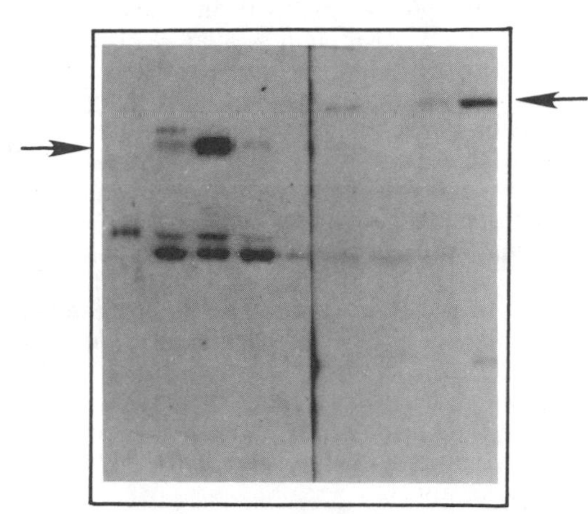

FIG. 5. Localization of LamB-PhoA hybrid proteins. Cells producing LamB-PhoA fusion proteins were labeled and fractionated as described in the Fig. 1 and 3 legends. Lanes: 1, wild-type PhoA protein from the periplasm of AW1046 (pHI1); 2–5, cytoplasmic, periplasmic, inner membrane, and outer membrane proteins from AW1061 (pHF1; encoding the LamB signal sequence plus 112 residues of mature LamB fused to PhoA); 6–9, the same fractions in order from AW1061 (pCH75; encoding the LamB signal sequence plus the first 176 residues of mature LamB fused to PhoA). The arrow on the left indicates the position of the pHF1-encoded hybrid protein. The arrow on the right indicates the position of the pCH75-encoded hybrid protein. The lower bands in lanes 2–5 are breakdown products seen in cells expressing some fusion proteins but not others.

Fusion proteins composed of the LamB signal sequence plus the first 176 or more residues of mature LamB fused to PhoA appeared to be largely localized in the outer membrane (Fig. 5, lane 9). Similar results were obtained when membrane-associated proteins were fractionated by Triton extraction (13; Fig. 5) or by density in sucrose gradients (4; data not shown).

Fusion proteins composed of the LamB signal sequence plus up to the first 112 amino acids of LamB fused to PhoA were localized almost solely in the periplasm (Fig. 5, lane 3). Benson et al. (1, 2) have observed that a fraction of the hybrid proteins containing a similar number of LamB residues fused to LacZ appear to be localized in the outer membrane. This may indicate that the LamB-PhoA fusion proteins lack sequences which would anchor the proteins to a membrane. Such sequences may be present in the longer LamB-PhoA fusion proteins described in the previous paragraph. Further studies are clearly necessary to define the sequences within the LamB protein that target it to the outer membrane.

Concluding Remarks

Two characteristics of alkaline phosphatase make it particularly well suited for the study of protein secretion. One is that the amino terminus is not required for activity and can be readily replaced by other protein sequences. The other is the property that its activity is absolutely dependent on its secretion from the cytoplasm. These properties allow, through the construction of gene fusions, the detection of secreted proteins and the characterization of sequences within them that are responsible for targeting the proteins to an extracytoplasmic location. The *phoA* system has been used successfully for studies of protein secretion and localization in a wide range of organisms including *Bacillus subtilis* (J. Pero, personal communication), *Y. enterocolitica* (R. Isberg, personal communication), *Vibrio cholerae* (J. Mekalanos, personal communication), and *Haemophilus influenzae* (this laboratory).

This work was initiated while one of us (A.W.) was on sabbatical in Jon Beckwith's laboratory. The initial stimulus for the work came from interactions with Hiroshi Inouye, and we dedicate this paper to his memory. We thank members of the Beckwith laboratory for advice and encouragement.

Our research was supported by Public Health Service grant RO1-AI20337 from the National Institute of Allergy and Infectious Diseases.

LITERATURE CITED

1. **Benson, S. A., E. Bremer, and T. J. Silhavy.** 1984. Intragenic regions required for LamB export. Proc. Natl. Acad. Sci. USA **81:**3830–3834.
2. **Benson, S. A., and T. J. Silhavy.** 1983. Information within the mature LamB protein necessary for localization of the outer membrane of *E. coli* K12. Cell **32:**1325–1335.
3. **Chang, C. N., W.-J. Kuang, and E. Y. Chen.** 1986. Nucleotide sequence of the alkaline phosphatase gene of *Escherichia coli* K-12. Gene **44:**121–125.
4. **Crowlesmith, I., M. Schindler, and M. J. Osborn.** 1978. Bacteriophage P22 is not a likely probe for zones of adhesion between the inner and outer membranes of *Salmonella typhimurium*. J. Bacteriol. **135:**259–269.
5. **Hoffman, C. S., and A. Wright.** 1985. Fusions of secreted proteins to alkaline phosphatase: an approach for studying protein secretion. Proc. Natl. Acad. Sci. USA **82:**5107–5111.
6. **Inouye, H., S. Michaelis, A. Wright, and J. Beckwith.** 1981. Cloning and restriction mapping of the alkaline phosphatase structural gene (*phoA*) of *Escherichia coli* and generation of deletion mutants in vitro. J. Bacteriol. **146:**668–675.
7. **Kelley, P. M., P. A. Neumann, K. Shriefer, F. Cancedda, M. J. Schlesinger, and R. A. Bradshaw.** 1973. Amino acid sequence of *Escherichia coli* alkaline phosphatase. Amino- and carboxyl-terminal sequences and variations between two isozymes. Biochemistry **12:**3499–3503.
8. **Manoil, C., and J. Beckwith.** 1985. Tn*phoA*: a transposon probe for protein export signals. Proc. Natl. Acad. Sci. USA **82:**8129–8133.
9. **Michaelis, S., and J. Beckwith.** 1982. Mechanism of incorporation of cell envelope proteins in *Escherichia coli* K12. Annu. Rev. Microbiol. **36:**435–464.
10. **Michaelis, S., H. Inouye, D. Oliver, and J. Beckwith.** 1983. Mutations that alter the signal sequence of alkaline phosphatase in *Escherichia coli*. J. Bacteriol. **154:**366–374.
11. **Roberts, C. H., and J. F. Chlebowski.** 1984. Trypsin modification of *Escherichia coli* alkaline phosphatase. J. Biol. Chem. **259:**729–733.
12. **Rothman, F., and R. Byrne.** 1963. Fingerprint analysis of alkaline phosphatase of *Escherichia coli* K12. J. Mol. Biol. **6:**330–340.
13. **Schnaitman, C. A.** 1971. Solubilization of the cytoplasmic membrane of *Escherichia coli* by Triton X-100. J. Bacteriol. **108:**545–552.
14. **Talmadge, K., S. Stahl, and W. Gilbert.** 1980. Eukaryotic signal sequence transports insulin antigen in *Escherichia coli*. Proc. Natl. Acad. Sci. USA **77:**3369–3373.
15. **Tommassen, J., H. van Tol, and B. Lugtenberg.** 1983. The ultimate localization of an outer membrane protein of *Escherichia coli* K-12 is not determined by the signal sequence. EMBO J. **2:**1275–1279.
16. **Wyckoff, H. W., M. Handschumacher, H. M. Krishna Murthy, and J. M. Sowadski.** 1983. The three dimensional structure of alkaline phosphatase of *E. coli*. Adv. Enzymol. Relat. Areas Mol. Biol. **55:**453–480.

Role of Phospholipids in the Energetics of Secretion of Proteins

MARINA A. NESMEYANOVA AND MICHAYL V. BOGDANOV

Institute of Biochemistry and Physiology of Microorganisms, USSR Academy of Sciences, Puschino, Moscow Region, 142292, USSR

Significant progress has been made toward understanding the molecular basis of bacterial exoprotein secretion. However, some important aspects of this problem are still obscure, such as the mechanisms of translocation of hydrophilic proteins through hydrophobic bilayer membranes and the energetics of this process. To understand these mechanisms, it is necessary to establish the role of membrane components, primarily phospholipids, in protein secretion.

Investigations of the role of phospholipids in the biosynthesis and secretion of alkaline phosphatase by *Escherichia coli* carried out recently in our laboratory have revealed a link between protein secretion and the metabolism, composition, and physicochemical state of membrane phospholipids (3–5, 14–16).

Participation of Phospholipids in the Translocation of Secreted Proteins through the Bacterial Cytoplasmic Membrane

We made a number of observations which indicate that synthesis and secretion of alkaline phosphatase in *E. coli* are correlated with increased de novo synthesis of lipids (10). Higher levels of a radioactive precursor of phospholipids are incorporated into phospholipids of cells synthesizing alkaline phosphatase than into phospholipids of cells unable to produce the enzyme. The biosynthesis of alkaline phosphatase correlates with an increase in the relative content of phosphatidylglycerol in these cells (16). This phospholipid is less accessible to the action of phospholipases A2 and C in cells producing high levels of alkaline phosphatase (16). These results led us to suggest an interaction between alkaline phosphatase and phosphatidylglycerol during biosynthesis of the enzyme. This hypothesis is supported by our observation that cells treated during growth with the lipotropic antibiotic polymyxin B, which interacts with acidic phospholipids, show a 50% decrease in the level of alkaline phosphatase (15).

The biosynthesis of secreted proteins also appears to depend on membrane fluidity and the content of unsaturated fatty acids (14). Incubation of bacterial cells at low temperature or in the presence of ethanol increased the quantity of *cis*-vaccenic acid and resulted in increased synthesis and secretion of alkaline phosphatase. On the other hand, a decrease in the level of *cis*-vaccenic acid, produced by addition of hexanol, resulted in a lower level of alkaline phosphatase synthesis (14).

These observations suggest that lipid molecules might play an active role in the process of secretion. Analysis of the dynamic properties of membrane lipids showed that conditions necessary to maintain transbilayer movement of lipids are the same as those required to promote protein secretion, suggesting coupling of the transmembrane movement of phospholipids and protein secretion (14).

Such a secretion mechanism assumes interactions between the signal peptide of the newly synthesized protein and the acid phospholipids of membranes. The phospholipids, having lost their negative charge during interactions with protein, acquire the ability to move to the hydrophobic region of the membrane, leading to movement of associated polypeptide. The newly synthesized protein and phospholipid simultaneously promote the movement of each other. It was suggested, on the basis of ideas of the metamorphic-mosaic structure of membranes (8), that interactions of the signal peptide and acid phospholipids modulate a nonbilayer configuration of phospholipids. A hydrophilic lipid channel (inverted short cylinder) is formed as an intermediary form in flip-flop of phospholipids, through which the major hydrophilic part of a secretory protein is linearly translocated during its synthesis, pulled by moving phospholipids with the signal peptide anchored in them.

To study the coupling of protein secretion and translocation of phospholipids through membranes, we analyzed the translocation and turnover of phospholipids in *E. coli* cells with different levels of alkaline phosphatase synthesis (Table 1). Phospholipids are synthesized on the inner side of the cytoplasmic membrane of *E. coli* and are translocated to its outer side and then into the outer membrane (9). Therefore, translocation of phospholipids was determined from the distribution of labeled individual phospholipids between cytoplasmic and outer membranes as well as from the dynamics of the radioactivity of individual phospholipids in the

TABLE 1. Level of phospholipids, IMPs, and alkaline phosphatase in membranes of *E. coli* in exponential and stationary phases of growth[a]

E. coli strain	Growth phase	AP (mU/mg of protein)	Source of phospholipids	Phospholipids (cpm/μg)			IMPs/μm²
				CL	PG	PEA	
K10 WT	Exponential	4.0	IM	170	200	2,800	3,200 ± 280
			OM	90	120	1,900	
C90 *phoU*	Exponential	95.0	IM	300	150	2,070	4,100 ± 290
			OM	260	180	2,070	
K10 WT	Stationary $-P_i$	83.0	IM	110	116	1,200	4,200 ± 320
			OM	110	174	1,800	
LEP1 *phoB*	Stationary $-P_i$	2.0	IM	60	80	1,800	3,000 ± 240
			OM	40	240	1,200	
K10 WT + 50 μM CCCP	Stationary $-P_i$	12	IM	20	6	280	3,200 ± 290
			OM	10	14	550	

[a] AP, Alkaline phosphatase; CL, cardiolipin; PG, phosphatidylglycerol; PEA, phosphatidylethanolamine; IM, inner membrane; OM, outer membrane; WT, wild type.

cytoplasmic membrane after pulse-labeling of cells (4).

It can be seen in Table 1 that cells with high rates of alkaline phosphatase synthesis show high levels of phospholipid translocation. There is an increase in the total and relative quantity of phospholipid in the outer membrane of such cells. In contrast, in a *phoB* mutant of *E. coli* unable to express *phoA*, the quantity of cardiolipin and phosphatidylethanolamine in the outer membrane is low while the phosphatidylglycerol content is high. Phosphatidylglycerol is a precursor of membrane-derived oligosaccharides present in the *E. coli* periplasm (12). It is probable that translocation of this phospholipid is coupled with its role as a precursor of these oligosaccharides. We have shown that cells producing alkaline phosphatase contain more oligosaccharide than those (*phoB* mutants) unable to produce the enzyme. Thus, accumulation of phosphatidylglycerol in the outer membrane of the *phoB* mutant (Table 1) is apparently due to its failure to enter the membrane-derived oligosaccharide biosynthetic pathway.

Acceleration of phospholipid translocation in cells producing alkaline phosphatase was revealed by analysis of radioactivity of pulse-labeled phospholipids in isolated cytoplasmic membranes. For instance, the quantity of radioactive cardiolipin in the cytoplasmic membrane of strain C90 (constitutive) in the presence of P_i decreased twice as rapidly as did cardiolipin in wild-type membranes.

The turnover of individual phospholipids in cells with different levels of alkaline phosphatase synthesis was studied in pulse-chase experiments. The turnover rate of acid phospholipids was two- to fourfold higher in cells producing alkaline phosphatase (Fig. 1). The half-time of turnover in wild-type *E. coli* starved for P_i (inducing condition) was 40 min for phosphatidylglycerol and 60 min for cardiolipin. Both lipids had a turnover half-time of 25 min in *E. coli* C90, a phosphatase-constitutive strain. In the phosphatase-negative strain LEP1, the phosphatidylglycerol half-life was 200 min, while the amount of cardiolipin actually increased over the course of the experiment. Phosphatidylethanolamine turnover, with a half-life of about 24 h, was not influenced by synthesis of secreted protein. These observations suggest a coupling of protein secretion and the turnover of acid phospholipids. The increased turnover is very likely due to increased translocation of phospholipids during protein secretion which requires resynthesis of phospholipids.

It is a peculiar feature of the ultrastructure of cytoplasmic membranes to contain the so-called intramembrane particles (IMPs) revealed by electron microscopic freeze-etching. The analysis of cytoplasmic membrane sections of *E. coli* cells producing different amounts of alkaline phosphatase (4) showed a correlation between secretion and the quantity of IMPs (Table 1). Cells producing high levels of alkaline phosphatase had 20 to 30% more IMPs than cells not producing the enzyme. Membranes from cells treated with the protonophore carbonyl cyanide *m*-chlorophenylhydrazone (CCCP), which inhibits alkaline phosphatase synthesis and secretion, were found to have areas devoid of IMPs. CCCP also inhibits synthesis and translocation of phospholipids (3). Removal of CCCP restores protein and lipid synthesis, and zones free from IMPs disappear. One may suppose that IMPs, the quantity of which correlates with the secretion of alkaline phosphatase, may represent a complex of protein and lipid being simultaneously transported across the membrane. It is quite possible that IMPs are a morphological reflection of lipid

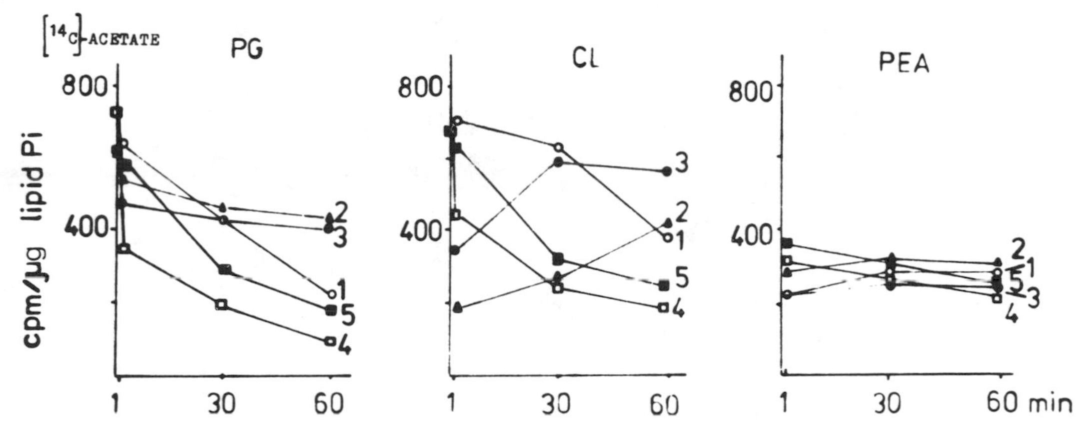

FIG. 1. Turnover of phosphatidylglycerol (PG), cardiolipin (CL), and phosphatidylethanolamine (PEA) in *E. coli* cells with different levels of biosynthesis and secretion of alkaline phosphatase. Curves: 1, strain K10 (− P_i); 2, strain LEP1 (− P_i); 3, strain K10 (− Pi + 50 μM CCCP); 4, strain C90 (+ Pi); 5, strain K10 (+ P_i). The times shown are minutes after pulse-labeling.

micelles or "canals" containing proteins being translocated through the membranes.

We have used immunoprecipitation to study interactions between phospholipid and newly synthesized protein (5). Immunoprecipitated membrane complexes, containing alkaline phosphatase, were found to be significantly enriched for phosphatidylglycerol. We have also studied changes in the availability of phospholipids to the action of phospholipase C after the removal of ribosomes from isolated cytoplasmic membranes. Ribosomes were removed by incubating membranes with puromycin in the presence of GTP and elongation factor G. After about 30% release of ribosomes, measured as release of labeled RNA, 30% of the phosphatidylglycerol in the membranes became sensitive to phospholipase C. Prior to removal of ribosomes, the phosphatidylglycerol was totally resistant to this enzyme.

The signal peptide of the *lamB* gene product interacts with monolayers of phosphatidylglycerol at a water surface (6). It is quite possible that the interaction of proteins with acid phospholipids is a widespread phenomenon which has a definite role in the biogenesis of membranes and secreted proteins.

Taken together, the above results suggest that protein secretion requires and is coupled to the transmembrane movement of acid phospholipids.

Role of Phospholipids in the Energetics of Protein Translocation through Bacterial Membranes

The proton ionophore CCCP, which removes the electrochemical potential of the cell membrane, inhibits the processing and secretion of proteins, leading to accumulation of their precursors in the cytoplasm or the membrane (17). Protein export in *E. coli* requires proton motive forces (both of which are components of the electrochemical potential gradient of protons: $\Delta\psi$ and ΔpH). It is unknown, however, which components of membranes are involved in the coupling of the membrane energy with protein translocation and how the membrane energy facilitates protein export.

Studies of the action of the protonophore CCCP on alkaline phosphatase synthesis and secretion showed that it suppresses not only these processes but also the metabolism and translocation of phospholipids, particularly cardiolipin (3). CCCP transports protons through the membrane (2), probably competing with phospholipids for this process, thus impairing their translocation and hence blocking protein secretion. Of particular interest is the change, in the presence of the protonophore, in the ratio of acid phospholipids (cardiolipin/phosphatidylglycerol), which is correlated with the secretion of proteins (10). There are data suggesting a correlation between the ratio and interconversion of acid phospholipids as well as the phosphorylating ability of cells (7) and the maintenance, in steady state, of bilayer and nonbilayer structures of membrane lipids (11). Besides, acid phospholipids may participate in energy coupling (13). For the above reasons, we have hypothesized (3, 14) that the translocation of acid phospholipids and their interconversion induced by interaction with a secretory protein generate the energy which may be used for the unidirectional movement of this protein through the membrane.

We have checked this hypothesis by comparative studies of the basic energetic parameters of *E. coli* cells which differ in their level of biosynthesis and secretion of alkaline phosphatase.

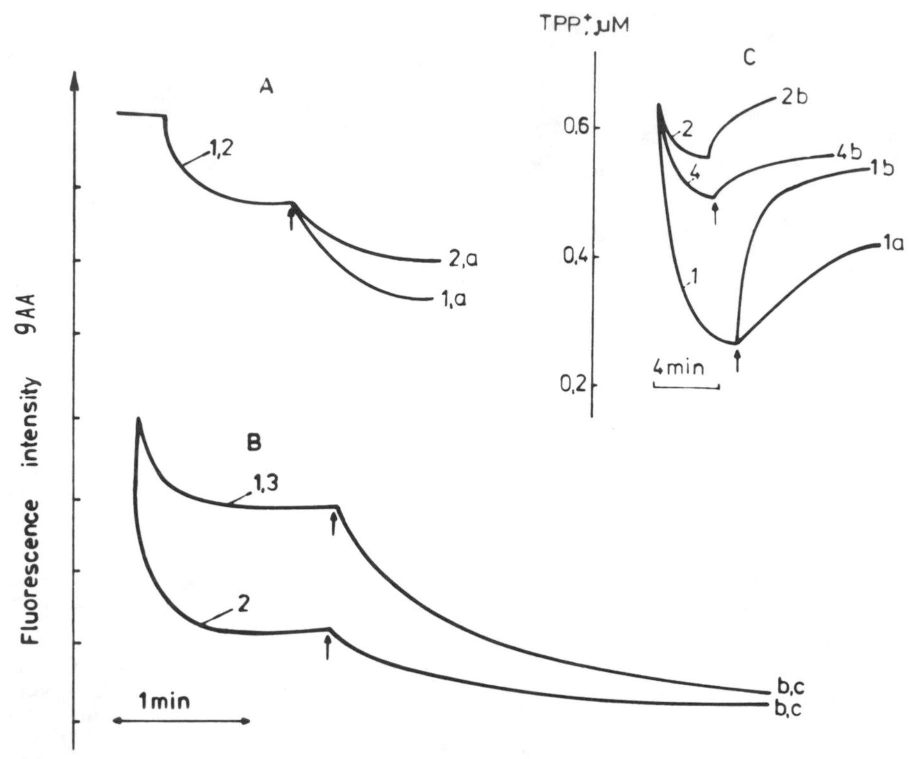

FIG. 2. Dynamics of 9-aminoacridine (9AA) (A, B) and tetraphenylphosphonium chloride (TPP^+) (C) uptake by *E. coli* cells with different levels of biosynthesis and secretion of alkaline phosphatase. (1) Strain K10, (2) strain LEP1, and (3) strain C90 all grown under phosphate starvation; (4) grown in the presence of 50 μM CCCP. Cell concentration was 0.5 mg of protein per ml of incubation medium. (A) Intact cells in cultivation medium, pH 7.8; (B and C) EDTA/lysozyme-treated cells in 25 mM Tris hydrochloride, pH 7.8, with 500 mM sucrose. Arrows indicate the time of addition of 1 mM P_i (a), 0.2 μg of nigericin per ml (b), and 5 μM CCCP (c). Intracellular water volume, 6.34 ± 0.23 μl/mg of protein, in all the strains under study, was determined with the use of ^3H_2O and [^{14}C]inulin as markers.

Simultaneously, we studied the biosynthesis of phospholipids and periplasmic membrane-derived oligosaccharides. We observed that in permeabilized *E. coli* K10 cells showing a high level of secreted alkaline phosphatase (P_i^- conditions) the percentage of tetraphenylphosphonium chloride ion uptake is higher than by a *phoB* mutant (LEP1) unable to synthesize the enzyme under similar conditions or by the wild-type strain grown in the presence of the protonophore (Fig. 2C). The increase in the membrane potential of secreting cells, estimated from the uptake of tetraphenylphosphonium chloride, correlates with a decrease in the quenching of 9-aminoacridine fluorescence, i.e., with the alkalization of the intracellular space. The addition of nigericin and CCCP (Fig. 2A and B) or of P_i results in the depletion of the membrane potential and the acidification of the cytoplasm, which is more intensive in the secreting cells (Fig. 2A).

The proton motive force of bacterial cells can be generated by respiration or by hydrolysis of intracellular ATP. We have established (Table 2) that the endogenous respiration rate (estimated polarographically) is the same in cells with different levels of synthesis and secretion of alkaline phosphatase. Table 2 also shows that the level of intracellular ATP is higher in cells with a high level of alkaline phosphatase than in cells without the enzyme. Moreover, ATP is accumulated in cells with active alkaline phosphatase secretion (e.g., in the constitutive strain *E. coli* C90 [*phoU35*]). In cells unable to synthesize the enzyme, e.g., the *phoB E. coli* mutant (LEP1), the ATP pool is gradually depleted (3a). However, the ATPase activity is equal in all the cells under study (Table 2).

Membrane-Derived Oligosaccharides: Their Role in Protein Secretion

Membrane-derived oligosaccharides are a major constituent of the periplasm of gram-negative organisms (12). They may play an active role in osmoregulation and, because of their polyanion nature, may be proton acceptors. We have

TABLE 2. Energetic parameters of *E. coli* cells with different levels of biosynthesis and secretion of alkaline phosphatase under phosphate starvation

Strains under phosphate starvation	Alkaline phosphatase activity (mU/mg of protein)	ATP[a] (nmol/mg of protein)	Membrane ATPase (mU/mg of protein)	Oxygen uptake (natom of O/min per mg of protein)	MDO[b] (cpm/mg of protein)	Half-life of acid phospholipids (min)[c]	
						PG	CL
K10	83	10.2	59	304	4,520	−40	−60
C90	355	16.5	58	300	4,710	−40	−35
LEP1	1.0	6.0	59	311	2,550	−200	+50
K10 + 50 μM CCCP	17.8	5.7	55	303	550	−180	+50

[a] The ATP concentration was determined by the luciferin-luciferase method.

[b] MDO, Membrane-derived oligosaccharides, estimated by the incorporation of [^{14}C]glycerol in cells with the preinduced *glp* operon.

[c] Half-life of acid phospholipids was calculated by using the formula for the calculation of kinetic process of the first order, $t_{1/2} = \ln 2/K$, where $K = \ln[It/Io]/t$ characterizes radioactivity of phospholipids after pulse labeling (t_0) and after a 60-min chase (t).

found that cells with high levels of protein secretion and high membrane potential contain more periplasmic oligosaccharide. It is known (12) that phosphatidylglycerol is the precursor of these oligosaccharides and that the phosphatidylglycerol content in the cytoplasmic membrane as well as its metabolism and translocation are correlated with the secretion process.

Generation of Membrane Potential during Protein Secretion and Formation of Additional ATP

We suppose (Fig. 3) that secretion itself, by facilitating transmembrane movement of protonated phospholipids, is an energy-generating process which produces a membrane potential. Both protonation and movement of phospholipids would be induced as the result of the electrostatic interaction between the de novo synthesized protein and negatively charged phospholipids, primarily cardiolipin. This results in the thermodynamically disadvantageous transfer of H^+ from H_2O and the transformation of cardiolipin to phosphatidylglycerol. The phosphatidylglycerol translocates across the membrane transferring the proton and is involved in the formation of periplasmic oligosaccharides which are coupled with secretion. This is followed by generation of a proton motive force, while synthesized oligosaccharides participate in the regulation of proton transport and stabilization of ΔpH. It is possible that the molecule of phosphatidic acid released by the splitting of cardiolipin, which is the second protonated acid

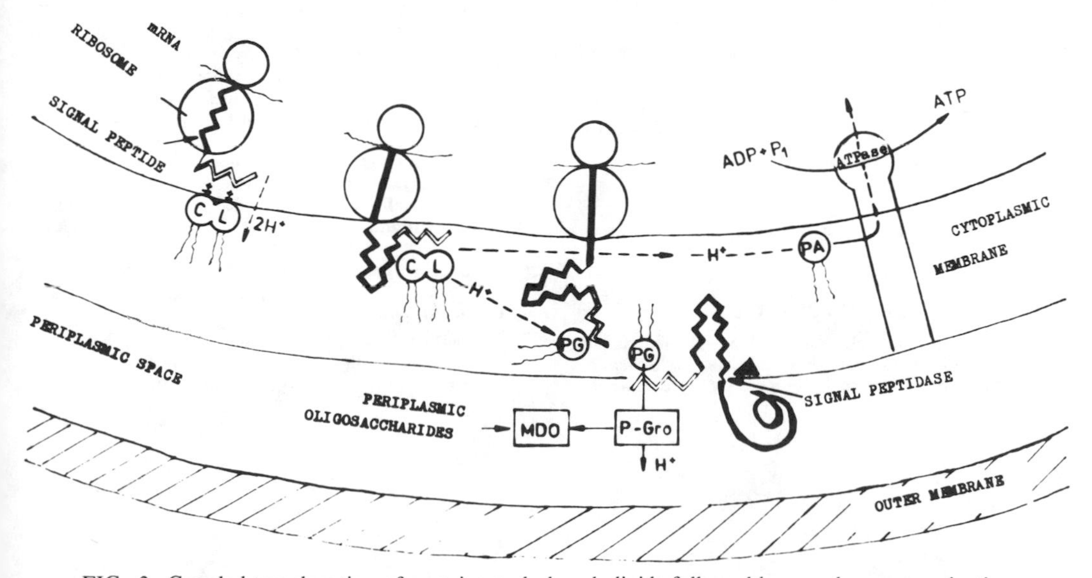

FIG. 3. Coupled translocation of proteins and phospholipids followed by membrane energization.

phospholipid, carries out a lateral transfer of the proton to the membrane ATPase, which utilizes it during the ATP synthesis.

A cardiolipin molecule, or a derivative, may participate in energy coupling as a hydrophobic highly energized compound, dephosphorylation of which is coupled with phosphorylation of ADP. It is quite possible that the ATP formed during secretion is used as an effector of the cardiolipin-specific phospholipase D, participating in cardiolipin metabolism.

We propose that translocation of proteins and phospholipids coupled with membrane-derived oligosaccharide metabolism generates membrane energy and that secretion of proteins is a self-organized process. The energization of secretion in gram-negative bacteria is determined by the local structural and metabolic rearrangement of the membrane, i.e., enhancement of translocation and metabolism of acid phospholipids and their interconversion, initiation of proton flows (in phospholipid-bound form) coupled with the unidirectional movement of the polypeptide chain, and synthesis of periplasmic oligosaccharides.

Conclusion

On the strength of the results obtained, we conclude that there is a relationship between protein secretion and metabolism, translocation, and structure of phospholipids. Thus, the protein secretion process is coupled to the transmembrane transfer of phospholipids. The electrostatic interactions of the secreted protein with phospholipids result in the energization of the membrane, promoting the unidirectional movement of the polypeptide chain through the membrane. The mechanism of membrane energization during secretion, indicated by our experiments, is important for the understanding of cell energetics as a whole. The relationship between protein secretion and lipid metabolism makes it possible to control this process by affecting the bacterial membranes.

We thank A. Torriani-Gorini for mutants of *E. coli* which we used in our work, and A. Wright for efforts in rewriting this paper.

LITERATURE CITED

1. **Bakker, E. P., and L. L. Randall.** 1984. The requirement for energy during export of β-lactamase in *Escherichia coli* is fulfilled by the total protonmotive force. EMBO J. **3:**895–900.
2. **Benz, R., and S. McLaughlin.** 1983. The molecular mechanism of action of the proton ionophore FCCP. Biophys. J. **41:**381–398.
3. **Bogdanov, M. V., I. S. Kulaev, and M. A. Nesmeyanova.** 1984. Study of lipid-protein interactions during protein transfer across bacterial membranes. I. Coordinated inhibition of the exoprotein synthesis, secretion and phospholipid metabolism by a protonophore. Biologicheskie Membrany **1:**495–502.

3a. **Bogdanov, M. V., and M. A. Nesmeyanova.** 1986. Energization of membrane during the protein secretion in gram-negative bacteria. Dokl. Acad. Nauk. SSSR **288:**1247–1250.

4. **Bogdanov, M. V., N. E. Suzina, and M. A. Nesmeyanova.** 1985. Peculiarities of phospholipid metabolism and ultrastructural organization of the *E. coli* cytoplasmic membrane in the alkaline phosphatase secretion. Biologicheskie Membrany **2:**367–375.
5. **Bogdanov, M. V., I. M. Tsfasman, and M. A. Nesmeyanova.** 1985. Study of lipid-protein interactions during protein transfer across bacterial membranes. II. Phospholipids in the translocation site of the alkaline phosphatase across *E. coli* membrane. Biologicheskie Membrany **2:**623–629.
6. **Briggs, M. S., L. M. Gierash, A. Zlotnick, J. D. Lear, and W. F. De Grado.** 1985. *In vivo* function and membrane binding properties are correlated for *Escherichia coli* LamB signal peptides. Science **228:**1096–1099.
7. **Cronan, J. E., Jr., and P. R. Vagelos.** 1972. Metabolism and function of the membrane phospholipids of *Escherichia coli*. Biochim. Biophys. Acta **265:**25–60.
8. **Cullis, P. R., B. De Kruijff, M. J. Hope, R. Nayar, and S. L. Schmid.** 1980. Phospholipids and membrane transport. Can. J. Biochem. **58:**1091–1099.
9. **Donohue-Rolfe, A. M., and M. Schaechter.** 1980. Translocation of phospholipids from the inner to the outer membrane of *Escherichia coli*. Proc. Natl. Acad. Sci. USA **77:** 1867–1871.
10. **Evdokimova, O. A., and M. A. Nesmeyanova.** 1977. Phospholipid composition of *E. coli* cells and membrane under repression and derepression of alkaline phosphatase biosynthesis. Biokhimia **42:**1791–1799.
11. **Gross, Z., S. Rottem, and R. Bittman.** 1982. Phospholipid interconversion in *Mycoplasma capricolum*. Eur. J. Biochem. **122:**169–174.
12. **Kennedy, E. P.** 1982. Osmotic regulation and the biosynthesis of membrane-derived oligosaccharides in *Escherichia coli*. Proc. Natl. Acad. Sci. USA **79:**1092–1095.
13. **Mikelsaar, X., I. I. Severina, and V. P. Sculachev.** 1974. Phospholipids and oxidative phosphorylation. Usp. Sovrem. Biol. **78:**348–370.
14. **Nesmeyanova, M. A.** 1982. On the possible participation of acid phospholipids in the translocation of secreted proteins through bacterial cytoplasmic membrane. FEBS Lett. **142:**189–193.
15. **Nesmeyanova, M. A., M. V. Bogdanov, M. N. Kolot, O. A. Zemlyanukhina, and I. S. Kulaev.** 1982. Interaction between alkaline phosphatase and acid phospholipids in *E. coli* cells and artificial membranes. Biokhimia **47:**671–677.
16. **Nesmeyanova, M. A., and O. A. Evdokimova.** 1979. Phospholipids of *E. coli* and activity of alkaline phosphatase. Biokhimia **44:**1512–1519.
17. **Zimmerman, R., and W. Wickner.** 1983. Energetics and intermediates of the assembly of protein OmpA into the outer membrane of *Escherichia coli*. J. Biol. Chem. **258:** 3920–3925.

Enzymatic Activity of Alkaline Phosphatase Precursor Depends on Its Cellular Location

DANA BOYD, CHU-DI GUAN,[†] SHERI WILLARD, WILL WRIGHT, KATHRYN STRAUCH, AND JON BECKWITH

Department of Microbiology and Molecular Genetics, Harvard Medical School, Boston, Massachusetts 02115

The assembly of the *Escherichia coli* enzyme alkaline phosphatase involves numerous steps. Since the enzyme is located in the periplasm (9), it is synthesized initially with an amino-terminal signal sequence (4, 5, 7, 20). Thus, as part of the assembly process, the newly synthesized alkaline phosphatase must pass through the cytoplasmic membrane, and cleavage of the signal peptide must occur. Further steps in assembly include the dimerization of identical monomeric subunits (12, 18, 22, 23), the formation of intrachain disulfide bonds (8), and the incorporation of Zn ion (17).

Signal sequence mutations, which prevent the efficient export of alkaline phosphatase to the periplasm, result in the accumulation in the cytoplasm of precursor to the protein, with its signal sequence intact (13, 14). While the periplasmic enzyme is quite stable, the cytoplasmically located precursor is unstable, being degraded with a half-life of 5 min at 37°C. Studies on strains carrying these signal sequence mutations suggested that the precursor found inside the cell was not able to assume an enzymatically active conformation. However, this conclusion was somewhat complicated by the instability of the cytoplasmic precursor. Studies with gene fusions of alkaline phosphatase have provided additional evidence that internalized alkaline phoshatase is unable to assume an active conformation (3, 10; Manoil, Boyd, and Beckwith, this volume).

Two lines of evidence suggest that the lack of activity of internalized alkaline phosphatase precursor is not due simply to the presence of the amino-terminal signal sequence. First, precursor alkaline phosphatase synthesized in vitro appears to exhibit enzymatic activity (5). Second, a number of gene fusions produce hybrid proteins in which alkaline phosphatase contains long amino-terminal extensions; these hybrid proteins exhibit high enzymatic activity when the alkaline phosphatase is on the periplasmic side of the cytoplasmic membrane (Manoil et al., this volume).

In this paper, we describe additional studies on the enzymatic activity of alkaline phosphatase precursor found in the cytoplasm. We also present evidence that alkaline phosphatase with a signal sequence attached can assemble into active enzyme when it is localized to the correct compartment, the periplasm.

Results

Stabilizing cytoplasmic precursor of alkaline phosphatase. The signal sequence mutation *phoA61* replaces a leucine with an arginine at position 14 of the alkaline phosphatase signal sequence (14). The mutation results in a severe defect in alkaline phosphatase export; only about 1% of the wild-type amount of the enzyme is found in the periplasm. The rest of the protein is found inside the cell as precursor. However, the cytoplasmic precursor is unstable.

We constructed a strain which carried the *phoA61* mutation and was defective in proteolysis. This strain carried both a *lon* mutation (11) and an amber mutation in the *htpR* gene that is weakly suppressed by a temperature-sensitive suppressor (1). At 30°C, this strain exhibited increased stability of the precursor found in *phoA61* (Fig. 1). Whereas wild-type alkaline phosphatase is completely stable during a 47-min chase following a 1-min pulse, the *phoA61* alkaline phosphatase has a half-life of approximately 20 min in the wild-type background. In the *lon htpR* background this half-life is increased by about fivefold. Despite the accumulation of stable precursor in this strain at 30°C, there was no substantial increase in alkaline phosphatase activity (Table 1).

Activity of mutant precursors of alkaline phosphatase. Using localized mutagenesis, we have isolated three new mutants with alterations in the early portion of the *phoA* gene. One of these mutations is in the *phoA* promoter region, and the other two are in the region of the gene coding for the signal sequence (Fig. 2). The mutation *phoA10* is a deletion of the first base of the arginine codon at the amino terminus of mature alkaline phosphatase which results in a translational frameshift.

We have isolated revertants of *phoA10* by selecting for the ability to grow on the chromo-

†Present address: New England BioLabs, Inc., Beverly, MA 01915.

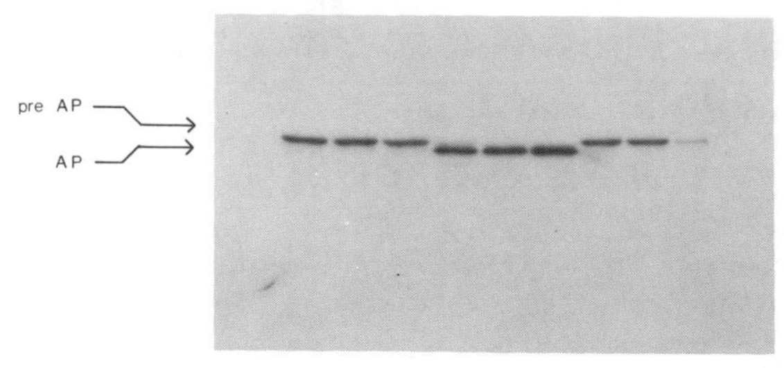

FIG. 1. Stabilization of cytoplasmic pre-alkaline phosphatase. Exponentially grown cultures in M63 minimal glucose medium supplemented with 19 amino acids at 30°C were pulse-labeled with [^{35}S]methionine (60 μCi/ml, 1,250 Ci/mmol) for 1 min and chased with 1 mM cold methionine for 1, 9, or 47 min. The chase was terminated by trichloroacetic acid precipitation, and samples were analyzed by immunoprecipitation with anti-alkaline phosphatase antibody and sodium dodecyl sulfate-polyacrylamide gel electrophoresis (6). All strains are derived from CAG456 *htpR*. Each set of three lanes includes 1-, 9-, and 47-min chases in that order. Lanes: a–c, *phoA61 lon htpR*; d–f, *phoA$^+$ lon$^+$ htpR$^+$*; g–i, *phoA61 lon$^+$ htpR$^+$*.

TABLE 1. Alkaline phosphatase activity of *phoA61*, *htpR*, and *lon* strains[a]

Strain genotype	Alkaline phosphatase[b]
phoA$^+$ lon$^+$ htpR$^+$	1,754
phoA61 lon$^+$ htpR$^+$	32
phoA61 lon htpR	11

[a] Cultures incubated in parallel with those of Fig. 1 were chilled on ice just before the 9-min chase point, washed in phosphate-free medium, and assayed for alkaline phosphatase activity (14). In other assays of alkaline phosphatase activity in *phoA$^+$* and *phoA61* strains done in minimal medium, no significant effects of the *lon* or *htpR* mutations either alone or in combination were seen.

[b] Expressed as units per milliliter per unit of optical density at 600 nm.

genic indicator 5-bromo-4-chloro-3-indolyl phosphate as sole phosphate source (21). The selection was performed in a strain carrying the mutation *prlA4* (2), which suppresses many signal sequence mutations. This was done so that revertants with slightly altered signal sequences, which might be detected only in the *prlA* background, could be obtained.

We chose for study five revertants of *phoA10*, only one of which showed a strong dependence on the *prlA4* mutation (Table 2). DNA sequence analysis showed that one of these revertants, R5, is a deletion of 5 base pairs restoring the original reading frame 15 amino acids upstream from the parental frameshift. This should result in synthesis of a protein with a wild-type signal sequence but with the first 16 amino acids of mature alkaline phosphatase replaced by a new sequence of 14 amino acids (Fig. 3). Sodium dodecyl sulfate-polyacrylamide gel electrophoresis of alkaline phosphatase immunoprecipitates from this strain showed that it makes an alkaline phosphatase that is somewhat slowly processed into a mature form (Fig. 4). Its electrophoretic mobility is slightly greater than that of the wild type (unpublished data).

The other four revertants have second-site frameshifts clustered in the middle of the hydrophobic core region of the signal sequence (Fig. 3). They share a common peptide preceding the normal cleavage site which is different from wild type. This amino acid sequence does not fit with the rules for processing which have been proposed (15, 25). Therefore, we have studied both the export of alkaline phosphatase to the periplasm and the processing of pre-alkaline phosphatase in these strains.

In Fig. 4 it can be seen that R7 and R8 make a precursor-sized protein that is somewhat unstable, but no mature form is visible. It seems likely that the portion of pre-alkaline phosphatase that is exported to the periplasm is relatively stable,

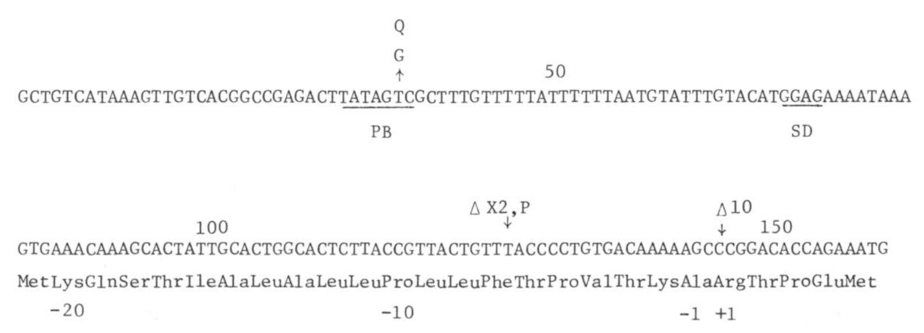

FIG. 2. Sequence of mutations in the *phoA* gene. PB stands for Pribnow box (of the promoter), and SD stands for Shine-Dalgarno sequence of the ribosome binding site (both putative). The amino acid sequence from −21 to −1 is the signal sequence. Mutation Q is a base change in the putative Pribnow box; mutations X2 and P are the same single-base-pair deletion in a codon for phenylalanine; mutation 10 is a single-base-pair deletion of the first base of the first codon of the mature protein.

while that remaining in the cytoplasm is unstable. This instability has been seen in other mutants which result in a cytoplasmic alkaline phosphatase (see above). In this view, the relative proportions of phosphatase in the two compartments would determine the shape of a biphasic decay curve. No label runs at the position of mature alkaline phosphatase in extracts from R7 and R8. It appears that the pre-alkaline phosphatase in these two revertants cannot be processed by signal peptidase.

To our surprise, in pulse-label experiments, the revertants R12 and R15 make alkaline phosphatases that are about 50% mature sized and 50% precursor sized despite the fact that their signal sequences are quite similar to those of R7 and R8 (Fig. 3). In particular, the sequences preceding the cleavage site are identical. In the *prlA*$^+$ background both mature and precursor forms of alkaline phosphatase are seen at early times. However, after a 60-min chase most of the precursor has either been converted to mature form or has been degraded. The alkaline phosphatase proteins of all four of these revertants comigrate with their wild-type precursor and mature counterparts on sodium dodecyl sulfate-polyacrylamide gel electrophoresis (data not shown). The alkaline phosphatase activity of these strains also comigrates with that of wild-type alkaline phosphatase in nondenaturing polyacrylamide gel electrophoresis. The isozyme pattern observed with these mutants is identical to that of the wild type in minimal medium with and without arginine in both log and stationary phases (data not shown). Therefore, it appears that the processing of the proteins in R12 and R15 is at the same site as in the

TABLE 2. Alkaline phosphatase activity of pseudorevertants of *phoA10*[a]

phoA allele	Alkaline phosphatase[b]	
	prlA4	*prlA*$^+$
R5	441	443
R7	472	253
R8	175	21
R12	689	491
R15	697	772
+	807	1,235
61	46	14

[a] Overnight cultures grown in M63 supplemented with 0.4% glycerol and 19 amino acids without methionine were assayed as described (14). All strains are MC1000 (14) derivatives that are *leu*$^+$ *phoR* Spc^r *lamB60*.

[b] Expressed as units per milliliter per unit of optical density at 600 nm.

wild-type pre-alkaline phosphatase (see Discussion).

Using the osmotic shock technique for separating the periplasm from cells (Fig. 5), we found that substantial amounts of the R12 and R15 alkaline phosphatase precursor-sized proteins are present in the periplasm in addition to the mature form. The R7 and R8 proteins which are present only as precursor forms are found mainly in the periplasm, with some in the membrane and less in the cytoplasm fractions. Thus, in all four strains, either most or a substantial amount of the precursor of alkaline phosphatase is found in the periplasm. In two strains, R7 and R8, this precursor must account for the observed alkaline phosphatase activity, since the small amounts of mature protein are not enough

```
+       GTGAAACAAAGCACTATTGCACTGGCACTCTTACCGTTACTGTTTACCCCTGTGACAAAAGCCCGGACACCAGAAATGCCTGTTCTGGAAAACCGGGCTGCTCAGGGCGATATT
        MetLysGlnSerThrIleAlaLeuAlaLeuLeuProLeuLeuPheThrProValThrLysAlaArgThrProGluMetProValLeuGluAsnArgAlaAlaGlnGlyAspIle

phoA10  GTGAAACAAAGCACTATTGCACTGGCACTCTTACCGTTACTGTTTACCCCTGTGACAAAAGCC GGACACCAGAAATGCCTGTTCTGGAAAACCGGGCTGCTCAGGGCGATATT
        MetLysGlnSerThrIleAlaLeuAlaLeuLeuProLeuLeuPheThrProValThrLysAla GlyHisGlnLysCysLeuPheTrpLysThrGlyLeuLeuArgAlaIleLe

R5      GTAAAACAAAGCACTATTGCACTGGCACTCTTACCGTTACTGTTTACCCCTGTGACAAAAGCC GGACACCAGAAATGCCTGTTCTGGAAAACCGGGCTGCTCA      ATATT
        MetLysGlnSerThrIleAlaLeuAlaLeuLeuProLeuLeuPheThrProValThrLysAla GlyHisGlnLysCysLeuPheTrpLysThrGlyLeuLeuA      snIle

R7      GTGAAACAAAGCACTATTGCACTGGCACT   TACCGTTACTGTTTACCCCTGTGACAAAAGCC GGACACCAGAAATGCCTGTTCTGGAAAACCGGGCTGCTCAGGGCGATATT
        MetLysGlnSerThrIleAlaLeuAlaLe   uThrValThrValTyrProCysAspLysSerA rgThrProGluMetProValLeuGluAsnArgAlaAlaGlnGlyAspIle

R8      GTGAAACAAAGCACTATTGCACT   CACTCTTACCGTTACTGTTTACCCCTGTGACAAAAGCC GGACACCAGAAATGCCTGTTCTGGAAAACCGGGCTGCTCAGGGCGATATT
        MetLysGlnSerThrIleAlaLe   uThrLeuThrValThrValTyrProCysAspLysSerA rgThrProGluMetProValLeuGluAsnArgAlaAlaGlnGlyAspIle

                                G
R12     GTGAAACAAAGCACTATTGCACTGGCACTCTTACCGTTACTGTTTACCCCTGTGACAAAAGCC GGACACCAGAAATGCCTGTTCTGGAAAACCGGGCTGCTCAGGGCGATATT
        MetLysGlnSerThrIleAlaLeuGyThrLeuThrValThrValTyrProCysAspLysSerA rgThrProGluMetProValLeuGluAsnArgAlaAlaGlnGlyAspIle
                                l

R15     GTGAAACAAAGCACTATTGCACTG       CTTACCGTTACTGTTTACCCCTGTGACAAAAGCC GGACACCAGAAATGCCTGTTCTGGAAAACCGGGCTGCTGAGGGCGATATT
        MetLysGlnSerThrIleAlaLeu       LeuThrValThrValTyrProCysAspLysSerA rgThrProGluMetProValLeuGluAsnArgAlaAlaGlnGlyAspIle
```

FIG. 3. Sequence of *phoA10* revertants. Deletions are indicated by gaps in the sequence, and the insertion in R12 is indicated by a raised letter. Translated amino acid sequences that are different from wild type are underlined. The frameshift in *phoA10* results in translation until a termination codon (not shown) 25 codons downstream from the processing site.

A

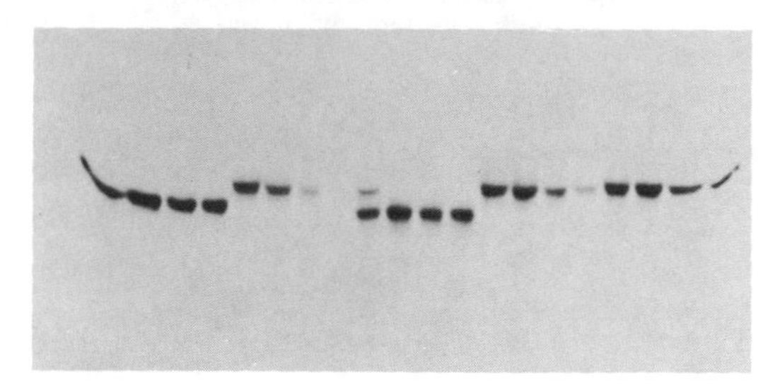

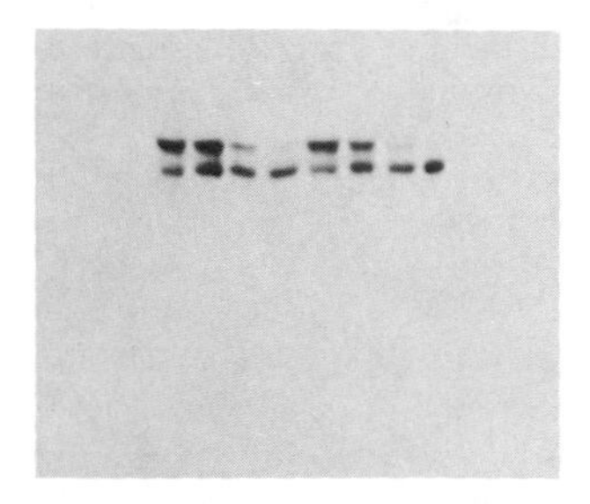

FIG. 4. Pulse-chase labeling of alkaline phosphatase in frameshift revertants. Exponentially grown cultures at 30°C in 2.0 ml of M63-glycerol medium supplemented with 19 amino acids were labeled with 80 μCi of [^{35}S]methionine for 30 s followed by addition of 1 mg of unlabeled methionine per ml. Samples were trichloroacetic acid precipitated and analyzed as in Fig. 1. These strains are *prlA*$^+$ *phoR*. Each set of four lanes includes 30-s and 15-, 30-, and 60-min chase in that order. The position of AP is indicated by lanes a–d in panel A and that of pre-AP is indicated by lanes e–h. Panel A: lanes a–d, *phoA*$^+$; lanes e–h, *phoA61*; lanes i–l, *phoA*R5; lanes m–p, *phoA*R7; lanes q–t, *phoA*R8. Panel B: lanes a–d, *phoA*R12; lanes e–h, *phoA*R15.

to account for the high levels of activity (Table 2).

Discussion

The precursor of alkaline phosphatase which accumulates in the cytoplasm of *phoA* signal sequence mutants is unstable. We have combined one of these mutations, *phoA61*, with the mutations *lon* and *htpR*, which reduce proteolysis. This proteolysis-defective strain exhibits high stability of the internalized precursor of alkaline phosphatase.

Despite the substantially increased stability of cytoplasmic alkaline phosphatase precursor, there is little or no increase in alkaline phosphatase activity. This finding has two implications. First, it provides stronger support for the conclusion that cytoplasmic precursor cannot assume an enzymatically active conformation. Previous analysis of this question was complicated by the instability of the precursor. Second, even though the small amount of alkaline phosphatase found in the periplasm of signal sequence mutants is exported posttranslationally from the internal precursor pool, increasing the stability of that precursor does not increase the amount of protein exported. This finding is consistent with the proposal that there are two pools of precursors of exported proteins in such cases, an export-competent and an export-incompetent pool (7a, 19). In the case of *phoA61*, it is possible that the two pools are determined at a very early stage and that the precursor destined to be exported is protected from proteolysis. Increasing the stability of cytoplasmic precursor is merely stabilizing a pool of that protein which has no possibility of being exported. We have been unable to activate this precursor and have not yet investigated its structure.

Studies on the frameshifted alkaline phosphatase proteins also add to our understanding of the assembly of alkaline phosphatase. The finding that strains carrying the R7 and R8 alleles are not processed but are highly active is a confirmation that the presence of an uncleaved signal sequence is not sufficient to inhibit alkaline phosphatase activity.

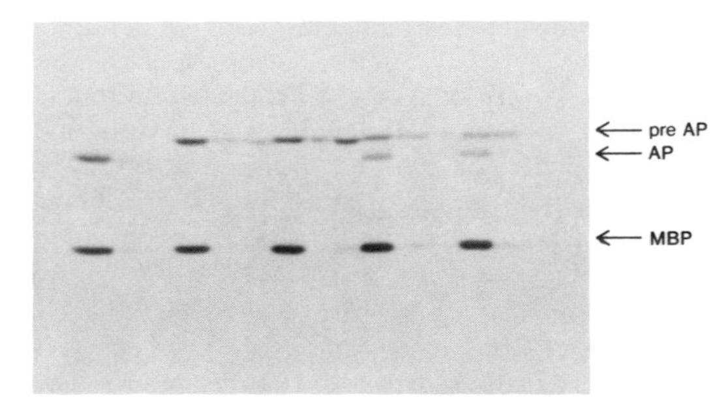

FIG. 5. Cells labeled as in Fig. 4 but pulsed for 5 min and chased for 15 min were spun and then suspended in 18% sucrose, 100 mM Tris hydrochloride (pH 8), 0.5 mM EDTA, and 40 μg of phenylmethylsulfonyl fluoride per ml. The mixture was set on ice for 5 min and centrifuged, after which the pellet was warmed to room temperature and suspended in cold water. $MgCl_2$ was added to 6 mM after 1 min, and the suspensions were centrifuged. The supernatant is the periplasm. The pellets were resuspended in the sucrose solution as above, lysed with lysozyme treatment and three cycles of freezing and thawing, and fractionated into membrane (pellet) and cytoplasm (supernatant). The samples were then trichloroacetic acid precipitated and analyzed by immune precipitation as in Fig. 1, using antibodies to both alkaline phosphatase (AP) and maltose-binding protein (MBP). In each set of three lanes the first is periplasm, the second is cytoplasm, and the third is the membrane fraction. Lanes: a–c, *phoA*$^+$; d–f, *phoA*R7; g–i, *phoA*R8; j–l, *phoA*R12; m–o, *phoA*R15.

Studies on the frameshift revertants also raise questions about specificity in the processing of signal sequence. The requirements for signal peptidase recognition deduced from a consensus of cleaved leader peptides are Ala, Ser, Gly, Cys, Thr, or Gln in position −1 and no aromatic, charged, or large polar groups at −3 (15, 25). The four mutant signal sequences considered here have an aspartic acid at position −3, thus violating the consensus rule. The fact that two of them are partially cleaved while two are not may indicate that when the fit of the substrate to the cleavage site of the peptidase is poor other factors can determine the rate at which the leader is cleaved. That the R12 and R15 alkaline phosphatases are, in fact, correctly processed is indicated by the preservation of the isozyme pattern of the wild type under several different growth conditions. Since this pattern results from cleavage of the N-terminal arginine from mature alkaline phosphatase (24) and is modified by inclusion of arginine in the medium (16), our finding of similar isozyme patterns with and without arginine suggests processing at the correct site.

This work was supported by an American Cancer Society Research Grant to J.B.

We thank Ann McIntosh for excellent assistance in the preparation of this manuscript and Elaine Garrett for lettering.

LITERATURE CITED

1. **Baker, T. A., A. D. Grossman, and C. A. Gross.** 1984. A gene regulating the heat shock response in *Escherichia coli* also affects proteolysis. Proc. Natl. Acad. Sci. USA **81:**6779–6783.
2. **Emr, S. D., S. Hanley-Way, and T. J. Silhavy.** 1981. Suppressor mutations that restore export of a protein with a defective signal sequence. Cell **23:**79–88.
3. **Hoffman, C., and A. Wright.** 1985. Fusions of secreted proteins to alkaline phosphatase: an approach for studying protein secretion. Proc. Natl. Acad. Sci. USA **82:**5107–5111.
4. **Inouye, H., W. Barnes, and J. Beckwith.** 1982. The signal sequence of alkaline phosphatase of *Escherichia coli*. J. Bacteriol. **149:**434–439.
5. **Inouye, H., and J. Beckwith.** 1977. Synthesis and processing of alkaline phosphatase precursor *in vitro*. Proc. Natl. Acad. Sci. USA **74:**1440–1444.
6. **Ito, K., P. Bassford, and J. Beckwith.** 1981. Protein localization in *E. coli*: is there a common step in the secretion of periplasmic and outer membrane proteins? Cell **24:**707–717.
7. **Kikuchi, Y., K. Yoda, M. Yamasaki, and G. Tamura.** 1981. The nucleotide sequence of the promoter and the amino-terminal region of alkaline phosphatase structural gene (*phoA*) of *Escherichia coli*. Nucleic Acids Res. **9:** 5671–5678.

7a. **Lee, C., and J. Beckwith.** 1986. Cotranslational and post-translational protein translocation in prokaryotic systems. Annu. Rev. Cell Biol. **2:**315–336.

8. **Levinthal, C., E. R. Signer, and K. Fetherolf.** 1962. Reactivation and hybridization of reduced alkaline phosphatase. Proc. Natl. Acad. Sci. USA **48:**1230–1237.
9. **Malamy, M., and B. Horecker.** 1961. The localization of alkaline phosphatase in *E. coli* K12. Biochem. Biophys. Res. Commun. **5:**104–108.
10. **Manoil, C., and J. Beckwith.** 1985. TnphoA: a transposon probe for protein export signals. Proc. Natl. Acad. Sci. USA **82:**8129–8133.
11. **Maurizi, M. R., P. Trisler, and S. Gottesman.** 1985. Insertional mutagenesis of the *lon* gene in *Escherichia coli*: *lon* is dispensable. J. Bacteriol. **164:**1124–1135.
12. **McCraeken, S., and B. Meighen.** 1980. Function and structural properties of immobilized subunits of *Escherichia coli* alkaline phosphatase. J. Biol. Chem. **255:**2396–2404.
13. **Michaelis, S., J. Hunt, and J. Beckwith.** 1986. Effects of signal sequence mutations on the kinetics of alkaline phosphatase export to the periplasm in *Escherichia coli*. J. Bacteriol. **167:**160–167.
14. **Michaelis, S., H. Inouye, D. Oliver, and J. Beckwith.** 1983. Mutations that alter the signal sequence of alkaline phosphatase of *Escherichia coli*. J. Bacteriol. **154:**366–374.
15. **Perlman, D., and H. O. Halvorson.** 1983. A putative signal peptidase recognition site and sequence in eukaryotic and prokaryotic signal peptides. J. Mol. Biol. **167:**391–409.
16. **Piggott, P. J., M. D. Sklar, and L. Gorini.** 1972. Ribosomal alterations controlling alkaline phosphatase isozymes in *Escherichia coli*. J. Bacteriol. **110:**291–299.
17. **Plocke, D. J., C. Levinthal, and B. L. Vallee.** 1962. Alkaline phosphatase of *Escherichia coli*: a zinc metalloenzyme. Biochemistry **1:**373–378.
18. **Rothman, F., and R. Byrne.** 1963. Fingerprint analysis of alkaline phosphatase of *Escherichia coli* K12. J. Mol. Biol. **6:**330–340.
19. **Ryan, J. P., and P. J. Bassford, Jr.** 1985. Post-translational export of maltose-binding protein in *Escherichia coli* strains harboring *malE* signal sequence mutations and either *prl*$^+$ or *prl* suppressor alleles. J. Biol. Chem. **260:** 14832–14837.
20. **Sarthy, A., A. Fowler, I. Zabin, and J. Beckwith.** 1979. Use of gene fusions to determine a partial signal sequence of alkaline phosphatase. J. Bacteriol. **139:**932–939.
21. **Sarthy, A., S. Michaelis, and J. Beckwith.** 1981. Deletion map of the *Escherichia coli* structural gene for alkaline phosphatase. J. Bacteriol. **145:**288–292.
22. **Schlesinger, M. J.** 1965. The reversible dissociation of the alkaline phosphatase of *Escherichia coli*. II. Properties of the subunit. J. Biol. Chem. **240:**4293–4298.
23. **Schlesinger, M. J., and K. Barrett.** 1965. The reversible dissociation of the alkaline phosphatase of *Escherichia coli*. I. Formation and reactivation of subunits. J. Biol. Chem. **24:**4284–4292.
24. **Schlesinger, M., W. Bloch, and P. Kelley.** 1975. Differences in the structure, function and formation of two isozymes of *E. coli* alkaline phosphatase, p. 333–342. *In* C. L. Markert (ed.), Isozymes, vol. 1. Molecular structure. Academic Press, Inc., New York.
25. **von Heijne, G.** 1983. Patterns of amino acids near signal-sequence cleavage sites. Eur. J. Biochem. **133:**17–21.

Using Fusions to Alkaline Phosphatase to Study Protein Insertion into the Cytoplasmic Membrane

COLIN MANOIL,† DANA BOYD, AND JON BECKWITH

Department of Microbiology and Molecular Genetics, Harvard Medical School, Boston, Massachusetts 02115

Complex transmembrane proteins, those whose polypeptide chains cross the lipid bilayer more than once, are ubiquitous in procaryotic and eucaryotic membranes and are generally responsible for the transport of matter (such as nutrient molecules) into the cell. Despite the importance of complex membrane proteins in membrane function, little is known of the mechanisms by which such proteins acquire their functional structures.

We have studied the insertion of complex membrane proteins in *Escherichia coli*, using fusions to the normally periplasmic protein alkaline phosphatase. It appears that alkaline phosphatase must be exported from the cytoplasm to be active (4, 7, 9), a property making it complementary to β-galactosidase, which shows low enzymatic activity if the cell attempts to export it (14). To generate alkaline phosphatase fusions, we constructed a transposon called Tn*phoA* (6). This paper describes general properties of the Tn*phoA* fusion system and the use of Tn*phoA* to analyze the topology of cytoplasmic membrane proteins.

Tn*phoA* Function

Tn*phoA* was constructed by inserting a fragment of DNA encoding most of alkaline phosphatase near the left end of transposon Tn*5* in such a way that transposition function was not disturbed. The DNA fragment inserted (*'phoA*) encodes all of alkaline phosphatase except its signal sequence and an additional five amino acid residues of the mature protein (Fig. 1).

The function of Tn*phoA* in generating protein fusions is illustrated in Fig. 2. When Tn*phoA* transposes into a gene (gene *X*) in the appropriate orientation and translational reading frame, it generates a gene fusion which encodes a hybrid protein with carboxyl-terminal sequences of the target gene product replaced by alkaline phosphatase (*X-phoA*). At the junction of every fusion (between *X* and *phoA* in Fig. 2) are 17 amino acid residues encoded mainly by sequences from the very left end of Tn*5* (Fig. 1), sequences necessary for transposition (5, 13).

†Present address: Department of Genetics, University of Washington, Seattle, WA 98195.

β-Lactamase–Alkaline Phosphatase Fusions

To examine Tn*phoA* function, we isolated a set of β-lactamase–alkaline phosphatase fusions (*bla-phoA* fusions) by transposition of Tn*phoA* into the small, high-copy-number plasmids pBR-322 and pBR325 (7). These studies showed that Tn*phoA* [and a derivative of it, Tn*phoA*(op)] fused *phoA* at a number of different places in *bla* (12 positions were distinguished for 14 insertions). Tn*phoA* thus retains the low DNA sequence specificity of insertion of its parent transposon Tn*5* (1).

Different *bla-phoA* hybrids showed the same alkaline phosphatase activity (measured by the rate of *p*-nitrophenyl phosphate hydrolysis by permeabilized whole cells). For example, cells producing fusions containing 7, 11, 22, or 27 kilodaltons of the 29-kilodalton β-lactamase showed enzyme activities within 20% of one another. Thus, there is no direct effect of the amount of amino-terminal hybrid protein sequence present on the activity observed.

We also recovered *bla-phoA* fusion proteins from periplasmic fractions, showing that they are secreted through the cytoplasmic membrane. These fusion proteins were relatively stable (with half-lives of 30 to 60 min), although most eventually broke down to give a product about the size of alkaline phosphatase itself (47 kilodaltons). A pulse-chase experiment illustrating this stability behavior is shown in Fig. 3 (SS^+ lancs).

Tn*phoA* Functions as a Probe for Protein Export Signals

Studies with alkaline phosphatase signal sequence missense and deletion mutants had indicated that alkaline phosphatase must be exported to give active enzyme (4, 9). Since Tn*phoA* does not encode the alkaline phosphatase signal sequence, we anticipated that, for a fusion generated by Tn*phoA* insertion into a gene to give activity, the gene product would have to contribute sequences able to promote export of the hybrid. Two lines of evidence indicate that this is true (7).

(i) It was not possible to isolate enzymatically active fusions of alkaline phosphatase to the cytoplasmic protein β-galactosidase after inser-

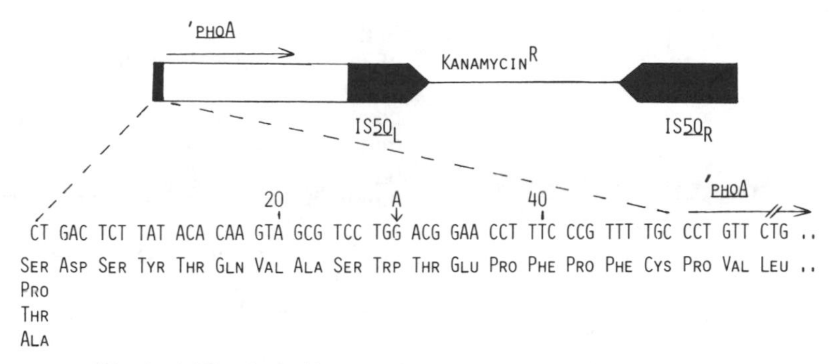

FIG. 1. Structure of Tn*phoA*. The shaded boxes represent sequences of the IS*50* elements of transposon Tn*5*. The open box represents a fragment containing most of the gene for alkaline phosphatase, a fragment derived from plasmid pCH39 (4).

tion of Tn*phoA* into a plasmid encoding β-galactosidase. However, when a collection of inserts into the β-galactosidase gene (*lacZ*) was screened for fusion protein production by using antibody to alkaline phosphatase, 4 of 20 were found to produce hybrids (Fig. 4). These *lacZ-phoA* fusion proteins fractionated to the cytoplasm and showed less than 2% of the level of alkaline phosphatase activity expected if they had the same specific activity and stability as alkaline phosphatase itself (unpublished data).

(ii) When one of the active *bla-phoA* fusions was deleted of nearly all of its signal sequence coding region (leaving that for only the two amino-terminal amino acids in its place), the hybrid protein no longer fractionated to the periplasm and showed less than 1% of its original

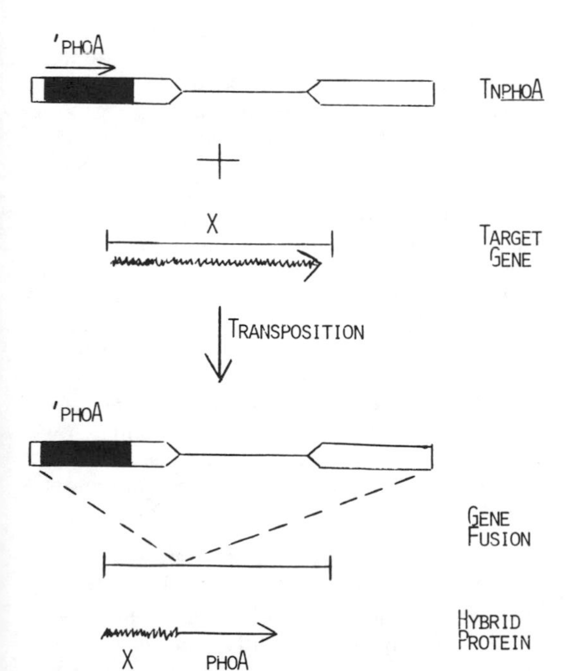

FIG. 2. Use of Tn*phoA* to generate hybrid proteins.

FIG. 3. Analysis of *bla-phoA* hybrid protein stability. Cells carrying plasmids encoding either a long *bla-phoA* hybrid protein with the signal sequence (SS^+) or the same hybrid deleted for most of the signal sequence (ΔSS) were treated for 1 min with [^{35}S]methionine before excess nonradioactive methionine was added. Cells were harvested immediately or after 10 or 60 min of further incubation at 37°C. Hybrid proteins were precipitated with antibody to alkaline phosphatase, the proteins were separated by sodium dodecyl sulfate-polyacrylamide gel electrophoresis, and the dried gels were used to expose X-ray films. Further experimental details are presented in reference 7.

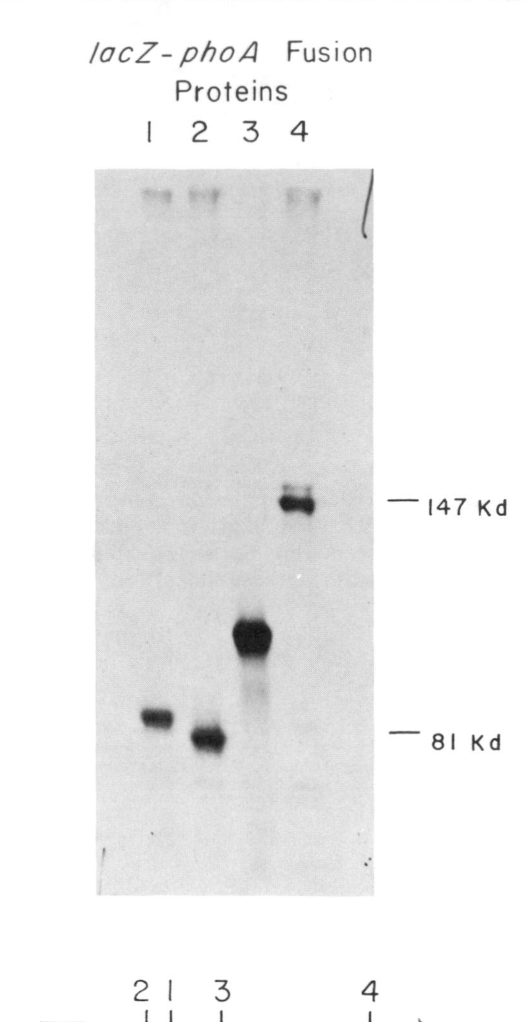

FIG. 4. β-Galactosidase–alkaline phosphatase hybrid proteins. Four *lacZ-phoA* hybrid proteins separated by sodium dodecyl sulfate-polyacrylamide electrophoresis after precipitation by antibody to alkaline phosphatase are shown. The figure also shows the approximate positions of Tn*phoA* insertions in *lacZ* for each of the hybrid proteins, as determined by restriction mapping.

nal alkaline phosphatase activity (7). In addition, the hybrid protein lacking its signal sequence was found to be much less stable than the corresponding hybrid synthesized with a signal sequence, with a half-life of less than 6 min (Fig. 3, ΔSS lanes). Since the alkaline phosphatase encoded by Tn*phoA* requires signals promoting export to be provided by the protein it is fused to for activity, Tn*phoA* functions as a probe for such protein export signals.

Why does alkaline phosphatase confined to the cytoplasm fail to give activity? The answer to this question is not yet known. The instability of most cytoplasmic hybrid proteins could account for their low activities; however, it is also possible that alkaline phosphatase cannot acquire an active conformation in the cytoplasm (e.g., because of an inability to form its disulfide bonds in the reducing environment of the cytoplasm [11]) and that this nonnative structure is both enzymatically inactive and a substrate for cytoplasmic proteases. We are currently attempting to distinguish these possibilities by stabilizing cytoplasmic alkaline phosphatase.

Fusions of Alkaline Phosphatase to Cytoplasmic Membrane Proteins Can Give Enzymatic Activity

In the process of isolating *bla-phoA* fusions using plasmids pBR322 and pBR325, we made the unexpected observation that Tn*phoA* insertion into the plasmid gene encoding the cytoplasmic membrane protein conferring tetracycline resistance could generate hybrids giving alkaline phosphatase activity (7). We have since found it possible to isolate active alkaline phosphatase fusions to every integral cytoplasmic membrane protein that we have tested, which includes seven additional proteins. Most of these proteins are thought to lack cleavable signal sequences (8), and we assume that internal signal sequences in them function to export alkaline phosphatase in these hybrids.

Analysis of Membrane Protein Topology Using Tn*phoA*

We have sought to determine whether Tn*phoA* can be used to analyze the topology of cytoplasmic membrane proteins. The idea underlying the approach we are testing is illustrated in Fig. 5 for a hypothetical complex membrane protein which spans the membrane six times. Since Tn*phoA* has a low specificity of insertion, it can fuse alkaline phosphatase at many positions in such a protein's structure. However, since alkaline phosphatase requires export for activity, it might be expected that fusions to periplasmic domains of the protein (e.g., position 1) would generate a periplasmic alkaline phosphatase moiety and therefore yield high enzyme activity, whereas fusions to cytoplasmic domains (e.g., position 2) would generate cytoplasmic, low-activity hybrids. Thus, the activities of fusions at different positions would help reveal the normal topology of the membrane protein.

Fusions of Alkaline Phosphatase to Methyl-Accepting Chemotaxis Proteins

To test whether the positions of alkaline phosphatase fusions can identify periplasmic domains of membrane proteins, we examined fusions to a group of proteins of relatively well-

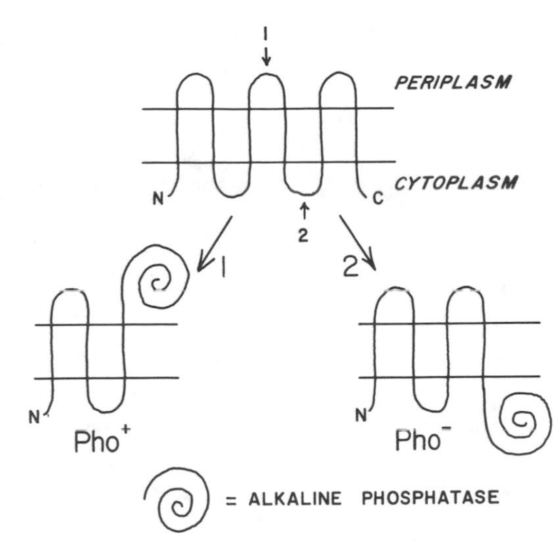

FIG. 5. Strategy for using alkaline phosphatase fusions to analyze cytoplasmic membrane protein topology.

defined transmembrane topology, the methyl-accepting chemotaxis proteins (MCPs). The four MCPs (encoded by *tsr, tar, tap*, and *trg*) are homologous in sequence and appear to have the transmembrane disposition shown in Fig. 6a, with two transmembrane helices defining an amino-terminal periplasmic domain (comprising about 30% of the protein) and a carboxyl-terminal cytoplasmic domain (comprising about 60% of the protein) (2, 6, 10, 12).

Fusions of alkaline phosphatase to three different MCPs were isolated by Tn*phoA* insertion into plasmids carrying their genes. Fusion plasmids yielding high and low alkaline phosphatase activities could be distinguished on media containing an alkaline phosphatase indicator (5-bromo-4-chloro-3-indolyl phosphate), and plasmids of each activity class were analyzed by restriction mapping and DNA sequencing. The sites of Tn*phoA* insertion giving high alkaline phosphatase activity are shown in Fig. 6b. All of nine different insertions giving high enzyme activity are positioned in or near the MCP periplasmic domain. This result is consistent with the scheme shown in Fig. 5.

DNA sequence analysis showed that low-activity *tsr-phoA* fusions could result from Tn*phoA* insertions which placed the *tsr* and *phoA* sequences in the same or different translational reading frames relative to each other (C. Manoil and J. Beckwith, manuscript in preparation). The two in-frame insertions giving low activity which we analyzed both fused alkaline phosphatase to the cytoplasmic domain of Tsr protein. Furthermore, it was possible to construct in-frame fusions of alkaline phosphatase to the cytoplasmic domain of Tsr protein by using an indirect method (Manoil and Beckwith, in preparation). First, we constructed a plasmid containing a derivative of *tsr* in which sequences encoding much of its periplasmic domain and all of its second transmembrane segment were deleted. The protein encoded by this derivative was expected to have normally cytoplasmic sequences of Tsr protein translocated to the periplasm as a result of the loss of the transmembrane sequence. High-activity fusions were isolated by Tn*phoA* insertion into the plasmid, and five insertions fused alkaline phosphatase at positions in the formerly cytoplasmic domain of Tsr protein. In a final step, the missing *tsr* sequences were added back to the five fusions. Cells carrying the *tsr-phoA* fusions generated by this method showed low alkaline phosphatase activity, just like the cytoplasmic domain fusions isolated directly by Tn*phoA* insertion into wild-type *tsr*. The positions of the seven in-frame, low-activity *tsr-phoA* fusions we have isolated are shown in Fig. 6c. Thus, as required by the approach diagrammed in Fig. 5, fusions of alkaline phosphatase to the cytoplasmic domain of Tsr protein showed low activity.

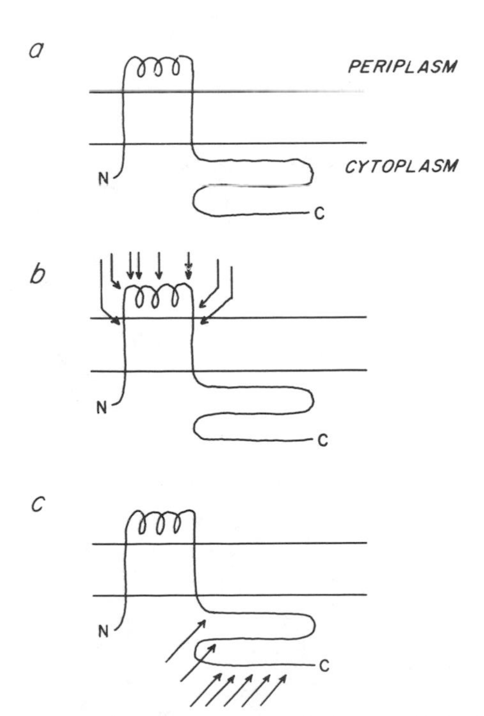

FIG. 6. Alkaline phosphatase fusions to MCPs. (a) Membrane topology of MCPs. (b) Positions of high-activity alkaline phosphatase fusions. Four *tsr*, four *tar*, and one *tap* fusion were analyzed. (c) Positions of in-frame, low-activity *tsr-phoA* fusions.

Conclusion

Our studies using the Tsr protein as a model indicate that an analysis of alkaline phosphatase fusions can help reveal the topology of a cytoplasmic membrane protein. However, Tsr protein has a relatively simple membrane topology, and we wish to determine whether this type of analysis can be used for proteins with more complex membrane configurations. To test this, we are currently engaged in a detailed analysis of alkaline phosphatase fusions to MalF protein, a cytoplasmic membrane protein involved in maltose uptake thought to have a complex topology in which the polypeptide chain spans the membrane eight times (3).

We are grateful to C. Lee for creative contributions.

C.M. was a Fellow of the Arthritis Foundation. This work was supported by a research grant from the American Cancer Society.

LITERATURE CITED

1. **Berg, D. E., M. A. Schmandt, and J. B. Lowe.** 1983. Specificity of transposon Tn*5* insertion. Genetics **105:**813–828.
2. **Bollinger, J., C. Park, H. Harayama, and G. L. Hazelbauer.** 1984. Structure of the Trg protein: homologies with and differences from other sensory transducers of *Escherichia coli*. Proc. Natl. Acad. Sci. USA **81:**3287–3291.
3. **Froshauer, S., and J. Beckwith.** 1984. The nucleotide sequence of the gene for *malF* protein, an inner membrane component of the maltose transport system of *Escherichia coli*. J. Biol. Chem. **259:**10896–10903.
4. **Hoffman, C. S., and A. Wright.** 1985. Fusions of secreted proteins to alkaline phosphatase: an approach for studying protein secretion. Proc. Natl. Acad. Sci. USA **82:**5107–5111.
5. **Johnson, R. C., and W. S. Reznikoff.** 1983. DNA sequences at the ends of transposon Tn*5* required for transposition. Nature (London) **304:**280–282.
6. **Krikos, A., M. P. Conley, A. Boyd, H. Berg, and M. I. Simon.** 1985. Chimeric chemosensory transducers of *Escherichia coli*. Proc. Natl. Acad. Sci. USA **82:**1326–1330.
7. **Manoil, C., and J. Beckwith.** 1985. Tn*phoA*: a transposon probe for protein export signals. Proc. Natl. Acad. Sci. USA **82:**8129–8133.
8. **Michaelis, S., and J. Beckwith.** 1982. Mechanism of incorporation of cell envelope proteins in *Escherichia coli*. Annu. Rev. Microbiol. **36:**435–464.
9. **Michaelis, S., H. Inouye, D. Oliver, and J. Beckwith.** 1983. Mutations that alter the signal sequence of alkaline phosphatase in *Escherichia coli* K-12. J. Bacteriol. **154:**366–374.
10. **Mowbray, S. L., D. Foster, and D. E. Koshland.** 1985. Proteolytic fragments identified with domains of the aspartate chemoreceptor. J. Biol. Chem. **260:**11711–11718.
11. **Pollit, S., and H. Zalkin.** 1983. Role of primary structure and disulfide bond formation in beta-lactamase secretion. J. Bacteriol. **153:**27–32.
12. **Russo, A. F., and D. E. Koshland.** 1983. Separation of signal transduction and adaptation functions of the aspartate receptor in bacterial sensing. Science **220:**1016–1020.
13. **Sasakawa, C., G. F. Carle, and D. E. Berg.** 1983. Sequences essential for transposition at the termini of IS*50*. Proc. Natl. Acad. Sci. USA **80:**7293–7297.
14. **Silhavy, T. J., and J. Beckwith.** 1985. Uses of *lac* fusions for the study of biological problems. Microbiol. Rev. **49:**398–418.

Secretion to Periplasm of Foreign Proteins in *Escherichia coli* by Aid of the *phoA*-Derived Secretion Vector Psi

KOJI YODA,[1] KOH-ICHI TACHIBANA,[1] SATORI WATANABE,[1] KUNIO YAMANE,[2] MAKARI YAMASAKI,[1] AND GAKUZO TAMURA[1]†

Department of Agricultural Chemistry, The University of Tokyo, Bunkyo-ku, Tokyo 113,[1] *and Institute of Biological Sciences, University of Tsukuba, Sakura, Ibaraki 305,*[2] *Japan*

Alkaline phosphatase of *Escherichia coli* is one of the well-known periplasmic (secretory) enzymes. The enzyme is synthesized as a precursor monomer form with a signal peptide at the amino-terminal end. The signal peptide of the precursor is cleaved during secretion to give the mature monomer. In the periplasmic space, 2 mol of monomer and 2 mol of Zn^{2+} coordinate to form an active alkaline phosphatase molecule (15). The synthesis of alkaline phosphatase is positively and negatively regulated by *phoB* and *phoR* genes, respectively. Under phosphate-limited conditions, alkaline phosphatase represents 6% of newly synthesized protein (16). The *phoA* gene is, therefore, equipped with a quite efficient and regulatable promoter.

We aimed to construct a vector containing the *phoA* promoter and signal peptide coding region that would be useful for secretion of foreign proteins into the periplasm of *E. coli*. For that purpose, we first obtained *phoA* transducing λ phage and subcloned the *phoA* gene on plasmid pBR322 (19). The nucleotide sequence of a part of the *phoA* gene including promoter and signal peptide coding regions has been determined (5). The signal peptide of alkaline phosphatase is deduced to be composed of 21 amino acids from the determined nucleotide sequence (5, 6). We used the *phoA*-containing plasmid to construct a *phoA*-derived secretion vector. During this work, it became known that the efficient expression of foreign genes in *E. coli* is apt to result in formation of insoluble inclusion bodies. Therefore, the efficient secretion of useful foreign gene products from *E. coli* has now become a more urgent and important problem than when we started the work. In this article, we report (i) the construction of a *phoA*-derived secretion vector, Psi, and improved versions of this vector, (ii) the efficient secretory production and secretion of α-amylase of *Bacillus subtilis* and human epidermal growth factor (EGF), and (iii) the isolation of *E. coli* mutants with increased secretory ability.

†Present address: Department of Applied Biological Science, Science University of Tokyo, Noda, Chiba 218, Japan.

Results

Construction of *phoA*-derived secretion vector Psi. We have already determined the nucleotide sequence of a part of the *phoA* gene including the 5′-flanking region and signal peptide coding region (5). As outlined in Fig. 1, an *Hpa*II site is present at the end of the signal peptide coding region. Therefore, all of the elements necessary for the construction of a *phoA*-derived secretion vector are contained on a 143-base-pair (bp) *Pvu*II-*Hpa*II fragment as shown in Fig. 1. We subcloned the fragment on pBR328. By introducing other cloning sites just after the signal peptide coding region and providing three different reading frames in relation to cloning sites, we constructed the *phoA*-derived secretion vector Psi (promoter-signal peptide). As shown in Fig. 2, the Psi vector (−1, 0, +1 coding frame, 3.14 kilobases) has three cloning sites, *Sma*I, *Eco*RI, and *Ava*I. Structural genes inserted at these sites were expected to be efficiently expressed under conditions of phosphate limitation.

Trials for secretion of several cytoplasmic proteins of *E. coli*. We first attempted to obtain secretion of typical cytoplasmic proteins, β-galactosidase, elongation factor Tu, and chloramphenicol acetyltransferase, using the Psi secretion vector. For that purpose we constructed plasmids containing in-frame fusions Psi-*lacZ*, Psi-*tufB*, and Psi-*cat*, respectively. Under phosphate-limited conditions, these cytoplasmic proteins were efficiently synthesized at levels ranging from 20 to 50% of total cellular protein (Table 1). These results indicate that the *phoA* promoter is very efficient. None of the proteins, however, was exported to the *E. coli* periplasm (Table 1). Moreover, all of them were present as inclusion bodies. We purified the inclusion bodies in all three cases by sucrose density gradient centrifugation and determined the sequences of the 25 amino-terminal amino acids for the protein in each inclusion body. We confirmed that the intact signal peptide was still attached to the cytoplasmic proteins (data not shown). This is the first direct determination of the amino acid sequence of the signal peptide of alkaline phosphatase.

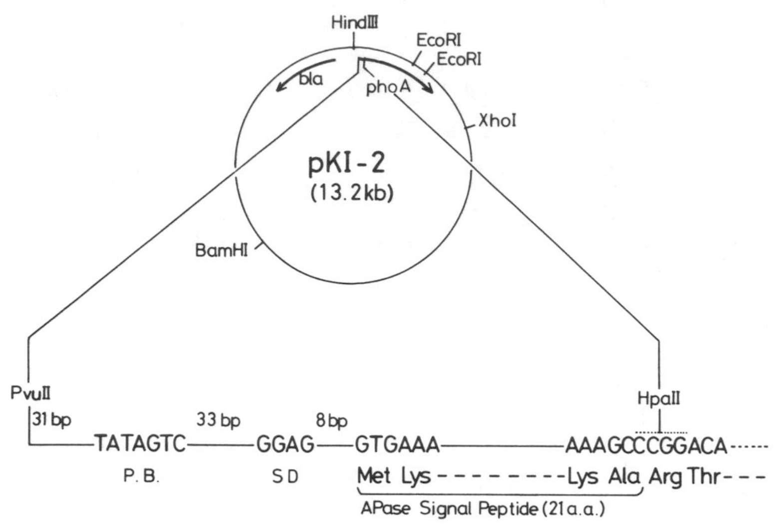

FIG. 1. Structure of $phoA^+$ plasmid pKI-2 with special reference to the promoter and signal peptide coding regions of *phoA*. P.B., Pribnow box; SD, Shine-Dalgarno sequence.

Improved Psi vector. Cloning of DNA restriction fragments in an appropriate Psi vector should result in in-frame fusion of the *phoA* signal sequence and the specific gene to be expressed. However, appropriate restriction sites are not always located near the open reading frame. Use of available restriction sites for cloning frequently results in unnecessary codons being located between the *phoA* signal peptide and the amino terminus of the mature protein.

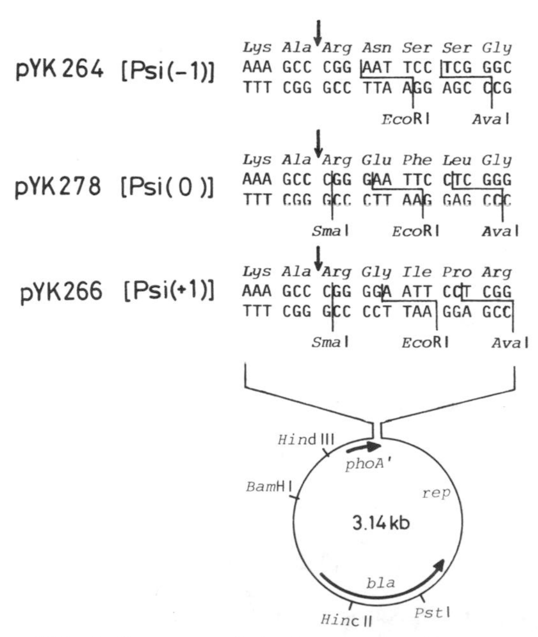

FIG. 2. Structure of the Psi vector with special reference to cloning sites in relation to coding frame.

We have found that oligonucleotide-directed site-specific deletion is a convenient method for making a direct connection (17). We introduced a multiple cloning site downstream of the *phoA* signal sequence coding region to facilitate DNA insertion at this site. A 514-bp *Rsa*I fragment of f1 phage replicative-form DNA which contains the intergenic region of this single-stranded phage was also introduced into a Psi vector to form pYK331 and pYK332. These plasmids differ in the orientation of inserted 514-bp *Rsa*I fragment. Single-stranded circular-form DNAs of both plasmids can be produced after infecting the cell with single-stranded phage f1, fd, or M13 (20). The availability of single-stranded circular DNA facilitates the oligonucleotide-directed site-specific deletion to remove unnecessary codons between the signal sequence and mature amino terminus.

Secretion of α-amylase of *B. subtilis* by the improved Psi vector. α-Amylase of *B. subtilis* is a well-characterized extracellular enzyme. The structural gene of the enzyme, *amyE*, has already been cloned on pUB110 (18). A 4.2-kilobase *Bgl*II-*Xba*I fragment which contains the *amyE* gene was cloned at the *Bam*HI site of the multiple cloning site of pYK331. Deletion of a 1-kilobase *Hin*dIII-*Hpa*II fragment produced a plasmid with the *phoA* signal peptide coding region fused through a 17-bp intervening sequence to the pro-α-amylase structural gene. The intervening 17-bp sequence was deleted by site-directed deletion, using a chemically synthesized 21mer oligonucleotide as outlined in Fig. 3. Under phosphate-limited conditions, α-amylase synthesis was about 5% of that of total cellular protein. Furthermore, most of the α-

TABLE 1. Expression of several genes by the Psi vector and secretion of some gene products

Gene	Secretion	Location of gene product	Signal peptide processing	Inclusion body formation	% Cellular protein	% Secretion
lacZ	−	Cytoplasm	−	+	10	
tufB[a]	−	Cytoplasm	−	+	50	
cat	−	Cytoplasm	−	+	50	
bla	+	Periplasm	+	(+)	10	>95
amyE	+	Periplasm	+	−	6	>95
EGF gene[b]	+	Periplasm	+	−	(2 mg/liter)	>95

[a] The *tufB* gene was subcloned from pTUB1 (10).
[b] T. Oka et al. (14).

amylase was detected in the periplasmic space of *E. coli*, as judged by α-amylase activity (Table 1) and immunoprecipitation (Fig. 4). We purified the periplasmic α-amylase and determined the sequence of its first five amino-terminal amino acids. The sequence Glu-Thr-Ala-Asn-Lys exactly coincides with the amino-terminal sequence of pro-α-amylase. This fact clearly shows that the *E. coli* alkaline phosphatase signal peptide was cleaved at the correct position.

Secretion of other secretory proteins by the improved Psi vector. We used the approach described for the *amyE* gene of *B. subtilis* to insert the *bla* gene into the multiple cloning site of the improved Psi vector. Through site-directed deletion we constructed a Psi-*bla* plasmid in which the *phoA* signal peptide coding region was fused directly to the mature β-lactamase structural gene. β-Lactamase was synthesized, under phosphate-limited conditions, to a level of 10% of total cellular protein and secreted to the periplasm (Table 1; H. Kadokura et al., manuscript in preparation). Miyake et al. have succeeded in obtaining secretion of human interferon-α in *E. coli* using the Psi vector (11). Oka et al. chemically synthesized the gene for human EGF and inserted it into the Psi vector. Under phosphate-limited conditions, human EGF was synthesized at a level of 2 mg/liter of culture and secreted in periplasm (14). More recently,

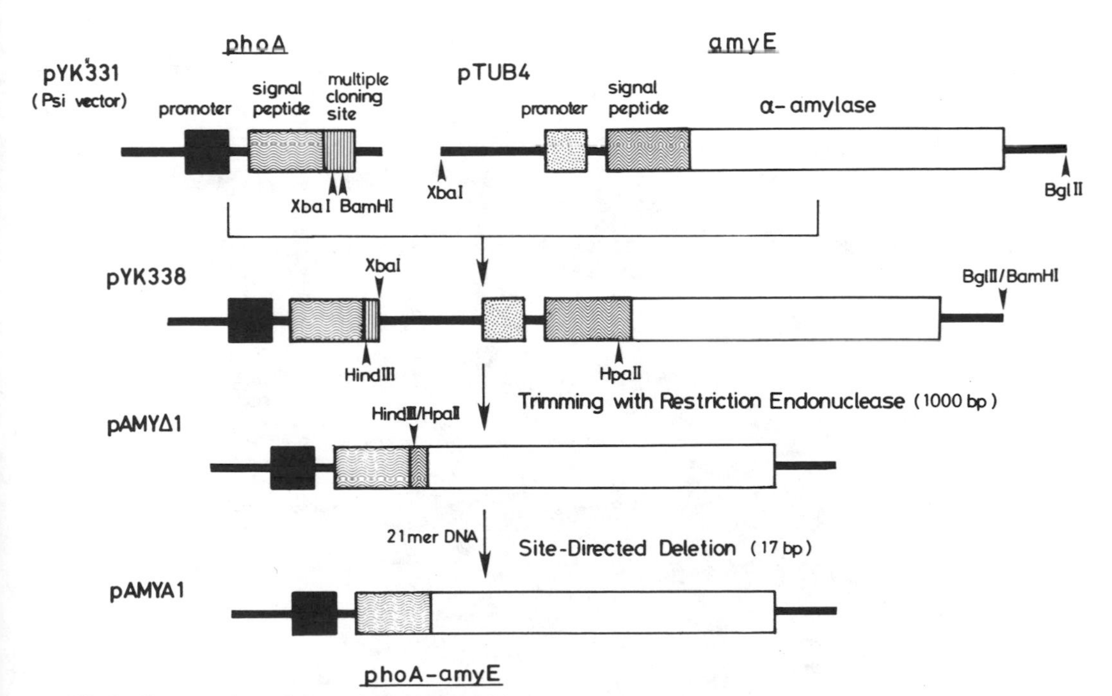

FIG. 3. Construction of the Psi-*amyE* plasmid pAMYA1. Plasmid DNAs were prepared by the method of Birnboim and Doly (1). Transformation and plasmid construction were done as described (9). Oligonucleotides were prepared by a manual solid-phase phosphotriester method (12). The site-specific deletion was performed as described (17) except that the templates were prepared from phage particles.

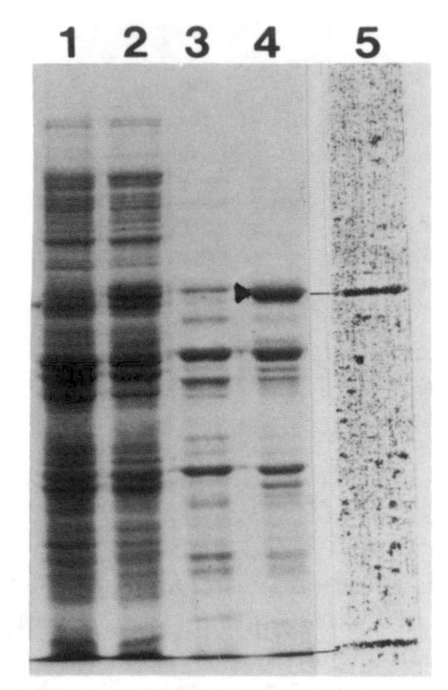

FIG. 4. Electrophoretic patterns of α-amylase produced by *E. coli* carrying pAMYA1 (lanes 1–4) and immunoprecipitates (lane 5). The media were as described (2). Periplasmic proteins were obtained by the osmotic shock procedure (13). Immunoprecipitates were prepared as described (3). Proteins were separated by SDS-polyacrylamide gel electrophoresis (7). Lane 1, Plus phosphate (P_i), whole cellular proteins; lane 2, minus P_i, whole cellular proteins; lane 3, plus P_i, periplasmic proteins; lanes 4 and 5, minus P_i, periplasmic proteins.

Tai et al. found that most of the EGF was present in the culture medium and the yield was greatly improved to the level of 40 mg/liter (K. Tai, I. Nishimoto, K. Moriyoshi, T. Miyake, S. Sakamoto, and T. Fuwa, Abstr. Annu. Meet. Agric. Chem. Soc. Japan 1986, p. 489).

***E. coli* mutants with increased secretory ability.** During a study of the expression of *phoA* on the multicopy plasmid pBR322 under phosphate-limited conditions, we observed that the level of alkaline phosphatase activity in the periplasm was much lower than expected, roughly twice that of the wild-type *E. coli*, which has a single copy of the *phoA* gene. About 80% of alkaline phosphatase failed to be secreted and formed insoluble inclusion bodies in the cytoplasm. These cells gradually lost colony-forming ability under phosphate-limited conditions. Addition of 1% sodium dodecyl sulfate (SDS) to the induction medium caused *E. coli* carrying the $phoA^+$ plasmid to rapidly lose viability (Y. Kikuchi et al., unpublished data).

The low yield of alkaline phosphatase in the periplasm and the loss of viability under phosphate-limited conditions are postulated to be due to saturation of the cellular secretory apparatus by the high level of *phoA* expression. We therefore aimed to isolate mutants of increased secretory ability among cells resistant to SDS. *E. coli* SW1036, SW1033, SW1038, and SW1043 thus obtained were resistant to 1% SDS and gave a 10-fold hyperproduction of alkaline phosphatase under phosphate-limited conditions. The SDS-polyacrylamide gel shown in Fig. 5 indicates the high efficiency of secretion of alkaline phospha-

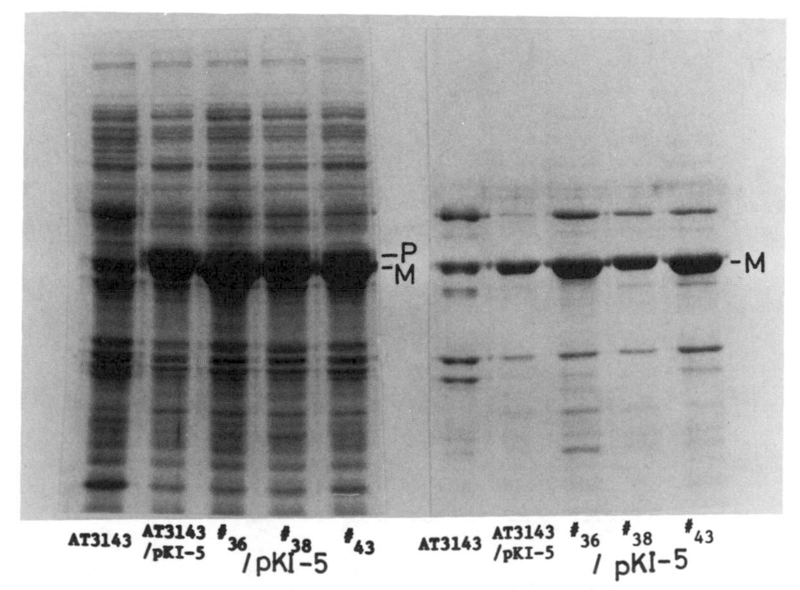

FIG. 5. Electrophoretic patterns of whole cellular proteins (left half) and of periplasmic proteins (right half) of mutants SW1036 (#36), SW1038 (#38), and SW1043 (#43). P, Precursor form of alkaline phosphatase monomer; M, mature form of alkaline phosphatase monomer. *E. coli* AT3143 has one copy of the *phoA* gene on the chromosome. Plasmid pKI-5 is a derivative of $phoA^+$ plasmid pKI-2 (19) (see Fig. 1).

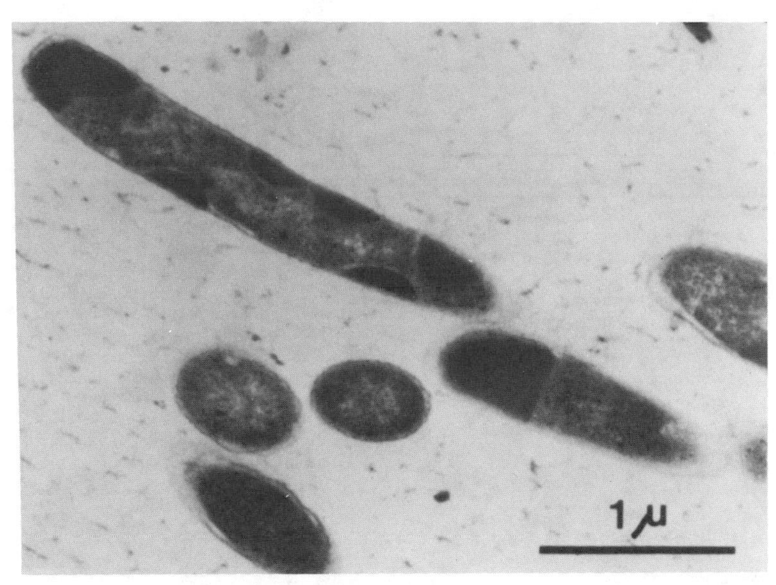

FIG. 6. Ultrathin section of *E. coli* SW1033 carrying *phoA'-'bla* fused gene plasmid pKY230 grown under phosphate-limited conditions.

tase in these strains as judged by an increased amount of mature alkaline phosphatase in the periplasm. An electron microscopic study further confirmed the enhanced secretion of alkaline phosphatase or *phoA'-'bla* fusion product by these mutants. Figure 6 shows that the *phoA'-'bla* fusion product accumulated as an electron-dense material in the periplasm of SW1033, which reached a level of about 30% of the total cellular protein. This will be the first visualization of the periplasm of *E. coli* without plasmolysis.

The secretion of alkaline phosphatase or the *phoA'-'bla* fusion product was enhanced in the mutants described above; however, the production of α-amylase by the multicopy *phoA-amyE* plasmid was not increased. This result may suggest that the level of expression of *amyE* was below that required to saturate the secretory apparatus. Mapping of the mutational sites of these mutants has not been completed.

Discussion

No one has succeeded in obtaining secretion of cytoplasmic proteins by adding a signal peptide to them with the exception of Lingappa et al., who reported (8) that addition of the *E. coli* β-lactamase signal peptide to α-globulin resulted in secretion of α-globin to the cisternae of the endoplasmic reticulum. A substantial fraction of elongation factor Tu is known to be released from normal *E. coli* strains by osmotic shock (4). This fact suggests that elongation factor Tu has a boundary nature between cytoplasmic and secretory proteins. Elongation factor Tu, however, was not secreted by the *phoA*-derived secretion vector Psi. Our present results suggest that cytoplasmic proteins are never exported across the cytoplasmic membrane by simple addition of a signal peptide.

On the other hand, secretory proteins such as α-amylase, β-lactamase, and human EGF were shown to be efficiently secreted when their own signal peptides were replaced by that of alkaline phosphatase. Moreover, in all cases the signal peptide was processed at the correct position. The reason these secretory proteins were expressed at relatively low levels by the Psi vectors compared with the amounts of cytoplasmic proteins produced by Psi vectors is still unknown. The relative instability of mRNA would be the most probable cause of the low yield of secretory proteins. If we can raise the yield of secretory proteins by any device, a higher yield of these proteins in the *E. coli* periplasm can be anticipated through the use of the bacterial mutants described here which show increased secretory ability.

We thank Y. Kikuchi, H. Kadokura, and Y. Katayama for their cooperation. We also express our gratitude to Y. Kajiro of the Institute of Medical Research, The University of Tokyo, and Y. Nagata of Nichirei Co. Ltd. for the generous gifts of *tufB* gene plasmid pTUB1 and rabbit anti-α-amylase antiserum, respectively.

This work was supported in part by a Grant-in-Aid of Special Scientific Research from the Ministry of Education, Science and Culture, Japan, and in part by a grant from the Institute of Physical and Chemical Research, Japan.

LITERATURE CITED

1. **Birnboim, H. C., and J. Doly.** 1979. A rapid alkaline extraction procedure for screening recombinant plasmid DNA. Nucleic Acids Res. **7:**1513–1523.
2. **Brickman, E., and J. Beckwith.** 1975. Analysis of the regulation of *Escherichia coli* alkaline phosphatase synthesis using deletion and ϕ80 transducing phages. J. Mol. Biol. **96:**307–316.
3. **Ito, K., P. J. Bassford, and J. Beckwith.** 1981. Protein localization in *E. coli*: is there a common step in the secretion of periplasmic and outer-membrane proteins? Cell **24:**707–717.
4. **Jacobson, G. R., and J. P. Rosenbusch.** 1976. Abundance and membrane association of elongation factor Tu in *E. coli*. Nature (London) **261:**23–26.
5. **Kikuchi, Y., K. Yoda, M. Yamasaki, and G. Tamura.** 1981. The nucleotide sequence of the promoter and the amino-terminal region of alkaline phosphatase structural gene (*phoA*) of *Escherichia coli*. Nucleic Acids Res. **9:** 5671–5678.
6. **Kikuchi, Y., K. Yoda, M. Yamasaki, and G. Tamura.** 1981. Sequence of the signal peptide of *E. coli* alkaline phosphatase. Agric. Biol. chem. **45:**2401–2402.
7. **Laemmli, U. K.** 1970. Cleavage of structural proteins during the assembly of the head of bacteriophage T4. Nature (London) **227:**680–685.
8. **Lingappa, V. D., J. Chaidez, C. S. Yost, and J. Hedgpeth.** 1984. Determinants for protein localization: β-lactamase signal sequence directs globin across microsomal membranes. Proc. Natl. Acad. Sci. USA **81:**456–460.
9. **Maniatis, T. E., E. F. Fritsch, and J. Sambrook.** 1982. Molecular cloning: a laboratory manual. Cold Spring Harbor Laboratory, Cold Spring Harbor, N.Y.
10. **Miyajima, A., M. Shibuya, and Y. Kaziro.** 1979. Construction and characterization of the two hybrid ColE1 plasmids carrying *Escherichia coli tufB* gene. FEBS Lett. **102:** 207–210.
11. **Miyake, T., T. Oka, T. Nishizawa, F. Misoka, T. Fuwa, K. Yoda, M. Yamasaki, and G. Tamura.** 1985. Secretion of human interferon by using secretion vectors containing a promoter and signal sequence of alkaline phosphatase gene of *Escherichia coli*. J. Biochem. **97:**1429–1436.
12. **Miyoshi, K., T. Miyake, T. Hozumi, and K. Itakura.** 1980. Solid-phase synthesis of polynucleotides. 2. Synthesis of polythymidylic acids by the block coupling phosphotriester methods. Nucleic Acids Res. **8:**5473–5489.
13. **Neu, H. C., and L. A. Heppel.** 1965. The release of enzymes from *Escherichia coli* by osmotic shock and during the formation of spheroplast. J. Biol. Chem. **240:** 3685–3692.
14. **Oka, T., S. Sakamoto, K. Miyoshi, T. Fuwa, K. Yoda, M. Yamasaki, G. Tamura, and T. Miyake.** 1985. Synthesis and secretion of human epidermal growth factor by *Escherichia coli*. Proc. Natl. Acad. Sci. USA **82:**7212–7216.
15. **Reid, T. W., and I. B. Wilson.** 1971. *E. coli* alkaline phosphatase, p. 373–415. *In* P. D. Boyer (ed.), The enzymes, vol. 4. Academic Press, Inc., New York.
16. **Torriani, A.** 1960. Influence of inorganic phosphate in the formation of phosphatases by *Escherichia coli*. Biochim. Biophys. Acta **38:**460–469.
17. **Wallace, R. B., P. F. Johnson, S. Tanaka, M. Schold, K. Itakura, and J. Abelson.** 1980. Directed deletion of a yeast transfer RNA intervening sequence. Science **209:**1396–1400.
18. **Yamazaki, H., K. Ohmura, A. Nakayama, Y. Takeichi, K. Otozai, M. Yamasaki, G. Tamura, and K. Yamane.** 1983. α-Amylase genes (*amyR2* and *amyE*$^+$) from an α-amylase-hyperproducing *Bacillus subtilis* strain: molecular cloning and nucleotide sequences. J. Bacteriol. **156:**327–337.
19. **Yoda, K., Y. Kikuchi, M. Yamasaki, and G. Tamura.** 1980. Cloning of the structural gene of alkaline phosphatase of Escherichia coli. Agric. Biol. Chem. **44:**1213–1214.
20. **Zagursky, R. J., and M. L. Berman.** 1984. Cloning vectors that yield high levels of single-stranded DNA for rapid DNA sequencing. Gene **27:**183–191.

Use of *lacZ* Fusions to Reveal Intragenic Export Signals of *Escherichia coli* LamB Protein

THOMAS J. SILHAVY, JOAN A. STADER, BETH A. RASMUSSEN, AND SPENCER A. BENSON

Department of Molecular Biology, Princeton University, Princeton, New Jersey 08544

In the past 15 years, it has become clear that the information that directs the export of proteins from their site of synthesis in the cytoplasm and governs correct cellular localization is contained within the structural gene. Often this export information is found in small linear units and it is expressed at the level of the amino acid sequence. For example, the signal sequence, a 15- to 30-amino acid peptide, is commonly found at the amino terminus of proteins destined for export. Usually, this signal is present in the precursor form of the protein only and it is removed proteolytically during the localization process. Similarly, membrane anchor sequences are small stretches of hydrophobic amino acids that span the lipid bilayer in many integral membrane proteins. An important goal for the understanding of protein localization is the identification and characterization of these and other intragenic export signals. Knowledge of their location, nature, and function will provide insights into the overall mechanism(s) of this complex and universal cellular process.

A variety of different experimental approaches have been brought to bear on the identification and characterization of intragenic export signals. In eucaryotic cells, cell biology and biochemistry have been successfully employed. The route followed by proteins destined for a variety of subcellular locations has been mapped, and several common mechanistic groupings have been assigned (7). In the case of membrane-bound and secreted glycoproteins, the system responsible for recognition has been elaborated in some detail (9). Key in this system are the amino-terminal signal sequence and a large, cytoplasmic, ribonucleoprotein particle termed SRP (signal recognition particle). SRP binds the signal sequence as it emerges from the ribosome and serves as a pilot to lead the translation complex to the endoplasmic reticular membrane, the site of protein translocation across the lipid bilayer. The component which identifies the endoplasmic reticulum and binds the SRP-ribosome-mRNA complex is called docking protein or SRP receptor. In addition to their role in recognition, SRP and docking protein may act to couple the processes of translation and export, serving as a type of safety mechanism to ensure that synthesis occurs at the opportune moment. Despite this progress, however, important mechanistic details and the critical structural requirements of the signals remain obscure.

In procaryotes, especially *Escherichia coli*, and in the lower eucaryotes, genetics has proved to be a powerful tool for the analysis of protein localization (2). A variety of mutants that are altered in protein export have been obtained and characterized. Through analysis of these mutants, intragenic information has been identified and genes whose products appear to function in protein export have been discovered. In a general sense, at least, results of this work support a conservation of export mechanism throughout biology. More specifically, the availability of mutants has permitted a detailed probe of export signals and their molecular interactions with the participating cellular machinery. These studies reveal a complex and perplexing pattern, and they demonstrate graphically how much is yet to be learned.

Export Signals and Chimeric Genetics

Genetic analysis of protein export was stimulated greatly by the development and application of gene fusion technology. As conceived originally, this technique was straightforward in a logical sense. If genes that specify exported proteins contain signals that control this process, then it should be possible to graft these signals genetically onto genes that specify nonexported proteins so as to alter the cellular location of the resulting product. This hypothesis could be tested by fusing the gene for an exported protein to a gene for a cytoplasmic protein and determining the location of the hybrid protein product. This has now been done with a variety of different genes in a number of different experimental systems. In general, the logic holds. However, there are some exceptions, and certain proteins behave in a manner that is not currently understood (8).

Much of our work has focused on the gene *lamB*, which specifies an *E. coli* outer membrane protein (LamB) that functions in maltose transport and serves as the receptor for bacteriophage λ. As a reporter gene, we have used *lacZ*, which specifies the cytoplasmic enzyme β-galactosidase. This protein offers a number of advan-

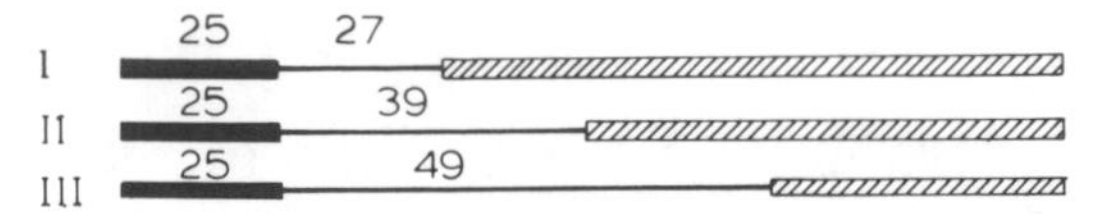

FIG. 1. Genetic structure of the relevant *lamB-lacZ* gene fusions (1). Localization of the hybrid protein product is as follows: I, cytoplasmic; II, inner membrane associated; III, outer membrane (30%). Full-size LamB contains a 25-amino acid signal sequence plus 421 amino acids in the mature protein. Symbols: line, mature LamB; solid box, signal sequence; hatched box, β-galactosidase.

tages. For example, its amino terminus can be replaced with any amino acid sequence without loss of function. In addition, its presence can be detected easily by biochemical or genetic methods.

To gain insights about the nature and location of intragenic export information within *lamB*, a number of different *lamB-lacZ* gene fusions have been constructed. Each of these fusions contains a different 5′ portion of *lamB* fused, in the correct reading frame, to a large and essentially constant 3′ portion of *lacZ*. We reasoned that by determining the cellular location of the hybrid protein produced by different fusion strains and comparing and contrasting these results such information could be obtained. For example, if a particular fusion specifies a hybrid protein that is localized to the cell envelope while another does not, it seems reasonable to conclude that the localization difference relates to the amount of *lamB* DNA present within the hybrid gene since this is the only relevant difference between the two.

Many different *lamB-lacZ* fusions have been constructed and characterized. Three of these provide particularly striking results with respect to the cellular location of the hybrid protein produced and the amount of *lamB* DNA present (1). The genetic structure of these fusions is shown in Fig. 1, and the results of cellular fractionation studies are summarized in the legend. Fusion I produces a hybrid protein with the signal sequence and the first 27 amino acids of mature LamB fused to β-galactosidase. Nevertheless, it remains in the cytoplasm. If the next 12 amino acids of LamB are added (fusion II), export occurs, but the hybrid does not appear to reach the outer membrane. For this to occur, an additional 10 amino acids of LamB are required (the signal sequence plus 49 amino acids of mature LamB, fusion III).

These results make several important points. First, they indicate that overall conformation of the LamB precursor is not required for export from the cytoplasm or for transport to the outer membrane. Second, since the largest of these three fusions contains only the first 17% of the *lamB* gene, it would appear that export information is confined in rather small regions and may, in fact, be arranged in a linear fashion. In addition, the possibility is raised that this information may be present in individual units each specific for a particular function. Viewed in this manner, the information that directs export from the cytoplasm would be contained in the signal sequence plus the first 39 amino acids of mature LamB, while information governing localization to the outer membrane would be contained, at least in part, within the region bounded by amino acids 39 to 49.

It must be stressed that, although the studies described above are informative, they cannot be offered as proof. Hybrid genes and proteins are man-made and decidedly unnatural, and the analysis of fusion strains can be fraught with a multitude of artifacts, such as interference with the cell physiology and insolubility or improper localization of the hybrid protein. Nevertheless, fusions provide a means to simplify an otherwise complex situation, and they suggest experimentally testable hypotheses. With respect to protein localization, intragenic regions identified with fusions must be altered systematically in an otherwise wild-type gene. Only here can the physiological effects of a particular mutation be analyzed reliably.

Fringe Benefits of *lacZ* Fusions

Long ago, it was realized that strains in which *lacZ* is fused to a gene coding for an exported protein often exhibit new phenotypes. For the geneticist this is a fortunate and useful development because such phenotypes can be exploited to select or screen for mutants which alter the phenotype conferred by the gene fusion. In several cases, such mutants have been shown to be defective in protein export. Since these phenotypes have been reviewed and discussed extensively (2, 8), we will simply restate current thought so that the arguments inherent in subsequent discussions will be clarified.

Apparently, the export machinery of *E. coli* cannot deal effectively with sequences of a cytoplasmic protein such as β-galactosidase. Accordingly, high-level synthesis of an exported LacZ hybrid is lethal as a result of an ultimate jamming of this essential cellular function. In the case of *lamB-lacZ* fusions, this overproduction lethality is evidenced as inducer sensitivity because the presence of inducer causes high-level synthesis of the harmful chimera. Mutations that block hybrid protein export can be recognized since they will relieve this inducer sensitivity.

To function catalytically, β-galactosidase must form a tetramer. Apparently, this oligomerization is inhibited when the LacZ monomer is directed to a membrane environment as a consequence of gene fusion. Thus, fusion strains in which the LacZ hybrid protein is localized to a membrane exhibit a lactose-negative phenotype. Mutations that block export and prevent the hybrid protein from reaching the membrane can be recognized because they restore the ability to utilize lactose efficiently.

What Is a Signal Sequence Anyway?

Because the signal sequence is small and ubiquitous in exported proteins, it was commonly suspected that the secrets of its functional role would be revealed by sequence analysis and homology comparisons. In analogy with other successful paradigms, it was thought that highly conserved amino acid residues would signify critical function and suggest possible mechanistic roles. Well, this is not the case. After a decade of sequence comparison, we are forced to conclude that at the primary amino acid sequence level, at least, signal sequences do not share significant homologies. The essential and identifying characteristics of a signal sequence must be subtle (10).

To identify the important molecular components of the LamB signal sequence, we identified a collection of mutations that altered function. By using this direct and more systematic approach, we hoped to reduce the background noise that is inherent in the more global analysis outlined above. To this end, a variety of selection procedures based on the unusual phenotypes of *lamB-lacZ* fusion strains cited in the preceding section have been designed and implemented. A remarkably high proportion of the mutants that survive these selections are altered within the portion of the *lamB* gene corresponding to the signal sequence. As expected, these mutations prevent hybrid protein export, causing the chimera to remain in the cytoplasm. A summary of these mutations is presented in Fig. 2.

The use of fusion strains to identify signal sequence mutations provides a now classic example of the utility of *lacZ* gene fusions for the analysis of protein export. To verify the predicted effect of the mutations on protein export, they were recombined into an otherwise wild-type *lamB* gene, thus permitting biochemical analysis under normal physiological conditions. Invariably, these mutations were found to block LamB export as well. Although quantitative differences were observed, fusions and their associated phenotypes proved to be a reliable measure of export (for review, see reference 2).

What insights into signal sequence function can be gained from the *lamB* mutations? To address this question, we have adopted an ex-

M M I T L R K L P L A V A V A A G V M A A Q A M A

10 - 13

12D 14D 16E 19R

19K

6S 13D 15E 17R

17D

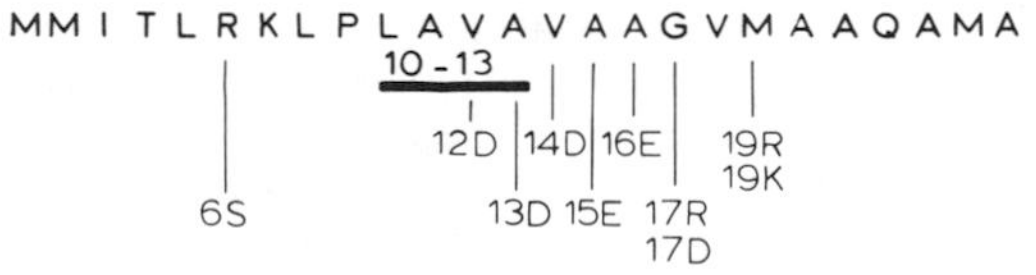

FIG. 2. LamB signal sequence and our collection of mutations. Numbers correspond to position, letters refer to amino acids, and vertical lines indicate point mutations. Heavy line (10-13), Deletion mutation.

port assay that allows quantitative measurements of export as a function of time (8a). Since the LamB protein is processed by a signal peptidase only after it is secreted, the kinetics of protein localization can be measured by pulse-chase experiments. When this is done with mutated forms of LamB, a ratio of the processed to total LamB, which can be separated on polyacrylamide gels, measures the efficiency of export. Results of such studies are summarized in Table 1.

Data from these experiments allow the assignment of *lamB* signal sequence mutations to five different classes (Table 1). The first two cause a near total export defect. Class 1 does so without inserting any charged residues into the hydrophobic core but alters secondary structure. The deletion mutations, such as 10-13, are members of class 1. Class 2 mutations include most of the charged amino acid substitutions. Other charge substitutions, however, behave differently in that they appear to slow the export process instead of stopping it altogether (class 3). Apparently, it is not simply the presence of a charged residue within the hydrophobic core of the signal sequence but rather its position that is important. The remaining two classes result in reduced synthesis of LamB, either without (class 4) or with (class 5) an accompanying export defect.

The deletion mutation 10-13 appears to alter secondary structure by preventing the formation

TABLE 1. Kinetic analysis of signal sequence mutations[a]

Mutant allele	Synthesis (%)[b]	Export[c]	Class
Wild type	100	100	
10-13	75–100	<15	1
12D, 13D, 14D, 15E, 19R, 19K	75–100	<15	2
17R, 17D	90–110	40–80	3
6S	20–30	100	4
16E	5–10	5	5

[a] Synthesis and export measurements were made after a 20-s pulse-labeling and a 4-min chase. See text for description of classes.

[b] Relative to wild-type levels.

[c] (LamB/pLamB + LamB) × 100.

of an α-helix in a critical region of the hydrophobic core. Previously reported genetic analysis of pseudorevertants of this deletion and biophysical characterization of synthetic peptides support this view (4, 5). Although this mutation has a defect similar to class 2 mutations as measured by our kinetic analysis, we think it likely that this mutation is quite different. The best evidence available to support this difference is the effect of *prlA*, a suppressor of signal sequence mutations (for review, see reference 3). This suppressor restores near-normal function to the truncated signal sequence of the 10-13 deletion. No class 2 signal sequence mutation is affected to this degree by *prlA*. Because of these pronounced differences between the deletion mutation and the other signal sequence alterations with respect to *prlA*, and since the former alters secondary structure while the point mutations probably do not, we propose that the functional defects caused by the deletion are distinct from those caused by the point mutations.

The remaining signal sequence mutations, 6S and 16E, are clearly different from those mentioned above because these mutations affect synthesis of LamB. The first, 6S (class 4), changes one of the basic residues (arginine) commonly found near the amino terminus of signal sequences to a serine (6). Our assay reveals that this mutation is not export defective at all; the protein that is made is exported with normal kinetics. Rather, it causes a four- to fivefold decrease in net synthesis, and previous work has shown that this synthesis defect is manifest at the level of translation. The other mutation, 16E (class 5), causes both an export defect and a decrease in synthesis. In terms of export, this mutation resembles most of the mutations which introduce a charged residue in the hydrophobic core. However, it causes, in addition, a 10-fold decrease in synthesis. It should be noted that *prlA* has no effect on the synthesis defects caused by these two signal sequence mutations. Apparently, this defect(s) occurs at a step(s) that does not involve PrlA.

In summary, by using a relatively simple but quantitative export assay, and by analyzing the effects of the *prlA* suppressor, we can separate our collection of *lamB* signal sequence mutations into at least four and probably five different classes. This complexity is surprising, and it forces us to reconsider the role of the signal sequence in protein export. It seems unlikely that this export signal performs a single function. If this were true, the various mutations would behave in a qualitatively similar manner, and this is clearly not the case. Accordingly, we suggest that the signal sequence may perform multiple functions. Viewed in this manner, the mutations would cause different effects because they interfere with different functions. To further complicate matters, some of the signal sequence mutations may be multiply defective. This could explain why *prlA* causes only limited suppression in most cases. Under suppression conditions, only one of several export defects is alleviated.

If our suggestion of multiple signal sequence function is correct, then more sophisticated biochemical assays are required. With these tools, it may be possible to separate the various mutations into clear mechanistic classes. If this could be done, the essential components of the signal sequence might become apparent, and thus would provide a framework for dissecting the important molecular elements of this remarkable and seemingly heterogeneous peptide.

Is There a Mechanism to Couple Synthesis and Export?

As mentioned in the introduction, there is biochemical evidence from studies with eucaryotes for a mechanism to couple translation and export. Although no biochemical data to indicate such a mechanism have yet been obtained with procaryotic systems, the 6S substitution in the LamB signal sequence suggests the presence of a similar phenomenon here as well. Results presented in Table 1 indicate that the mutation causes a 70 to 80% decrease in net synthesis of LamB, and studies with *lamB-lacZ* fusions show that this decrease is due to altered translation and that it is strictly tied to export. To account for these results, it has been proposed that 6S interferes with the mechanism that normally couples export and synthesis (6). However, this conclusion has been the subject of some debate for the following reasons: (i) complexities have been encountered in the analysis of 6S, (ii) a critical piece of data rests solely on gene fusions, and (iii) there is only a single mutation of this type.

We have found yet another *lamB* signal sequence mutation that causes a decrease in synthesis, 16E. In fact, the synthesis defect associated with 16E seems to be even more severe than that caused by 6S (Table 1). The nature of this defect has not yet been determined. However, the possibility of a transcriptional defect seems remote, and the lesion is located at some distance from the site of translation initiation. Accordingly, the possibility that 16E decreases synthesis in a manner related to export looms large, and if this can be demonstrated directly, it will greatly strengthen the proposal that synthesis and export are coupled in bacteria.

Closer examination of the kinetic data obtained in the analysis of signal sequence mutations reveals an apparent effect on LamB synthesis (J. Stader and T. Silhavy, unpublished data). With wild-type strains, after a 20-s pulse of labeled methionine we found that full-length LamB molecules which contain label continue to appear for more than 1 min following the chase. Presumably, this reflects the time required to complete synthesis of this protein. However, with strains that carry certain signal sequence mutations, the time required for LamB to appear is greatly reduced. For example, consider the deletion mutation 10-13. To avoid potential variation between strains and experiments, we standardize values relative to the amount of LamB present after a 2-min chase during each assay. This corresponds to the maximum amount of protein that is labeled over the course of a particular experiment. As a proportionate measure of the time required to complete synthesis of the protein, the amount of full-length molecules present after 10 s of chase is measured and the result is expressed as a fraction of the 2-min maximum total. This value is typically 10 to 15% for wild-type strains. With the *lamB* deletion 10-13, however, the value is increased three- to fivefold. This suggests that this signal sequence mutation increases the rate of LamB synthesis substantially. Since mRNA should be at steady-state levels in all strains during the experiment and because we are monitoring amino acid incorporation, this indicates that the increased rate of synthesis is the result of an increased rate of translation.

We have considered two models to account for the effect of the *lamB* signal sequence mutations on the rate of translation. The first derives from studies with SRP in eucaryotic cells. Under certain circumstances in vitro, SRP can bind the signal sequence while the protein is still nascent and cause a stable arrest in translation. This arrest is maintained until the polysome reaches the membrane and docking protein intervenes. If a similar mechanism operates in *E. coli*, then we might expect the synthesis of exported proteins to be slowed somewhat as a result of the transient arrest caused by an SRP-like complex. The signal sequence mutation would cause an increase in the rate of synthesis by preventing recognition by the SRP-like complex and thus the imposition of the resulting translation arrest. A second model raises the possibility that the translation of exported proteins is, in general, slower than that of nonexported proteins. This could be caused by physical constraints such as the availability of charged tRNA, etc., that create a different rate-limiting step. Viewed in this manner, the signal sequence mutations would increase rates simply by preventing export. The data available at present do not allow us to distinguish these possibilities.

Regardless of which, if any, of the models presented in the paragraph above is correct, the results argue that under normal physiological conditions, LamB export occurs while translation is ongoing, i.e., export must be cotranslational. If synthesis was complete before export initiated, i.e., posttranslational, then signal sequence mutations could not possibly affect the rate of translation.

In summary, we find that many signal sequence mutations affect translation in one way or another. Some cause a net decrease in synthesis while others cause an increase in the rate of translation. We have argued that, under normal conditions in vivo, LamB export occurs in a cotranslational manner. If this is true, then it would not be surprising if the cell has evolved a mechanism to ensure that export and synthesis are coupled. This remains, however, a matter for speculation.

Is There Export Information within Mature Sequences of *lamB*?

Results obtained with *lamB-lacZ* fusions predicted that information required for export from the cytoplasm was located beyond the region of the gene specifying the signal sequence but prior to codon 39 of the mature protein. In addition, these studies predicted that the region between codons 39 and 49 contained information required for proper routing to the outer membrane (Fig. 1). Nevertheless, the selections designed for the isolation of export-defective mutants based on phenotypes of fusion strains yielded no point mutations or deletions, which were internal to the *lamB* portion of the hybrid gene, that blocked export and did not alter the LamB signal sequence. Accordingly, to test these predictions, other techniques were employed.

To selectively alter regions of *lamB*, we adapted a biochemical technique to create deletions in vitro (3). By starting with a plasmid carrying a large *lamB-lacZ* gene fusion, we could screen simply for deletions that were internal to *lamB* and which did not alter translational reading frame by demanding that the lesion not prevent the production of β-galactosidase activity. Initially, we utilized an existing restriction enzyme cleavage site located at codon 183 to produce a series of deletions that collectively removed amino acids 70 to 220 of mature LamB. Results show that the deletions do not alter the export of the hybrid protein. Moreover, when recombined genetically into an otherwise wild-type *lamB* gene, the truncated LamB protein is localized to the outer membrane in an apparently normal fashion as well. Here again, fusions proved to be a reliable indicator for export.

Owing to a paucity of useful restriction enzyme cleavage sites in the early, i.e., 5′, portion of *lamB*, a similar analysis of the region promoter proximal to the codon for amino acid 70 was not possible directly. To circumvent this problem, we have engineered an insertion mutation which introduces simultaneously an in-frame nonsense codon and a unique restriction enzyme cleavage site at codon 45 of *lamB* (B. Rasmussen and T. Silhavy, Genes Dev., in press). By using techniques of in vitro mutagenesis, the insertion mutation was incorporated onto a plasmid which carries a large *lamB-lacZ* fusion. This facilitates identification of the desired mutant plasmid because the nonsense codon will prevent production of β-galactosidase activity. Moreover, this activity will be restored upon introduction of a nonsense suppressor. Once constructed, this plasmid was employed to create a series of internal in-frame deletions using an experimental approach similar to that described in the preceding paragraph. In total, more than 60 deletions have been constructed, and by physical methods we estimate that these deletions range in size from approximately 6 to 270 base pairs.

As an initial step in the characterization of the internal, in-frame deletions, they were recombined onto a λ *lamB-lacZ* fusion transducing phage. These phage were used to make lysogens to permit phenotypic analysis. In a previous section, we reviewed the characteristic phenotypes conferred by *lamB-lacZ* fusions. One of these is inducer (maltose) sensitivity, and previous work has demonstrated a direct correlation between this phenotype and attempts by the cell to export the hybrid protein. Using a zone of inhibition assay, we determined the inducer sensitivity caused by the various fusions (Table 2). On the basis of these data, the deletions can be divided into three distinct classes. The first does not alter sensitivity, the second causes intermediate sensitivity, and the third causes complete resistance. Not surprisingly, perhaps, the approximate size of the deletion correlates with the level of sensitivity, the smallest having no apparent effect. On the basis of previous work with fusions, we predict that deletions that do not alter sensitivity also do not affect export. In contrast, those that abolish sensitivity are expected to block export. Deletions that result in an intermediate phenotype most likely block export as well, but to a lesser degree. Representative examples from each phenotypic class were chosen, and the extent of the various deletions was determined by DNA sequence analysis.

TABLE 2. Inducer sensitivity conferred by *lamB-lacZ* fusions with internal *lamB* deletions[a]

Regions removed	Zone of inhibition (mm) with:	
	0.5% inducer	5% inducer
None	18	27
2	16	25
1 and 2	0	10
1, 2, and 3	0	0

[a] Sensitivity was measured by spotting a paper disk saturated with a solution (0.5 or 5%) of inducer onto a lawn of the fusion strain and measuring the diameter of the zone of inhibition.

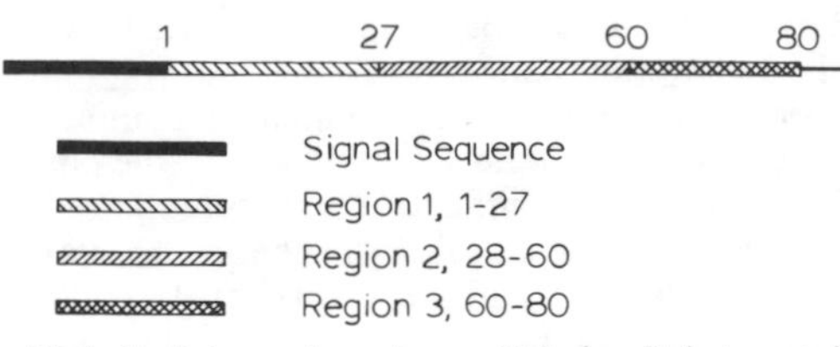

FIG. 3. Intragenic regions within *lamB* that contain export-related information.

To test the predictions concerning export based on the inducer sensitivity conferred by the truncated fusions, the internal deletions were recombined genetically onto an otherwise wild-type chromosomal *lamB* gene. Under these physiologically normal circumstances, export can be measured in a quantitative fashion using our pulse-chase assay. The results of these experiments (summarized in Fig. 3 and Table 3), in combination with the DNA sequence analysis, allow the identification of three regions prior to codon 80 but distal to the signal sequence. The first region contains codons 1 to 28 of mature LamB. The second region is bounded by codons 28 and 60, and because this region contains the restriction enzyme cleavage site from which all deletions were generated, this region is altered in all cases. Deletions that are contained within region 2 have no effect on LamB export from the cytoplasm. Deletions that extend into region 1, which is bound by codons 1 and 28, alter export but do not block it completely. Rather, they appear to slow the export process. Deletions that extend from region 2 into both regions 1 and 3 block export almost completely.

The results obtained with these internal in-frame *lamB* deletions provide the first clear demonstration that information which specifies export from the cytoplasm is contained within

TABLE 3. Kinetic analysis of internal *lamB* deletions[a]

Regions removed	Export (%)
None	100
2	100
1 and 2	40–60
1, 2, and 3	<10

[a] Export measurements were made after a 20-s pulse-labeling and a 2-min chase.

sequences that correspond to the mature portion of a noncytoplasmic protein, and they show that, for LamB at least, an intact signal sequence by itself is not sufficient to initiate the localization process. In addition, they suggest that this important export-related information is located in at least two physically distinct sites, regions 1 and 3.

Here again, *lacZ* fusions have proven to be a reliable indicator for the export process. Intragenic regions that are required for hybrid protein export are required for the export of the corresponding wild-type protein as well. Some differences can be seen upon careful scrutiny. However, these are slight and qualitative in nature.

Two general models have been considered to explain the requirement for regions 1 and 3 in the export process. The first proposes specific sites located in these regions that are required for recognition by some component(s) of the cellular export machinery. Alternatively, it may be that these regions participate in a more indirect fashion by allowing correct conformational presentation of the signal sequence. Experiments are currently in progress to distinguish these alternatives.

On the basis of our analysis of gene fusions, we have predicted that information required to direct LamB to its normal outer membrane location is contained within a region of the gene bounded by codons 39 and 49. This region, of course, is contained within region 2. The data currently available allow us to conclude that there is no information located in this region that is required for export from the cytoplasm, and this is consistent with the fusion data. We do not yet know, however, whether the alterations of this region affect correct localization of the truncated product to the outer membrane. These studies are complicated by the relative instability of the mutant protein product, and accordingly, the answer to this important question must await further analysis and perhaps the development of more sensitive biochemical techniques.

Summary

We have used *lacZ* gene fusions to identify regions within a gene, *lamB*, that specify protein localization to the outer membrane. By determining the cellular location of the various hybrid proteins produced and by using a combination of classical and modern genetic approaches to exploit the phenotypes exhibited by these fusion strains, we have found that most of the essential export information appears to be located in the 5′ portion of the gene at a position preceding the codon for the 80th amino acid of mature LamB. On the basis of mutant analysis, we have divided this portion of the gene into four regions. The first, and most extensively studied, corresponds to the signal sequence. Mutants altered in this region exhibit a variety of defects suggesting that this signal may in fact perform multiple functions. In light of the observed effect of many of these mutations on translation, one or more of these functions may relate to a cellular mechanism that couples synthesis and export. Two additional regions, bounded by codons 1 and 28 and 60 and 80 of the mature protein, contain information that is required for export of the protein from the cytoplasm. Without these regions, the signal sequence is rendered nonfunctional. A final region bounded by codons 29 to 59 may contain information required for proper routing to the outer membrane. As predicted, lesions in this region do not prevent export from the cytoplasm; however, more work is required to clarify its possible role in routing.

This work was supported by fellowships to B.A.R. and J.A.S. from the American Cancer Society and by Public Health Service grant GM34821 to T.J.S. from the National Institutes of Health.

LITERATURE CITED

1. **Benson, S. A., E. Bremer, and T. J. Silhavy.** 1984. Intragenic regions required for LamB export. Proc. Natl. Acad. Sci. USA **81:**3830–3834.
2. **Benson, S. A., M. N. Hall, and T. J. Silhavy.** 1985. Genetic analysis of protein export in *Escherichia coli* K 12. Annu. Rev. Biochem. **54:**101–134.
3. **Benson, S. A., and T. J. Silhavy.** 1983. Information within the mature LamB protein necessary for localization to the outer membrane of *E. coli* K 12. Cell **32:**1325–1335.
4. **Briggs, M. S., and L. M. Gierasch.** 1984. Exploring the conformational roles of signal sequences: synthesis and conformational analysis of a λ receptor protein wild-type and mutant signal peptides. Biochemistry **23:**3111–3114.
5. **Emr, S. D., and T. J. Silhavy.** 1983. Importance of secondary structure in the signal sequence for protein secretion. Proc. Natl. Acad. Sci. USA **80:**4599–4603.
6. **Hall, M. N., J. Gabay, and M. Schwartz.** 1983. Evidence for a coupling of synthesis and export of an outer membrane protein in *Escherichia coli.* EMBO J. **2:**15–19.
7. **Palade, G.** 1975. Intracellular aspects of the process of protein synthesis. Science **189:**347–358.
8. **Silhavy, T. J., and J. R. Beckwith.** 1985. Uses of *lac* fusions for the study of biological problems. Microbiol. Rev. **49:**398–418.

8a. **Stader, J., S. A. Benson, and T. J. Silhavy.** 1986. Kinetic analysis of *lamB* mutants suggests the signal sequence plays multiple roles in protein export. J. Biol. Chem. **261:**15075–15080.

9. **Walter, P., R. Gilmore, and G. Blobel.** 1984. Protein translocation across the endoplasmic reticulum. Cell **38:**5–8.
10. **Watson, M. E. E.** 1984. Compilation of published signal sequences. Nucleic Acids Res. **12:**5145–5164.

IV. STRUCTURE AND FUNCTION OF ALKALINE PHOSPHATASE

Introduction

JOSEPH E. COLEMAN

Department of Molecular Biophysics and Biochemistry, Yale University, New Haven, Connecticut 06510

Alkaline phosphatase needs only a short introduction since it is an enzyme that has received an enormous amount of investigation; over 6,000 citations in the review by McComb et al. in 1968 (8). Part of this, of course, stems from the fact that changes in the "activity" of this enzyme in various human tissues have been widely used as a signal for various pathological processes.

With the discovery in the late 1950s and early 1960s (10) that *Escherichia coli* possessed an alkaline phosphatase that was derepressible by phosphate starvation and remarkably heat stable, molecular studies of alkaline phosphatase rapidly shifted to the bacterial prototype. Engström (2) observed that the mammalian intestinal enzyme formed a phosphoseryl intermediate believed to be on the reaction path. This same phosphoseryl intermediate was identified in the *E. coli* enzyme, and the view is now widely held that most of the molecular features of the bacterial enzyme can in all probability be applied to the structure of the membrane-attached mammalian enzymes. The recently determined sequence of the cDNA of the human placental enzyme (7) strongly supports the assumption that the structure of the active center and most of the details of the mechanism of action must be nearly identical.

With the discovery that *E. coli* alkaline phosphatase is a protein secreted into the periplasmic space by a "signal" peptide mechanism, alkaline phosphatase became a model for the synthesis and transport of secretory proteins (6), a subject which is discussed in the papers on the preenzyme by Boyd and his colleagues.

In 1962, the *E. coli* alkaline phosphatase joined the class of zinc metalloenzymes with the demonstration by Plocke et al. that the enzyme contained stoichiometric amounts of zinc (9). During the 1960s and 1970s, the enzyme played an important part in elucidating the mechanisms of structure-function relationships involved in the control of protein synthesis and gene expression; this work came principally from the laboratory of Garen and his colleagues (4). The gene *phoA* was shown to be part of a cluster of genes whose products modulated the transcription of the alkaline phosphatase gene. The existence of suppressor tRNAs was also demonstrated with mutations in the structural gene of alkaline phosphatase, as well as a large series of amino acid substitutions resulting in catalytic changes, alterations in metal binding, and mutations affecting the monomer-monomer interactions in the alkaline phosphatase dimer.

Crystals of the *E. coli* alkaline phosphatase suitable for X-ray diffraction were obtained (1) in 1968, and a high-resolution crystal structure is now available from the laboratory of Wyckoff and his colleagues. We were thus able to compare the three-dimensional molecular structure in the crystal with the solution studies on the structure-function relationships in this enzyme; principally, a large body of information was accumulated on the function of the metal ions as well as on the structure and function of the enzyme by nuclear magnetic resonance studies. An active program of site-directed mutagenesis has recently been initiated in the laboratories of Tom Kaiser. Kendall and Kaiser's paper covers the probing of the alkaline phosphatase mechanism by this approach.

We are thus able to cover a remarkable spectrum within the biochemistry of this enzyme, from crystal structure to detailed descriptions of the control of its synthesis and its catalytic mechanism. One area, however, that we cannot tread with such surety is perhaps the most important of all, its general physiological function, other than to say that in bacteria, when they found themselves in an environment consisting of principally organic phosphate esters as a source of phosphate (perhaps as a carbon source as well), alkaline phosphatase could be derepressed to protect the organism from phosphate starvation. This obvious statement begs the question of the reason for the phosphoseryl intermediate and its ability to transfer the phosphate to a large number of acceptor alcohols. It also does not address the reasons for the later evolutionary attachment of this enzyme to a variety of cell membranes in higher animals and its apparent association with phosphate transport, bone formation, and induction by vitamin D.

LITERATURE CITED

1. **Applebury, M. L., B. P. Johnson, and J. E. Coleman.** 1970. Phosphate binding to alkaline phosphatase. J. Biol. Chem. **245:**4968–4975.
2. **Engström, L.** 1961. Studies on calf-intestinal alkaline phosphatase. II. Incorporation of inorganic phosphate into a highly purified preparation. Biochim. Biophys. Acta **52:**49–59.
3. **Gallucci, E., and A. Garen.** 1966. Suppressor genes for nonsense mutations. II. The Su-4 and Su-5 suppressor genes of *Escherichia coli*. J. Mol. Biol. **15:**193–200.
4. **Garen, A., S. Garen, and R. C. Wilhelm.** 1965. Suppressor genes for nonsense mutations. I. The Su-1, Su-2 and Su-3 genes of *Escherichia coli*. J. Mol. Biol. **14:**167–178.
5. **Hanson, A. W., M. L. Applebury, J. E. Coleman, and H. W. Wyckoff.** 1970. X-ray studies on single crystals of E. coli alkaline phosphatase. J. Biol. Chem. **245:**4975.
6. **Inouye, H., C. Pratt, J. Beckwith, and A. Torriani.** 1977. Alkaline phosphatase synthesis in a cell-free system using DNA and RNA templates. J. Mol. Biol. **110:**75–87.
7. **Kam, W., E. Clauser, Y. S. Kim, Y. W. Kan, and W. J. Rutter.** 1985. Cloning, sequencing, and chromosomal localization of human term placental alkaline phosphatase cDNA. Proc. Natl. Acad. Sci. USA **82:**8715–8719.
8. **McComb, R. B., G. N. Bowers, and S. Pozen.** 1979. Alkaline phosphatase. Plenum Publishing Corp., New York.
9. **Plocke, D. J., C. Levinthal, and B. L. Vallee.** 1962. Alkaline phosphatase of E. coli: a zinc metalloenzyme. Biochemistry **1:**373–378.
10. **Torriani, A.** 1960. Influence of inorganic phosphate in the formation of phosphatases by E. coli. Biochim. Biophys. Acta **38:**460–469.

Site-Directed Mutagenesis in the Redesign of the Active Site and Signal Peptide Regions of *Escherichia coli* Alkaline Phosphatase

DEBRA A. KENDALL AND E. T. KAISER

Laboratory of Bioorganic Chemistry and Biochemistry, The Rockefeller University, New York, New York 10021

Active Site Mutation

As part of our effort to redesign active site and structural regions of enzymes, we have carried out site-directed mutagenesis both on the active site and on the signal region of *Escherichia coli* alkaline phosphatase (5, 7). Our observations on mutations in the active site region are most complete for the conversion of the nucleophilic residue Ser-102 to a Cys residue. We found that near pH 7.5 to 8.0 the Cys mutant catalyzes the hydrolysis of phosphate monoesters with high efficiency when the alcohol product is a good leaving group, but much less effectively when the leaving group is poor. This contrasts with the wild-type enzyme, where the k_{cat} values for a wide range of phosphate monoesters show no significant dependency on the leaving group tendency of the product alcohol.

With Tris used as a nucleophilic trapping agent, the phosphoryl group of *p*-nitrophenyl phosphate was diverted from the formation of P_i in the Cys-102 alkaline phosphatase-catalyzed hydrolysis of this phosphate monoester. This interception of the phosphoryl group by Tris is similar to what is seen with the wild-type enzyme, where Tris has been shown to trap the phosphoryl-Ser-102 intermediate formed transiently in phosphate monoester hydrolysis. While addition of Tris increased the apparent rate of hydrolysis of 10.0 mM *p*-nitrophenyl phosphate at pH 8.0 by the wild-type enzyme, a similar increase was not seen for the reaction of the Cys-102 mutant. These observations have been interpreted to suggest that in phosphate monoester hydrolysis by thiol alkaline phosphatase (in the pH range near pH 7.5 to 8.0), the rate-limiting step is formation of the phosphoryl-Cys-102 intermediate. This differs from the wild-type enzyme where, depending on the pH, under typical reaction conditions either dephosphorylation of the phosphoryl enzyme or release of product P_i is rate limiting.

Signal Sequence Mutation

In the other phase of our work on alkaline phosphatase, we have studied the signal region of the enzyme, employing site-directed mutagenesis. In *E. coli*, alkaline phosphatase is synthesized as a precursor with a 21-amino acid signal sequence. This signal peptide is necessary for transport of the enzyme from its cytoplasmic site of synthesis to its final periplasmic location (6, 8). We are investigating the structural elements of this signal peptide which are involved in the transport process. Our approach involves constructing mutants with signal sequences optimized for formation of a particular structural feature. The efficacy with which these mutants complement the natural translocation machinery is then tested by transport studies with *E. coli* harboring the mutant alkaline phosphatase gene.

Previous studies suggested that the hydrophobicity (11, 12) and helix-forming potential (3, 4, 10) of the residues in the signal peptide core region are important for function. To examine the involvement of these features in the alkaline phosphatase signal peptide, our goal has been to produce a mutant containing a polyleucine core segment. Leucine was chosen since it is a hydrophobic residue and synthetic peptides incorporating polyleucine regions exhibit strong α-helical character (1). Rather than make all the necessary residue changes in one step, however, we first chose to study individually some potentially problematic substitutions to assess the limitations of this approach. These include two separate Pro to Leu substitutions since proline disrupts helical structures, and in one case this residue is located near the site of signal peptide cleavage. Also, a single Phe to Leu substitution was studied since this is the only residue in the natural sequence with a large, aromatic side chain and therefore could be specifically required for activity. For this purpose, site-specific mutagenesis was used to produce the signal sequences shown in Fig. 1. Cell fractionation studies with mutants containing the single residue changes indicated that mature alkaline phosphatase is transported to the periplasm in each case. A comparison of whole-cell samples, however, showed some accumulation of precursor. The relative amount of precursor observed was mutant 3 > mutant 2 > mutant 1 > wild type. These results were supported by the rate of

Wild Type:	MKQSTIALALLPLLFTPVTKA
Mutant 1 (Pro to Leu):	MKQSTIALALLLLLFTPVTKA
Mutant 2 (Phe to Leu):	MKQSTIALALLPLLLTPVTKA
Mutant 3 (Pro to Leu):	MKQSTIALALLPLLFTLVTKA
Mutant 4 (9 x Leu):	MKQSTILLLLLLLLLTPVTKA

FIG. 1. Signal sequences produced by site-specific mutagenesis.

precursor processing observed in pulse-chase studies.

Although two of the mutants shown in Fig. 1 involved separate Pro to Leu substitutions, each had a markedly different effect on transport of alkaline phosphatase. It has been suggested that the presence of a proline near the cleavage site might serve to nucleate a β-turn conformation in that region (2, 9, 12). Replacing this residue with leucine (mutant 3) would diminish the potential for β-turn formation and consequently might reduce recognition by the signal peptidase. This may explain the substantial accumulation of precursor observed with mutant 3, but since some mature enzyme was transported, the data also indicate that this β-turn structure is not absolutely required for the cleavage process. More surprising was the finding that replacement of the proline at the center of the hydrophobic core region (mutant 1) produced only a subtle change in transport properties. Prior to these experiments, we considered the possibility that some structural irregularity might serve as an important trigger for membrane perturbation and subsequent translocation of the enzyme. The presence of a proline at the center of the core region is an ideal candidate for distorting the structure of this region, yet substitution of this residue had little effect. Conversely, the Phe to Leu substitution (mutant 2), which should not greatly affect the helix-forming potential of this region, produced a more notable reduction in precursor processing. These results exemplify the difficulty in correlating clearly the structural features and transport properties of mutants with single amino acid changes.

On the basis of the above results, mutant 4 was designed to retain the proline in the cleavage region. Construction of this mutant did, however, involve replacement of two alanines, the phenylalanine, and the core region proline with leucine. This resulted in a signal sequence comprising nine consecutive leucines in the core region (mutant 4). Cell fractionation studies with this mutant indicated that mature alkaline phosphatase is transported and correctly localized within the periplasm. No precursor accumulation was observed in a whole-cell sample from this mutant. Pulse-chase studies further indicated that maturation of the mutant 4 precursor is accelerated relative to wild type. In parallel experiments, some wild-type precursor was evident up to 1 min after chase initiation but no precursor was observed from mutant 4. These results indicate that a signal peptide core segment, optimized for hydrophobic α-helix formation, functions extremely efficiently in the transport process and that specific residues are not required in the hydrophobic domain of the signal sequence.

Construction of mutant 4 involved four amino acid changes, including the replacement of the phenylalanine and one proline each with leucine. Incorporation of these substitutions in mutant 4 did not, however, have an additive influence on transport and processing. Instead, the resultant signal peptide functioned more efficiently than mutants containing any single substitution. This indicates that the overall structure of the signal peptide core region, as opposed to primary sequence specificity, is a critical feature for transport activity. The negative effect of single residue changes was eliminated when the structure was optimized over the entire region.

These results further underline the value of making global mutations in one defined region of the signal peptide. By studying mutants containing simple sequences, such as polyleucine, clear correlations can be made with regard to the importance of a given structural parameter and transport activity. The outstanding feature of polyleucine lies in its strong propensity to adopt a hydrophobic α-helix conformation. Mutant 4 incorporates this sequence and functions extremely well in the transport process. This result can be attributed to the physical properties of the core region and suggests that formation of a hydrophobic α-helix may be a key event during translocation of alkaline phosphatase.

LITERATURE CITED

1. **Arfmann, H.-A., R. Labitzke, and K. G. Wagner.** 1977. Conformational properties of L-leucine, L-isoleucine, and L-norleucine side chains in L-lysine copolymers. Biopolymers **16:**1815–1826.
2. **Austen, B. M.** 1979. Predicted secondary structures of amino-terminal extension sequences of secreted proteins. FEBS Lett. **103:**308–313.
3. **Bankaitis, V. A., B. A. Rasmussen, and P. J. Bassford, Jr.** 1984. Intragenic suppressor mutations that restore export of maltose binding protein with a truncated signal peptide. Cell **37:**243–252.
4. **Briggs, M. S., and L. M. Gierasch.** 1984. Exploring the conformational roles of signal sequences: synthesis and conformational analysis of λ receptor protein wild-type and mutant signal peptides. Biochemistry **23:**3111–3114.
5. **Ghosh, S. S., S. C. Bock, S. E. Rokita, and E. T. Kaiser.** 1986. Modification of the active site of alkaline phosphatase by site-directed mutagenesis. Science **231:**145–148.
6. **Inouye, H., W. Barnes, and J. Beckwith.** 1982. Signal sequence of alkaline phosphatase of *Escherichia coli*. J. Bacteriol. **149:**434–439.
7. **Kendall, D. A., S. C. Bock, and E. T. Kaiser.** 1986. Ideal-

ization of the hydrophobic segment of the alkaline phosphatase signal peptide. Nature (London) **321:**706–708.
8. **Michaelis, A., and J. Beckwith.** 1982. Mechanism of incorporation of cell envelope proteins in *Escherichia coli.* Annu. Rev. Microbiol. **36:**435–465.
9. **Perlman, D., and H. O. Halvorson.** 1983. A putative signal peptidase recognition site and sequence in eukaryotic and prokaryotic signal peptides. J. Mol. Biol. **167:**391–409.
10. **Shinnar, A. E., and E. T. Kaiser.** 1984. Physical and conformational properties of a synthetic leader peptide from M13 coat protein. J. Am. Chem. Soc. **106:**5006–5007.
11. **von Heijne, G.** 1981. On the hydrophobic nature of signal sequences. Eur. J. Biochem. **116:**419–422.
12. **von Heijne, G.** 1985. Signal sequences. The limits of variation. J. Mol. Biol. **184:**99–105.

Structure of *Escherichia coli* Alkaline Phosphatase Determined by X-Ray Diffraction

HAROLD W. WYCKOFF

Department of Molecular Biophysics and Biochemistry, Yale University, New Haven, Connecticut 06511

The X-ray diffraction studies on which this paper is based were begun in 1969 when Applebury and Coleman succeeded in producing suitable crystals of a trigonal form of alkaline phosphatase from *Escherichia coli* (1). Hanson et al. took the first diffraction patterns and determined the first space group (4). Knox and Wyckoff solved that crystal form at 7.7-Å (0.77-nm) resolution (8), and W. D. Carlson (Ph.D. thesis, Yale University, New Haven, Conn., 1975) continued these studies. Technical problems led to abandonment of the trigonal form in favor of an orthorhombic form which had the full dimer in the asymmetric unit. Sowadski et al. (9) solved this form at 0.6-nm resolution, revealing the same principal features as seen previously and allowing progress toward an interpretable higher-resolution map. M. D. Handschumacher did most of the later chain tracing and detailed fitting. H. M. K. Murthy did the computer refinements that led to better and more extended phases, while Handshumacher did the interactive graphics human intervention and J. M. Sowadski collected further data and did most of the illustrations used herein. S. Katti has engaged in the most recent work. These studies have been reported by Wyckoff et al. (13) and Sowadski et al. (10, 11). Coleman has provided protein and crystals and intellectual support throughout, along with his steady stream of interesting studies that make the efforts in combination much more than either would be in isolation. References to all of the X-ray studies are in this paragraph and will not be cited repeatedly throughout the text.

Basic Structure

The three-dimensional structure of the alkaline phosphatase dimer from *E. coli*, determined by X-ray diffraction, is basically symmetric and globular, with two active sites, each containing a metal ion triple, located 3.2 nm from each other. The residues providing ligands for the metal ion in each active-site region derive from one subunit, but the base of residue 412 in particular lies at the intersubunit interface. The twofold symmetry is generally good, but specific differences may appear in the future analysis and the active sites may be different in detail. We cannot tell at the present time.

Each subunit has a major, 10-stranded beta sheet with a curled corner. The strands are parallel except for one antiparallel insertion. The butterfly figure (Fig. 1) shows the relationship of the principal relatively flat segments of these sheets in the two monomers. Helices of various lengths flank the sheets in a typical alpha-beta pattern characteristic of parallel sheet protein domains. Figure 1 shows the helices behind the sheet only. The remaining helices can be imagined by flipping the diagram over around the twofold axis which is vertical in the figure. Only one half of the residues are in these sheets and helices; the other half are in defined but hard-to-describe loops, coils, and connecting segments. Figure 2 shows the complete alpha-carbon trace in the same view as Fig. 1. Figure 3 is a schematic representation of one monomer as seen from the subunit interface. Some of the complex region at the top of the molecule is poorly defined in the electron density map, and in particular residues from 390 to 409 are essentially undefined as a result of disorder. In the apo enzyme residues 322 to 333 become disordered also.

There are three metal ions, designated A, B, and C, in each of the active-site regions. The A and B metals are in a pocket which just accommodates the phosphate moiety of the substrate, located at the bottom of a broad valley. Coleman (this volume) has discussed and cited the references to extensive spectroscopic investigations of these metal sites and their catalytic significance. The A-site zinc ion is held in place by three histidine imidazole nitrogens from residues 331, 372, and 412 and one carboxyl oxygen from Asp-327. The B-site metal, which can be either zinc or magnesium in a fully functional molecule, is liganded by His-370, Asp-369, Asp-51 (which bridges to the C-site metal), and Ser-102, which is the active phosphorylatable serine. The C-site ligands are all oxygens, contributed by Asp-51, Asp-153, Glu-322, and Thr-155. Caveats to these statements are needed. Either His-372 or Asp-327 may exclusively be directly liganded to metal A, or both may be simultaneous ligands, or each may be liganded in different subunits or even at different times during a catalytic cycle. The metal-binding sites were not, in fact, fully occupied in many of our experiments, in spite of

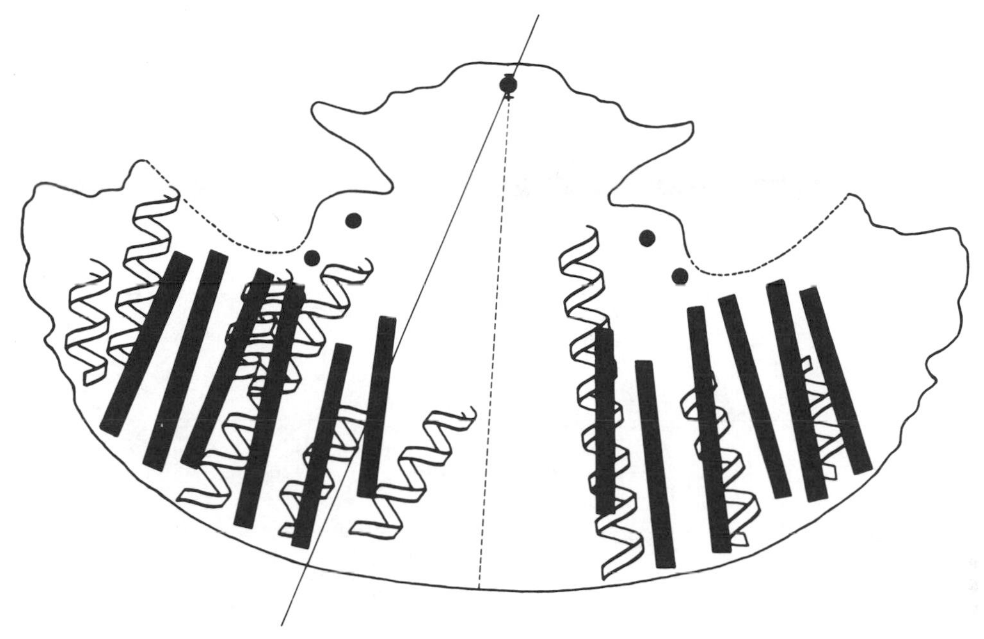

FIG. 1. Schematic view of the alkaline phosphatase dimer. The molecular twofold axis is vertical. The crystal axis is at 20 degrees. Helices in front of the beta sheet have been omitted. Taken from Sowadski et al. (9).

tests that at first indicated saturation. In addition, the hydroxyl oxygen of Ser-102 may be only indirectly connected to the B-site metal. Some of the C-site ligands might be indirect also; the metal-oxygen distances seem too long for direct ligation.

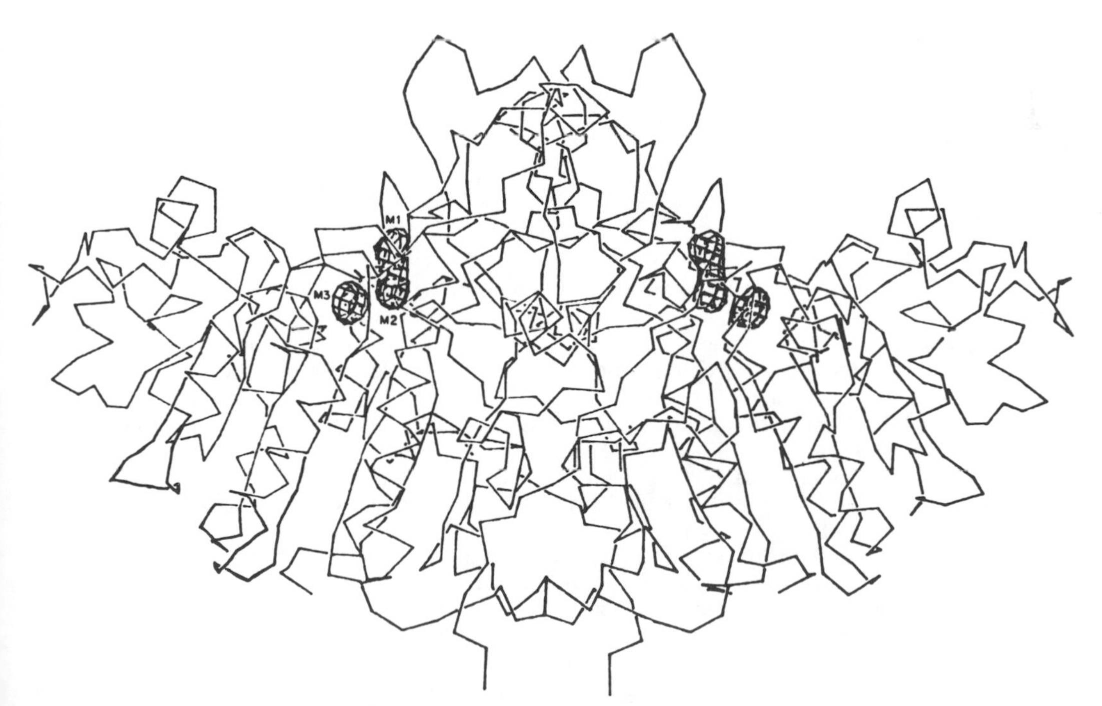

FIG. 2. Chain trace (alpha-carbon) of the complete dimer with the cadmium anomalous dispersion map superimposed showing the three metal sites in each active center. M1, M2, M3, Sites designated A, B, and C, respectively, in the text. Taken from Wyckoff et al. (13) by permission of John Wiley & Sons, Inc.

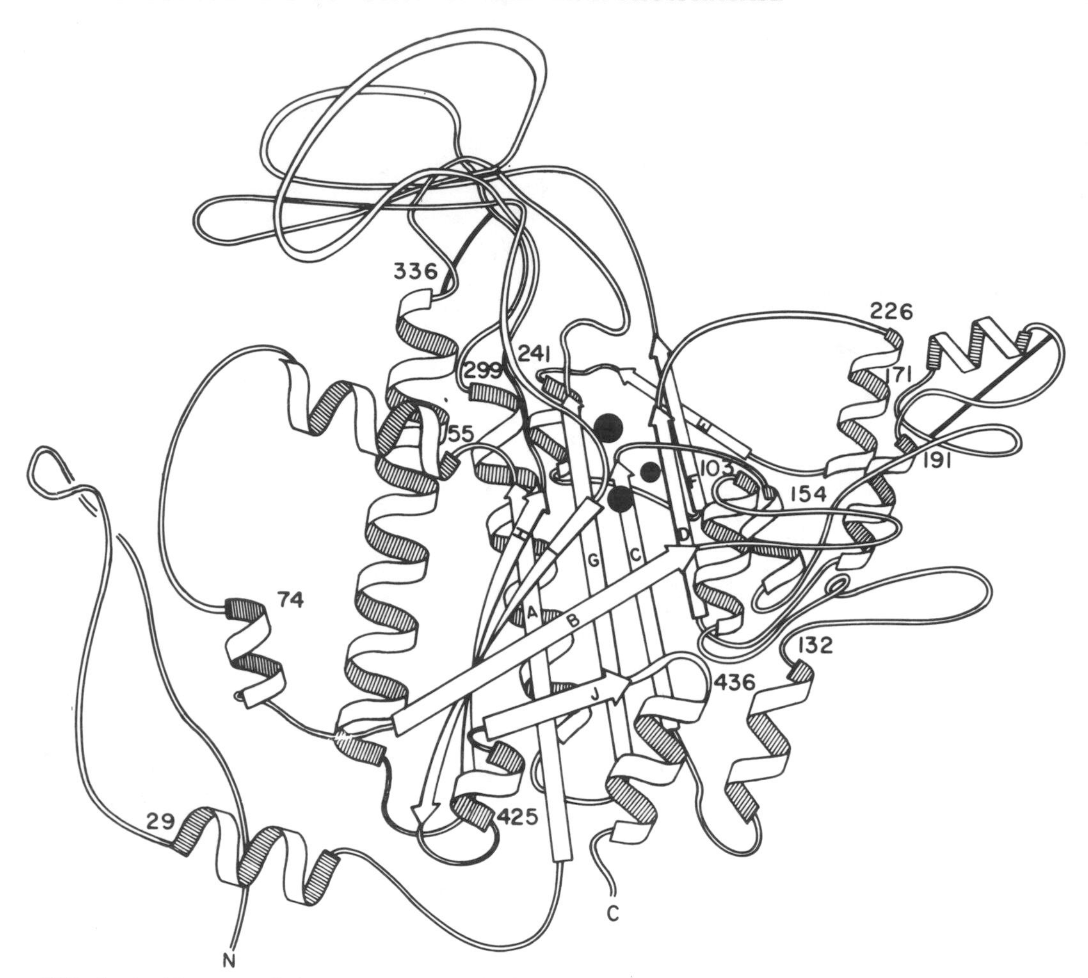

FIG. 3. Architecture of the monomer of alkaline phosphatase as viewed from the monomer-monomer interface. The three functional metal sites are drawn as black circles: the upper circle is A, the lower is B, and C is farther away with a smaller radius. The sequence numbers on the drawing indicate N termini of the corresponding helices. Taken from Sowadski et al. (10).

The distance from each alpha-carbon in one subunit to the B-site metal is presented in a two-dimensional polar plot in Fig. 4. The inward-pointing spikes focus attention on the residues within 1 nm, spaced approximately 50 residues apart in the first third and last third of the chain. These residues are Asp-51, Ser-102, Thr-155, His-331, Asp-369, and His-412. Much of the structure is far from the active site, far from the subunit interface, and not obviously involved in any important structural feature. The two disulfide bridges do not seem to have any important role. Many surface loops could apparently be shortened. Residues 390 to 409 are so disorganized that the corresponding electron density is absent or untraceable. The helix at residue 171 in the upper right corner of Fig. 3 could be dispensed with. A whole segment at the outer end of the sheet, residues 206 to 299, seems unnecessary. The molecule is much larger than RNase, for example, which performs a similar catalytic function and is almost as stable. The explanation may be largely evolutionary rather than functional. The facts give rise to many interesting molecular engineering possibilities.

How the monomer structure develops is particularly puzzling since the beta sheet appears to be a dominant and directing structural feature and yet it derives from disparate chain segments in a disjointed way. Figure 5 shows the sheet connectedness schematically and details of the layout of the sheet residues. Figure 6 shows a top view of the same.

The first 86 amino acids provide two of the strands of the beta sheet, A and B, one in the middle and one on the left side. A is flanked by rather late strands G and H; B is flanked by very

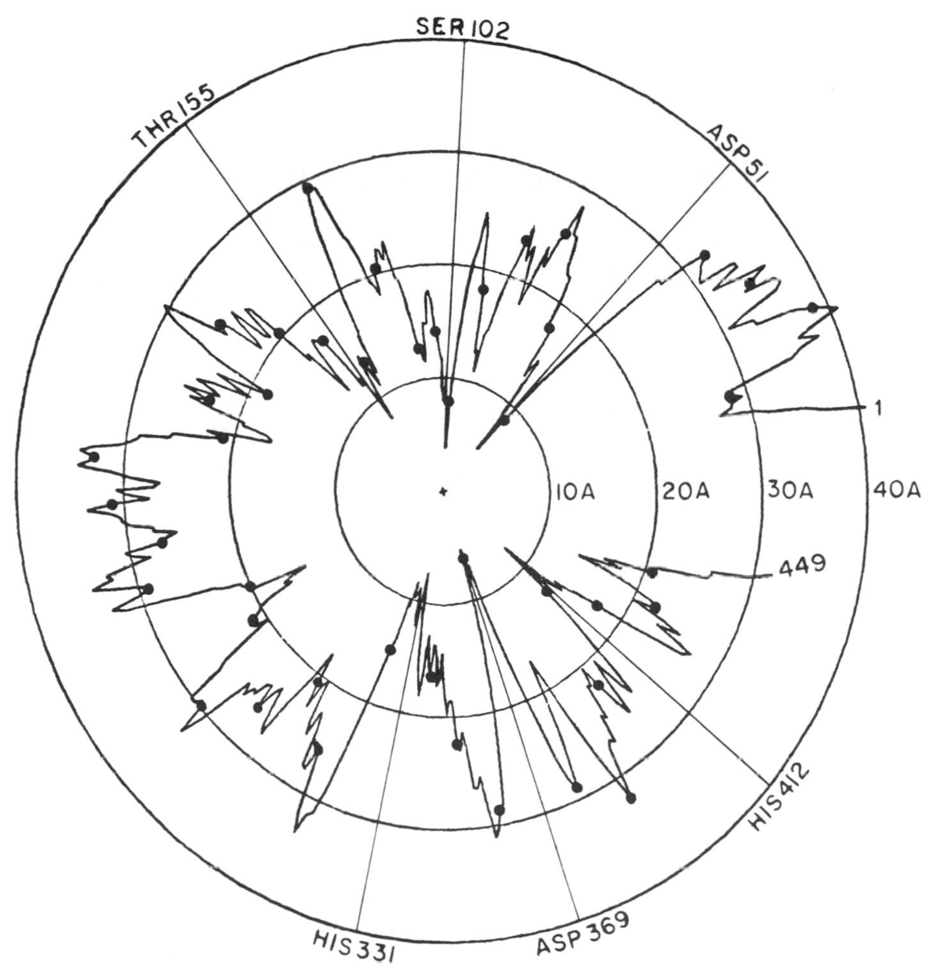

FIG. 4. Polar plot of distances of alpha-carbon atoms from the central metal of one subunit. Every tenth residue is marked with a ●. Taken from Wyckoff et al. (13) by permission of John Wiley & Sons, Inc.

late strands I and J. The segment of chain from residues 86 to 200, from the top of strand B to the bottom of strand D, contains only one strand of the beta sheet and otherwise lies entirely on one side of the sheet. This segment contains five short helical sections and long stretches of nondescript strands, loops, and bends. Strand D, composed of residues 201 to 206, completes the first pair of adjacent strands. The segment from 299 to 380 contains the two longest helices and two central strands in the sheet, G and H, which bracket strand A. Five of the ligands to the A- and B-site metals are contained in this stretch. Following the disorganized region from 390 to 409, the rest of the sequence contains two strands of the sheet, carries one A-site ligand, and is heavily involved in the subunit interface. The carboxyl-terminal residues are helical as in many other proteins. They are far from the active site and relatively near the amino terminus.

Helices, Sheet, and Subunit Interface

The helix and sheet portion of the molecule are somewhat wedge shaped, narrow at the bottom and broad at the level of the active site. The sheet is largely "classical" in that it twists in a right-handed sense as we proceed from one strand to the next, the segments between strands contain helices, and these connections proceed in a right-handed sense. These connections do not necessarily proceed to physically adjacent strands. In the alkaline phosphate dimer from *E. coli*, the central portion of the sheet is flatter than in many proteins, there are fewer physically adjacent, sequentially adjacent strands, and one surface contains more hydrophilic groups than normal. The nature of the sheet surfaces will be discussed below in connection with the sequence homologies of human placental alkaline phosphatase.

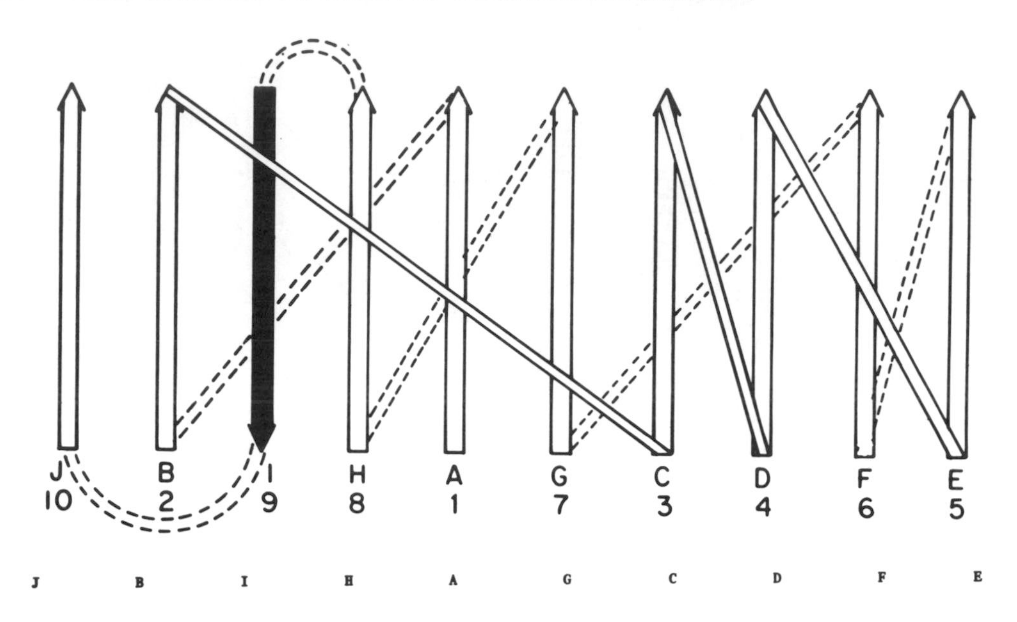

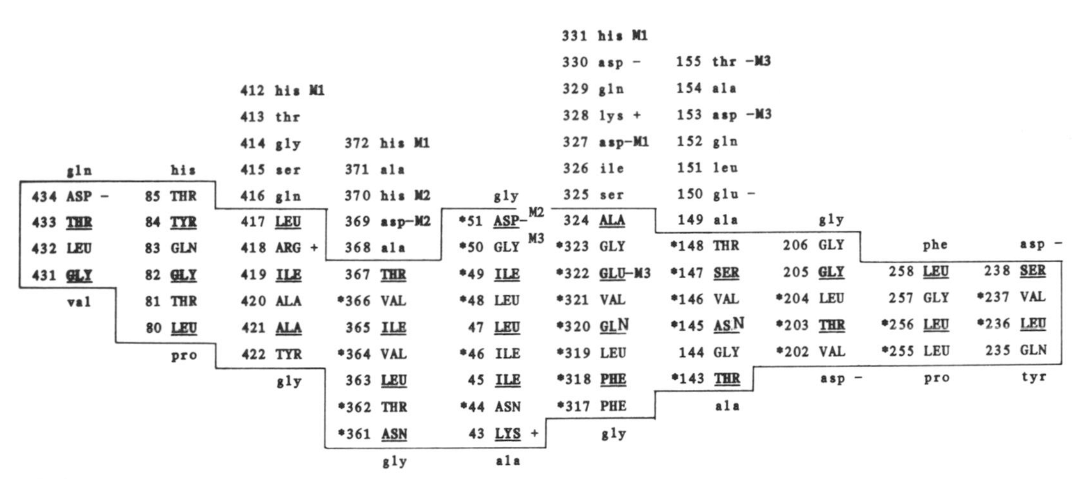

FIG. 5. Connectivity diagram and amino acid sequence of the beta sheet. Antiparallel strand I(9) is indicated in black. Only two strands, A(1) and G(7), provide ligands to the metal sites. Ligands are, however, contained in short extensions to the strands I(9), H(8), G(7), and C(3). Asterisks indicate residues that are in contact with helices. Amino acids underlined and in boldface are on the front side of the sheet. Amino acids given in lowercase are outside the sheet. Taken from Sowadski et al. (10).

Looking along the top edge of the sheet from the outer edge of the dimer, we see that the connections go right, right, left, left, right, right, up, down, left, left. This pattern facilitates the involvement of six sequentially well-separated chain segments that are physically close together in the metal binding and the active center.

Formation of stable globular subunits which then associate into a dimer (or higher) is conceptually a tricky business, assumed to occur in *E. coli* alkaline phosphatase and other multisubunit structures. The interface must be hydrophobic enough to induce association but not so hydrophobic as to induce inside-out folding. Within the first 129 residues in the sequence, approximately 49 are interfacial, as are 37 in the last 77 residues, with none in between. The surface area of positively charged groups buried by dimerization is 3.15 nm^2, and the negative surface area buried is 2.00 nm^2. On the surface of each subunit there is one Arg/Asp pair and one Lys/Asp pair, and one lone Asp is buried in the interfacial region. There is appreciable unaccounted-for electron density between the monomers which must represent trapped water and possibly inorganic or even organic ions. The

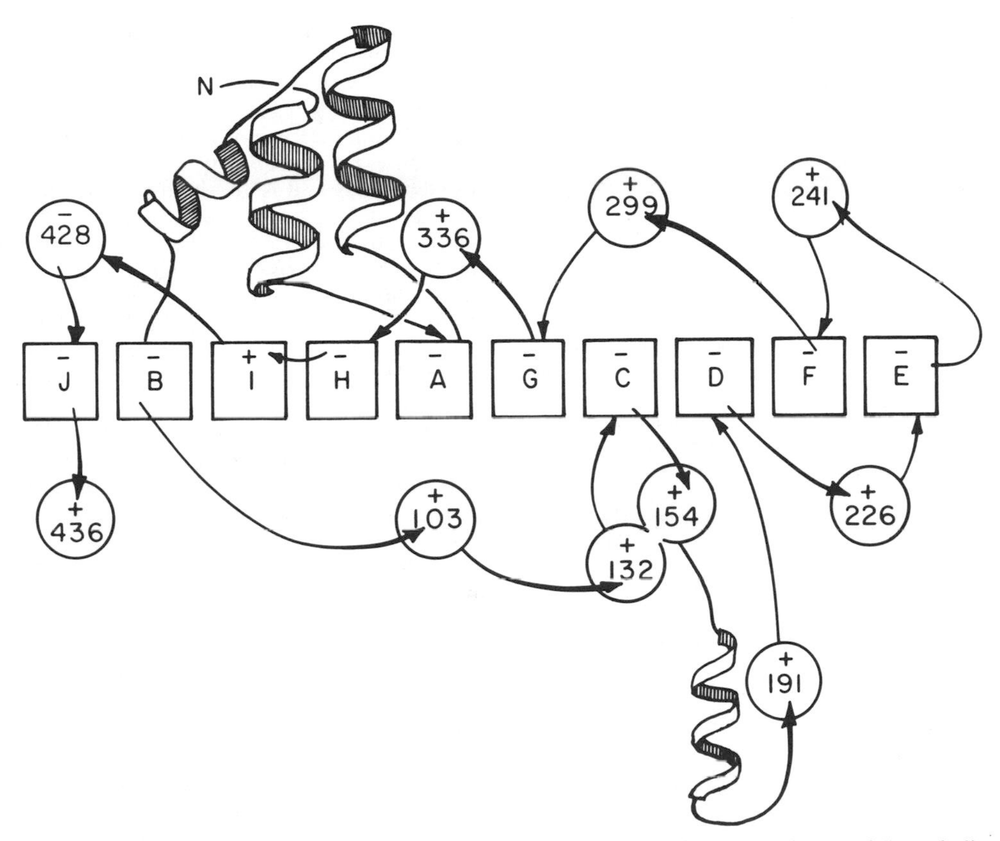

FIG. 6. Connectivity diagram of alkaline phosphatase. Carboxyl ends of beta strands (A and J) are indicated by − and N termini of helices are indicated by +, to indicate the equivalent partial charges due to dipole moments. Taken from Sowadski et al. (10).

interfacial area is large. Nondescript strands are involved as well as three or four helices and the beta sheet.

Enzymatic Mechanism

Alkaline phosphatase is a phosphohydrolase and phosphotransferase which operates through phosphorylation and dephosphorylation of the Ser-102 hydroxyl group in *E. coli* alkaline phosphatase. The overall reaction scheme is depicted in Fig. 7. In each step the tetrahedral phosphoryl ester group is putatively attacked on the face distal to the linking oxygen by a nucleophile in an in-line mechanism (7). The tetrahedron is inverted as the original linking oxygen is displaced and a new bond is formed. At the midpoint in this reaction, five oxygens surround the

$$R_1OP + E \rightleftharpoons R_1OP \cdot E \overset{R_1OH}{\rightleftharpoons} E-P \overset{H_2O}{\rightleftharpoons} E \cdot P \rightleftharpoons E + P_i$$
$$E-P \xrightarrow{R_2OH} R_2OP + E$$

FIG. 7. Alkaline phosphatase reaction (see the text).

phosphorus atom in a trigonal bipyramid configuration. Details of the intermediate states are conjectural. Jones et al. (5), with an isotopically chiral substrate, have shown that the complete transfer reaction yields overall retention of chirality, implying two in-line reactions. In the second step the phosphorylserine is attacked by water or an alcohol to complete the hydrolysis or transfer, respectively. In each step two oxygens are being activated, one to attack and the other to leave.

Alkaline phosphatase is relatively nonspecific. Almost any phosphomonoester is cleaved, and the K_ms and maximum velocities are not very variable. This is a consequence of the fact that the rate-limiting step is departure of the P_i produced. Physically, the active-site pocket contains the phosphate with little extra space. The leaving group is largely dangling in open solvent. Conversely, any accepting group in the transfer reaction can freely approach the phosphoryl serine. Nitrogen functions on the beta- or gamma-carbons of acceptors do enhance transfer activity in examples such as Tris buffer, beta-amino ethanol, and gamma-amino propanol. X-

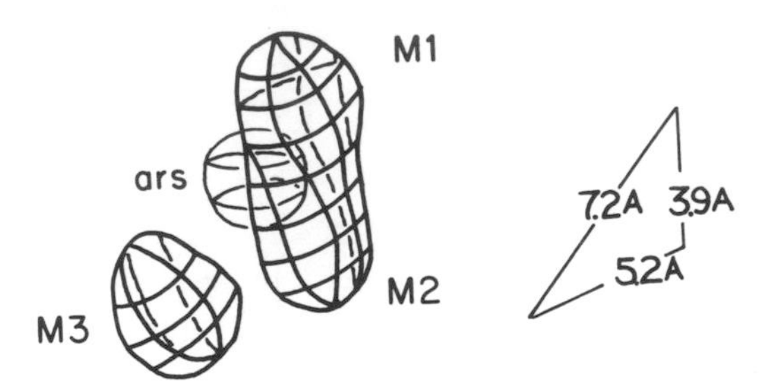

FIG. 8. Electron density map of Cd in alkaline phosphatase with Cd (50 mM + 2 mM P_i) replacing Zn and Mg, based on 0.34-nm resolution anomalous dispersion data and multiple isomorphous replacement phases. The orientation is as in the left side of Fig. 2. Arsenate electron density (from anomalous dispersion) is also represented as found in a Zn/Mg/AsO_4^{2-}/ alkaline phosphatase crystal (As, 0.2 mM). The arsenate density is behind the plane of the metals. M1, M2, and M3 are A, B, and C sites, respectively. The Ser-102 oxygen would be behind or within the arsenate density. Taken from Sowadski et al. (11).

ray studies of appropriate substrate analog complexes are planned.

J. E. Coleman has suggested for many years that the final hydrolysis step is an attack by a hydroxyl group formed from water by direct interaction with the A-site zinc ion. He has presented and discussed this idea (Coleman, this volume), and I will not try to document it here. I agree with his basic idea.

The three metals in each active-site region, A, B, and C, are illustrated in Fig. 8 in terms of the electron density map of cadmium replacements, as explained in the legend. The histidyl ligands would be in front of these metal ions. The arsenate as indicated is behind the plane of the metals, and Ser-102 oxygen is behind or within the arsenate density as contoured.

It is reasonable to propose that the leaving oxyanion in the original transfer of the phosphoryl to the protein is stabilized during the bond-breaking events by contact with the A-site zinc, just as water is activated in the final hydrolysis step. Likewise, the hydroxyl group of the incoming alcohol in the overall transfer reaction could be activated by the A-site metal. Equivalent events can be proposed for activation of the serine oxygen during the first-step attack on the phosphorus or the second-step breakage of the phosphorus-serine-oxygen bond using the B-site metal. Either zinc or magnesium can apparently function in the B site. Perhaps this is related to water being one of the reactants at the A site whereas the serine alcoholic oxygen is always the reactant at the B site. In detail, the B-site environment is different from the A site, with only one histidine ligand at B and a different degree of solvent accessibility. Alcohols ionize in the same pH range as water. Represenative ionization pKs tabulated by Streitwieser and Heathcock (12) are 15.7 for water, 15.5 for methanol, 10.1 for phenol, and 7.2 for *p*-nitrophenol. Note that in this context the pK_a of water refers to a first-order process and K_a is the ionization product 10 exp(−14) divided by the concentration of water, which is 55 M. The serine activation mechanism I proposed here may be either direct metal coordination or more distant via electrostatic effects moderated by the local dielectric environment. It is noteworthy that there is no second nucleophile such as the carboxyl-histidine couple in the serine proteases to activate Ser-102 in *E. coli* alkaline phosphatase. Arsenate has been found to bind near both the A and B sites and Ser-102. A substrate analog, phenylphosphonate, has been located in this region in one low-resolution experiment (unpublished data), but the details of the crucial orientation of the in-line attack axis cannot be inferred at present.

The guanidinium group of Arg-166 is in position to provide stabilization of charged intermediates or transition states along the reaction pathway. The carboxyl group of Asp-101 is positioned away from direct contact with the serine or the phosphate but serves to position Arg-166. The seven residues following Asp-101 and Ser-102 form a helical stem that is buried against the beta sheet by other helices and loops, providing an anchor for these residues.

Ghosh et al. (3) have shown that cysteine can substitute for Ser-102 and yield an active enzyme. If glycine were put there instead, perhaps a water molecule would take the place of the serine hydroxyl and be activated to produce direct hydrolysis with inversion. Transfer would then be replaced by direct phosphorylation of an incoming alcohol by P_i with probable racemization rather than retention of chirality or inversion.

Site-directed mutagenic replacement of Arg-166 with a neutral amino acid, which is being investigated by E. R. Kantrowitz (private communication), might affect the maximum velocity positively or negatively. Departure of the product P_i should be enhanced considerably, but the positive charge on Arg-166 may be very significant in the bond reactions, and they could become rate limiting with a substantial slowing of the overall reaction. Hydrolysis might be speeded up and transferase activity might be slowed down.

The involvement of the C-site metal in catalysis is apparently not prominent. It is not clear why it should be where it is. Specific mutations to decrease binding at the C site might help resolve this quandary.

```
Outer surface residues

    432       83      418                             48
Asp Leu|Thr Gln Thr|Arg Ala Tyr|    Val     |    Leu
       |           |           |Val     Thr|Gly     Ile Asn|
Glu Val|Pro Arg Met|Ala Phe Arg|    Ser     |    Phe

                                              255
                                          |Gly Leu|    Gln|
Gly Val Leu Phe|Thr Val Gly|Gly Leu Val|       |Val    |
                                          |Asn Val|    Asn|

Inner surface residues

         84             417 419           365        361  51   49  47  45
Thr    |Tyr Gly Leu|Leu Ile    |    Ile      Asn|    Ile Leu Ile
    Gly|           |        Ala|Thr     Leu    |Asp
Gln    |Phe Asp Ala|Val Val    |    Leu      Asp|    Leu Ile Leu

       322 320        147 145 143     203
   |Ala     Gln    |Ser Asn Thr|    Thr|Leu Leu|Ser    |
Lys|    Glu     Phe|           |Gly    |       |    Leu|
   |Gly     Phe    |Thr Val Val|    Ile|Arg Trp|Gln    |
```

FIG. 9. Comparison of the outer and inner surface residues of the beta sheet.

Placental Alkaline Phosphatase: Positive Comments on the Putative Physical Homology

The sequence of human term placental alkaline phosphatase has been determined by Kam et al. (6) by sequencing the cloned cDNA, and a clear homology with *E. coli* alkaline phosphatase has been established by the authors. With appropriate deletions and insertions they establish a 30% strict homology and cite many functionally relevant identities and similarities. Coleman (this volume) has commented on their findings relative to the metal binding sites in particular. The placental sequence is indeed compatible with all important aspects of the three-dimensional structure. I will comment on the surfaces of the beta sheet, the intersubunit interface, and several other aspects relevant to interspecies comparisons and molecular engineering.

The beta sheet can be described as having an inner surface and an outer surface relative to the center of gravity of the monomer. The curled corner wraps around the center of the monomer, and the helix supporting Ser-102 is near this center. A comparison of the residues on each sheet surface with the equivalent residues in the placental sequence is tabulated in Fig. 9, with the *E. coli* amino acids in the upper row. The strands are in their physical order, and succeeding strands are separated with a vertical line. The residues are in physically descending order in the orientation of the molecule used consistently in this paper.

The outer surface list of 29 residues has 16 identities (53%). Leu to Val changes at residues 432 and 255 and the Leu to Phe change at 48 are conservative. In the physical context the changes at 83 and 418 can also be considered pairwise to be conservative in that these residues are adjacent, on the same side of the sheet, and the positively charged guanidinium group is attached to one or the other. The physical position of the charge could be the same. The Asp to Glu change at 434 is conservative. Adding these six residues to the conserved category raises the fraction to 88%.

Of the 30 inner surface residues only 10 are identical. The metal ligands in this list, Asp-51 and Glu-322, are strict homologies. The changes at 84, 417, 419, 365, 361, 49, 47, and 45 are clearly conservative in context. This increases the conserved list to 18 (60%). The clusters of changes at 320, 145, 143, and 203 are particularly interesting. The somewhat anomalous hydrophilic patch on the *E. coli* sheet is replaced with a more classical hydrophobic area. Including the Ser to Thr change at 147, the volume of the residues in this pocket is increased slightly, but this presents no problem. There is at least one water molecule internal to the protein here in *E. coli*, and there are additional unexplained electron density and free volume. These changes might be considered constructive instead of conservative in that they would likely increase the stability and facilitate folding. It can also be inferred that the somewhat special situation in *E. coli* is not functionally important.

As noted by Kam et al. (6) and by Coleman (this volume), the metal ligands are conserved except for fully acceptable replacements at the relatively unimportant C site, where histidine replaces Asp-153 and serine replaces Thr-155. The Asp and His at 327 and 331 are conserved, so no clue is provided to help resolve the difficulties we have with assigning ligand status to these residues in *E. coli*.

The carbohydrate attachment sites in the placental enzyme and the insertions present no physical problem. The 16-residue insertion between 404 and 405 of *E. coli* occurs in an undefined surface region of the structure. The 12-residue insertion between 411 and 412 can easily be accommodated. Residues through 409 are in the undefined region.

The deletion of 10 residues between 31 and 42 is a particularly interesting case. This includes part of the 29 to 36 helix of *E. coli* alkaline phosphatase which is interfacial and would force the remainder of this helix to be moved to the bottom of the important, central 43 to 51 strand of the sheet and well away from the interface. The five-residue connection 37 to 42 is fully extended in *E. coli*. The problem is resolved by examining the helix prediction values calculated for us by J. Ponder with the algorithm of Garnier et al. (2). The homologies listed by Kam et al. equate placental 34 with *E. coli* 43 and correspondingly 33 with 32. The strong helical prediction for the placental sequence points to residues 17 to 22 instead of to the residues homologous with the *E. coli* alkaline phosphatase helix. Nine residues of weakly helical character between these and residue 34 allow the predicted helix to be placed physically where the *E. coli* helix is at the expense of the amino acid homology. The now longer connecting piece would loop out a bit.

The four cysteines of the two disulfide links in *E. coli* alkaline phosphatase are all replaced in the placental alkaline phosphatase. As noted above, neither of these disulfides seems to be important in the structure, a view confirmed by these differences. In addition, part of the apparently dispensable helix between Cys-168 and Cys-178 is in fact deleted.

All told, there is little doubt that these structures are quite homologous and that the structure presented here based on *E. coli* alkaline phosphatase is in essence the structure of many alkaline phosphatases. The structure provides a rational basis for evolutionary comparisons, for understanding fusion constructs, for molecular modification directed at a better understanding of structure and function, and for molecular engineering for practical ends.

This work was supported by Public Health Service grant GM22778 from the National Institute of General Medical Sciences.

LITERATURE CITED

1. **Applebury, M. L., and J. E. Coleman.** 1969. *Escherichia coli* alkaline phosphatase metal binding, protein conformation, and quaternary structure. J. Biol. Chem. **244:**308–318.
2. **Garnier, J., D. J. Osguthorpe, and B. Robson.** 1978. Analysis of the accuracy and implications of simple methods for predicting the secondary structure of globular proteins. J. Mol. Biol. **120:**97–120.
3. **Ghosh, S. S., S. C. Bock, S. E. Rokita, and E. T. Kaiser.** 1986. Modification of the active site of alkaline phosphatase by site-directed mutagenesis. Science **231:**145–148.
4. **Hanson, A. W., M. L. Applebury, J. E. Coleman, and H. W. Wyckoff.** 1970. X-ray studies on single crystals of *Escherichia coli* alkaline phosphatase. J. Biol. Chem. **245:** 4975–4976.
5. **Jones, S. R., L. A. Kindman, and J. R. Knowles.** 1978. Stereochemistry of phosphoryl group transfer using a chiral [^{16}O, ^{17}O, ^{18}O,] stereochemical course of alkaline phosphatase. Nature (London) **275:**564–565.
6. **Kam, W., E. Clauser, Y. S. Kim, Y. W. Kan, and W. J. Rutter.** 1985. Cloning, sequencing, and chromosomal localization of human term placental alkaline phosphatase cDNA. Proc. Natl. Acad. Sci. USA **82:**8715–8719.
7. **Knowles, J. R.** 1980. Enzyme-catalysed phosphoryl transfer reactions. Annu. Rev. Biochem. **49:**877–919.
8. **Knox, J. R., and H. W. Wyckoff.** 1973. A crystallographic study of alkaline phosphatase at 7.7 Å resolution. J. Mol. Biol. **74:**533–545.
9. **Sowadski, J. M., B. A. Foster, and H. W. Wyckoff.** 1981. Structure of alkaline phosphatase with zinc/magnesium cobalt or cadmium in the functional metal sites. J. Mol. Biol. **150:**245–272.
10. **Sowadski, J. M., M. D. Handschumacher, H. M. K. Murthy, B. A. Foster, and H. W. Wyckoff.** 1985. Refined structure of alkaline phosphatase from *Escherichia coli* at 2.8 Å resolution. J. Mol. Biol. **186:**417–433.
11. **Sowadski, J. M., M. D. Handschumacher, H. M. K. Murthy, C. E. Kundrot, and H. W. Wyckoff.** 1983. Crystallographic observations of the metal ion triple in the active site region of alkaline phosphatase. J. Mol. Biol. **170:**575–581.
12. **Streitwieser, A., Jr., and C. H. Heathcock.** 1985. Introduction to organic chemistry, 3rd ed., p. 195, 813. Macmillan Publishing Co., New York.
13. **Wyckoff, H. W., M. D. Handschumacher, H. M. K. Murthy, and J. M. Sowadski.** 1983. The three dimensional structure of alkaline phosphatase from *E. coli*. Adv. Enzymol. Relat. Areas Mol. Biol. **55:**453–480.

Multinuclear Nuclear Magnetic Resonance Approaches to the Structure and Mechanism of Alkaline Phosphatase

JOSEPH E. COLEMAN

Department of Molecular Biophysics and Biochemistry, Yale University, New Haven, Connecticut 06510

Alkaline phosphatase is a nonspecific phosphomonoesterase that hydrolyzes all phosphate monoesters, ROP, at approximately the same rate regardless of the nature of the R group or the pK_a of the leaving group, ROH (7–9). The enzyme is a Zn(II) metalloenzyme, widely distributed in nature from bacteria to humans (intestinal brush border, placenta, and bone) (8, 9). The enzyme from *Escherichia coli* has been most extensively studied and forms the prototype for this class of enzyme (8, 30). The *E. coli* enzyme is a dimer (molecular weight, 94,000) of identical subunits (47,000) containing 449 amino acid residues (3, 23). The enzyme catalyzes phosphate monoester hydrolysis via a phosphoseryl (Ser-102) covalent intermediate (E-P). The mechanism of action of the mammalian enzymes appears to be identical to that of the bacterial enzyme (the phosphoseryl intermediate was first identified in the calf intestinal enzyme) (11), although the molecules are larger and appear to include a membrane-soluble domain that anchors the enzyme to the brush border membrane or the membrane of the osteoblast cell (24).

The active center of the enzyme from *E. coli* contains three metal-binding sites (8, 30). One of these sites consists of three histidyl residues as ligands and preferentially binds Zn(II). The Zn(II) at this site participates directly in the mechanism by coordinating the incoming phosphate in the noncovalent Michaelis complex with either P_i (E·P) or ester (E·ROP) (8, 30). In *E. coli* the enzyme is synthesized on membrane-bound ribosomes as a preenzyme including a "signal" peptide of 22 residues (Fig. 1) and is secreted through the inner cytoplasmic membrane to the periplasmic space, where it is processed to the mature enzyme by a signal peptidase located on the inner membrane (4, 5, 13, 25). The properties of the *E. coli* enzyme are summarized in Table 1.

Catalytic Reactions of Alkaline Phosphatase

Alkaline phosphatase catalyzes the hydrolysis of phosphate monoesters by the steps illustrated in Fig. 2, where E-P is a phosphoseryl group formed with Ser-102 of the enzyme, and E·ROP and E·P are the noncovalent Michaelis complexes with substrate and product, respectively. At low pH (6 and below), dephosphorylation of E-P (k_3) is rate limiting, whereas at pH 8 to 9, where the enzyme is maximally active, product dissociation (k_4) is rate limiting (8). For phosphate-free enzymes, pre-steady-state bursts of RO^- are seen at both acid and alkaline pH, in agreement with the above assignment of rate-limiting steps (7, 8, 27).

Alkaline phosphatase also catalyzes a phosphoryl transfer reaction to alcohols. Amino alcohols like Tris are most effective, although aliphatic alcohols like glycerol are also reasonably efficient. Both increased ionic strength and Tris buffer (participating in transphosphorylation) activate alkaline phosphatase (15). The usual minimal representation of the transferase pathway is illustrated in Fig. 3. The mechanism is probably symmetrical about E-P, and thus transferase activity must involve formation of an E·R′OP complex and dissociation of the phosphoalcohol as separate steps. This mechanism would parallel normal hydrolysis, with R′OH taking the place of H_2O as the species forming the attacking nucleophile.

^{113}Cd NMR and Cd X-Ray Diffraction Applied to Alkaline Phosphatase

The Cd(II)-substituted alkaline phosphatases (see Table 1) have been very useful in structural and mechanistic studies of the enzyme in solution by virtue of the fact that natural-abundance Cd contains two spin ½ isotopes, ^{113}Cd (12.34%) and ^{111}Cd (12.86%), which have reasonable nuclear magnetic resonance (NMR) sensitivity (1, 12). Using ^{113}Cd, it is possible to obtain reasonable NMR signals from ^{113}Cd(II) bound at the active sites of metalloproteins (1). The example of alkaline phosphatase, substituted with ^{113}Cd(II) at the three metal-binding sites of each monomer of the symmetrical dimer, is shown in Fig. 4E. Each site gives rise to a separate resonance, labeled A, B, and C, with chemical shifts of 153, 70, and 2 ppm relative to the standard, $Cd(ClO_4)_2$ (Fig. 4E). Each resonance integrates to two Cd(II) ions, confirming the symemtrical nature of the dimer. The rest of Fig. 4 (parts A to D) illustrates the sequential occupancy of A, B, and C sites by progressive addition of ^{113}Cd(II) ions. It is possible to occupy A and B sites in two different orders. If $^{113}Cd(II)_2$ alkaline phosphatase is phosphoryl-

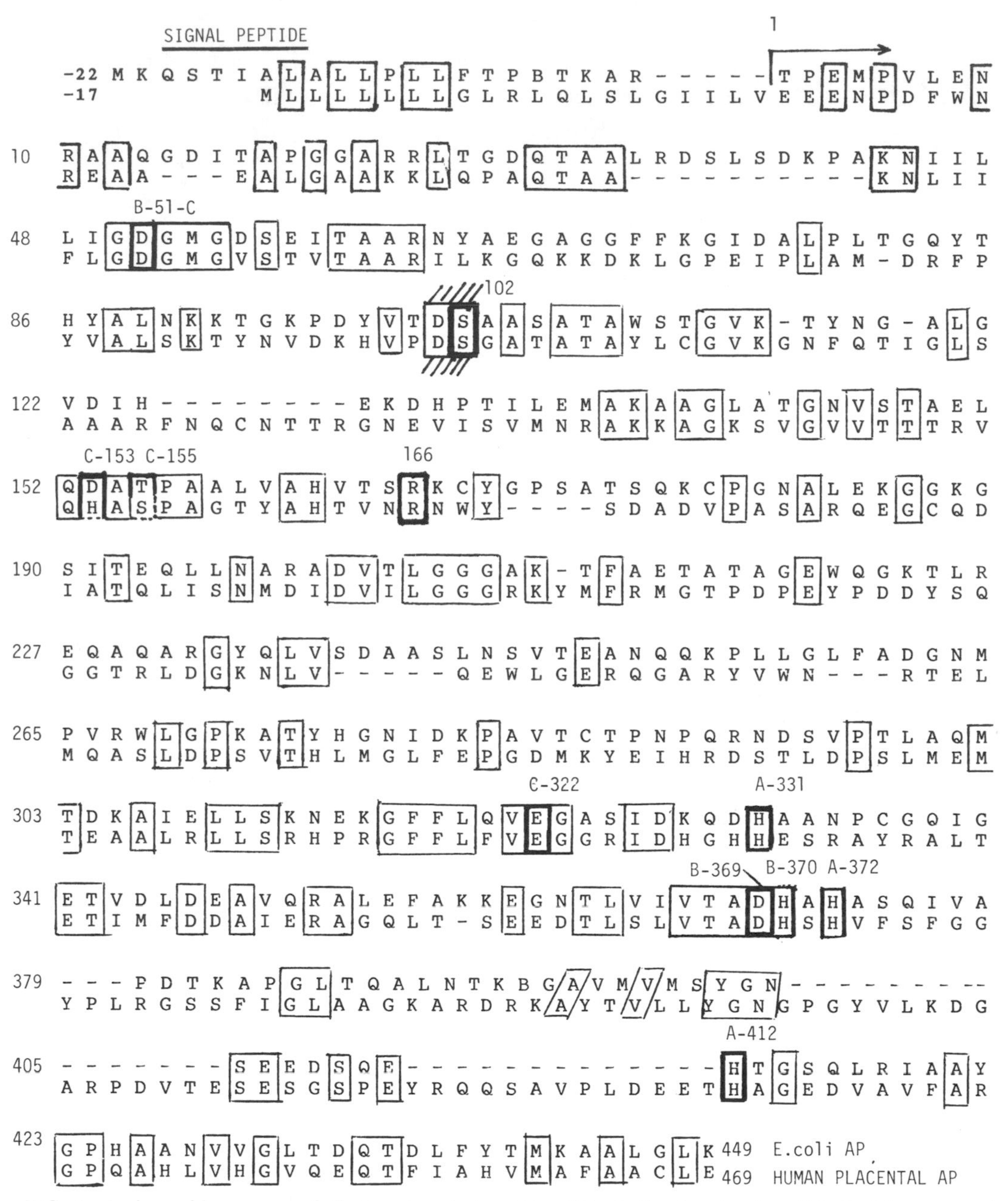

FIG. 1. Amino acid sequence of alkaline phosphatase (*E. coli*) including the signal peptide (3, 23), compared with the amino acid sequence of alkaline phosphatase from human placenta (22).

ated, only one phosphoseryl residue forms, and the original symmetrical distribution of the two ^{113}Cd(II) ions in each of the two A sites, giving a single resonance (Fig. 4A), shifts to two equal resonances, one ^{113}Cd(II) in the A site and one ^{113}Cd(II) in the adjacent B site of the phosphorylated monomer (Fig. 4B). Formation of the phosphoryl enzyme, E-P, so stabilizes the Cd_A-Ser P-Cd_B triad that the Cd(II) ion migrates from one A site to the B site on the phosphorylated monomer, thus leaving one monomer devoid of metal ions (12).

All four metal-binding sites, two A and two B sites per dimer, can be filled first (Fig. 4C). In this case, however, a chemical exchange broadening of the ^{113}Cd NMR signals due to conformational flux at the active center broadens the A-site resonance and completely prevents observation of the B-site resonances (8, 12) (Fig. 4C). This exchange broadening appears to be

TABLE 1. Molecular properties of the alkaline phosphatase from *E. coli*

1. Molecular weight 94,000.
2. Two identical subunits (molecular weight 47,000).
3. 449 amino acid residues.
4. Ser-102 is phosphorylated to form the covalent phosphoenzyme, E-P, during turnover.
5. Dephosphorylation of E-P is rate limiting at low pH.
6. Dissociation of product, P_i, from the noncovalent complex, E·P, is rate limiting at alkaline pH, $k_{cat} = 10^2\ s^{-1}$.
7. From the crystal structure there are three metal-binding sites, A, B, and C, at each active center.
   - Ligands to A: His-331, His-372, His-412, and one or two H_2O molecules.
   - Ligands to B: His-370, Asp-51, Asp-369.
   - Ligands to C: Asp-153, Glu-322, Thr-155, ?sharing of Asp-51 with site C.
8. Alkaline phosphatases of the following metal ion compositions and distributions between sites have been prepared. When site C is not indicated, it is not filled. Both active centers of the symmetrical dimer are indicated.
   - $(Zn_AZn_B)\ (Zn_AZn_B)$
   - $(Zn_AMg_B)\ (Zn_AMg_B)$
   - $(Cd_AZn_B)\ (Cd_AZn_B)$
   - $(Cd_A—_B)\ (Cd_A—_B)$
   - $(Cd_ACd_B)\ (—_A—_B)$
   - $(Zn_ACd_B)\ (Zn_ACd_B)$
   - $(Cd_ACd_B)\ (Cd_ACd_B)$
   - $(Cd_ACd_B)\ (Zn_AZn_B)$
   - $(Zn_AZn_BMg_C)\ (Zn_AZn_BMg_C)$
   - $(Zn_AMg_BMg_C)\ (Zn_AMg_BMg_C)$
   - $(Cd_ACd_BCd_C)\ (Cd_ACd_BCd_C)$
9. ^{113}Cd ^{31}P coupling on the ^{31}P signal of E·P shows that the A-site metal ion is directly coordinated to phosphate in the E·P complex.
10. The general structure of each monomer is an α-β-α core with 10 major β ribbons (both parallel and antiparallel) running through the core of the molecule.

due to exchange of the $^{113}Cd(II)$ at the B site with free $^{113}Cd(II)$ at a rate corresponding to intermediate exchange on the NMR time scale. If the $Cd(II)_4$ enzyme is phosphorylated, 2 mol of phosphoserine forms and the ^{113}Cd NMR spectrum becomes identical to that in part B except that each resonance represents two $^{113}Cd(II)$ ions of the symmetrical dimer. Thus, phosphorylation stabilizes the structure and slows down the exchange process.

The turnover number of the Cd(II) enzyme is $\sim 10^3$ times smaller than that of the Zn(II) enzyme, although it can be shown both by $H^{32}PO_4^{2-}$ labeling of the enzyme and by ^{31}P NMR that the Cd(II) enzyme forms both noncovalent (E·P) and covalent (E-P) phosphoenzymes (8, 9). Thus, the Cd(II) enzyme is a model in slow motion for the native enzyme. While the

$$E + ROP \underset{k_{-1}}{\overset{k_1}{\rightleftharpoons}} E \cdot ROP \underset{k_{-2}}{\overset{k_2,\ \uparrow RO^-}{\rightleftharpoons}} E\text{-}P \underset{k_{-3}}{\overset{H_2O \downarrow,\ k_3}{\rightleftharpoons}} E \cdot P \underset{k_{-4}}{\overset{k_4}{\rightleftharpoons}} E + P_i$$

FIG. 2. Hydrolysis of phosphate monoesters, by catalyzed alkaline phosphatase.

$$E + ROP \rightleftharpoons E \cdot ROP \overset{\uparrow RO^-}{\rightleftharpoons} E\text{-}P \overset{R'OH \downarrow}{\rightleftharpoons} R'OP + E$$

FIG. 3. The transferase pathway.

^{113}Cd NMR signals identify three separate metal-binding sites in each monomer, they do not establish the spatial relationships between them. That has been done by X-ray diffraction studies of the crystalline Cd(II) enzyme.

Cadmium shows significant anomalous dispersion of X-rays such that pairs of reflections are not of the same intensity and thus yield phase information. Thus, an electron density map of the atoms participating in anomalous dispersion can be calculated directly; the electron density map of the Cd(II) ions in $Cd(II)_6$ alkaline phosphatase is shown in Fig. 5 (29, 30). The Cd(II) density is embedded in the polypeptide structure of the native enzyme determined by the more usual isomorphous replacement method (29). This technique avoids the problems associated with the Fourier difference maps between metallo- and apoenzymes in which amino acid side-chain motion associated with metal ion removal is also included (30). The striking feature of the metal-binding sites in alkaline phosphatase is that all three are spaced very close to one another at each active center, occupying a triangle 3.9 by 4.9 by 7 Å (0.39 by 0.49 by 0.7 nm) (Fig. 5). The interpretation of this map is shown in Fig. 6. Seryl residue 102, known to be the residue phosphorylated in E-P (3), is located between metal ions A and B and slightly to one side of the axis connecting them. Since it is constructed by embedding the Cd(II) electron density in the map of the native zinc enzyme, the precise molecular structure must await a map of the Cd_6 alkaline phosphatase itself.

Assignment of the ^{113}Cd NMR Signals

Given the nature of the ligands to Cd(II) at each of the three Cd(II) sites (Fig. 6), a reasonable assignment of the three $^{113}Cd(II)$ resonances can be made. Sites containing all oxygen ligands show $^{113}Cd(II)$ NMR signals near 0 ppm or at an even higher field of the Cd perchlorate signal (the standard at 0 ppm). Thus, the 2-ppm signal is easily assigned to the C site. Cd(II) sites containing three N ligands tend to resonate between 150 and 250 ppm (8); hence, the ^{113}Cd resonance at 153 is assigned to the A site consisting of three His as ligands. The intermediate signal at 70 ppm is then assigned to the B site consisting of one His nitrogen and three oxygen atoms as ligands, a chemical shift reasonable for this ligand composition. Hence, from low field to high field the ^{113}Cd resonances are assigned in the order A site (153 ppm), B site (70 ppm), and

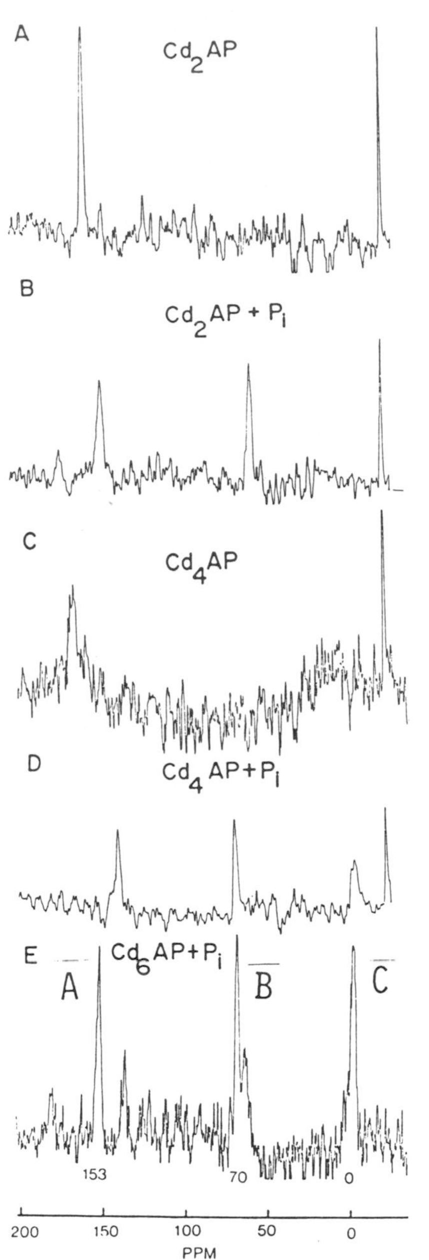

FIG. 4. ^{113}Cd NMR spectra of ^{113}Cd-substituted alkaline phosphatase (AP) as functions of Cd(II) and phosphate stoichiometry. (A) $^{113}Cd_2AP$, pH 6, no phosphate. (B) Sample A + phosphate. (C) $^{113}Cd_4AP$, pH 6, no phosphate. (D) Sample C + phosphate, pH 6.5. (E) $^{113}Cd_6AP$ + phosphate, pH 8.3. ^{113}Cd spectra were obtained on a Bruker CXP-200 at 44.37 MHz, and chemical shifts are expressed relative to $Cd(ClO_4)_2$. The standard at the far right is cadmium acetate resonating under these conditions at −17 ppm. Data from references 8, 9, and 12.

C site (2 ppm). The characteristics of these signals can thus be used to probe the structure and function of these three separate metal ion sites.

Comparative Biochemistry of the Alkaline Phosphatases

Alkaline phosphatases are of great importance to mammalian physiology. These membrane-attached proteins participate in Ca(II) and HPO_4^{2-} uptake at the intestinal brush border, are essential for the formation of hydroxyapatite, and have less well determined functions in kidney and placenta. Although the molecular events that insert these enzymes (whose currently known function is the nonspecific hydrolysis of phosphate monoesters) into these complex processes are unknown, the periplasmic alkaline phosphatase from *E. coli* has proved to be a valid prototype for the catalytic domains of the more complex mammalian enzymes. These prototypic features include the Zn(II) ions and the formation of a phosphoserine intermediate. Very recently, the complete amino acid sequence of the human placental alkaline phosphatase has been obtained from a clone of the cDNA for the placental enzyme (22). The sequence of the placental enzyme is compared with that of the *E. coli* enzyme in Fig. 1. Additions and deletions have been made in both enzymes to given maximum conservation of sequence. Not surprisingly, there appear to have been a number of inserts in the larger mammalian enzyme. While there has been a good deal of drift in the sequence up the evolutionary tree, if one examines the critical residues found in the *E. coli* enzyme around the three-metal site, there has been remarkable conservation of these residues.

Ser-102 and the positive charge at the active center contributed by Arg-166 are both preserved. Among the ligands, the three His ligands to A-site Zn(II) are preserved, as is His-370, the N ligand to B site. Asp-51, bridging B and C sites, is preserved, as are the other ligands to the B site contributed by Glu-369. The two additional ligands to the C site have changed, but maintain the metal liganding properties, Glu-153 to His-153 and Thr-155 to Ser-155. Thus, the structure of the active center and much of the mechanism of action of alkaline phosphatase have been preserved from *E. coli* to humans.

What Does the Combination of ^{31}P and ^{113}Cd NMR Say about the Precise Relationships between Substrate and Metal Ions at Each Active Center?

It is possible, by combining ^{113}Cd and ^{31}P NMR, to follow the progressive phosphorylation of Ser-102 in the NMR tube because catalysis by the Cd(II) enzyme is sufficiently slow. An example is shown in Fig. 7 for the phosphorylation of

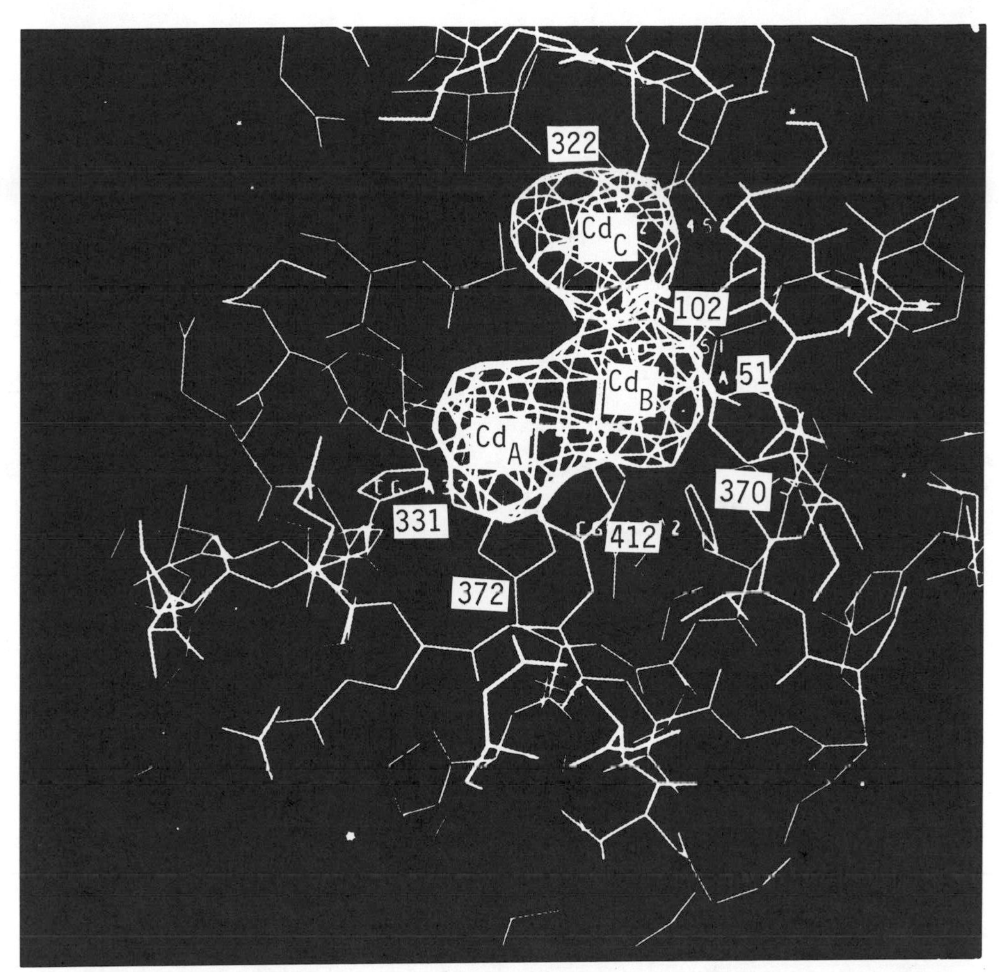

FIG. 5. Computer graphics representation of the immediate vicinity of the active center of alkaline phosphatase. The metal-binding sites, A, B, and C, at each active center are represented by the electron densities of the Cd(II) ions in the crystalline $Cd(II)_6$ alkaline phosphatase calculated from anomalous dispersion data (30).

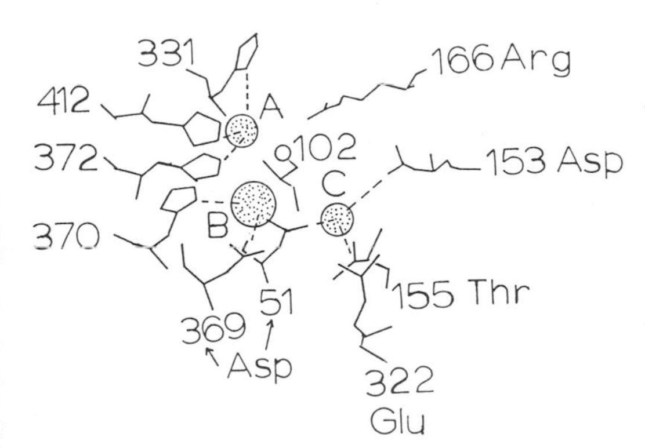

FIG. 6. Diagram of the residues immediately surrounding the three metal ions at the active center of alkaline phosphatase, drawn from a computer graphics fitting of the electron density map (data from the laboratory of H. W. Wyckoff). Residue 102 is the serine phosphorylated in E-P.

the $Cd(II)_4$ alkaline phosphatase (A and B sites fully occupied) (13). Addition of P_i at pH 9.0 results in the immediate binding of phosphate, giving rise to a ^{31}P resonance at ~13 ppm (Fig. 7a). We have assigned this resonance to E·P, since magnetization transfer studies show this resonance to be in faster exchange with free phosphate than the resonance assigned to the phosphoseryl residue (see below for comments on the chemical shift). The E·P ^{31}P resonance shows a 30-Hz $^{113}Cd\text{-}O\text{-}^{31}P$ coupling to a single $^{113}Cd(II)$ ion. Hence phosphate in E·P is directly coordinated to one, but not both, Cd(II) ions at the active center. Heteronuclear decoupling, i.e., irradiating the ^{113}Cd resonances (A′ and B′, Fig. 7a) and observing phosphorus resonances, shows that irradiation of A′ removes the coupling. Irradiation of B′ has no effect. Hence, the phosphate of E·P is coordinated to the A-site

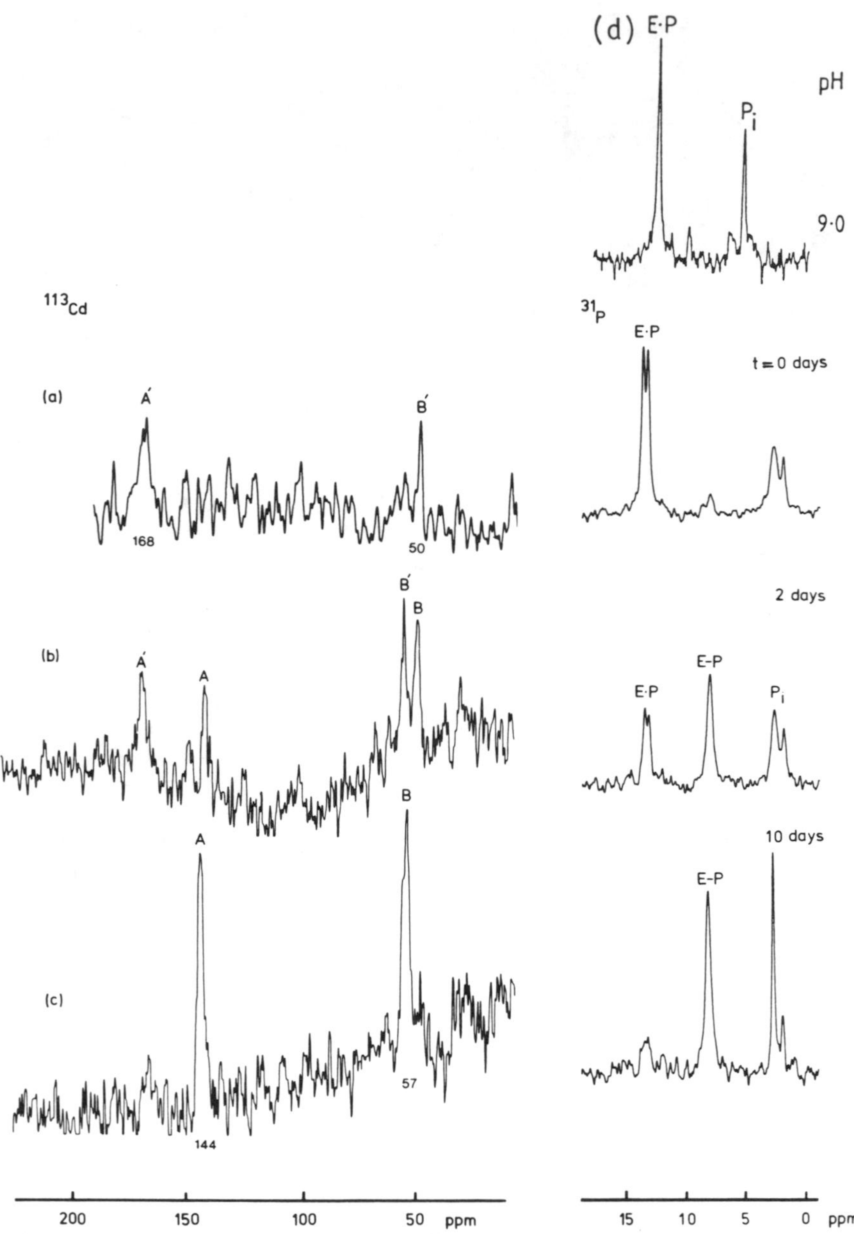

FIG. 7. Time course of the slow phosphorylation of $^{113}Cd(II)_2$ alkaline phosphatase (2.76 mM) containing two equivalents of P_i at pH 9.0. The spectra on the right are ^{31}P NMR (80.9 MHz) and on the left are ^{113}Cd NMR (44.3 MHz, a; 19.96 MHz, b and c) (13). (d) ^{31}P spectrum of the phospho-$Zn(II)_ACd(II)_B$ alkaline phosphatase hybrid at pH 9.0.

Cd(II) ion, but is not close enough to coordinate B.

After 2 days, half the phosphate has phosphorylated Ser-102, giving rise to a new ^{31}P resonance at 8.3 ppm (assigned to E-P) (13). This resonance is a singlet; hence, the phosphoseryl group is coordinated to neither A- nor B-site metal ion, a fact confirmed by the finding that irradiation of none of the ^{113}Cd resonances narrows this ^{31}P line. When an E·P to E-P ratio of 50:50 is reached (Fig. 7b), the ^{113}Cd signals from both A and B sites have split into pairs, A′ and B′, reflecting active centers in the E·P form, and A and B, reflecting active centers in the E-P form. Hence, the proximity of the two phosphoenzyme intermediates to the A- and B-site metal ions is clearly evident. The ^{113}Cd chemical shift of the A site is most sensitive to the E·P →

E-P change, moving upfield by ~25 ppm, consistent with the phosphate ion leaving its coordination sphere as it moves on to the serine hydroxyl.

$[(Zn(II)_A Cd(II)_B)]_2$ Alkaline Phosphatase

In a preparation where Zn(II) is slowly removed from $Zn(II)_4$ alkaline phosphatase by the chelating agent 1,10-phenanthroline, it is possible to remove Zn(II) preferentially from the B sites, leaving a nearly homogeneous $Zn(II)_{2(A)}$ alkaline phosphatase to which $^{113}Cd(II)$ ions can be added to form $[Zn(II)_{2A}Cd(II)_{2B})]_2$ alkaline phosphatase (14). The ^{113}Cd NMR spectrum of this hybrid shows a single major resonance at 30 ppm which corresponds to $^{113}Cd(II)$ in the B sites. Thus the ^{113}Cd NMR confirms the picture of this hybrid as one in which Cd(II) occupies B sites and Zn(II) occupies the A sites. At pH 9.0 the ^{31}P spectrum of the $Zn(II)_A{}^{113}Cd(II)_B$ alkaline phosphatase shows a prominent singlet at 12.66 ppm, assignable to E·P of the hybrid (Fig. 7d). Thus, replacement of $^{113}Cd(II)$ at the A site by Zn(II) removes the 30-Hz coupling on E·P, as expected if the phosphate of E·P is coordinated to the A-site Zn(II), but not to the B-site metal ion which remains $^{113}Cd(II)$. The unusual downfield shift of the E·P resonance [12.66 ppm compared with 4.3 ppm for the native Zn(II) enzyme] remains (Fig. 7d). This suggests that the Cd(II) in the B site is primarily responsible for the low-field shift. The adjacent noncoordinated metal ion in the B site thus seems to be primarily responsible for the conformational change causing the large change in chemical shift in the phosphorus of E·P.

$[(Zn(II)_A Mg(II)_B)]_2$ Alkaline Phosphatase

Another hybrid of considerable interest, which may be of some physiological significance, is the $[(Zn(II)_A Mg(II)_B)]_2$ alkaline phosphatase. Zn(II) has reasonably high affinity for both A and B sites, such that if only two Zn(II) ions are added back to the apoenzyme, there is heterogeneous occupancy of both A and B sites (10, 13). It is also easy to isolate alkaline phosphatase containing three to four Zn(II) per molecule, i.e., with most of both A and B sites filled (2, 17, 28). In contrast, Mg(II) has little affinity for the A site and will not compete with Zn(II) for the A site (2, 10). Mg(II) will bind to the B site and successfully competes with Zn(II) for the B site at high enough concentrations (10). Hence, if two Zn(II) ions are added to the apoenzyme plus 10 mM Mg(II), one forms a homogeneous $[(Zn(II)_A Mg(II)_B Mg(II)_C)]_2$ alkaline phosphatase from which Mg(II) can be dialyzed to yield $[(Zn(II)_A Mg(II)_B)]_2$ alkaline phosphatase. When phosphate is bound to this species, the major ^{31}P resonance appearing from enzyme-bound phosphate is at 1.8 ppm (14). This is an E·P species that occurs unusually far upfield, compared with 3.4 ppm and 4.3 ppm found for $[(Zn(II)_A Zn(II)_B Mg(II)_C)]_2$ and $[(Zn(II)_A Zn(II)_B)]_2$ alkaline phosphatase species. As will be detailed below, this hybrid shows a significant shift in the pH profile describing the E·P $\rightleftharpoons$ E-P equilibrium. Under the standard assay conditions which include 1 M Tris, this hybrid shows full activity (10, 15). Close examination reveals, however, that in the absence of Tris as an acceptor alcohol, the hydrolysis rate of this enzyme is only about 10% that of the native protein (15). Ninety percent of the observed activity in the normal 1 M Tris buffer is due to the phosphotransferase reaction forming Tris phosphate, a feature confirmed by the ^{31}P NMR assay of transphosphorylation (15). The molecular basis for this shift from hydrolysis to transphosphorylation induced by Mg(II) at the B site is currently not understood.

Functional Consequences of Metal Ion Substitutions at A and B Sites

Native alkaline phosphatase shows a sigmoid pH-rate profile with an apparent pK_a from 7.5 to 8, with maximum activity above pH 9 (7). Detailed metal-binding studies defining the properties of the three closely spaced metal-binding sites at each active center show that it is difficult to form homogeneous species unless one adds six of one species of metal ion per dimer (7, 13, 14). Hence, caution must be exercised in the interpretation of data on function unless one is certain which sites are occupied by which metal ions. Procedures have now been developed to uniquely occupy each set of sites in the dimer with a specific type of metal ion (14), as illustrated by the hybrids discussed above. One major functional consequence of metal ion substitution at the active center of alkaline phosphatase has already been mentioned; i.e., Cd(II) substitution for Zn(II) drops the catalytic activity observed at pH 8 by $\sim 10^3$ (8).

The apparent pK_a of activity is a major reflection of the proton dissociation behavior of one or more groups involved in the catalytic mechanism. Arguments based on the minimal mechanism for this enzyme suggest that this pK_a directly controls the equilibrium between E·P and E-P, i.e., the overall rates of phosphorylation and dephosphorylation of Ser-102, i.e., k_3/k_{-3} (13, 14). At low pH, $k_{-3} >> k_3$ and hence E-P is the major equilibrium species, while at alkaline pH, $k_3 >> k_{-3}$ and E·P is the major equilibrium species, even though in the alkaline pH range the reaction is still proceeding through E-P (21). Hence the E-P/E·P ratio as a function of metal ion species and pH implies much about the catalytic steps. ^{31}P NMR methods (exchange,

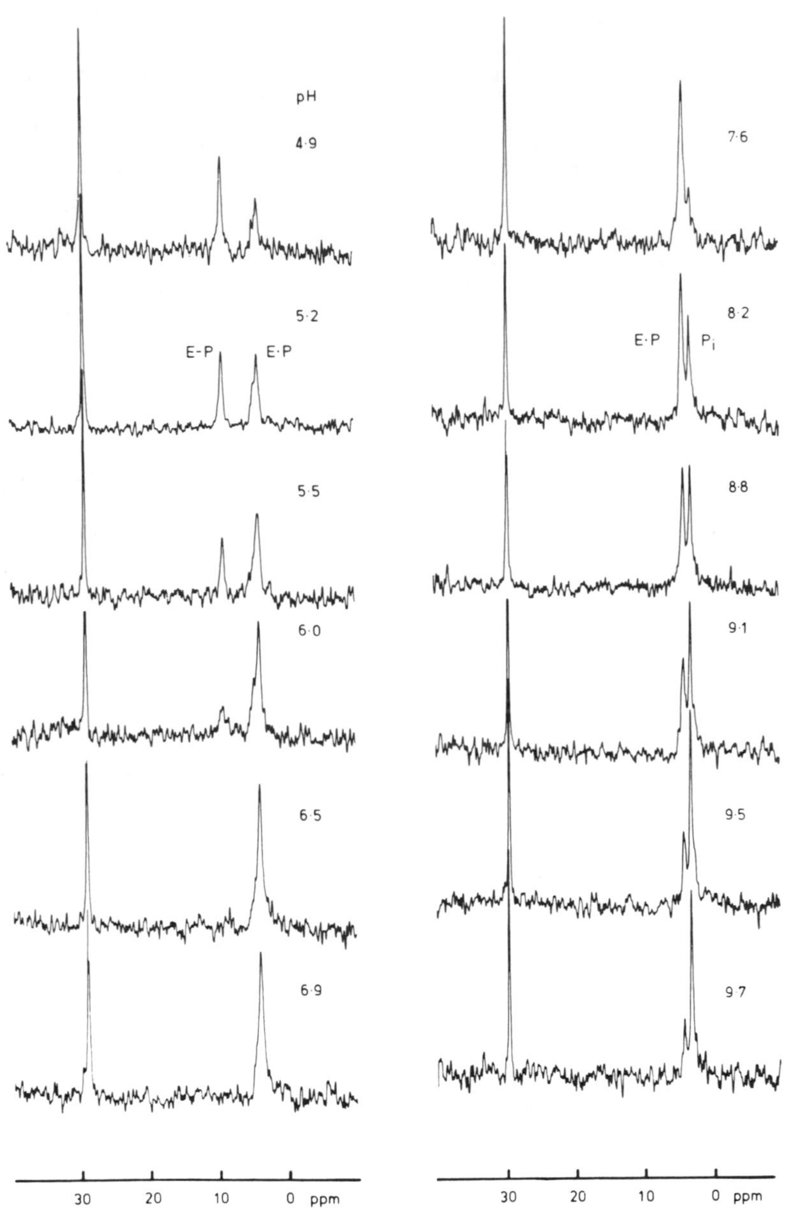

FIG. 8. ^{31}P NMR spectra (80.9 MHz) of $Zn(II)_4Mg(II)_2$ alkaline phosphatase, containing one equivalent of phosphate, as a function of pH. The enzyme concentration was 2.13 mM. Spectra are the average of 14,000 transients. The most down-field resonance is external methyl phosphonate (13).

magnetization transfer [26], and inversion transfer [15]) have been very valuable in measuring product (phosphate) dissociation rates, the rate-limiting step when the enzyme is most active (6, 8, 18).

The pH dependency of the E-P/E·P ratio is a direct reflection of the pK_a of the activity-linked group, which we will identify as the pK_a of a water molecule coordinated to the A-site metal ion. The best way of following the pH dependency of the ratio of phospho intermediates is to follow the ^{31}P NMR signals of the two intermediates as a function of pH, as illustrated in Fig. 8 for the $[(Zn(II)_AZn(II)_BMg(II)_C)]_2$ alkaline phosphatase, where the phosphoenzyme intermediate shifts from primarily E-P at pH 4.9 (resonance, 8.6 ppm) to exclusively E·P at pH 7 (resonance, 4.2 ppm) and finally to dissociation of a fraction of P_i from E·P at pH values above 8 (resonance, 3.5 ppm), where the phosphate dis-

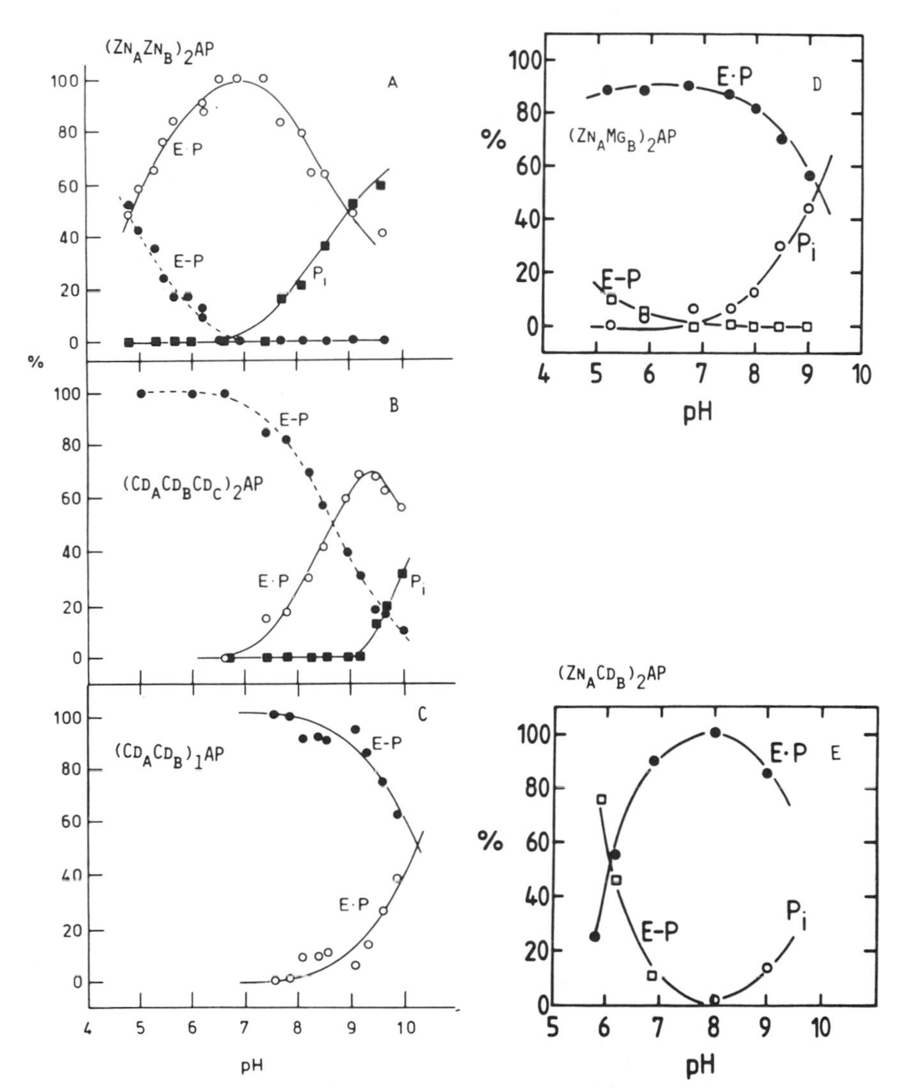

FIG. 9. Relative equilibrium concentrations of phosphorus species (E·P, E-P, P_i) for zinc- and cadmium-containing alkaline phosphatases (AP) as a function of pH. (A) 1.98 mM $Zn(II)_4AP$ plus one equivalent of phosphate; (B) 1.83 mM $Cd(II)_2AP$ plus two equivalents of phosphate; (C) 2.00 mM $Cd(II)_6AP$ plus two equivalents of phosphate. Phosphate was added at pH 6.5 and equilibrium was attained before the pH was adjusted. Only one equivalent of phosphate remained bound throughout the titration, and therefore the percentages of E·P and E-P are normalized to this concentration rather than to two equivalents. Solid lines are for visual aid. Broken lines represent E-P concentrations determined earlier from ^{32}P-labeling measurements (13). (D) Percentage of E-P (□), E·P (●), and phosphate (○) in the presence of $Zn(II)_2Mg(II)_2AP$ versus pH (14). Two equivalents of total phosphate per dimer are present. (E) Percentage of E-P (□), E·P (●), and phosphate (○) in the presence of $[Zn(II)_ACD(II)_B]_2AP$ versus pH (14). Two equivalents of total phosphate per dimer are present.

sociation constant becomes larger. The pH profiles for the [E-P] $\rightleftharpoons$ [E·P] $\rightleftharpoons$ [P_i] + [E] equilibrium are collected in Fig. 9 for various metalloalkaline phosphatases, including those with homogenous metal ion composition and those with different metal ions in each of the several sites (13, 14). The pH at which [E-P] = [E·P] is listed in Table 2 along with the ^{31}P chemical shifts of the two species.

Two conclusions can be drawn. The metal ion at the A site appears to have the major influence on the pK_a of the group controlling this equilibrium. Secondly, Zn(II) in A must induce a relatively low pK_a such that the midpoint of the E-P $\rightleftharpoons$ E·P equilibrium occurs near pH 5 (Table 2). In contrast, Cd(II) at the A site results in a pK_a at least 3 pH units higher, between 8 and 10 (Table 2). The actual pK_a value is considerably higher

TABLE 2. Effect of metal ions on the E-P ⇌ E·P equilibrium of alkaline phosphatase (AP)

Metal site occupancy at the active center	pH at which [E-P] = [E·P]	Chemical shift	
		E-P	E·P
$[Zn(II)_A Zn(II)_B]_2AP$	5	8.6	4.3
$[Zn(II)_A Zn(II)_B Mg(II)_C]_2AP$	5	8.6	3.4
$[Zn(II)_A Mg(II)_B]_2AP$	4.5	8.5	1.8
$[Zn(II)_A Cd(II)_B]_2AP$	6	8.0	12.6
$[Cd(II)_A Cd(II)_B]_1AP$	10	8.4	13.4
$[Cd(II)_A Cd(II)_B Cd(II)_C]_2AP$	8.7	8.7	13.0
$[Cd(II)_A Zn(II)_B]_2AP$ (unstable)	7	9.3	

than the pH of the midpoint of the E·P ⇌ E-P equilibrium (see below). The most plausible explanation for this phenomenon associated with the Zn(II)-to-Cd(II) substitution is that the activity-linked pK_a represents that of a coordinated solvent on the A-site metal ion (13, 14). Above the pK_a this becomes a coordinated hydroxide ion and is likely to be the nucleophile attacking the phosphorus atom of E-P in the dephosphorylation reaction. Within the dramatic changes in pH profile induced by changing Zn(II) to Cd(II) in the A site, the nature of the B-site metal ion has a secondary and much smaller effect on the pK_a; e.g., Mg(II) in B lowers the pK_a, while Cd(II) in B raises the pK_a.

Mechanism of Action Incorporating Conclusions from the NMR Data

^{31}P NMR data suggest that the A-site metal ion, initially coordinating phosphate in the E·P complex, develops an open coordination site after formation of E-P (Fig. 10). This movement would allow a solvent molecule to coordinate the A-site metal ion. Thus, in the phosphoryl native enzyme, Zn(II) in the A site is postulated to coordinate a solvent molecule with a relatively low pK_a (7 to 8). The midpoint of the E-P ⇌ E·P equilibrium is several pH units lower than the actual pK_a because of the ratio of kinetic constants, $k_{phosphorylation}/k_{dephosphorylation}$. The pK_a would be expected to be much higher for a solvent (H_2O) coordinated to Cd(II), a softer metal with less polarizing power than Zn(II), thus making the coordinated solvent a poorer acid. The proximity of B site to A site, only 0.4 nm apart, would explain why there are secondary modifications of the pK_a of the coordinated solvent by the metal ion occupying the B site.

Influence of the B-Site Metal Ion on the Conformation of the E·P and E-P Intermediates

The chemical shifts of the noncovalent E·P complexes of the three enzyme species containing Zn(II) in A sites and either Mg(II), Zn(II), or Cd(II) in B sites show large variation, reflecting the different metal in the B site. E·P resonates at 1.8, 4.3, and 12.6 ppm for species containing Mg(II), Zn(II), or Cd(II), respectively, at the B site with A occupied by Zn(II) in all cases (Table 2). One might attribute such a range of chemical shifts to direct coordination of the phosphate in E·P to the different species of metal ion occupying the B site. There are, however, no model system data suggesting such variation in chemical shifts in the case of coordination compounds containing phosphate groups. Chemical shift changes on coordination are generally small, are similar for several metals, and reflect displacement of the proton (20). Current data clearly support coordination of phosphate only to the A-site metal ion, and at no step in the mechanism does it appear to coordinate B (Fig. 10).

Phosphate of E·P is anchored by coordination to the A-site metal ion and possibly also by interaction with Arg-166, which swings down toward the A-site metal ion (30). Thus, one may picture E·P as a complex in which phosphate is relatively fixed, possibly allowing small conformational changes induced by the adjacent B-site metal ion to cause significant distortion of phosphate bond angles, a feature known to induce large down-field ^{31}P shifts (16). On the other hand, the noncoordinated phosphate group of the phosphoserine may not be rigidly held. Compatible with the postulate that the phosphoseryl group is not coordinated is the observation that the ^{31}P shift of E-P is relatively insensitive to the nature of the metal ion at either A or B sites, being 8.0, 8.6, 8.4, and 8.5 ppm for $[(Zn(II)_A Cd(II)_B)]_2$, $[(Zn(II)_A Zn(II)_B)]_2$, $[(Cd(II)_A Cd(II)_B)]_2$,

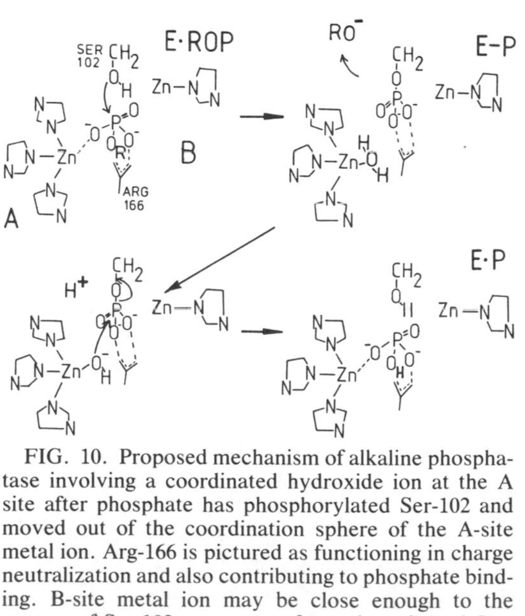

FIG. 10. Proposed mechanism of alkaline phosphatase involving a coordinated hydroxide ion at the A site after phosphate has phosphorylated Ser-102 and moved out of the coordination sphere of the A-site metal ion. Arg-166 is pictured as functioning in charge neutralization and also contributing to phosphate binding. B-site metal ion may be close enough to the oxygen of Ser-102 to promote formation of seryl-O^-.

and $[(Zn(II)_A Mg(II)_B)]_2$ alkaline phosphatases, respectively.

Influence of B-Site Metal Ion on Dissociation of P_i from the Active Center

The rate-limiting step in the hydrolysis of phosphate monoesters by alkaline phosphatase acid pH is dephosphorylation of E-P, while at alkaline pH where the enzyme is maximally active, dissociation of product, P_i, is rate limiting (8, 15, 26). The substitution of Cd(II) for Zn(II) in both A and B sites not only shifts the midpoint of the E-P $\rightleftharpoons$ E·P equilibrium to alkaline pH, but also dramatically lowers the phosphate dissociation rate (15), while enzyme turnover is reduced by two to three orders of magnitude (8). At pH 9 the dissociation rate of P_i from the Cd(II) enzyme is $\sim 1\ s^{-1}$ (26), which is still too fast for dissociation to be rate limiting and suggests that dephosphorylation remains rate limiting for $Cd(II)_6$ alkaline phosphatase, even at pH 9. If B-site Zn(II) is replaced by Cd(II) to form the $[Zn(II)_A Cd(II)_B]_2$ alkaline phosphatase hybrid, P_i dissociation falls dramatically, from a value of 20 to 50 s^{-1} calculated for $Zn(II)_4 Mg(II)_2$ alkaline phosphatase at alkaline pH (10, 14, 28) to 2 to 6 s^{-1}, as determined by saturation transfer (15). Slowing of product dissociation is accompanied by the unusual downfield shift in the E·P resonance, from the near-free value of 3.4 ppm for $Zn(II)_4 Mg(II)_2$ alkaline phosphatase to values between 12 and 13.4 ppm when cadmium occupies the B site (14). An altered phosphate conformation and a reduced phosphate dissociation rate may both be manifestations of small changes in the active site brought about by replacement of the native B-site Zn(II) or Mg(II) with the larger Cd(II) ion.

A diagram of the mechanism which it is possible to synthesize from the current data is shown in Fig. 10. The characteristics of the hybrid metal phosphatases provide a much clearer picture of the functions of the separate metal ions. The metal ion in the A site, zinc in the native enzyme, serves to coordinate the phosphate group in the noncovalent complex. As the phosphate is transferred to Ser-102, its coordination site on the A-site metal ion originally occupied by phosphate (or ROP) would become available to solvent. Phosphorylation of Ser-102 can proceed slowly in the absence of B-site metal ion, but is accelerated by its presence (10, 12). Acceleration of phosphorylation is probably due to additional charge neutralization of the negative phosphate group as it moves toward Ser-102 and therefore closer to the B-site metal ion. The metal ion at the B site could be close enough to the seryl oxygen to potentiate the formation of seryl-O^-, though there is no evidence as yet for direct coordination. Presumably, the noncovalent complex with a phosphomonoester as substrate, E·ROP, would be analogous. Dephosphorylation of the phosphoserine is inefficient and rate limiting at low pH where the concentration of $Zn\text{-}OH^-$, formed by ionization of a water molecule bound to A-site zinc, would be small. At alkaline pH dephosphorylation is greatly accelerated, paralleled by a shift in the E-P $\rightleftharpoons$ E·P equilibirum to the right and a large increase in the concentration of the nucleophile. Charge neutralization at the phosphoserine by the metal ion at the B site would also potentiate the nucleophilic attack of the coordinated ^-OH at the second stage of the mechanism. Dissociation of product phosphate from the highly positively charged active center then apparently becomes rate limiting. The pK_a of the bound water is primarily determined by the A-site metal ion and thus differs most between $Zn(II)_4$ and $Cd(II)_6$ alkaline phosphatases (Fig. 10).

This work was supported by Public Health Service grants AM09070 and AM32067 from the National Institutes of Health.

LITERATURE CITED

1. **Armitage, I. M., A. J. M. Schoot-Uiterkamp, J. F. Chlebowski, and J. E. Coleman.** 1978. ^{113}Cd NMR as a probe of the active sites of metalloenzymes. J. Mag. Res. **29:**375–392.
2. **Bosron, W. F., R. A. Anderson, M. C. Falk, F. S. Kennedy, and B. L. Vallee.** 1977. Effect of magnesium on the properties of zinc alkaline phosphotase. Biochemistry **16:**610–614.
3. **Bradshaw, R. A., F. Cancedda, L. H. Ericsson, P. A. Neuman, S. P. Piccoli, M. J. Schlesinger, K. Shrifer, and K. A. Walsh.** 1981. Amino acid sequence of *Escherichia coli* alkaline phosphatase. Proc. Natl. Acad. Sci. USA **78:** 3473–3477.
4. **Brockman, R. W., and L. A. Heppel.** 1968. On the localization of alkaline phosphatase and cyclic phosphodiesterase in *Escherichia coli*. Biochemistry **7:**2554–2562.
5. **Chang, C. N., H. Inouye, P. Model, and J. Beckwith.** 1980. Processing of alkaline phosphatase precursor to the mature enzyme by an *Escherichia coli* inner membrane preparation. J. Bacteriol. **142:**726–728.
6. **Chlebowski, J. F., I. M. Armitage, P. Tusa, and J. E. Coleman.** 1976. ^{31}P NMR of phosphate and phosphonate complexes of metalloalkaline phosphatases. J. Biol. Chem. **151:**1207–1216.
7. **Chlebowski, J. F., and J. E. Coleman.** 1974. Mechanisms of hydrolysis of O-phosphorothioates and inorganic thiophosphate by *Escherichia coli* alkaline phosphatase. J. Biol. Chem. **249:**7192–7202.
8. **Coleman, J. E., and P. Gettins.** 1983. Alkaline phosphatase, solution structure, and mechanism. Adv. Enzymol. Relat. Areas Mol. Biol. **55:**381–452.
9. **Coleman, J. E., and P. Gettins.** 1983. Molecular properties and mechanism of alkaline phosphatase, p. 153–217. *In* T. G. Sprio (ed.), Zinc enzymes. John Wiley and Sons, Inc., New York.
10. **Coleman, J. E., K.-I. Nakamura, and J. F. Chlebowski.** 1983. $^{65}Zn(II)$, $^{115m}Cd(II)$, $^{60}Co(II)$, and Mg(II) binding to alkaline phosphatase of *Escherichia coli*. J. Biol. Chem. **258:**386–395.
11. **Engstrom, L.** 1961. Studies on calf-intestinal alkaline phosphatase. Biochim. Biophys. Acta **52:**49–59.

12. **Gettins, P., and J. E. Coleman.** 1983. ^{113}Cd nuclear magnetic resonance of Cd(II) alkaline phosphatases. J. Biol. Chem. **258:**396–407.
13. **Gettins, P., and J. E. Coleman.** 1983. ^{13}P nuclear magnetic resonance of phosphoenzyme intermediates of alkaline phosphatase. J. Biol. Chem. **258:**408–416.
14. **Gettins, P., and J. E. Coleman.** 1984. Zn(II)-^{113}Cd(II) and Zn(II)-Mg(II) hybrids of alkaline phosphatase. J. Biol. Chem. **259:**4991–4997.
15. **Gettins, P., M. Metzler, and J. E. Coleman.** 1985. Alkaline phosphatase. ^{31}P NMR probes of the mechanism. J. Biol. Chem. **260:**2875–2883.
16. **Gorenstein, D. G.** 1981. Nucleotide conformational analysis by ^{31}P nuclear magnetic resonance spectroscopy. Annu. Rev. Biophys. Bioeng. **10:**355–386.
17. **Harris, M. L., and J. E. Coleman.** 1968. The biosynthesis of apo- and metalloalkaline phosphatases of *Escherichia coli*. J. Biol. Chem. **243:**5063–5073.
18. **Hull, W. E., S. E. Halford, H. Gutfreund, and B. D. Sykes.** 1976. ^{31}P nuclear magnetic resonance study of alkaline phosphatase: the role of inorganic phosphate in limiting the enzyme turnover rate at alkaline pH. Biochemistry **15:** 1547–1561.
19. **Inouye, H., C. Pratt, J. Beckwith, and A. Torriani.** 1977. Alkaline phosphatase synthesis in cell-free system using DNA and RNA templates. J. Mol. Biol. **110:**75–87.
20. **Jaffe, E. K., and M. Cohn.** 1978. ^{31}P nuclear magnetic resonance spectra of the thiophosphate analogues of adenine nucleotides; effects of pH and Mg^{2+} binding. Biochemistry **17:**652–657.
21. **Jones, S. R., L. A. Kindman, and J. R. Knowles.** 1978. Stereochemistry of phosphoryl group transfer using a chiral [^{16}O, ^{17}O, ^{18}O]: stereochemical course of alkaline phosphatase. Nature (London) **257:**564–565.
22. **Kam, W., E. Clauser, Y. S. Kim, Y. W. Kan, and W. J. Rutter.** 1985. Cloning, sequencing, and chromosomal localization of human term placental alkaline phosphatase cDNA. Proc. Natl. Acad. Sci. USA **82:**8715–8719.
23. **Kikuchi, Y., K. Yoda, M. Yamasaki, and G. Tamura.** 1981. The nucleotide sequence of the promoter and the amino-terminal region of alkaline phosphatase structural gene (phoA) of *Escherichia coli*. Nucleic Acids Res. **9:** 5671–5678.
24. **McComb, R. B., G. N. Bowers, and S. Posen.** 1979. Alkaline phosphatase. Plenum Publishing Co., Inc., New York.
25. **Neu, H. C., and L. A. Heppel.** 1965. The release of enzymes from *Escherichia coli* by osmotic shock and during the formation of spherolasts. J. Biol. Chem. **240:** 3685–3692.
26. **Otvos, J. D., J. R. Alger, J. E. Coleman, and I. M. Armitage.** 1979. ^{31}P NMR of alkaline phosphatase. J. Biol. Chem. **254:**1778–1780.
27. **Reid, T. W., and I. B. Wilson.** 1971. *E. coli* alkaline phosphatase. Enzymes **4:**373–415.
28. **Simpson, R. T., and B. L. Vallee.** 1968. Two differentiable classes of metal atoms in alkaline phosphatase of *Escherichia coli*. Biochemistry **7:**4343–4350.
29. **Sowadski, J. M., M. Handschumacher, H. M. K. Murthy, B. Z. Foster, and H. W. Wyckoff.** 1985. Refined structure of alkaline phosphatase from *Escherichia coli* at 2.8 Å resolution. J. Mol. Biol. **186:**417–433.
30. **Wyckoff, H. W., M. Handschumacher, K. Murthy, and J. M. Sowadski.** 1983. The three dimensional structure of alkaline phosphatase from *E. coli*. Adv. Enzymol. Relat. Areas Mol. Biol. **55:**453–479.

Molecular Mechanism of Isozyme Formation of Alkaline Phosphatase in *Escherichia coli*

ATSUO NAKATA,[1] HIDEO SHINAGAWA,[1] AND FRANK G. ROTHMAN[2]

Department of Experimental Chemotherapy, The Research Institute for Microbial Diseases, Osaka University, 3–1, Yamadaoka, Suita, Osaka, Japan 565,[1] and Division of Biology and Medicine, Brown University, Providence, Rhode Island 02912[2]

Analysis of highly purified alkaline phosphatase from *Escherichia coli* by starch gel electrophoresis reveals a pattern of several enzymatically active isozymes (24; M. L. Bach, E. R. Signer, C. Levinthal, and I. W. Sizer, Fed. Proc. **20**:255, 1961). The isozymes are all coded for by the single *phoA* gene (3) since the entire pattern is shifted by a point mutation in this gene (Bach et al., Fed. Proc. **20**:255, 1961). E. Signer (Ph.D. thesis, Massachusetts Institute of Technology, Cambridge, 1963) showed that the isozymes were not artifacts generated during the isolation, purification, or analysis of the enzyme, and that the relative amounts of the different isozymes synthesized were influenced by the conditions of bacterial growth.

Although more than three isozyme bands have been observed in extracts of cells grown under various conditions (9, 16, 18, 20), structural studies have focused on the three prevalent isozymes produced in Tris-glucose medium (6, 11) (Fig. 1). These isozymes are numbered 1, 2, and 3 in the order of electrophoretic mobility, 1 being the slowest. Better resolution of these three isozymes by electrophoresis is obtained in polyacrylamide gels, using 10-fold diluted buffer, than in starch gels (9, 15). Individual isozymes can be purified by DEAE-cellulose chromatography (6, 8, 20, 23). Isozymes 1 and 3 are identical or similar in optical rotatory dispersion, heat stability, substrate specificity, pH optimum, and a number of catalytic properties measured at pH values above 7.4 (8, 20). However, the steady-state rate of substrate hydrolysis at pH 5.5 catalyzed by isozyme 1 is almost twice that catalyzed by isozyme 3 (2, 21).

E. coli alkaline phosphatase is a dimer of two identical subunits (19). When monomers of isozymes 1 and 3 are mixed and allowed to reassociate, phosphatase with the mobility of isozyme 2 as well as 1 and 3 is formed (8, 20). This result indicates that isozyme 2 is a hybrid dimer containing one monomer each of isozymes 1 and 3.

Schlesinger and co-workers elucidated the chemical difference between isozymes 1 and 3 by sequence analysis of the amino-terminal peptides obtained after cyanogen bromide cleavage (6). Isozyme 1 was found to have an extra amino acid residue at the amino terminus, namely, arginine.

This result suggested that the origin of isozymes 2 and 3 is the sequential cleavage of the N-terminal arginine residues of isozyme 1 by a protease, consistent with results of in vivo labeling experiments which had indicated that isozyme 1 is a precursor to isozymes 2 and 3 (20, 21). Addition of arginine to the culture medium inhibits the conversion of isozyme 1 to isozyme 3 (18, 21). The proteolytic nature of the isozyme conversion was also suggested (13) by its inhibition by some serine-protease (endopeptidase) inhibitors, such as antipain and leupeptin. Bestatin, an inhibitor of aminopeptidase B, did not inhibit isozyme conversion.

An alternative explanation for conversion of isozyme 1 to isozymes 2 and 3, namely, that arginine is added posttranslationally to the amino terminus of isozyme 3, is not likely. DNA sequence analysis shows a codon for arginine at the amino terminus of the mature enzyme, following a sequence encoding a signal peptide 21 residues long (4, 7).

Piggott et al. (18) suggested that translational ambiguity contributes to the formation of phosphatase isozymes. They observed effects of *strA* and *ram* mutations as well as of streptomycin on the isozyme patterns. The spread or reduction of isozyme patterns paralleled the restriction or enhancement of translational ambiguity. This mechanism may contribute to heterogeneity within each band at a level not detected in amino acid sequence studies and may account for the origin of minor isozyme bands.

We have used a genetic approach to explore the mechanism of conversion of isozyme 1 to isozymes 2 and 3. Mutant strains defective in this conversion were isolated (9, 15) and designated *iap* (isozyme of alkaline phosphatase). The mutants mapped at 59 min on the *E. coli* standard genetic map (1), with map order *srl-iap-cys-argA-thyA-lysA* (16). The wild-type *iap* gene was found to be dominant (10).

Since an isozyme pattern similar to that in *E. coli* was observed in *Salmonella typhimurium* and *Serratia marcescens* carrying the *E. coli*

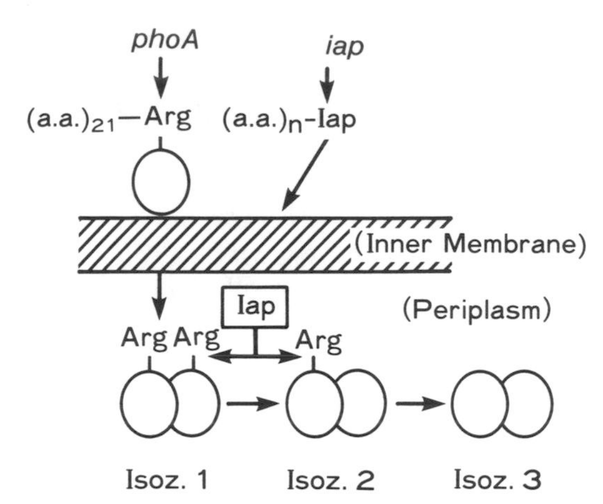

FIG. 1. Schematic illustration of alkaline phosphatase isozyme conversion.

phoA gene (22, 24), an *iap* gene was presumably active in each of these species. The *S. typhimurium iap* gene has been identified and maps at a similar locus as in *E. coli* (8).

A chromosomal fragment carrying the *E. coli iap* gene has been cloned on a multicopy plasmid (12). Cells carrying the multicopy plasmid containing *iap* converted isozyme 1 to isozymes 2 and 3 more efficiently than Iap^+ cells without the plasmid, indicating that the product of the *iap* gene responsible for isozyme conversion is overproduced in these cells.

In vitro studies with extracts of cells carrying many copies of the cloned *iap* gene have confirmed a gene dosage effect and have provided a convenient system for further characterization of the isozyme-converting reaction(s) (14). Leupeptin was found to inhibit at lower concentrations than in vivo, and no inhibition was observed with ovomucoid trypsin inhibitor. Inhibition by arginine and antipain was observed in vitro, but only at higher concentrations than in vivo. This observation raises the possibility that synthesis of the enzyme responsible for isozyme conversion is repressed by arginine in the growth medium. Extracts from cells grown under phosphate-deprived conditions had only marginally more conversion activity than those of cells grown in excess phosphate, suggesting that synthesis of the conversion enzyme is not induced by phosphate starvation.

```
    5         10        15        20
MetPheSerAlaLeuArgHisArgThrAlaAlaLeuAlaLeuGlyValCysPheIleLeu

        25
ProValHisAlaSerSerPro.....
```

FIG. 2. Amino acid sequence of the Iap protein deduced from DNA sequence. A putative signal peptide is underlined.

The nucleotide sequence of the *iap* gene has recently been determined (unpublished data). It contains an open reading frame coding for a protein of 345 amino acid residues (molecular weight, 37,920). The amino-terminal sequence of this deduced protein (Fig. 2) contains the characteristic features of a signal sequence (17): 1 to 3 positively charged residues near the amino terminus (at positions 6 to 8); following these, a stretch of 14 to 20 neutral, primarily hydrophobic residues (at positions 9 to 22); and a putative cutting site preceded by a preferred sequence (Val-X-Ala at residues 22 to 24). If the Iap protein does indeed contain a signal sequence, we would expect it to be secreted into the periplasm, where alkaline phosphatase is located and isozyme conversion is therefore presumed to take place. The cellular localization of isozyme-converting activity is under investigation.

The biological role of the Iap protein is still unclear. The similarity of the enzymatic parameters of isozymes 1 and 3 at physiological pH, and the presence of the *iap* gene in *S. typhimurium* which does not synthesize alkaline phosphatase, suggest that the *iap* gene has not solely evolved for alkaline phosphatase isozyme conversion. The Iap protein may work on other substrates as well and may function primarily to replenish arginine when the pool is depleted.

LITERATURE CITED

1. **Bachmann, B. J.** 1983. Linkage map of *Escherichia coli* K-12, edition 7. Microbiol. Rev. **47:**180–230.
2. **Bloch, W., and M. J. Schlesinger.** 1974. Kinetics of substrate hydrolysis by molecular variants of *E. coli* alkaline phosphatase. J. Biol. Chem. **249:**1760–1768.
3. **Garen, A., and S. Garen.** 1963. Complementation *in vivo* between structural mutants of alkaline phosphatase from *E. coli*. J. Mol. Biol. **7:**13–22.
4. **Inouye, H., W. Barnes, and J. Beckwith.** 1982. Signal sequence of alkaline phosphatase of *Escherichia coli*. J. Bacteriol. **149:**434–439.
5. **Ishizu, J., T. Uematsu, S. Takaichi, T. Seki, and S. Fukuda.** 1984. Existence of an "*iap*" gene in *Salmonella typhimurium* and its location on the chromosome. Bull. Liber. Arts Sci. NMS **5:**9–16.
6. **Kelley, P. M., P. A. Neumann, K. Shriefer, F. Cancedda, M. J. Schlesinger, and R. A. Bradshaw.** 1973. Amino acid sequence of *Escherichia coli* alkaline phosphatase. Amino- and carboxyl-terminal sequences and variations between two isozymes. Biochemistry **12:**3499–3503.
7. **Kikuchi, Y., K. Yoda, M. Yamasaki, and G. Tamura.** 1981. The nucleotide sequence of the promoter and the amino-terminal region of alkaline phosphatase structural gene (phoA) of Escherichia coli. Nucleic Acids Res. **21:** 5671–5678.
8. **Lazdunski, C., and M. Lazdunski.** 1967. Les isophosphatases alcalines d'*Escherichia coli*. Separation, proprietés cinetiques et structurales. Biochim. Biophys. Acta **147:** 280–288.
9. **Nakata, A., M. Amemura, M. Yamaguchi, and K. Izutani.** 1977. Factors affecting the formation of alkaline phosphatase isozymes in *Escherichia coli*. Biken J. **20:**47–55.
10. **Nakata, A., and J. Kawamata.** 1980. Dominance of alkaline phosphatase isozyme gene (*iap*) of *Escherichia coli*. Biken J. **23:**49–51.

11. **Nakata, A., G. R. Peterson, E. L. Brooks, and F. G. Rothman.** 1971. Location and orientation of the *phoA* locus on the *Escherichia coli* K-12 linkage map. J. Bacteriol. **107:**683–689.
12. **Nakata, A., H. Shinagawa, and M. Amemura.** 1982. Cloning of alkaline phosphatase isozyme gene (*iap*) of *Escherichia coli*. Gene **19:**313–319.
13. **Nakata, A., H. Shinagawa, and J. Kawamata.** 1979. Inhibition of alkaline phosphatase isozyme conversion by protease inhibitors in *Escherichia coli* K-12. FEBS Lett. **105:**147–1150.
14. **Nakata, A., H. Shinagawa, and H. Shima.** 1984. Alkaline phosphatase isozyme conversion by cell-free extract of *Escherichia coli*. FEBS Lett. **175:**343–348.
15. **Nakata, A., M. Yamaguchi, K. Izutani, and M. Amemura.** 1978. *Escherichia coli* mutants deficient in the production of alkaline phosphatase isozymes. J. Bacteriol. **134:**287–294.
16. **Nesmeyanova, M. A., O. B. Motlokh, M. N. Kolot, and I. S. Kulaev.** 1981. Multiple forms of alkaline phosphatase from *Escherichia coli* cells with repressed and derepressed biosynthesis of the enzyme. J. Bacteriol. **146:**453–459.
17. **Oliver, D.** 1985. Protein secretion in *Escherichia coli*. Annu. Rev. Microbiol. **39:**615–648.
18. **Piggot, P. J., M. D. Sklar, and L. Gorini.** 1972. Ribosomal alterations controlling alkaline phosphatase isozymes in *Escherichia coli*. J. Bacteriol. **110:**291–299.
19. **Rothman, F., and R. Byrne.** 1963. Fingerprint analysis of alkaline phosphatase of *E. coli* K12. J. Mol. Biol. **6:**330–340.
20. **Schlesinger, M. J., and L. Andersen.** 1968. Multiple molecular forms of the alkaline phosphatase of *Escherichia coli*. Ann. N.Y. Acad. Sci. **151:**159–170.
21. **Schlesinger, M. J., W. Bloch, and P. M. Kelley.** 1975. Differences in the structure, function, and formation of two isozymes of *Escherichia coli* alkaline phosphatase, p. 333–342. *In* C. L. Markert (ed.), Isozymes I. Molecular structure. Academic Press, Inc., New York.
22. **Schlesinger, M. J., and R. Olsen.** 1968. Expression and localization of *Escherichia coli* alkaline phosphatase synthesized in *Salmonella typhimurium* cytoplasm. J. Bacteriol. **96:**1601–1605.
23. **Schlesinger, M. J., and R. Olsen.** 1970. A new, simple, rapid procedure for purification of *Escherichia coli* alkaline phosphatase. Anal. Biochem. **36:**86–90.
24. **Signer, E., A. Torriani, and C. Levinthal.** 1961. Gene expression in intergeneric merozygotes. Cold Spring Harbor Symp. Quant. Biol. **26:**31–34.

V. TRANSPORT OF PHOSPHATE AND PHOSPHORYLATED COMPOUNDS IN *ESCHERICHIA COLI*

Introduction

EZRA YAGIL

Department of Biochemistry, The George S. Wise Faculty of Life Sciences, Tel Aviv University, Tel Aviv 69978, Israel

The importance of P_i and some of its organic carbohydrate esters in the metabolism of *Escherichia coli* is reflected by the number of active transport systems for these compounds which the organism has acquired. There exist two systems for the active transport of P_i, two for *sn*-glycerol 3-phosphate (Gly3P), and one for hexose phosphate (1, 4, 4a, 8, 9, 11, 13–18, 20; for a comprehensive review, see H. Rosenberg, *in* S. Silver and B. P. Rosen, ed., *Ion Transport in Prokaryotes*, in press).

Among these systems, it is convenient to distinguish between the ones which operate under "normal" growth conditions, i.e., when there is ample P_i in the growth medium (over 5 μM), and those which become the major ones when the cells are starved for P_i (Table 1). When P_i is abundant, the Pit (P_i transport) system operates as the major active transport system for P_i, GlpT (Gly3P transport) operates for Gly3P, and Uhp (uptake of hexose phosphate) operates for hexose phosphates (Table 1A). These three systems belong to the group of systems energized by the proton motive force (3). They are considered to be relatively simple and less efficient because they consist of only one membrane component and their transport constant (K_t) toward their substrates is relatively high, in the range of 20 to 100 μM. Indeed, under conditions of abundant P_i the cells grow well and do not require any overspecialized means to scavenge scarce sources of phosphates. The Pit system is constitutive, whereas GlpT and Uhp are induced by their corresponding substrates (Gly3P and hexose phosphate, respectively). The transport of the organic phosphoesters in the latter two systems is linked with an efflux of P_i; i.e., there is an exchange of one P_i for each molecule of Gly3P or hexose phosphate which enters the cell (4–6, 8, 10, 12).

When, on the other hand, the concentration of P_i in the medium drops below 5 μM, cell growth stops, and the *pho* regulon is turned on by a circuit of signals which is discussed by Lugtenberg, Torriani-Gorini, Shinagawa et al., and Surin et al. (this volume), but is not yet fully understood (19). Along with alkaline phosphatase and several other components of the *pho* regulon, two additional, more efficient, but more complex transport systems are turned on: Pst (P_i-specific transport), which is specific to P_i, and Ugp (uptake of Gly3P), which is specific to Gly3P (Table 1B). These two systems can transport much lesser amounts of substrate because their K_ts are 150-fold lower (for P_i) and 10-fold lower (for Gly3P) when compared with the parallel systems, Pit and GlpT, which operate under excess P_i. Indeed, their structure is more complex since it is composed of at least three different membrane-associated proteins and one specific periplasmic binding protein. The energy of Pst and probably also of Ugp is supplied (at least in part) by phosphate-rich energy bonds. The genes involved and their location are summarized in Table 1(B). Notably, Gly3P transported by Ugp can be used only as a source of P_i, and no parallel system is known for hexose phosphate (1, 12–14, 18).

Along with Pst and Ugp, the *pho* regulon turns on the synthesis of PhoE, a unique porin which serves as an outer membrane-specific funnel which facilitates the entry of P_i, small phosphorylated compounds, and other anions into the periplasmic space (2, 7). Thus, Pst and Ugp have evolved as complex transport systems composed of specific components in the outer membrane, in the periplasm, and in the cytoplasmic membrane, serving as scavengers of phosphate residues which may have been left in the P_i-meager environment.

In the following articles, the systems mentioned above and their structure, function, and regulation are discussed in greater detail.

TABLE 1. Transport systems of P_i, Gly3P, and hexose phosphate in *E. coli*

Conditions	Substrate transported	Transport system	Membrane transport genes	Genetic location (min)	Transport constant, K_t (μM)	Regulation
A. In excess P_i	P_i	Pit	*pit*	77	25	No
	Gly3P	GlpT	*glpT*	48	20	Yes
	Hexose phosphate	Uhp	*uhpT*	80	100	Yes
B. Under P_i starvation	P_i	Pst	*pstS*,[a] *pstC*, *pstB*, *pstA*	83	0.16	Yes
	Gly3P	Ugp	*ugpD*, *ugpB*,[a] *ugpA*, *ugpC*	73.5	2	Yes

[a] Gene coding for periplasmic binding protein.

LITERATURE CITED

1. **Amemura, M., K. Makino, H. Shinagawa, A. Kobayashi, and A. Nakata.** 1985. Nucleotide sequence of the genes involved in phosphate transport and regulation of the phosphate regulon in *Escherichia coli*. J. Mol. Biol. **184:** 241–250.
2. **Benz, R., R. P. Darveau, and R. E. W. Hancock.** 1984. Outer membrane protein PhoE from *Escherichia coli* forms anion-selective pores in lipid-bilayer membranes. Eur. J. Biochem. **140:**319–324.
3. **Berger, E. A.** 1973. Different mechanisms of energy coupling for the active transport of proline and gluthamine in *Escherichia coli*. Proc. Natl. Acad. Sci. USA **70:**1514–1518.
4. **Dietz, G. W.** 1976. The hexose phosphate transport system of *Escherichia coli*. *In* A. Meisner (ed.), Advances in enzymology, p. 237–259. John Wiley & Sons, New York.

4a. **Elvin, C. M., N. E. Dixon, and H. Rosenberg.** 1986. Molecular cloning of the phosphate (inorganic) transport (*pit*) gene of *Escherichia coli* K-12: identification of the *pit* gene product and physical mapping of the *pit-gor* region. Mol. Gen. Genet. **204:**477–484.

5. **Elvin, C. M., C. M. Hardy, and H. Rosenberg.** 1985. Inorganic phosphate exchange mediated by the GlpT-dependent *sn*-glycerol-3-phosphate transport system in *Escherichia coli*. J Bacteriol. **161:**1054–1058.
6. **Essenberg, R. C., and H. L. Kornberg.** 1975. Energy coupling in the uptake of hexose phosphates by *Escherichia coli*. J. Biol. Chem. **250:**939–945.
7. **Korteland, J., P. de Graaf, and B. Lugtenberg.** 1984. PhoE protein pores in the outer membrane of *Escherichia coli* K-12 not only have a preference for Pi and Pi-containing solutes but are general anion preferring channels. Biochim. Biophys. Acta **778:**311–316.
8. **Larson, T. J., M. Ehrmann, and W. Boos.** 1983. Periplasmic glycerophosphodiester phosphodiesterase of *Escherichia coli*, a new enzyme of the *glp* regulon. J. Biol. Chem. **258:**5428–5432.
9. **Larson, T. J., G. Schumacher, and W. Boos.** 1982. Identification of the *glpT*-encoded *sn*-glycerol-3-phosphate permease of *Escherichia coli*, an oligometric integral membrane protein. J. Bacteriol. **152:**1008–1021.
10. **Ludke, D., T. J. Larson, C. Beck, and W. Boos.** 1982. Only one gene is required for the *glpT*-dependent transport of *sn*-glycerol-3-phosphate in *Escherichia coli*. Mol. Gen. Genet. **186:**540–547.
11. **Rosenberg, H., R. G. Gerdes, and K. Chegwidden.** 1977. Two systems for the uptake of phosphate in *Escherichia coli*. J. Bacteriol. **131:**505–511.
12. **Rosenberg, H., R. G. Gerdes, and F. M. Harold.** 1979. Energy coupling to the transport of inorganic phosphate in *Escherichia coli*. Biochem. J. **178:**133–137.
13. **Schweizer, H., M. Argast, and W. Boos.** 1982. Characteristics of a binding protein-dependent transport system for *sn*-glycerol-3-phosphate in *Escherichia coli* that is part of the *pho* regulon. J. Bacteriol. **150:**1154–1163.
14. **Schweizer, H., and W. Boos.** 1985. Regulation of *ugp*, the *sn*-glycerol-3-phosphate transport system of *Escherichia coli* K-12 that is a part of the *pho* regulon. J. Bacteriol. **163:**392–394.
15. **Schweizer, H., W. Boos, and T. J. Larson.** 1985. Repressor for the *sn*-glycerol-3-phosphate regulon of *Escherichia coli* K-12: cloning of the *glpR* gene and identification of its product. J. Bacteriol. **161:**536–566.
16. **Shattuck-Eidens, D. M., and R. J. Kadner.** 1981. Exogenous induction of the *Escherichia coli* hexose phosphate transport system defined by *uhp-lac* operon fusions. J. Bacteriol. **148:**203–209.
17. **Shattuck-Eidens, D. M., and R. J. Kadner.** 1983. Molecular cloning of the *uhp* region and evidence for a positive activator for expression of the hexose phosphate transport system of *Escherichia coli*. J. Bacteriol. **155:**1062–1070.
18. **Surin, B. P., H. Rosenberg, and G. B. Cox.** 1985. Phosphate-specific transport system of *Escherichia coli*: nucleotide sequence and gene-polypeptide relationships. J. Bacteriol. **161:**189–198.
19. **Torriani, A., and D. N. Ludtke.** 1985. The *pho* regulon of *Escherichia coli*, p. 224–242. *In* M. Schaechter, F. C. Neidhart, J. Ingraham, and O. Kjeldgaard (eds.), The molecular biology of bacterial growth. Jones and Bartlett Publishers, Boston.
20. **Willsky, G. R., and M. H. Malamy.** 1980. Characterization of two genetically separable inorganic phosphate transport systems in *Escherichia coli*. J. Bacteriol. **144:**356–365.

Molecular Studies on the Phosphate-Specific Transport System of *Escherichia coli*

BRIAN P. SURIN, GRAEME B. COX, AND HARRY ROSENBERG

Department of Biochemistry, John Curtin School of Medical Research, Australian National University, Canberra City, A.C.T., Australia 2601

The phosphate-specific transport (Pst) system of *Escherichia coli* K-12 is a multicomponent system consisting of several membrane proteins and a soluble periplasmic binding protein. Originally described by Medveczky and Rosenberg (11, 12), this system has been the subject of further study in several laboratories (2, 5, 6, 12, 22, 23), culminating in a detailed description (14, 25) of its properties, as well as its comparison with a companion system, the phosphate inorganic transport system, described elsewhere in this book (Elvin, Hardy, and Rosenberg).

The Pst system is repressible by phosphate in the medium at concentrations between 10^{-4} and 10^{-3} M (24). It is a high-affinity system (K_t = 0.16 μM), energized by an unidentified source of phosphate bond energy as well as by some component(s) of the proton motive force (15, 16). In this system phosphate uptake is accompanied by the entry of H^+, which is rapidly followed by an exchange of H^+ for K^+. The net result is the entry of two K^+ ions with one of phosphate (17), representing essentially an electroneutral process.

Recent work resulted in the sequencing of the entire *pst* region (1, 9, 19, 20), revealing the existence of five genes: *pstS, pstC, pstA, pstB,* and *phoU*, transcribed anticlockwise on the *E. coli* chromosome. The products of the first four genes are required for the transport of P_i, and together with the fifth (*phoU*) they are involved in the regulation of the "phosphate regulon" (Fig. 1). The PhoU protein is not required for transport. The information available that bears on function varies with each protein. The best studied is the *pstS* product, the phosphate-binding protein. This was purified in the past (5, 10) and shown to reconstitute phosphate transport in spheroplasts of wild-type *E. coli* but not in *pstS* mutants available at that time (6). The protein has also been crystallized, and its crystal structure was determined from X-ray diffraction studies (7a).

The existence of the *pstC* gene became known only as a result of the sequencing work. None of the mutants previously isolated in the *pst* region were *pstC*. The *pstC* gene encodes a hydrophobic protein that is extractable from the membranes by 1 M salt, suggesting that it is a peripheral membrane protein (20). We have not studied this protein further. The presence of an intact *pstC* gene is essential for phosphate transport through the Pst system. This was demonstrated by the construction of a *pstC* mutant by the deletion of 345 base pairs from the gene which had been cloned into a plasmid. Subsequently, the mutation was transferred to the chromosome, and the resultant *pstC* mutant was shown to lack phosphate transport through the Pst system and to synthesize alkaline phosphatase constitutively (Table 1). Transformation with a variety of plasmids carrying the *pstC* gene resulted in the restoration of the transport and repressibility of the phosphatase (not shown).

The *pstA* product is also hydrophobic. Its failure to focus during isoelectric gel electrophoresis is reminiscent of subunits *a* and *c* of the membrane ATPase and indicative of tightly bound lipid. Its secondary and tertiary structure, derived theoretically (3), indicates the presence of six helical sections capable of traversing the membrane (Fig. 2).

The *pstB* protein has features similar to those of a specific class of proteins, some of which are components of other binding protein-dependent transport systems. These membrane proteins are all distinguished by the presence of highly conserved amino acid sequences (Fig. 3) which can form a nucleotide-binding fold found also in ATP-requiring enzymes (7, 21). In view of the findings (15, 16) that energy coupling to the Pst system involves a source (as yet unidentified) of phosphate bond energy in close metabolic relation to ATP, we have suggested that the PstB component may be involved in energy coupling to the transport process (H. Rosenberg, *in* S. Silver and B. P. Rosen, ed., *Ion Transport in Prokaryotes*, in press).

The PhoU protein has been purified in our laboratory (18) from a strain which overproduced this protein under conditions of thermally induced transcription, resulting in yields of PhoU of about 5% of the total cell protein. In induced cells the protein formed aggregates which were isolated by differential centrifugation, affording a considerable degree of purification in this step. The purified protein was analyzed for amino acid composition and N-termi-

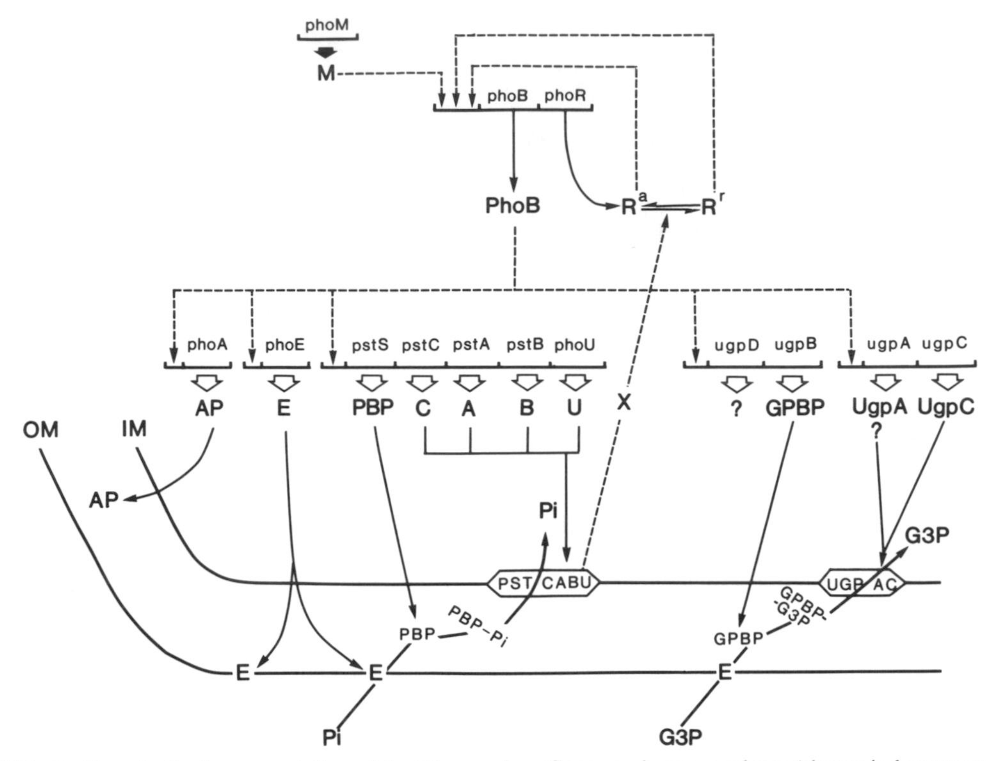

FIG. 1. Hypothetical representation of the Pho regulon. Genes and gene products (shown in lowercase and uppercase, respectively) are connected by solid-line arrows to their locations in the cell envelope. Dotted lines with arrows represent proposed gene control mechanisms. Question marks represent gene products that have not been identified. OM and IM are outer and inner membranes, respectively. AP, Alkaline phosphatase; PBP, phosphate-binding protein; GPBP, glycerol 3-phosphate–binding protein (also shown in complex with their substrates). E, Outer membrane PhoE pore. R^a and R^r are the *phoR* gene product in its activator and repressor forms, respectively. X is a putative hypothetical intermediate. The elements of the Pst and Ugp systems depicted in the inner membrane do not represent any real arrangement of these proteins. Alternative designations have been used in the literature for the Pst genes. These were *phoS* for *pstS*, *phoT* for *pstA*, and *phoW* for *pstC*. The nomenclature used here is based on the premise that the four genes coding for the proteins of the Pst system should carry the *pst* mnemonic.

nal sequence, both of which confirmed the reading frame previously established for the *phoU* gene (1, 20).

The PhoU protein plays a regulatory role in the *pho* regulon and is suggested (Fig. 1) to be involved in the provision of an intermediary signal between the Pst system and the other regulatory genes of the regulon. A possible involvement of certain nucleotides in this process was suggested in a recent report (13). The *phoU* product is not required for phosphate transport: *phoU* mutants take up phosphate at normal rates (20).

With the suggested involvement of the PstB protein in energy coupling, and the known role of the PstS protein as the periplasmic phosphate-binding protein, the PstA and the PstC proteins remain the most likely candidates for the actual transport channel. They are the two most hydrophobic proteins of the Pst system. At present,

TABLE 1. Alkaline phosphatase activity and phosphate transport rates in *E. coli* mutants and heterodiploids

Strain	Relative genotype	Alkaline phosphatase[a]			Transport[b]		
		Control	+pAN273	+pSN5083	Control	+pAN273	+pSN5083
HR314	Δ*pstS*	0.36	0.0	0.0	2.4	35.0	40.1
HR317	*pstS164*	0.49	0.0	0.016	0.7	2.8	17.0
AN1696	*pstA*	0.36	0.0	0.49	1.9	0.6	2.0
AN2537	Δ*pstC*	1.41			0.5		

[a] Expressed as micromoles of *p*-nitrophenyl phosphate cleaved per minute per milligram of protein.
[b] Expressed as nanomoles of P_i per minute per milligram of dry weight.

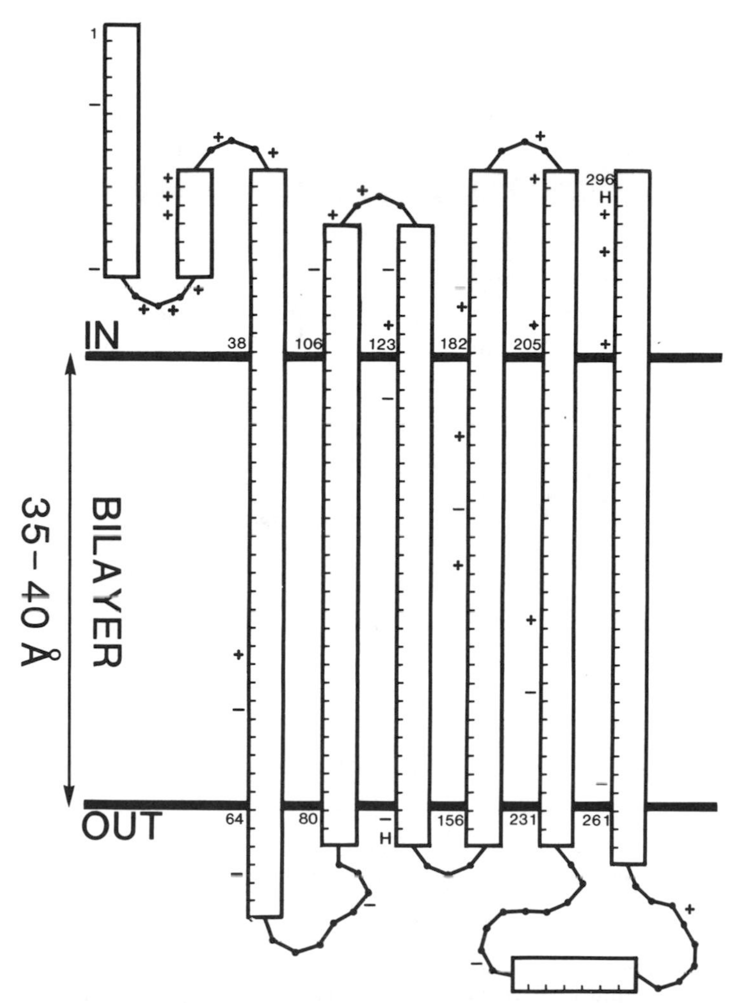

FIG. 2. Secondary and tertiary structure of the PstA protein predicted according to a procedure developed for membrane proteins (3). Boxed sections represent α helices (84% of the molecule). Curved sections denote β turns. Charged side chains are as shown. H, Histidine.

however, there is no proof for their proposed function as a channel.

In an independent study of the Pst system, Nakata and colleagues (this volume) report their findings and discuss the possible functions of the Pst gene products. Their views and ours are in excellent agreement, both on the structure of the operon and on the proposed functions of its components, both in phosphate transport and in the regulation of the *pho* regulon.

We have attempted to clear up an ambiguity concerning the *pstS* gene, following a report that the *phoS64* allele was in a cistron distinct from other *phoS*(*pstS*) mutations and that it was located distal to *phoU* (8). Our sequencing data exclude the possibility of another gene existing within the 83-min *pst* locus, since it is closely limited by the neighboring genes: *glmS*, upstream of *pstS*, with an intergenic region of 310 nucleotides, and *bgl*, downstream from *phoU*, separated by 150 nucleotides (Fig. 4). It is now known that two species of the *phoS64* allele exist, derived from the original C75a and C75b strains (4). That derived from C75a is actually a *phoR* type mutation and is to be known as *pho-64*. The other will be known as *pstS164* (B. Bachmann, personal communication; see also Rosenberg, in press).

An unambiguous *pstS* mutant was constructed in this laboratory by the deletion of 156 base pairs from the gene carried on a plasmid and the subsequent transfer of this mutation to the genome. The mutant was constitutive for alkaline phosphatase synthesis and could not transport phosphate via the Pst system. Transformation with (multicopy) plasmid pAN273 (carrying the

(A)

PROTEIN	RESIDUES	SEQUENCE
Bovine ATPase β	149-168	K G G K I G L F - G G A G V G K T - V F I M
E. coli ATPase β	142-161	K G G K V G L F - G G A G V G K T - V N M M
E. coli ATPase α	161-180	R G Q R E L I I - G D R G T G K T - A L A I
Adenylate Kinase	6- 26	K K S K I I F V V G G P G S G K G - T Q C E
OppD protein	46- 66	A G E T L G I V - G E S G S G K S Q S R L R
HisP protein	31- 50	A G D V I S I I - G S S G S G K S - T F L R
MalK protein	28- 47	E G G F V V F V - G P S G C G K S - T L L R
PstB protein	35- 54	K N Q V T A F I - G P S G C G K S - T L L R
CONSENSUS		G G G K S/T

(B)

PROTEIN	RESIDUES	SEQUENCE
Bovine ATPase β	241-267	V A E Y F R D Q E G Q D V L L F I D N I F R F T Q A G
E. coli ATPase β	227-252	M A E K F R D - E G R D V L L F V D N I Y R Y T L A G
E. coli ATPase α	265-290	M G E Y F R D - R G E D A L I I Y D D L S K Q A V A Y
Adenylate Kinase	102-127	G E E F E R K - I G Q P T L L L Y V D A G P E T M T K
OppD protein	173-200	Q R V M I A M A L L C R P K L L I A D E P T T A L D V T
HisP protein	160-187	Q R V S I A R A L A M E P D V L L F D E P T S A L D P E
MalK protein	140-167	Q R V A I G R T L V A E P S V F L L D E P L S N L D A A
PstB protein	160-187	Q R L C I A R G I A I R P E V L L L D E P C S A L D P I
CONSENSUS		R * * * * D D/E

FIG. 3. Consensus nucleotide-binding sequences in several ATP-utilizing enzymes and in one membrane component of four shock-sensitive bacterial transport systems. The sequences of three ATPases and adenylate kinase and the consensus nucleotide-binding sequence are as described by Walker et al. (21). The OppD, HisP, and MalK sequences are from Higgins et al. (7). The sequences of PstB have been aligned by C. F. Higgins (personal communication). The asterisks in the consensus line denote hydrophobic amino acids.

pstS gene) restored both phosphate transport and repressibility of the phosphatase. By comparison, repressibility of phosphatase only, but not phosphate transport, was restored when the strain carrying the allele *pstS164* was likewise transformed (Table 1). However, when these strains were transformed with the single-copy plasmid pSN5083, carrying $pstS^+$ (a gift from H. Shinagawa), both transport and phosphatase repressibility were restored in each strain (Table 1). We believe that in the multicopy transformant the large number of copies of genes produced may lead to the "mopping up" of the *phoB* gene product, thus making it unavailable to the promoter regions of the chromosomal *phoA* gene and *pst* genes, preventing both alkaline phosphatase synthesis and P_i transport. This was confirmed by the observation that an apparent restoration of repressible alkaline phosphatase synthesis (but not of P_i transport) could be demonstrated in a *pstA* mutant transformed with the multicopy plasmid pAN273 carrying only the *pstS* gene. On the other hand, the same *pstA*

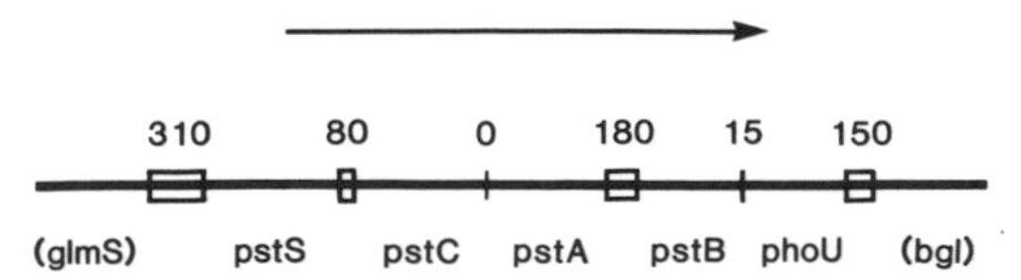

FIG. 4. Organization of the genes in the *pst* region. The five genes (including regulatory regions) are shown flanked by *glmS* and *bgl*. The length of the intergenic regions (boxes) is indicated by numbers of base pairs. The arrow shows the direction of transcription which is anticlockwise on the *E. coli* genetic map.

mutant strain, when transformed with the single-copy plasmid pSN5083, still showed constitutive alkaline phosphatase synthesis (Table 1).

The reconstitution of transport in the Δ*pstS* mutant transformed with pAN273, and the higher rates observed, may be due to the effect of the particular mutation, which is an in-frame deletion within the gene. The nature and exact location of *pstS164* are not known.

LITERATURE CITED

1. **Amemura, M., K. Makino, H. Shinagawa, A. Kobayashi, and A. Nakata.** 1985. Nucleotide sequence of the genes involved in phosphate transport and regulation of the phosphate regulon in *Escherichia coli*. J. Mol. Biol. **184:** 241–250.
2. **Bennett, R. L., and M. H. Malamy.** 1970. Arsenate resistant mutants of *Escherichia coli* and phosphate transport. Biochem. Biophys. Res. Commun. **40:**496–503.
3. **Cox, G. B., D. A. Jans, A. L. Fimmel, F. Gibson, and L. Hatch.** 1984. The mechanism of ATP synthase. Conformational change by rotation of the β-subunit. Biochim. Biophys. Acta **768:**201–208.
4. **Garen, A., and N. Otsuji.** 1964. Isolation of a protein specified by a regulator gene. J. Mol. Biol. **8:**841–852.
5. **Gerdes, R. G., and H. Rosenberg.** 1974. The relationship between the phosphate-binding protein and a regulator gene product from *Escherichia coli*. Biochim. Biophys. Acta **351:**77–86.
6. **Gerdes, R. G., K. P. Strickland, and H. Rosenberg.** 1977. Restoration of phosphate transport by the phosphate-binding protein of *Escherichia coli*. J. Bacteriol. **131:**512–518.
7. **Higgins, C. F., P. D. Haag, K. Nikaido, F. Ardeshir, G. Garcia, and G. F. Ames.** 1982. Complete nucleotide sequence and the identification of membrane components of the histidine transport operon of *S. typhimurium*. Nature (London) **298:**723–727.

7a. **Kubena, B. D., H. Luecke, H. Rosenberg, and F. A. Quiocho.** 1986. Crystallization and x-ray diffraction studies on a phosphate-binding protein involved in active transport in *Escherichia coli*. J. Biol. Chem. **261:** 7995–7996.

8. **Levitz, R., A. Klar, N. Sar, and E. Yagil.** 1984. A new locus in the phosphate specific transport (PST) region of *Escherichia coli*. Mol. Gen. Genet. **197:**98–103.
9. **Magota, K., N. Otsuji, T. Miki, T. Horiuchi, S. Tsunasawa, J. Kondo, F. Sakiyama, M. Amemura, T. Morita, H. Shinagawa, and A. Nakata.** 1984. Nucleotide sequence of the *phoS* gene, the structural gene for the phosphate-binding protein of *Escherichia coli*. J. Bacteriol. **157:**909–917.
10. **Medveczky, N., and H. Rosenberg.** 1970. The phosphate-binding protein of *Escherichia coli*. Biochim. Biophys. Acta **211:**158–168.
11. **Medveczky, N., and H. Rosenberg.** 1971. Phosphate transport in *Escherichia coli*. Biochim. Biophys. Acta **241:** 494–506.
12. **Rae, A. S., K. P. Strickland, N. Medveczky, and H. Rosenberg.** 1976. Studies on phosphate transport in *Escherichia coli*. I. Reexamination of the effect of osmotic and cold shock on phosphate uptake and some attempts to restore uptake with phosphate binding protein. Biochim. Biophys. Acta **433:**555–563.
13. **Rao, N. N., E. Wang, J. Yashphe, and A. Torriani.** 1986. Nucleotide pool in *pho* regulon mutants and alkaline phosphatase synthesis in *Escherichia coli*. J. Bacteriol. **166:**205–211.
14. **Rosenberg, H., R. G. Gerdes, and K. Chegwidden.** 1977. Two systems for the uptake of phosphate in *Escherichia coli*. J. Bacteriol. **131:**505–511.
15. **Rosenberg, H., R. G. Gerdes, and F. M. Harold.** 1979. Energy coupling to the transport of inorganic phosphate in *Escherichia coli*. Biochem. J. **178:**133–137.
16. **Rosenberg, H., C. M. Hardy, and B. P. Surin.** 1984. Energy coupling to phosphate transport in *Escherichia coli*, p. 50–52. *In* L. Leive and D. Schlessinger (ed.), Microbiology—1984. American Society for Microbiology, Washington, D.C.
17. **Russell, L. M., and H. Rosenberg.** 1980. The nature of the link between potassium transport and phosphate transport in *Escherichia coli*. Biochem. J. **188:**715–723.
18. **Surin, B. P., N. E. Dixon, and H. Rosenberg.** 1986. Purification of the PhoU protein, a negative regulator of the *pho* regulon of *Escherichia coli* K-12. J. Bacteriol. **168:** 631–635.
19. **Surin, B. P., D. A. Jans, A. L. Fimmel, D. C. Shaw, G. B. Cox, and H. Rosenberg.** 1984. Structural gene for the phosphate-repressible phosphate-binding protein of *Escherichia coli* has its own promoter: complete nucleotide sequence of the *phoS* gene. J. Bacteriol. **157:**772–778.
20. **Surin, B. P., H. Rosenberg, and G. B. Cox.** 1985. Phosphate-specific transport system of *Escherichia coli*: nucleotide sequence and gene-polypeptide relationships. J. Bacteriol. **161:**189–198.
21. **Walker, J. E., M. Saraste, M. J. Runswick, and N. J. Gay.** 1982. Distantly related sequences in the α- and β-subunits of ATP synthase, myosin, kinases and other ATP-requiring enzymes and a common nucleotide-binding fold. EMBO J. **1:**945–951.
22. **Willsky, G. R., R. L. Bennett, and M. H. Malamy.** 1973. Inorganic phosphate transport in *Escherichia coli*: involvement of two genes which play a role in alkaline phosphatase regulation. J. Bacteriol. **113:**529–539.
23. **Willsky, G. R., and M. H. Malamy.** 1974. The loss of the *phoS* periplasmic protein leads to a change in the specificity of a constitutive inorganic phosphate transport system in *Escherichia coli*. Biochem. Biophys. Res. Commun. **60:**226–233.
24. **Willsky, G. R., and M. H. Malamy.** 1976. Control of the synthesis of alkaline phosphatase and the phosphate-binding protein in *Escherichia coli*. J. Bacteriol. **127:**595–609.
25. **Willsky, G. R., and M. H. Malamy.** 1980. Characterization of two genetically separable inorganic phosphate transport systems in *Escherichia coli*. J. Bacteriol. **144:** 356–365.

Genetic and Biochemical Analysis of the Phosphate-Specific Transport System in *Escherichia coli*

ATSUO NAKATA, MITSUKO AMEMURA, KOZO MAKINO, AND HIDEO SHINAGAWA

Department of Experimental Chemotherapy, The Research Institute for Microbial Diseases, Osaka University, 3-1 Yamadaoka, Suita, Osaka, Japan 565

In the early genetic studies on the regulation of alkaline phosphatase synthesis in *Escherichia coli*, two unlinked negative regulatory mutations were isolated and mapped at R1 and R2 loci (6). More recent studies have shown that *E. coli* has two major systems for incorporation of P_i, the phosphate inorganic transport system and the phosphate-specific transport (Pst) system, the genes located at the R2 locus constituting the Pst system (5; for review, see references 24 and 25). Thus, some of the negative regulatory genes for alkaline phosphatase synthesis which mapped at the R2 position (83.5 min on the *E. coli* map [3]) were shown to code for the components of the Pst system.

We attempted to analyze the genetic components in this locus to reveal the organization and regulation of genes involved in the regulation of the phosphate (*pho*) regulon and the Pst system.

Pst Operon

Structure of the Pst operon. We cloned an *E. coli* chromosomal DNA fragment encompassing the R2 locus on a vector plasmid and carried out complementation tests of strains with mutations at the R2 region by constructing various deletion plasmids (2). We (1, 14) and Surin et al. (22, 23) determined the nucleotide sequence of the DNA fragment and found five cistrons, *pstS*(*phoS*)-*pstC*(*phoW*)-*pstA*(*phoT*)-*pstB*-*phoU*, arranged counterclockwise on the *E. coli* standard genetic map (3). Different nomenclatures have been used for the genes in the Pst operon. To avoid confusion, the participants in this symposium have agreed hereafter to use the nomenclature shown above for the genes in the Pst operon. (Alternative designations are indicated in parentheses.) In these studies, *pstC*(*phoW*) was discovered as a new open reading frame. Since alkaline phosphatase constitutive mutations of two strains in our collection were complemented by the plasmid with this open reading frame, the gene corresponding to the open reading frame is involved in the negative regulation of the *pho* regulon. We have not examined whether this gene is related to transport of P_i. We designated it the *phoW* gene (1), whereas Surin et al. (23) called it *pstC*.

Each cistron is preceded by a putative ribosome binding site at an appropriate distance from the start codon. Since no putative promoter sequences (20) were identified in the flanking regions of each cistron, except in the upstream flanking region of the *pstS* gene, these five cistrons may constitute a multicistronic operon, with the *pstS* promoter proximal (Fig. 1).

The *pstS* gene product is the periplasmic phosphate-binding protein (PBP), and like alkaline phosphatase, it is synthesized in a phosphate-limited medium. To examine the expression of other *pho*-related genes in this region under the same conditions, we constructed a multicopy plasmid containing the *pstS*-*pstC*-*pstA'*-*'lacZ* fusion gene by recloning an appropriate DNA fragment into a regulation probe vector, pMC1403 (4). The expression of the *pstA* gene was measured by monitoring the β-galactosidase activity. A *lac*-deletion strain carrying the recombinant plasmid synthesized 10-fold more β-galactosidase in a phosphate-limited medium than in an excess-phosphate medium. The expression of *pstA* is induced by phosphate deprivation, like that of *phoA* and *pstS*, although the expression of *pstA* was much lower than that of *pstS*.

We also constructed a multicopy plasmid with a deletion from the *Hpa*I site in *pstS* (nucleotide at 666) to the *Mlu*I site immediately upstream of *pstC* (nucleotide at 1104), where we found a putative transcriptional attenuator (Fig. 1 and 2). Overproduction of the PstB and PhoU proteins was observed in a strain carrying this plasmid, and the production of these proteins was induced by phosphate limitation. The molecular weights of these proteins were consistent with those deduced from DNA sequence and the maxicell experiment (17). This result suggests that transcription of *pstB* and *phoU* was initiated from a phosphate-regulated promoter, probably the *pstS* promoter, and it was enhanced by removing the putative attenuator located in the *pstS*-*pstC* intercistronic region (see below).

Promoter structure of the operon. We found a consensus sequence, "the phosphate box," shared with the regulatory regions of *phoA* (13), *phoB* (15), *phoE* (18), and *pstS* (14, 22) upstream of the *pstS* coding region. A well-conserved

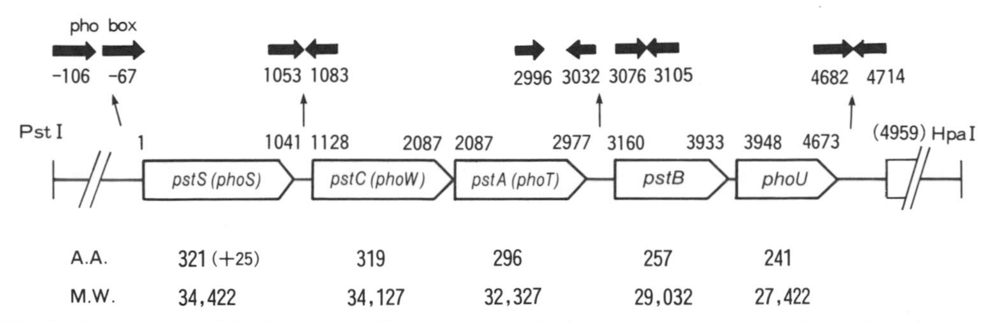

FIG. 1. Organization of the Pst operon. The open arrows indicate the arrangement and direction of the genes. The numbers above each gene are the nucleotide numbers, with the first letter of the translational initiation codon of the *pstS* gene taken as +1. The thick arrows indicate the size and positions of the phosphate box and the inverted repeat sequences. The number of amino acid residues and the molecular weights of the gene products deduced from the nucleotide sequences are shown below the corresponding genes. Abbreviations: A.A., number of amino acid residues; M.W., molecular weight.

18-nucleotide sequence, CTG/TTCATAA/TAA/TCTGTCAC/T, was found 10 or 11 nucleotides upstream of the putative Pribnow box in all cases. We recently analyzed the transcription initiation site of *pstS* both in vivo and in vitro, and found the transcription to be initiated 8 nucleotides distal to the Pribnow box of *pstS* (TATttT). We showed that the PhoB protein bound to the phosphate box and activated transcription of the genes in the *pho* regulon in vitro, including *pstS* (Shinagawa et al., this volume). The phosphate box is tandemly repeated in the regulatory region of *pstS*, and this may be the reason for the very high expression of *pstS*.

Nucleotide structures in the intercistronic spaces. In the intercistronic regions of the Pst operon, there are 89 bases between *pstS* and *pstC*, 185 bases between *pstA* and *pstB*, and 17 bases between *pstB* and *phoU*. The last letter of the termination codon of *pstC* overlaps with the first letter of the initiation codon of *pstA*, so they may be coordinately regulated.

There are three palindromic nucleotide sequences: one between *pstS* and *pstC*, and two between *pstA* and *pstB* (Fig. 1). The sequence immediately downstream of *pstS* (nucleotides 1053 to 1088) (Fig. 2) is homologous to the consensus sequence of the REP (repetitive extracistronic palindromic) sequence (7, 21) (82% homology). A similar nucleotide sequence immediately upstream of *pstB* (nucleotides 3076 to 3109) is more homologous to the consensus REP sequence, 97% (Fig. 2). Two sets of inverted repeats of 7 and 15 bases are included in these two REP sequences (nucleotides 1051 to 1057 and 1070 to 1084 and nucleotides 3107 to 3101 and 3089 to 3075, respectively; underlined in Fig. 2).

Immediately downstream of *pstA*, the transcript of nucleotides from 2996 to 3032 can form three alternative, and mutually exclusive, stem-and-loop structures.

The biological function of these sequences is not known, although it was assumed that they modulate the expression of the immediately upstream or downstream cistrons (21). As mentioned above, we constructed a deletion plasmid which lacked two-thirds of the 3′-terminal part of *pstS* and most of the intercistronic space between *pstS* and *pstC*, where we had found an REP-like sequence. We observed overproduction of the PstB and PhoU proteins in the strain carrying the deletion plasmid in a phosphate-limited medium. Therefore, it is likely that the REP nucleotide sequence distal to *pstS* reduces the expression of the genes downstream of *pstS* probably by terminating transcription. This assumption is also consistent with the finding that the induced level of *pstA* expression is much lower than that of *pstS* expression. Similarly,

(pstS)--agGCCGGgTa CGGtGTTTT ACGCCgcATCCGGCatTAC--(pstC)

(pstA)---tGCCGGATG CGGCGT GA ACGCCTgATCCGGCC TAC--(pstB)

Consensus: $GCC^{G}_{T}GATG.CG^{G}_{A}CG^{C}_{T}....^{G}_{A}CG^{C}_{T}CTTATC^{C}_{A}GGCC$ TAC

FIG. 2. Nucleotide sequences that are homologous to the consensus REP sequence (21) found in the intercistronic spaces between the *pstS* and *pstC* genes (nucleotides 1053 to 1088), and between the *pstA* and *pstB* genes (nucleotides 3076 to 3109) are shown in capital boldface letters. Two inverted repeat sequences with 7 and 15 bases (underlined) are also found in these sequences.

the other REP sequence might also affect the expression of the *pstB* and *phoU* genes. Wanner (26) suggested that these REP-like sequences might be involved in phase variation by gene inversion.

Transcriptional terminator. Eight nucleotides distal to the translational termination codon of *phoU*, there is a sequence whose transcript may form a stable stem-and-loop structure with $\Delta G = -23.0$ kcal/mol (nucleotides 4682 to 4714). This nucleotide sequence may be the transcriptional terminator of the Pst operon.

Taking these results into consideration, we conclude that these five cistrons constitute a multicistronic operon, and we propose to designate it the Pst operon, although the *phoU* gene is not involved in the Pst system (Fig. 1). Surin, Cox, and Rosenberg (this volume) examined DNA sequences of the flanking regions of the Pst operon and concluded that there is no other gene related to phosphate transport or *pho* regulation in this region. Since the products of the Pst operon are involved in the reception and transmission of the phosphate signal as well as in the transport of phosphate, and since this operon is regulated by phosphate in the medium and by the products of the operon, this system constitutes an autonomous circular regulatory network. Expression of the Pst system constitutes an adaptive response in *E. coli* to the condition of phosphate deprivation.

Components of the Periplasmic Binding Protein-Mediated Transport System

Composition of the Pst system. In *E. coli* and *Salmonella typhimurium*, several transport systems mediated by periplasmic binding protein for low-molecular-weight materials, such as maltose, oligopeptides, and histidine, have been extensively investigated (7–9, 11, 12). These transport systems consist of four protein components, a binding protein localized in periplasm, two highly hydrophobic proteins presumably localized in inner membrane, and a protein with lower hydrophobicity.

The protein components of the Pst system are very similar to these systems, although the arrangement of the corresponding genes is not always the same. We examined the hydrophilicity profiles of the PstC, PstA, and PstB proteins deduced from the DNA sequence and found that both the PstC and PstA proteins contain long stretches of hydrophobic amino acids. Therefore, these proteins may be located in the inner membrane. The PstB protein is not hydrophobic, suggesting that it is not embedded in the membrane. Thus, the Pst system also consists of a periplasmic PBP (PstS), two highly hydrophobic proteins (PstC and PstA), and a less hydrophobic protein (PstB).

Homology among the PstB, OppD, MalK, and HisP proteins. We found that the PstB protein is homologous to the MalK (9) and OppD (12) proteins of *E. coli* and to the HisP protein (11) of *S. typhimurium* (Fig. 3). These four proteins are more homologous in the regions assumed to be involved in nucleotide binding (12) (underlined in Fig. 3). This high similarity suggests that they evolved from a common ancestor, and they still retain similar functions such as ATP-binding or ATPase activity that are required for active transport.

Model for the Functions of the Pst Operon in Transport of Phosphate and Regulation of the Phosphate Regulon

Recent studies on the Pst operon have probably revealed all the components that participate in the Pst system and also most or all of the regulatory proteins, coded for by the genes in this operon, for the *pho* regulon, in addition to the previously known PhoB, PhoR, and PhoM. It may be useful to propose more refined models that can be tested experimentally. According to our current model (Fig. 4), P_i or organic phosphates passively pass through the outer membrane pores, such as the PhoE porin, into the periplasm. P_i is also released in the periplasm from organic phosphate compounds by the action of alkaline phosphatase and is subsequently concentrated in the periplasm by binding to the PBP, the product of *pstS*. PBP functions as a scavenger of phosphate from the environment where it is scarce. The phosphate is transferred from the PBP into the cytoplasm through pores which are made by a specific interaction of PstC, PstA, and PstB. This process requires energy that may be provided by ATP hydrolysis catalyzed by PstB since PstB contains a consensus sequence for an ATP-binding domain. PstC and PstA span the inner membrane, and PstB interacts with either one or both of them on the cytoplasmic side of the complex. The pore gate will be opened by allosteric effects initiated by interaction of phosphate-bound PBP with the complex on one side and ATP hydrolysis by PstB on the other side of the membrane.

Consistent with this model, we recently identified the PstB protein in the membrane fraction from which it was solubilized by nonionic detergents (unpublished data). This suggests that the protein is associated with the membrane.

PhoU may also interact directly with this Pst complex or catalyze formation of an effector molecule from a phosphate derivative that is produced from P_i by the function of the Pst complex. PhoU that has received the phosphate signal or the effector molecule modifies PhoR to

```
PstB:            msmvetapsKiQVrNLnfyYGkfhALKnINLDIaknqVt
HisP:                mmsEnKlhVidLhKrYGgheVLKgvsLqaRAGdVi
MalK:                   masvQlqNvtKawGeVvVsKDINLDIheGEfv
OppD: mslsetatqapqpanvllEvndlrvtfatpdGdVtAvnDlNftlRAGEtl

PstB: aFIGPSGCGKS TLLRtfNkmfelypeqrae GEilldGdnILTnsqdia
HisP: sIIGsSGSGKS TfLRcINfLEkpseGaIivnGqniNlvRdkDgqLkvAd
MalK: vFVGPSGCGKS TLLRmIaGLEtitsGdlfi GEk    RmnDTP P AE
OppD: gIVGeSGSGKSqsrLR lmGLlatn GrIggsatf N GReIL PLPerE

PstB:      LLRakVGMVFQkPtpfpmSiydniafgvrlfeKlsrAdmdErvQw
HisP: kNqlRLLRtrltMVFQhfn  LwSHMtVlENvMEapiqvlGlSKhdArER
MalK:     R   g VGMVFQsya  LyPHlsVaEN MsfglKpaGAkKevinQR
OppD: lNtrRa  eqisMiFQdPmtsLnPyMrVgEqlMEvlmlhkGmSKaEAfEe

PstB: ALtk  AalwnEtkdkLhqsgysLSGGQQQRlcIARgiAiRPeVLLLDEP
HisP: ALkyL A KVgideraqgkYPvhLSGGQQQRVsIARALAmEPdVLLfDEP
MalK: vnq v A eVlqlahlLdrkPkaLSGGQRQRVaIgRtLvaEPsVfLLDEP
OppD: svrmLdAvKmpEarkrmkmYPhefSGGmRQRVmIAmALlcRPklLiaDEP

PstB: cSALDPistG   RIEe LitELKqdy TvVI VTHnMqqAArcSDHtaF
HisP: TSALDPeLvGevlRI MqqLaEe Gk  TMVv VTHeMGfArhvSsHViF
MalK: lSnLDaaLrVQ mRIEisrLhkrlGr  TMIy VTHDqveAmtlaDKivV
OppD: TtALDvt  VQ aqI MtLLnELKrefnTaIImiTHDlGvvAgicDKVlV

PstB: MYlGeliEfsnt DdlFtkpakkqtedyitgryg
HisP: LhqGkieEeGdPeqV F GnPqsPRlqqFLKgsLk
MalK: LdAGRvaqvGKPlaVplsGrPfcrRiyrFaKdeLl (242-370)
OppD: MYAGRtmEyGKarDV F yqPvhPysiglLnavpr (273-335)
```

FIG. 3. Homologous amino acid sequences (capital letters) of the PstB, HisP, MalK, and OppD proteins. The amino acid sequences of these proteins deduced from the nucleotide sequences (8, 9, 11) are aligned to give maximum matching. The underlined sequences are the consensus sequence for the nucleotide-binding domain (12).

a form that inactivates the transcriptional activator function of PhoB (Shinagawa et al., this volume). The PhoU protein is envisaged as a cytoplasmic protein, since we identified it in the soluble fraction but not in the membrane fraction of the cell lysate (unpublished data). In this respect, it is possible that the specific nucleotides detected in the cells grown under either excess or limited phosphate conditions are the signal nucleotides, or effectors, involved in the *pho* regulation (19).

Elucidation of the PstS Function by Mutant Analysis

We designed an experiment to test the validity of our model. According to the model, PBP should possess at least two domains, one for specific binding of phosphate and the other for specific interaction with a component of the PstC-PstA complex in the inner membrane. The *pstS* mutants that become constitutive for alkaline phosphatase synthesis should include both

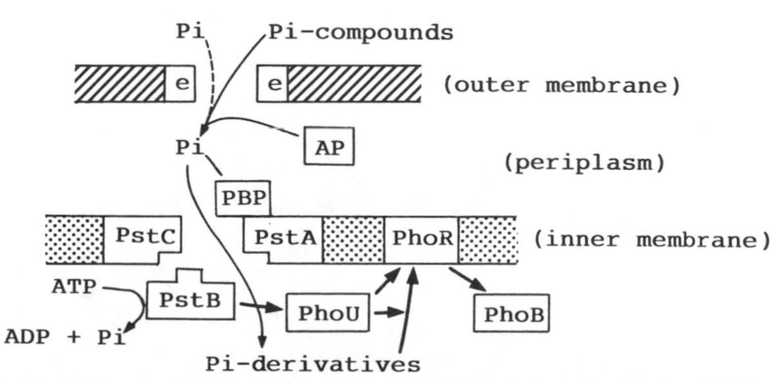

FIG. 4. Model for the Pst system. The inner membrane gate (consisting of the PstC and PstA proteins) for the phosphate bound to the PBP will be opened by the PstB protein upon hydrolysis of ATP. Abbreviations: e, porin e; AP, alkaline phosphatase.

types of mutants, since any defect in these domains will result in a defect in transfer of the phosphate signal. We mutagenized a plasmid DNA carrying the *pstS* gene with hydroxylamine in vitro (10) and introduced the plasmid into a *pstS* mutant strain by transformation, and alkaline phosphatase constitutive transformants were selected. Among them, we selected mutants that produced and secreted mutated PBPs into the periplasm which were indistinguishable from the intact PBP as examined by sodium dodecyl sulfate-polyacrylamide gel electrophoresis (16), since we wanted to eliminate the *pstS* mutants with gross morphological changes or nonsense mutations.

So far, we have determined the nucleotide sequences of 12 *pstS* mutants. Eight mutants contained a single amino acid change of Thr-10 to Ile-10, and two contained a single amino acid change of Ser-254 to Phe-254. One mutant contained two changed amino acids, Thr-10 to Ile-10 and Gly-140 to Glu-140, and one contained three changed amino acids, Thr-10 to Ile-10, Thr-253 to Ile-253, and Ser-254 to Phe-254. Extraordinarily high and moderately high frequencies of mutations that resulted in changes of Thr-10 to Ile-10 and Ser-254 to Phe-154, respectively, are very suggestive. We found that all of these mutant PBPs, except for the mutant with two changed amino acids, can bind phosphate. These results may suggest that Thr-10 and Ser-254 are involved in the interaction with the membrane components of the Pst system, whereas Gly-140 is involved in the binding of phosphate. Alternatively, there may be more than one phosphate-binding domain in PBP, and Thr-10 or Ser-254 may also be involved in phosphate binding.

Studies on the Pst operon are also presented by Surin et al. (this volume), and there are no contradictions between the results shown by them and by us.

This study was supported by a Grant-in-Aid for Scientific Research from the Ministry of Education, Science and Culture of Japan.

We also thank H. Rosenberg for correcting the English of this manuscript.

ADDENDUM IN PROOF

The amino acid sequence of the GlnQ protein of *E. coli* involved in transport of glutamine has been reported to be highly homologous to those of the PstB, HisP, MalK, and OppD proteins (T. Nohno, T. Saito, and J.-S. Hong, Mol. Gen. Genet. **205:**260–269, 1986).

LITERATURE CITED

1. **Amemura, M., K. Makino, H. Shinagawa, A. Kobayashi, and A. Nakata.** 1985. Nucleotide sequence of the genes involved in phosphate transport and regulation of the phosphate regulon in *Escherichia coli*. J. Mol. Biol. **184:** 241–250.
2. **Amemura, M., H. Shinagawa, K. Makino, N. Otsuji, and A. Nakata.** 1982. Cloning of and complementation tests with alkaline phosphatase regulatory genes (*phoS* and *phoT*) of *Escherichia coli*. J. Bacteriol. **152:**692–701.
3. **Bachmann, B. J.** 1983. Linkage map of *Escherichia coli* K-12, edition 7. Microbiol. Rev. **47:**180–230.
4. **Casadaban, M. J., J. Chou, and S. N. Cohen.** 1980. In vitro gene fusions that join an enzymatically active β-galactosidase segment of amino-terminal fragments of exogenous proteins: *Escherichia coli* plasmid vectors for the detection and cloning of translational initiation signals. J. Bacteriol. **143:**971–980.
5. **Cox, G. B., H. Rosenberg, J. A. Downie, and S. Silver.** 1981. Genetic analysis of mutants affected in the Pst inorganic phosphate transport system. J. Bacteriol. **148:**1–9.
6. **Echols, H., A. Garen, S. Garen, and A. Torriani.** 1961. Genetic control of repression of alkaline phosphatase in *E. coli*. J. Mol. Biol. **3:**425.
7. **Gilson, E., J.-M. Clement, D. Brutlag, and M. Hofnung.** 1984. A family of dispersed repetitive extragenic palindromic DNA sequences in *E. coli*. EMBO J. **3:**1417–1421.
8. **Gilson, E., C. F. Higgins, M. Hofnung, G. F.-L. Ames, and H. Nikaido.** 1982. Extensive homology between membrane-associated components of histidine and maltose transport systems of *Salmonella typhimurium* and *Escherichia coli*. J. Biol. Chem. **257:**9915–9918.
9. **Gilson, E., H. Nikaido, and M. Hofnung.** 1982. Sequence of the *malK* gene in *E. coli* K12. Nucleic Acids Res. **22:** 7449–7458.
10. **Hashimoto, T., and M. Sekiguchi.** 1976. Isolation of temperature-sensitive mutants of R plasmid by in vitro mutagenesis with hydroxylamine. J. Bacteriol. **127:**1561–1563.
11. **Higgins, C. F., P. D. Haag, K. Nikaido, F. Ardeshir, G. Garcia, and G. F.-L. Ames.** 1982. Complete nucleotide sequence and identification of membrane components of the histidine transport operon of *S. typhimurium*. Nature (London) **298:**723–727.
12. **Higgins, C. F., I. D. Hiles, K. Whalley, and D. J. Jamieson.** 1985. Nucleotide binding by membrane components of bacterial periplasmic binding protein-dependent transport systems. EMBO J. **4:**1033–1040.
13. **Kikuchi, Y., K. Yoda, M. Yamasaki, and G. Tamura.** 1981. The nucleotide sequence of the promoter and the amino-terminal region of alkaline phosphatase structural gene (*phoA*) of *Escherichia coli*. Nucleic Acids Res. **9:** 5671–5678.
14. **Magota, K., N. Otsuji, T. Miki, T. Horiuchi, S. Tsunasawa, J. Kondo, F. Sakiyama, M. Amemura, T. Morita, H. Shinagawa, and A. Nakata.** 1984. Nucleotide sequence of the *phoS* gene, the structural gene for the phosphate-binding protein of *Escherichia coli*. J. Bacteriol. **157:**909–917.
15. **Makino, K., H. Shinagawa, M. Amemura, and A. Nakata.** 1986. Nucleotide sequence of the *phoB* gene, the positive regulatory gene for the phosphate regulon of *Escherichia coli* K-12. J. Mol. Biol. **190:**37–44.
16. **Morita, T., M. Amemura, K. Makino, H. Shinagawa, K. Magota, N. Otsuji, and A. Nakata.** 1983. Hyperproduction of phosphate-binding protein, *phoS*, and pre-*phoS* proteins in *Escherichia coli* carrying a cloned *phoS* gene. Eur. J. Biochem. **130:**427–435.
17. **Nakata, A., M. Amemura, and H. Shinagawa.** 1984. Regulation of the phosphate regulon in *Escherichia coli* K-12: regulation of the negative regulatory gene *phoU* and identification of the gene product. J. Bacteriol. **159:**979–985.
18. **Overbeeke, N., H. Bergmans, F. von Mansfield, and B. Lugtenberg.** 1983. Complete nucleotide sequence of *phoE*, the structural gene for the phosphate limitation inducible outer membrane pore protein of *Escherichia coli* K12. J. Mol. Biol. **163:**513–532.
19. **Rao, N. N., E. Wang, J. Yashphe, and A. Torriani.** 1986. Nucleotide pool in *pho* regulon mutants and alkaline phosphatase synthesis in *Escherichia coli*. J. Bacteriol. **166:**205–211.
20. **Rosenberg, M., and D. Court.** 1979. Regulatory sequences

involved in the promotion and termination of RNA transcription. Annu. Rev. Genet. **13**:319–353.

21. **Stern, M. J., G. F.-L. Ames, N. H. Smith, E. C. Robinson, and C. F. Higgins.** 1984. Repetitive extragenic palindromic sequences: a major component of the bacterial genome. Cell **37**:1015–1026.
22. **Surin, B. P., D. A. Jans, A. L. Fimmel, D. C. Shaw, G. B. Cox, and H. Rosenberg.** 1984. Structural gene for the phosphate-responsible phosphate-binding protein of *Escherichia coli* has its own promoter: complete nucleotide sequence of the *phoS* gene. J. Bacteriol. **157**:772–778.
23. **Surin, B. P., H. Rosenberg, and G. B. Cox.** 1985. Phosphate-specific transport system of *Escherichia coli*: nucleotide sequence and gene-polypeptide relationships. J. Bacteriol. **161**:189–198.
24. **Tommassen, J., and B. Lugtenberg.** 1982. *Pho*-regulon of *Escherichia coli* K12: a minireview. Ann. Microbiol. (Paris) **133A**:243–249.
25. **Torriani, A., and D. N. Ludtke.** 1985. The *pho* regulon of *Escherichia coli*, p. 224–242. *In* M. Schaechter, F. C. Neidhardt, J. Ingraham, and N. O. Kjeldgaard (ed.), The molecular biology of bacterial growth. Jones and Bartlett Publishers, Boston.
26. **Wanner, B. L.** 1986. Novel regulatory mutants of the phosphate regulon in *Escherichia coli* K-12. J. Mol. Biol. **190**:900–920.

Molecular Studies on the Phosphate Inorganic Transport System of *Escherichia coli*

C. M. ELVIN, C. M. HARDY, AND H. ROSENBERG

Department of Biochemistry, John Curtin School of Medical Research, Australian National University, Canberra City, A.C.T., Australia 2601

The low-affinity phosphate inorganic transport (Pit) system of *Escherichia coli* belongs to the group of systems which are energized by the proton motive force. The existence of this system in *E. coli* was originally reported in the 1970s (2, 10). Its properties were subsequently studied and reported in greater detail (5, 9).

The Pit system is constitutive, and although its affinity for phosphate (K_t = 25 μM) is two orders of magnitude lower than that of the inducible phosphate-specific transport system, its maximal velocity, about 70 nmol of phosphate per min per mg (dry weight), is the same. Phosphate transport through the Pit system is coupled exclusively to the proton motive force

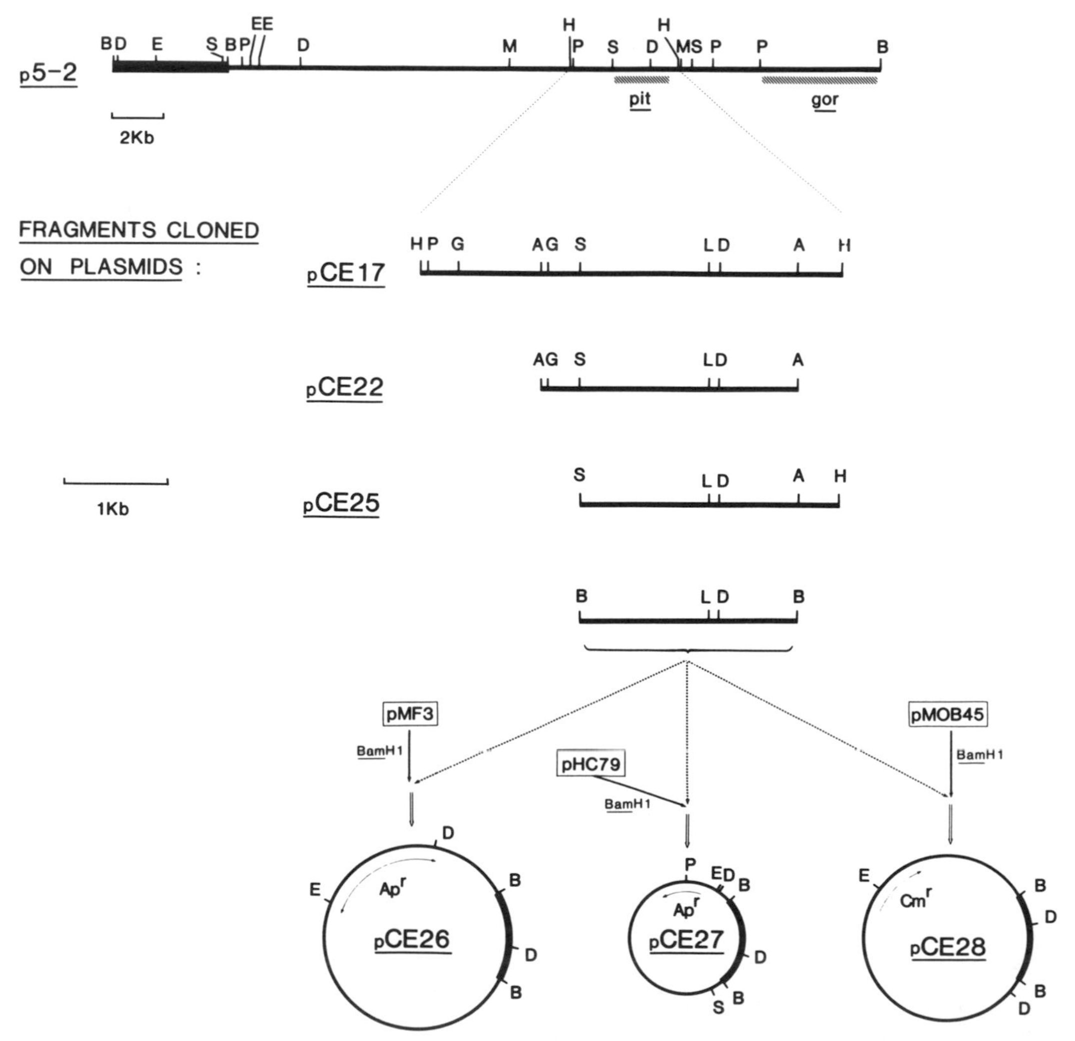

FIG. 1. Cloning of the *pit* gene into three plasmids differing in copy number.

and is completely abolished by uncouplers (6, 7). As in the case of the phosphate-specific transport system, phosphate is symported with protons, but electroneutrality is maintained by a concomitant K^+/H^+ exchange, so that the observed ion movement appears as the symport of two K^+ ions for one of phosphate (8). Unlike the phosphate-specific transport system, the Pit system does not possess a periplasmic component. Its function has been demonstrated both in spheroplasts (5) and in membrane vesicles energized by ascorbate-phenazine methosulfate (4).

Transport systems energized by the proton motive force are generally believed to involve a single membrane component. In the case of the lactose carrier of *E. coli*, which has been extensively studied, it has been shown that a single membrane protein is sufficient to carry out lactose transport in reconstituted vesicles (3). While such direct proof is not currently available for the Pit system, circumstantial evidence supports this statement. Thus, all known *pit* mutations map in the same locus of the *E. coli* chromosome and, as shown below, are complemented by transformation with a plasmid carrying a small fragment of DNA which directs the synthesis of a single protein. Furthermore, *pit* mutants transformed with a multicopy plasmid carrying the same DNA fragment transport P_i 8 to 10 times faster than wild strains (see Fig. 2 below). If the Pit protein was not the sole component of the Pit system, such amplification would not be expected.

Cloning of the *pit*$^+$ Gene

Recent work in this laboratory (2a) resulted in the precise mapping of the *pit* gene with respect to *gor*, positioning these at about 77.1 and 77.3 min, respectively, on the *E. coli* linkage map (1). The *pit*$^+$ gene was isolated from a library of *E. coli* genes inserted in the cosmid vector pHC79. Subcloning and deletion analysis indicated that the entire *pit*$^+$ gene was located within a 2.2-kilobase *Sal*I-*Ava*I fragment. *Bam*HI oligonucleotide linkers were attached to this fragment to facilitate cloning into different plasmid vectors.

Restoration of P_i Transport in *pit* Mutants

The 2.2-kilobase *Bam*HI fragment was eventually inserted into three different vectors, pMF3, pHC79, and pMOB45 (Fig. 1). When the first two of these were used to transform a *pit* strain incapable of transporting P_i, normal transport was restored in transformants carrying plasmid pCE26 (vector pMF3, single-copy *pit*$^+$), while transport rates 8 to 10 times those observed in wild strains were seen in transformants carrying plasmid pCE27, where the DNA fragment was cloned in the multicopy vector pHC79

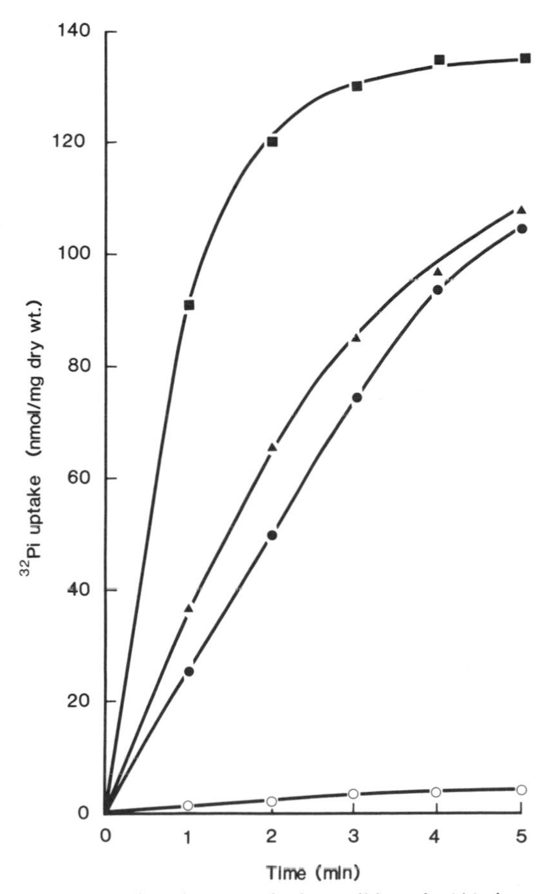

FIG. 2. Phosphate uptake in a wild strain (▲), in a *pit* mutant (○), and in the mutant carrying the single-copy plasmid pCE26 (●) or the multicopy plasmid pCE27 (■).

(Fig. 2). The steady-state intracellular concentration of P_i was, however, the same in both cases (not shown in Fig. 2). All transformants displayed the low affinity characteristic of the Pit system, showed 98 to 99% inhibition by carbonyl cyanide-*m*-chlorophenyl hydrazone, and transported phosphate constitutively.

Identification of the *pit*$^+$ Gene Product

In a minicell strain (DS410) transformed with plasmid pCE27, the 2.2-kilobase DNA fragment directed the synthesis of a single [^{35}S]methionine-labeled polypeptide that was not produced in the corresponding cells carrying the vector (pHC79) only. The polypeptide showed an M_r of 39,000 on sodium dodecyl sulfate-polyacrylamide gel electrophoresis and was found exclusively in the cell membrane. The same polypeptide, again located in the membrane, was synthesized in larger amounts by cells carrying the multicopy "runaway" plasmid pCE28 (Fig. 1), in which the *pit*$^+$ fragment was cloned into

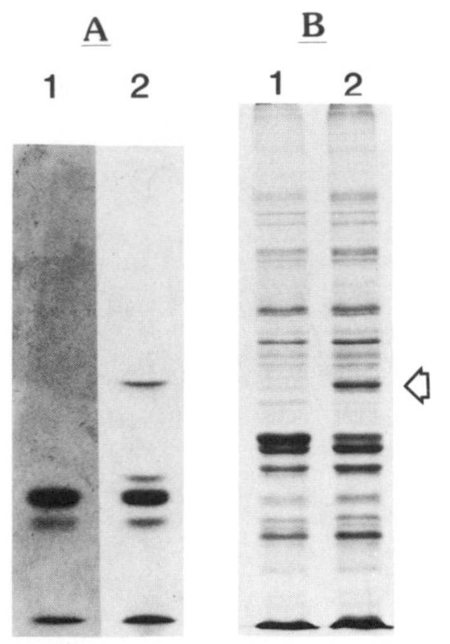

FIG. 3. Plasmid-directed synthesis of the Pit protein. (A) Radioautograph of [^{35}S]methionine-labeled proteins from minicells carrying vector pHC79 alone (1) or pCE27 (2). (B) Coomassie-stained gels of cell envelope fractions from cells carrying vector pMOB45 alone (1) or plasmid pCE28 (2). The position of the Pit protein is shown by the arrow. All gels were 10% polyacrylamide with sodium dodecyl sulfate.

the vector pMOB45. Again, the 39-kilodalton polypeptide was not seen in transformants carrying the vector alone (Fig. 3). The Pit protein

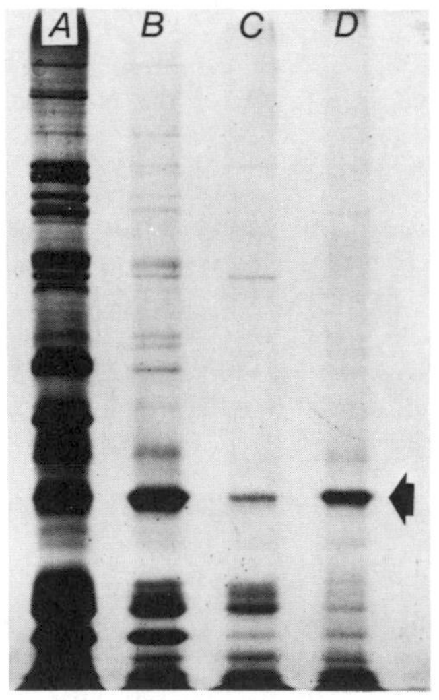

FIG. 4. Separation by sodium dodecyl sulfate-polyacrylamide gel electrophoresis of membrane fractions in progressive stages of purification. The membranes were from a strain carrying the plasmid pCE28. Lanes: A, crude membrane vesicles; B, vesicles after wash with 5 M urea followed by 5% sodium cholate; C, *N*-octyl-glucoside (1.25%) extract of B; D, centrifuged aggregate (see text). The position of the Pit protein is shown by the arrow.

cannot be discerned in gels prepared from wild strains, probably because of the low number of carriers normally present in the cells. The amplification seen in transformants carrying pCE28 is therefore considerable.

Purification of the Pit Protein

Only partial purification of the Pit protein has been achieved to date. One of the difficulties has been the propensity of the detergent-extracted *pit* product to aggregate upon concentration and freezing. While this interfered with chromatography because the material remained on top of columns, the property was used successfully to effect some degree of purification (Fig. 4).

Sequencing of the *pit* Gene

The entire 2.2-kilobase fragment was sequenced. An open reading frame comprising 1,287 nucleotides was derived. The two alternative reading frames were not open for more than 400 nucleotides. The sequence codes for a polypeptide of 429 amino acids with an M_r of 46,200. The sequence has not been confirmed because purification of the Pit protein is as yet incomplete. The complete sequence and its analysis will be published as soon as it is confirmed by N-terminal amino acid analysis of the purified protein.

LITERATURE CITED

1. **Bachmann, B. J.** 1983. Linkage map of *Escherichia coli* K-12, edition 7. Microbiol. Rev. **47:**180–230.
2. **Bennett, R. L., and M. H. Malamy.** 1970. Arsenate resistant mutants of *Escherichia coli* and phosphate transport. Biochem. Biophys. Res. Commun. **40:**496–503.

2a. **Elvin, C. M., N. E. Dixon, and H. Rosenberg.** 1986. Molecular cloning of the phosphate (inorganic) transport (*pit*) gene of *Escherichia coli* K-12: identification of the *pit*$^+$ gene product and physical mapping of the *pit-gor* region. Mol. Gen. Genet. **204:**477–484.

3. **Foster, D. L., M. L. Garcia, M. J. Newman, L. Patel, and H. R. Kaback.** 1982. Lactose-proton symport by purified *lac* carrier protein. Biochemistry **21:**5634–5638.
4. **Konings, W. N., and H. Rosenberg.** 1978. Phosphate transport in membrane vesicles from *Escherichia coli*. Biochim. Biophys. Acta **508:370–378.**
5. **Rosenberg, H., R. G. Gerdes, and K. Chegwidden.** 1977. Two systems for the uptake of phosphate in *Escherichia coli*. J. Bacteriol. **131:**505–511.
6. **Rosenberg, H., R. G. Gerdes, and F. M. Harold.** 1979. Energy coupling to the transport of inorganic phosphate in *Escherichia coli*. Biochem. J. **178:**133 137.
7. **Rosenberg, H., C. M. Hardy, and B. P. Surin.** 1984. Energy coupling to phosphate transport in *Escherichia coli*, p. 50–52. *In* L. Leive and D. Schlessinger (ed.), Microbiology—1984. American Society for Microbiology, Washington, D.C.
8. **Russell, L. M., and H. Rosenberg.** 1980. The nature of the link between potassium transport and phosphate transport in *Escherichia coli*. Biochem. J. **188:**715–723.
9. **Willsky, G. R., and M. H. Malamy.** 1980. Characterization of two genetically separable inorganic phosphate transport systems in *Escherichia coli*. J. Bacteriol. **144:**356–365.
10. **Willsky, G. R., R. L. Bennett, and M. H. Malamy.** 1973. Inorganic phosphate transport in *Escherichia coli*: involvement of two genes which play a role in alkaline phosphatase regulation. J. Bacteriol. **113:**529–539.

PhoE Protein Structure and Function

PETER VAN DER LEY AND JAN TOMMASSEN

Department of Molecular Cell Biology and Institute for Molecular Biology, State University of Utrecht, 3584 CH Utrecht, The Netherlands

The cell envelope of gram-negative bacteria consists of an inner and an outer membrane, separated by the peptidoglycan layer and the periplasmic space. The outer membrane functions as a molecular sieve by the presence of pore-forming proteins (11, 16). These proteins form transmembrane channels, which allow the passage of small, hydrophilic solutes by a diffusionlike process. Under standard laboratory conditions, *Escherichia coli* K-12 synthesizes two general pore-forming proteins, OmpF protein and OmpC protein. When cells are grown under phosphate limitation, the synthesis of another pore protein, i.e., PhoE protein, is induced (18). The *phoE* gene is part of the *pho* regulon (19). In contrast to the cation-selective OmpF and OmpC pores, PhoE pores are anion selective (1, 9).

Nucleotide sequence analysis of the *ompF*, *ompC*, and *phoE* genes (7, 12, 17) has revealed an extensive homology of approximately 60% in the primary structures of the three proteins (Fig. 1). The porins are unusual among membrane proteins in their lack of hydrophobic sequences long enough to span the membrane. The functional unit is a trimer with a compact structure, lacking large extramembranous domains (5).

To study the structure-function relationships of PhoE protein and the topology of the protein in the membrane, we have employed three distinct approaches, namely, the construction of OmpC-PhoE and PhoE-OmpC hybrid genes, the analysis of PhoE proteins from different enterobacterial species, and the isolation of mutant PhoE proteins with single amino acid substitutions.

Hybrid Pore Proteins

The OmpC and PhoE proteins are very homologous. To study the structure-function relationship of these proteins, we have constructed a series of *ompC-phoE* hybrid genes in which the DNA encoding part of one protein is replaced by the corresponding part of the other gene (22). These hybrid genes were obtained by in vivo recombination. The fusion sites in the hybrid genes were mapped by using restriction enzymes which cleave either in *ompC* or in *phoE*. In this way, the fusion joints in the different hybrid genes could be allocated to 1 of the 10 regions, indicated as I to X in Table 1, defined by the positions of differing restriction sites.

The proteins encoded by the hybrid genes were normally expressed and transported to the outer membrane. They were characterized for functions and properties in which the native PhoE and OmpC proteins differ (Table 1). Probably as a result of the PhoE anion selectivity, the negatively charged β-lactam antibiotic cefsulodin permeates much faster through PhoE pores than through OmpC pores (15). The rate of uptake of cefsulodin, when measured in cells producing the hybrid porins as the only pores, was reduced in three steps from the level of PhoE protein pores in class I hybrids to a level comparable with OmpC protein pores in class VI hybrids. These results suggest that at least three distinct regions in the N-terminal 141 amino acid residues of PhoE protein contribute to its anion specificity. They are in agreement with the results obtained with an *ompF-phoE* hybrid gene which was constructed with the aid of restriction enzymes (21). The resulting OmpF-PhoE hybrid protein, in which the N-terminal 74 amino acid residues of PhoE are replaced by the corresponding part of OmpF, was found to have a reduced anion selectivity.

All class I to III hybrid OmpC-PhoE hybrid proteins, and some class IV hybrid proteins, functioned as receptor for the PhoE-specific phage TC45 (3) and its host-range derivative TC45 *hr*N3. Sensitivity for OmpC-specific phages was found only with class VIII to X hybrid proteins. Thus, at least part of the receptor site for the PhoE-specific phages is located on region IV of PhoE protein, and at least part of the receptor site for the OmpC-specific phages is located in the region of OmpC between the fusion joints of the class VII and VIII proteins. Finally, the ability of 10 monoclonal antibodies, which recognize the cell surface-exposed part of PhoE protein (23), to bind to cells producing the OmpC-PhoE hybrid proteins was tested. All antibodies could bind to cells producing the class I to VII hybrid proteins, but none could bind to cells with class VIII to X hybrids. Thus, the antigenic determinants for these antibodies are at least partly located in the region of PhoE protein between the fusion joints of the class VII and VIII hybrid proteins.

In a similar way, reversed hybrids have also been isolated, i.e., hybrid proteins with an N-terminal part of PhoE and a C-terminal part of

```
OmpF           V L   AVGL  F KGNGENSYGGN  MT A L       45
PhoE  AEIYNKDGNKLDVYGKVKAMHYMSDNASKD*****GDQSYIRFGF     40
OmpC   V         L    DGL  F   KDV          T M L       40

          SD      Q  YN Q  NSEG DAQ GN*                 87
      KGETQINDQLTGYGRWEAEFAGN***KAESDTAQQKTRLAFAG       80
         VT       Q  YQIQ     S  NENNS*W                79

        A V         Y VV  ALGY   L      *T YS          126
      LKYKDLGSFDYGRNLGALYDVEAWTDMFPEFGGDSSAQTD         120
       FQ V         Y VV   TS   VL       *TYGS          118

      D FVG VG V     SN   LV    FAV  L    *          TA 166
      NFMTKRASGLATYRNTDFFGVIDGLNLTLQYQGKNEN**************RD*V 160
        QQ GN F          LV    FAV     G PSGEGFTSGVTNNG  A 172

      RRS    V G IS EYE ** G V   GAA    L E A P        204
      KKQNGDGFGTSLTYDFGGSDFAISGAYTNSDRTNEQN*LQS        200
      LR     V G I   YE ** G G  ISS K  DA  TAAY        211

      L N  K  Q               AN G   NA    NKFTNTS      250
      RGTGKRAEAWATGLKYDANNIYLATFYSETRKMTPIT******GGF   240
      I N D   TYTG            AQ TQ YNA RVG     SLGW    252

         DVLL                IA TK  A  V       V       290
      ANKTQNFEAVAQYQFDFGLRPSLGYVLSKGKDIE*GIGDED        280
        A                    A LQ     NLGR YD          293

        FE              TY    I    I      GVGS ***     330
      LVNYIDVGATYYFNKNMSAFVDYKINQLDSDNKLN****INNDD     320
      ILK V        P    TY        L  *  QFTRDAG  T N   336

      T   IV                                           340
      IVAVGMTYQF                                       330
        L LV                                           346
```

FIG. 1. Amino acid sequence of PhoE protein (middle line) compared with OmpF (upper line) and OmpC (bottom line). Amino acids in OmpF and OmpC are indicated only when they differ from PhoE. The proteins are aligned to give maximal homology. Only the sequences of the mature proteins are shown. An asterisk indicates a deletion of one amino acid residue.

OmpC (P. Van der Ley et al., submitted for publication). Analysis of these hybrids showed that the antigenic determinants for 9 of the 10 monoclonal antibodies are split into at least two regions separated in the primary structure: in addition to the region between the fusion joints of the class VII and VIII OmpC-PhoE hybrid proteins, binding of these antibodies is also dependent on part of the regions IX and X. This result is consistent with the fact that the antibodies recognize only the native trimeric PhoE protein and not the sodium dodecyl sulfate-denatured polypeptide; in other words, they bind to conformational rather than sequential determinants.

TABLE 1. Properties of OmpC-PhoE hybrid proteins

Class[a]	Cefsulodin uptake[b]	PhoE phages[c]	OmpC phages[c]	Monoclonal anti-PhoE[d]
I (1–11)	11	S	R	+
	5	S	R	+
II (12–49)	5	S	R	+
III (50–74)	5	S	R	+
IV (75–110)	5	S	R	+
	2	R	R	+
V (111–131)	2	R	R	+
VI (132–141)	<1	R	R	+
VII (142–173)	<1	R	R	+
VIII (174–267)	<1	R	S	−
	<1	R	S	−
IX (268–279)	<1	R	S	−
X (280–330)	<1	R	S	−
PhoE$^+$	12	S	R	+
OmpC$^+$	<1	R	S	−

[a] The fusion sites in the hybrid genes were localized in 10 regions; the corresponding amino acid segments in PhoE are indicated.

[b] The rate of uptake of cefsulodin is expressed as nanomoles per minute per 10^8 cells.

[c] Resistance (R) or sensitivity (S) to the phages is indicated.

[d] Presence (+) or absence (−) of the epitopes for 10 PhoE-specific monoclonal antibodies is indicated.

PhoE Proteins from Different Enterobacterial Species

The *phoE* genes from *Enterobacter cloacae* (24) and *Klebsiella pneumoniae* (Van der Ley et al., submitted) were cloned in *E. coli* K-12 by selecting for the closely linked *proA* and *proB* genes. Expression and regulation of both genes in *E. coli* K-12 are similar to those of the *E. coli phoE* gene. Functionally, the products of these *phoE* genes behave very similarly since they form anion-selective pores (R. Benz and J. Korteland, personal communications) and since they serve as receptors for phage TC45. In addition, antigenic determinants of the proteins exposed at the cell surface appear to be well conserved, since differences in binding were found for only 2 of 10 PhoE-specific monoclonal antibodies (23; Van der Ley et al., submitted).

Preliminary sequencing data on both *phoE* genes show a strong conservation of the amino acid sequence and, to a lesser extent, of the nucleotide sequence. Many of the amino acid substitutions are clustered in a few variable regions. A comparative study on the *ompA* gene in five enterobacterial species has demonstrated that the most variable regions correspond with external domains (2). Therefore, the variable regions in PhoE are likely to represent cell surface-exposed parts too where constraints on the amino acid sequence are less pronounced. They correspond to variable regions from the OmpF/OmpC/PhoE comparison, but in the latter case the number of regions with major sequence divergence is higher.

Mutant PhoE Proteins with Amino Acid Substitutions in Cell Surface-Exposed Regions

The PhoE-specific phage TC45 was used to isolate *phoE* missense mutations conferring resistance to this phage (10). Five independently isolated mutants were found to differ from the wild type by the same single amino acid substitution in PhoE protein: Arg-158 → His. This mutant pore protein has lost its receptor function for phage TC45 and part of its preference for negatively charged solutes, but is still able to

TABLE 2. Properties of mutant PhoE proteins

Amino acid substitution	Selection[a]	Phage TC45[b]	Epitopes present[c]
Arg-158 → His	TC45	R	10
Arg-201 → His	mAb/C	S	6
Arg-201 → Cys	mAb/C	S	3
Gly-238 → Ser	mAb/C	S	9
Gly-275 → Ser	mAb/C	S	2
Gly-275 → Asp	mAb/C	S	1

[a] Mutants were selected with phage TC45 or with monoclonal antibody plus complement (mAb/C).

[b] Sensitivity (S) or resistance (R) to phage TC45 is indicated.

[c] Number of monoclonal antibodies of a total of 10 that can still bind to the mutant proteins.

bind the 10 PhoE-specific monoclonal antibodies (Table 2).

Monoclonal antibodies which recognize the cell surface-exposed part of PhoE protein have been used to select for mutants with altered antibody-binding properties (23a). The selection procedure was based on the antibody-dependent bactericidal action of the complement system. Using two distinct PhoE-specific monoclonal antibodies in combination with complement, seven independent mutants with an altered PhoE protein were isolated, in addition to a large majority of mutants entirely lacking PhoE protein. Among these seven mutants, five distinct binding patterns were observed with a panel of 10 monoclonal antibodies. DNA sequence analysis revealed the following substitutions: Arg-201 → His (three isolates), Arg-201 → Cys, Gly-238 → Ser, Gly-275 → Ser, and Gly-275 → Asp. Binding of phage TC45 is not affected by these substitutions (Table 2).

It is expected that the amino acid residues Arg-158, Arg-201, Gly-238, and Gly-275 are directly involved in binding of phage or monoclonal antibody and that the alterations in the mutants do not affect these properties through some long-range conformational alterations, for the following reasons. First, the mutations have very specific effects. None of the changes affects all cell surface-associated PhoE functions (Table 2), as would be expected for mutants with gross conformational alterations. In addition, pore function is retained in all cases. Second, the conclusion that residues 201, 238, and 275 are involved in binding of monoclonal antibodies is in good agreement with the results obtained with the hybrid pore proteins which showed that the antigenic specificity of PhoE protein is located in the C-terminal half of the protein.

Topology of PhoE Protein

Physical studies with OmpF protein trimers have shown that approximately two-thirds of the polypeptide chain is in the form of an antiparallel β-pleated sheet, with many 10- to 12-residue-long membrane-spanning segments which run approximately normal to the plane of the membrane (8). Given the close homology of the OmpF, OmpC, and PhoE proteins, a similar structure for all three pore proteins is to be expected.

The four amino acid residues of PhoE protein that are changed in phage-resistant or monoclonal antibody/complement-resistant mutants, i.e., Arg-158, Arg-201, Gly-238, and Gly-275 (Table 2), show a remarkably regular spacing, being approximately 40 residues apart. Inspection of the PhoE amino acid sequence reveals that these four areas correspond to hydrophilic maxima (17) and to regions with pronounced differences in the PhoE, OmpF, and OmpC proteins (Fig. 1). These properties are to be expected for cell surface-exposed regions. In fact, in PhoE protein eight such regions can be found, all of which are separated by approximately 40 amino acid residues (Fig. 1). Mizuno et al. (12) identified six regions where the largest differences in the OmpF, OmpC, and PhoE proteins can be found. In addition, we include the regions 115 to 121 and 272 to 279, where the differences are somewhat less pronounced but still considerable. A distance of approximately 40 amino acid residues between the variable regions corresponds with the expected length of two membrane-spanning β-sheet strands plus turns at the outside and inside of the outer membrane. We therefore propose a model for the topology of the *E. coli* porins in which the polypeptide repeatedly traverses the outer membrane in β-sheet structure, exposing the eight hydrophilic/variable regions at the cell surface (Fig. 2).

Secondary structure predictions (6) show that regions likely to form reverse turns are present in parts of the protein which, according to the model, should be located at the periplasmic side of the outer membrane (Fig. 2). In addition, the model predicts location of the N and C termini at the periplasmic side. This has already been demonstrated to be the case for the N terminus (20).

Further support for the proposed model has come from linker insertion mutagenesis of the *phoE* gene (D. Bosch et al., submitted for publication; M. Agterberg et al., manuscript in preparation). Insertions of four amino acid residues in regions predicted to be cell surface exposed gave the following results. Insertion in region 64 to 73 resulted in TC45 resistance, whereas insertions in regions 156 to 163 and 272 to 279 caused the loss of the epitopes for one and nine monoclonal antibodies, respectively. An insertion at residue 174, which according to our model should be located at or close to the periplasmic side of the outer membrane, did not

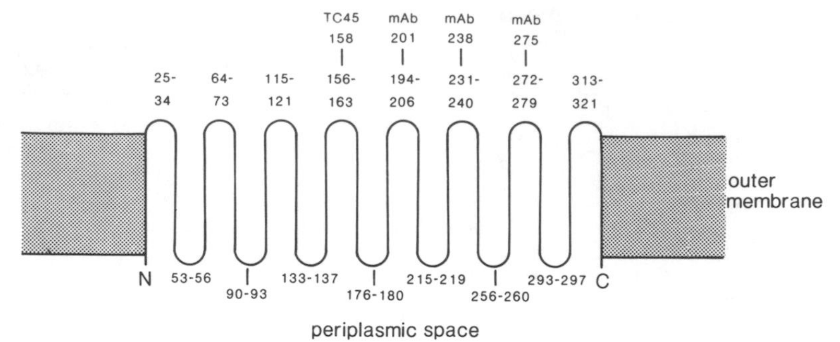

FIG. 2. Tentative model for the topology of PhoE protein in the outer membrane. The numbers refer to amino acid positions in the mature PhoE protein. The regions predicted to be surface exposed are shown at the top of the figure and are arbitrarily delineated by the first and last residue which is not conserved among OmpF, OmpC, and PhoE proteins in that region. Positions changed in phage TC45-resistant and monoclonal antibody/complement-resistant mutants are indicated. At the periplasmic side, regions predicted to form reverse turns are indicated.

interfere with any of these surface-associated PhoE functions.

In a β-sheet structure, the amino acid side chains alternate between the two surfaces of the sheet. Since in the porins approximately two-thirds of the polypeptide chain is in the form of an antiparallel β-pleated sheet, the parts of the sequence that contact the hydrophobic part of the outer membrane might be identified by the presence of exclusively hydrophobic residues at one side of a potential β sheet. The segments of PhoE predicted to be membrane spanning were inspected for this property (Table 3). Remarkably, for 14 of the 16 segments it was found that the residues at one side of a potential sheet are invariably hydrophobic, whereas the residues at the other side are on average less hydrophobic. For 12 of these 14 segments, the high hydrophobicity of one side is conserved among the PhoE/OmpF/OmpC proteins and the PhoE proteins of three different enterobacterial species. It therefore seems likely that these regions have their hydrophobic residues facing outward to the lipid bilayer and the less hydrophobic side facing inward to the inside of the pore or other parts of the protein. In this way, the porin surface that

TABLE 3. Potentially membrane-spanning segments of PhoE protein

Region	Amino acid sequence[a]	Avg hydrophobicity[b]	
10–19	Lys *Leu* Asp *Val* Tyr *Gly* Lys *Val* Lys *Ala*	−1.03	*+0.88*
35–43	Tyr *Ile* Arg *Phe* Gly *Phe* Lys *Gly* Glu	−0.80	*+1.07*
55–62	Arg *Trp* Glu *Ala* Glu *Phe* Ala *Gly*	−0.84	*+0.78*
76–84	Leu *Ala* Phe *Ala* Gly *Leu* Lys *Tyr* Lys	−0.04	*+0.65*
95–104	*Leu* Gly *Ala* Leu *Tyr* Asp *Val* Glu *Ala* Trp	+0.15	*+0.74*
126–133[c]	Arg *Ala* Ser *Gly* Leu *Ala* Thr *Tyr*	−0.41	*+0.50*
144–153	Gly *Leu* Asn *Leu* Thr *Leu* Gln *Tyr* Gln *Gly*	−0.41	*+0.81*
164–173[c]	Asn *Gly* Asp *Gly* Phe *Gly* Thr *Ser* Leu *Thr*	+0.11	*+0.24*
181–188	*Phe* Ala *Ile* Ser *Gly* Ala *Tyr* Thr	+0.25	*+0.84*
206–214[d]	Arg *Ala* Glu *Ala* Trp *Ala* Thr *Gly* Leu	−0.28	*+0.59*
220–228	Asn *Ile* Tyr *Leu* Ala *Thr* Phe *Tyr* Ser	+0.22	*+0.68*
246–255	Asn *Phe* Glu *Ala* Val *Ala* Gln *Tyr* Gln *Phe*	−0.42	*+0.78*
260–268[d]	Arg *Pro* Ser *Leu* Gly *Tyr* Val *Leu* Ser	−0.26	*+0.65*
284–293	Tyr *Ile* Asp *Val* Gly *Ala* Thr *Tyr* Tyr *Phe*	+0.01	*+0.92*
298–306	Ser *Ala* Phe *Val* Asp *Tyr* Lys *Ile* Asn	−0.43	*+0.85*
321–330	Ile *Val* Ala *Val* Gly *Met* Thr *Tyr* Gln *Phe*	+0.32	*+0.86*

[a] Amino acid residues on the most hydrophobic side of a potential β-sheet strand are shown in italics.
[b] Average hydrophobicity of each side of a potential β-sheet strand. The values in italics correspond with the residues shown in italics. The hydrophobicity values used are those of the normalized consensus scale of Eisenberg (4).
[c] Region without a pronounced hydrophobic side.
[d] Region where the hydrophobic side does not retain its hydrophobicity in OmpF and OmpC protein and the PhoE proteins from two other enterobacterial species.

contacts the lipids of the outer membrane can be very hydrophobic, in spite of the absence of unbroken hydrophobic sequences long enough to span the membrane.

Interestingly, a model similar to the one we present for the topology of PhoE protein has been proposed for the 180-residue-long outer membrane-located N-terminal domain of OmpA protein (13). It was suggested that this part of the OmpA polypeptide traverses the outer membrane eight times in cross-β structure, exposing four areas to the outside. This model was based on the observation that all amino acid changes found in phage-resistant *ompA* mutants are clustered in four regions which are evenly spaced with about 40 residues between the regions (14). The fact that amino acid substitutions affecting cell surface-associated functions in two different outer membrane proteins (which have no significant sequence homology and are also functionally very different) all show this 40-amino acid spacing is very remarkable and suggests that it reflects a common folding pattern for outer membrane proteins. This view is supported by our finding that the eight potentially membrane-spanning β-sheet segments of OmpA protein each have one side with exclusively hydrophobic residues, just as is shown for the porins in Table 3. This pattern is conserved in the OmpA sequences of five different enterobacterial species (2).

The contributions of Ben Lugtenberg, Marja van Zeijl, Marja Agterberg, August Bekkers, and Marlies Struyvé are gratefully acknowledged.

This work was supported by the Netherlands Foundation for Chemical Research (S.O.N.) with financial aid from the Netherlands Organization for the Advancement of Pure Research (Z.W.O.).

LITERATURE CITED

1. **Benz, R., A. Schmid, and R. Hancock.** 1985. Ion selectivity of gram-negative bacterial porins. J. Bacteriol. **162:**722–727.
2. **Braun, G., and S. T. Cole.** 1984. DNA sequence analysis of the *Serratia marcescens ompA* gene: implications for the organisation of an enterobacterial outer membrane protein. Mol. Gen. Genet. **195:**321–328.
3. **Chai, T., and J. Foulds.** 1978. Two bacteriophages which utilize a new *Escherichia coli* major outer membrane protein as part of their receptor. J. Bacteriol. **135:**164–170.
4. **Eisenberg, D.** 1984. Three-dimensional structure of membrane and surface proteins. Annu. Rev. Biochem. **53:**595–623.
5. **Garavito, R. M., J. Jenkins, J. N. Jansonius, R. Karlsson, and J. P. Rosenbusch.** 1983. X-ray diffraction analysis of matrix porin, an integral membrane protein from *Escherichia coli* outer membranes. J. Mol. Biol. **164:**313–327.
6. **Garnier, J., D. J. Osguthorpe, and B. Robson.** 1978. Analysis of the accuracy and implications of simple methods for predicting the secondary structure of globular proteins. J. Mol. Biol. **120:**97–120.
7. **Inokuchi, K., N. Mutoh, S. Matsuyama, and S. Mizushima.** 1982. Primary structure of the *ompF* gene that codes for a major outer membrane protein of Escherichia coli K-12. Nucleic Acids Res. **10:**6957–6968.
8. **Kleffel, B., R. M. Garavito, W. Baumeister, and J. P. Rosenbusch.** 1985. Secondary structure of a channel-forming protein: porin from *E. coli* outer membranes. EMBO J. **4:**1589–1592.
9. **Korteland, J., P. de Graaff, and B. Lugtenberg.** 1984. PhoE protein pores in the outer membrane of *Escherichia coli* K-12 not only have a preference for Pi and Pi-containing solutes but are general anion-preferring channels. Biochim. Biophys. Acta **778:**311–316.
10. **Korteland, J., N. Overbeeke, P. de Graaff, P. Overduin, and B. Lugtenberg.** 1985. Role of the Arg[158] residue of the outer membrane PhoE pore protein of *Escherichia coli* in bacteriophage TC45 recognition and in channel characteristics. Eur. J. Biochem. **152:**691–697.
11. **Lugtenberg, B., and L. Van Alphen.** 1983. Molecular architecture and functioning of the outer membrane of *Escherichia coli* and other Gram-negative bacteria. Biochim. Biophys. Acta **737:**51–115.
12. **Mizuno, T., M.-Y. Chou, and M. Inouye.** 1983. A comparative study on the genes for three porins of the *Escherichia coli* outer membrane. DNA sequence of the osmoregulated *ompC* gene. J. Biol. Chem. **258:**6932–6940.
13. **Morona, R., M. Klose, and U. Henning.** 1984. *Escherichia coli* K-12 outer membrane protein (OmpA) as a bacteriophage receptor: analysis of mutant genes expressing altered proteins. J. Bacteriol. **159:**570–578.
14. **Morona, R., C. Krämer, and U. Henning.** 1985. Bacteriophage receptor area of outer membrane protein OmpA of *Escherichia coli* K-12. J. Bacteriol. **164:**539–543.
15. **Nikaido, H., E. Y. Rosenberg, and J. Foulds.** 1983. Porin channels in *Escherichia coli*: studies with β-lactams in intact cells. J. Bacteriol. **153:**232–240.
16. **Nikaido, H., and M. Vaara.** 1985. Molecular basis of bacterial outer membrane permeability. Microbiol. Rev. **49:**1–32.
17. **Overbeeke, N., H. Bergmans, F. Van Mansfeld, and B. Lugtenberg.** 1983. Complete nucleotide sequence of *phoE*, the structural gene for the phosphate limitation inducible outer membrane pore protein of *Escherichia coli* K-12. J. Mol. Biol. **163:**513–532.
18. **Overbeeke, N., and B. Lugtenberg.** 1980. Expression of outer membrane protein e of *Escherichia coli* K12 by phosphate limitation. FEBS Lett. **112:**229–232.
19. **Tommassen, J., and B. Lugtenberg.** 1980. Outer membrane protein e of *Escherichia coli* K-12 is co-regulated with alkaline phosphatase. J. Bacteriol. **143:**151–157.
20. **Tommassen, J., and B. Lugtenberg.** 1984. Amino terminus of outer membrane PhoE protein: localization by use of a *bla-phoE* hybrid gene. J. Bacteriol. **157:**327–329.
21. **Tommassen, J., A. P. Pugsley, J. Korteland, J. Verbakel, and B. Lugtenberg.** 1984. Gene encoding a hybrid OmpF-PhoE pore protein in the outer membrane of *Escherichia coli* K12. Mol. Gen. Genet. **197:**503–508.
22. **Tommassen, J., P. Van der Ley, M. Van Zeijl, and M. Agterberg.** 1985. Localization of functional domains in *E. coli* outer membrane porins. EMBO J. **4:**1583–1587.
23. **Van der Ley, P., H. Amesz, J. Tommassen, and B. Lugtenberg.** 1985. Monoclonal antibodies directed against the cell surface-exposed part of PhoE pore protein of the *Escherichia coli* K-12 outer membrane. Eur. J. Biochem. **147:**401–407.

23a. **Van der Ley, P., M. Struyvé, and J. Tommassen.** 1986. Topology of outer membrane pore protein PhoE of *Escherichia coli*. Identification of cell surface-exposed amino acids with the aid of monoclonal antibodies. J. Biol. Chem. **261:**12222–12225.

24. **Verhoef, C., C. Van Koppen, P. Overduin, B. Lugtenberg, J. Korteland, and J. Tommassen.** 1984. Cloning and expression in *Escherichia coli* K-12 of the structural gene for outer membrane PhoE protein from *Enterobacter cloacae*. Gene **32:**107–115.

glpT-Dependent Transport of *sn*-Glycerol 3-Phosphate in *Escherichia coli* K-12

TIMOTHY J. LARSON†

Department of Biochemistry and Molecular Biology, University of North Dakota School of Medicine, Grand Forks, North Dakota 58202

sn-Glycerol 3-phosphate (G3P) is a key metabolic intermediate having both anabolic and catabolic fates. Acylated derivatives of G3P, the glycerophospholipids, are present in virtually all biological membranes. Thus, G3P is a direct precursor for phospholipid synthesis. G3P also serves as a source of carbon and energy for a variety of microorganisms. It is not surprising that organisms have developed transport systems for this compound.

The pathways for the metabolism of G3P in *Escherichia coli* are shown in Fig. 1. The proteins catalyzing the catabolic steps are encoded by the genes of the *glp* regulon (10). The *glp* genes constitute a regulon because they are organized in several operons which are all negatively regulated by a common repressor encoded by *glpR*. The inducer for the *glp* operons is G3P.

Extracellular G3P is actively accumulated by the cell via a cytoplasmic membrane-associated carrier encoded by *glpT* (9). Besides G3P, glycerol and glycerophosphodiesters serve as carbon and energy sources and as precursors to phospholipid (Fig. 1). Extracellular glycerophosphodiesters are converted to G3P plus an alcohol by action of a periplasmic phosphodiesterase (*glpQ* encoded) specific for these compounds (8). Glycerol enters the cell via the *glpF*-encoded glycerol diffusion facilitator and is phosphorylated by glycerol kinase (encoded by *glpK* [10]). Cytoplasmic G3P is oxidized to dihydroxyacetone phosphate. This reaction is catalyzed by the *glpD*-encoded G3P dehydrogenase during aerobic growth (10) or by the *glpACB* operon-encoded G3P dehydrogenase during anaerobic growth with fumarate or nitrate as electron acceptor (7, 10).

In the absence of exogenous sources of G3P, the G3P required for phospholipid synthesis is derived from dihydroxyacetone phosphate by action of G3P synthase encoded by *gpsA*. The cellular concentration of G3P is stringently regulated by feedback inhibition of this enzyme by the product G3P. During conditions of phosphate deprivation, the *pho* regulon is derepressed, and exogenous G3P can be actively accumulated via the *pho* regulon-dependent high-affinity active transport system encoded by *ugp* (13).

The genes encoding the components of the *glp* regulon have been mapped to three regions on the linkage map of *E. coli* (Fig. 2). Mutations in *glpD* affect aerobic G3P dehydrogenase activity and map adjacent to the *glpR* locus near min 75 (10, 14). The *glpE* and *glpG* genes (14) map between the divergently transcribed *glpD* and *glpR* genes. The functions and directions of transcription of *glpE* and *glpG* are unknown. The *glpK* and *glpF* loci map at min 88 (10).

The *glpTQ* and the *glpACB* operons are located near min 49. Genetic analysis of this region indicates that *glpT* is promoter proximal to *glpQ* in an operon transcribed counterclockwise on the chromosome (9, 11). The catalytic components of the anaerobic dehydrogenase are encoded by *glpA* and *glpC* located promoter proximal to *glpB*, which is thought to encode a membrane anchor subunit of the enzyme (3a, 7). The *glpACB* operon is directly adjacent to and is transcribed divergently from the *glpTQ* operon (see next section).

Genetic Analysis of the glpT-Dependent Transport System

After the initial characterization of the *glpT* system by Lin and co-workers (5, 10), its characterization continued in the laboratory of W. Boos. Comparison of the *glpT* wild-type and mutant strains revealed a close correlation between G3P active transport function and the presence of a 40,000-M_r periplasmic protein (15). Isolation of recombinant plasmids which carry $glpT^+$, but are deleted for the gene encoding the periplasmic protein, showed that this protein is not needed for transport of G3P. Finding of a single genetic complementation group in a collection of *glpT* mutations (11) is consistent with the idea that a single membrane carrier protein (encoded by a promoter-proximal gene, *glpT*) should suffice for transport of G3P. The close correlation with the 40,000-M_r periplasmic protein was due to the fact that the gene, named *glpQ*, encoding the periplasmic protein occurs in the same transcriptional unit. Some of the

†Present address: Department of Biochemistry and Nutrition, Virginia Polytechnic Institute and State University, Blacksburg, VA 24061.

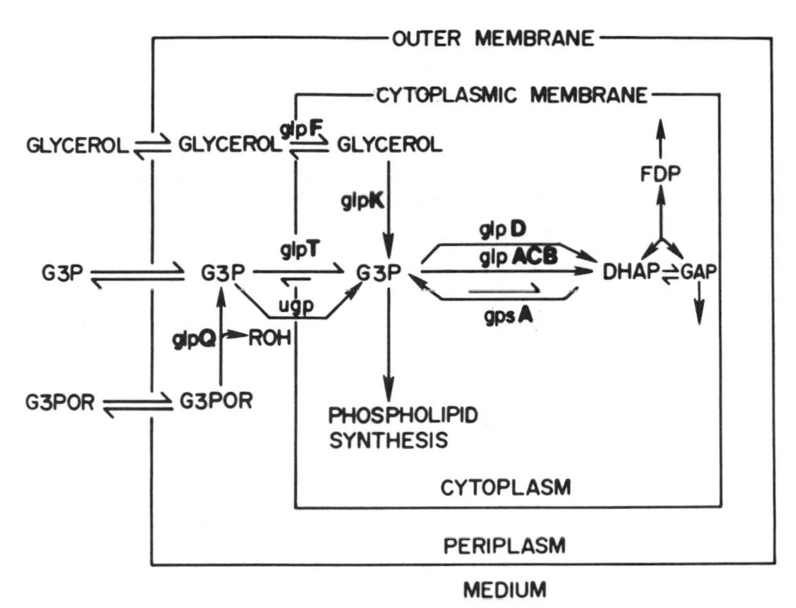

FIG. 1. G3P metabolism in *E. coli*. Gene designations are shown adjacent to the function catalyzed by the respective gene product. The designations are *glpF*, glycerol diffusion facilitator; *glpK*, glycerol kinase; *glpD*, aerobic G3P dehydrogenase; *glpACB*, anaerobic G3P dehydrogenase; *glpT*, G3P active transport; *glpQ*, glycerophosphodiester phosphodiesterase; *ugp*, *pho* regulon-dependent G3P active transport; *gpsA*, G3P synthase. DHAP, dihydroxyacetone phosphate; GAP, glyceraldehyde 3-phosphate; G3POR, glycerophosphodiester; FDP, fructose 1,6-diphosphate.

strains analyzed contained polar mutations in *glpT* which caused decreased expression of *glpQ*. Later, we (8) found that the periplasmic protein is an enzyme which hydrolyzes glycerophosphodiesters such as glycerophosphoethanolamine, glycerophosphocholine, glycerophosphoglycerol, and bis(glycerophospho)glycerol. Thus, the *glpQ* gene product expands the catabolic capability of the cells to include a variety of glycerophosphodiesters. The phosphodiesterase encoded by *glpQ* is specific for these compounds because it did not hydrolyze bis(*p*-nitrophenyl)phosphate, a general phosphodiesterase substrate (8).

The availability of strains harboring phage λ integrated at *glpT-lac*, *glpA-lac*, and *glpB-lac* fusions (3a) facilitated isolation of specialized phages carrying portions of *glpT*. Genetic crosses between these phages and the collection of *glpT* mutant strains were carried out. It became apparent that *glpT* is transcribed divergently from both *glpA* and *glpB* (Fig. 3). The promoter-proximal end of *glpT* is on the right of Fig. 3 because phages derived from *glpT-lac* fusions carried this material (11). The *glpA* and *glpB* genes are transcribed divergently from *glpT* because independent phage lines derived from fusions in *glpA* and *glpB* carried various amounts of *glpT* material beginning with the promoter-proximal end of *glpT*. Earlier experiments (11) indicated that *glpT* is transcribed in the counterclockwise direction on the linkage map. All of these results demonstrate that this region of the chromosome is organized as shown in Fig. 2. In *Salmonella typhimurium*, *glpA* is

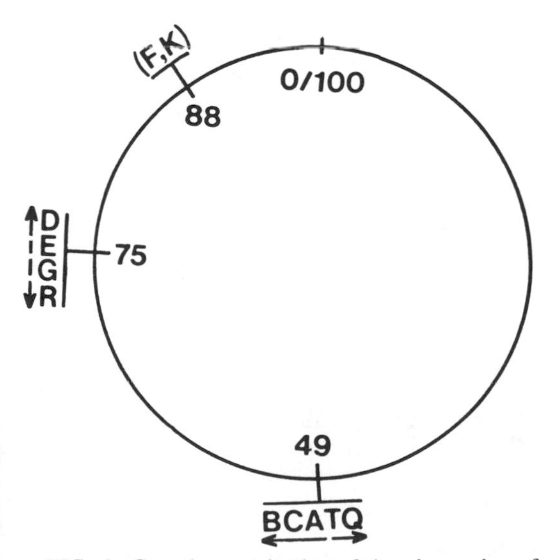

FIG. 2. Genetic organization of the *glp* regulon of *E. coli*. The positions on the linkage map and the directions of transcription (if known) are indicated. Gene designations not defined in the Fig. 1 legend are *glpR*, repressor for the *glp* regulon; *glpE* and *glpG*, two genes of unknown function.

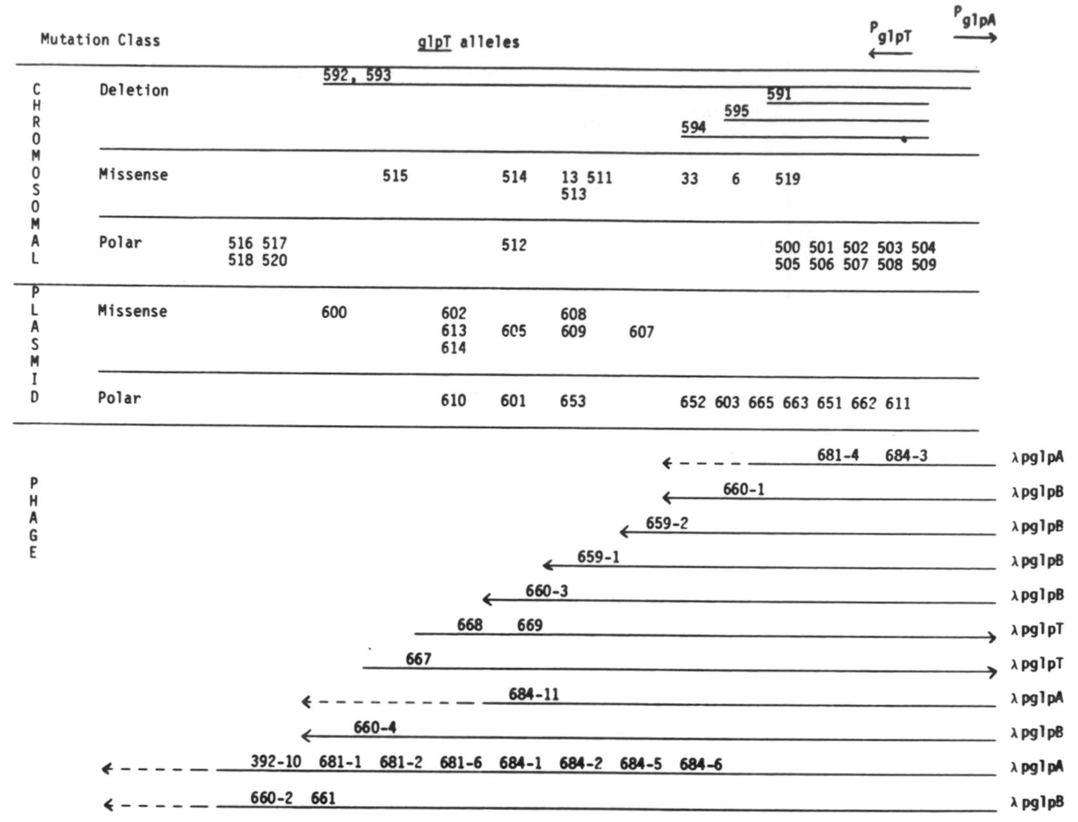

FIG. 3. Genetic map of *glpT*. Mutant strains harboring the indicated chromosomal or plasmid-borne *glpT* alleles were crossed with λ phages derived from strains with λ integrated at *glpA-lac, glpB-lac*, or *glpT-lac* fusions as indicated on the right of the figure. Phages carried chromosomal DNA shown as a solid line; the dashed extensions of lines indicate uncertainty about endpoints. Arrows point away from *glp-lac* fusion joints. The first three numbers for a phage isolate indicate the bacterial strain from which it was derived; numbers after the hyphens are isolate numbers. The chromosomal and plasmid-borne mutations mapped were those described earlier (9, 11). The chromosomal deletions are indicated by the solid lines at the top of the drawing. Point mutations were divided into two classes; they were considered missense mutations if they had no polar effect on expression of *glpQ* and were not suppressed by ochre or amber suppressors.

positioned in the same way with respect to the *glpTQ* operon (6).

In addition to providing information about the directions of transcription of the *glpT, glpA*, and *glpB* genes, the phages described in Fig. 3 were useful for mapping of plasmid-borne mutations generated by using hydroxylamine mutagenesis in vitro (9). The order of the amber mutations correlated well with the sizes of the amber fragments synthesized in vitro (9), providing additional evidence in support of the proposed direction of transcription of *glpT*. Mapping of these amber mutations to a given deletion interval allowed prediction of the physical locations of other mutations. For example, most of the plasmid-borne missense mutations clustered near the carboxyl-terminal end of *glpT* (Fig. 3). All of these missense mutations except *glpT607* mapped in the same or in a more promoter-distal deletion interval than *glpT653*(Am), which gives rise to a 27,000-M_r amber fragment (9). The G3P carrier has an apparent molecular weight of 33,000 (9; see next section).

Identification and Characterization of the G3P Carrier Protein

Construction of a collection of recombinant plasmids carrying either $glpT^+$ or mutated *glpT* (9) proved very useful for the following studies. Plasmids in which portions of the *glpTQ* operon had been deleted were constructed by using restriction endonucleases and DNA ligase. Plasmids with point mutation *glpT* were isolated by using hydroxylamine mutagenesis in vitro. The plasmid-borne missense and polar mutations mapped in Fig. 3 were among those isolated in this way.

The G3P carrier was identified by sodium

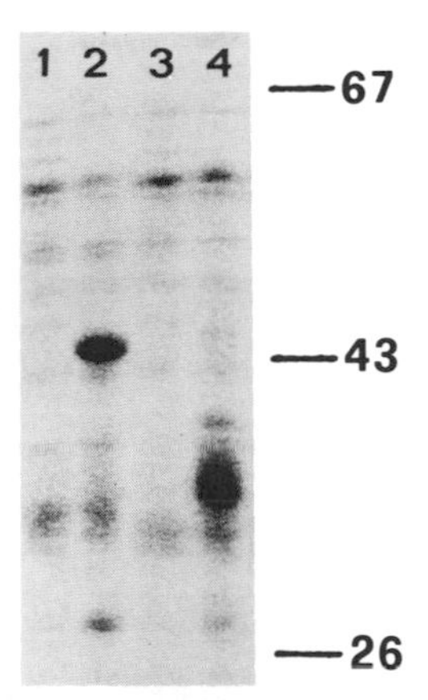

FIG. 4. Identification of the G3P carrier protein. Cytoplasmic membrane fractions from strain DL291 (pGS3103) (lanes 1 and 3) and DL291(pGS31) (lanes 2 and 4) were analyzed by sodium dodecyl sulfate-polyacrylamide gel electrophoresis followed by Coomassie blue staining (9). Strain DL291 carries Δ*glpT*; pGS31 carries $glpT^+$ cloned into pBR322; pGS3103 is identical to pGS31 but carries *glpT603*(Am) (9). Samples containing 40 μg of protein, 0.8% sodium dodecyl sulfate and 10 mM dithiothreitol were heated at 50°C for 10 min (lanes 1 and 2) or at 50°C for 10 min followed by treatment at 95°C for 5 min (lanes 3 and 4). The molecular weights of protein standards are indicated (10^3) on the right.

dodecyl sulfate-polyacrylamide gel electrophoretic analysis of cytoplasmic membrane fractions isolated from strains harboring plasmids with either wild-type or mutated *glpT* (9). Cytoplasmic membrane fractions from the $glpT^+$ overproducing strain DL291(pGS31) contained large amounts of a protein exhibiting an apparent M_r of 44,000 (Fig. 4, lane 2). This protein was missing from *glpT*(Am) (lane 1) and Δ*glpT* (not shown) mutant strains. Visualization of the G3P carrier required comparison of the cytoplasmic membrane fractions. No differences between protein profiles displayed on sodium dodecyl sulfate gels were apparent when crude extracts or crude membrane fractions from wild-type and mutant strains were compared. The protocol for sample treatment prior to electrophoresis was also critical for identification of the *glpT* product. Treatment at 50°C was used to visualize the carrier. If samples were heated at 100°C prior to electrophoresis, no apparent differences between wild-type and mutant strains were observed. This may have been due to aggregation of the carrier, resulting in failure of the protein to enter the gel. Such behavior is reminiscent of that exhibited by the lactose carrier of *E. coli* (16).

Exploration of different protocols for sample treatment prior to electrophoresis led to the discovery that the 44,000-M_r form of the G3P carrier was modified by treatment of samples at higher temperature (9). When samples were first treated at 50°C for 10 min and then at 95°C for 5 min, the apparent M_r of the carrier decreased to 33,000. A possible interpretation is that the 44,000-M_r protein is in a solubilized but incompletely denatured oligomeric form. The lower-M_r form could be the completely denatured and solubilized subunit. An alternative explanation is that both forms are monomeric, and the lower-M_r form binds more sodium dodecyl sulfate or is more completely denatured, or both.

The thermal modifiability of the G3P carrier suggested that the protein may be a dimer or oligomer in the membrane. If this is so, then some missense mutations in *glpT* might be negatively dominant over wild-type *glpT* function. To test this possibility, eight recombinant plasmids carrying nonpolar mutations in *glpT* were introduced by transformation into $glpT^+$ *recA* strains (9) and tested for growth on minimal G3P medium. Seven of these eight strains were unable to grow on G3P. These mutations are negatively dominant, indicating that the active form of G3P carrier is oligomeric. Control plasmids with *glpT*(Am) or Δ*glpT* did not produce a $G3P^-$ phenotype. Similar results have been observed with the lactose carrier (12).

Reconstitution of the G3P Carrier

Elvin, Hardy, and Rosenberg (4) have shown that the *glpT* transport system mediates the efflux of cellular P_i in response to externally added P_i or G3P. They proposed that the G3P carrier mediates G3P:P_i and P_i:P_i antiport. To study the molecular basis for transport via the *glpT* system, movement of these anions was analyzed in intact cells and in reconstituted proteoliposomes. Transport in strain DL291 (pGS31), which overproduces the G3P carrier, was compared with transport in the isogenic Δ*glpT* strain DL291(pGS34).

Transport of both [^{14}C]G3P and $^{31}P_i$ was dependent upon the presence of the G3P carrier (reference 1 and additional data not shown). Transport of $^{32}P_i$ was abolished when G3P was included in the incubation mixture. $^{32}P_i$ accumulated via the G3P carrier was released in response to externally added P_i. These results show that the G3P transporter recognizes both G3P and P_i, as proposed earlier (4, 5).

The results indicated that transport of G3P and P_i via the G3P carrier may be analogous to transport via the hexose-phosphate carrier of *Streptococcus lactis*, which catalyzes both P_i:P_i and hexose-phosphate:P_i exchange (3). If this is so, the methods used for reconstitution of antiport from *S. lactis* (2) might be useful for reconstitution of the G3P carrier from *E. coli*. Indeed, it was possible to demonstrate both P_i:P_i and

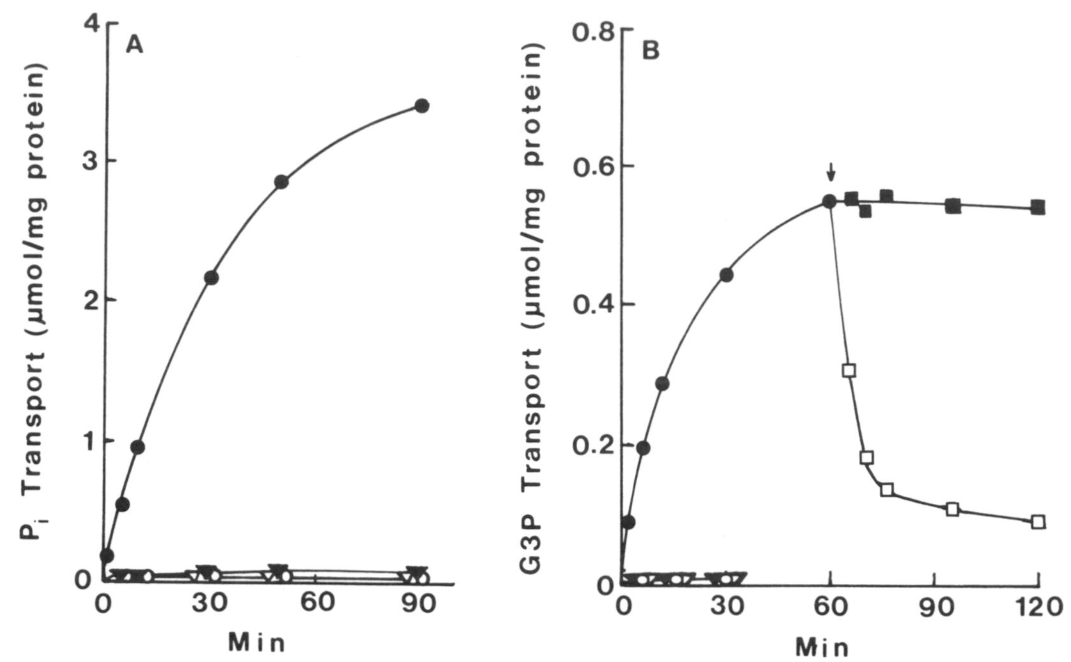

FIG. 5. Reconstitution of the G3P carrier. Proteoliposomes were prepared (1) by dilution of DL291(pGS31) (●, ○) or DL291(pGS34) (▼, ▽) into either 0.1 M KP_i (●, ▲) or 0.1 M K-MOPS (○, △). Liposomes were collected by centrifugation, washed in 20 mM K-MOPS–75 mM K_2SO_4–2.5 mM $MgSO_4$, pH 7, and resuspended in this buffer prior to the following transport measurements. (A) Samples of proteoliposomes incubated in the presence of 250 μM $^{32}P_i$ were filtered at the indicated time points, and the radioactivity was determined. (B) An experiment similar to that shown in A was done using 30 μM [^{14}C]G3P. KP_i (30 mM) (□) or assay buffer (■) was added to a portion of proteoliposomes at the time indicated by the arrow. Strain DL291(pGS31) is described in the Fig. 4 legend; pGS34 is a Δ*glpT* derivative of pGS31 (9).

G3P:P_i exchange in proteoliposomes reconstituted from octylglucoside extracts of membranes from the *glpT*$^+$ overproducing strain (1; Fig. 5). Figure 5A shows that $^{32}P_i$ uptake into proteoliposome was dependent upon preloading the proteoliposomes with KP_i. No uptake occurred when the liposomes were preloaded with potassium-morpholinepropranesulfonic acid (K-MOPS). Furthermore, $^{32}P_i$ uptake was dependent upon the G3P transport system because no transport was observed in proteoliposomes prepared from the Δ*glpT* strain. Other evidence indicating that transport of P_i occurred via GlpT was that $^{32}P_i$ transport was inhibited by G3P. Half-maximal inhibition of $^{32}P_i$ transport occurred at 40 μM G3P, which is near the Michaelis constant for transport of G3P (see below).

Similar results were observed when [^{14}C]G3P transport into proteoliposomes was assessed (1). Again, transport was dependent upon preloading of proteoliposomes with KP_i and was observed only in proteoliposomes prepared from the *glpT*$^+$ overproducing strain (Fig. 5B). Accumulated G3P rapidly exited the proteoliposomes in response to 30 mM KP_i added externally (Fig. 5B, arrow). Other experiments (not shown) indicated that the G3P:P_i exchange reaction was saturable. The Michaelis constant for transport of G3P was 20 μM, which is similar to the values of 12 μM (5) and 20 μM (13) observed for transport in intact cells. The maximal velocity for transport of G3P was 127 nmol/min per mg of protein.

The results indicate that the reconstituted G3P carrier encoded by *glpT* catalyzes both P_i:P_i and G3P:P_i exchange. The exchanges catalyzed by and the kinetic properties of the reconstituted G3P carrier appear to be the same as those of the G3P carrier characterized in intact cells. All of the results support the proposition that anion exchange is the molecular basis for transport via the *glpT*-encoded carrier of *E. coli*.

I thank W. Boos, who provided a stimulating environment for initiation of this work. Members of his group who contributed to this project included D. Ludtke, G. Schumacher, M. Ehrmann, R. Hengge, and H. Schweizer. I am also grateful to P. Maloney and S. Ambudkar, who provided the laboratory space, expertise, and assistance necessary for the reconstitution studies.

I was supported by a fellowship from the Alexander von Humboldt Foundation and by a travel award from the University of North Dakota Faculty Research Committee. Some of the work was supported by grant DMB-8502166 from the National Science Foundation.

LITERATURE CITED

1. **Ambudkar, S. V., T. J. Larson, and P. C. Maloney.** 1986. Reconstitution of sugar phosphate transport systems of *Escherichia coli*. J. Biol. Chem. **261:**9083–9086.
2. **Ambudkar, S. V., and P. C. Maloney.** 1986. Anion exchange in bacteria. Reconstitution of phosphate:hexose 6-phosphate antiport from *Streptococcus lactis*. Methods Enzymol. **125:**558–563.
3. **Ambudkar, S. V., L. A. Sonna, and P. C. Maloney.** 1986. Variable stoichiometry of phosphate-linked anion exchange in *Streptococcus lactis*. Implications for the mechanism of sugar phosphate transport by bacteria. Proc. Natl. Acad. Sci. USA **83:**280–284

3a. **Ehrmann, M., W. Boos, E. Ormseth, H. Schweizer, and T. J. Larson.** 1987. Divergent transcription of the *sn*-glycerol-3-phosphate active transport (*glpT*) and anaerobic *sn*-glycerol-3-phosphate dehydrogenase (*glpA glpC glpB*) genes of *Escherichia coli* K-12. J. Bacteriol. **169:**526–532.

4. **Elvin, C. M., C. M. Hardy, and H. Rosenberg.** 1985. P_i exchange mediated by the GlpT-dependent *sn*-glycerol-3-phosphate transport system in *Escherichia coli*. J. Bacteriol. **161:**1054–1058.
5. **Hayashi, S., J. P. Koch, and E. C. C. Lin.** 1964. Active transport of L-α-glycerophosphate in *Escherichia coli*. J. Biol. Chem. **239:**3098–3105.
6. **Hengge, R., T. J. Larson, and W. Boos.** 1983. *sn*-Glycerol-3-phosphate transport in *Salmonella typhimurium*. J. Bacteriol. **155:**186–195.
7. **Kuritzkes, D. R., X.-Y. Zhang, and E. C. C. Lin.** 1984. Use of Φ(*glp-lac*) in studies of respiratory regulation of the *Escherichia coli* anaerobic *sn*-glycerol-3-phosphate dehydrogenase genes (*glpAB*). J. Bacteriol. **157:**591–598.
8. **Larson, T. J., M. Ehrmann, and W. Boos.** 1983. Periplasmic glycerophosphodiester phosphodiesterase of *Escherichia coli*, a new enzyme of the *glp* regulon. J. Biol. Chem. **258:**5428–5432.
9. **Larson, T. J., G. Schumacher, and W. Boos.** 1982. Identification of the *glpT*-encoded *sn*-glycerol-3-phosphate permease of *Escherichia coli*, an oligomeric integral membrane protein. J. Bacteriol. **152:**1008–1021.
10. **Lin, E. C. C.** 1976. Glycerol dissimilation and its regulation in bacteria. Annu. Rev. Microbiol. **30:**535–578.
11. **Ludtke, D., T. J. Larson, C. Beck, and W. Boos.** 1982. Only one gene is necessary for the *glpT*-dependent transport of *sn*-glycerol-3-phosphate in *Escherichia coli*. Mol. Gen. Genet. **186:**540–547.
12. **Mieschendahl, M., D. Buchel, H. Blockage, and B. Müller-Hill.** 1981. Mutations in the *lacY* gene of *Escherichia coli* define functional organization of lactose permease. Proc. Natl. Acad. Sci. USA **78:**7652–7656.
13. **Schweizer, H., M. Argast, and W. Boos.** 1982. Characteristics of a binding protein-dependent transport system for *sn*-glycerol-3-phosphate in *Escherichia coli* that is part of the *pho* regulon. J. Bacteriol. **150:**1154–1163.
14. **Schweizer, H., G. Sweet, and T. J. Larson.** 1986. Physical and genetic structure of the *glpD-malT* interval of the *Escherichia coli* K-12 chromosome. Identification of two new structural genes of the *glp*-regulon. Mol. Gen. Genet. **202:**488–492.
15. **Silhavy, T. J., I. Hartig-Beecken, and W. Boos.** 1976. Periplasmic protein related to the *sn*-glycerol-3-phosphate transport system of *Escherichia coli*. J. Bacteriol. **126:**951–958.
16. **Teather, R. M., B. Müller-Hill, U. Abrutsch, G. Aichele, and P. Overath.** 1978. Amplification of the lactose carrier protein in *Escherichia coli* using a plasmid vector. Mol. Gen. Genet. **159:**239–248.

ugp-Dependent Transport System for *sn*-Glycerol 3-Phosphate of *Escherichia coli*

PIUS BRZOSKA, HERBERT SCHWEIZER, MANFRED ARGAST, AND WINFRIED BOOS

Department of Biology, University of Konstanz, Konstanz, Federal Republic of Germany

The uptake and metabolism of *sn*-glycerol 3-phosphate (G3P) and glycerophosphoryldiesters in *Escherichia coli* wild-type cells growing on minimal media of high (>10 mM) phosphate content are mediated by the proteins encoded by the *glp* genes (10). The expression of the *glp* genes is under the control of *glpR*, coding for a repressor that is inactivated by G3P (18). The *glpT* gene codes for a G3P transport (GlpT) system with an apparent K_m of 12 μM (7). It is dependent on the proton motive force, it is composed of more than one identical polypeptide chain (9), and it functions essentially as a G3P/P_i exchange system (6). It does not recognize glycerophosphoryldiesters (8). However, there is a periplasmic glycerophosphoryldiester phosphodiesterase encoded by *glpQ* on the same operon as *glpT*. This enzyme hydrolyzes these compounds to G3P, which can subsequently be taken up by the transport system (8). Once inside the cytoplasm, G3P is oxidized to dihydroxyacetone phosphate by an aerobic or anaerobic dehydrogenase (10), enzymes that are both encoded by genes *glpD* and *glpA*. Thus, the *glp*-encoded enzymes and transport system are geared for the utilization of G3P and the glycerophosphoryldiesters as carbon source. However, cells containing only the GlpT system can use G3P as sole source of carbon as well as phosphate (14).

During transport, the GlpT protein exchanges external G3P with internal phosphate (6). The observed G3P/Pi exchange activity not only provides a favorable chemiosmotic energy balance but also creates an exit pathway for the excess of cellular phosphate that results from the utilization of G3P as the sole source of carbon, energy, and phosphate.

Mutants defective in *glpT* are unable to grow on G3P as the sole source of carbon when the concentration of P_i in the growth medium is high enough to repress alkaline phosphatase synthesis. These mutants "revert" to G3P$^+$ by secondary mutations in genes that cause constitutivity of the *pho* regulon (4). Growth on G3P is then due to the action of constitutive alkaline phosphatase that hydrolyzes G3P in the periplasm to glycerol, which ultimately is used as carbon source. However, simultaneously with the appearance of constitutive alkaline phosphatase, a second transport system for G3P, the Ugp (uptake of G3P) system, is synthesized (3).

Physiologically, the Ugp system is geared for the utilization of phosphate, but apparently not of carbon. Strains that express Ugp as the only G3P-transporting system, as well as the G3P dehydrogenase constitutively, are able to use G3P as sole source of phosphate but not of carbon. However, G3P is not chemically altered during transport and can be used as precursor for phospholipid biosynthesis. In contrast, after transport, glycerophosphorylethanolamine (GPE) is not accumulated as such but is hydrolyzed to G3P that accumulates and ethanolamine that diffuses back into the medium. The enzyme responsible for this phosphodiesterase activity is most likely located at the inner face of the cytoplasmic membrane.

We isolated mutants that are deficient in this glycerophosphoryldiester phosphodiesterase. These mutants still transport G3P normally and are now able to accumulate GPE. The mutation maps near or within the *ugp* region. When plasmids containing part of or the entire *ugp* region were used in a complementation assay, it became clear that the diesterase is coded for by *ugpC* or by an as yet unidentified gene distal to *ugpC* on the same operon.

In this article we describe the different aspects of the GlpT and Ugp systems.

Ugp Transport System

Substrate specificity and regulation. The Ugp system can be measured conveniently in strains that are *glpT* and, therefore, deficient in the major G3P uptake system in *E. coli*. Strains that are *glpT ugp*$^+$ are resistant to the antibiotic fosfomycin, while *ugp glpT*$^+$ strains are not; *glpT ugp* strains are resistant to the antibiotic 3,4-dihydroxybutylphosphonate, while *glpT ugp*$^+$ strains are not. In this way one can easily distinguish the occurrence of these systems in the cell. For maximal expression of Ugp, the cells have to be induced by phosphate starvation or mutants have to be used that are expressing the *pho* regulon constitutively (*phoR* or *pst*). The observation that mutations in *phoB* as well as double mutants in *phoR* and *phoM* prevent expression identified the Ugp transport system as a member of the *pho* regulon (17). The apparent K_m of uptake of the Ugp system for its major substrate G3P is around 2 μM; thus, the Ugp system has about a 10-fold higher affinity for G3P than the GlpT system (14). Also, the range of substrate specificity is different with the two systems. While the GlpT system recognizes (in addition to G3P) glyceraldehyde 3-phosphate and the toxic G3P analogs fosfomycin and 3,4-dihydroxybutyl phosphonate, as well as phosphate (10), the Ugp system recognizes (in addition to G3P) glycerophosphoryldiesters with affinities similar to those for G3P but not phosphate, glyceraldehyde phosphate, or fosfomycin. These glycerolphosphoryldiesters are hydrolyzed very quickly during or after transport; consequently, one can detect inside the cells only the split products, namely, G3P and the corresponding alcohol. Besides G3P, the Ugp system recognizes 3,4-dihydroxybutyl phosphonate, an observation that can be exploited to isolate mutants in *ugp* (19). The substrate specificity of Ugp is determined by a periplasmic binding protein (2), and membrane vesicles are not active in Ugp-mediated transport (14).

Ugp paradox. Ugp was discovered by the analysis of *glpT* mutants that had reverted to G3P$^+$ on media containing G3P as the only source of carbon and a P_i concentration below 1 mM (4). Surprisingly, these G3P$^+$ revertants no longer grew on G3P when the P_i concentration in the medium was raised to 100 mM. The explanation for this phenomenon was that these revertants became *pho* constitutive by a spontaneous mutation in the *pho* regulatory genes (*pst, phoR*); consequently, growth on G3P at low P_i concentration was due to the hydrolysis of G3P by alkaline phosphatase to P_i and glycerol, which subsequently was used as a carbon source. High P_i concentration inhibits alkaline phosphatase by feedback inhibition (12), and thus hydrolysis of G3P no longer occurs. However, these revertants showed, in addition to

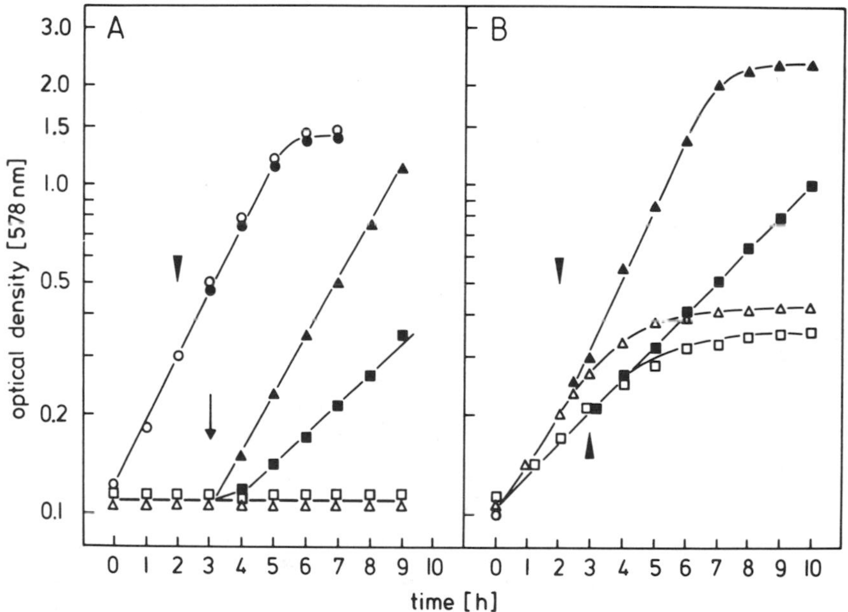

FIG. 1. Growth of different *ugp*$^+$ strains on G3P as the sole source of carbon or phosphate. G3P transported exclusively by Ugp can serve as a sole source of phosphate but not of carbon. (A) A *ugp*$^+$ strain pregrown overnight in minimal medium A (11) containing 0.2% glucose and 100 mM P_i was divided into two fractions. One fraction was diluted into fresh minimal medium A containing 0.2% glucose as the carbon source (○). At the time indicated by an arrowhead, the culture was divided, and to one part 0.2% G3P was added (●). The remaining fraction of the overnight culture was washed twice in Tris medium without phosphate and suspended in the same medium containing 60 μM P_i and 0.2% G3P (△). At the time indicated by an arrow, the culture was divided, and to one part 0.2% glucose was added (▲). (■, □) Growth curves of a *ugp*$^+$ *glpR* strain which was pregrown overnight in minimal medium containing 0.4% neutralized succinate as the carbon source. The cells were treated as described above and suspended in Tris medium containing 0.2% G3P (□). At the time indicated by an arrow, the culture was divided, and to one part 0.4% succinate was added (■). (B) The *ugp* strains were pregrown and pretreated as described under A. The *ugp*$^+$ strain was grown in Tris medium containing 60 μM P_i and 0.2% glucose as the carbon source (△). The *ugp*$^+$ glpR strain was grown in the same medium, except that 0.4% succinate was the carbon source (□). At the time points indicated by an arrowhead, each culture was divided, and to one part of each 0.2% G3P was added (▲, ■). (Taken from reference 14.)

constitutive alkaline phosphatase, a high G3P-transport activity (mediated by Ugp) which was independent of the phosphate concentration in the assay. After introduction of a *phoA* mutation into these revertants, they could no longer grow on G3P, despite the fact that they still transported G3P and still synthesized constitutive levels of G3P-metabolizing enzymes. However, in the presence of an alternative carbon source, G3P entering the cell exclusively via Ugp could be used as the sole source of phosphate (14) (Fig. 1). Here, a strain that lacks alkaline phosphatase (*phoA*), glycerolkinase (*glpK*), and the *glpT*-dependent G3P-transport system, but that is constitutive for the Ugp system as well as the G3P dehydrogenase, was tested for growth on different carbon sources. Mutations in *phoA* and *glpK* were introduced to avoid growth on glycerol which could emerge by the action of alkaline phosphatase or another phosphatase hydrolyzing G3P. As can be seen, the mutant did not grow on G3P, but readily grew in the additional presence of glucose or succinate. The experiment shows that G3P entering the cell through Ugp can provide enough cellular phosphate but not enough carbon for growth. At present, there is no satisfactory explanation for this paradox, but some additional information is available. (i) By thin-layer chromatography (TLC) analysis of cellular extracts, it could be shown that G3P is transported as such; ^{14}C as well as ^{32}P label in G3P could readily be found in phospholipids. The formation of lipids from G3P is consistent with the observation that G3P transported exclusively via Ugp can be used as supplement for the G3P auxotrophy in a *plsB* strain (15). (ii) In the presence of an alternative carbon source, ^{14}C label provided by G3P was also incorporated into amino acids. (iii) When the radioactively labeled compounds were compared by TLC after incubating cells with [^{14}C]G3P for a short period of time, the same spots were observed whether a *ugp*$^+$ *glpT* or a *glpT*$^+$ *ugp* strain was used.

From these observations one can conclude that G3P, transported exclusively via the Ugp system, does not form a toxic compound or a compound that is no longer metabolizable. Our working hypothesis to explain this phenomenon considers the missing function of *glpT* in a strain that transports G3P exclusively and unidirectionally via Ugp. With G3P as a sole carbon source, too much phosphate is accumulated, the excess of which should be excreted. Since GlpT appears to be the only exit system for P_i available in these strains (1, 5), P_i may accumulate and exert a *trans*-inhibition on the further income of G3P via the Ugp system.

Mapping and cloning of *ugp*. The finding that Ugp does transport the toxic G3P analog 3,4-

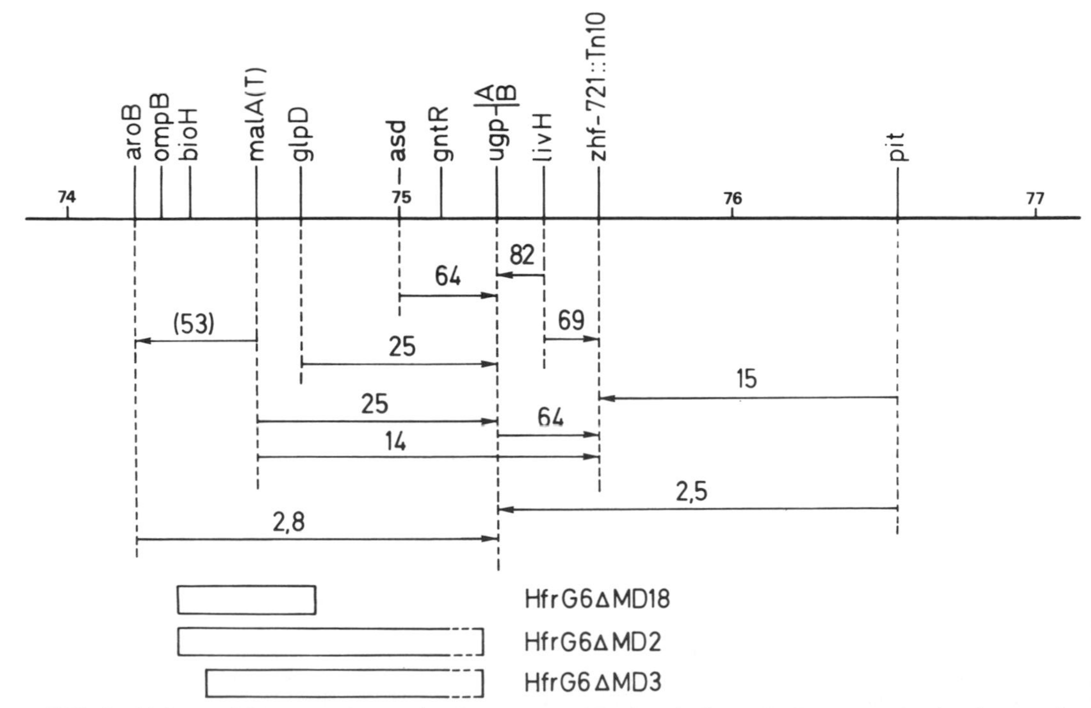

FIG. 2. Linkage of the *ugp* region on the chromosome. Numbers indicate the P1 cotransduction frequencies in percent between the indicated genetic markers. Bars indicate deletions on the chromosome. (Taken from reference 19.)

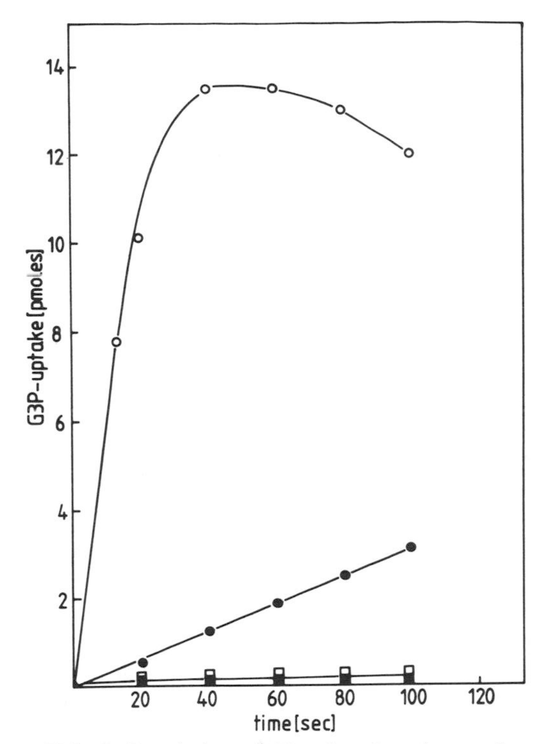

FIG. 3. Regulation of Ugp by the *pho* regulon. [^{14}C]G3P uptake assays were done with a *ugp* strain carrying plasmid pSH12 containing *ugp*$^+$ (○, ●) or the vector plasmid pBR322 (□, ■). Open symbols indicate growth of the strains under *ugp* derepressing conditions (0.1 mM P_i); closed symbols indicate growth of the strains under *ugp* repressing conditions (1 mM P_i). (Taken from reference 16.)

dihydroxybutyl phosphonate made it possible to isolate a variety of mutants in *ugp* including Tn*10* insertions and *lacZ* fusions (19). The mapping of these mutations by P1 transduction located them at 76 min on the *E. coli* linkage map (Fig. 2).

The cloning of *ugp* was initiated by lysogenic complementation using a gene library of *E. coli Eco*RI fragments in phage λgt7 (5). As a positive selection for Ugp$^+$ in a *glpT* strain, we used the ability of G3P, transported through Ugp, to fulfill the G3P requirement of a *plsB* mutation (15). We isolated a phage that carried at least part of the *ugp* region. The cloned fragment was subcloned into plasmid pACYC184, forming pSH100. The analysis of the DNA insert in pSH100 revealed the presence of two *Eco*RI fragments, of which only one carried structural information of *ugp*, even though the expression of Ugp transport activity was not under *pho* control. Thus, during cloning in λgt7 one *Eco*RI site had separated the structural genes from their correct *phoB*-dependent promoter. The second *Eco*RI fragment that had been carried along during the selection simply provided promoter activity for the *ugp* region (15) but was not located next to *ugp* in the chromosome. The correct promoter fragment was subsequently isolated from a lambda phage carrying a *ugpA-lacZ* fusion that exhibited PhoB-dependent expression of β-galactosidase activity. The religation of this promoter fragment with the *ugp*-carrying fragment of pSH100 yielded a plasmid, pSH12, that expressed Ugp transport activity under the correct and *pho*-dependent regulatory scheme (16; Fig. 3). The restriction nuclease analysis of pSH12 as well as other plasmids containing fragments of *ugp* is shown in Fig. 4.

Organization and gene products of the *ugp* region. Tn*5* insertions were isolated in the *ugp*$^+$ plasmid pSH100 and screened for their loss in Ugp transport activity. These plasmids were analyzed for their ability to complement different chromosomal *ugp* mutations, for the position of insertion, and, using in vivo protein-synthesizing systems, for their ability to synthesize *ugp*-dependent proteins. From the combined results of these experiments we have drawn the following conclusions (Fig. 4). There are at least four genes organized in two operons that are located adjacent to each other and are transcribed in the same direction (counterclockwise on the *E. coli* linkage map). We named these genes *ugpD, ugpB, ugpA*, and *ugpC*. So far, two gene products have been identified; *ugpB* codes for the periplasmic binding protein for G3P (molecular weight, 42,000), and *ugpC* codes for a protein of 40,000 molecular weight of unknown location.

A glycerophosphoryldiester phosphodiesterase is connected to the *ugp* system. As mentioned above, glycerophosphoryldiesters, in particular GPE, are excellent substrates of the Ugp transport system. Table 1 shows that G3P transport is effectively inhibited by GPE. Also, periplasmic protein preparations isolated from *ugp*$^+$ strains, but not from strains lacking the G3P-binding protein, are able to bind GPE that is ^{14}C labeled in the ethanolamine portion (Table 2). Binding is not accompanied by hydrolysis, nor is ethanolamine bound by the preparations containing the binding protein.

However, when [^{14}C]GPE labeled in the ethanolamine portion is used as substrate in the transport assay, the accumulation of radioactivity is temporary and can be observed only when the GPE concentration is above the K_m for this substrate. Analyzing the supernatant fluids of cells incubated with [^{14}C]GPE by TLC, one can detect the formation of ethanolamine. The rate at which ethanolamine is formed in different strains depends on the expression of the Ugp transport system (Fig. 5). The question arises as

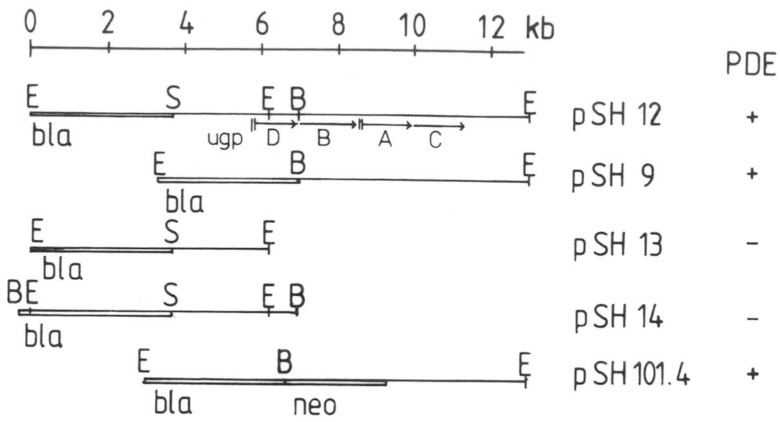

FIG. 4. Restriction nuclease analysis of plasmid pSH12 containing *ugp*$^+$ and its subclones. Symbols above the lines indicate restriction nuclease sites (E, *Eco*RI; B, *Bgl*II; S, *Sal*I); *bla* and *neo* indicate the genes for β-lactamase and neomycin resistance. Double lines indicate the vector portion of the plasmid, and single lines indicate the portion of the insert. Lettering below the line in pSH12 indicates the *ugp* genes D, B, A, and C. Vertical double lines indicate promoters, and arrows indicate the direction of transcription. PDE indicates the presence of the *ugp*-dependent glycerophosphoryl phosphodiesterase, checked by complementation assays with chromosomal mutations.

to where in the cell envelope the hydrolysis of GPE takes place and whether it is part of the transport mechanism itself. To answer this question, we chose a strain that is constitutive for *ugp* and lacks *glpT*. In addition, it lacks *glpD* and, therefore, cannot metabolize G3P. We incubated this strain with different ^{14}C-labeled substrates and analyzed its cellular extracts as well as medium supernatants by TLC. The results obtained are summarized in Fig. 6. With G3P as substrate (Fig. 6A), the medium was quickly depleted and intact G3P appeared within the cytoplasm. Simultaneously, additional products appeared that most likely represent lipid intermediates. With GPE labeled in the ethanolamine portion (Fig. 6B), the substrate disappeared from the medium within 1 min, but no GPE could be seen within the cytoplasm. Instead, free ethanolamine was the only product that could be detected outside. With [^{14}C]GPE labeled in the G3P portion of the molecule (Fig. 6D), again no cytoplasmic GPE accumulated, but the internal formation of G3P could easily be recognized. In this experiment, for technical reasons (low specific radioactivity), the substrate concentration had to be much higher. As a result, GPE in the medium disappeared only slowly, and from the internal G3P only traces of lipid precursors were being formed. Similar data were obtained with [^{14}C]glycerophosphorylglycerol labeled in both glycerol molecules (Fig. 6C). Here again, no accumulation of glycerophosphorylglycerol could be observed. Instead, free glycerol was detected outside and G3P inside the cell.

All attempts to measure the glycerophosphoryldiester phosphodiesterase activity in cellular extracts, membranes, or a combination thereof

TABLE 1. Inhibition of *ugp*-dependent G3P transport by glycerophosphoryldiesters[a]

Inhibitor	Concn (μM)	% Inhibition of G3P transport
GPE	10	96
GPG	100	65
GPC	10	78
G3P	10	57

[a] Transport of [^{14}C]G3P was measured at 0.1 μM concentration in cells of a *glpT ugp*$^+$ strain at a cell density of 8×10^8 cells per ml. At 15 to 60 s, 250 μl was sampled by filter assay and washed with 5 ml of minimal medium A. Uptake of G3P was linear during this time. The uptake was repeated in the presence of unlabeled glycerophosphorylethanolamine (GPE), glycerophosphorylglycerol (GPG), and glycerophosphorylcholine (GPC). Percent inhibition is given as inhibition of the initial rate of transport.

TABLE 2. Binding of [^{14}C]GPE by periplasmic protein preparations[a]

Genotype	GPE bound (nmol/mg of protein)
ugp$^+$	3.04
phoB	0.19
ugp$^+$ on multicopy plasmid	11.2
ugpB	0.03

[a] The binding assay was done by the ammonium sulfate precipitation technique (13) at a free substrate concentration of 5 μM and in a total volume of 30 μl of periplasmic protein solution at 0.35 mg/ml.

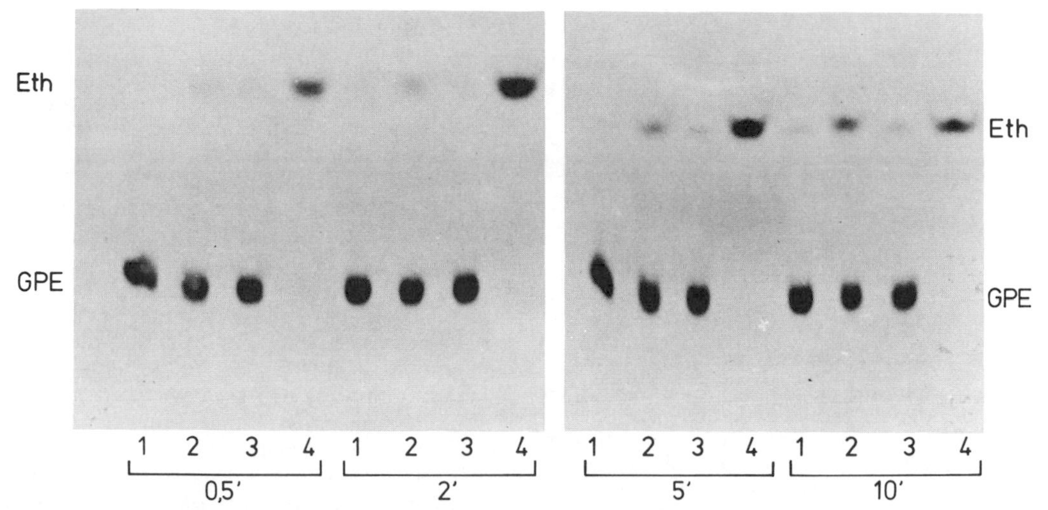

FIG. 5. Hydrolysis of glycerophosphoryldiesters by Ugp. Strains were grown under limiting phosphate conditions and were then incubated with [^{14}C]GPE labeled in the ethanolamine portion for 0.5, 2, 5, and 10 min. The supernatant fluids of the cells were chromatographed on TLC plates. (1) A *ugpA glpT glpQ* strain; (2) same strain as in 1, but carrying plasmid pSH12 (see Fig. 4) coding for *ugp*; (3) *phoB glpT glpQ phoS*; (4) *ugp*^{c} *glpT glpQ phoS*. Eth, Ethanolamine.

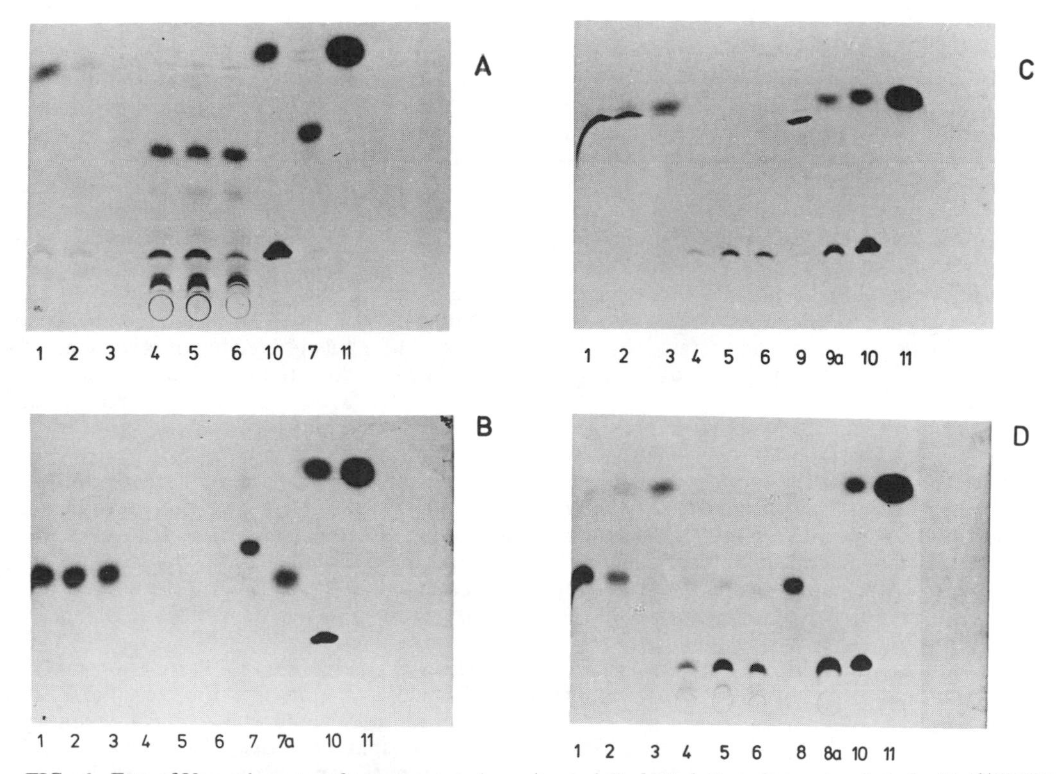

FIG. 6. Fate of Ugp substrates after transport. A *ugp*$^+$ *pst glpT glpQ glpD* strain was incubated with [^{14}C]G3P (A), [^{14}C]glycerophosphorylglycerol (C), [^{14}C]GPE (B) labeled in the ethanolamine portion, and [^{14}C]GPE (D) labeled in the glycerol portion. Supernatant fluids of the cells and soluble fraction were chromatographed on TLC plates. Lanes: (1, 2, 3) Supernatant fluids of cells after 0.5, 2, and 5 min; (4, 5, 6) soluble fraction of cells after 0.5, 2, and 5 min. Controls (lanes) are GPE (7), ethanolamine (7a), GPE (8), G3P (8a), glycerophosphorylglycerol (9), G3P and glycerol (9a), G3P and glycerol (10), and glycerol (11).

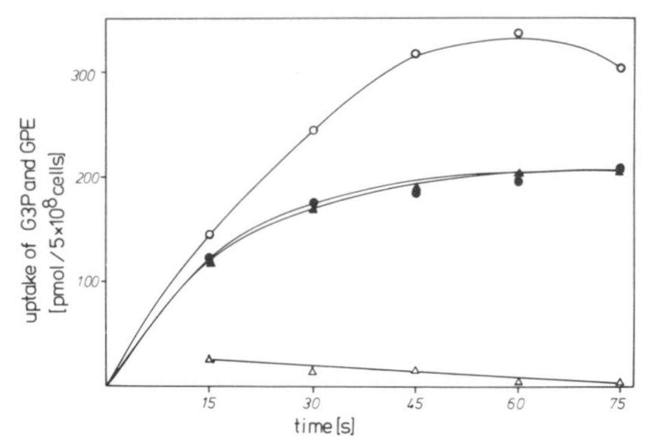

FIG. 7. Transport of G3P and GPE in a strain lacking the Ugp-dependent glycerophosphoryldiester phosphodiesterase. Transport assays were done with 0.1 μM [^{14}C]GPE, labeled in the ethanolamine portion (○, △), and 0.1 μM [^{14}C]G3P (●, ▲). (△, ▲) Result of a *ugp*$^+$ *glpT* strain; (○) result of the same strain carrying a defect in the Ugp-dependent phosphodiesterase.

have failed so far. Apparently, only the transported diester is recognized as substrate.

To obtain mutants in the Ugp-related phosphodiesterase, we mutagenized a *plsB glpT glpQ ugp*$^+$ strain. In this strain the requirement for G3P can be supplied by G3P itself or by GPE, both of which are entering the cell only via the Ugp system. The mutants obtained were screened for the use of G3P but not of GPE as supplement. Of 12,000 colonies we found 3 with this phenotype. By P1 transduction the mutation was found closely linked to or within the *ugp* region. Transport assays using G3P and GPE (labeled in the ethanolamine portion) as substrates with one of these mutants and its parent are shown in Fig. 7. As can be seen, in contrast to the wild type, the mutant strain can now accumulate GPE, while both strains transport G3P equally well.

To determine whether the structural gene for glycerophosphoryldiester phosphodiesterase maps within the *ugp* region, we transformed this strain with several plasmids that contain different portions of the *ugp* region. As can be seen in Fig. 4, only plasmids carrying *ugpC* were able to complement the G3P auxotrophy with GPE. Therefore, the structural gene of glycerophosphoryldiester phosphodiesterase is either identical with *ugpC* or represents an as yet unidentified gene distal to *ugpC* on the same operon.

Discussion

The Ugp system transports G3P and glycerophosphoryldiesters. However, during the transport process the diesters do not appear as intact molecules in the cytoplasm but are hydrolyzed to G3P and the corresponding alcohol. G3P remains in the cytoplasm, while the alcohol diffuses into the medium.

The phosphodiesterase activity is not present in the periplasm, and it cannot be demonstrated in cytoplasmic extracts, but it appears to be connected to the intact transport system. However, it is clear that the diesterase is not an essential part of the transport machinery. We were able to isolate mutants that still transport G3P as well as GPE but were unable to hydrolyze GPE. In fact, these mutants accumulate GPE, in contrast to the parent strain; the mutation responsible for the loss of diesterase activity maps within the *ugp* region.

From these data we propose that the diesterase function is intimately associated with the transport system. It releases the incoming diester from the transport machinery only in hydrolyzed form, without being involved in the transport step itself. It is clear that hydrolysis takes place at the internal face of the membrane since the diffusion of the corresponding alcohol into the medium is dependent on its chemical nature. It is fast for ethanolamine and glycerol but much slower for choline. Neither isolated membranes nor cellular extracts retain this phosphodiesterase activity.

The Ugp system is a member of the *pho* regulon; its synthesis is derepressed when the P_i concentration of the medium is low. The membership in the *pho* regulon becomes apparent when Ugp is the only G3P-transporting system available to the cell. In such a situation, both G3P and the glycerophosphoryldiesters can serve as a sole source of phosphate and as a supplement for the G3P auxotrophy in lipid biosynthesis, but not as a sole source of carbon. The reason for this phenomenon is unclear, but it might be related to the imbalance of phosphate that is created when the cells are forced to use G3P as the only source of carbon.

It is interesting to compare the two major G3P-transporting systems of *E. coli*, the *pho*-dependent Ugp system and the *pho*-independent GlpT system. While the Ugp system is derepressed by P_i starvation, the GlpT system is induced by G3P and geared for carbon utilization. Here, the problem of P_i imbalance is elegantly solved by the ability of the GlpT carrier to exchange internal P_i for external G3P (1, 5; see also Larson, this volume). Both systems are connected to a glycerophosphoryldiester phosphodiesterase. In the case of the Ugp system the enzyme is located at the internal face of the membrane, hydrolyzing the substrate appearing in the cytoplasm. In the case of GlpT the diesters are hydrolyzed in the periplasm, producing G3P, the substrate for the GlpT system.

LITERATURE CITED

1. **Ambudkar, S. V., T. J. Larson, and P. C. Maloney.** 1986. Reconstitution of sugar phosphate transport systems of *E. coli*. J. Biol. Chem. **261:**9083–9086.
2. **Argast, M., and W. Boos.** 1979. Purification and properties of the *sn*-glycerol-3-phosphate binding protein in *Escherichia coli*. J. Biol. Chem. **254:**10931–10935.
3. **Argast, M., and W. Boos.** 1980. Co-regulation in *Escherichia coli* of a novel transport system for *sn*-glycerol-3-phosphate and outer membrane protein Ic (e, E) with alkaline phosphatase and phosphate binding protein. J. Bacteriol. **143:**142–150.
4. **Argast, M., D. Ludtke, T. U. Silhavy, and W. Boos.** 1978. A second transport system for *sn*-glycerol-3-phosphate in *Escherichia coli*. J. Bacteriol. **136:**1070–1083.
5. **Davis, R. W., D. Botstein, and J. D. Roth.** 1980. Advanced bacterial genetics. Cold Spring Harbor Laboratory, Cold Spring Harbor, N.Y.
6. **Elvin, C. M., C. M. Hardy, and H. Rosenberg.** 1985. P_i exchange mediated by the GlpT-dependent *sn*-glycerol-3-phosphate transport system in *Escherichia coli*. J. Bacteriol. **161:**1054–1058.
7. **Hayashi, S.-I., J. P. Koch, and E. C. C. Lin.** 1964. Active transport of L-α-glycerophosphate in *Escherichia coli*. J. Biol. Chem. **239:**3098–3105.
8. **Larson, T. U., M. Ehrmann, and W. Boos.** 1983. Periplasmic glycerophosphodiester phosphodiesterase of *Escherichia coli*, a new enzyme of the *glp* regulon. J. Biol. Chem. **258:**5428–5432.
9. **Larson, T. U., G. Schumacher, and W. Boos.** 1982. Identification of the *glpT*-encoded *sn*-glycerol-3-phosphate permease of *Escherichia coli*, an oligomeric integral membrane protein. J. Bacteriol. **152:**1008–1021.
10. **Lin, E. C. C.** 1976. Glycerol dissimilation and its regulation in bacteria. Annu. Rev. Microbiol. **30:**535–575
11. **Miller, J. H.** 1972. Experiments in molecular genetics. Cold Spring Harbor Laboratory, Cold Spring Harbor, N.Y.
12. **Reid, T., and J. Wilson.** 1971. E. coli alkaline phosphatase, p. 373–415. *In* P. Boyer (ed.), The enzymes, vol. 4. Academic Press, Inc., New York.
13. **Richarme, G., and A. Kepes.** 1983. Study of binding protein-ligand interaction by ammonium sulfate-assisted adsorbtion on cellulose esters filters. Biochim. Biophys. Acta **742:**16–24.
14. **Schweizer, H., M. Argast, and W. Boos.** 1982. Characteristics of a binding protein-dependent transport system for *sn*-glycerol-3-phosphate in *Escherichia coli* that is part of the *pho* regulon. J. Bacteriol. **150:**1154–1163.
15. **Schweizer, H., and W. Boos.** 1983. Cloning of the *ugp* region containing the structural genes for the *pho* regulon dependent *sn*-glycerol-3-phosphate transport system of *Escherichia coli*. Mol. Gen. Genet. **192:**177–186.
16. **Schweizer, H., and W. Boos.** 1984. Characterization of the *ugp* region containing the genes for the *phoB* dependent *sn*-glycerol-3-phosphate transport system of *Escherichia coli*. Mol. Gen. Genet. **197:**161–168.
17. **Schweizer, H., and W. Boos.** 1985. Regulation of *ugp*, the *sn*-glycerol-3-phosphate transport system of *Escherichia coli* K-12. J. Bacteriol. **163:**392–394.
18. **Schweizer, H., W. Boos, and T. Larson.** 1985. Repressor for the *sn*-glycerol-3-phosphate regulon of *Escherichia coli* K-12: cloning of the *glpR* gene and identification of its product. J. Bacteriol. **161:**563–566.
19. **Schweizer, H., T. Grussenmeyer, and W. Boos.** 1982. Mapping of two *ugp* genes coding for the *pho* regulon dependent *sn*-glycerol-3-phosphate transport system of *Escherichia coli*. J. Bacteriol. **150:**1164–1171.
20. **Tomassen, J., and B. Lugtenberg.** 1982. *pho* regulon of *Escherichia coli* K-12: a minireview. Ann. Microbiol. (Paris) **133A:**243–249.

Exogenous Induction of the *uhp* Sugar Phosphate Transport System

ROBERT J. KADNER, DONNA M. SHATTUCK-EIDENS,† AND LUCY A. WESTON

Department of Microbiology and the Molecular Biology Institute, School of Medicine, The University of Virginia, Charlottesville, Virginia 22908

Cells of *Escherichia coli* possess an inducible active transport system responsible for the accumulation in unaltered form of a variety of sugar phosphates (5, 13). Substrates for this transport system range from three to seven carbon atoms in size and include amino sugars and sugar alcohols. Recent studies have indicated that this transport system operates by a phosphate-antiport process, in which each sugar phosphate enters in exchange for one or more molecules of P_i (S. V. Ambudkar, T. J. Larson, and P. J. Maloney, Abstr. Annu. Meet. Am. Soc. Microbiol. 1986, K68, p. 204). This is a very interesting finding, but perhaps the most intriguing feature of this transport system involves the regulation of its synthesis. Although the substrate specificity of the transporter is relatively broad, its induction is quite specific, requiring the presence of external glucose 6-phosphate (G6P) or 2-deoxy-glucose 6-phosphate (3, 13, 17). Internally generated G6P does not provide an inducing signal. The term exogenous induction was coined to describe regulatory systems that respond exclusively to an external effector.

Exogenous induction is expected to be a preferred mechanism for controlling genes responsible for the transport or catabolism of substrates that are present at appreciable levels during normal cellular metabolism. Systems known to be regulated in this manner include, besides the *uhp* sugar phosphate transport system in *E. coli*, the *pgt* phosphoglycerate and the tricarboxylic acid transport systems of *Salmonella typhimurium*, systems for the transport and degradation of proline and for citrate-mediated iron uptake, and components of the phosphoenolpyruvate:sugar phosphotransferase system in *E. coli* (4, 7, 10, 12, 14). Most of the models to explain the lack of response to the inducer that is normally present inside the cell invoke the existence of a vectorial regulatory system, one component of which must be located in the cytoplasmic membrane and must possess an externally oriented inducer-binding site. The fundamental question is how the presence of external inducer is signaled to the transcription system. Transmembrane signaling is frequently encountered in eucaryotic cells, and the properties of a number of membrane-related regulatory processes in bacteria have been recently reviewed (6).

The exogenous induction of the *uhp*-encoded sugar phosphate transport system has been recognized for 20 years (13). Previous studies from this laboratory employed *uhpT-lac* operon fusions to demonstrate that the induction of *uhpT*, the structural gene for the sugar phosphate transporter, operated at the transcriptional level and that this regulation occurred normally even in strains that lacked any detectable uptake of the inducer (15). Genetic studies showed that all mutations that specifically affected the sugar phosphate transport system were located in the *uhp* region, which is at min 82 of the genetic map (1, 8). Fine-structure mapping and preliminary analysis of the cloned *uhp* region led to a model that invoked the presence of three regulatory genes controlling the expression of *uhpT* (9, 16). One of these, *uhpA*, was proposed to be an activator of *uhpT* transcription because mutations in *uhpA* prevented expression of the transport system and because strains carrying *uhpA* on multicopy plasmids exhibited greatly elevated uninduced levels of *uhpT* expression. In this communication we report recent results that allowed the identification of the regulatory genes and their products, and we discuss the revisions that must be made to our previous model of *uhp* regulation.

uhp Gene Organization

Transposon insertions. The intact *uhp* region has been cloned as a 6.5-kilobase-pair DNA fragment from plasmid pLC17-47 of the Clarke-Carbon collection (16). A restriction map of this fragment is portrayed in Fig. 1. To correlate this physical map with the genetic maps we previously presented (9), we isolated a collection of transposon Tn*10* and Tn*1000* insertions in the cloned fragment. Their locations relative to the restriction sites were mapped (Fig. 1). Plasmids carrying the insertions were introduced by transformation into host strains that carried deletions that entered into *uhpA* from the left end (Δ*uhpA*) or into *uhpT* from the right (Δ*uhpT*) or which

†Present address: NPI, Inc., Salt Lake City, UT 84108.

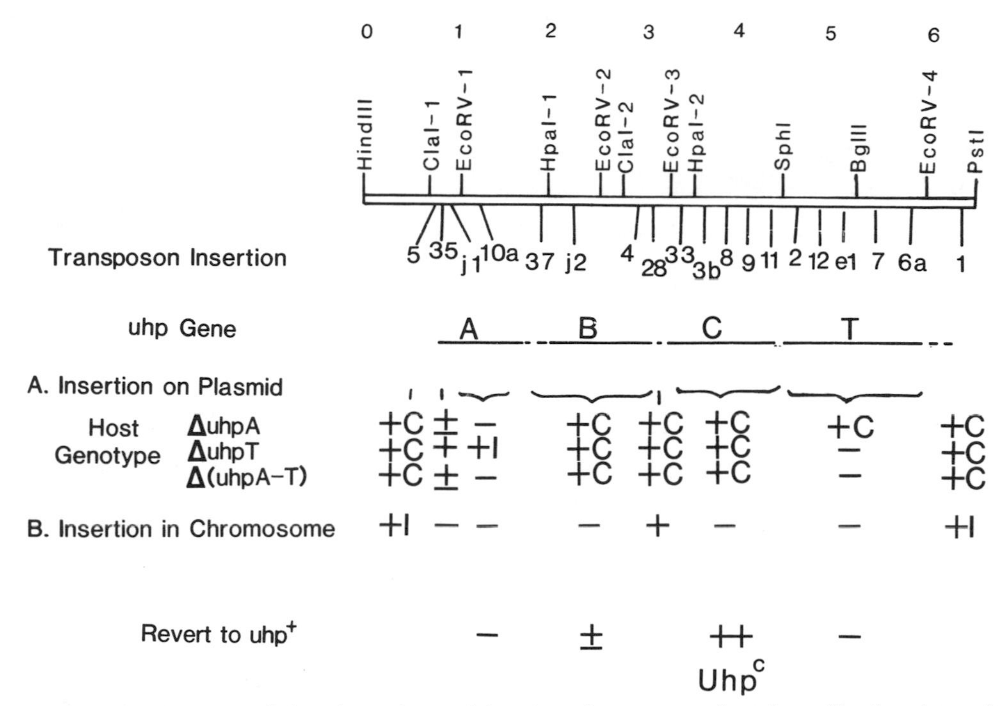

FIG. 1. Restriction map of the *uhp* region and location of transposon insertions. The locations of the restriction endonuclease cleavage sites were determined by combinations of enzymes and were confirmed by the nucleotide sequence. The transposon insertions are identified by the isolation numbers and include Tn*10* derivatives, designated by letters, and Tn*1000* insertions, indicated by numbers alone. The Uhp phenotypes are shown in the column under each insertion, with insertions giving identical phenotypes clustered by brackets. Section A presents the Uhp phenotypes when the insertions are present on multicopy plasmids and the chromosomal *uhp* region carries the indicated deletions. Section B shows the phenotype of strains in which the transposon insertion has been crossed onto the chromosome and is now present in the haploid state. Growth on fructose 6-phosphate as the sole carbon source is indicated by +, −, or ±; the regulatory phenotypes were constitutive or inducible, indicated by C and I, respectively. The bottom line indicates the ability of the strains carrying the indicated *uhp*::Tn insertions to give rise to Uhp$^+$ revertants: −, less than 10^{-8} revertants; ++, around 10^{-6} revertants; ±, around 10^{-6} revertants, but only small colonies on selection plates.

removed the entire *uhp* region (Δ*uhpA–T*). These plasmid-bearing strains were tested for their expression and regulation of the Uhp transport system.

Insertions into *uhpA* and *uhpT* were defined by their inability to complement the corresponding chromosomal deletions. Plasmids carrying the intact *uhpA* gene conferred constitutive expression of the *uhpT* gene present on either the plasmid or the chromosome, as previously observed (16). Insertions in *uhpA* rendered *uhpT* expression dependent on the chromosomal copy of *uhpA*$^+$. This expression was inducible, although lower than normal as a result of competition for the limiting amount of activator. Thus, the derangement in regulation seen in the multicopy situation is due solely to the elevated dosage of *uhpA*. Insertion 35, on the left end of *uhpA*, only partially complemented the *uhpA* mutant for growth (the ± response in Fig. 1), and this expression was inducible. Maxicell expression showed that this insertion resulted in decreased production of a normal-sized *uhpA* product.

Insertions into the plasmid in the region between *uhpA* and *uhpT* did not interfere either with the expression of *uhpT* or with the constitutive expression in response to multiple copies of the *uhpA* gene.

Each transposon insertion was then transferred onto the bacterial chromosome by a homologous recombination process. In this haploid situation, all except three of the insertions resulted in the Uhp$^-$ phenotype, showing that not only *uhpA* and *uhpT*, but also the entire region between them, are required for proper Uhp expression (Fig. 1B). The requirement for these other genes is circumvented when *uhpA* is supplied on a multicopy plasmid. The insertions on the two ends help to define the extent of the *uhp* region. We presume that the reason insertion 28 did not confer the Uhp$^-$ phenotype is that it lies

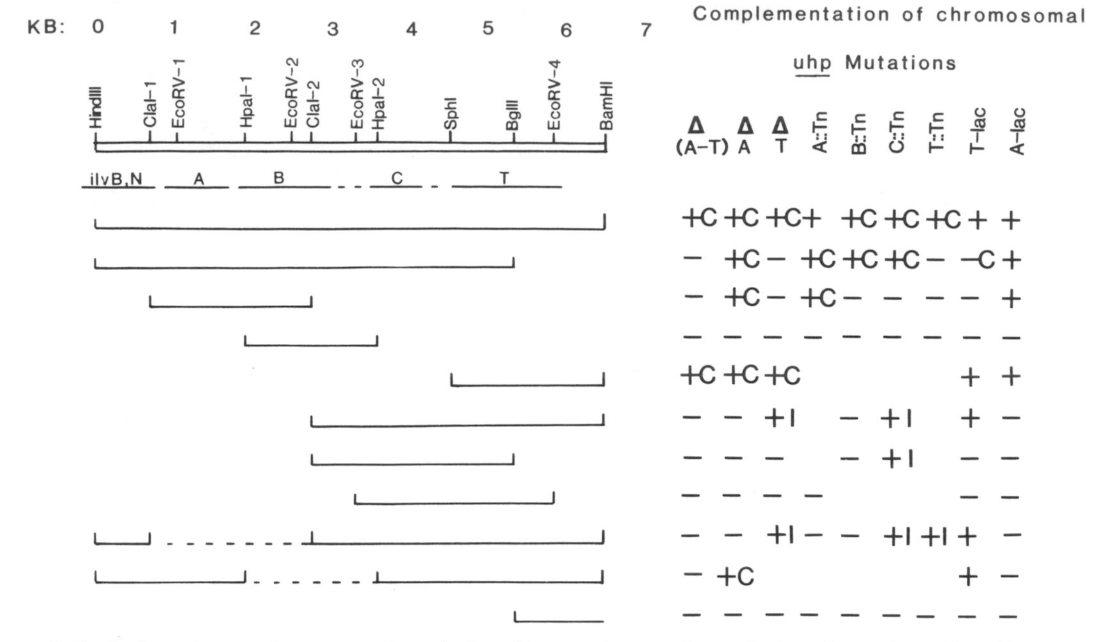

FIG. 2. Complementation properties of plasmids carrying portions of the *uhp* region. Plasmids were constructed by cloning restriction fragments into the pUC18 or 19 vectors. Complementation of the Uhp phenotype in host strains carrying the indicated *uhp* mutations is designated as in Fig. 1. In the case of the strain carrying the chromosomal *uhpT-lac* operon fusion, the plasmid carrying the *Hin*dIII-*Bgl*II fragment (second plasmid) did not confer the Uhp$^+$ phenotype, as a result of the absence of an intact *uhpT* gene, but did confer constitutive expression of β-galactosidase.

in an intergenic region, which suggests that there are at least two *uhp* genes lying between *uhpA* and *uhpT*.

Another indication for the existence of these two genes came from examination of the ability of the chromosomal insertion mutants to give rise to Uhp$^+$ revertants (Fig. 1, bottom line). Insertions in *uhpA* and *uhpT* did not give rise to cells able to utilize fructose 6-phosphate as carbon source at any appreciable frequency; the few revertants that arose appeared to have undergone precise excision of the transposon. Insertions in the gene termed *uhpB*, lying between *uhpA* and insertion 28, also did not give rise to Uhp$^+$ revertants, but only to very small colonies which lacked appreciable Uhp transport activity. In contrast, insertions in *uhpC*, between insertion 28 and *uhpT*, did revert at a relatively high frequency. These revertants still carried the transposon, and many of them displayed high constitutive expression of the transport system. Thus, these results pointed to the existence of three regulatory genes, each of which is necessary for *uhpT* expression when present at normal copy number.

Subclones of the *uhp* region. These conclusions were corroborated and extended by the construction and analysis of subclones carrying defined portions of the *uhp* region in various plasmid vectors. The extent of the material carried on these subclones and their complementation properties are summarized in Fig. 2. Plasmids carrying the region between the *Cla*I-1 and *Hpa*I-1 sites complemented all *uhpA* deletion and insertion mutations. Presence of these plasmids resulted in constitutive expression of *uhpT* in all *uhpT*$^+$ hosts, confirming the observation that overexpression of *uhpA* overcomes the Uhp$^-$ phenotype that results from mutations in *uhpB* or *uhpC*.

Plasmids carrying the region from *Cla*I-2 to or beyond the *Bgl*II site were able to complement *uhpC* but not *uhpB* insertions. The expression in these cases, where *uhpA* is not in multicopy, is inducible by G6P. No plasmids have been obtained yet that complement only *uhpB* mutations.

Some plasmids with the sequences rightward from the *Sph*I site to the end of the insert carry and express the *uhpT* gene, but in a manner independent of *uhpA* or of the presence of inducer. This transcription is presumably initiated from a vector promoter. This result does indicate that *uhpT* is the only gene in the region whose presence is required for transport activity and that all of the other genes are involved only in regulation. Preliminary transcript mapping studies have shown that normal transcription of *uhpT* is initiated very near the *Sph*I site, and thus these plasmid constructions must lack the normal *uhpT* regulatory region. This result does not prove that UhpT is the only polypeptide required for the transport system, because it is possible that another gene encoded elsewhere on the chromosome might be involved. However, fosfomycin-resistant mutants that are specifically defective in G6P uptake have never been found to carry mutations outside the *uhp* region (8).

uhp Polypeptides

Maxicell analysis. The polypeptide products of the *uhp* region were investigated by use of maxicells carrying some of the plasmids described above (Fig. 3). The *uhpT* product was identified as a diffuse band migrating during sodium dodecyl sulfate-polyacrylamide gel electrophoresis with an apparent molecular weight of 38,000. Production of this polypeptide was blocked by insertions in *uhpT* (lane 15) and was dependent on *uhpA*$^+$ function provided on either the plasmid or the chromosome. When *uhpA*$^+$ was present only on the chromosome, the production of UhpT in maxicells was seen to be induced by the addition of G6P during the labeling period (lanes 3–8). When *uhpA*$^+$ was carried on the plasmid, production of the UhpT polypeptide was constitutive (lanes 1, 2, 9–14), just as is the transport activity in normal cells. The appearance of UhpT as a diffuse band that migrates faster than expected based on its actual molecular weight is reminiscent of the behavior of LacY, the lactose permease, and other nonpolar cytoplasmic membrane transport proteins (2, 11).

The UhpA polypeptide migrated as a discrete band corresponding to a molecular weight of 26,000. Its synthesis was evoked by all of the plasmids with transposon insertions except those in *uhpA* (Fig. 3, lanes 3–8) and only by those subclones able to complement *uhpA* mutations (not shown). Its synthesis in maxicells was not influenced by the presence of inducer.

The *uhpB* product has a molecular weight of 45,000. Its synthesis was blocked by insertions in *uhpA* or *uhpB* (Fig. 3, lanes 3–11). This result means either that *uhpB* is a distal gene in an operon with *uhpA* or that UhpA is also a positive activator of UhpB transcription.

The UhpC protein is not readily detected in maxicells as a result of its migration with nonspecific polypeptides. Recent experiments in which *uhpC*-containing restriction fragments were transcribed from a phage T7 promoter suggest that UhpC has a molecular weight around 21,000 (not shown). Confirmation of this result and detection of the transcription of this gene from its own promoter and of the requirement for the other *uhp* products remain to be determined.

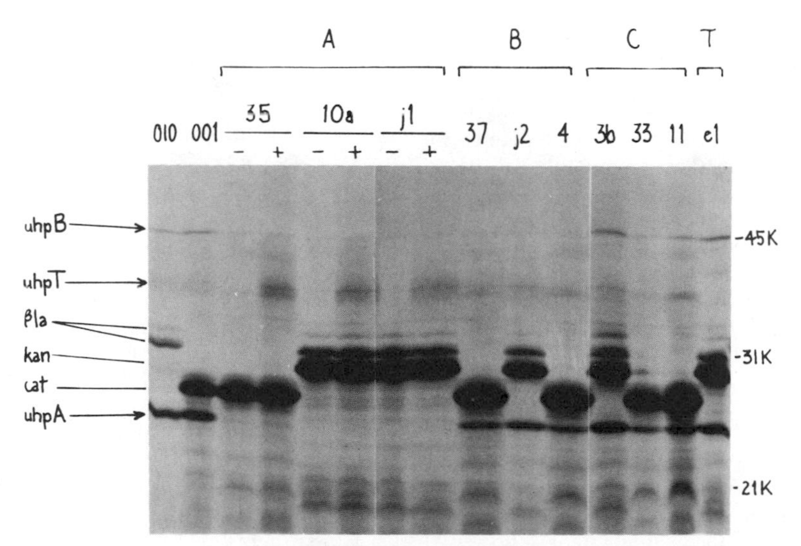

FIG. 3. Polypeptides produced in maxicells in response to plasmids carrying transposon insertions and subcloned portions of the *uhp* region. Identification of the polypeptides is given on the left, and the mobility of molecular weight standards is on the right. The host strain was RK4991 (Δ*uhpT recA*). The first two lanes were from maxicells carrying plasmids containing the complete *uhp* region (6.5-kilobase *Hin*dIII-*Bam*HI) cloned in different vectors: pBR322 in lane 1 (pRJK010) and pACYC184 in lane 2 (pLAW001). Plasmids represented in the subsequent lanes carried transposon insertions identified by their isolation number and defined in Fig. 1. The lanes indicated by + and − represent the presence or absence of G6P during the period of labeling with [^{35}S]methionine; in the other lanes, the maxicells were labeled in the absence of inducer.

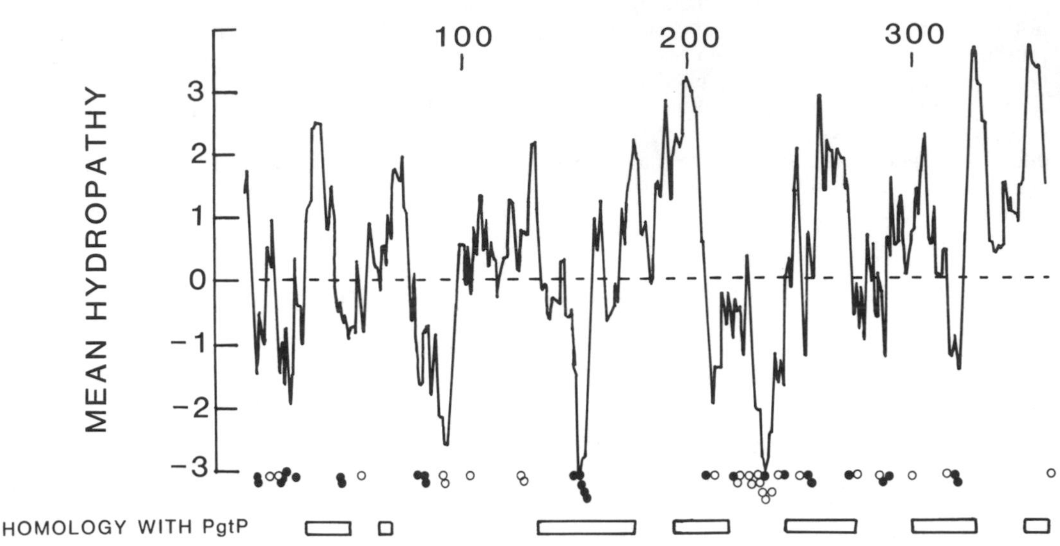

FIG. 4. Properties of the UhpT polypeptide. Results are presented for the first 360 amino acids of the protein sequence deduced from the nucleotide sequence. The hydropathy profile calculated according to Kyte and Doolittle is portrayed, with positive values representing hydrophobic segments; a span of seven residues, with negatively charged residues shown by the open circles. The boxes at the bottom indicate regions of UhpT that have substantial homology with PgtP, determined by J.-S. Hong.

Fractionation of labeled maxicells showed that both UhpB and UhpT reside in the membrane fraction and are released by treatment with Triton X-100. Most of the UhpA polypeptides are soluble, although a fraction is membrane associated. The location of UhpC is not known.

Structure of UhpT. The determination of the nucleotide sequence of the *uhp* region is nearing completion. Information is available on the sequence of *uhpT* extending for 360 amino acids. Figure 4 presents some features of the predicted polypeptide. Kyte-Doolittle hydropathy plots show that the protein is relatively nonpolar (14% charged amino acids) and possesses up to nine segments of sufficient length and hydrophobic character to constitute α-helical transmembrane regions. The charged residues are strongly clustered, and these probably are in portions of the protein that are exposed on one side or the other of the membrane. Extensive topological studies will be required to elucidate the transmembrane orientation of this protein.

Comparison of the sequence of UhpT with that of the phosphoglycerate transporter PgtP of *S. typhimurium* determined by J.-S. Hong (unpublished data) revealed the presence of numerous areas of moderate conservation of amino acid sequence. These conserved areas included both polar and nonpolar regions. This finding strongly suggests that these two proteins are related and that both carry out a similar mechanism of energy coupling, namely, phosphate/organophosphate exchange, although this has not been experimentally determined. The two proteins are very similar in size and mobility during sodium dodecyl sulfate-polyacrylamide gel electrophoresis.

Models for Regulation

The model of *uhp* regulation that we previously presented must be modified somewhat (16). Results presented here provide strong evidence for the existence of three regulatory proteins, which have been identified and whose coding regions have been defined. Pending completion of the nucleotide sequence, comparison of the apparent sizes of the regulatory proteins (26, 45, and 21 kilodaltons) with the maximum extent of their coding region between the end of the adjacent *ilvBN* operon and the start of *uhpT* (3.5 kilobases) shows that there is sufficient coding capacity for the three proteins, but not for any more.

Experiments with the *uhpT-lac* fusions indicated that induction affects the transcription of *uhpT*. It is possible that the induction process actually acts initially to increase the level of the UhpA activator and that this results directly in increased transcription of *uhpT*. The constitutive expression of *uhpT* seen in strains with elevated *uhpA* gene dosage could be explained in this way. However, this possibility has been discounted by study of a *uhpA-lac* fusion. In this strain, the level of β-galactosidase, which serves as an indicator of *uhpA* transcription, was not affected by the presence of inducer, even when the $uhpA^+$ gene was provided in *trans*. This result is also consistent with the observations that in maxicells the levels of UhpA are not affected by the presence of inducer.

If the level of UhpA is not changed during regulation and if it is the factor that controls the extent of *uhpT* transcription, several simple models can be proposed to account for the control by external G6P of the ability of the activator protein to bind to the *uhpT* promoter. One model suggests that UhpA is normally bound in a complex with the membrane-localized regulatory proteins and is released from that complex only upon binding of inducer to external domains of UhpB or UhpC. Overproduction of UhpA could overwhelm the capacity of the UhpBC proteins to bind and inactivate it, resulting in constitutive expression. This model predicts that mutants lacking both UhpB and UhpC might exhibit constitutive expression, which is not the case.

Another possibility is that the UhpBC proteins, upon binding of G6P to an external domain, catalyze the synthesis of some low-molecular-weight effector which diffuses through the cytoplasm, binds to UhpA, and stabilizes a conformation of UhpA that is able to bind to the *uhpT* promoter. It is clear that neither UhpB nor UhpC is necessary to allow elevated levels of UhpA to stimulate transcription, but perhaps the elevated concentration of UhpA under those conditions results in an increased concentration of the activated species through equilibrium. Neither of these models explains all the experimental results, but their key features are accessible to testing.

The research summarized here was supported by National Science Foundation grant PCM-82-15915.

LITERATURE CITED

1. **Bachman, B. J.** 1983. Linkage map of *Escherichia coli* K-12, edition 7. Microbiol. Rev. **47:**180–230.
2. **Buchel, D. E., B. Gronnenborn, and B. Muller-Hill.** 1980. Sequence of the lactose permease gene. Nature (London) **283:**541–545.
3. **Dietz, G. W., and L. A. Heppel.** 1971. Studies in the uptake of hexose phosphates. II The induction of the glucose-6-phosphate transport system by exogenous but not by endogenously formed glucose-6-phosphate. J. Biol. Chem. **246:**2885–2890.
4. **Dills, S. S., A. Apperson, M. R. Schmidt, and M. H. Saier.** 1980. Carbohydrate transport in bacteria. Microbiol. Rev. **44:**385–418.
5. **Eidels, L., P. D. Rick, N. P. Stimler, and M. J. Osborn.** 1974. Transport of D-arabinose-5-phosphate and sedoheptulose-7-phosphate by the hexose phosphate transport system of *Salmonella typhimurium*. J. Bacteriol. **119:**138–143.
6. **Epstein, W.** 1983. Membrane-mediated regulation of gene expression in bacteria, p. 281–292. *In* J. Beckwith, J. Davies, and J. A. Gallant (ed.), Gene function in prokaryotes. Cold Spring Harbor Laboratory, Cold Spring Harbor, N.Y.
7. **Hussein, S., K. Hantke, and V. Braun.** 1981. Citrate-dependent iron transport system in *Escherichia coli* K-12. Eur. J. Biochem. **117:**431–437.
8. **Kadner, R. J.** 1973. Genetic control of the transport of hexose phosphates in *Escherichia coli*: mapping of the *uhp* locus. J. Bacteriol. **116:**764–770.
9. **Kadner, R. J., and D. M. Shattuck-Eidens.** 1983. Genetic control of the hexose phosphate transport system of *Escherichia coli*: mapping of deletion and insertion mutations in the *uhp* region. J. Bacteriol. **155:**1052–1061.
10. **Kay, W. W., and M. Cameron.** 1978. Citrate transport in *Salmonella typhimurium*. Arch. Biochem. Biophys. **190:**270–280.
11. **Larson, T. J., G. Schumacher, and W. Boos.** 1982. Identification of the *uhpT*-encoded *sn*-glycerol-3-phosphate permease of *Escherichia coli*, an oligomeroic integral membrane protein. J. Bacteriol. **152:**1008–1021.
12. **Maloy, S., and J. Roth.** 1983. Regulation of proline utilization in *Salmonella typhimurium*: characterization of *put*::Mu *d*(Ap *lac*) operon fusions. J. Bacteriol. **154:**561–568.
13. **Pogell, B. M., B. R. Maity, S. Frumkin, and S. Shapiro.** 1966. Induction of an active transport system for glucose-6-phosphate in *Escherichia coli*. Arch. Biochem. Biophys. **116:**406–415.
14. **Saier, M. H., D. L. Wentzel, B. U. Feucht, and J. J. Judice.** 1975. A transport system for phosphoenolpyruvate, 2-phosphoglycerate, and 3-phosphoglycerate in *Salmonella typhimurium*. J. Biol. Chem. **250:**5089–5096.
15. **Shattuck-Eidens, D. M., and R. J. Kadner.** 1981. Exogenous induction of the *Escherichia coli* hexose phosphate transport system defined by *uhp-lac* operon fusions. J. Bacteriol. **148:**203–209.
16. **Shattuck-Eidens, D. M., and R. J. Kadner.** 1983. Molecular cloning of the *uhp* region and evidence for a positive activator for expression of the hexose phosphate transport system of *Escherichia coli*. J. Bacteriol. **155:**1062–1070.
17. **Winkler, H. H.** 1970. Compartmentation in the induction of the hexose-6-phosphate transport system of *Escherichia coli*. J. Bacteriol. **101:**470–475.

VI. MECHANISMS AND ENERGETICS OF PHOSPHATE TRANSPORT IN OTHER ORGANISMS

Introduction

WINFRIED BOOS

Fakultät für Biologie, Universität Konstanz, D7750 Konstanz, Federal Republic of Germany

The proper maintenance of phosphate is of fundamental importance to any living cell. On one hand, the synthesis of phosphate-containing macromolecules, such as DNA and RNA, has to continue at certain minimal rates; on the other hand, an adequate pool of phosphate-containing intermediates has to be maintained to allow the production of energy for the synthesis of macromolecules and cell growth. The gate to the outside world is established by transport systems that bring in either P_i or phosphate covalently linked to organic molecules such as glucose or glycerol. Several factors are relevant in the task of balancing the internal phosphate account: (i) the regulation of transport on the level of expression, (ii) avoiding an overload of P_i when cells are forced to use organic phosphates as carbon sources (e.g., the use of glycerolphosphate as carbon source brings in 20 times more phosphate than is needed); (iii) regulating the activity of phosphate transport systems (e.g., by pH), and (iv) the energy coupling needed for the concentration of phosphate against its concentration gradient.

In this section Hancock et al. describe the properties of a phosphate-specific channel in the outer membrane of *Pseudomonas aeruginosa* which has many similarities with the system in *E. coli*. The synthesis of this phosphate channel is under the same regulation as the inner membrane phosphate transport system as well as additional proteins, such as the periplasmic alkaline phosphatase, that are needed for scavenging phosphate. The signal for turning on the corresponding genes is a drop in the phosphate concentration of the medium.

Maloney et al. elucidate in detail the mechanism and energetics of sugar phosphate transport systems in bacteria. Two aspects are relevant: first, sugar phosphate is taken up in exchange for phosphate, thus avoiding an overaccumulation of phosphate; second, the circulation of phosphate can drive (energetically) the uptake of sugar phosphate. Two modes of the carrier can be envisioned. In the electroneutral exchange one monovalent phosphate is exchanged against one monovalent sugar phosphate. In the accumulation mode (in the presence of a pH gradient) two monovalent anions exchange with single divalent anion, equivalent to a proton-anion symport mechanism.

The mechanisms of energy coupling to solute transport are discussed on a broader basis by Konings and Poolman. Most systems termed secondary transport systems are driven by electrochemical ion (mostly proton) gradients. In membrane vesicles from organisms which possess cytochrome-linked electron-transfer systems a proton motive force can be generated by electron transfer. Another way to generate a proton motive force is by incorporating a proton pump (bacterial rhodopsin or cytochrome oxidase) by fusion of the membrane vesicles with liposomes containing the proton pumps. The results indicate that the general view of energy coupling of secondary transport systems via proton substrate symport might be too simplified. The demonstration of a direct interaction of electron transfer systems and secondary transport indicates the participation of redox reactions directly with the carrier protein. The participation of dithiol/disulfide groups is indicated by Konings and Poolman.

The last article by Borst-Pauwels and Peters deals with phosphate uptake in yeasts. Several systems are present in these organisms, some of them are derepressed when the phosphate concentration is falling below a certain threshold, and others are constitutive.

Kinetic analysis of these transport systems reveals that genetic regulation is not the only way to control the flow of phosphate into the cell. Transport is dependent on phosphate concentration, on internal pH, and on surface potential. Particularly the latter may be rather complicated since changes in the surface potential have opposite effects on the interfacial concentration of the negatively charged phosphate anion and the positively charged proton.

Phosphate-Binding Site of *Pseudomonas aeruginosa* Outer Membrane Protein P

ROBERT E. W. HANCOCK,[1] E. A. WOROBEC,[1] K. POOLE,[1] AND ROLAND BENZ[2]

Department of Microbiology, University of British Columbia, Vancouver, British Columbia, Canada, V6T 1W5,[1] and Lehrstuhl für Biotechnogie, Universität Würzburg, D8700 Würzburg, Federal Republic of Germany[2]

When *Pseudomonas aeruginosa* cells are grown in phosphate-deficient medium, a high-affinity phosphate transport system is induced. Although some features of this transport system are quite similar to the well-studied phosphate-specific transport system of *Escherichia coli* (20), the outer membrane transport component, protein P, shows substantial differences from the equivalent component, the PhoE porin (2), of the phosphate-specific transport system. For example, the PhoE protein is a typical porin demonstrating large, weakly selective channels (although it has been suggested that PhoE is polyphosphate selective [12]). In contrast, protein P forms constricted, anion-specific channels (3, 9) which contain a strong phosphate-binding site (8). This paper reviews the attempts of our laboratories to relate the structure of protein P to its function in phosphate transport.

Coregulation of Protein P with a Phosphate Starvation-Inducible Regulon

The syntheses of a number of proteins are derepressed in *P. aeruginosa* upon growth in phosphate-limiting (0.2 mM P_i) medium. These include outer membrane protein P, a periplasmic phosphate-binding protein, and both periplasmic and excreted alkaline phosphatase and phospholipase C (16, 17). Coincidentally, a high-affinity phosphate transport system is derepressed (16, 17). A pleiotropic negative mutant strain was isolated which failed to respond to phosphate starvation by increasing levels of the above proteins. In addition, a mutant constitutive for alkaline phosphatase was additionally constitutive for phospholipase C, protein P, and the phosphate-binding protein (17). These data strongly support the existence of a phosphate regulon in *P. aeruginosa*, with protein P actively involved in phosphate transport across the outer membrane.

Involvement of Protein P and a Phosphate-Binding Protein in Phosphate Transport In Vivo

Mutants lacking protein P and a phosphate-binding protein were isolated. In the case of protein P, a Tn*501* insertion mutant was produced and confirmed as protein P deficient by its inability to interact with a protein P-specific antiserum (19). This mutant exhibited a K_m for high-affinity P_i transport 10 times greater than that of the parent strain (19).

Using the procedure of Brinkman and Beckwith (6), mutants constitutive for alkaline phosphatase and deficient in phosphate-binding protein were obtained (17). These mutants lacked a high-affinity phosphate transport system. Both of the above classes of mutants were deficient in their abilities to grow in phosphate-limiting medium. These data demonstrate that both protein P and the phosphate-binding protein are components of a high-affinity phosphate transport system.

Immunochemistry of Protein P

Protein P was demonstrated to form sodium dodecyl sulfate (SDS)-stable trimers (1) which dissociate into monomers of 47,000 daltons upon SDS solubilization at high temperatures (9). Immunoblot assays revealed that polyclonal antibodies raised against the native trimer form of protein P reacted exclusively with the trimer and not the monomer form. Conversely, P monomer-specific antiserum interacted only with the monomer (18; E. Worobec, unpublished data). This suggested that the major epitopes on the trimer are conformational.

Using a protein P trimer-specific polyclonal antiserum, it was possible to show by immunoblot assays a cross-reactivity of the oligomeric forms of phosphate starvation-inducible outer membrane proteins from bacteria representing 11 different genera of the families *Enterobacteriaceae* and *Pseudomonadaceae* (18). This antiserum did not react with the monomer form of these proteins, and P monomer-specific antiserum did not react with either the monomer or oligomer forms of any of these proteins (18). This demonstrated conserved conformational epitopes which may reflect the related porin functions of these proteins.

Protein P, in the native trimer conformation (purified by electroelution as described by Parr et al. [14]), was found to be resistant to all proteases tested except for trypsin, proteinase

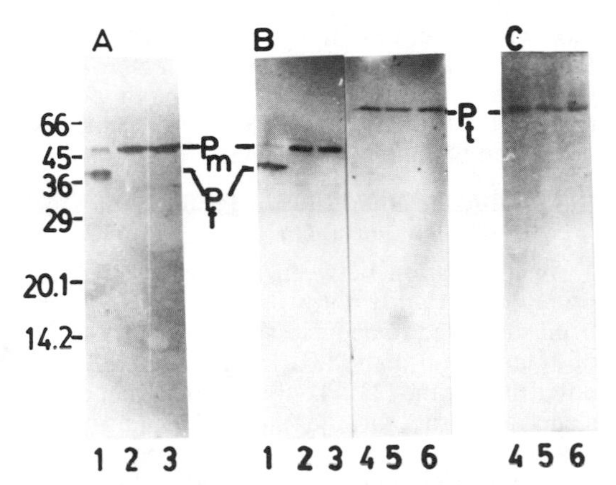

FIG. 1. SDS-PAGE and corresponding immunoblots of proteolytically digested protein P trimer. P_t, P trimer; P_m, P monomer; P_f, proteolytic fragment of P. Purified protein P trimer was mixed at a 50:1 weight ratio with a variety of proteases. In all cases digestion was for 24 h at 37°C. The digestion products were run on SDS-PAGE, either unheated (lanes 4–6) or heated at 100°C for 10 min prior to electrophoresis (lanes 1–3). SDS-PAGE and immunoblots were performed as previously described (13). (A) Immunoblot using polyclonal P monomer-specific antiserum at a 1:200 dilution. (B) Coomassie blue-stained SDS-PAGE. (C) Immunoblot using polyclonal P trimer-specific antiserum at a 1:200 dilution. Lanes 1 and 4, Protein P digested with pronase (similar data were obtained with trypsin and proteinase K); lanes 2 and 5, protein P digested with papain (similar data were obtained with chymotrypsin, carboxypeptidase A, and *Staphylococcus aureus* V8 protease); lanes 3 and 6, untreated protein P. Molecular weights of protein standards ($\times 10^3$) are indicated on the left.

K, and pronase. These enzymes did not alter the mobility of the trimer on SDS-polyacrylamide gel electrophoresis (PAGE), and proteolysis of protein P trimers with proteinase K had no effect on its pore-forming function as assayed by the black lipid model membrane system (Worobec, unpublished data). Heating the trimers in SDS resulted in the appearance of monomers with molecular weights of 37,000 (cf. 47,000 for the native protein) (Fig. 1B). P monomer-specific antiserum recognized these digestion products (Fig. 1A). Thus, there appears to be a protease-susceptible site on protein P which can be cleaved without grossly perturbing the functional trimer. This attests to the importance of tertiary and quarternary interactions in the formation of protein P trimers.

Black Lipid Bilayer Studies on Protein P

When purified and reconstituted into black lipid bilayer membranes (9), protein P trimers form well-defined channels with a considerably smaller single-channel conductance (0.25 nS in 1 M KCl) than other porin proteins (usually larger than 1.5 nS) (4, 5). Zero current potential measurements of ion selectivity demonstrated that this protein is more than 100-fold selective for anions over cations (3). This high selectivity for anions allowed an estimation of the effective diameter of the channel, 0.6 nm, based on the permeability of the channel for anions of different sizes (3). The basis of anion specificity is the presence of an anion-binding site within the channel with a binding constant, K_d, of, for example, 40 mM for Cl^-.

Lipid bilayer experiments in the presence of phosphate ions indicated that protein P channels were permeable to phosphate. Macroscopic conductance experiments revealed that phosphate was capable of inhibiting chloride flux in a dose- (Fig. 2) and pH-dependent fashion (8). From this information, I_{50} values (concentration of inhibitor giving 50% inhibition) were calculated (e.g., I_{50} = 0.38 mM P_i at pH 8.0). These values suggested that the affinity of protein P for phosphate was 60- to 100-fold greater than the affinity for other ions such as Cl^-. Phosphate analogs, such as pyrophosphate and arsenate, also inhibited the transport but not to the extent seen with phosphate (I_{50} at pH 7.0 = 4.9 mM PP_i and 1.3 mM AsO_4^{2-}, respectively).

Chemical Modification Studies on Protein P

To probe the nature of the phosphate-binding site of protein P, the ε-amino groups of the available lysine residues in the trimer were chem-

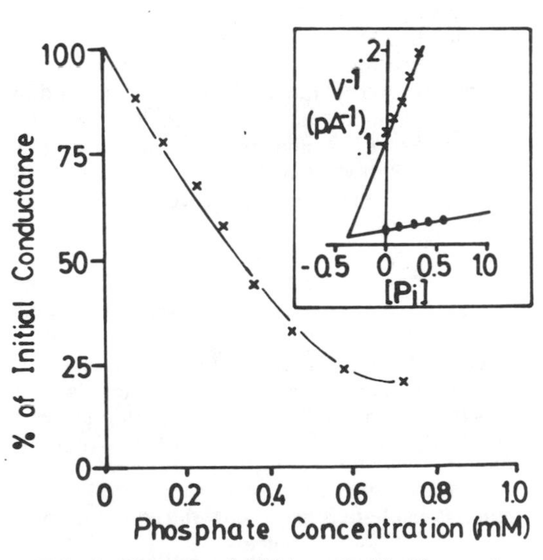

FIG. 2. Phosphate inhibition of chloride conductance at pH 8.0. In the insert, a Dixon plot is presented in which the inverse of the measured membrane current (at 20 mV applied voltage) was graphed against the cumulative phosphate concentration (millimolar) for two experiments in which different numbers of protein P channels were inserted into the membrane.

TABLE 1. Effect of chemical modification of the lysines of protein P on single-channel conductance and affinity for chloride and phosphate

Lysine modification	Single-channel conductance in 1 M KCl (pS)[a]	Affinity (mM) for:	
		Cl^{-}[b]	Phosphate[c]
None	260	40	0.46
Trinitrophenylation	290	ND[d]	ND
Methylation	140	1,000	1.25
Acetylation	25	>3,000	>90[e]
Carbamylation	43	>3,000	>90[e]

[a] Averaged for >200 single-channel events.
[b] Estimated as the K_d for Cl^- from experiments testing the effect of Cl^- concentration on single-channel conductance.
[c] Estimated as the I_{50} from experiments like that depicted in Fig. 2.
[d] ND, Not determined, but the channel remained strongly anion selective.
[e] These data were extrapolated from experiments like that in Fig. 2 but performed in the presence of 1 M KCl.

ically modified, and the protein was then studied for its ability to transport anions in the lipid bilayer system (Table 1). Modification of all available lysines with the bulky reagent trinitrobenzene sulfonate, by the method of Fields (7), had little effect on protein P channel properties, indicating that the critical lysine groups are located in a constriction of the channel which was not available for modification (10a). To confirm this hypothesis, the lysine ε-amino groups were modified by use of less bulky reagents (8). Acetylation (10) and carbamylation (15, 22) produced uncharged residues and destroyed the anion- and phosphate-binding sites (Table 1). Methylation (11) resulted in the modification of lysines but still allowed the retention of the positive charge. Methylated protein P channels were still anion specific but had an altered affinity for chloride and phosphate ions (Table 1). In all cases, the modified proteins retained their trimer conformation as determined by SDS-PAGE. These data strongly suggest that lysines are the key residues involved in both the anion- and phosphate-binding sites of protein P, and suggest that these two sites involve the same lysine residues.

Other Phosphate Starvation-Induced Membrane Proteins

A number of bacterial strains were shown to synthesize phosphate starvation-inducible outer membrane proteins (18). Four members of the family *Pseudomonadaceae* (*Pseudomonas fluorescens* branch) produced proteins which had several functional characteristics in common with protein P. Those examined were found to form small (180–297 pS) anion-selective channels which possessed phosphate-binding sites (K. Poole, T. R. Parr, and R. E. W. Hancock, submitted for publication).

Physical Association between Protein P and the Periplasmic Phosphate-Binding Protein

An association between the maltose-binding protein and LamB porin protein of *E. coli* has been suggested to be necessary for the transport of maltose and maltodextrins across the *E. coli* outer membrane (23). To determine whether the same was true for the high-affinity transport system involving protein P and the periplasmic phosphate-binding protein, a variety of methods were used to examine the association in vitro. By use of a modified enzyme-linked immunosorbent assay, the phosphate-binding protein (17) was immobilized in the bottom of wells of microtiter plates and examined for its ability to interact with protein P. The association of protein P with the immobilized phosphate-binding protein was indicated by a 56% increase over background of an enzyme-linked immunosorbent assay using protein P-specific antibodies (K. Poole, Ph.D. thesis, University of British Columbia, Vancouver, British Columbia, Canada, 1986). This assay was reproducible but of low sensitivity. Numerous attempts to demonstrate interaction, using either protein P or phosphate-binding protein immobilized on an affinity column matrix, were unsuccessful.

Black lipid bilayer experiments were performed in which protein P and a 300-fold molar excess of phosphate-binding protein were mixed prior to the measurement of macroscopic conductance in the presence of 1 M KCl (Fig. 3). The presence of the phosphate-binding protein caused a 10-fold decrease in the final conductance level achieved. However, we were unable to distinguish between the possibilities that the binding protein caused blockage of channels or that it prevented incorporation of protein P into the membrane. These data, while by no means conclusive, indicate a physical association of these proteins.

Discussion

Protein P provides an excellent model protein for examination of structure-function relationships in facilitated diffusion proteins. The presence within the channel of a defined binding site(s) for phosphate (K_d, approximately 0.3 mM P_i) and other anions (e.g., K_d = 40 mM Cl^-) provides this protein with a set of easily studied functional properties including anion specificity, saturation at higher salt concentrations, and competitive blockage of anion (e.g., Cl^-) movement by the preferred anionic substrate phos-

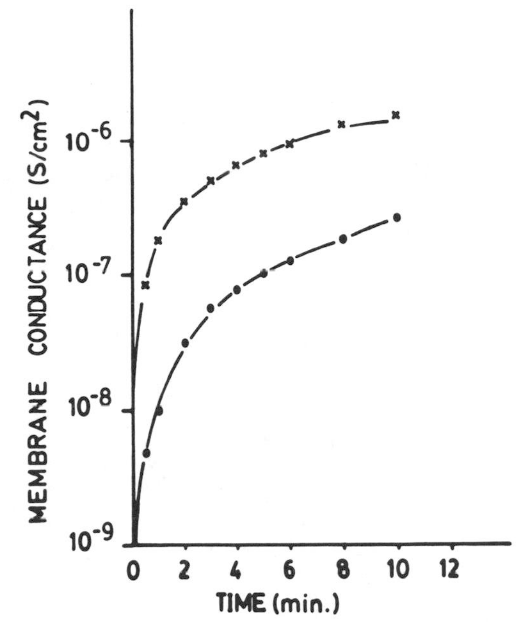

FIG. 3. Inhibition of chloride conduction of protein P by the periplasmic phosphate-binding protein. Protein P was incubated with a 300-fold molar excess of the phosphate-binding protein prior to lipid bilayer experiments. Both proteins were present in soluble forms in 20 mM Tris hydrochloride (pH 6.8) plus 0.05% (vol/vol) Triton X-100. The experiments were carried out in 1 M KCl and 20 mV was the applied voltage. Results from an average of five to eight separate experiments are presented. A conductance of 10^{-7} S/cm^2 is equivalent to 400 protein P channels per cm^2. Symbols: ×, protein P alone; ●, protein P plus the phosphate-binding protein.

phate. This has allowed us to provide a mathematical model for anion movement through the protein P channel (R. Benz and R. E. W. Hancock, submitted for publication). The simplest form of this model is a two-barrier, one-site model. In this model there is a binding site within the channel. Binding of an anion to this site is preceded by an energy barrier (presumably partial dehydration of the anion) and is followed by a second energy barrier (probably due to dissociation of the anion from the binding site). The net energy level of the anion approaching the site and departing from the site would be identical since the anion would almost certainly be hydrated at both times. Nevertheless, these energy barriers must be overcome, and presumably the "driving force" for this would be the concentration gradient. In this regard, the presence of a stronger binding site in the periplasm, like the high-affinity binding site (K_d = 0.3 μM P_i) of the phosphate-binding protein, would be valuable in maintaining the concentration gradient. The possible interaction of protein P and the phosphate-binding protein may favor this mechanism (i.e., maintenance of the concentration gradient), although it is not obligatory.

Protein P, like other porins, has certain disadvantages in structural studies. The apparent lack of linear epitopes on the native trimer, and the maintenance of a stable trimer form even after protease digestion, suggest that protein P has extensive tertiary and quaternary structure (possibly β-sheet folding like other porins [21]). Therefore, we anticipate some difficulty in determining its three-dimensional configuration. Nevertheless, its production in large amounts, ease of purification, and extreme stability to detergents and chemical reagents have allowed us to provide a realistic model for the phosphate/anion-binding site. Given the data described above showing that the protein is a trimer (1), contains its phosphate/anion-binding site in a channel constriction of approximately 0.6 nm (3), and contains lysines as critical residues in the binding site (10), we have suggested (8) that three lysines (one per monomer subunit) are symmetrically arranged in a circle of approximately 0.6 nm diameter within the channel. The HPO_4^{2-} anion is about 0.58 nm in diameter and contains three symmetrically arranged charges represented by two negatively charged oxygens and a partial negative charge on the double-bonded oxygen. Thus, in this model the symmetrical positive charges on the lysine ϵ-amino groups would "coordinate" the symmetrical negative charges on these oxygens. Our current research is directed at determining the protein structure in the neighborhood of this binding site with a view to testing and expanding on this model.

This work was supported by grants from the Natural Sciences and Engineering Research Council of Canada and the Medical Research Council (for the immunochemistry experiments) to R.H., the Deutsche Forschungsgemeinschaft (Be865/1-2 and Be865/3-1) to R.B., and the Alberta Heritage Foundation for Medical Research in the form of a postdoctoral fellowship to E.W.

LITERATURE CITED

1. **Angus, B. L., and R. E. W. Hancock.** 1983. Outer membrane porin proteins F, P and D1 of *Pseudomonas aeruginosa* and PhoE of *Escherichia coli*: chemical cross-linking to reveal native oligomers. J. Bacteriol. **155:**1042–1051.
2. **Benz, R., R. P. Darveau, and R. E. W. Hancock.** 1984. Outer membrane protein PhoE from *Escherichia coli* forms anion selective pores in lipid bilayer membrane. Eur. J. Biochem. **140:**319–324.
3. **Benz, R., M. Gimple, K. Poole, and R. E. W. Hancock.** 1983. An anion-selective channel from the *Pseudomonas aeruginosa* outer membrane. Biochim. Biophys. Acta **730:** 387–390.
4. **Benz, R., and R. E. W. Hancock.** 1981. Properties of the large ion-permeable pores formed from protein F of *Pseudomonas aeruginosa* in lipid bilayer membranes. Bio-

chim. Biophys. Acta **646:**298–308.

5. **Benz, R., A. Schmid, and R. E. W. Hancock.** 1985. Ion selectivity of gram-negative bacterial pores. J. Bacteriol. **162:**722–727.
6. **Brinkman, E., and J. Beckwith.** 1975. Analysis of regulation of *Escherichia coli* alkaline phosphatase synthesis using deletions and ϕ80-transducing phages. J. Mol. Biol. **96:**307–316.
7. **Fields, R.** 1972. The rapid determination of amino groups with TNBS. Methods Enzymol. **25:**464–469.
8. **Hancock, R. E. W., and R. Benz.** 1986. Demonstration and chemical modification of a specific phosphate binding site in the phosphate-starvation inducible outer membrane porin protein P of *Pseudomonas aeruginosa*. Biochim. Biophys. Acta **860:**699–707.
9. **Hancock, R. E. W., K. Poole, and R. Benz.** 1982. Outer membrane protein P of *Pseudomonas aeruginosa*: formation of small anion specific channels in lipid bilayer membrane. J. Bacteriol. **150:**730–738.
10. **Hancock, R. E. W., K. Poole, M. Gimple, and R. Benz.** 1983. Modification of the conductance, selectivity and concentration-dependent-saturation of *Pseudomonas aeruginosa* protein P channels by chemical acetylation. Biochim. Biophys. Acta **735:**137–144.

10a. **Hancock, R. E. W., A. Schmid, K. Bauer, and R. Benz.** 1986. Role of lysines in ion selectivity of bacterial outer membrane porins. Biochim. Biophys. Acta **860:**263–267.

11. **Jentoft, N., and D. G. Dearborn.** 1980. Protein labeling by reductive methylation with sodium cyanoborohydride: effect of cyanide and metal ions on the reaction. Anal. Biochem. **106:**186–190.
12. **Korteland, J., J. Tommassen, and B. Lugtenberg.** 1982. PhoE protein pore of the outer membrane of *Escherichia coli* K12 is a particularly efficient channel for organic and inorganic phosphate. Biochim. Biophys. Acta **690:**282–289.
13. **Mutharia, L. M., and R. E. W. Hancock.** 1983. Surface localization of *Pseudomonas aeruginosa* porin protein F by using monoclonal antibodies. Infect. Immun. **42:**1027–1033.
14. **Parr, T. B., Jr., K. Poole, G. W. K. Crockford, and R. E. W. Hancock.** 1986. Lipopolysaccharide-free *Escherichia coli* OmpF and *Pseudomonas aeruginosa* protein P porins are functionally active in lipid bilayer membranes. J. Bacteriol. **165:**523–526.
15. **Plapp, B. V., S. Moore, and W. H. Stein.** 1971. Activity of bovine pancreatic deoxyribonuclease A with modified amino groups. J. Biol. Chem. **246:**939–945.
16. **Poole, K., and R. E. W. Hancock.** 1983. Secretion of alkaline phosphatase and phospholipase C in *Pseudomonas aeruginosa* is specific and does not involve an increase in outer membrane permeability. FEMS Microbiol. Lett. **16:**25–29.
17. **Poole, K., and R. E. W. Hancock.** 1984. Phosphate transport in *Pseudomonas aeruginosa*: involvement of a periplasmic phosphate-binding protein. Eur. J. Biochem. **144:** 607–612.
18. **Poole, K., and R. E. W. Hancock.** 1986. Phosphate-starvation-induced outer membrane proteins of members of the families *Enterobacteriaceae* and *Pseudomonadaceae*: demonstration of immunological cross-reactivity with an antiserum specific for porin protein P of *Pseudomonas aeruginosa*. J. Bacteriol. **165:**987–993.
19. **Poole, K., and R. E. W. Hancock.** 1986. Isolation of a Tn*501* insertion mutant lacking porin protein P of *Pseudomonas aeruginosa*. Mol. Gen. Genet. **202:**403–409.
20. **Rosenberg, H., R. G. Gerdes, and K. Chegwidden.** 1977. Two systems for the uptake of phosphate in *Escherichia coli*. J. Bacteriol. **131:**505–511.
21. **Rosenbusch, J. P.** 1974. Characterization of the major envelope protein from *Escherichia coli*. Regular arrangement on the peptidoglycan and unusual dodecylsulfate binding. J. Biol. Chem. **249:**8019–8029.
22. **Stark, G.** 1972. Use of cyanate for determining NH_2-terminal residues in protein. Methods Enzymol. **25:**579–584.
23. **Wandersmann, C., M. Schwartz, and F. Ferenci.** 1979. *Escherichia coli* mutants impaired in maltodextrin transport. J. Bacteriol. **140:**1–13.

Anion Exchange as the Molecular Basis of Sugar Phosphate Transport by Bacteria

PETER C. MALONEY, SURESH V. AMBUDKAR, AND LARRY A. SONNA

Department of Physiology, The Johns Hopkins University School of Medicine, Baltimore, Maryland 21205

Chemiosmotic Circulations and Bacterial Transport

Bacteria, like other cells, expend metabolic energy in the accumulation of essential materials and in the extrusion of toxic or unwanted substances. To a significant degree, such events are now understood in the context of chemiosmotic theory (13, 23, 26), a set of ideas which suggests that ionic "circulations" underlie phenomena as distinct as ATP synthesis, solute transport, and (in bacteria) even flagellar rotation. Such arrangements are usually discussed with specific reference to the proton (H^+) as a circulating ion and often with the tacit assumption that chemiosmotic porters use only a cation(s) to circulate between driving and driven reactions. Anions are not usually assigned independent roles (halorhodopsin [36] notwithstanding). In part, this consensus has been justified by experience, since, when anion movement is observed, it is most often coupled to the flux of some driving cation H^+,Mg^{2+}/citrate cotransport (37) or H^+/lactate symport (15, 21), for example. However, more recent work suggests that there is an important class of bacterial porters whose operation involves the movement of anions alone (22).

The idea that bacteria harbor anion-exchange systems was required to account for a novel sugar 6-phosphate transport system in the gram-positive anaerobe *Streptococcus lactis* (4, 5, 22). Further study has made it clear that similar P_i-linked exchange systems are found in many cell types and that presently described examples reflect a family of exchangers spread widely among biomembranes (5, 11, 20). For this reason, while the study of such reactions in bacteria has intrinsic value, it must also be viewed as an opportunity to make special contributions to the more general field of transport. Accordingly, the purpose of this report is twofold: to summarize the current status and to emphasize areas in which such work might prompt conclusions of wider significance.

Phosphate Transport: Net Movement and Exchange

It is convenient to introduce the topic of P_i-linked exchange by distinguishing the various P_i transport systems available to bacteria, since the historical record has some ambiguity regarding activities devoted to net P_i transport and those which participate only in an exchange. In the mid-1950s, Mitchell and Moyle (25, 27) reported their studies of resting cells of *Micrococcus pyogenes* (now *Staphylococcus aureus*). They carefully characterized a rapid P_i:P_i exchange and showed that it was based on the selective movement of monovalent phosphate, $H_2PO_4^{1-}$. It was reasonably presumed that such P_i:P_i exchange reflected operation of a system normally directed to net P_i accumulation. Roughly a decade later, Harold et al. (14) made quite different observations using another gram-positive cell, *Streptococcus faecalis*. In that case, an exchange reaction was not found, and when net flux was characterized, it was shown to have an absolute dependence on concurrent metabolism (ATP) with selectivity favoring the divalent form, HPO_4^{2-}, not the monoanion.

These striking differences (*S. aureus* versus *S. faecalis*) were early hints of a diversity appreciated later, when studies of *Escherichia coli* outlined a set of (at least) four systems, each of which carried isotopically labeled P_i into the cell (6, 32, 38). Two of these, designated Pst (phosphate-specific transport) and Pit (phosphate inorganic transport), proved highly specific for P_i itself, while two others seemed to accept P_i as an analog of the natural substrate, an organophosphate. The latter have been designated as GlpT (*sn*-glycerol 3-phosphate transport [16]) and UhpT (hexose phosphate uptake [8]).

For purposes of classification, Pst is placed with those systems that use ATP ("phosphate bond energy") to drive solute transport, while Pit seems to be a chemiosmotic carrier that displays nH^+/P_i symport (33). Pst and Pit represent common solutions to P_i transport by bacteria. Pst-like energetics would accommodate Harold's (14) findings in *S. faecalis*, and Pit-like behavior could account for P_i transport driven by respiration in *Micrococcus lysodeikticus* (12), *Paracoccus denitrificans* (7), or *S. aureus* (L. A. Sonna and P. C. Maloney, unpublished data). The remaining $^{32}P_i$ transporters, GlpT and UhpT, are usually placed, along with Pit, in the category of nH^+/anion symport (10, 31, 39), but as outlined below, this assignment is probably not correct. Instead, the simplest interpretation of current evidence is that systems which incorporate $^{32}P_i$ come in two varieties: those designed for net P_i movement (Pit, Pst, and others) and those which mediate exchange reactions involving P_i or phosphorylated substrates, or both (GlpT, UhpT, and others). It is likely that Harold et al. (14) described one of the former (dedicated) category, while Mitchell's discovery (25, 27) belongs to the second (exchange) class.

Phosphate-Linked Exchange in *Streptococcus lactis*

Since the observations suggesting anion exchange arose during studies of $^{32}P_i$ transport by *S. lactis*, it is useful to summarize relevant information from that work. For example, the exchange phenomenon was demonstrated by an experiment (Fig. 1A) in which resting (starving) cells of *S. lactis* were exposed to 20 μM $^{32}P_i$. The added $^{32}P_i$ was readily incorporated into the free intracellular P_i pool (ca. 50 mM), but incorporation of $^{32}P_i$ was unaffected by inhibitors of energy metabolism (carbonyl cyanide *p*-trifluoromethoxy phenylhydrazone, *N,N'*-dicyclohexylcarbodiimide), a finding which posed the fundamental paradox: the membrane was "permeable" to $^{32}P_i$, yet an outward P_i chemical gradient (2,500-fold) was maintained indefinitely (>180 min), with no dependence on energy. This paradox was resolved only when other work showed that $^{32}P_i$ transfer was accompanied by a 1:1 exchange with internal (nonradioactive) P_i. Clearly, $^{32}P_i$ incorporation reflected isotope exchange, and the role of metabolic energy was moot. Further study showed that P_i:P_i exchange used monovalent P_i ($H_2PO_4^{1-}$), and this suggested an important similarity to the exchange found in *S. aureus*. Subsequently, however, the streptococcal example was set apart from that early precedent by the finding that $^{32}P_i$ entry was specifically blocked by low concentrations of certain sugar 6-phosphates (Fig. 1B). It seemed likely, therefore, that P_i:P_i exchange was a manifestation of an anion-exchange carrier normally directed to the net transport of sugar 6-phosphate.

This proposal was tested directly by studies with membrane vesicles (4), and as well as documenting the heterologous reaction, the vesicle work provided quantitative estimates of exchange stoichiometry. Thus, for every 2 mol of P_i which moved (in or out), 1 mol of 2-deoxyglucose 6-phosphate moved in exchange. Because those assays were at pH 7, where divalent sugar phosphate was enriched (pK_2 = 6.1), the 2:1 ratio was to be expected if the reaction was a neutral exchange involving two monovalent P_i anions and a single, divalent 2-deoxyglucose 6-phosphate. That interpretation was consistent with a selectivity for monovalent P_i during P_i:P_i exchange (see above) and also with the failure of several ion-conductive ionophores to affect the rates of either homologous (P_i:P_i) or heterologous (P_i:2-deoxyglucose 6-phosphate) exchange.

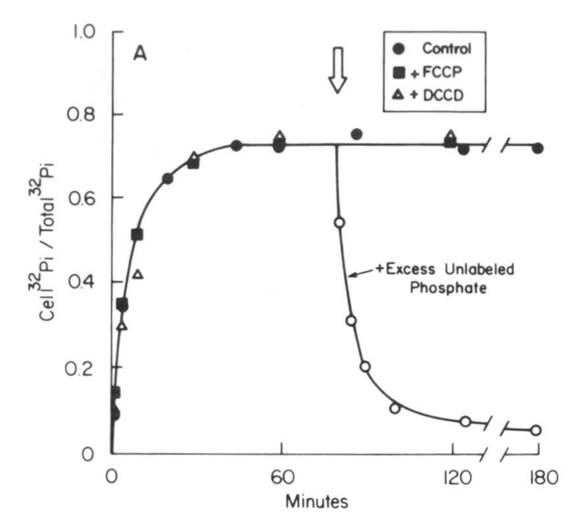

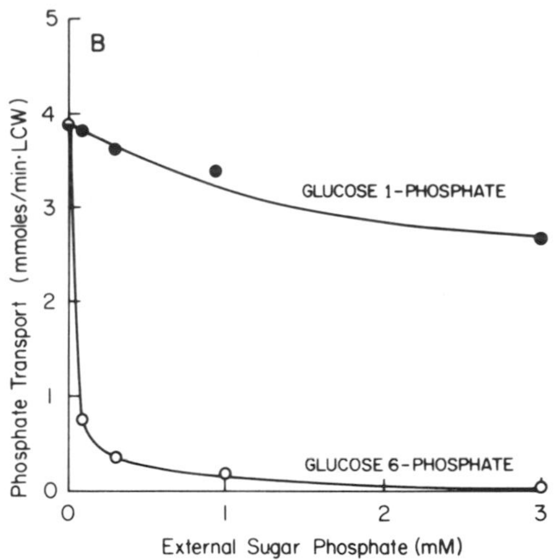

FIG. 1. Phosphate exchange in *S. lactis*. (A) Cells in 300 mM KCl–20 mM MOPS/K^+ at pH 7 were given 20 μM $^{32}P_i$ at zero time. Entry and exchange of $^{32}P_i$ were determined at intervals thereafter. Pretreatments with 1 mM *N,N'*-dicyclohexylcarbodiimide (DCCD) or 1 μM carbonyl cyanide *p*-trifluoromethoxy phenylhydrazone (FCCP) were as described (22). (B) In a similar experiment using 60 μM $^{32}P_i$, $^{32}P_i$ entry was estimated after 1 min of incubation with glucose 1-phosphate or glucose 6-phosphate. LCW, Liters of cell water. (Data from reference 22.)

Such observations are conveniently summarized by the physiological model of Fig. 2, in which this exchange participates in a chemios-

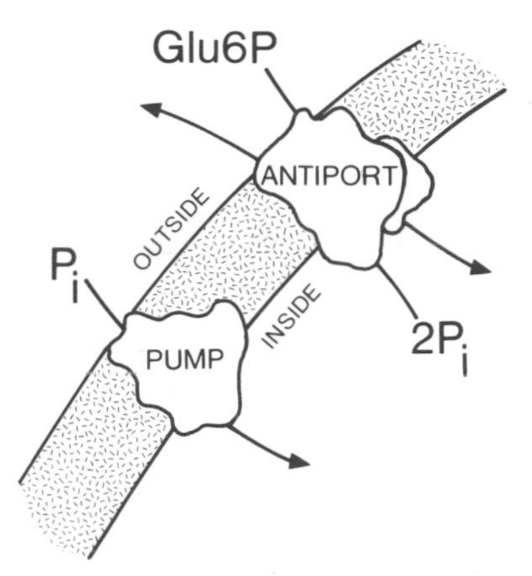

FIG. 2. Accumulation of glucose 6-phosphate (Glu6P) by a P_i circulation. See text for discussion. The term pump is used in the broadest context to indicate any active transport system which accumulates P_i from the medium.

motic circulation. Note, however, that circulation of an anion (P_i) links the driving and driven reactions. One presumes that a P_i pump (streptococcal Pst- or Pit-like systems) accumulates P_i and that internal (monovalent) P_i leaves in a neutral 2:1 exchange with external (divalent) glucose 6-phosphate. The net reaction, then, has a zero sum in regard to P_i and a gain of 1 glucose for each unit of energy (ATP, nH^+, etc.) expended during P_i accumulation by the pump.

It seems likely that this cycle (Fig. 2) operates in at least two situations. For example, many cells tend to accumulate P_i as they enter stationary phase (for reasons not entirely clear [21, 24]), and chemiosmotic exchange based on P_i would sensibly exploit such elevated internal P_i. It is also probable that this cycle operates in vitro, when sugar phosphate transport is studied in the presence of P_i buffers (see below). As noted later, a more subtle exchange is invoked to account for events during active growth.

Tests of Anion Exchange as a General Mechanism in Bacterial Transport

Experiments of the sort outlined above (4, 5, 22) have made a strong case for the operation of P_i-linked exchange in *S. lactis* and in its relative, *S. cremoris* (strain E8) (unpublished data). But this exchange reaction is not found in several other streptococcal species (22) (as expected from Harold et al. [14]), so that it has become necessary to question the general significance of anion exchange.

Recently, the impact of findings in *S. lactis* has been reinforced by experiments using *S. aureus*. Our new experiments have confirmed the presence of P_i:P_i exchange in *S. aureus* (25, 27) and have indicated that such exchange is associated with a sugar phosphate carrier of broader specificity than that found in *S. lactis*. Perhaps the most important experiments with *S. aureus* have been those which discriminate between the two likely mechanisms of transport. For example, the idea of nH^+/anion symport has been supported by both direct (10) and indirect (31) evidence gathered in *E. coli*, yet the experiments with *S. lactis* (4, 5, 22) point to anion: anion exchange as the molecular basis of transport.

The experiment illustrated in Fig. 3 presents a crucial observation in this context. Membrane vesicles were prepared by osmotic lysis in the presence of morpholinepropanesulfonic acid (MOPS) (no P_i was present) and then tested for P_i and 2-deoxyglucose 6-phosphate transport in response to the proton motive force developed by oxidation of DL-lactate. The MOPS-loaded vesicles did accumulate P_i (Fig. 3A), and a

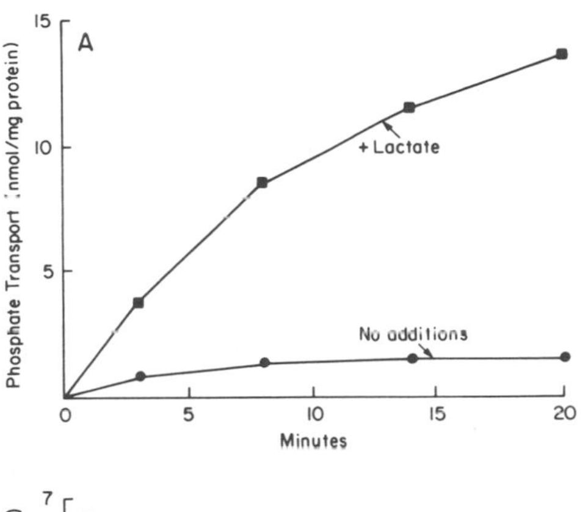

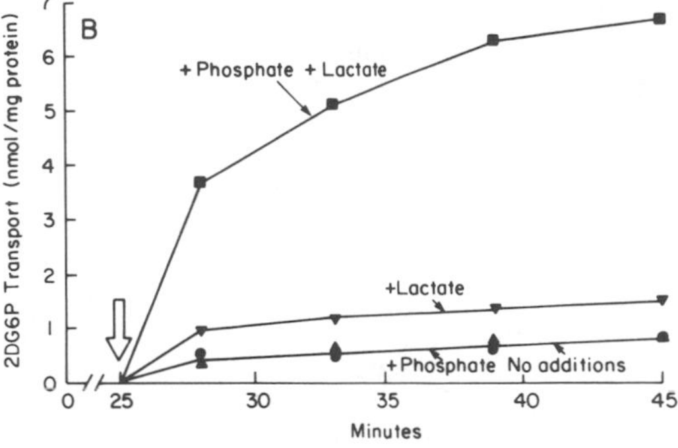

FIG. 3. Sugar phosphate transport by *S. aureus*. (A) Transport of $^{32}P_i$. Membrane vesicles (4) were loaded at pH 7 with 70 mM K^+/MOPS, 10 mM K_2SO_4, 10 mM $MgSO_4$, and 0.25 mM Na_3VO_4. Vesicles were assayed in 125 mM K_2SO_4, 20 mM MOPS/K, and 0.25 mM Na_3VO_4 (pH 7), with DL-lactate (33 mM) or nonradioactive (100 μM) P_i, if indicated. (B) In a parallel experiment, vesicles were preincubated (25 min) with the indicated substrates before tests of [^{14}C]2-deoxyglucose 6-phosphate (100 μM) (2DG6P) transport.

sensitivity to a protonophore (not shown) argued that this was the result of nH^+/P_i symport, as suggested for Pit in *E. coli* (33). By contrast, the transport of 2-deoxyglucose 6-phosphate could not be observed for these conditions. Instead, sugar phosphate transport required the presence of internal P_i, which could be provided only by suitable preincubation (Fig. 3B). Thus, pretreatment with P_i and DL-lactate led to later 2-deoxyglucose 6-phosphate accumulation, whereas the pretreatments with P_i or DL-lactate alone were ineffective. Such findings are not easily interpreted to indicate electrogenic nH^+/2-deoxyglucose 6-phosphate symport as proposed for *E. coli* (31), but they are readily understood if 2-deoxyglucose 6-phosphate movement is based on an exchange which can use P_i (Fig. 2).

Reconstitution of P_i-Linked Exchange from *E. coli*

One implication of the interpretation above is that carriers thought of as n(cation)/anion symport might be constructed as anion exchange at the molecular level, and this is one factor prompting a reexamination of the GlpT and UhpT transporters of *E. coli* (1). In addition, recent studies of whole cells suggest that P_i-linked exchanges might be attributed to GlpT (9, 34). The literature relevant to UhpT transport has a similar uncertainty with regard to mechanism since, unlike H^+/lactose and other symport reactions, the movement of sugar phosphate via UhpT persists to a significant degree in "energy-starved" cells (10, 39).

We have reconstituted P_i-linked activities from *E. coli* strains of the appropriate genotype (1), using a method first introduced by Racker et al. (30) and by Newman and Wilson (28) and then modified in our laboratory (2, 3) to optimize recovery of fully active protein. That work shows that both homologous and heterologous P_i-linked exchanges can be mediated by the GlpT and UhpT proteins in reconstituted proteoliposomes. The data (1) are consistent with the idea that both the GlpT and UhpT transporters operate using anion exchange. It will now be feasible to conduct more direct tests using reconstituted material.

Such results encourage the view that anion exchange subserves nutrient transport in both gram-positive and gram-negative cells, and in the development of this argument, it will be necessary to satisfy physiological demands with a biochemical mechanism based on neutral exchange. In particular, it will be important to understand how an exchange can operate to give the net entry of carbon and phosphorus, since it is common experience (*E. coli*) that either *sn*-glycerol 3-phosphate (via GlpT) or glucose 6-phosphate (via UhpT) can serve as a sole P_i source, without extracellular hydrolysis. It is not obvious how this is accomplished by antiport of the kind discussed above.

Variable Stoichiometry of Antiport: Implications for Mechanism

One goal of work with *S. lactis* has been to identify the ionic forms of P_i and 2-deoxyglucose 6-phosphate which serve as substrates. The participation of both monovalent P_i and divalent sugar phosphate was clearly indicated by the 2:1 stoichiometry of exchange at pH 7 (4), but to examine the role of monovalent sugar phosphate, it was necessary to examine both kinetic behavior and exchange stoichiometry as the pH was lowered from pH 7 (5).

A series of kinetic studies of P_i:2-deoxyglucose 6-phosphate antiport revealed that assay pH (pH 5.2 to 7) had little effect on the affinity for 2-deoxyglucose 6-phosphate, suggesting that the exchange selected randomly among available mono- and divalent forms of 2-deoxyglucose 6-phosphate. By contrast, quite specific changes were observed with regard to exchange stoichiometry, and as pH fell from 8 to 5.2, the exchange ratio shifted from 2:1 (P_i:2-deoxyglucose 6-phosphate) to about 1:1 (5). The pattern of response suggested that overall stoichiometry was determined by titration of the 2-deoxyglucose 6-phosphate substrate, rather than by effects on the protein, a proposal which may clarify the general issue of "variable" stoichiometry (18). But of more significance in the present context, this changing stoichiometry directs attention to a simple biochemical model.

Figure 4A shows the biochemical model which most simply describes the variable stoichiometry found during exchange. The model proposes that exchange is always electrically neutral, that the active site contains two binding regions which together alternate between membrane surfaces, and that each of these binding sites can accept a monoanion substrate (of P_i, HG6P [monovalent glucose 6-phosphate], 2-deoxyglucose 6-phosphate, etc.). It is also assumed that organic substrates (e.g., sugar 6-phosphates, not P_i) can trigger a change of conformation which brings the separate binding sites in close enough proximity to accept a single divalent substrate, precluding occupancy by any monovalent species. If reorientation is rate limiting (a common assumption), one would then expect (as found) a variation of exchange stoichiometry which parallels the pH-dependent titration of the complex substrate.

Even as it stands, this biochemical model has an important consequence for physiological function. Thus, a direct prediction is that when the organophosphate substrate is used, there will be net flux, but only in the presence of a pH

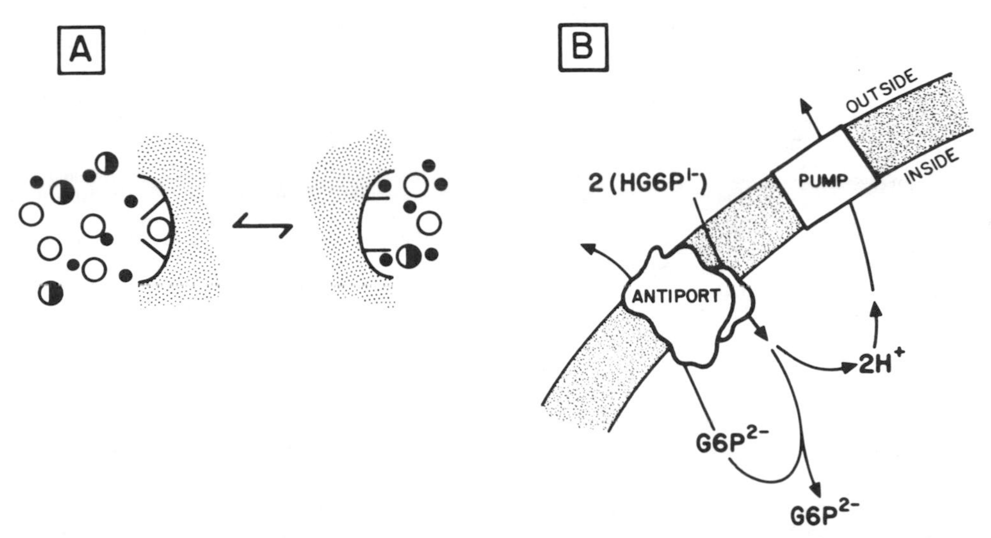

FIG. 4. Models of anion exchange. (A) A model of exchange involving monovalent P_i (●), monovalent sugar phosphate (◑), or divalent sugar phosphate (○). As shown, binding is to divalent sugar phosphate (left) or to a pair of monovalent P_i ions (right). (Taken from reference 5.) (B) A consequence of the model in part A is that net glucose 6-phosphate accumulation will occur in the presence of a pH gradient (see text).

gradient. For example, there should be antiport of a single divalent $G6P^{2-}$ against two monovalent $HG6P^-$ anions, with the result that net sugar phosphate accumulation is driven (indirectly) by a pH gradient (Fig. 4B). In this instance, then, the energetics would implicate H^+/anion symport, yet the mechanism would rest on an obligatory anion exchange. This explains how an exchange reaction can provide for the net influx of phosphorus, and this is a feature essential to any realistic view of GlpT or UhpT function. This arrangement is a novel one (compare with Fig. 2); it is a perspective which shows how a conflict between thermodynamics (H^+/symport) and mechanism (anion exchange) might be resolved, and why systems classified as H^+/symport (GlpT, UhpT, and others) might be anion exchange at a molecular level.

Physiological Position of Anion Exchange in Microorganisms

The experiments discussed above argue in favor of antiport as a general mechanism for the transport of certain anionic nutrients. The idea is most strongly and directly supported by studies of gram-positive cells (*S. lactis, S. cremoris, S. aureus*), but the status of work with *E. coli* (GlpT, UhpT) clearly points to extension of the same principle to gram-negative organisms. This view (Fig. 2, Fig. 4B) implies that some porters now classified as H^+/anion symport are, in fact, anion-exchange systems, and along with the cases discussed above one might cite as possible examples an *E. coli* cyclic AMP extrusion mechanism (29) and a system which mediates the accumulation of phosphoenolpyruvate in *Salmonella typhimurium* (35). It seems useful to direct attention to examples of P_i-linked exchange in eucaryote organelles. There, the finding of P_i:malate exchange (mitochondria [20]) and P_i:phosphoenolpyruvate or P_i:triose-P_i antiport (chloroplasts [11]) immediately suggests that P_i-linked reactions constitute a widespread family with representatives in many types of cells and organelles. Adenine nucleotide exchange, found in both mitochondria (17, 20) and intracellular rickettsiae (19), offers still another example that emphasizes the similarity of bacterial and organelle transport. Such comparisons suggest in addition that anion exchange in bacteria may include the transport of Krebs cycle intermediates; mitochondria handle these substrates with great facility using exchange reactions (20).

The juxtaposition of bacterial transporters with their (presumed) relatives in eucaryote compartments poses a challenge for the near future: to what degree are the reactions similar, and how do they differ? Discoveries in bacteria may have already contributed new information to this general field (5). Certainly, the continued exploration with microorganisms will be essential for progress beyond the stage of kinetic description.

We thank T. J. Larson and J. Thompson for providing bacterial strains.

Work in this laboratory has been supported by Public Health Service grant GM24195 from the National Institutes of Health.

LITERATURE CITED

1. **Ambudkar, S. V., T. J. Larson, and P. C. Maloney.** 1986. Reconstitution of sugar phosphate transport systems of *Escherichia coli*. J. Biol. Chem. **261:**9083–9086.
2. **Ambudkar, S. V., and P. C. Maloney.** 1986. Anion exchange in bacteria: reconstitution of phosphate:hexose 6-phosphate antiport from *Streptococcus lactis*. Methods Enzymol. **125:**558–563.
3. **Ambudkar, S. V., and P. C. Maloney.** 1986. Bacterial anion exchange: use of osmolytes during solubilization and reconstitution of phosphate-linked antiport from *Streptococcus lactis*. J. Biol. Chem. **261:**10079–10086.
4. **Ambudkar, S. V., and P. C. Maloney.** 1984. Characterization of phosphate:hexose 6-phosphate antiport in membrane vesicles of *Streptococcus lactis*. J. Biol. Chem. **259:** 12576–12585.
5. **Ambudkar, S. V., L. A. Sonna, and P. C. Maloney.** 1986. Variable stoichiometry of phosphate-linked anion exchange in *Streptococcus lactis*: implications for the mechanism of sugar phosphate transport by bacteria. Proc. Natl. Acad. Sci. USA **83:**280–284.
6. **Bennett, R. L., and M. H. Malamy.** 1970. Arsenate resistant mutants of *Escherichia coli* and phosphate transport. Biochem. Biophys. Res. Commun. **40:**496–503.
7. **Burnell, J. N., P. John, and F. R. Whatley.** 1975. Phosphate transport in membrane vesicles of *Paracoccus denitrificans*. FEBS Lett. **58:**215–218.
8. **Deitz, G.** 1976. The hexose phosphate transport system of *Escherichia coli*. Adv. Enzymol. **44:**237–259.
9. **Elvin, C. M., C. M. Hardy, and H. Rosenberg.** 1985. P_i exchange mediated by the GlpT-dependent *sn*-glycerol-3-phosphate transport system in *Escherichia coli*. J. Bacteriol. **161:**1054–1058.
10. **Essenberg, R. C., and H. L. Kornberg.** 1975. Energy coupling in the uptake of hexose phosphates by *Escherichia coli*. J. Biol. Chem. **250:**939–945.
11. **Fliege, R., U.-F. Flugge, K. Werdan, and H. W. Heldt.** 1978. Specific transport of inorganic phosphate, 3-phosphoglycerate and triose-phosphates across the inner membrane of the envelope of spinach chloroplasts. Biochim. Biophys. Acta **502:**232–247.
12. **Friedberg, I.** 1977. The effect of ionophores on phosphate and arsenate transport in *Micrococcus lysodeikticus*. FEBS Lett. **81:**264–266.
13. **Harold, F. M.** 1977. Membranes and energy transduction in bacteria. Curr. Top. Bioenerg. **6:**83–149.
14. **Harold, F. M., R. L. Harold, and A. Abrams.** 1965. A mutant of *Streptococcus faecalis* defective in phosphate transport. J. Biol. Chem. **240:**3145–3153.
15. **Harold, F. M., and E. Levin.** 1974. Lactic acid translocation: terminal step in glycolysis by *Streptococcus faecalis*. J. Bacteriol. **117:**1141–1148.
16. **Hayashi, S.-I., J. P. Koch, and E. C. C. Lin.** 1964. Active transport of L-a-glycerophosphate in *Escherichia coli*. J. Biol. Chem. **239:**3098–3105.
17. **Klingenberg, M., H. Hackenberg, R. Kramer, C. S. Lin, and H. Aquila.** 1980. Two transport proteins from mitochondria. I. Mechanistic aspects of asymmetry of the ADP/ATP translocator. II. The uncoupling protein of brown adipose tissue mitochondria. Ann. N.Y. Acad. Sci. **358:**83–95.
18. **Konings, W. N., and I. R. Booth.** 1981. Do the stoichiometries of ion-linked transport systems vary? Trends Biochem. Sci. **6:**257–262.
19. **Krause, D. C., H. H. Winkler, and D. O. Wood.** 1985. Cloning and expression of the *Rickettsia prowazekii* ADP/ATP translocator in *Escherichia coli*. Proc. Natl. Acad. Sci. USA **82:**3015–3019.
20. **LaNoue, K. F., and A. C. Schoolwerth.** 1979. Metabolite transport in mitochondria. Annu. Rev. Biochem. **48:**871–922.
21. **Maloney, P. C.** 1983. Relationship between phosphorylation potential and electrochemical H^+ gradient during glycolysis in *Streptococcus lactis*. J. Bacteriol. **153:**1461–1470.
22. **Maloney, P. C., S. V. Ambudkar, J. Thomas, and L. Schiller.** 1984. Phosphate/hexose 6-phosphate antiport in *Streptococcus lactis*. J. Bacteriol. **158:**238–245.
23. **Maloney, P. C., and T. H. Wilson.** 1985. Evolution of ion pumps. Bioscience **35:**43–48.
24. **Mason, P. W., D. P. Carbone, R. A. Cushman, and A. S. Waggoner.** 1981. The importance of inorganic phosphate in regulation of energy metabolism of *Streptococcus faecalis*. J. Biol. Chem. **256:**1861–1866.
25. **Mitchell, P.** 1954. Transport of phosphate across the osmotic barrier of *Micrococcus pyogenes*: specificity and kinetics. J. Gen. Microbiol. **11:**73–82.
26. **Mitchell, P.** 1979. Compartmentation and communication in living systems. Ligand conduction: a general catalytic principle in chemical, osmotic and chemiosmotic reaction systems. Eur. J. Biochem. **95:**1–20.
27. **Mitchell, P., and J. M. Moyle.** 1953. Paths of phosphate transfer in *Micrococcus pyogenes*: phosphate turnover in nucleic acids and other fractions. J. Gen. Microbiol. **9:** 257–272.
28. **Newman, M. J., and T. H. Wilson.** 1980. Solubilization and reconstitution of the lactose transport system from *Escherichia coli*. J. Biol. Chem. **255:**10583–10586.
29. **Potter, K., G. Chaloner-Larsson, and H. Yamazaki.** 1974. Abnormally high rate of cyclic AMP excretion from an *Escherichia coli* mutant deficient in cyclic AMP receptor protein. Biochem. Biophys. Res. Commun. **57:**379–385.
30. **Racker, E., B. Violand, S. O'Neal, M. Alfonzo, and J. Telford.** 1979. Reconstitution, a way of biochemical research; some new approaches to membrane-bound enzymes. Arch. Biochem. Biophys. **198:**470–477.
31. **Ramos, S., and H. R. Kaback.** 1977. pH-dependent changes in proton:substrate stoichiometries during active transport in *Escherichia coli* membrane vesicles. Biochemistry **16:**4271–4275.
32. **Rosenberg, H., R. G. Gerdes, and K. Chegwidden.** 1977. Two systems for the uptake of phosphate by *Escherichia coli*. J. Bacteriol. **131:**505–511.
33. **Rosenberg, H., R. G. Gerdes, and F. M. Harold.** 1979. Energy coupling to the transport of inorganic phosphate in *Escherichia coli* K12. Biochem. J. **178:**133–137.
34. **Rosenberg, H., L. M. Russel, P. A. Jacomb, and K. Chegwidden.** 1982. Phosphate exchange in the *Pit* system in *Escherichia coli*. J. Bacteriol. **149:**123–130.
35. **Saier, M. H., Jr., D. L. Wentzel, B. U. Feucht, and J. J. Judice.** 1975. A transport system for phosphoenolpyruvate, 2-phosphoglycerate and 3-phosphoglycerate in *Salmonella typhimurium*. J. Biol. Chem. **250:**5089–5096.
36. **Schobert, B., and J. K. Lanyi.** 1982. Halorhodopsin is a light-driven chloride pump. J. Biol. Chem. **257:**10306–10313.
37. **Willecke, K., E.-M. Grier, and P. Oehr.** 1973. Coupled transport of citrate and magnesium in *Bacillus subtilis*. J. Biol. Chem. **248:**807–814.
38. **Willsky, G. R., and M. H. Malamy.** 1980. Characterization of two genetically separable inorganic phosphate transport systems in *Escherichia coli*. J. Bacteriol. **144:** 366–374.
39. **Winkler, H. H.** 1973. Energy coupling to the hexose phosphate transport system in *Escherichia coli*. J. Bacteriol. **116:**203–209.

Solute and Ion Transport across Bacterial Membranes

W. N. KONINGS AND B. POOLMAN

Department of Microbiology, University of Groningen, 9751 NN Haren, The Netherlands

Bacteria require organic and inorganic solutes which have to be transported across their cytoplasmic membranes for growth and survival. The mechanisms and the energy requirement of the translocation processes have been investigated extensively since the pioneering studies of Cohen and Monod (1). With respect to the energy requirement of solute transport, several schools of thought have dominated over the years. Since the postulation of Lipman in 1941 (20) that ATP functioned as the molecular energy currency in living cells, it was generally assumed that transport of solutes across the cytoplasmic membrane against their concentration gradients required the hydrolysis of ATP (14). It eventually became clear that bacterial cells have distinct energy coupling mechanisms at their disposal for solute translocation. In 1964, Kundig et al. (19) found that a number of sugars can be transported by phosphoenolpyruvate-dependent group translocation systems. In gram-negative bacteria transport of several solutes can be mediated by systems which are energized by ATP or a phosphorylated intermediate and in which periplasmic binding proteins play an essential role (29). Solid evidence for a direct role of ATP as energy source has also been presented for the transport of several ions (K^+, Na^+, Ca^{2+}) via ion-ATPases (10). In some anaerobes indirect evidence has been obtained for the involvement of ATP in the translocation of several solutes such as glutamate and phosphate (11).

The majority of solutes, however, appear to be translocated by diffusion processes. For membrane-permeable or lipophilic solutes (e.g., weak acids and bases, methanol) no specific membrane proteins are required for the translocation processes, and transport can occur by passive diffusion. However, the translocation across the cytoplasmic membrane of membrane-impermeable solutes (e.g., sugars, amino acids, most organic acids, inorganic cations and anions) and the rapid translocation of membrane-permeant solutes require the involvement of specific transport proteins (carriers). These facilitated diffusion processes have been studied most extensively. Kaback's pioneering studies in isolated cytoplasmic membrane vesicles unequivocally demonstrated that neither ATP nor any other phosphorylated intermediate functions as the direct energy source for facilitated solute transport (13). In the past 15 years, solid evidence has been presented for the involvement of the electrochemical proton gradient between the bulk phases at both sides of the cytoplasmic membrane as the driving force for several facilitated diffusion systems (chemiosmotic transport systems) in a wide variety of bacteria (16). These transport systems are termed secondary transport systems since they convert the electrochemical energy of protons (the proton motive force) into the electrochemical energy of solute (16). The systems which convert (photo)chemical energy into electrochemical energy of protons or other solutes are primary transport systems. The proton pumps such as electron transfer systems and membrane-bound ATPases which generate a proton motive force (Δp) are primary transport systems. The proton motive force is composed of two components, the electrical potential ($\Delta\psi$) and the chemical gradient of protons across the cytoplasmic membrane (ΔpH) according to the equation $\Delta p = \Delta\psi - Z\Delta\text{pH}$ (mV), in which Z equals 2.3 RT/F (R is the gas constant, T is the absolute temperature, and F is the Faraday constant).

The electrochemical driving force for solute translocation is composed of the chemical potentials of the translocated solute(s) ($\Delta\mu A/F$ and $-Z\Delta\text{pH}$) and of the electrical potential ($\Delta\psi$) if net charge is translocated. The following equation holds for secondary transport processes (17): $\Delta\mu_A = -(n + m)F\Delta\psi + n\,2.3\,RT\,\Delta\text{pH}$, in which A is the translocated solute with charge m, $\Delta\mu_A = 2.3\,RT \log [A_{in}]/[A_{out}]$ and n is the number of translocated protons.

In the past 5 years, results have been presented which appear to be at variance with a chemiosmotic mechanism of solute transport, and this has led to the postulation of other mechanisms of energy coupling and also to a revival of the localized chemiosmotic concept of Williams (28).

We have studied the mechanism of energy coupling to secondary transport of several solutes in *Rhodopseudomonas sphaeroides*, *Escherichia coli*, *Streptococcus lactis*, and *Streptococcus cremoris*. These studies demonstrate that rather complex interactions exist between the different energy transducing systems in bacteria. Conclusions about the mechanism of energy coupling often cannot be drawn from studies in intact cells, and studies in model systems such as membrane vesicles are required to obtain such information. However, such studies cannot supply information about the role of cytoplasmic components in regulatory processes. A combination of studies in membrane vesicles and intact cells will clearly be required for a full understanding of the transport processes.

Regulation by Electron Transfer

In the phototrophic bacterium *R. sphaeroides* the existence of a proton motive force alone is not sufficient for solute uptake; turnover of the cyclic electron transfer chain is also necessary (4). The initial rate of uptake of the amino acid alanine has been measured at a constant light intensity (= constant rate of cyclic electron transfer) but varying magnitudes of the proton motive force. Under these conditions the rate of transport increased exponentially with the proton motive force. However, when the proton motive force was kept constant and the rate of electron transfer was varied, the rate of transport increased linearly with the light intensity (Fig. 1). At low light intensities there was no uptake of alanine, even when the proton motive force was high. These results demonstrate that the electron transfer chain not only functions as a generator of a proton motive force but also directly influences the activity of the carrier.

Such an interaction between the electron transfer system and solute transport carriers is not specific for *R. sphaeroides*. In a strain of *R. sphaeroides* in which the *E. coli* transport protein for lactose (the M protein) was incorporated via genetic transformation, a similar relation between the rate of cyclic electron transfer and lactose transport was observed (5). Kinetic analysis of the changes in the initial rate of both alanine and lactose uptake indicated that the regulation is due to a light-dependent change of the activity of the carrier molecules in the membrane (5).

A very attractive experimental system for investigating the role of electron transfer and the proton motive force in solute transport was found in membrane vesicles of *E. coli* which contain pyrollo-quinoline quinone (PQQ)-dependent glucose dehydrogenase. Glucose dehydrogenase is found in *E. coli* as the apoenzyme which can be converted into the active holoenzyme by the addition of its prosthetic group PQQ (27). This system is ideal for transport studies because the enzyme is coupled to the respiratory chain and its activity can be increased by adding increasing amounts of PQQ to *E. coli* membrane vesicles. Therefore, the rate

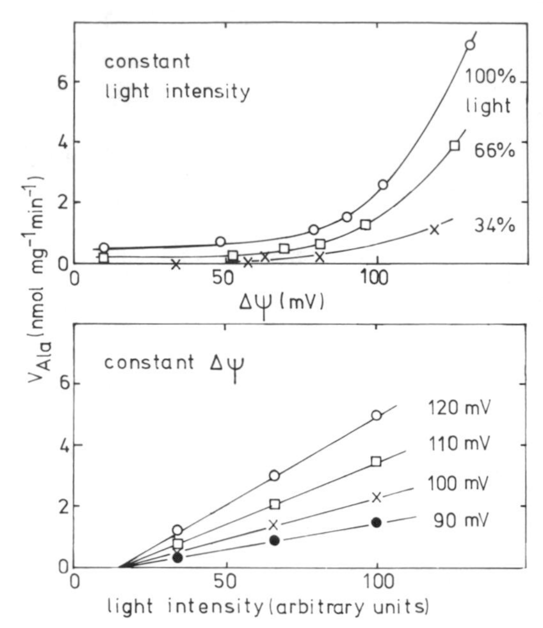

FIG. 1. Relation beween the rate of alanine uptake (V_{Ala}) and $\Delta\psi$ at constant light intensity (upper panel) and between V_{Ala} and light intensity at constant $\Delta\psi$ (lower panel) in cells of *R. sphaeroides*. Alanine uptake and $\Delta\psi$ were measured simultaneously in cells treated with 2 mM K^+-EDTA and 2 μM nigericin at different light intensities. The incubation medium contained 50 mM potassium phosphate (pH 8), 5 mM $MgSO_4$, 50 μM [^{14}C]alanine, 4 μM tetraphenylphosphonium, cells at 0.64 mg of protein per ml, and various concentrations of valinomycin. Data from the upper panel were replotted to obtain the lower panel. (From Elferink et al. [4], with permission.)

of electron flow and the redox state of the components of the electron transfer chain can be varied by adding various amounts of PQQ (Fig. 2). At constant $\Delta\psi$ value, the rate of lactose uptake decreased with the respiration rate, while the equilibrium $\Delta\mu_{lactose}$ remained constant. At lower respiration rates (lower PQQ concentrations) both the $\Delta\psi$ and $\Delta\mu_{lactose}$ decreased. This decrease of the uptake rate was steeper since the driving force ($\Delta\psi$) also decreased at lower rates of electron transfer (27).

Regulation by Dithiol-Disulfide Interchange

A large number of reports have appeared on the involvement of sulfhydryl groups in the function of membrane proteins such as carriers, energy transducing enzymes, and receptor proteins. Dithiol-disulfide interconversions have been reported to play an essential role in many solute transport and energy transducing systems in bacteria, mitochondria, and chloroplasts. Moreover, many reports have demonstrated that the sensitivity of various transport systems to

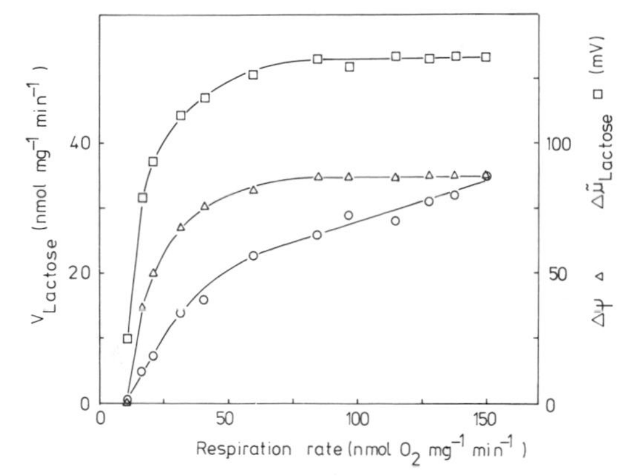

FIG. 2. Initial rate of lactose uptake, steady-state accumulation level, and $\Delta\psi$ as a function of the respiration rate in *E. coli* membrane vesicles energized by PQQ-dependent glucose oxidation. Lactose uptake, $\Delta\psi$, and respiration rate were measured simultaneously at 30°C. The incubation medium contained 50 mM potassium phosphate (pH 7.5), 5 mM $MgSO_4$, 3 μM tetraphenylphosphonium, and the membrane vesicles at 0.7 mg of protein per ml. The membrane vesicles were pretreated with 1 μM nigericin. After 5 min of incubation with increasing amounts of PQQ (0 to 12.5 μM), 20 mM glucose was added. After 2 min, when the $\Delta\psi$ was maximal, the uptake experiment was started by adding 200 μM ^{14}C-lactose to the membrane vesicle suspension. Symbols: △, electrical potential; ○, initial rate of lactose uptake; □, steady-state level of lactose accumulation (J. M. van Dijl, M. G. L. Elferink and K. J. Hellingwerf, unpublished data).

SH reagents is changed either by addition of substrates or by energization of the membrane or by both (for review, see reference 6). Such sulfhydryl reagent sensitivity has been reported for the phosphoenolpyruvate-dependent glucose transport system (7) and the lactose transport system of *E. coli* (2). These two transport systems operate by completely different mechanisms. The lactose transport system is a secondary transport system which translocates lactose in symport with protons. The phosphoenolpyruvate-dependent glucose transport system couples the transport of glucose to the hydrolysis of phosphoenolpyruvate via a series of phosphoryl group transfer reactions (23). Evidence has been presented that redox-sensitive dithiol groups in the membrane-bound enzyme II of the phosphoenolpyruvate-dependent glucose phosphate transport system are involved and that the redox state of these groups controls the activity of the enzyme (25).

The activity of the lactose carrier has also been shown to be subject to redox control of dithiol-disulfide groups (18), as is indicated by the following experiments. The accumulation of L-proline and lactose in right-side-out membrane vesicles of *E. coli* can be energized by ascorbate-phenazine methosulfate, D-lactate, or an artificially imposed chlorate or potassium diffusion potential. This uptake can be inhibited to the uptake levels observed under nonenergized conditions by lipophilic oxidants such as plumbagin, phenazine methosulfate, menadione, and 1,2-naphthoquinone (18). Strong inhibitions of uptake were also observed with polar oxidants such as ferricyanide, provided that the membrane vesicles were preincubated for several minutes with these oxidants (18, 22). The inhibitions exerted by these oxidants could not be explained by a decreased flow of electrons to oxygen or a decreased proton motive force. In all cases the addition of dithiothreitol to the membrane vesicle suspension restored the transport activity fully (Fig. 3). These data demonstrate that the activity of the carriers for L-proline and lactose of *E. coli* can be regulated by altering the redox potential in the membrane in a

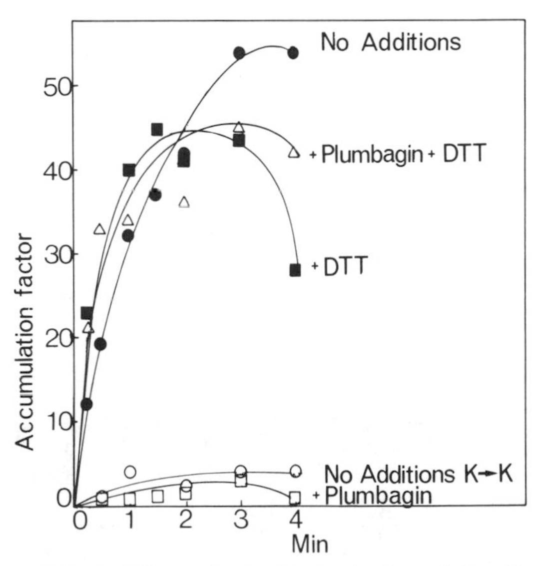

FIG. 3. Effects of 0.5 mM plumbagin and 10 mM dithiothreitol (DTT) on uptake of L-proline by membrane vesicles of *E. coli* energized by a valinomycin-induced potassium diffusion potential. Membrane vesicles (35 mg of protein per ml) in 50 mM potassium phosphate (pH 8) were incubated with 10 μM valinomycin for 30 min and with the redox reagents for 5 min each prior to 100-fold dilution into 50 mM Tris maleate (pH 8) containing 10 mM $MgSO_4$ and 3.5 μM L-[^{14}C]proline or 50 mM potassium phosphate (pH 8) containing 10 mM $MgSO_4$ and 3.5 μM L-[^{14}C]proline ($K^+ \rightarrow K^+$). Experiments were performed at room temperature. Symbols: ●, no additions, with a potassium gradient; △, plumbagin plus DTT, with a potassium gradient, incubated with 0.5 mM plumbagin and then with 10 mM DTT; ■, DTT, with a potassium gradient, incubated with 10 mM DTT; ○, no additions $K^+ \rightarrow K^+$, no potassium gradient; □, + plumbagin, with a potassium gradient, incubated with 0.5 mM plumbagin. (From reference 18 with permission.)

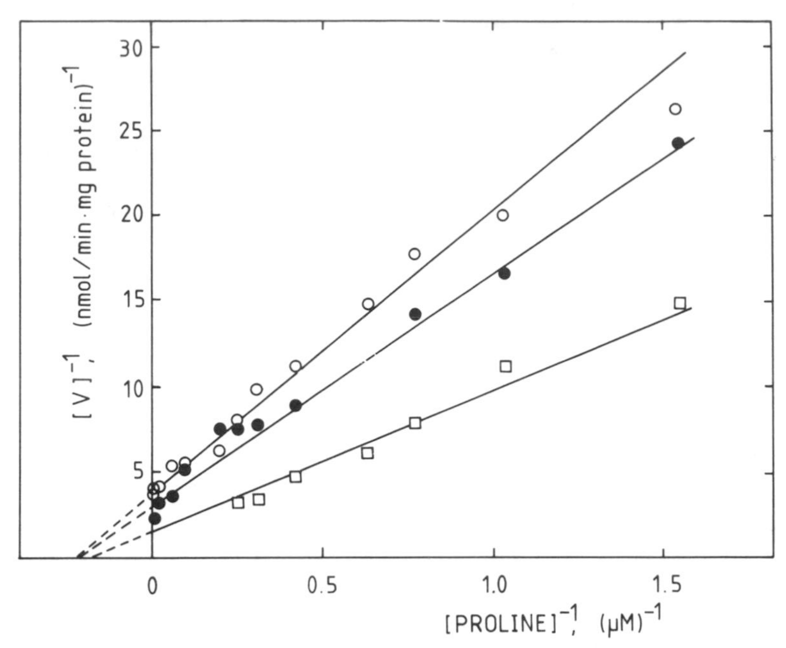

FIG. 4. Effect of plumbagin on the kinetics of proline transport in membrane vesicles of *E. coli* ML 308-225. Membrane vesicles were incubated in the presence of 0 (□), 50 (●), or 100 (○) μM plumbagin for 30 min on ice prior to the uptake assay. Proline transport was driven by a chlorate diffusion potential at pH 8.0 as described in reference 22.

way similar to that demonstrated for the phosphoenolpyruvate-dependent glucose transport system. Kinetic studies of L-proline transport energized by a chlorate diffusion potential in the presence of different concentrations of plumbagin demonstrate that oxidation of the carriers leads to a decrease of the V_{max} and thus to inactivated carriers (Fig. 4). A similar conclusion has been drawn from studies on the binding of galactosides to the *lac* carrier of *E. coli* in the presence of plumbagin (21).

The redox process involves the conversion of dithiols to disulfide or vice versa in the carrier proteins. In the proline carrier two sets of dithiol/disulfides appear to play a role: one set located at the outer surface and the other at the inner surface of the cytoplasmic membrane (22). This conclusion is based on the following observations. In right-side-out membrane vesicles, electron transfer in the respirator chain leads to the generation of a proton motive force (inside, negative, and alkaline) and to the conversion at the outer surface of a disulfide to a dithiol, as is shown by the increased inhibition of proline transport by the membrane-impermeable dithiol reagent thorin (22). The inhibition exerted by thorin was completely reversed by dithiothreitol. A similar but irreversible inhibition was observed with the membrane-impermeable SH reagent glutathione hexane maleimide (22). Pretreatment of the membrane vesicles with ferricyanide or with thorin protected against glutathione hexane maleimide inhibition since the transport activity of L-proline was fully restored after dithiothreitol treatment. Both SH reagents therefore appear to react with the same SH groups of a dithiol located at the outer surface.

Evidence for the involvement of dithiols located at the inner surface of the cytoplasmic membrane was obtained from studies with inside-out membrane vesicles. These inside-out membrane vesicles accumulate L-proline if a chlorate diffusion potential (inside negative) is imposed. Ferricyanide reversibly inhibits this proline uptake, and this appears to be the result of oxidation of SH groups since ferricyanide also protects against the irreversible inhibition by glutathione hexane maleimide. A similar protection against glutathione hexane maleimide inactivation in these inside-out membrane vesicles could be achieved by imposition of a proton motive force by electron flow, inside positive and acid, indicating that a dithiol is converted to a disulfide upon energization. These results strongly indicate that two redox-sensitive dithiol groups play a role, at least in the carrier of proline.

The redox control of the transport proteins combined with the control exerted by electron transfer suggests the existence of a redox interaction of the transport proteins with components of the electron transfer system. Such an interaction could possibly operate as shown schematically in Fig. 5. According to this scheme electrons can be transferred from a yet unknown electron transfer intermediate to a redox center located at the outer surface, thereby reducing this redox center (Fig. 5, top). This reduction causes an opening of the carrier and an exposure of the solute binding site to the outer surface. When solute and proton(s) are bound, electron transfer from the outer redox center to the inner redox center occurs which leads to an oxidation of the center at the outer surface and a reduction of the center at the inner surface (Fig. 5, bottom). This results in a closing of the carrier at the outer surface and an exposure of the binding site to the inner surface. When solute and proton(s) are released in the cytoplasm, electrons are transferred from the inner redox center to a second unknown intermediate of the electron transfer system with redox potential higher than the first intermediate. In solute transport systems which are not regulated by electron transfer, such redox transitions require intramolecular electron transfer between the redox centers only (see also reference 26). The redox sensitivity of the carrier proteins suggests that in intact cells a redox control coupled to glutathione or thioredoxine is very likely to occur.

The results presented above for secondary transport systems reveal that a complex set of factors influences the activity of these systems. In the following section some properties of ATP-dependent transport systems are described which indicate that these systems are even more influenced by various parameters.

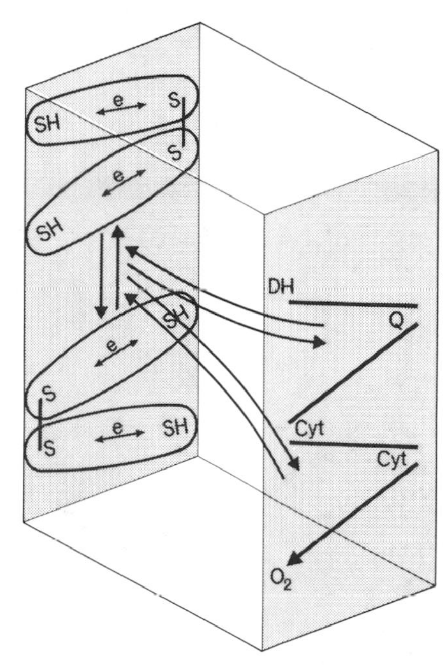

FIG. 5. Hypothetical scheme of the redox interaction between the respiratory chain and a carrier protein as described in text. DH, Membrane-bound dehydrogenase; Q, quinone pool; Cyt, cytochromes. The carrier is at the outside in the dithiol form at the top of the figure and in the disulfide form at the bottom of the figure.

Role of ATP in Solute Transport

Glycolyzing cells of *S. cremoris* and *S. lactis* can accumulate anions such as glutamate and phosphate to concentration gradients that exceed the thermodynamic equilibrium level with the proton motive force (assuming an H^+/solute stoichiometry of 1). Furthermore, transport of glutamate and phosphate is essentially unidirectional. Efflux of these anions from starved cells is extremely slow; the $t_{1/2}$ values of efflux at pH 6 are greater than 4 h. These observations, which are not compatible with a reversible secondary transport mechanism, prompted us to investigate the energetics of these systems in more detail.

To elucidate the mechanism of energy coupling to the transport systems, the initial rate of glutamate and phosphate uptake was measured between pH 5 and 8. At pH 5 the proton motive force of *S. lactis* cells was largely composed of a $Z\Delta pH$ (Fig. 6). With increasing external pH, the $Z\Delta pH$ decreased as the internal pH remained essentially constant, i.e., between 7.2 and 7.4. The $\Delta\psi$ component of the proton motive force increased, whereas the intracellular ATP concentration was not significantly affected by the increase in the extracellular pH. At a glutamate concentration of 100 μM the initial rates of glutamate uptake decreased with increasing external pH from pH 5 to 7.8. However, when the initial rates of glutamate uptake were measured under V_{max} conditions (see below), hardly any variation of the V_{max} was observed at different external pH values, despite large changes in the magnitude and composition of the components of the proton motive force (Table 1). Glutamate transport also appeared to be fully insensitive to dissipation of the membrane potential by the ionophore valinomycin, indicating that glutamate is translocated across the cytoplasmic membrane independent of the proton motive force.

Instead of varying the $Z\Delta pH$ by changing the external pH, the magnitude of $Z\Delta pH$ can be altered by titration with nigericin at a fixed extracellular pH value. Under these conditions the internal pH varies at constant external pH. Simultaneous recording of the $Z\Delta pH$ and the rate of glutamate uptake under these experimental conditions showed that glutamate transport was heavily dependent on the absolute value of the internal pH. The rate of glutamate uptake

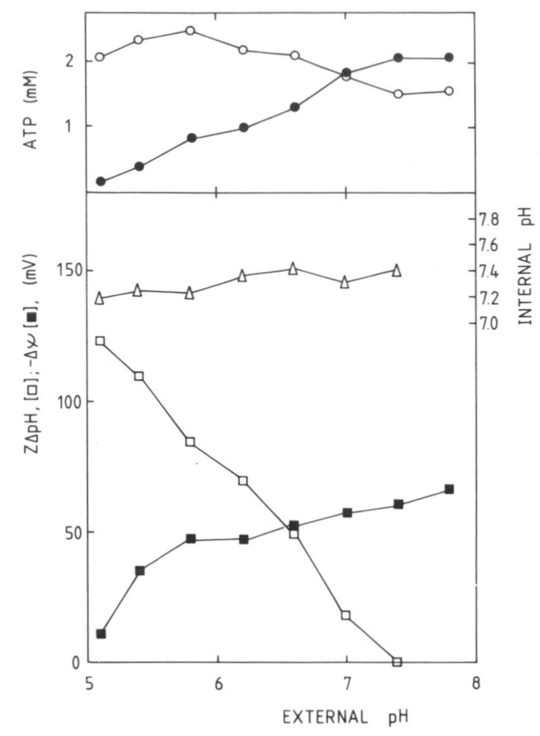

FIG. 6. Effect of external pH on the components of the proton motive force, the internal pH, and the intracellular ATP concentration in *S. lactis* ML3. Cells were suspended in 30 mM K^+-morpholineethane-sulfonic acid–30 mM K^+-PIPES [piperazine-*N,N'*-bis(2-ethanesulfonic acid)]–30 mM K^+-HEPES (*N*-2- hydroxyethylpiperazine-*N'*-2-ethanesulfonic acid)–5 mM $MgCl_2$ at the indicated pH at 30°C to a final protein concentration of 2.1 mg/ml. The $\Delta\psi$ (■) and the $Z\Delta pH$ (□) were determined from the distribution of tetraphenylphosphonium (4 μM) and salicylate (100 μM), respectively, using ion-selective electrodes. Samples (100 μl) were withdrawn for the determination of the intracellular ATP concentration 5 min after the addition of 10 mM lactose. Following extraction, ATP was analyzed by the luciferin/luciferase asay. Internal ATP concentrations were determined in the absence (○) and presence (●) of 1.0 μM nigericin; △, internal pH. (B. Poolman, unpublished data.)

TABLE 1. Effect of external pH on the kinetic parameters of glutamate transport in *S. lactis* ML3[a]

pH	K_m (μM)	V_{max} (nmol/mg of protein·min)
5.1	11.2	20.2
6.0	77	23.0
7.0	>500	21.0

[a] Cells were suspended to a final protein concentration of 1.0 mg/ml as described in the legend to Fig. 6. [U-^{14}C]Glutamate was added 3 min after the addition of lactose to final concentrations ranging from 1.8 μM to 1.0 mM. Initial rates of glutamate uptake were determined after 10 s by separating the cells from the external medium by filtration (B. Poolman, unpublished data).

was maximal at an internal pH of 7.4, whereas it was nearly completely inhibited at pH 6.0.

Attempts to induce glutamate uptake in starved cells in response to an artificially induced pH gradient were repeatedly unsuccessful (Fig. 7). Treatment of streptococci with the Ca^{2+}, Mg^{2+}-ATPase inhibitor *N,N'*-dicyclohexylcarbodiimide enables glycolyzing cells to synthesize ATP without effectively generating a proton motive force. Under these conditions and at an external pH of 6.0, a slow rate of glutamate uptake was observed (Fig. 7). However, a high rate of glutamate uptake was observed upon the imposition of an outwardly directed acetate diffusion potential, resulting in the transient increase of the internal pH, together with metabolism of lactose in cells treated with *N,N'*-dicyclohexylcarbodiimide (Fig. 7). The initial lag phase of transport reflects the lag in ATP synthesis. Autoradiography of extracted cell material revealed that glutamate was not metabolized

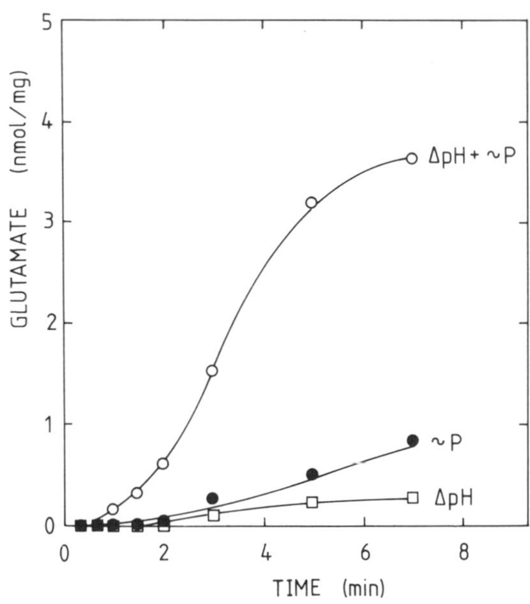

FIG. 7. Accumulation of glutamate in response to an artificial pH gradient. *S. lactis* ML3 cells, suspended in 100 mM K^+-PIPES (pH 6.0)–5 mM $MgSO_4$, were treated for 30 min at room temperature with 200 μM *N,N'*-dicyclohexylcarbodiimide. Subsequently, the cells were washed and suspended in 20 mM K^+-PIPES–100 mM K^+-acetate–5 mM $MgSO_4$ (pH 6.0), to a final protein concentration of approximately 70 mg/ml. Uptake of 42 μM [U-^{14}C]glutamate was initiated by diluting the cells 100-fold into the same buffer containing 5 mM lactose (●, ~P) or into buffer in which acetate was replaced by gluconate and in the presence (○, ΔpH + ~P) or absence (□, ΔpH) of 5 mM lactose. (B. Poolman, unpublished data.)

during the course of these experiments. These results indicate that glutamate transport is driven by ATP or another phosphorylated intermediate and regulated by the intracellular pH. The results obtained for glutamate transport in *S. lactis* and also *S. cremoris* differ with respect to the regulation by the cytoplasmic pH from those in *S. faecalis* (11). In *S. faecalis* uptake of glutamate and aspartate is hardly affected by *N,N'*-dicyclohexylcarbodiimide, nigericin, or uncouplers that dissipate the components of the proton motive force at both pH 6.2 and pH 7.5. It was concluded that glutamate and aspartate uptake in *S. faecalis* does not require a proton motive force but that transport of these amino acids is driven by ATP directly. Another difference between *S. cremoris, S. lactis*, and *S. faecalis* with respect to transport of these anions is related to the substrate specificity of the transport systems. In *S. faecalis* glutamate and aspartate appear to be taken up by the same system(s) (24), whereas glutamate uptake in *S. cremoris* and *S. lactis* is catalyzed at pH 6 by a single kinetically distinguishable transport system with a high affinity for glutamine and very low affinity for aspartate. Interestingly, the affinity constant (K_m) for glutamate increased severely with increasing pH (Table 1), whereas the K_m for glutamine was approximately 1.5 μM at pH 5.1, 6.0, and 7.0. This indicates that the protonation state of the γ-carboxyl group of glutamate ($pK_a = 4.25$) strongly affects the binding of the substrate to the transport system. In fact, at pH 4.0 the K_m for glutamate uptake was 3.5 μM, very similar to that of glutamine transport.

The uptake of phosphate was studied under similar conditions (Fig. 8). The initial rate of phosphate uptake was maximal at pH 5.8 and above pH 7.8, suggesting that in *S. lactis* (as in *S. faecalis* [8, 9]) two distinct but overlapping transport systems for phosphate are present. In the presence of nigericin a virtually complete inhibition of phosphate transport was observed between pH 5 and 6, whereas above pH 7 the rate of uptake was hardly affected. In *S. lactis* phosphate transport was not inhibited by valinomycin at any pH value. Therefore, at alkaline pH transport of phosphate is independent of the proton motive force. Possibly ATP or another energy-rich phosphorylated compound is involved in the energization of phosphate transport in a way similar to that proposed for *S. faecalis* (11).

The decrease of this transport activity below pH 7 in the presence of nigericin might then reflect the internal pH sensitivity of the system(s) (see also reference 11) or might be caused by the drop in the intracellular ATP concentration (Fig. 6).

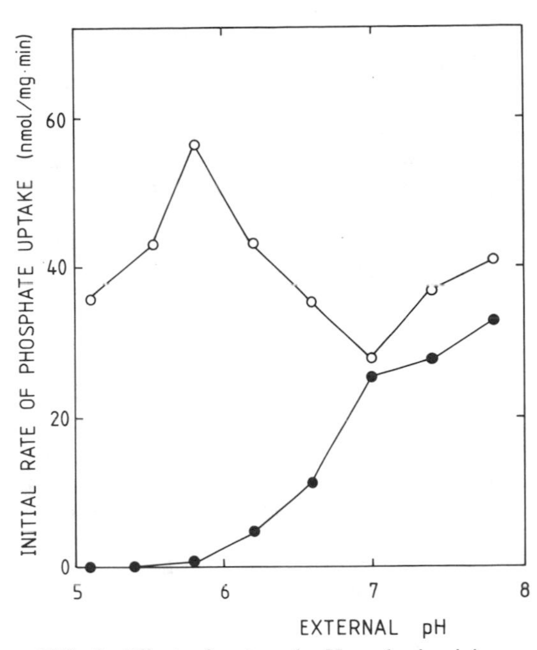

FIG. 8. Effect of external pH and nigericin on phosphate transport in *S. lactis* ML3. Conditions were as described in the footnote to Table 1 except that 100 μM $^{32}P_i$ was added instead of [U-^{14}C]glutamate. Uptake was determined in the absence (○) and presence (●) of 1.0 μM nigericin. (B. Poolman, unpublished data.)

Arginine is an excellent source of energy for ATP synthesis in cells of *S. lactis* induced for the arginine dihydrolase pathway (3). At pH 6 arginine-dependent uptake of phosphate by *S. lactis* was not observed despite the presence of a ZΔpH of approximately 60 mV. At pH 7.4 the rate of phosphate uptake in the presence of arginine was 30 to 40% of the rate in the presence of a glycolytic substrate. This inability of arginine to support transport of phosphate (or its analog arsenate) has also been found in *S. faecalis* and has been explained by the existence of a large cytoplasmic pool of P_i maintained in the presence of arginine (11) which exerts a negative feedback control on phosphate uptake.

Concluding Remarks

The results presented in this brief review show that many environmental and (intra)cellular factors modulate the properties of solute transport systems. The rate of solute transport by secondary transport systems is directly determined by the proton motive force. The composition and magnitude of this proton motive force is a complex function of the activities of primary transport systems, secondary transport systems, other processes driven by the proton motive force, external parameters (ionic composition, pH) and internal parameters (12). In previous studies we demonstrated that the number of protons which are translocated with a solute can vary with the external pH or external solute concentration, or both (15), showing that the role of the proton motive force as driving force in transport of one particular solute can vary.

The activities of the secondary transport systems appear to be affected by the activity of electron transfer systems and to be subject to redox control. Recently, the internal pH has been shown to control the activity of a proton motive force-driven amino acid transport system in *S. cremoris* (A. J. M. Driessen, unpublished data). The results presented for the ATP-driven transport systems for glutamate and phosphate in *S. lactis* show that both the internal and the external pH affect the activity of these systems. The differential effects of internal and external pH on transport make it difficult to discriminate between regulation and a role of ΔpH as a driving force of solute transport.

The known and unknown effects on solute transport systems limit the conclusions that can be drawn about the mechanism or energetics of solute transport from studies in intact cells. The advantages of model systems, such as membrane vesicles, for the studies of solute transport are obvious. Unfortunately, model systems for studies on ATP-driven solute transport have not yet been developed. This severely limits the information that can be obtained from the ATP-driven transport systems.

LITERATURE CITED

1. **Cohen, G. N., and J. Monod.** 1957. Bacterial permeases. Bacteriol. Rev. **21:**169–194.
2. **Cohn, D. E., G. J. Kaczorowski, and H. R. Kaback.** 1981. Effect of the proton electrochemical gradient on maleimide inactivation of active transport in *Escherichia coli* membrane vesicles. Biochemistry **20:**3308–3313.
3. **Crow, V. L., and T. D. Thomas.** 1982. Arginine metabolism in lactic streptococci. J. Bacteriol. **150:**1024–1032.
4. **Elferink, M. G. L., I. Friedberg, K. J. Hellingwerf, and W. N. Konings.** 1983. The role of the proton motive force and electron flow in light-driven solute transport in *Rhodopseudomonas sphaeroides*. Eur. J. Biochem. **129:**583–587.
5. **Elferink, M. G. L., K. J. Hellingwerf, F. E. Nano, S. Kaplan, and W. N. Konings.** 1983. The lactose carrier of *Escherichia coli* functionally incorporated in *Rhodopseudomonas sphaeroides* obeys the regulatory conditions of the phototrophic bacterium. FEBS Lett. **164:**185–190.
6. **Fonyo, A.** 1978. SH-group reagents as tools in the study of mitochondrial anion transport. J. Bioenerg. Biomembr. **10:**171–194.
7. **Haguenauer-Tsapis, R., and A. Kepes.** 1977. Unmasking of an essential thiol during function of the membrane bound enzyme II of the phosphoenolpyruvate glucose phosphotransferase system of *Escherichia coli*. Biochim. Biophys. Acta **465:**118–130.
8. **Harold, F. M., and J. R. Baarda.** 1966. Interaction of arsenate with phosphate transport systems in wild-type and mutant *Streptococcus faecalis*. J. Bacteriol. **91:**2257–2262.
9. **Harold, F. M., R. L. Harold, and A. Abrams.** 1965. A mutant of *Streptococcus faecalis* defective in phosphate uptake. J. Biol. Chem. **240:**3145–3153.
10. **Harold, F. M., and Y. Kakinuma.** 1985. Primary and secondary transport of cations in bacteria, p. 375–383. *In* G. Semenza and R. Kinne (ed.), Membrane transport driven by ion gradients. Ann. N. Y. Acad. Sci. **456:**375–383.
11. **Harold, F. M., and E. Spitz.** 1975. Accumulation of arsenate, phosphate, and aspartate by *Streptococcus faecalis*. J. Bacteriol. **122:**266–277.
12. **Hellingwerf, K. J., and W. N. Konings.** 1985. The energy flow in bacteria: the main free energy intermediates and their regulatory role. Adv. Microb. Physiol. **26:**125–154.
13. **Kaback, H. R., and L. S. Milner.** 1970. Relationship of a membrane bound D-(−)-lactic dehydrogenase to amino acid transport in isolated bacterial membrane preparations. Proc. Natl. Acad. Sci. USA **66:**1008–1015.
14. **Kepes, A.** 1970. Galactoside permease in *Escherichia coli*. Curr. Top. Membr. Transp. **1:**101–134.
15. **Konings, W. N.** 1985. Generation of metabolic energy by end-product efflux. Trends Biochem. Sci. **10:**317–319.
16. **Konings, W. N., K. J. Hellingwerf, and G. T. Robillard.** 1981. Transport across bacterial membranes, p. 257–283. *In* S. L. Bonting and J. J. H. H. M. de Pont (ed.), Membrane transport. Elsevier/North Holland Publishing Co., Amsterdam.
17. **Konings, W. N., and P. A. M. Michels.** 1980. Electron transfer driven solute accumulation across bacterial membranes, p. 33–86. *In* C. J. Knowles (ed.), Diversity of bacterial respiratory systems. CRC Press, Inc., Boca Raton, Fla.
18. **Konings, W. N., and G. T. Robillard.** 1982. The physical mechanism for regulation of protein solute transport in *Escherichia coli*. Proc. Natl. Acad. Sci. USA **79:**5480–5484.
19. **Kundig, W., S. Ghosh, and S. Roseman.** 1964. Phosphate bound to histidine in a protein as an intermediate in a novel phosphotransferase system. Proc. Natl. Acad. Sci. USA **52:**1067–1074.
20. **Lipman, F.** 1941. Metabolic generation and utilization of phosphate bond energy. Adv. Enzymol. **1:**99–162.
21. **Neuhaus, J. M., and J. K. Wright.** 1983. Chemical modification of the lactose carrier of *Escherichia coli* by plumbagin, phenylarsine oxide or diethylpyrocarbonate affects the binding of galactoside. Eur. J. Biochem. **127:**597–604.
22. **Poolman, B., W. N. Konings, and G. T. Robillard.** 1983. The location of redox-sensitive groups in the carrier protein of proline at the outer and inner surface of the membrane in *Escherichia coli*. Eur. J. Biochem. **135:**41–46.
23. **Postma, P. W., and S. Roseman.** 1976. The bacterial phosphoenol pyruvate:sugar phosphotransferase system. Biochim. Biophys. Acta **457:**213–257.
24. **Reid, K. G., N. M. Utech, and J. T. Holden.** 1970. Multiple transport components for dicarboxylic acids in *Streptococcus faecalis*. J. Biol. Chem. **245:**5261–5272.
25. **Robillard, G. T., and W. N. Konings.** 1981. Physical mechanism for regulation of phosphoenol pyruvate-dependent glucose transport activity in *Escherichia coli*. Biochemistry **20:**5025–5032.
26. **Robillard, G. T., and W. N. Konings.** 1982. The role of dithiol-disulphide interchange in solute transport and energy transduction processes. Eur. J. Biochem. **127:**597–604.
27. **van Schie, B. J., K. J. Hellingwerf, J. P. van Dijken, M. G. L. Elferink, J. M. van Dijl, J. G. Kuenen, and W. N. Konings.** 1985. Energy transduction by electron transfer via a pyrrolo-quinoline quinone-dependent glucose dehydrogenase in *Escherichia coli, Pseudomonas aeruginosa*, and *Acinetobacter calcoaceticus* (var. *lwoffi*). J. Bacteriol. **163:**493–499.
28. **Williams, R. J. P.** 1978. The history and the hypothesis concerning ATP-formation by energized protons. FEBS Lett. **85:**9–19.
29. **Wilson, D. B.** 1978. Cellular transport mechanisms. Annu. Rev. Biochem. **47:**933–965.

Phosphate Uptake in *Saccharomyces cerevisiae*

G. W. F. H. BORST-PAUWELS AND P. H. J. PETERS

Laboratory of Chemical Cytology, 6525 ED Nijmegen, The Netherlands

Phosphate uptake in *Saccharomyces cerevisiae* cells is a very complicated process. There are three different ways by which phosphate can enter into the cells. Goodman and Rothstein (8) have shown that phosphate can be translocated by means of a low-affinity transport system with a K_m of approximately 1 mM. Besides this low-affinity process, a high-affinity transport process with a K_m of 1 to 15 μM can also be operative in the cells (2, 9). This system becomes derepressed when the cells are incubated in the absence of phosphate with a suitable metabolizable substrate (9). It consists of a proton-phosphate cotransport system (7) which cotransports two or three protons with one monovalent anion of phosphate. The third system involved in phosphate transport consists of a Na^+-phosphate cotransport system (6, 10) which also becomes derepressed on incubating the cells in the absence of phosphate.

This article deals with the effects of cell pH upon phosphate transport by means of both the high-affinity proton-phosphate cotransport system and the low-affinity transport system. Effects of changes in the surface potential upon the rate of phosphate uptake by means of both the proton-phosphate and the Na^+-phosphate cotransport system are discussed (13).

Effect of External and Cellular pH upon Phosphate Uptake

The maximal rate of phosphate uptake by means of the high-affinity proton-phosphate cotransport system depends not only upon the pH of the medium (5), but also upon the type of buffer applied. At low and medium pH, phosphate uptake is low and increases with increasing pH to a maximum value. On increasing the pH of the medium further, the maximal rate of uptake decreases again. The pH at which the maximal rate has an optimal value is far lower in Tris citrate buffer than in Tris succinate (see Table 1). Furthermore, the optimum pH depends upon the metabolic conditions. On shifting from anaerobic conditions to aerobic conditions and replacing glucose by ethanol, the optimum pH is increased. It appears, however, that the effects of the medium pH upon the maximal rate of uptake are mainly due to changes in cell pH accompanying changes in external pH. As a matter of fact, a single relationship exists between the maximal rate of phosphate uptake and the cell pH irrespective of the buffer system applied (see also Table 1).

The view that the cell pH is the main factor determining the maximal rate of phosphate uptake is further supported by the fact that decreasing the cell pH by incubating the cells at a fixed external pH with various amounts of butyrate, thereby acidifying the cells, leads to a similar dependence of the maximal rate of uptake upon the cell pH, as is found on varying the external pH (5).

The dependence of the K_m upon the external pH is still more complicated than that of the maximal rate of uptake (5). At low pH the K_m shows a dependence upon the external pH similar to that found for the maximal rate of phosphate uptake. This effect may be ascribed to a change in cell pH. Accordingly, decreasing the cell pH at fixed medium pH leads to a change in K_m. At external pH values above 5.5, however, the K_m increases, whereas the maximal rate of uptake decreases. It appears that two effects are involved, namely, an effect of the cell pH which, because the cell pH is higher than the optimal pH 6.86, leads to a simultaneous decrease in both K_m and maximal rate of uptake, and an effect of the external pH which leads to an increase in K_m. The latter effect may be ascribed to either a competitive inhibition of hydroxyl anions with phosphate for the binding sites on the phosphate carrier or to a change in the apparent affinity of phosphate for the carrier due to the fact that the concentration of the co-ions, the protons, is decreased. On correcting the K_m for the apparent competitive effect of hydroxyl anions of the medium, a relationship between the K_m and the cell pH is found that is similar to that for the maximal rate of phosphate uptake and the cell pH (see Fig. 1). As a matter of fact, a linear relationship exists between the K_m and the maximal rate of uptake according to the equation (4, 5): $K_m = K_m' + qV_{max}$. Neither K_m' nor q depends upon the cell pH. On the other hand, the coefficient q still depends upon the medium pH and increases with increasing medium pH.

The dependence of the K_m and the maximal rate of uptake upon the cell pH can be explained if the transport of phosphate through the yeast cell membrane is mediated by a so-called mobile carrier, i.e., by a carrier which can move from one side of the plasma membrane to the other. In

TABLE 1. Optimal pH for phosphate uptake[a]

Buffer applied	Medium pH	Cell pH
100 mM Tris citrate + 10 mM KCl + 3% glucose	4.5	6.86
100 mM Tris succinate + 3% glucose	5.2	6.86
100 mM Tris succinate + 1% ethanol	5.6	6.86

[a] External pH and cellular pH at which the rate of phosphate uptake by means of the high-affinity proton phosphate cotransport system is optimal (5). Experiments with ethanol were carried out with air in the gas phase, and those with glucose were carried out with N_2 in the gas phase. The cell pH was determined as described previously (3). The cells were washed, frozen in liquid nitrogen, and thawed in a small amount of water; the cells were then boiled. The pH was determined by means of a glass electrode in the yeast suspension after cooling to room temperature. Therefore, the pH is an average of the contributions of the various compartments of the cell.

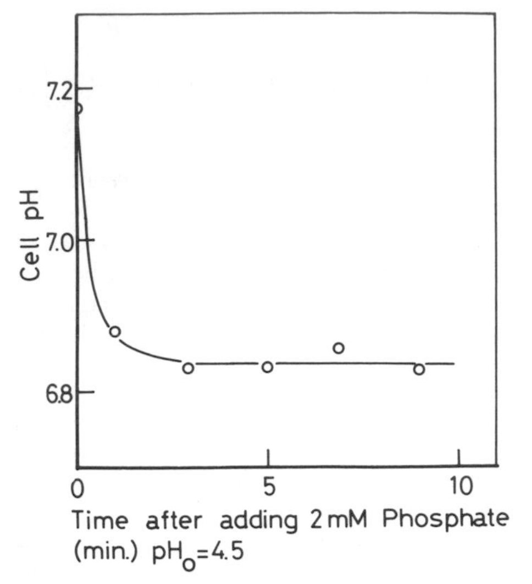

FIG. 2. Acidification of metabolizing yeast cells during uptake of 0.1 mM phosphate by means of the high-affinity proton-phosphate cotransport system at pH 4.5.

such a case the kinetic constants depend upon the concentrations of intracellular solutes which show affinity to the carrier (1, 14). It should be noted that it is not necessary for the whole carrier to be moving through the cell membrane. Transport kinetics of the mobile carrier type will also apply if only the phosphate-binding groups on the carrier can switch from an orientation toward the inner side to an orientation toward the external side of the cell membrane. Since the high-affinity transport system consists of a proton-phosphate cotransport system, protons are also substrates of the carrier, and changes in cell pH are expected to affect the kinetic constants of the transport process. For an apparently mobile transport system with a single binding site for the substrate ion, in this case phosphate, and one or more binding sites for a co-ion (protons), the K_m assigned to the substrate ion will be related to the maximal rate of uptake by an equation of the form of the experimentally found relation given above (1). This supports our view that an apparently mobile carrier is involved.

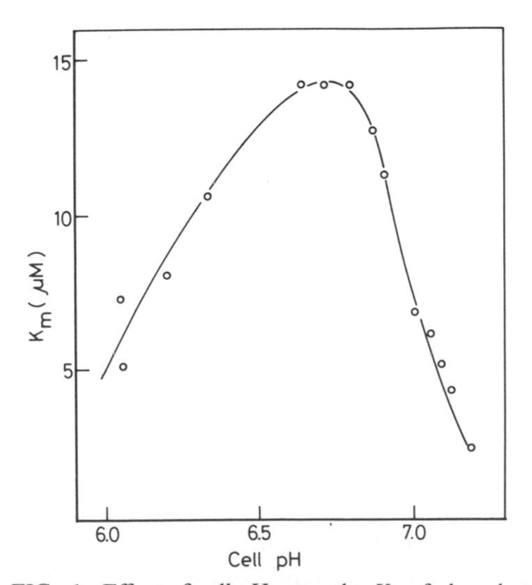

FIG. 1. Effect of cell pH upon the K_m of phosphate uptake mediated by the high-affinity proton-phosphate cotransport system. The K_m values are referred to monovalent phosphate and are corrected for the effect of external OH^- or H^+ as described previously (5).

Phosphate uptake leads to uptake of protons into the cells (7). Figure 2 shows that the cells can be considerably acidified during uptake of phosphate by means of the proton-phosphate cotransport system. Since the rate of phosphate uptake depends upon the cell pH, one would expect that during phosphate uptake the influx rate will change. This is especially important at high medium pH, when the cell pH is much higher than the optimum pH for phosphate uptake. Figure 3 shows that at an external pH of 7.0 the uptake of phosphate into metabolizing cells increases initially more than proportionally with the time. On preincubating the cells with butyrate, the cells are acidified and the cell pH approaches the optimum pH. Figure 4 shows that in this case the rate of phosphate uptake does not increase during uptake of phosphate. Furthermore, Fig. 4 shows that a measurable increase in the rate of phosphate uptake has already occurred within 20 s. In studies of the kinetics of phosphate transport in *S. cerevisiae* it is therefore very important to take the cell

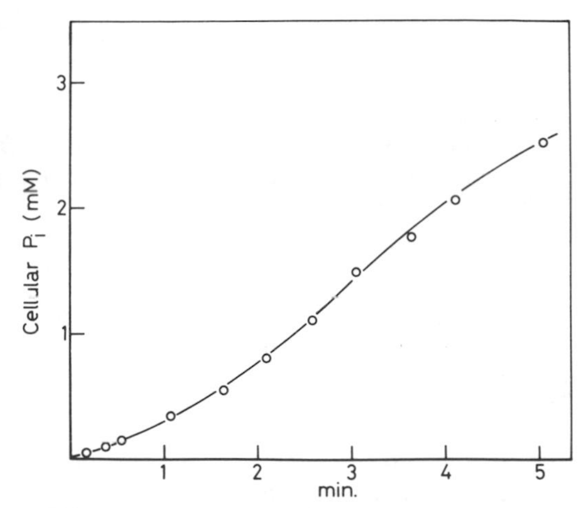

FIG. 3. Time course of uptake of phosphate at pH 7.0 by means of the high-affinity proton-phosphate cotransport system. The initial extracellular concentration was 0.1 mM.

samples as soon as possible to be sure that changes in cell pH caused by phosphate uptake into the cells have not affected the influx rate. If one is not aware of the effects of changes in cell pH upon the influx rate of phosphate, serious mistakes may be made in the interpretation of the experimental results.

The low-affinity phosphate uptake system also depends upon the cell pH. The maximal rate of phosphate uptake is decreased on acidifying the cells. At an external pH of 4.5, changing the cell pH from 6.76 to 6.39 causes a 50% decrease in V_{max}. The K_m, on the other hand, remains virtually constant (1.65 mM) under these conditions

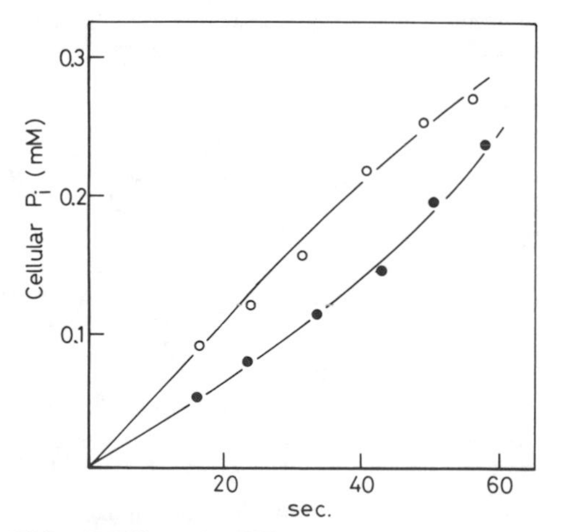

FIG. 4. Effect of acidification of the cells upon the time course of phosphate uptake at pH 6.5. Symbols: ○, 6 min prior to 0.2 mM phosphate, 50 mM butyrate was added, by which the cells were acidified; ●, control without butyrate.

(B. J. W. M. Nieuwenhuis, Thesis, Nijmegen, The Netherlands, 1983).

In studies of the repression or derepression of the high-affinity systems and the contribution of the low-affinity system to phosphate transport, one should be aware that the observed changes in the translocation rate of phosphate by these systems are not necessarily due to changes in carrier concentration, but may be due to changes in cell pH. If one is not accounting for these indirect effects, serious mistakes may be made in the interpretation of the experimental results.

Na^+-Phosphate Cotransport System

Besides the low-affinity system for phosphate uptake and the high-affinity proton-phosphate cotransport system, a Na^+-phosphate cotransport system is also operative in yeast cells. Phosphate uptake at pH 7 is increased greatly by Na^+ (6, 10). At pH 7.2 and in the presence of Na^+, a nonlinear Hofstee plot (v against v/s) is found which is typical for the involvement of a dual mechanism consisting of two independent, simultaneously operating transport systems. One system has a K_m, referring to monovalent phosphate, of 30 μM, and the other has a K_m of 0.6 μM. The maximal rate of the second process is increased by increasing the Na^+ concentration, whereas the maximum rate of the first process is not affected at all by Na^+. The first process should be ascribed to proton-phosphate cotransport, and the second, to Na^+-phosphate cotransport. The occurrence of a dual mechanism excludes the possibility that the enhancement of phosphate uptake by Na^+ is due to occupation of the proton binding sites of the proton-phosphate cotransport carrier by Na^+ ions. Also, Na^+ uptake is increased by phosphate, whereas Rb^+ uptake is decreased (as a result of the depolarization caused by proton-phosphate cotransport [11]). On raising the external pH from 7 to 9, the relative contribution of the Na^+-phosphate cotransport system to the total phosphate uptake increases, as shown in Table 2. This is due to both a decrease in the proton-phosphate and an increase in the Na^+-phosphate cotransport with increasing pH. The cell pH remains constant despite the changes in external pH.

Effect of the Surface Potential

To describe the kinetics of phosphate transport properly, one also has to account for the effect of the surface potential as well as of external pH and cell pH upon phosphate uptake. A proper description of ion transport kinetics needs replacement of the bulk aqueous phase concentrations S by the interfacial concentrations S' near the plasma membrane (15). These

TABLE 2. Effects of medium pH on [^{32}P]phosphate transport[a]

Medium pH	Cell pH	Na-P (%)	H-P (%)
7.1	6.75	50	50
7.6	6.75	98	12
8.2	6.74	140	7
8.8	6.75	172	16

[a] Effect of the pH of the medium upon carrier-free [^{32}P]phosphate transport by both proton-phosphate cotransport and Na^+-phosphate cotransport (Nieuwenhuis, thesis, 1983). The uptake rates are expressed as percentages of the total phosphate uptake mediated by the two transport systems at pH 7.1. We have accounted for decreases in the relative contribution of monovalent phosphate to the total phosphate with increasing medium pH. Na-P means Na^+-phosphate cotransport and H-P means proton-phosphate cotransport.

interfacial concentrations normally differ from those of the bulk aqueous phase, as a result of the fact that plasma membranes of most living cells bear a net negative charge which gives rise to a negative surface potential ψ near the plasma membranes. S' is related to S by the Boltzmann equation: $S' = S \exp(-zF\psi/RT)$, where z is the valency of the ion concerned, and F, R, and T have their usual meanings. For cations the interfacial concentration is increased and for anions the interfacial concentration is decreased. Since phosphate uptake by the two derepressible systems consists of a cotransport of the phosphate anion and one to three cations, rather complicated effects of changes in the surface potential may occur (12). A reduction of the surface potential will lead to an increase in interfacial phosphate concentration and concomitantly to a decrease in proton or Na^+ interfacial concentration. Whether phosphate uptake will be enhanced or decreased will depend upon the experimental conditions. For example, at high pH, where the proton concentration may be limiting for proton-phosphate cotransport, a reduction in the surface potential may lead to a decrease in the rate of phosphate uptake despite the fact that the interfacial phosphate concentration will increase. On the other hand, at low pH a reduction in the negative surface potential will mainly lead to an increase in phosphate uptake because the proton concentration is no longer limiting. As shown in Table 3, reduction of the surface potential by addition of Ca^{2+} at pH 7.2 leads to a decrease in phosphate transport by means of the proton-phosphate cotransport, whereas Na^+-phosphate cotransport is increased under these conditions. On the other hand Ca^{2+} increases the rate of phosphate uptake via the proton-phosphate cotransport system at pH 4.5.

The transport of Na^+ by means of the Na^+-phosphate cotransport system is influenced by changes in the surface potential. This uptake is reduced on increasing the Ca^{2+} concentration (13).

TABLE 3. Effects of Ca^{2+} on [^{32}P]phosphate uptake[a]

Medium pH	Mechanism	Phosphate uptake (%)	
		4 mM Ca^{2+}	25 mM Ca^{2+}
4.5	H-P	260	320
7.2	H-P	85	70
7.2	Na-P	380	850

[a] Effect of Ca^{2+} at concentrations which cause a reduction in the surface potential upon carrier-free [^{32}P]phosphate uptake under different experimental conditions (13). Corrections for complex formation between phosphate and Ca^{2+} have been made. H-P means proton-phosphate cotransport and Na-P means Na^+-phosphate cotransport. Uptake with no Ca^{2+} is taken as 100%.

Conclusions

In this article we have reviewed the current knowledge of the kinetics of phosphate uptake in *S. cerevisiae*. The kinetics of phosphate uptake are rather complicated for several reasons. (i) Three transport systems can be simultaneously operative. (ii) Uptake depends on cell pH; this dependence is not always recognized, but is important and must be taken into account in comparing kinetic parameters of phosphate uptake from data obtained under different conditions. (iii) The interfacial concentrations of both phosphate and cotransported cations (protons or Na^+) depend upon the surface potential of the yeast cell. The effects of polyvalent cations upon phosphate uptake may be traced to changes in the surface potential rather than to a direct interaction of these cations with the phosphate translocating system. Theoretically, carrier-mediated cotransport of phosphate and protons depends not only upon the surface potential and cell pH but also upon the electrical potential across the plasmalemma. It may be worthwhile to examine the effect of changes in the membrane potential upon the kinetics of phosphate uptake.

LITERATURE CITED

1. **Borst-Pauwels, G. W. F. H.** 1974. Multi-site carrier transport comparison with enzyme kinetics. J. Theor. Biol. **48:** 183–195.
2. **Borst-Pauwels, G. W. F. H.** 1981. Ion transport in yeast. Biochim. Biophys. Acta **650:**88–127.
3. **Borst-Pauwels, G. W. F. H., and J. Dobbelmann.** 1972. Determination of the yeast cell pH. Acta Bot. Neerl. **21:** 149–154.
4. **Borst-Pauwels, G. W. F. H., and S. Jager.** 1969. Inhibition of phosphate and arsenate uptake in yeast by

monoiodoacetate, fluoride, 2,4-dinitrophenol and acetate. Biochim. Biophys. Acta **172:**399–406.

5. **Borst-Pauwels, G. W. F. H., and P. H. J. Peters.** 1977. Effect of the medium pH and the cell pH upon the kinetical parameters of phosphate uptake by yeast. Biochim. Biophys. Acta **466:**488–495.
6. **Borst-Pauwels, G. W. F. H., A. P. R. Theuvenet, and P. H. J. Peters.** 1975. Uptake by yeast: interaction of Rb^+, Na^+ and phosphate. Physiol. Plant. **33:**8–12.
7. **Cockburn, M., P. Earnshaw, and A. A. Eddy.** 1975. The stoicheiometry of the absorption of protons with phosphate and L-glutamate by yeasts of the genus Saccharomyces. Biochem. J. **146:**705–712.
8. **Goodman, J., and A. Rothstein.** 1957. The active transport of phosphate into the yeast cell. J. Gen. Physiol. **40:**915–923.
9. **Leggett, J. E.** 1961. Entry of phosphate into yeast cells. Plant Physiol. **36:**277–284.
10. **Roomans, G. M., F. Blasco, and G. W. F. H. Borst-Pauwels.** 1977. Cotransport of phosphate and sodium by yeast. Biochim. Biophys. Acta **467:**65–71.
11. **Roomans, G. M., and G. W. F. H. Borst-Pauwels.** 1977. Interaction of phosphate with monovalent cation uptake in yeast. Biochim. Biophys. Acta **470:**84–91.
12. **Roomans, G. M., and G. W. F. H. Borst-Pauwels.** 1978. Co-transport of anions and neutral solutes with cations across charged biological membranes. Effect of surface potential on uptake kinetics. J. Theor. Biol. **73:**453–468.
13. **Roomans, G. M., and G. W. F. H. Borst-Pauwels.** 1979. Interaction of cations with phosphate uptake by Saccharomyces cerevisiae. Effects of surface potential. Biochem. J. **178:**521–527.
14. **Rosenberg, T., and W. Wilbrandt.** 1963. Carrier transport uphill. J. Theor. Biol. **5:**288–305.
15. **Theuvenet, A. P. R., and G. W. F. H. Borst-Pauwels.** 1976. The influence of surface charge on the kinetics of ion-translocation across biological membranes. J. Theor. Biol. **57:**313–329.

VII. PHOSPHATE RESERVES AND ENERGY STORAGE: POLYPHOSPHATES

Introduction

IGOR S. KULAEV

Institute of Biochemistry and Physiology of Microorganisms, USSR Academy of Sciences, Puschino, Moscow Region, 142292, USSR

Almost a century has passed since Liebermann (9) found high-polymer polyphosphates in yeasts. But it was only after the fundamental studies by Wiame (13), Ebel (2), Kornberg et al. (5), Belozersky (1), Lohmann (10), and others in the late 1940s through the 1950s that biochemists turned their attention to these compounds. That stage was summarized at the symposium in Strasbourg in 1961 (3), where the important role of polyphosphates in phosphate accumulation and energy storage in microorganisms was convincingly demonstrated. That period of research on microbial polyphosphates was associated primarily with studies of the so-called volutine or polyphosphate granules revealed in the cytoplasm of many microorganisms and with the conception of them as "microbial phosphagenes."

In the 25 years since the Strasbourg symposium, our ideas of the physiological role, metabolism, and localization of microbial polyphosphates have undergone significant changes. These changes are connected with the work of Weimberg and Orton (12), Szymona and Ostrowski (11), Harold (4), and a number of other investigators (6–8), who showed that polyphosphates are localized not only in the polyphosphate granules but also in other parts of microbial cells, specifically, on their surface. It was also established that polyphosphates, or at least some fraction of these compounds, are not microbial phosphagenes. The enzymes catalyzing the utilization and biosynthesis of polyphosphates without the participation of ATP and other adenyl nucleotides were recognized. In addition, the multifunctionality of polyphosphates was revealed. Specifically, their participation in the regulation of phosphorus metabolism was detected by many workers. Thus, it is not by chance that the problem of polyphosphates is at present discussed primarily from the point of view of their participation in the regulation of phosphorus and energy metabolism. Neither is it by chance that in recent years a considerable escalation of interest in polyphosphates and their physiological role has been observed. Finally, it is not by chance that the problems of metabolism, localization, and the physiological role of polyphosphates are presented in this section.

The papers that follow show new approaches to research into the structure and function of inorganic polyphosphates and present new pathways for their biosynthesis and utilization to meet requirements of microbial cells. New data on the important role of polyphosphates in the regulation of phosphorus and energy metabolism in microorganisms are presented. Finally, from the material in this section one can draw quite a definite conclusion about the significance of studying polyphosphate metabolism in microorganisms, not only to more correctly understand the basic principles of eucaryotic and procaryotic cell functioning, but also for solving practical problems. It was recently established that the knowledge of polyphosphate metabolism in microorganisms oversynthesizing these compounds and the revelation of conditions promoting such overproduction can be of significant help in solving the problem of sewage and industrial water purification from excessive phosphates. This is a problem which the biotechnologists of many countries throughout the world face at present.

LITERATURE CITED

1. **Belozersky, A. N.** 1958. The formation and functions of polyphosphates in the developmental processes of some lower organisms. Comm. and Reports to the Fourth Intern. Biochem. Congress, Vienna.
2. **Ebel, J. P.** 1948. Sur le dosage des metaphosphates dans les microorganismes par hydrolyse differentielle technique et application aux levures. C.R. Acad. Sci. **226:**2184–2186.
3. **Ebel, J. P., and M. Grunberg-Manago (ed.).** 1962. Acides ribonucléiques et polyphosphates. Structure, synthèse, et

fonctions. Colloq. Int. C.N.R.S., Paris.
4. **Harold, F. M.** 1966. Inorganic polyphosphates in biology: structure, metabolism, and functions. Bacteriol. Rev. **30:** 772–785.
5. **Kornberg, A., S. Kornberg, and E. Simms.** 1956. Metaphosphate synthesis by an enzyme from *Escherichia coli*. Biochim. Biophys. Acta **20:**215–227.
6. **Kulaev, I. S.** 1979. Biochemistry of inorganic polyphosphates. John Wiley & Sons, Inc., New York.
7. **Kulaev, I. S.** 1985. Some aspects of environmental regulation of microbial phosphorus metabolism, p. 1–25. *In* I. S. Kulaev, A. E. Dawes, and D. W. Tempest (ed.), Environmental regulation of microbial metabolism. Academic Press, Inc., Orlando, Fla.
8. **Kulaev, I. S., and V. M. Vagabov.** 1983. Polyphosphate metabolism in microorganisms. Adv. Microb. Physiol. **24:** 83–171.
9. **Liebermann, L.** 1888. Über das Nuclein der Hefe und künetliche Darstellung eines Nuclein's Eiweiss und Metaphosphorsäure. Ber. Dtsch. Chem. Ges. **21:**598–607.
10. **Lohmann, K.** 1958. Über das Vorkommen der kondensierten Phosphate in Lebewesen, p. 29–44. *In* Kondensierte Phosphate in Lebensmitteln. Springer Verlag, Berlin.
11. **Szymona, M., and W. Ostrowski.** 1964. Inorganic polyphosphate glucokinase of *Mycobacterium phlei*. Biochim. Biophys. Acta **85:**283–295.
12. **Weimberg, R., and W. L. Orton.** 1965. Synthesis and breakdown of the polyphosphate fraction and acid phosphomonesterase of *Saccharomyces mellis* and their location in the cell. J. Bacteriol. **89:**740–742.
13. **Wiame, J. M.** 1947. Etude d'une substance polyphosphorée basophile et metachromatique chez les levures. Biochim. Biophys. Acta **1:**234–255.

Polyphosphate Accumulation and Metabolism in *Escherichia coli*

N. N. RAO,[1] M. F. ROBERTS,[2] AND A. TORRIANI[1]

Departments of Biology[1] and Chemistry,[2] Massachusetts Institute of Technology, Cambridge, Massachusetts 02139

Inorganic polyphosphates ($polyP_is$) are widely distributed among many microorganisms, where large amounts of cytologically demonstrable $polyP_is$ are accumulated (12). As a phosphate source, they can be utilized by microorganisms from the medium. In this paper, we approach both problems: (i) how $polyP_is$ are accumulated and metabolized within the cell and (ii) how microorganisms utilize large $polyP_is$ from the medium as the sole source of P_i. The organism we chose was *Escherichia coli*.

Accumulation of $PolyP_is$ in *E. coli*

Many microorganisms accumulate $polyP_is$ when subjected to unfavorable growth conditions such as low pH, anaerobiosis, or sulfur starvation, or following a period of phosphate starvation (6, 27). These $polyP_is$ are present in cells in the form of an acid-soluble fraction ranging in chain length from two (PP_i) to several hundreds and an acid-insoluble fraction of higher molecular weight, complexed with RNA and proteins (12).

E. coli does not accumulate appreciable amounts of $polyP_is$, especially those of high-molecular-weight types (11). Yet, *E. coli* appears to be the ideal choice for the study of $polyP_i$ metabolism since in this organism the enzymes involved in the biosynthesis and degradation of $polyP_is$ have been described (8, 17).

The study of $polyP_i$ accumulation in *E. coli* was made possible by the finding of a mutant K10C4 (*phoU35*) which is constitutive for the synthesis of alkaline phosphatase and is able, under conditions of anaerobiosis, to synthesize more than four times the amount of $polyP_i$ produced by its wild-type parental strain (28). The amount of $polyP_i$ synthesized by other *pho* regulon mutants of *E. coli* was compared. It was observed that a mutation in the positive regulatory gene *phoB*, which codes for a transcriptional inducer, or in the alkaline phosphatase gene *phoA* (26) decreased the synthesis (Table 1). To study the accumulation of $polyP_is$ in *E. coli*, the *phoU35* mutant was grown anaerobically in Penassay broth (28); the $polyP_i$ was extracted and measured as acid-soluble and alkali-soluble fractions. $PolyP_i$ began to accumulate during exponential growth (between 2 and 4 h) and reached a maximum at 6 h when the cells were in early stationary phase. Subsequently, $polyP_i$ was rapidly degraded (23). The two fractions (acid-soluble and alkali-soluble $polyP_i$) appeared to have independent but possibly related metabolism since the alkali-soluble fraction was synthesized when more than 85% of the acid-soluble $polyP_i$ was already hydrolyzed.

The $polyP_is$ were further analyzed by gel filtration (Fig. 1). The extracts of the 6-h anaerobic cells contained a $polyP_i$ with an average chain length of 15 to 20 phosphate residues as the major species (83%) and another with an average chain length of 142. The extracts of the 24-h cells contained a higher ratio (25%) of high- to low-molecular-weight polymers. The $polyP_i$ from the 24-h culture was separated into five peaks with average chain lengths of 42, 132, 230, 750, and over 1,000 phosphate units (23). Evidence that the extraction procedure did not modify the cellular $polyP_i$ was obtained by a coextraction of $polyP_i$ from *E. coli* cells in the presence of chemically synthesized $[^{32}P]polyP_i$. There were no significant changes either in the amount or in the size of the indicator $[^{32}P]polyP_i$. Thus, these results should represent the nature of the cellular $polyP_i$.

These observations were tested by ^{31}P nuclear magnetic resonance (^{31}P NMR) spectroscopy. High-resolution ^{31}P NMR spectra at 109.3 MHz were recorded for aerobically and anaerobically grown cells (23). No $polyP_i$ resonances could be detected in either intact cells or cell extracts from aerobic cultures. In contrast, the spectra of *E. coli* cells grown anaerobically showed the presence of $polyP_i$ and confirmed the observation made by chemical analysis, namely, that 6-h cultures accumulated more soluble $polyP_i$ than did the 24-h cultures, and verified the results obtained by gel filtration about the chain length (23). It is possible to distinguish between surface-associated $polyP_is$ from cytoplasmic pools by treating cells with membrane-impermeable chelators such as EDTA and to compare the ^{31}P NMR spectra of intact cells with and without EDTA (19). Addition of 10 to 20 mM EDTA had no effect on the intensity of the terminal phosphate resonance of anaerobic *E. coli* cells, indicating the absence of surface pools of $polyP_i$.

TABLE 1. PolyP$_i$ accumulation in *E. coli* K10 during aerobic and anaerobic growth[a]

Strain	PolyP$_i$ after growth for 6 h	
	Aerobic	Anaerobic
K10 wild type	0.25	0.72
C4 *phoU35*	1.0	2.97
D15 *phoU35 phoB72*	0.40	0.90
E15 Δ*phoA8*	0.22	0.61
S3 *phoB63*	0.05	0.06

[a] Values are expressed in milligrams of polyP$_i$ per 100 mg of dry cell weight. All strains are derivatives of K10 Hfr *relA1 pit-10 tonA22* T2^r. Medium used was Difco antibiotic medium no. 3 (Penassay broth). After 6 h in aerobiosis the optical density at 540 nm was 3.5; anaerobic growth stopped when the optical density at 540 nm was 0.7, after 6 h. Extraction and estimation of polyP$_i$ were carried out as described previously (22).

A limitation of this analysis is that only the polyP$_i$s which are soluble inside the cell can be detected. PolyP$_i$s, especially high-molecular-weighttypes, are found to be associated with either proteins or nucleic acids and in granules or vacuoles (12), and hence may not be detected by solution-state NMR. Nevertheless, the above results suggest that *E. coli* cells synthesize and accumulate low-molecular-weight, acid-soluble polyP$_i$s during anaerobic growth. Once the growth ceases, high-molecular-weight polyP$_i$s are accumulated, possibly from the low-molecular-weight precursors.

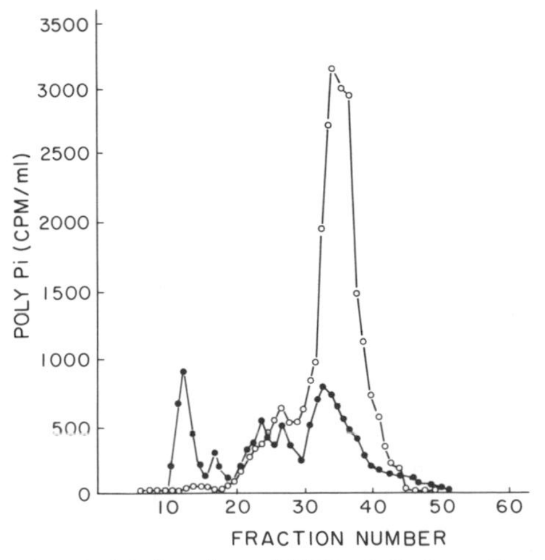

FIG. 1. Gel filtration of [^{32}P]polyP$_i$ from *E. coli* K10 *phoU35*. Cultures (750 ml of Penassay broth containing 50 μCi of $H_3{}^{32}PO_4$ per ml) were grown (optical density at 540 nm of 0.7) for 6 h (○) or incubated 18 h longer (●) at 37°C under anaerobiosis. [^{32}P]polyP$_i$ was extracted by using sodium hypochlorite at pH 9.8 (22). The extracts were purified and chromatographed on a Sepharose-4B column as described previously (23).

Metabolism of Cytoplasmic PolyP$_i$ in *E. coli*

Several enzymes have been implicated in microbial synthesis and degradation of polyP$_i$ (4, 11, 12). The enzymes involved in the synthesis are polyP$_i$ kinase, ATP-polyP$_i$ phosphotransferase (EC 2.7.4.1), and 1,3-diphosphoglycerate: polyP$_i$ phosphotransferase (EC 2.7.4.17). PolyP$_i$ kinase was purified and studied in *E. coli* (8, 14). The enzyme 1,3-diphosphoglycerate:polyP$_i$ phosphotransferase is involved in the transfer of the energy-rich phosphate residue from 1,3-diphosphoglycerate to polyP$_i$ (12, 17). The utilization and degradation of the polymer is catalyzed by polyphosphatases (17, 27) and by specific kinases, viz., polyP$_i$ glucokinase and polyP$_i$ fructokinase (4). The enzymes AMP-polyP$_i$ phosphotransferase (2) and polyP$_i$ kinase may also participate in the utilization of polyP$_i$ in *E. coli*. Mutants in polyP$_i$ metabolism have been isolated in *Klebsiella aerogenes* by Harold and Harold (5). They showed that in this organism polyP$_i$ kinase was the only route of polyP$_i$ biosynthesis and that its synthesis depended upon the P$_i$ concentration of the growth medium. The degradation of polyP$_i$ was brought about by polyphosphatase (4). Both polyP$_i$ kinase (8) and polyphosphatase (17, 27) are present in *E. coli*.

To explain the synthesis of polyP$_i$ by the *pho* regulon mutants, we measured polyP$_i$ kinase activity directed toward the formation of acid-insoluble polyP$_i$, in cells grown aerobically (18 h) and in cells grown anaerobically from 0 to 24 h (Fig. 2). The activity in the *phoU35* mutant C4 grown aerobically for 18 h increased threefold during anaerobiosis, and it was higher than in K10 (wild type), D15 (*phoU35 phoB72*), S3 (*phoB63*), or E15 (Δ*phoA8*) (Fig. 2). The maximum increase in activity under anaerobic growth was noticed at 24 h of incubation. However, we observed that the peak of polyP$_i$ synthesis was at 6 h (23).

Nesmeyanova et al. (17) observed, as we do here, that the increase in polyP$_i$ kinase activity coincided with the decrease in the intracellular polyP$_i$ concentration. This enzyme was shown in vitro to degrade polyP$_i$ by transferring phosphate from polyP$_i$ to ADP (9, 24). We observed that the activity of polyP$_i$ kinase extracted from *E. coli* was stimulated by P$_i$ and also that the synthesis in aerobically growing cells was enhanced by it. Experiments conducted in *Mycobacterium smegmatis* strongly suggested that polyP$_i$ would be synthesized and utilized by one and the same polyP$_i$ kinase (25). It was predicted that the action of this enzyme and the utilization of polyP$_i$ in the cells may be controlled by the concentration of a variety of phosphorylated compounds and by a cationic environment (13, 25). A similar situation may exist in *E. coli*, and

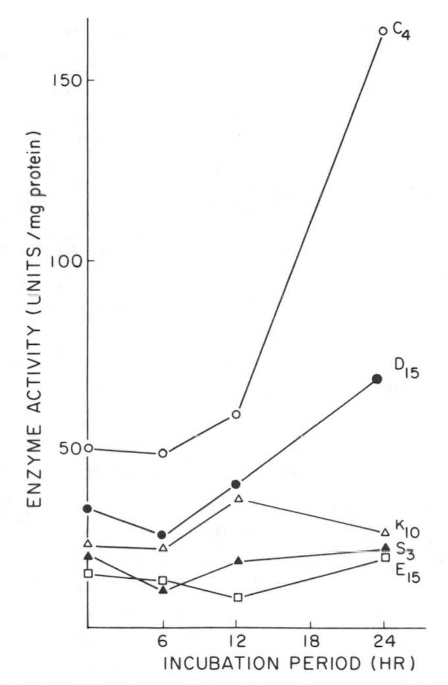

FIG. 2. Kinetics of polyP$_i$ kinase activity in anaerobically incubated *E. coli* K10 mutants. The values at time zero refer to enzyme activity of aerobic cultures grown for 18 h. One unit of enzyme is defined as the amount of enzyme producing 1 nmol of acid-insoluble P$_i$ in 30 min. Symbols: ○, C4 (*phoU35*); ●, D$_{15}$ (*phoU35 phoB72*); △, K10 (wild type); ▲, S3 (*phoB63*); □, E15 (Δ*phoA8*).

the activity of the enzyme may depend upon the ATP/ADP ratio. A need for short-chain polyP$_i$ as primer in the polyP$_i$ kinase reaction in *E. coli* is speculative (8). However, short-chain polyP$_i$ accumulated initially during anaerobic growth, followed by a period of synthesis of long-chain polyP$_i$ which may be initiated by polyP$_i$ kinase in *E. coli*.

Nesmeyanova and co-workers described a cytoplasmic phosphohydrolase-polyphosphatase in aerobically growing *E. coli* cells. Unexpectedly, they observed that the peak of activity of the polyphosphatase coincided with the peak of polyP$_i$ content in the cells (17). We measured polyphosphatase activity in the *E. coli* K10 wild type and mutants grown anaerobically. Since anaerobiosis induces polyP$_i$, these values were compared with the activity of cells grown anaerobically for 6 h (Table 2). It was observed that in anaerobiosis the polyphosphatase activity was rapidly lost in C4 *phoU35* (Table 2). All the mutants (S3 *phoB63*, D15 *phoU35 phoB72*, H_2 *phoB52*, and LEP_1 *phoB23*), incapable of synthesis of large amounts of polyP$_i$, showed no significant level of polyphosphatase activity in any growth condition. However, it should be pointed out that *phoB* mutants are alkaline phosphatase negative (cannot be induced). To check whether the lack of polyphosphatase activity was due to a lack of alkaline phosphatase activity, we used a Δ*phoA* mutant (E15) grown aerobically and found that it possessed a high level of polyphosphatase activity which was lost during anaerobiosis (Table 2). Mutations exhibiting pleiotropic positive or negative effects on the synthesis of alkaline phosphatase, polyphosphatase, and polyP$_i$ kinase have previously been identified (5, 18). Our results strongly suggest that *phoB* positively regulates the synthesis of polyphosphatase.

Utilization of External PolyP$_i$ by *E. coli*

Polyphosphate can replace P$_i$ as a sole source of phosphate in the growth medium (21). The phosphate available in the medium as P$_i$, glycerol 3-phosphate, or short linear polyP$_i$ (chain length, 15 phosphate residues) can enter the periplasm through pores, called porins, present in the outer membrane. When P$_i$ is present in the medium, it diffuses through the porins OmpF and OmpC (15). Growth of *E. coli* under conditions of phosphate limitation results in the synthesis of another porin, the product of the *phoE* gene (20). It was suggested previously (1) that PhoE porin may play a role in the uptake of polyP$_i$, teichoic acid, DNA, and RNA. Korteland et al. (10) and Overbeeke and Lugtenberg (21) found that cells deprived of P$_i$ can utilize large linear polyP$_i$ for growth, since the induced PhoE porin possesses a translocation site for negatively charged P$_i$-containing solutes. Thus, wild-type *E. coli* cells derepressed for *phoE* grew well in a medium containing polyP$_i$ of average chain length of 35 phosphate residues with a generation time of 90 min. In contrast, a strain containing a deletion of *phoA* grew quite poorly (generation time of 450 min). These results suggested that at least two genes of the *pho*

TABLE 2. Polyphosphatase activity in *E. coli* K10 mutants

Strain	Activity[a]	
	Aerobic[b]	Anaerobic[c]
K10 wild type	22.5	19.6
C4 *phoU35*	28.6	2.6
D15 *phoU35 phoB72*	0.3	0.1
E15 Δ*phoA8*	16.8	0.6
S3 *phoB63*	<0.1	<0.1

[a] Polyphosphatase activity is expressed as units per optical density at 540 nm. One unit is the amount of enzyme producing 1 nmol of P$_i$ per min. The substrate was a chemically synthesized (16) polyP$_i$ of about 200 phosphate residues.

[b] The cultures were grown for 18 h in aerobiosis.

[c] The cells were then transferred to a fresh medium under anaerobiosis and incubated for 6 h.

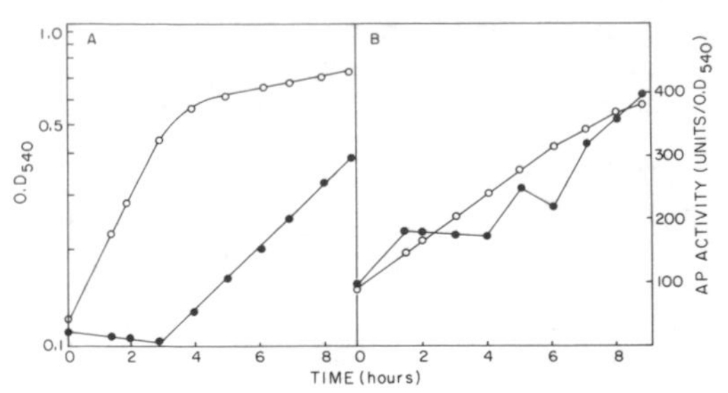

FIG. 3. Growth and alkaline phosphatase (AP) synthesis by *E. coli* K10. The culture was grown in MOPS medium with either 0.1 mM K_2HPO_4 (A) or 0.001 mM polyP100 (B) as sole phosphate source. Symbols: ○, growth; ●, alkaline phosphatase activity expressed as nanomoles of *p*-nitrophenol liberated per minute at 27°C per 4 × 10^8 cells per ml.

regulon, i.e., *phoA* and *phoE*, are required for the utilization of polyP$_i$ (21). Besides PhoE and alkaline phosphatase, an acid phosphatase (pH optimum of 2.5) located in the periplasm has been implicated in the utilization of linear polyP$_i$ in *E. coli* (3).

To further analyze the mechanism by which long-chain polyP$_i$s are utilized, we chose a linear polyP$_i$ with an average chain length of 100 phosphate residues as a sole source of P$_i$. This polymer (polyP100) did not hydrolyze spontaneously in the medium. In contrast, commercial polyP$_i$ ranging in average chain length from 5 to 45 phosphate residues provided a low level of P$_i$ caused by spontaneous hydrolysis; hence, we observed that a strain devoid of alkaline phosphatase activity could grow in this medium.

The question of polyP$_i$ utilization was addressed by two strategies. The first involved growing the *E. coli* mutants in the presence of polyP$_i$ to record the growth rate and to measure the activities of alkaline phosphatase and of acid phosphatase to determine which of these is essential for the utilization of exogenous polyP100. The second strategy used ^{31}P NMR spectroscopy to analyze the uptake and degradation of polyP100.

When *E. coli* K10 was grown in a medium containing 0.1 mM K_2HPO_4, the alkaline phosphatase activity was repressed until the end of the exponential phase of growth, followed by a steady increase in the enzyme activity after the medium became deficient in P$_i$ (Fig. 3A), as expected (26). The growth of *E. coli* in polyP$_i$-containing medium (0.001 mM polyP100) was exponential up to 6 h, and the synthesis of alkaline phosphatase appeared to be biphasic with bursts of repression and derepression (Fig. 3B). This can be explained if polyP$_i$ is degraded into P$_i$ which represses the enzyme synthesis. Once the available P$_i$ is utilized by the growing cells, the enzyme synthesis and the activity reappear, resulting in further degradation of polyP$_i$.

This supports the hypothesis that alkaline phosphatase hydrolyzes the polyP$_i$ into P$_i$ which enters the cytoplasm. It involves the function of a series of genes of the Pho regulon: genes for alkaline phosphatase (*phoA*) and its regulation (*phoB, phoU*), genes for porin E (*phoE*), and the gene for acid phosphatase (*appA*). We investigated the effect of mutations in these systems by measuring the doubling time and the activities of alkaline phosphatase and acid phosphatase in cells growing on polyP100 or high P$_i$ (2 mM) (Table 3). The *phoU* mutation provoked a constitutive synthesis of the two enzymes. However, the growth of the *phoU* mutant was twice as slow as that of the wild type. We have, at present, no explanation for this result. Both alkaline phosphatase and acid phosphatase activities were higher in polyP$_i$-grown wild-type *E. coli* W3110 as compared with cells grown in P$_i$-containing medium, suggesting a slow release of P$_i$ from polyP100, causing P$_i$ limitation. Mutants lacking alkaline phosphatase activity, viz., Δ*phoA8, phoB63, phoU35* Δ*phoA8*, and *phoU35 phoB63* mutants, did not utilize polyP100 (Table 3). These observations suggest that the functions of *phoB* and *phoA* are required. The presence of PhoE, which is constitutively synthesized in *phoU* and *phoU phoA* mutants, is insufficient without alkaline phosphatase activity for polyP100 utilization.

The experiments using *E. coli* K10 strains (Table 3) also included the enzyme acid phosphatase, which has a high specificity for low-molecular-weight polyP$_i$. A strain deleted for the acid phosphatase gene (Δ*appA*) and possessing a low level of the enzyme could utilize polyP100. Strains deleted either for *phoA* or for *phoA* and *appA* (as in Δ*phoA8* Δ*appA*) did not grow. It was concluded, therefore, that polyP100 could be utilized by *E. coli* cells possessing a low level of acid phosphatase activity, but the presence of alkaline phosphatase activity was absolutely essential for polyP$_i$ degradation, as previously stated by Overbeeke and Lugtenberg (21).

To corroborate the observations made on the utilization of polyP100 by growth experiments, ^{31}P NMR spectroscopy was employed to study the uptake and degradation of polyP100 by intact cells. High-resolution ^{31}P NMR spectra at 109.3 MHz were recorded for aerobically grown *E. coli* cells. Typically, internal phosphate resonates in the region of 2 and 0 ppm, depending on the intracellular pH from external 85% H_3PO_4, while a resonance for P$_i$ outside the cell is detected slightly upfield (0 to −1 ppm). The resonances for PP$_i$ linkages lie in the region of −10 to −13 ppm, and all except the terminal

TABLE 3. Growth, alkaline phosphatase activity, and acid phosphatase activity of *E. coli* mutants[a]

Strain	Growth on polyP100 (doubling time, min)	Phosphatase activity[b]			
		Alkaline		Acid	
		PolyP100	High P_i	PolyP100	High P_i
W3110					
Wild type	185	65.6	0.10	4.9	10.32
ΔphoA8	NG[c]		0.09		9.05
phoB63	NG		0.09		7.62
phoU35	240	419.3	90.6	8.0	18.04
phoU35 phoB63	NG		0.2		9.31
ΔphoA8 phoU35	NG		0.18		9.01
K10					
Wild type	90	440.5	0.8	26.7	10.77
ΔphoA8	NG		0.03		2.90
ΔappA	100	444.8	0.12	1.61	1.57
ΔphoA8 ΔappA	NG		0.03		2.1

[a] The enzyme activities were measured in cells growing in MOPS-buffered medium with a high (2 mM K_2HPO_4) level of P_i and in cultures utilizing polyP100 as the only source of P_i. No activities were recorded in cultures unable to utilize polyP100 for growth. The strain W3100 *phoU35* is high constitutive for alkaline phosphatase and low constitutive for acid phosphatase.

[b] Phosphatase activity: U/4 × 10^8 cells per ml.

[c] NG, No growth observed up to 24 h at 37°C.

phosphates of polyP_i resonate in the region of −20 to −22 ppm.

E. coli K10 *phoA8 ΔappA*, incapable of synthesizing both alkaline phosphatase and acid phosphatase, was grown overnight (ca. 18 h) in a low-P_i (0.2 mM K_2HPO_4) morpholinepropanesulfonic acid (MOPS) medium. The cells were washed free from medium and were incubated for 30 min at 25°C in PIPES buffer [piperazine-*N,N'*-bis(2-ethanesulfonic acid), 0.05 M, pH 7.2] containing 1.3 mM polyP100 and 1 mM NaF. The cells were then washed free from polyP100 and resuspended in NMR buffer (7). The ^{31}P NMR spectra were recorded as described previously (23). The spectrum in Fig. 4A shows the resonances due to sugar phosphates, internal P_i, external P_i, PP_i, and polyP_i in the intact cells. To learn whether the polyP_i is intracellular, trapped in the periplasm, or freely permeable to outer membrane, the cells from Fig. 4A were centrifuged, washed, and resuspended in NMR buffer. The spectra of the resulting supernatant (Fig. 4B) and the recentrifuged (Fig. 4C) cells were recorded. The supernatant showed the presence of only external P_i and polyP_i. The centrifuged cells (Fig. 4C) exhibited the same pattern as before the centrifugation.

These observations rule out the possibility that the observed polyP_i resonance is due to polyP_i bound to the surface of the cells or trapped within the pellet as free polyP_i. After two washings (23), the cell pellet could not contain more than 1 μM polyP100 free in the supernatant, a concentration too low to be detected by NMR in vivo. Thus, the polyP_i observed in our experiments must be either free in the periplasm and exchange with the medium or be in equilibrium with undetectable bound polyP100 internal or external to the cell wall. Our results favor the possibility that, in this *E. coli* strain, polyP_i of chain length 100 equilibrates between the medium and the periplasm.

The addition of a shift reagent, praseodymium, to the intact cells (Fig. 4D) broadened the polyP_i resonance so dramatically that it cannot be seen above the background. The gain of the spectrum has been increased threefold to emphasize this fact. Since *E. coli* cells are impermeable to praseodymium, this implies that the bulk of the polyP100 was not in the cytoplasm. The resonances of the phosphate esters and the PP_i were not broadened by praseodymium and were also not detected in the supernatant. Hence, these are solely in the cytoplasm.

The kinetics of polyP_i degradation by intact cells was monitored by ^{31}P NMR spectroscopy. For this purpose, *E. coli* possessing both acid and alkaline phosphatase activities (*E. coli* K10 *appA*$^+$ *phoA*$^+$) was grown overnight (ca. 18 h) in a low-P_i (0.2 mM K_2HPO_4) medium. The cells were washed and incubated with 2.0 mM polyP100 for 2 h. The spectra of washed cell suspensions showed that the levels of both intracellular and extracellular P_i increased and the levels of polyP100 associated with the cells decreased (data not shown).

These observations suggest that the cells are able to take polyP100 from the medium, hydrolyze it, and transport the resulting P_i into the cytoplasm. However, we cannot exclude the possibility of a small amount of leakage of alkaline phosphatase into the growth medium which

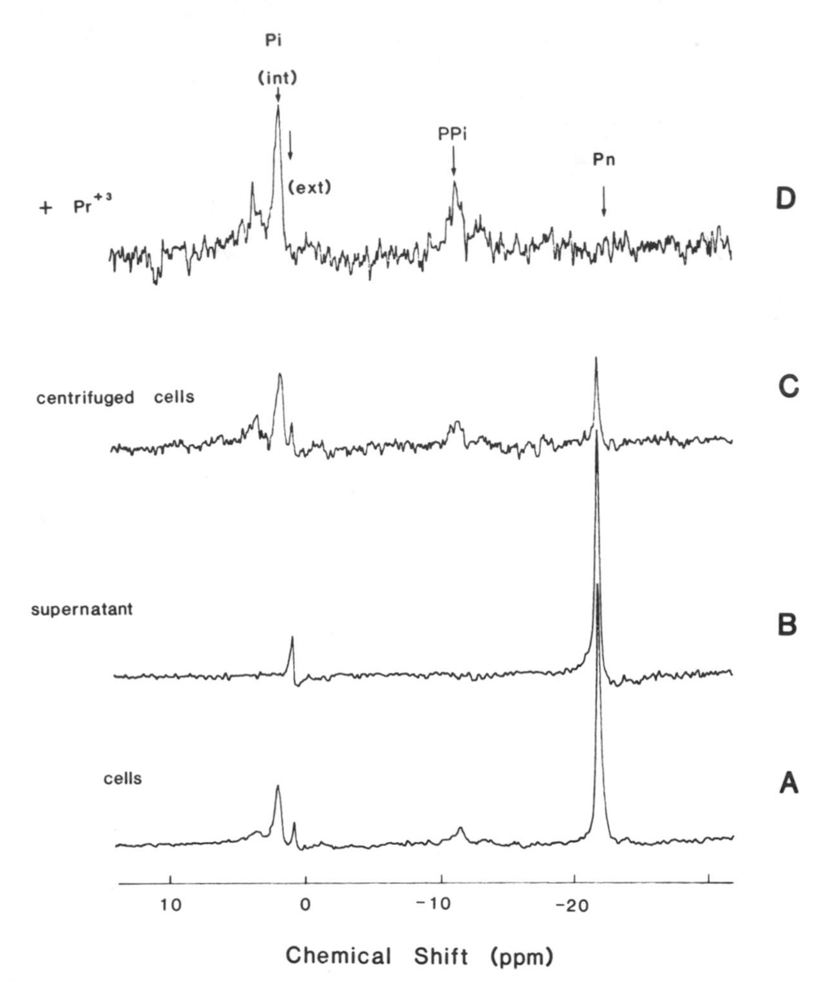

FIG. 4. ^{31}P NMR spectra depicting P_i, PP_i, and polyphosphates (Pn) of cell suspensions of K10 Δ*phoA8* Δ*appA* in NMR buffer (7) containing 1 mM NaF. The cells grown in 600 ml of MOPS low-P_i (0.2 mM K_2HPO_4) medium for 18 h were harvested and washed once with 0.85% (wt/vol) sodium chloride solution. The washed cells were incubated in 0.33 mM polyP100 in 0.05 M PIPES buffer, pH 7.2, at 25°C for 30 min. After the incubation, the cells were centrifuged at 10,000 × g for 10 min, suspended in 10 ml of PIPES buffer (pH 7.2, 0.05 M), and centrifuged as before. The washed cells were suspended in 1 ml of NMR buffer (A). The culture from A was recentrifuged, resulting in supernatant (B), and cells were resuspended in 1 ml of NMR buffer (C). Praseodymium (Pr^{3+}) was added to the centrifuged cells (D).

generates sufficient P_i for growth from polyP100 outside the cell, although no measurable activity was observed in the supernatant.

When a chemically synthesized, [^{32}P]polyP_i of high molecular weight (chain length, ca. 200) was incubated with a purified preparation of alkaline phosphatase from *E. coli*, a constant rate of hydrolysis, producing P_i, was observed up to 5 h of incubation. Thus, both the in vivo and in vitro studies indicate that alkaline phosphatase can act as a polyphosphatase.

The ^{31}P NMR studies complemented the information that alkaline phosphatase is essential for the utilization of external polyP100. Both PhoE and acid phosphatase are not essential, since their presence did not enhance the utilization of polyP100. The periplasmic polyphosphatase, which is distinct from alkaline phosphatase and derepressed under conditions of phosphate limitation (2), may not be involved in the utilization of polyP100, since *E. coli* lacking alkaline phosphatase only did not grow in a medium containing polyP100.

Concluding Remarks

An alkaline phosphatase-constitutive *phoU35* mutant of *E. coli* synthesized polyP_i in a phosphate-rich medium under anaerobiosis. All the polyP_is were cytoplasmic, and no surface pool was detected. The amount and types of polyP_i

accumulated were time dependent. The majority of $polyP_is$ extracted at 6 h had smaller chain lengths (20 ± 5 phosphate units), whereas the $polyP_is$ extracted from cells incubated for 24 h contained high-molecular-weight $polyP_i$ species (chain length > 200).

Our results suggest that the accumulation of $polyP_i$ in *E. coli* could be attributed to an equilibrium between $polyP_i$ kinase and polyphosphatase activities. These enzymes may be members of the *pho* regulon since the *phoU35* mutation enhanced $polyP_i$ kinase activity by approximately threefold and a mutation in the gene of positive control, *phoB*, caused a complete or a nearly complete loss of polyphosphatase activity and prevented $polyP_i$ accumulation.

The utilization of a high-molecular-weight $polyP_i$ (chain length, 100 phosphate residues) as a sole source of P_i in the medium depended upon the presence of alkaline phosphatase activity and the inducer of the regulon, PhoB. PhoE and acid phosphatase were not essential since their absence did not prevent the utilization of polyP100. The ^{31}P NMR studies with intact *E. coli* cells indicate that free polyP100 either entered the periplasmic space or was in equilibrium with polyP100 loosely bound to the outer membrane and hydrolyzed to P_i by the action of alkaline phosphatase.

LITERATURE CITED

1. **Argast, M., and W. Boos.** 1980. Co-regulation in *Escherichia coli* of a novel transport system for *sn*-glycerol-3-phosphate and outer membrane protein Ic (e,E) with alkaline phosphatase and phosphate-binding protein. J. Bacteriol. **143:**142–150.
2. **Chalykoff, P., and H. Yamazaki.** 1978. Phosphorylation of adenosine 5′-monophosphate by a dialyzed cell-free extract from *Escherichia coli*. Can. J. Biochem. **56:**839–841.
3. **Dassa, E., and P. L. Boquet.** 1981. Is the acid phosphatase of *Escherichia coli* with pH optimum of 2.5 a polyphosphate depolymerase? FEBS Lett. **135:**148–150.
4. **Harold, F. M.** 1966. Inorganic polyphosphates in biology: structure, metabolism, and function. Bacteriol. Rev. **30:** 772–794.
5. **Harold, F. M., and R. L. Harold.** 1965. Degradation of inorganic polyphosphate in mutants of *Aerobacter aerogenes*. J. Bacteriol. **89:**1262–1270.
6. **Harold, F. M., and S. Sylvan.** 1963. Accumulation of inorganic polyphosphate in *Aerobacter aerogenes*. II. Environmental control and the role of sulfur compounds. J. Bacteriol. **86:**222–231.
7. **Herrero, A. A., R. F. Gomez, and M. F. Roberts.** 1985. ^{31}P NMR studies of *Clostridium thermocellum*. Mechanism of end product inhibition by ethanol. J. Biol. Chem. **260:** 7442–7451.
8. **Kornberg, A., S. R. Kornberg, and E. S. Simms.** 1956. Metaphosphate synthesis by an enzyme from *Escherichia coli*. Biochim. Biophys. Acta **20:**215–227.
9. **Kornberg, S. R.** 1957. Adenosine triphosphate synthesis from polyphosphate by an enzyme from *Escherichia coli*. Biochim. Biophys. Acta **26:**294–300.
10. **Korteland, J., P. De Graaff, and B. Lugtenberg.** 1984. PhoE protein pores in the outer membrane of *Escherichia coli* K-12 not only have a preference for Pi and Pi-containing solutes but are general anion-preferring channels. Biochim. Biophys. Acta **778:**311–316.
11. **Kulaev, I. S.** 1975. Biochemistry of inorganic polyphosphates. Rev. Physiol. Biochem. Pharmacol. **73:**131–158.
12. **Kulaev, I. S., and V. M. Vagabov.** 1983. Polyphosphate metabolism in microorganisms. Adv. Microb. Physiol. **24:** 81–171.
13. **Levinson, S. L., L. H. Jacobs, T. A. Krulwich, and H. C. Li.** 1975. Purification and characterization of a polyphosphate kinase from *Arthrobacter atrocyaneus*. J. Gen. Microbiol. **88:**65–74.
14. **Li, H. C., and G. G. Brown.** 1973. Orthophosphate and histone dependent polyphosphate kinase from *E. coli*. Biochem. Biophys. Res. Commun. **53:**875–881.
15. **Lugtenberg, B., and L. Van Alphen.** 1983. Molecular architecture and functioning of the outer membrane of *Escherichia coli* and other gram-negative bacteria. Biochim. Biophys. Acta **737:**51–115.
16. **Muhammed, A., A. Rogers, and D. E. Hughes.** 1959. Purification and properties of a polymetaphosphatase from *Corynebacterium xerosis*. J. Gen. Microbiol. **20:**482–495.
17. **Nesmeyanova, M. A., A. D. Dmitriev, and I. S. Kulaev.** 1973. High molecular weight polyphosphates and enzymes of polyphosphate metabolism in the process of *Escherichia coli* growth. Mikrobiologiya **42:**213–219.
18. **Nesmeyanova, M. A., S. A. Gonina, and I. S. Kulaev.** 1975. Biosynthesis of polyphosphatases of *Escherichia coli* under the control of regulatory genes in common with alkaline phosphatase. Dokl. Acad. Nauk USSR **224:**710–712.
19. **Ostrovskii, D. N., N. F. Sepetov, V. I. Reshetnyak, and L. S. Siberl'dina.** 1980. Investigation of the localization of polyphosphates in cells of microorganisms by the method of high resolution ^{31}P-NMR-145.78 MHz. Biokhimiya **45:** 392–398.
20. **Overbeeke, N., and B. Lugtenberg.** 1980. Expression of outer membrane protein e of *Escherichia coli* K12 by phosphate limitation. FEBS Lett. **112:**229–232.
21. **Overbeeke, N., and B. Lugtenberg.** 1982. Recognition site for phosphorus-containing compounds and other negatively charged solutes on the phoE protein pore of the outer membrane of *Escherichia coli* K12. Eur. J. Biochem. **126:**113–118.
22. **Poindexter, J. S., and E. F. Eley.** 1983. Combined procedure for assays of poly-β-hydroxy butyric acid and inorganic polyphosphate. J. Microbiol. Methods **1:**1–17.
23. **Rao, N. N., M. F. Roberts, and A. Torriani.** 1985. Amount and chain length of polyphosphates in *Escherichia coli* depend on cell growth conditions. J. Bacteriol. **162:**242–247.
24. **Robinson, N. A., and H. G. Wood.** 1986. Polyphosphate kinase from *Propionibacterium shermanii*: demonstration that the synthesis and utilization of polyphosphate is by a processive mechanism. J. Biol. Chem. **261:**4481–4485.
25. **Suzuki, H., T. Kaneko, and Y. Ikeda.** 1972. Properties of polyphosphate kinase prepared from *Mycobacterium smegmatis*. Biochim. Biophys. Acta **268:**381–390.
26. **Torriani, A., and D. N. Ludke.** 1985. The pho regulon of *Escherichia coli*, p. 224–242. *In* M. Shaechter, F. C. Neidhart, J. Ingraham, and N. O. Kjeldgaard (ed.), The molecular biology of bacterial growth. Jones and Bartlett Publishers, Boston.
27. **Yagil, E.** 1975. Derepression of polyphosphatase in *Escherichia coli* by starvation for inorganic phosphate. FEBS Lett. **55:**124–127.
28. **Zuckier, G., E. Ingenito, and A. Torriani.** 1980. Pleiotropic effects of alkaline phosphatase regulatory mutations *phoB* and *phoT* on anaerobic growth of and polyphosphate synthesis in *Escherichia coli*. J. Bacteriol. **143:**934–941.

Metabolically Active Surface Polyphosphate Pool in *Acinetobacter lwoffi*

H. O. HALVORSON,[1] N. SURESH,[1] M. F. ROBERTS,[2] M. COCCIA,[1] AND H. M. CHIKARMANE[1]

Rosenstiel Basic Medical Sciences Research Center, Brandeis University, Waltham, Massachusetts 02254,[1] and Department of Chemistry, Massachusetts Institute of Technology, Cambridge, Massachusetts 02139[2]

Excess phosphate accumulation in microorganisms (10) has led to an increased interest in the role and the mechanisms of phosphate uptake and release. Polyphosphate (P_n), the form in which excess phosphate is frequently stored, has been detected in a wide variety of microorganisms. P_n may be present in the periplasmic space (20, 23) or inside the cell as long-chain cytoplasic reserves (volutin) (7, 9, 14). P_n-containing granules in bacteria have been unequivocally demonstrated by various techniques (6, 9, 24) including cytochemical methods (8). High levels of K^+ and low levels of Ca^{2+} and Mg^{2+} are associated with P_n granules. P_n is also detectable by ^{31}P nuclear magnetic resonance (^{31}P-NMR) spectroscopy, a noninvasive technique for monitoring soluble pools of phosphate-containing metabolites in cells (2, 22).

Fuhs and Chen (4), Deinema et al. (3), and M. Spector (U.S. Patent 4,056,465, November 1977) showed that effective phosphate removal from activated sludge requires alternate anaerobic and aerobic cycles which would be expected to enrich for bacterial populations capable of P_n accumulation under balanced nutritional conditions. In the anaerobic zone Timmerman (21) showed that approximately 1 mol of P_i is released to the aqueous phase per mol of glucose absorbed by the biomass from the aqueous phase. Under subsequent aerobic conditions, both the phosphate released in the anaerobic phase and that originally present in the feed to the system were taken up by the cells as the glucose was oxidized. The initial anaerobic feeding phase appears essential for efficient phosphate removal.

^{31}P-NMR studies of concentrated sludge samples have confirmed the decrease in P_n and the consequent increase in P_i after anaerobic treatment (18). The value obtained, however, did not quantitatively agree with that measured by dry weight, suggesting there are at least two P_n pools: one that is readily measured by NMR and is active physiologically and a second that is more permanent, spectroscopically invisible, and gradually converted to more accessible P_n. Further evidence comes from analysis of dried sludge by chemical methods or by scanning electron microscopy coupled with X-ray microprobe analysis. By both techniques, the measured phosphate content of the sludge remains constant whether the sample is taken oxidatively or anaerobically. The quantity of phosphate released anaerobically is only 10% of the total phosphate content of the sludge.

We have investigated the partitioning and utilization of P_n in *Acinetobacter lwoffi* JW11 to determine the role of these pools. This strain, isolated from a wastewater plant, is highly efficient in coupling phosphate removal and anaerobic release with nutrient uptake (18).

Accumulation of P_n in *A. lwoffi*

P_n content in microorganisms is known to be related to conditions of growth (7). *A. lwoffi* was grown aerobically in lactate-minimal medium, and the total P_n content was measured as a function of time. At the end of exponential growth, the concentration of P_n reached a value of 200 μmol/g (dry weight) of cells. The highest levels of P_n (24% of total cellular P) were obtained when cells were 20 h into the stationary phase.

Evidence for Multiple P_n Pools

Cytoplasmic P_n granules in *A. lwoffi* could be detected by polarized light microscopy and stained by Neisser's method. Surface P_n was detected by metachromatic spectral shifts with toluidine blue and ^{31}P-NMR spectroscopy.

Metachromatic shifts. At low concentrations, toluidine blue does not penetrate the cell membrane in microorganisms (5). It has an absorption maximum at 630 nm in aqueous solution. As shown in Fig. 1, when the dye was incubated with a negatively charged polyelectrolyte such as purified P_n, the absorption maximum shifted to 545 nm (trace B compared with trace A). When lactate-grown *A. lwoffi* cells were incubated with the dye, a similar metachromatic shift was observed (Fig. 1, trace C).

^{31}P-NMR spectroscopy. In vivo high-resolution ^{31}P-NMR spectroscopy selectively monitors mobile, soluble pools of P_n (2, 22). Cytoplasmic P_n complexes in volutin granules are not detected under normal spectral conditions. If, however, P_n complexes (presumably with divalent cations) are dispersed by treatment with metal ion chelators, a signal is observed. This characteristic

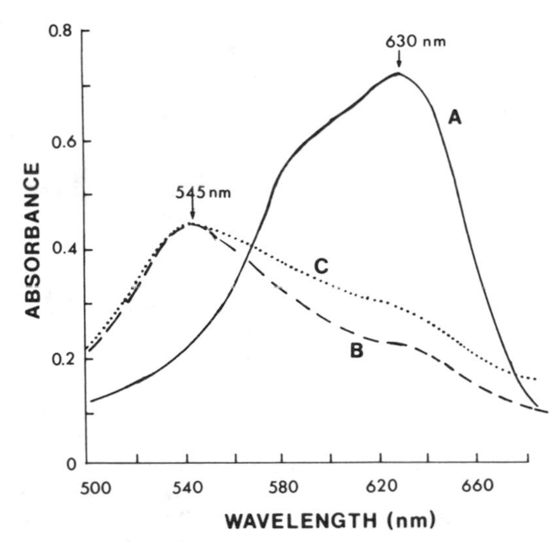

FIG. 1. Absorption spectra of 20 μM toluidine blue in 0.25 M sucrose. (A) Toluidine blue alone. (B) After the addition of 500 μM purified P_n. (C) With the addition of cells which were grown in lactate-minimal medium, washed, and resuspended in 0.25 M sucrose to an A_{600} of 0.2. An equivalent concentration of cells was also added to the reference cuvette.

allows one to distinguish between surface-associated P_n, which is accessible to membrane-impermeable chelators such as EDTA, and cytoplasmic pools of P_n. The failure to detect a P_n signal with *A. lwoffi* grown in Luria broth or lactate-minimal medium unless EDTA was included (Fig. 2A and B) supported the view that the NMR-detectable P_n in intact cells is from surface P_n that was initially immobile. If cells were lysed, increased P_n was initially detected (Fig. 2C), but this was rapidly hydrolyzed (no P_n was detected after 30 min at 4°C, whereas the NMR-observable P_n in starved cells was stable for at least 6 h). In *Mycobacterium smegmatis*, an increase in the intensity of the terminal phosphate resonance was observed when cells were treated with EDTA (15). Similarly, with *Saccharomyces fragilis*, nonpenetrating cations like UO_2^{2+} and Eu^{3+} reduced the intensity of the P_n signal (20); addition of EDTA restored it.

An estimate of total P_n present in cells requires near-quantitative extraction. Common acid extraction treatments tend to precipitate P_n-protein-ion complexes. A simple extraction procedure using NaOCl (16) selectively purifies P_n and poly-β-hydroxybutyrate. NMR spectra of this material can be used to determine the amount and the average chain length of P_n. These can then be compared with spectra of P_n observed in vivo.

In *A. lwoffi* both surface and cytoplasmic P_n were large polymers (>200 phosphate units) similar to those observed during anaerobic P_n accumulation in *Escherichia coli* (17). Surface P_n represented only a minor fraction (1.3%) of the total P_n in lactate-grown cells. This may be a minimal estimate of noncytoplasmic P_n since surface P_n inaccessible to EDTA would not be detected by NMR.

Effect of Energy Limitation on P_n Pools

P_n has been proposed as an alternative energy source when ATP is in short supply (10). Under such conditions, it is conceivable that surface P_n would be preferentially utilized. Two series of experiments suggest that surface P_n is indeed more mobile physiologically than is cytoplasmic P_n: (i) when dicyclohexylcarbodiimide (DCCD), an inhibitor of membrane ATPase, was added to cells grown under a variety of conditions, they

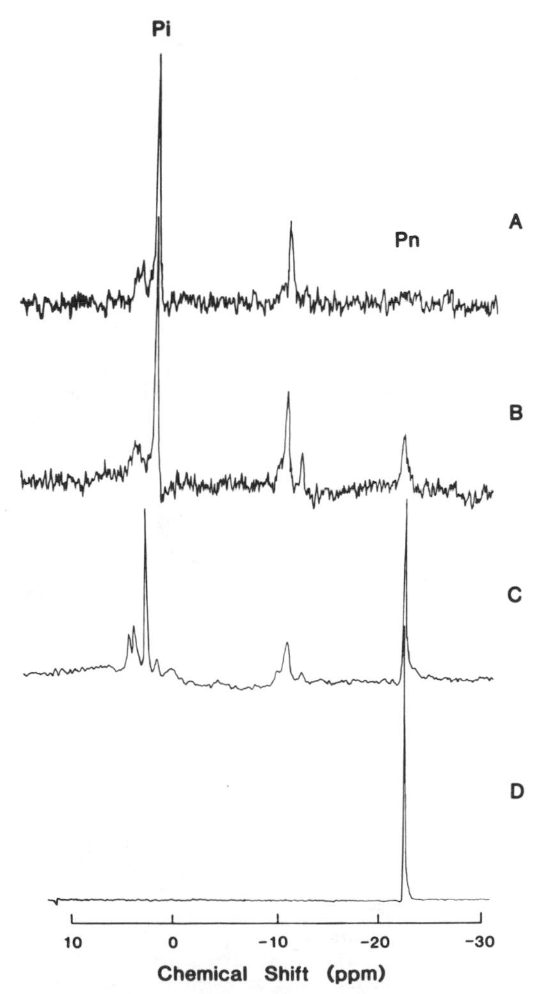

FIG. 2. ^{31}P-NMR (109.3 MHz) of the phosphate and P_n region. (A) Intact cells (~10^{11} cells per ml) resuspended in buffer. (B) EDTA (10 mM) added to intact cells. (C) EDTA (10 mM) added to cells disrupted with glass beads. (D) Hypochlorite extracts of cells. The peak at −10 ppm represents organic PP_i.

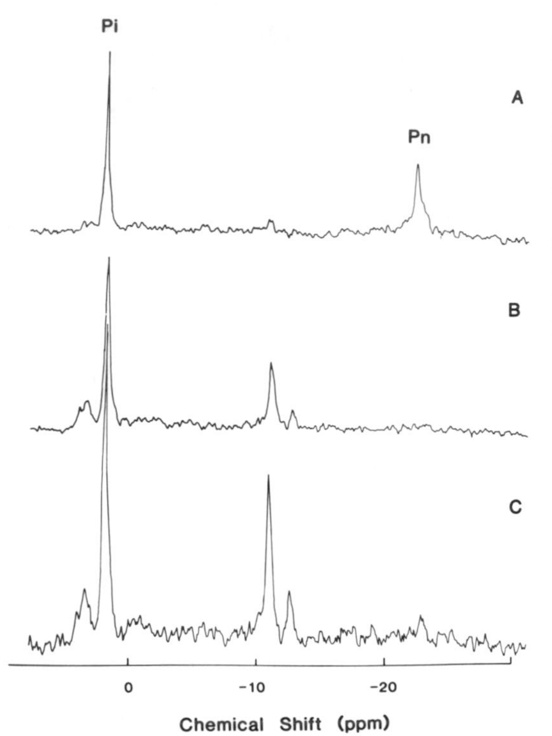

FIG. 3. ^{31}P-NMR spectra at 109.3 MHz of DCCD-treated *A. lwoffi*. Cells were grown to stationary phase and were then transferred to fresh medium containing 200 μM DCCD and aerated for 2 h. (A) Cells grown in lactate medium followed by a shift to lactate medium with DCCD. (B) Cells grown in Luria broth and then shifted to broth plus DCCD. (C) Luria broth-grown cells transferred to glucose-supplemented broth plus DCCD.

lost surface P_n, and (ii) when cells (*A. lwoffi* is a strict aerobe) were incubated under anaerobic conditions, surface P_n disappeared. In both cases, the level of cytoplasmic P_n did not change, and the loss of surface P_n was accompanied by release of P_i and accumulation of organic PP_i (Fig. 3B and C). It was shown earlier with *Acinetobacter* spp. (4, 13) that P_i is released to the medium by anaerobiosis or after addition of acetate to the medium. In addition, reduced compounds such as isocitrate, citrate, and butane-1,2 diol stimulate phosphate release during anaerobiosis (3).

Cadmium Induces Loss of Surface P_n

It was shown earlier that, under sulfate-limited conditions, *Klebsiella aerogenes* growing in the presence of Cd^{2+} loses P_n and has a substantial increase in P_i (1, 7). Cd^{2+} uptake in several bacteria occurs by either Mn^{2+} (11, 25) or Zn^{2+} (12) active transport systems. Since in neither case is substrate phosphorylation involved, Cd^{2+} active transport provides an additional system to test whether P_n can serve as an alternative energy source. The following experiments were done to see whether there is a similar effect in *A. lwoffi*.

Cd^{2+} accumulation in *A. lwoffi* was sensitive to low temperature (4°C) and to the addition of carbonylcyanide *m*-chlorophenyl-hydrazone, an uncoupling agent that abolishes proton symport (Fig. 4). Thus, Cd^{2+} is accumulated by active transport in *A. lwoffi*.

Figure 5 shows the effect of increasing concentrations of Cd^{2+} on the growth of *A. lwoffi*. It

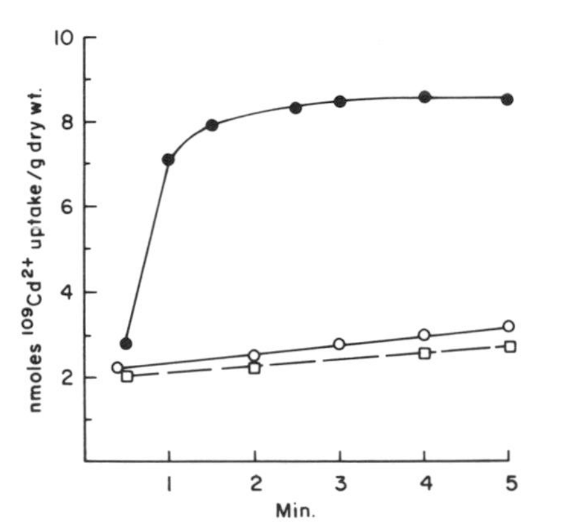

FIG. 4. Kinetics of cadmium uptake. *A. lwoffi* cells were grown to stationary phase in broth, centrifuged, resuspended in lactate medium to a density of 5×10^8 cells per ml, and incubated aerobically at 30°C. At zero time, 0.1 μM $^{109}Cd^{2+}$ was added to cultures aerated at 30°C (●) or transferred to 4°C (□), or aerated at 30°C after 5-min preincubation with 200 μM *m*-chlorophenyl-hydrazone (○). At intervals, 1-ml samples from each culture were filtered through 0.45-μm membrane filters (Millipore Corp.), washed, and counted.

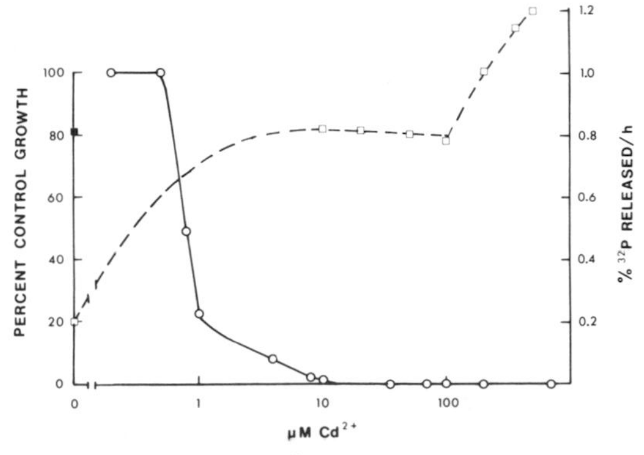

FIG. 5. Effect of Cd^{2+} on growth and phosphate release by *A. lwoffi*. Cultures were grown overnight at 30°C and diluted 1:100 into Luria broth containing Cd^{2+}. Cultures were aerated on roller drums. A_{600} was measured after 24 h (○). To measure the effect of Cd^{2+} on phosphate release, cells were grown to stationary phase in lactate-minimal medium containing 1 μCi of $^{32}P_i$ per ml, washed by centrifugation, and resuspended in one-fifth volume of minimal medium containing various concentrations of Cd^{2+}. After 2 h of aerobic (□) or anaerobic (■) incubation at room temperature (25°C), the cultures were centrifuged and a sample of the supernatant was removed to measure ^{32}P release.

is common experience that in complex medium (e.g., Luria broth) high concentrations of Cd^{2+} are required to inhibit growth. However, in synthetic medium growth is inhibited at Cd^{2+} above 0.5 μM. Cd^{2+} uptake in lactate-minimal medium was accompanied by phosphate release (Fig. 5). Between 0 and 10 μM Cd^{2+} there was a fourfold increase in ^{32}P release. This same rate was observed when *A. lwoffi* was incubated under anaerobic conditions in lactate-minimal medium. At Cd^{2+} concentrations above 200 μM the rate of P_i release was even higher.

When cells grown in lactate-minimal medium were incubated in the presence of Cd^{2+}, they showed a loss of the NMR-detectable surface P_n (Fig. 6A), while the total P_n pool was essentially unaltered (Fig. 6B). Direct Cd^{2+}-P_n interactions were not responsible, since 200 μM Cd^{2+} did not affect the line width or the intensity of high-molecular-weight P_n. The loss of surface P_n paralleled the loss of metachromasy with toluidine blue.

This report summarizes experiments with *A. lwoffi* to show that there are two pools of P_n, one surface associated and the other cytoplasmic. The surface P_n pool is the metabolically more active one and disappears under conditions of metabolic stress, e.g., ATP limitation. It is very likely that surface P_n is utilized for transport of nutrients when ATP is limiting. Tijssen et al. (19) earlier suggested that surface P_n is involved in sugar uptake and phosphorylation under anaerobic conditions in *Kluyveromyces marxianus*. This requires that the enzymes involved in P_n synthesis and degradation be associated with the cytoplasmic membrane. We have preliminary evidence in *A. lwoffi* that this might indeed be the case.

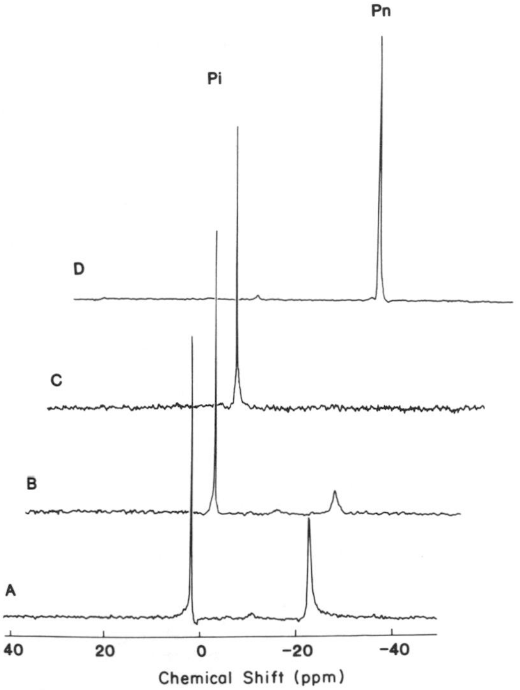

Pn
Pi
D
C
B
A
40 20 0 -20 -40
Chemicol Shift (ppm)

FIG. 6. NMR spectra of intact cells of *A. lwoffi*. (A) Grown in lactate-minimal medium. (B) As in A, incubated for 30 min with 200 μM $CdCl_2$. Traces are shown displaced for reasons of clarity.

This research was supported by grants from Air Products and Chemicals, Inc., Allentown, Pennsylvania, and International Minerals & Chemical Corporation, Northbrook, Illinois. M.F.R. acknowledges the Alfred P. Sloan Foundation for support.

LITERATURE CITED

1. **Aiking, H., A. Stijnman, C. van Garderen, H. van Heerikhuizen, and J. Van'T Riet.** 1984. Inorganic phosphate accumulation and cadmium detoxification in *Klebsiella aerogenes* NCTC 418 growing in continuous culture. Appl. Environ. Microbiol. **47:**374–377.
2. **Burt, C. T., and M. F. Roberts.** 1984. ^{31}Phosphorus nuclear magnetic resonance observation of less-expected phosphorus metabolites, p. 231–242. *In* T. L. James and L. Margulis (ed.), Biomedical magnetic resonance. Radiology Research and Education Foundation, San Francisco.
3. **Deinema, M. H., M. van Loosdrecht, and A. Scholten.** 1985. Some physiological characteristics of *Acinetobacter* spp. accumulating large amounts of phosphate. Water Sci. Technol. **17:**119–126.
4. **Fuhs, G. W., and M. Chen.** 1975. Microbiological basis of phosphate removal in the activated sludge process for the treatment of waste water. Microbol. Ecol. **2:**119–138.
5. **Griffin, J. B., N. M. Davidian, and R. Penniall.** 1965. Studies of phosphorus metabolism by isolated nuclei. VII. Identification of polyphosphate as a product. J. Biol. Chem. **240:**4427–4434.
6. **Guerrini, A. M., N. Barni, and P. Donini.** 1980. Chromatographic separation and identification of short-chain acid-soluble polyphosphates from *Saccharomyces cerevisiae*. J. Chromatogr. **189:**440–444.
7. **Harold, F. M.** 1966. Inorganic polyphosphates in biology: structure, metabolism and function. Bacteriol. Rev. **30:** 772–794.
8. **Keck, K., and H. Stich.** 1957. The widespread occurrrence of polyphosphate in lower plants. Ann. Bot. **21:**611–619.
9. **Kulaev, I. S.** 1979. The biochemistry of inorganic polyphosphates. John Wiley & Sons, Inc., New York.
10. **Kulaev, I. S., and V. M. Vagabov.** 1983. Polyphosphate metabolism in microorganisms. Adv. Microb. Physiol. **24:** 83–171.
11. **Laddaga, R. A., R. Bessen, and S. Silver.** 1985. Cadmium-resistant mutant of *Bacillus subtilis* 168 with reduced cadmium transport. J. Bacteriol. **162:**1106–1110.
12. **Laddaga, R. A., and S. Silver.** 1985. Cadmium uptake in *Escherichia coli* K-12. J. Bacteriol. **162:**1100–1105.
13. **Lotter, L. H.** 1985. The role of bacterial phosphate metabolism in enhanced phosphorus removal from the activated sludge process. Water Sci. Technol. **17:**127–138.
14. **Miller, J. J.** 1984. *In vitro* experiments concerning the state of polyphosphate in the yeast vacuole. Can. J. Microbiol. **30:**236–246.
15. **Ostrovskii, D. N., N. F. Sepetov, V. I. Reshetnyak, and L. S. Siberl'dina.** 1980. Investigation of the localization of polyphosphates in cells of microorganisms by the method of high-resolution ^{31}P-NMR-145·78 MHz. Biokhimiya **45:** 392–398.
16. **Poindexter, J. S., and L. F. Alley.** 1983. Combined proce-

dure for assays of poly-β-hydroxybutyric acid and inorganic polyphosphate. J. Microbiol. Methods **1**:1–17.

17. **Rao, N. N., M. F. Roberts, and A. Torriani.** 1985. Amount and chain length of polyphosphates in *Escherichia coli* depend on cell growth conditions. J. Bacteriol. **162**:242–247.
18. **Suresh, N., R. Warburg, M. Timmerman, J. Wells, M. Coccia, M. F. Roberts, and H. O. Halvorson.** 1985. New strategies for the isolation of microorganisms responsible for phosphate accumulation. Water Sci. Technol. **17**:99–111.
19. **Tijssen, J. P. F., P. J. A. van den Broek, and J. van Steveninck.** 1984. The involvement of high-energy phosphate in 2-deoxy-D-glucose transport in *Kluyveromyces marxianus*. Biochim. Biophys. Acta **778**:87–93.
20. **Tijssen, J. P. F., and J. van Steveninck.** 1984. Detection of a yeast polyphosphate fraction localized outside the plasma membrane by the method of phosphorus-31 nuclear magnetic resonance. Biochem. Biophys. Res. Commun. **119**:447–451.
21. **Timmerman, W. M.** 1979. Biological phosphate removal from domestic waste water using anerobic/aerobic treatment. Dev. Ind. Microbiol. **20**:285–298.
22. **Ugurbil, K., R. G. Shulman, and T. R. Brown.** 1979. High resolution ^{31}P and ^{13}C nuclear magnetic resonance studies of *Escherichia coli* cells *in vivo*, p. 537–589. *In* R. G. Shulman (ed.), Biological applications of magnetic resonance. Academic Press, Inc., New York.
23. **Umnov, A. M., A. G. Steblyak, S. Umnova, S. E. Mansurova, and I. S. Kulaev.** 1975. Possible physiological role of the high molecular weight polyphosphate and polyphosphate phosphohydrolase system in *Neurospora crassa*. Mikrobiologiya **44**:414–421.
24. **Varma, A. K., W. Rigsby, and D. C. Jordan.** 1983. A new inorganic pyrophosphate utilizing bacterium from a stagnant lake. Can. J. Microbiol. **29**:1470–1474.
25. **Weiss, A. A., S. Silver, and T. G. Kinscherf.** 1978. Cation transport alteration associated with plasmid-determined resistance to cadmium in *Staphylococcus aureus*. Antimicrob. Agents Chemother. **14**:856–865.

Polyphosphate Kinase and Polyphosphate Glucokinase of *Propionibacterium shermanii*

HARLAND G. WOOD, NANCY A. ROBINSON, CATHERINE A. PEPIN, AND JOAN E. CLARK

Department of Biochemistry, School of Medicine, Case Western Reserve University, Cleveland, Ohio 44106

Our investigations of polyphosphates [poly-(P)s] were stimulated by the reports of Kulaev and co-workers (4, 6, 18) that *Propionibacterium shermanii* contains enzymes which catalyze the following reactions:

$$\text{ATP} + \text{poly(P}_n) \xrightleftharpoons{\text{poly(P) kinase}} \text{ADP} + \text{poly(P}_{n+1})$$

$$\text{Glucose} + \text{poly(P}_n) \xrightarrow{\text{poly(P) glucokinase}} \text{glucose 6-phosphate} + \text{poly(P}_{n-1})$$

$$\text{1,3-diphosphoglycerate} + \text{poly(P}_n) \xrightleftharpoons{\text{poly(P) 3-phosphoglycerate kinase}} \text{3-phosphoglycerate} + \text{poly(P}_{n+1})$$

We wondered whether poly(P) is used by *P. shermanii* as a source of energy in addition to PP_i and ATP. Our earlier studies with *P. shermanii*, beginning with the discovery of carboxytransphosphorylase (13), followed by the discovery of other enzymes that use PP_i, had shown that PP_i is used as an energy source by this organism (19, 20). It is now clear that PP_i is utilized by numerous organisms (see the articles in this volume by Reeves, Baltscheffsky and Nyrén, and Black et al.). To begin the studies with poly(P), we surveyed the enzymes of *P. shermanii* and confirmed the presence of poly(P) kinase and poly(P) glucokinase but found insignificant amounts of poly(P) phosphoglycerate kinase present and an abundant amount of ATP 3-phosphoglycerate kinase (21).

Although poly(P)s occur in practically all forms of life and although there have been numerous publications during the past 35 years concerning their possible roles (see references 3 and 5 for comprehensive reviews), much uncertainty still remains as to their exact role in metabolism. This uncertainty arises in part (i) because no methods are available for isolating long-chain poly(P) from cells, (ii) because the methods for accurately determining the sizes of poly(P) are inadequate, especially if only a small amount of poly(P) is available, and (iii) because the properties of the enzymes of poly(P) metabolism have not been thoroughly characterized. It is clear that one must have information concerning the properties of the enzymes involved to evaluate the role of poly(P) in metabolism. Therefore, we have undertaken an investigation of the properties of poly(P) kinase and poly(P) glucokinase, which is reviewed here; the isolation and sizing of poly(P) are dealt with elsewhere (1, 1a, 10, 11).

Properties of Poly(P) Kinase

Two questions, important in evaluating the role of poly(P) kinase in metabolism, have existed for a long time: (i) is the catalytic mechanism of poly(P) kinase strictly processive, and (ii) is there an initiator or primer involved in the reaction? Kornberg et al. (2), in 1956, suggested that a primer might be involved in the reaction. They found that $^{32}PP_i$ was incorporated into poly(P) when included in the reaction along with ATP. However, PP_i was not required and even inhibited the rate of the reaction. It was concluded that PP_i is not the in vivo primer. Muhammed (7) found that addition of poly(P) to a reaction with the poly(P) kinase from *Corynebacterium xerosis* had no effect on the rate of synthesis and stressed that the involvement of a primer was not supported by the experimental data.

We have purified the poly(P) kinase from *P. shermanii* to about 70% homogeneity and have shown by activity staining of nondenaturing gels, by gel filtration, and by sodium dodecyl sulfate-gel electrophoresis that it is a monomeric enzyme of M_r 83,000 (N. A. Robinson, J. E. Clark, and H. G. Wood, J. Biol. Chem., in press).

We found that the addition of P_i or short-chain poly(P) stimulates the rate of synthesis of poly(P) from ATP about 10-fold (Robinson et al., in press). The question then becomes, is either P_i or short-chain poly(P) incorporated in the long-chain product and perhaps serving as a primer in the reaction, or is the stimulation of rate due solely to an allosteric effect? We found that $^{32}P_i$ was incorporated in the long-chain poly(P) product; however, we also found that our enzyme preparation (either the enzyme per se or a contaminant) catalyzed an exchange of $^{32}P_i$ with ATP, and thus further analysis of the direct incorporation of $^{32}P_i$ was not possible. However, short-chain [^{32}P]poly(P) was also incorporated in the long-chain product, and since

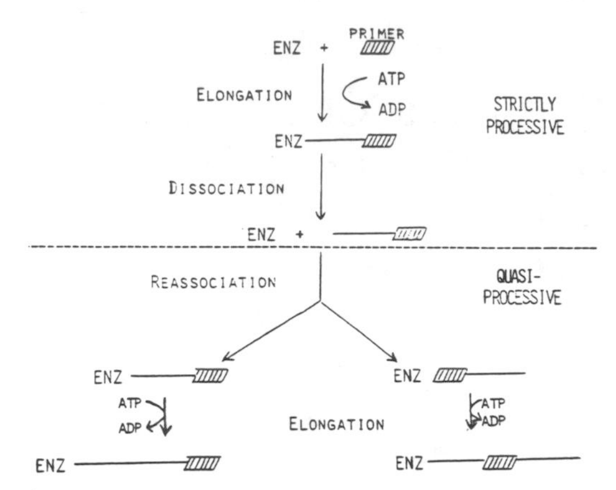

FIG. 1. Illustration of a strictly processive process and a quasi-processive process and the use of a ^{32}P-labeled primer to differentiate between them.

ATP was not labeled from this source, further analysis could be performed.

Our definition of a strictly processive process is illustrated in Fig. 1. It is assumed in this illustration that the reaction occurs with the utilization of a short-chain ^{32}P-labeled poly(P) primer and that elongation of the chain occurs by repeated addition of phosphate from unlabeled ATP, prior to dissociation of the chain from the enzyme. If the process is strictly processive the release of poly(P) from the enzyme terminates the reaction. There is no reassociation following dissociation in a strictly processive process, and the poly(P) is end labeled as illustrated in Fig. 1. In a quasi-processive process, by our definition, there is reassociation with the enzyme and further additions of phosphate residues from ATP to the chain. In this case, if poly(P) is a symmetrical molecule (which seems to be a reasonable assumption), the reassociation will be random with either end, and half of the molecules will have the ^{32}P of the primer internalized in the chain as illustrated in Fig. 1. Of course, if the process is nonprocessive, i.e., if there is release of the poly(P) from the enzyme after each addition of phosphate, the ^{32}P will likewise be internalized. The essential point is that, if the process is strictly processive, the poly(P) will be exclusively end labeled.

The procedure used in our experiment is outlined in Fig. 2. [^{32}P]poly(P) was synthesized with poly(P) kinase using [γ-^{32}P]ATP, and the resulting long-chain [^{32}P]poly(P) was hydrolyzed in boiling 1 N NaOH and then fractionated by gel filtration. Fractions containing chain lengths from 5 to 80, as determined by gel electrophoresis, were combined. Long-chain poly(P) was then synthesized with poly(P) kinase using unlabeled ATP in combination with this short-chain [^{32}P]poly(P). The resulting long-chain poly(P) was in turn separated from the unreacted short-chain [^{32}P]poly(P) and was then allowed to react with poly(P) glucokinase, which removes phosphates exclusively from the ends. Samples (1,000 cpm) were removed when 0, 20, 40, 60, 80, and 100% (lanes 1 through 6, respectively, of Fig. 3) of the poly(P) had been utilized by the poly(P) glucokinase and were electrophoresed on a 15% polyacrylamide gel. An autoradiogram of the gel is shown in Fig. 3C, and the gel stained with toluidine blue, for observation of the total residual poly(P), is shown in Fig. 3A. It is clear from lane 2, Fig. 3C, that all the label was removed from the poly(P) when only 20% of the poly(P) was utilized. The ^{32}P was present in the glucose 6-phosphate which migrated at the lower end of the gel (not shown). It is also clear from the toluidine blue stain (Fig. 3A) that a large amount of the poly(P) still remained unused after all the radioactivity had been removed from it. A control experiment with [γ-^{32}P]ATP and unlabeled short-chain poly(P) was done to obtain results with poly(P) labeled internally. The autoradiogram is shown in Fig. 3D, and the toluidine blue stain is shown in Fig. 3B. In this control, with the poly(P) labeled throughout the

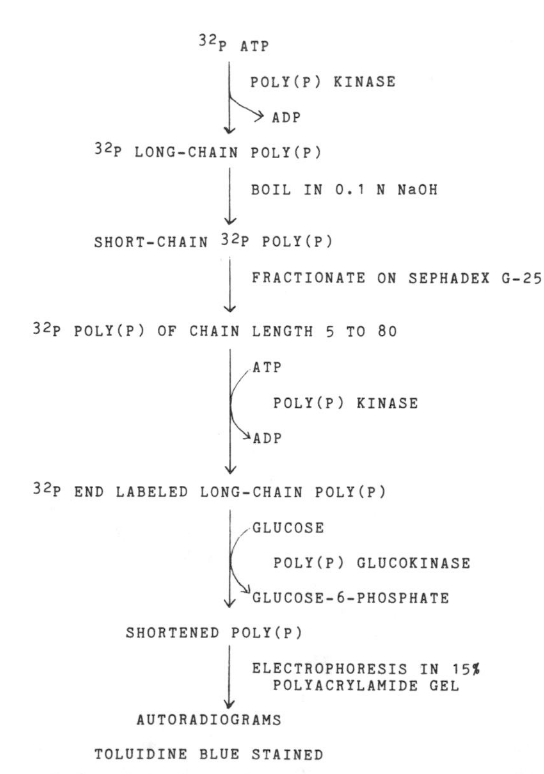

FIG. 2. Procedure for the preparation of short-chain ^{32}P-labeled primer and for proving that, when it is incorporated by poly(P) kinase into long chains, the resulting poly(P) is end labeled.

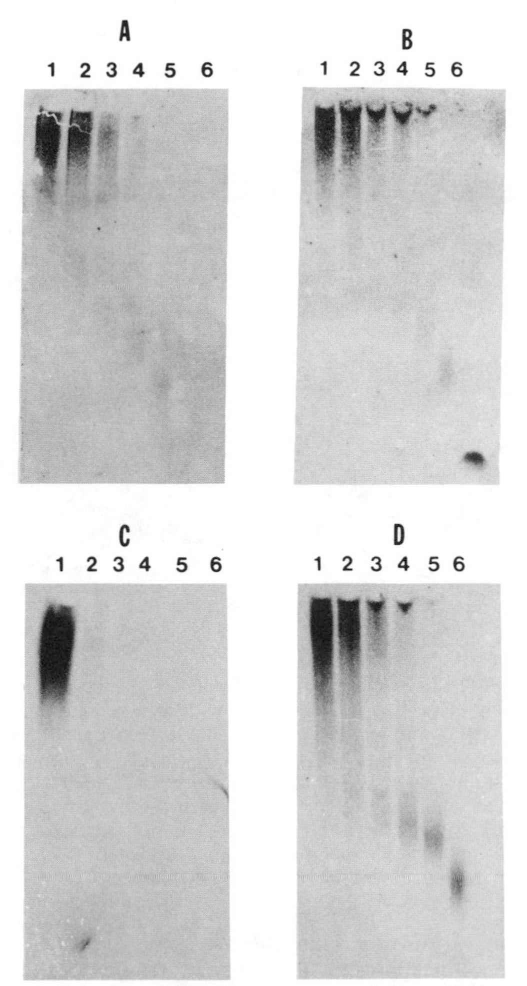

FIG. 3. Demonstration by use of poly(P) glucokinase and by electrophoresis (15% polyacrylamide) that poly(P), synthesized by poly(P) kinase wih ^{32}P-labeled short-chain poly(P) as a primer, is end labeled and that the mechanism is strictly processive (panels A and C). Panels B and D are controls with poly(P) synthesized with [^{32}P]ATP and unlabeled primer. A and B were stained with toluidine blue; C and D are autoradiograms. Samples in lanes 1 through 6 were taken after 0, 20, 40, 60, 80 and 100% of the poly(P) was utilized by poly(P) glucokinase. (From Robinson et al., in press).

chain, except at one end, the radioactivity remained present until all the poly(P) was utilized (Fig. 3D) and corresponded to the toluidine blue stain of the total phosphate (Fig. 3B).

These results clearly show that the short-chain [^{32}P]poly(P) was located only at the end of the long-chain poly(P) synthesized from ATP by poly(P) kinase. Therefore, this short-chain poly(P) serves as a primer, and the mechanism of the poly(P) kinase is strictly processive.

Although the mechanism of poly(P) kinase is strictly processive, the termination of synthesis does not yield a narrow range of sizes. The termination is influenced by the temperature of the reaction and the concentration of the primer. Figure 4 shows densitometry scans of lanes from a 1.5% agarose gel which had been stained with toluidine blue. The relative optical amplitude, plotted on the ordinate, is proportional to the total phosphate concentration. The chain lengths, indicated at the top, were determined as described elsewhere (1a). The poly(P) synthesis was conducted at 20°C (top panel) and 30°C (bottom panel). Experiments were done with ATP in the presence of 20 mM P_i, 0.1 mg of Sigma type 15 poly(P) per ml as primer, or 0.5 mg of type 15 poly(P) per ml as primer. It is obvious from Fig. 4 that the longest poly(P) was present in the largest proportion when formed from ATP and P_i and that the largest proportion of shorter poly(P) occurred when the primer concentration was the highest (0.5 mg/ml) and the temperature was 30°C. Apparently, the off rate of the poly(P) is increased with temperature more than the overall rate of elongation. Perhaps when the concentration of primer is high, it competes with and displaces the poly(P) from the enzyme more frequently, thus leading to the formation of shorter chains.

Although we have gained considerable information about poly(P) kinase, many puzzling features remain. For example, when calculations are made of chain length, based on the

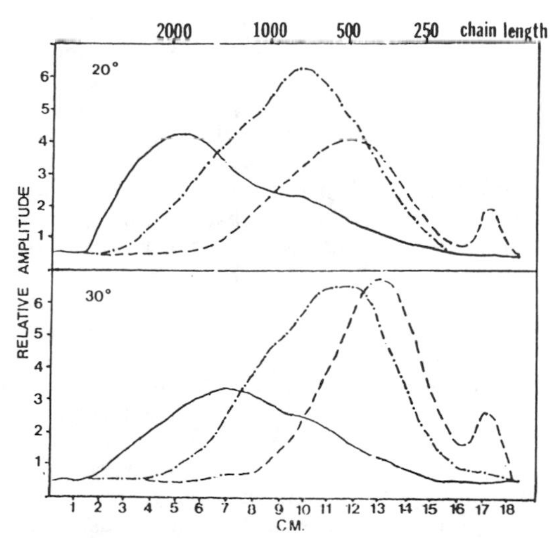

FIG. 4. Effect of temperature, concentration of primer, and phosphate on termination of synthesis of poly(P) by poly(P) kinase. The poly(P) from each reaction was electrophoresed in a 1.5% agarose gel. ——,With ATP and 20 mM P_i; —·—, with ATP and type 15 poly(P) (0.1 mg/ml); – – –, with ATP and type 15 poly(P) (0.5 mg/ml). With 0.5 mg of primer per ml, the excess primer is evident at 17 to 18 cm. The relative amplitude [proportional to phosphate in poly(P)] is plotted versus centimeters of migration into the gel. (From Robinson et al., in press.)

assumption that 1 mol of primer is incorporated per mol of synthesized poly(P), the values do not agree with those obtained by other methods. It appears that the synthesis of poly(P) may be initiated by two distinct pathways, one which is unknown and another which involves poly(P) as a primer (see Robinson et al. [in press] for further details). We have observed (11, 12) that basic proteins, such as histones or polylysine, form insoluble complexes with poly(P) kinase, and the rate of synthesis of poly(P) is about fourfold higher than in the absence of basic protein. The reverse reaction is inhibited by basic proteins (12). Perhaps basic proteins, such as histones, have some role in the regulation of poly(P) kinase in vivo.

Properties of Poly(P) Glucokinase

Poly(P) glucokinase was discovered by Szymona in 1956 (14) and occurs in a wide variety of microorganisms (5), but so far, it has not been reported in animals, plants, eubacteria, yeasts, or fungi. One of the curious features of poly(P) glucokinase is that, even with a homogeneous protein (8), some activity has been observed with ATP in the absence of poly(P). As yet, it is not certain whether the single protein has two activities or whether there are two glucokinases which have not been completely separated. In our case (9), we have purified the enzyme 960-fold, and the ratio of poly(P) glucokinase to ATP glucokinase activity has remained fairly constant throughout the purification at approximately 4:1. It seems likely that both activities are catalyzed by the same protein (9).

An interesting feature of the poly(P) glucokinase reaction with long-chain poly(P) is that the chain is reduced in size to about 100 residues and then the poly(P) accumulates with very little change in size until all the longer chains are utilized. In these initial experiments (9), Maddrell salt was used as the substrate, and samples were removed after reaction with poly(P) glucokinase and electrophoresed in 15% polyacrylamide gels. Since no intermediate sizes were observed prior to chain length 100, we considered the mechanism to be strictly processive. It was recognized, however, that the formation of intermediate sizes might not have been detected because of the limited resolution of large sizes on the 15% gel. We have now investigated this question in greater detail (C. A. Pepin and H. G. Wood, J. Biol. Chem., in press), and the results are shown in Fig. 5. In this experiment, two concentrations of gels were used to analyze the products formed by poly(P) glucokinase during reaction with substrate poly(P_{700}) plus poly(P_{210}). The 15% polyacrylamide gel (left panel) is appropriate for monitoring the shortening of the poly(P_{210}) substrate, and

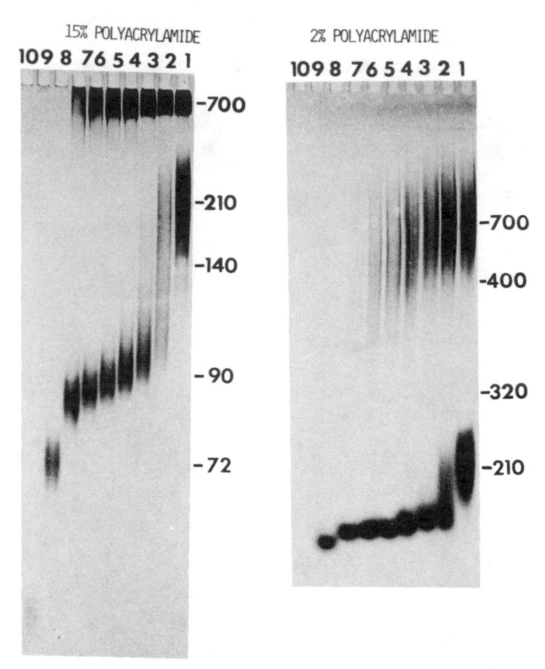

FIG. 5. Electrophoresis of poly(P) formed from a combination of poly(P)s, one with an average chain length of 210 and the other of 700. Lanes 1 to 10 are from the reaction at 0, 8, 16, 24, 30, 38, 46, 54, 62, and 71 min, respectively. Two concentrations of gels were used which permit resolution of the different sizes: 15% polyacrylamide (left) for poly(P) of chain length 210 and 2% polyacrylamide–0.5% agarose for chain length 700. (From Pepin and Wood, in press.)

the 2% polyacrylamide–0.5% agarose gel (right panel) is appropriate for the poly(P_{700}) substrate (1a). Samples initially containing 10 μg of each poly(P) substrate were removed after 0, 8, 16, 24, 30, 38, 46, 54, 62, and 71 min of reaction (lanes 1 through 10, respectively). With 15% polyacrylamide, the long-chain poly(P) remained stacked at the top of gel. The poly(P_{210}) migrated into the gel (Fig. 5, left, lane 1), and the resolution was sufficient with the 15% polyacrylamide gel to determine changes in this size of poly(P). It is evident at 8 min (lane 2) that a wide range of sizes were present, from about chain length 190 to 90, and by 24 min (lane 4), most of this poly(P) had been reduced to an average chain length of about 90. These results clearly show that the poly(P_{210}) was not used by a strictly processive process. From both sizes of poly(P), shortened poly(P) was formed that accumulated, and its chain length did not decrease substantially until all the long-chain poly(P), stacked at the top of the gel, was utilized (Fig. 5, lane 9). Only then was the accumulated short-chain poly(P) shortened further (lane 10).

For investigation of the changes occurring with the poly(P_{700}), the 2% polyacrylamide gel was used (Fig. 5, right). On this gel, both poly(P) fractions migrated into the gel. Again, it is evident that poly(P_{210}) was rapidly reduced to the size that accumulates. The 2% gel shows that a wide range of sizes were formed from poly(P_{700}). For example, in lanes 6 and 7 (38 and 46 min), some of the chains have been reduced in size to about 320 residues, and a wide span of sizes is present. The stain is very light since, at this time, a large part of the poly(P) has been converted to glucose 6-phosphate. Clearly, the longer chains are not used by a strictly processive process. Thus, these results confirm those obtained with the end-labeled poly(P) in the experiment of Fig. 3.

Experiments also were done with short-chain forms of poly(P) (9). These experiments are very informative because the individual bands are visible on 15% polyacrylamide gels. In Fig. 6, results are shown which were obtained with both poly(P) glucokinase (panel A) and poly(P) kinase (panel B). Two discrete sizes of poly(P) were included as substrates, and successive samples were removed and electrophoresed in a 15% polyacrylamide gel. Two differences are immediately apparent. First, the longer chain is used preferentially by the poly(P) glucokinase and the smaller size is not used until the longer size is reduced to the smaller size. Poly(P) kinase, on the other hand, uses both sizes simultaneously and, thus, does not differentiate between sizes. Second, there is a sequential decrease in the size of the poly(P) by poly(P) glucokinase. Clearly, this mechanism is not strictly processive. In contrast, no intermediate sizes are formed by poly(P) kinase, the mechanism being strictly processive.

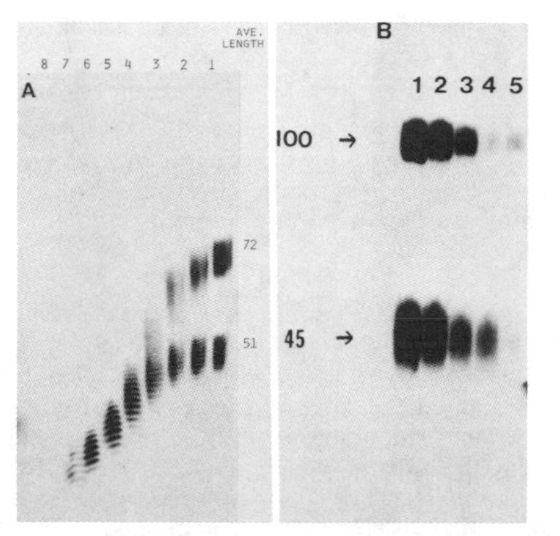

FIG. 6. Comparison of the utilization of short-chain poly(P) in the phosphorylation of glucose by poly(P) glucokinase (A) and in the phosphorylation of ADP by poly(P) kinase (B). Two discrete sizes as indicated were included as substrates in each case. Samples were removed at 0, 1, 2, 4, 6, 8, 10, and 12 min (lanes 1 through 8, respectively) in the experiment with poly(P) glucokinase and at 0, 10, 15, 20, and 25 min (lanes 1 through 5, respectively) with poly(P) kinase. Electrophoresis was in a 15% polyacrylamide gel. (A is from reference 9, and B is from reference 12.)

Initial velocity kinetics were examined with poly(P) glucokinase, using poly(P) of various chain lengths (9). The poly(P) samples contained discrete sizes of poly(P) and were isolated by preparative gel electrophoresis. The initial rate was determined with various concentrations of poly(P) of each size, and the K_m was estimated from a plot of $1/V$ versus $1/S$. Such studies would not have been possible with commercial sources of poly(P) because of the wide range of sizes that are present. The results of the experiments are summarized in Fig. 7, in which the chain length of poly(P) is plotted against the log of K_m and also against the log of V_{max}/K_m. The curve for K_m was quite similar when plotted on the basis of the total phosphate in the samples of poly(P). A striking feature was the remarkable decrease in K_m with increase in chain length. There was a 2,000-fold difference (4.3 μM/2.0 × 10^{-3} μM) in the K_m of poly(P_{30}) from that of poly(P_{724}). A distinct inflection of K_m occurred at about chain length 100, and the K_m rose rapidly thereafter as the chain length was shortened. This change coincided with the observed accumulation of poly(P) at a chain length of about 100 when longer chains were the substrate, suggesting that the K_m is indeed a reflection of the affinity of the enzyme for poly(P). The long chains with a high affinity (low K_m) apparently saturate the active site, and the short chains with a low affinity cannot compete until their concentration is much higher than that of the long chains.

It was also observed that, with long chains, the V_{max}/K_m was more or less constant, beginning at a chain length of about 200, and was larger than those observed with the shorter chains. This is in accord with the observation that long chains are used preferentially prior to the utilization of the shorter chains.

We have seen (Fig. 5 and 6) that the mechanism of poly(P) glucokinase is not strictly processive. The question then arises, is the mechanism nonprocessive and the poly(P) released from the enzyme after each phosphorylation or is it quasi-processive and several phosphorylations occur before the poly(P) is released from the enzyme? The experiment of Fig. 6 with short chains provides some clues. In lanes 6 and 7 of Fig. 6A, about seven bands [each band designating a poly(P) differing by one residue] are visible. If the mechanism were quasi-processive and involved, for example, 10 phosphorylations prior to release from the enzyme, there

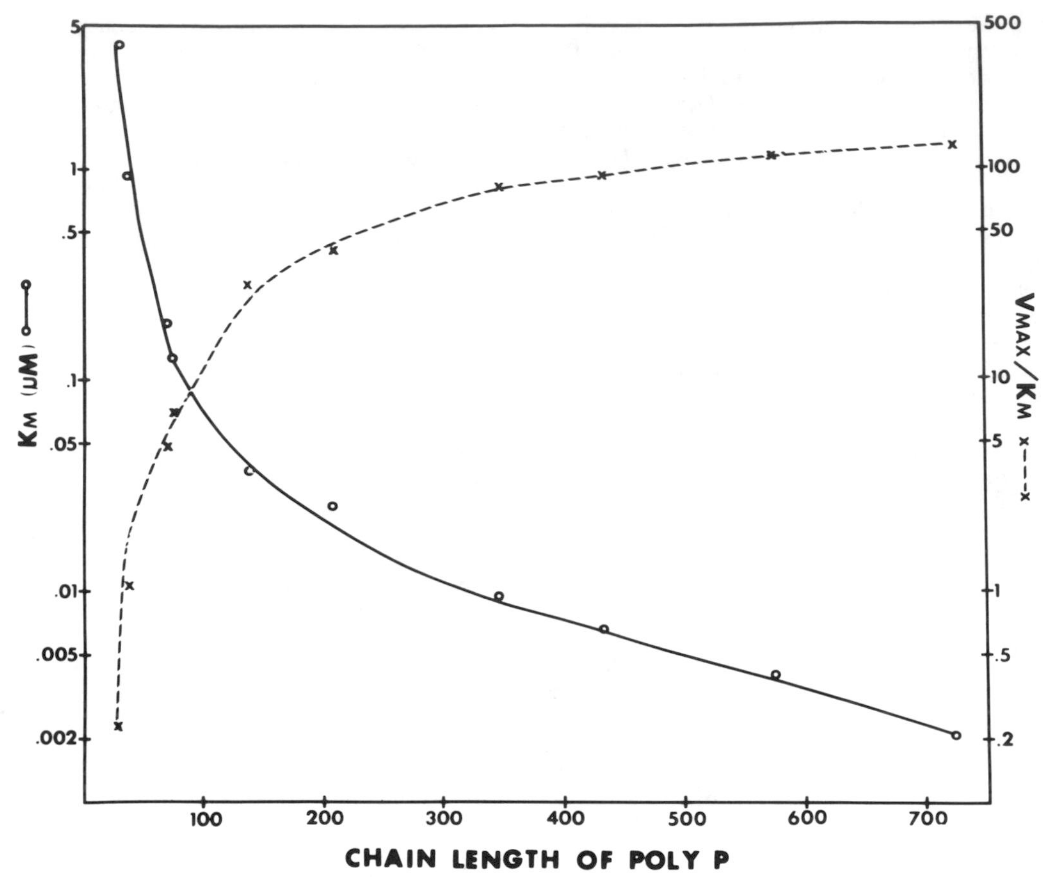

FIG. 7. Semilog plot of K_m and of V_{max}/K_m of poly(P) glucokinase for poly(P) versus chain length. (From reference 9.)

would be a gap equivalent to three or four bands between the previous group and the newly formed group. Thus, if the mechanism is quasi-processive with short chains, apparently no more than five or six phosphorylations are involved in succession prior to release from the enzyme. With long chains, the affinity is so high, compared with short chains, that it seems more likely that several phosphorylations would occur prior to release from the enzyme.

It is interesting that long chains consisting of a large span of sizes are converted to a very limited number of sizes of short chains (see Fig. 5 and reference 9). A possible explanation was that long chains are used preferentially, as occurs with the longer of the short chains (Fig. 6), therefore causing a reduction in the span of sizes. However, it is to be noted that ^{32}P was removed from end-labeled poly(P), which contained a large span of sizes (see Fig. 4), long before most of the poly(P) was utilized; thus, there was little if any differentiation in the utilization by size among the long chains. (See also Fig. 5, which shows there is no differentiation of sizes of chain lengths greater than 100.) Apparently, the clustering of sizes occurs when the chain lengths approach 100, and the longer chains are then used preferentially.

Concluding Remarks

One role poly(P) could have in the metabolism of the propionibacteria is in the phosphorylation of glucose by poly(P) glucokinase. We have investigated the rates of the ATP- and poly(P)-dependent glucokinases in these organisms (21). From the rate at which glucose was used by *Propionibacterium freudenreichi*, it was estimated that 1.8 U of ATP glucokinase would be required per g of cells (wet weight) to provide for the required conversion of glucose to glucose 6-phosphate. We found the activity of the ATP glucokinase to be 1.2 ± 0.4 U/g of cells and that with poly(P) to be 5.6 ± 2.1 U. The corresponding values for *P. shermanii* were 2.3 ± 0.6 U/g of cells with ATP and 7.1 ± 1.5 U/g of cells with poly(P). Clearly, with both species, the poly(P) activity was much greater than the ATP activity, and there is every reason to consider that poly(P) has a role in the phosphorylation of glucose.

One of the most convincing pieces of evidence that poly(P) is used as a source of energy for phosphorylation of sugars has been provided by Szymona and Szumilo (17). They found that, when *Mycobacterium phlei* is grown in fructose, poly(P) fructokinase is present, but it is absent when glucose is the source of carbon. It seems probable that the enzyme would be adaptive only if poly(P) were actually being used in the pathway. Two other adaptive enzymes have been observed (16), poly(P) mannokinase and poly(P) gluconatokinase.

The only means that we have found for formation of poly(P) by *P. shermanni* is via poly(P) kinase. Assuming that this is the case and that poly(P) is utilized by poly(P) glucokinase, what can we predict from the mechanisms of these two enzymes? On the one hand, the poly(P) glucokinase would shorten the poly(P) chain, and on the other hand, the poly(P) kinase would utilize the shortened poly(P) as a primer, lengthening the chain without release until it had made a long chain. The poly(P) glucokinase has a much lower K_m for long-chain poly(P) than for short chains. Thus, it would be predicted that the long chains would be used by the poly(P) glucokinase preferentially and short chains would accumulate, if the production of long chains by the poly(P) kinase were rate limiting.

In the lactate-grown cells, when no glucose is present, only long chains are found, presumably because in the absence of glucose there is no phosphorylation to promote its removal and turnover (1). We have not found a polyphosphatase which hydrolytically cleaves the poly(P) chain, although Kulaev et al. (6) have reported the presence of a polyphosphatase in *P. shermanii*. Glucose-grown cells contain 100-fold less poly(P) than lactate-grown cells (1), which may indicate that the production of poly(P) by poly(P) kinase is rate limiting. Both lactate-grown cells and glucose-grown cells contain poly(P) kinase (21), but since the assay is difficult in the crude extracts, exact activities are uncertain. Short-chain poly(P) is found in glucose-grown cells but not in lactate-grown cells, indicating that the long-chain poly(P) is shortened as a result of utilization by the poly(P) glucokinase.

Further evidence that poly(P) is utilized to phosphorylate glucose has been obtained by transferring lactate-grown cells, containing a high amount of poly(P), to fresh glucose media. During incubation with glucose, the poly(P) decreased in a manner consistent with the mechanism of poly(P) glucokinase. The extremely long-chain poly(P) decreased, and there was accumulation in 15% polyacrylamide gels of shorter poly(P) of about 100 residues. In similar studies with *M. phlei*, Szymona and Szymona (15) demonstrated that, on incubation with glucose, the volutin granules in the cells disappeared.

It is intriguing to consider how the long-chain poly(P) combines with poly(P) glucokinase. From crystallographic data, poly(P) forms a helical structure with a period of 0.7 nm (3). If it is assumed that there are three phosphoryl residues per period, then poly(P_{725}) would have a length of 170 nm. An enzyme of molecular weight 90,000 (enolase) has a diameter of about 7 nm; thus, for this size protein, the poly(P) is 25 times longer than the protein and 3 times longer than its circumference. The K_m of poly(P) with poly(P) glucokinase decreases with an increase in the chain length. It therefore appears that the entire chain is involved in binding to the enzyme. How this occurs is of considerable interest. Perhaps some understanding can be obtained by electron microscopy of the enzyme-poly(P) complex.

This work was supported by Public Health Service grant GM29569 from the National Institutes of Health.

LITERATURE CITED

1. **Clark, J. E., H. Beegen, and H. G. Wood.** 1986. Isolation of intact chains of polyphosphate from *Propionibacterium shermanii* grown on glucose or lactate. J. Bacteriol. **168:** 1212–1219.

1a. **Clark, J. E., and H. G. Wood.** 1987. Preparation of standards and determination of sizes of long-chain polyphosphates by gel electrophoresis. Anal. Biochem. **161:**280–290.

2. **Kornberg, A., S. R. Kornberg, and E. C. Simms.** 1956. Metaphosphate synthesis by an enzyme from *E. coli*. Biochim. Biophys. Acta **20:**215–227.

3. **Kulaev, I. S.** 1979. The biochemistry of inorganic polyphosphates. John Wiley & Sons, Inc., New York.

4. **Kulaev, I. S., M. A. Bobyk, N. M. Nikalaev, N. S. Serveev, and S. O. Uryson.** 1971. Polyphosphate synthesizing enzymes in some fungi and bacteria. Biochem. Russ. **36:**791–795. (English translation.)

5. **Kulaev, I. S., and V. M. Vagabov.** 1983. Polyphosphate metabolism in microorganisms. Adv. Microb. Physiol. **24:** 83–171.

6. **Kulaev, I. S., L. I. Vorob'eva, L. V. Konovalova, M. A. Bobyk, G. I. Konoshenko, and S. O. Uryson.** 1973. Enzymes of polyphosphate metabolism during the growth of *P. shermanii* under normal conditions and in the presence of polymyxin M. Biochem. Russ. **38:**595–599. (English translation.)

7. **Muhammed, A.** 1961. Studies on biosynthesis of polymetaphosphate by an enzyme from *Corynebacterium xerosis*. Biochim. Biophys. Acta **54:**121–123.

8. **Pastuszak, I., and M. Szymona.** 1980. Occurrence of a large molecular size form of polyphosphate-glucose phosphotransferase in extracts of *Mycobacterium tuberculosis*. Acta Microbiol. Pol. **29:**49–56.

9. **Pepin, C. A., and H. G. Wood.** 1986. Polyphosphate glucokinase from *Propionibacterium shermanii*. Kinetics and demonstration that the mechanism involves both processive and nonprocessive type reactions. J. Biol. Chem. **261:**4476–4480.

10. **Pepin, C. A., H. G. Wood, and N. A. Robinson.** 1986.

Determination of the size of polyphosphates with polyphosphate glucokinase. Biochem. Int. **12:**111–123.

11. **Robinson, N. A., N. H. Goss, and H. G. Wood.** 1984. Polyphosphate kinase from *Propionibacterium shermanii*. Formation of an enzymatically-active insoluble complex with basic proteins and characterization of synthesized polyphosphate. Biochem. Int. **8:**757–769.
12. **Robinson, N. A., and H. G. Wood.** 1986. Polyphosphate kinase from *Propionibacterium shermanii*. Demonstration that the synthesis and utilization of polyphosphate is by a processive mechanism. J. Biol. Chem. **261:**4481–4485.
13. **Siu, P. M., and H. G. Wood.** 1962. Phosphoenolpyruvic carboxytransphosphorylase, a CO_2 fixation enzyme from propionic acid bacteria. J. Biol. Chem. **237:**3044–4051.
14. **Szymona, M.** 1956. The phosphorylation of glucose and glucosamine by acetone-powder of *Mycobacterium phlei*. Bull. Acad. Pol. Sci. Cl. 2 **4:**121–124.
15. **Szymona, M., and O. Szymona.** 1961. Participation of volutin in the hexokinase reaction of *Corynebacterium diptheriae*. Bull. Acad. Pol. Sci. Cl. 2 **9:**371–374.
16. **Szymona, O., H. Kowalska, and M. Szymona.** 1969. Search for inducible sugar kinases in *Mycobacterium phlei*. Ann. Univ. Mariae Curie-Sklodowska Sect. D **24:**1–12.
17. **Szymona, O., and T. Szumilo.** 1966. Adenosine triphosphate and inorganic polyphosphate fructokinases of *Mycobacterium phlei*. Acta Biochim. Pol. **13:**129–143.
18. **Uryson, S. O., and I. S. Kulaev.** 1968. The presence of polyphosphate glucokinase in bacteria. Dokl. Biol. Sci. **183:**697–699. (English translation.)
19. **Wood, H. G.** 1977. Some reactions in which inorganic pyrophosphate replaces ATP and serves as a source of energy. Fed. Proc. **36:**2197–2205.
20. **Wood, H. G.** 1985. Inorganic pyrophosphate and polyphosphates as sources of energy. Curr. Top. Cell. Regul. **26:** 355–369.
21. **Wood, H. G., and N. H. Goss.** 1985. Phosphorylation enzymes of the propionic acid bacteria and the roles of ATP, inorganic pyrophosphate and polyphosphates. Proc. Natl. Acad. Sci. USA **82:**312–315.

New Data on Biosynthesis of Polyphosphates in Yeasts

IGOR S. KULAEV, VLADIMIR M. VAGABOV, AND YURI A. SHABALIN

Institute of Biochemistry and Physiology of Microorganisms, USSR Academy of Sciences, Pushchino, Moscow Region, 142292, USSR

Inorganic polyphosphates are linear polymers (Fig. 1) with orthophosphoric acid residues interconnected by phosphoanhydrite bonds. The hydrolysis of these bonds releases the same amount of energy as the splitting out of terminal ATP phosphate. Thus, polyphosphates are important for phosphorus and energy metabolism in microorganisms (4, 5).

Two systems for synthesis of low-polymeric polyphosphates have been found in fungi: (i) ATP:polyphosphate phosphotransferase (EC 2.7.4.1) catalyzing the synthesis of polyphosphate due to terminal phosphate ATP, and (ii) 1,3-diphosphoglycerate:polyphosphate phosphotransferase (EC 2.7.4.17) synthesizing the polymer due to energy-rich phosphate of 1,3-diphosphoglyceric acid (5).

A new system for the synthesis of high-polymeric polyphosphates localized outside the cytoplasmic membrane has been found and partially characterized in our laboratory (5).

We observed a correlation between the rates of accumulation of mannoproteins and the polyP4 polyphosphates localized in the yeast cell wall and consisting on the average of 200 to 300 phosphate residues (5) (Table 1). No such correlation between other polyphosphate fractions and polysaccharide-containing compounds in *Saccharomyces cerevisiae* was found. From these initial observations we focused our attention on the understanding of the coordination between the accumulation of cell wall polyphosphates and the mannoproteins.

The structure of yeast wall mannoproteins is rather complex (Fig. 2). Like other glycoproteins, they have short oligomannoside chains attached to protein (serine or threonine) via O-glycoside bonds, in addition to large polymannosyl blocks consisting of about 15 to 17 mannoses (inner core) and a high-molecular-weight outer chain. According to Nakajima and Ballou, the latter contains up to 250 mannose residues. The blocks are attached to asparagine of protein by *N*-glycoside bonds via a chitobiose bridge (11).

The synthesis of cell wall mannoproteins from GDP-mannose realized in a multistage manner proceeds inside the cell, first in the endoplasmic reticulum, then in membranes of Golgi apparatus, and, finally, in vesicles which bud off and move toward the cell surface (8). PP_i released during the synthesis of GDP-mannose from mannose 1-phosphate and GTP might be one of the coupling sites for the biosynthesis of polysaccharides and polyphosphates (6) (Fig. 3). However, the existence of this pathway has not been proved experimentally.

At the same time, with membrane preparations isolated from *S. carlsbergensis* it was possible to demonstrate the incorporation of radioactivity into polymeric products from both GDP-[^{14}C]mannose and GDP-[β-^{32}P]mannose (Table 2). The maximal incorporation occurs in the microsomal yeast fraction which is a peculiar localization of mannan-synthetase (11). Mn^{2+} ions necessary for the activation of mannan-synthetase also activate the incorporation of ^{32}P into a phosphorus-containing product. The latter was shown to be precipitated by Ba^{2+} ions at pH 4.5 and to be hydrolyzed up to 95% in 1 N HCl at 100°C; i.e., it exhibits the properties typical for polyphosphates. Finally, the analysis of the mobility of products by DEAE-cellulose chromatography, as well as the production of cyclic trimetaphosphate upon hydrolysis of a radioactive compound by the method of Thilo and Wieker (20), supports unambiguously the fact that high-molecular-weight polyphosphates are actually synthesized from GDP-[β-^{32}P]mannose. It was shown further that from GDP-[^{14}C]mannose the radioactive label is transferred to mannoproteins (15). How can two different polymers, polyphosphates and mannoproteins, be synthesized from one and the same precursor, GDP-mannose? Basic pathways for the biosynthesis of asparagine-bound carbohydrate components of yeast mannoprotein cores are schematically shown in Fig. 4. This process involves two important lipid intermediates, dolichyl monophosphate and dolichyl diphosphate (7).

The synthetic cycle of the mannoprotein carbohydrate part starts after the transformation of dolichyl diphosphate to dolichyl monophosphate. The fate of phosphate released in this reaction had been unknown and it was clarified in the following experiments.

After incubation of membrane preparations from *S. carlsbergensis* with GDP[β-^{32}P]mannose under conditions optimal for mannan biosynthesis, the lipid fraction was isolated and labeled with ^{32}P. The ^{32}P-labeled lipids were able to serve as phosphate donors for polyphosphate bio-

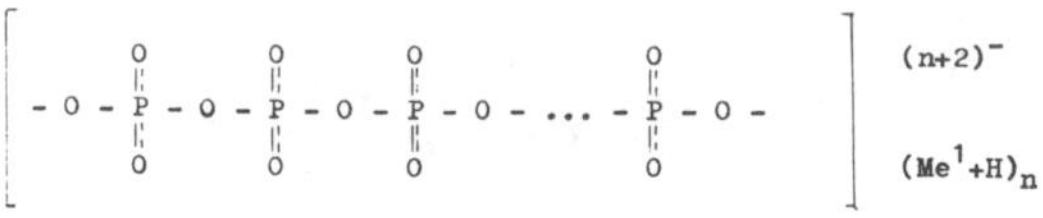

```
      O       O       O               O           (n+2)-
      ‖       ‖       ‖               ‖
 - O - P - O - P - O - P - O - ... - P - O -
      ‖       ‖       ‖               ‖
      O       O       O               O           (Me1+H)n
```

FIG. 1. Molecular structure of linear polyphosphate. Me^1, Monovalent cation; n, number of P_i residues in the chain.

synthesis. Up to 50% of the radioactive phosphorus of lipids was converted to high-molecular-weight polyphosphate. If, however, radioactive phosphorus of GDP-[β-^{32}P]mannose was used as a donor, no more than 3 to 5% of the initial label was incorporated into these compounds. One can infer from these data that phosphorylated lipids are immediate donors of phosphate for the synthesis of polyphosphates (16).

The analysis of the labeled lipid fraction showed that yeast membrane preparations may synthesize dolichyl diphosphate mannose in addition to dolichyl monophosphate mannose. Thus, we propose the mechanism of the coupled biosynthesis of mannoproteins and polyphosphates as shown in Fig. 5.

TABLE 1. Correlation factors (*r*) of the accumulation rates of different polysaccharides and polyphosphate fractions in *S. carlsbergensis* cells (6)[a]

Polysaccharides	Polyphosphates	*r*
(Polysaccharides)	(Polyphosphates)	0.806 ± 0.68
Glycogen	polyP$_1$	0.077 ± 0.02
Glycogen	polyP$_2$	0.141 ± 0.008
Glycogen	(polyP$_2$, polyP$_3$, polyP$_4$, polyP$_5$)	0.173 ± 0.018
Glucan + mannoproteins	(polyP$_2$, polyP$_3$, polyP$_4$, polyP$_5$)	0.750 ± 0.087
Glucan	polyP$_2$	0.291 ± 0.180
Glucan	polyP$_3$	0.615 ± 0.122
Mannoproteins	polyP$_2$	0.136 ± 0.192
Mannoproteins	polyP$_3$	0.035 ± 0.196
Mannoproteins	polyP$_4$	0.813 ± 0.098

[a] *S. carlsbergensis* IBPhM Y-366 was grown on a Reader (13) mineral medium with 2% glucose and 0.2% yeast extract. The cells were grown to the exhaustion of glucose in the medium, and then were separated into two parts, transferred on fresh nutrient- and nitrogen-deficient media, and grown for 10 h more. During growth, samples were withdrawn, washed in a medium without phosphorus and glucose at 0°C, and frozen by liquid nitrogen. The content of polysaccharides, by the method of Trevelyan and Harrison (21), and of inorganic polyphosphates was determined for each sample. Five polyphosphate fractions were isolated: polyP$_1$, acid soluble, extracted with 0.5 N $HClO_4$ at 0°C; polyP$_2$, acid soluble, extracted with $NaClO_4$ saturated solution at 0°C; polyP$_3$, extracted with NaOH solution at pH 9 to 10 at 0°C; polyP$_4$, extracted with NaOH solution at pH 12 at 0°C; polyP$_5$, defined by hydrolysis of the residue left in 1 N $HClO_4$ at 100°C for 10 min after all the other fractions had been extracted. Data obtained in 36 experiments were used to calculate correlation factors (*r*) of accumulation rates of various polyphosphate fractions and polysaccharides by the method of Bailey (1).

According to this hypothesis, dolichyl diphosphate mannose may be a donor of mannosyl residues during the synthesis of yeast mannoprotein, while the resulting dolichyl diphosphate may serve as a phosphate donor for polyphosphate biosynthesis.

To prove the correctness of this scheme, one should (i) demonstrate the existence of an enzymatic system using dolichyl diphosphate for biosynthesis of polyphosphates and (ii) show the participation of dolichyl diphosphate mannose in mannan biosynthesis.

Biosynthesis of Mannoprotein from Dolichyl Diphosphate Mannose

Dolichol used to synthesize dolichyl diphosphate mannose was isolated from pig liver. Labeled dolichyl diphosphate mannose was synthesized as described by Shibaev et al. (19). Dolichyl monophosphate was preliminarily prepared by the method of Danilov and Chojnacki, using phosphorus chlorine oxide as a phosphorylating agent (2).

The dolichyl monophosphate was purified by ion-exchange chromatography and activated with sulfanyldiimidazole. The resulting dolichyl-monophosphorylimidazole reacted with labeled [^{14}C]mannoso-1-phosphate, yielding dolichyl diphosphate [^{14}C]mannose (14).

The membrane fraction of the yeast *S. carlsbergensis* was used as the enzyme preparation and was incubated with the labeled substrate in the presence of Triton X-100 and magnesium ions. Table 3 shows the results of the incorporation of radioactive mannose from dolichyl diphosphate [^{14}C]mannose into mannoprotein and the formation of a lipid-oligosaccharide fraction which was extracted with chloroform-methanol-water (10:10:3). The incorporation of the label into both products was more than doubled in the presence of Mn^{2+} ions. Essential components for glycoprotein biosynthesis such as GDP-mannose and UDP-*N*-acetylglucosamine stimulated even further mannose transfer from dolichyl diphosphate mannose to lipid-oligosaccharide and mannoproteins. The analysis of a lipid fraction of the chloroform-methanol (3:2) extract obtained after the cessation of the reaction showed the presence, alongside nonreacted dolichyl diphosphate mannose, of another, more hydrophilic compound, probably dolichyl diphosphate mannobiose, as judged by silica gel chromatography. The data obtained show that yeast dolichyl diphosphate mannose participates in the biosynthesis of mannoproteins (17). The

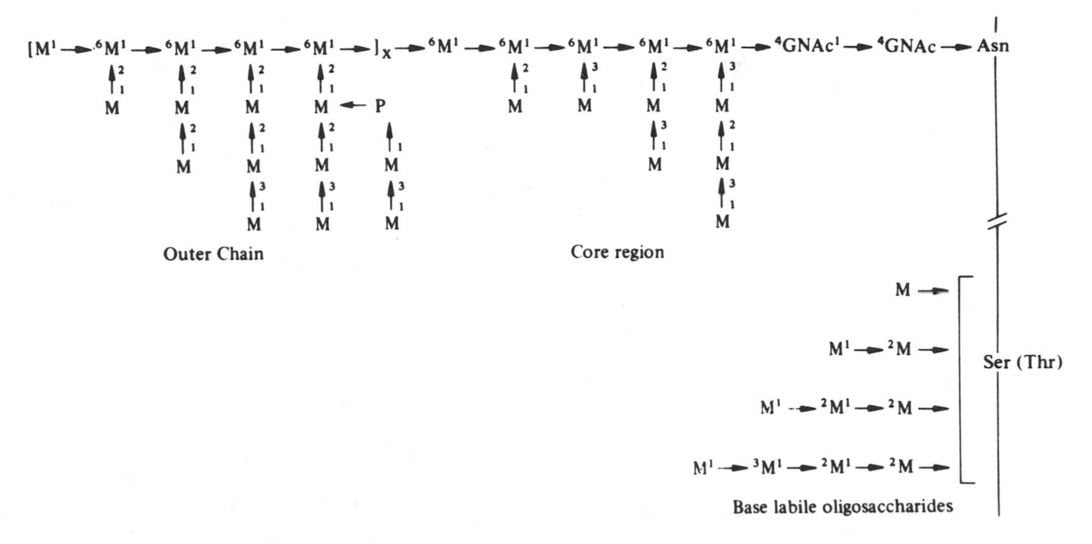

FIG. 2. Structure of the carbohydrate portion of *S. cerevisiae* mannoproteins (11). M, Mannopyranose residue; GNac, *N*-acetylglucosamine residue; P, P_i residue; Asn, Ser, and Thr, amino acids asparagine, serine, and threonine, respectively.

presence in the dolichyl diphosphate compound of two mannose residues allows one to predict that mannoprotein synthesis may also yield an oligosaccharide block on this lipid intermediate. Evidently, the presence of a PP_i group in the dolichyl diphosphate molecule can provide it with a new property, the ability to serve as a carrier of oligosaccharide blocks. The pathways for the synthesis of the so-called inner core of the glycoprotein carbohydrate part have been thoroughly investigated; however, much is still unknown about the synthesis of the outer branched polysaccharide chains.

As already mentioned, this branched molecule consists of a central chain, in which mannose residues are connected by (1-6) bonds, and side chains with (1-2, 1-3) bonds. Theoretically, the synthesis of such polysaccharides may proceed in two ways: (i) by formation of a linear chain

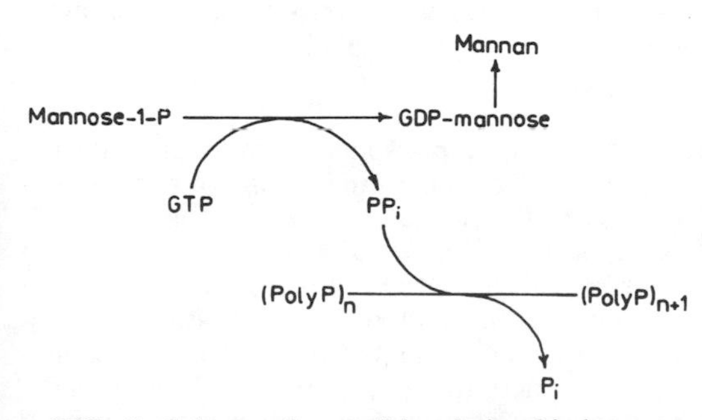

FIG. 3. Scheme of a possible relationship between biosynthesis of mannoproteins and the high-molecular-weight polyphosphate fraction (6). $(Poly\ P)_n$, High-molecular-weight polyphosphates.

with (1-6) bonds to which side chains are attached later, and (ii) by synthesis of side chains

TABLE 2. Incorporation of ^{14}C and ^{32}P from GDP-[^{14}C]mannose and [β-^{32}P]GDP-mannose into mannoproteins and polyphosphate[a]

Fraction	Radioactivity (cpm/mg of protein)	
	Mannoproteins	Polyphosphate
Lysate	1,380	600
Vacuoles	920	0
Microsomes	4,810	3,400

[a] Cells of *S. carlsbergensis* IBPhM Y 366 grown to the middle of the logarithmic growth phase were used to obtain protoplasts by snail enzyme. Vacuoles from protoplasts were isolated by flotation in a Ficoll density gradient (18). Lysate was obtained after the osmotic shock of protoplasts by centrifugation at 3,000 × *g* for 20 min. The supernatant was successively centrifuged at 10,000 × *g* for 20 min (the residue was removed) and 70,000 × *g* for 60 min. The pellet was the "microsomal" fraction. Biosynthesis of mannoproteins and polyphosphate was judged by the inclusion in $HClO_4$ precipitate of radioactivity from GDP-[^{14}C]mannose or GDP-[β-^{32}P]mannose (15), respectively. Incubation medium contained $MnCl_2$ (15 mM), Tris hydrochloride (0.025 M, pH 7.4), GDP-[β-^{32}P]mannose (0.09 μM, 1.3 μCi/μmol) or GDP-[^{14}C]mannose (1 mM, 3 μCi/μmol), and membrane preparation of 30 to 100 μg of protein in the total volume of 100 μl. Incubation time was 30 min at 30°C. The reaction was stopped by adding 100 μl of 1 N $HClO_4$ on ice. The acid-insoluble residue was transferred to Whatman GF/F glass-fiber paper filters and washed several times with a cold solution of 0.5 N $HClO_4$. Radioactivity of the residues was counted with an SL-30 Intertechnique liquid scintillation spectrophotometer.

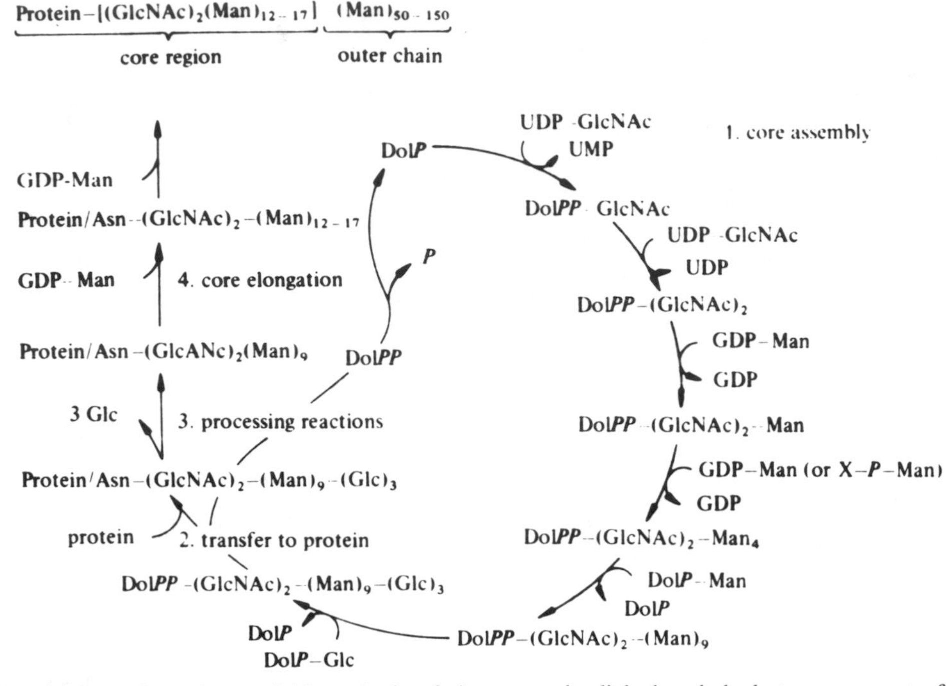

FIG. 4. Schematic pathway of biosynthesis of the asparagine-linked carbohydrate component of yeast mannoproteins (7). Dol*P*, Dolichyl monophosphate; Dol*PP*, dolichyl diphosphate; $(GlcNAc)_2$, chitobiose.

on any precursor followed by the interconnection of these chains by (1-6) bonds. Our data suggest the participation of dolichyl diphosphate mannose in the second of the two mechanisms of mannoprotein synthesis (Fig. 6).

A no less important role is attributed to dolichyl diphosphate mannose which serves as a carrier of mannosyl groups across the membrane, as in the case of dolichyl diphosphate-*N*-acetylglucosamine (9) and dolichyl monophosphate mannose (3).

Dolichyl diphosphate mannose can transfer mannose residues from the GDP-mannose in the cytosol either to GDP in the inner space of the membrane structure or, as mentioned above, directly to the growing chain of mannoprotein

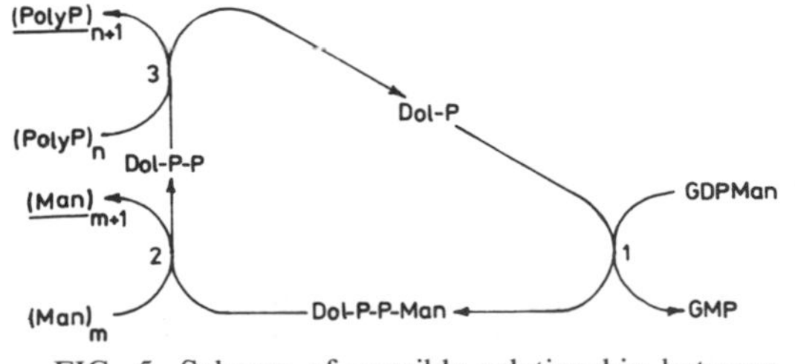

FIG. 5. Scheme of possible relationship between biosynthesis of mannoproteins and polyphosphate on the level of dolichyl diphosphate mannose. Dol-P-P-Man, Dolichyl diphosphate mannose; $(Man)_m$, mannose-oligosaccharides of mannoproteins.

(22). Further investigations in the field will clarify the role of dolichyl diphosphate mannose in the biosynthesis of yeast glycoproteins.

Dolichyl Diphosphate as a Precursor of Polyphosphate Synthesis

After we had demonstrated the participation of dolichyl diphosphate mannose in the synthesis of mannoproteins, it was important to establish whether dolichyl diphosphate yielded by this process might be a precursor of polyphosphates synthesized simultaneously, i.e., to elucidate whether the membrane preparations of yeasts might possess a dolichyl diphosphate:polyphosphate phosphotransferase activity. To do this, we synthesized, jointly with A. Naumov, dolichyl diphosphate labeled by ^{33}P in the β position.

The technique used for the synthesis of labeled dolichyl diphosphate is identical with the synthesis of dolichyl diphosphate mannose, except for the replacement of [^{14}C]mannoso-1-phosphate by $^{33}P_i$. Table 4 shows that the yeast microsomal membrane fraction exhibits both dolichyl diphosphate:polyphosphate phosphotransferase and dolichyl diphosphate-phosphohydrolase activities (12). The resulting polyphosphates were identified as a characteristic product of mild acid hydrolysis, namely, cyclic trimetaphosphate. The transferase activity requires the presence of Mg^{2+} or Mn^{2+} ions,

TABLE 3. Incorporation of ^{14}C from dolichyl diphosphate-[^{14}C]mannose into oligosaccharide-lipid fraction and mannoproteins[a]

Incubation mixture	Radioactivity (cpm/100 mg of protein)	
	Oligosaccharide-lipid	Mannoproteins
Dolichyl diphosphate-[^{14}C]mannose	240	156
Dolichyl diphosphate-[^{14}C]mannose + $MnCl_2$ (10 mM)	1,042	318
Dolichyl diphosphate-[^{14}C]mannose + $MnCl_2$ (10 nM) + UDP-*N*-acetylglucosamine (20 nM) + GDP-mannose (30 nM)	820	264
Control	76	42

[a] *S. carlsbergensis* IBPhM Y-366 was grown on a medium containing 2% glucose, 1% peptone, and 0.5% yeast extract to the middle of the logarithmic growth phase. The membrane fraction from protoplasts was obtained as described in the footnote to Table 2. Dolichyl diphosphate-[^{14}C]mannose was obtained as described previously (14). The incubation mixture contained dolichyl diphosphate-[^{14}C]mannose (35,000 cpm), Tris hydrochloride (0.05 M, pH 7.4), and Triton X-100, plus the additions listed and membrane fraction (100 μg of protein). Incubation was for 45 min at 30°C (17). The reaction was stopped by adding 1.5 ml of chloroform-methanol (3:2). To obtain the oligosaccharide-lipid fraction, the residue was extracted with chloroform-methanol-H_2O (10:10:3). The residue obtained was transferred to CF/A Whatman filters, and its radioactivity was determined. In the control the membrane preparation was preinactivated by heating at 100°C for 5 min.

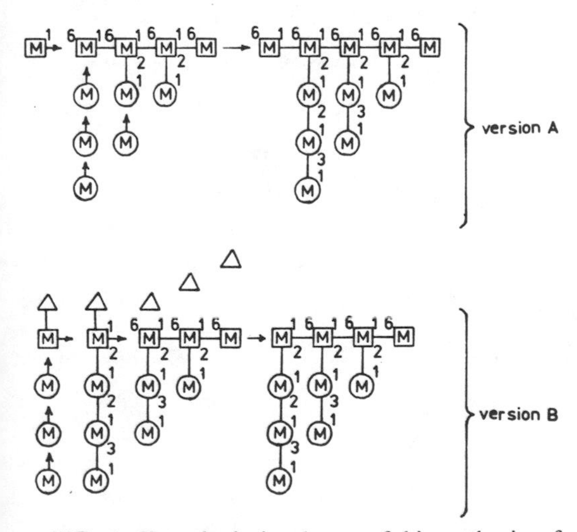

FIG. 6. Hypothetical scheme of biosynthesis of branched outer chain of yeast mannoproteins. △, Dolichyl diphosphate; □, mannose residue forming α-1,6-linked backbone chains; ○, mannose residues forming α-1,2- and α-1,3-linked side chains.

TABLE 4. Incorporation of ^{33}P from dolichyl [β-^{33}P]diphosphate into polyphosphate and hydrolysis to P_i[a]

Addition	Radioactivity (cpm/mg of protein)	
	Polyphosphate	P_i
No cations	15,130	39,380
$MgCl_2$, 10 mM	23,700	27,270
$MnCl_2$, 10 mM	17,860	5,510
EDTA, 1.5 mM	3,020	77,470

[a] Conditions of growth of *S. carlsbergensis* IBPhM Y-366 and isolation of membrane preparation were as in Table 3. Dolichyl [β-^{33}P]diphosphate was synthesized as described previously (14). Incubation mixture contained dolichyl [β-^{33}P]diphosphate (14,000 cpm), Triton X-100 (0.1%), Tris hydrochloride (0.5 M, pH 7.2), membrane preparation (60 mg of protein), cations as listed in the table, and EDTA (12). Incubation time was 30 min at 30°C.

whereas the phosphohydrolase activity is maximal in the absence of bivalent cations.

Heating of membrane preparations decreased both transferase and hydrolase activities. This confirms the enzymatic nature of the processes under study. It is unclear yet whether we witness the activity of one and the same enzyme which operates, depending on conditions (e.g., the content of bivalent cations), as transferase or hydrolase, or two different enzymes from the yeast membrane preparation which attack one and the same substrate, dolichyl diphosphate.

One may regard these facts as proof that the central link coupling the biosynthesis of mannoproteins and polyphosphates in yeasts is dolichyl diphosphate. The revealed dolichyl diphosphate:polyphosphate phosphotransferase activity catalyzes the transfer of the terminal phosphate to polyphosphates, thus promoting the involvement of dolichyl monophosphate formed therewith in a new cycle of reactions of mannan biosynthesis. The resulting P_i accumulates as inorganic polyphosphates.

The above results suggest that the mechanism of formation of polyphosphates, localized in the yeast envelope, differs from other known biosynthetic mechanisms of these compounds. It is coupled with the conversion of dolichyl diphosphate to dolichyl monophosphate which, according to our data, participates in mannoprotein biosynthesis in *Saccharomyces* spp.

One can suggest that dephosphorylation of dolichyl diphosphate, either by transferring β-phosphate to polyphosphate or by splitting off during hydrolysis to P_i by a dolichyl diphosphatase, is a necessary stage for this lipid carrier to go across the Golgi apparatus membrane. This appears to be the most probable mechanism because the reaction of dolichyl monophosphate with GDP-mannose occurs at the cytozole side and the formation of the carbohydrate chain in mannoproteins is carried out in the lumen of the Golgi apparatus.

The splitting-out of β-phosphate from dolichyl diphosphate appears to be a prerequisite for bringing the lipid carrier back to the cytozole side and its involvement in a new cycle. A similar mechanism was revealed in the synthesis of surface glycoproteins of *Halobacterium salivarium* (10).

Thus, one of the reaction products, dolichyl monophosphate, leaves the sphere of mannoprotein biosynthesis by going across the membrane. Depending on the conditions under which the reaction occurs, specifically on the presence or absence in the Golgi lumina of sufficient concentrations of bivalent cations (Mn^{2+}, Mg^{2+}, etc.), the conversion of dolichyl diphosphate to dolichyl monophosphate can be coupled with either the formation of polyphosphates or the accumulation of P_i. However, further experiments are required to confirm this hypothesis.

LITERATURE CITED

1. **Bailey, N. T. J.** 1959. Statistical methods in biology. The English Universities Press Ltd., Oxford.
2. **Danilov, L. L., and T. Chojnacki.** 1981. A simple procedure for preparing dolichylmonophosphate by the use of $POCl_3$. FEBS Lett. **131**:310–312.
3. **Haselbeck, A., and W. Tanner.** 1984. Further evidence for dolichylphosphate mediated glycosyl translocation through membrane. FEMS Microbiol. Lett. **21**:305–308.
4. **Kulaev, I. S.** 1979. The biochemistry of inorganic polyphosphates. John Wiley & Sons, Inc., New York.
5. **Kulaev, I. S., and V. M. Vagabov.** 1983. Polyphosphate metabolism in microorganisms. Adv. Microb. Physiol. **24**: 83–171.
6. **Kulaev, I. S., V. M. Vagabov, and A. B. Tsiomenko.** 1972. On the correlation of accumulation of polysaccharides and condensed inorganic phosphates of the yeast, p. 229–238. *In* A. Kockova-Kratochvilova and E. Minarik (ed.), Yeast models in science and technics. Publishing House of the Slovak Academy of Sciences, Bratislava, Czechoslovakia.
7. **Lehle, L.** 1980. Biosynthesis of the core regions of yeast mannoproteins. Formation of a glycosilated dolichol-bond oligosaccharide precursor, its transfer to protein and subsequent modification. Eur. J. Biochem. **109**:589–601.
8. **Lehle, L., F. Bauer, and W. Tanner.** 1977. The formation of glycosidic bonds in yeast glycoproteins. Intracellular localisation of the reactions. Arch. Microbiol. **114**:77–81.
9. **Lennarz, W. J.** 1982. Topological and regulatory aspects of dolichyl phosphate mediated glycosylation of proteins. Phil. Trans. R. Soc. London Ser. B **300**:129–144.
10. **Mescher, M. F.** 1981. Glycoproteins as cell-surface structural components. TJBS **6**:97–99.
11. **Nakajima, T., and C. E. Ballou.** 1974. Structure of the linkage region between the polysaccharide and protein parts of Saccharomyces cerevisiae mannan. J. Biol. Chem. **249**:7685–7694.
12. **Naumov, A. V., Y. A. Shabalin, V. M. Vagabov, and I. S. Kulaev.** 1985. Two pathways of dolichyldiphosphate dephosphorylation in yeasts. Biokhimiya **50**:651–658.
13. **Reader, V.** 1927. The relation of the growth of certain microorganisms to the composition of the medium. I. The synthetic culture medium. Biochem. J. **21**:901.
14. **Shabalin, Y. A., A. V. Naumov, V. M. Vagabov, I. S. Kulaev, L. L. Danilov, and V. N. Shibaev.** 1985. Synthesis of dolichyl-(β,^{33}P)-pyrophosphate and dolichylpyrophosphate-(^{14}C)-mannose. Bioorg. Khim. **11**:651–654.
15. **Shabalin, Y. A., V. M. Vagabov, and I. S. Kulaev.** 1978. The biosynthesis of high-molecular polyphosphates from β,^{32}P GDP-mannose by membrane fraction of *Saccharomyces carlsbergensis* yeasts. Dokl. Akad. Nauk SSSR **239**:490–492.
16. **Shabalin, Y. A., V. M. Vagabov, and I. S. Kulaev.** 1979. On the coupling mechanism of biosynthesis of high-molecular polyphosphates and mannan in *Saccharomyces carlsbergensis* yeasts. Dokl. Akad. Nauk SSSR **239**:490–492.
17. **Shabalin, Y. A., V. M. Vagabov, and I. S. Kulaev.** 1985. Dolichyldiphosphatemannose—an intermediate of glycoprotein biosynthesis in yeasts? Dokl. Akad. Nauk SSSR **283**:720–724.
18. **Shabalin, Y. A., V. M. Vagabov, A. B. Tsiomenko, O. A. Zemlenukhina, and I. S. Kulaev.** 1977. Study of the polyphosphate kinase activity in yeast vacuoles. Biokhimiya **42**:1642–1648.
19. **Shibaev, V. N., L. L. Danilov, V. N. Chekunchikov, Y. Y. Kusov, and N. K. Kochetkov.** 1979. A new synthesis of polyprenyl glycosyl pyrophosphates. Bioorg. Chem. (Bioorg. Khim.) **5**:308–310.
20. **Thilo, E., and W. Wieker.** 1957. Die anionenhydrolyse kondensierten Phosphate in verdünter wasseriger Lösung. Zeitschr. Anorg. Allgem. Chem. **291**:164–185.
21. **Trevelyan, W. E., and I. S. Harrison.** 1952. Studies on yeast metabolism. I. Fractionation and microdetermination of cell carbohydrates. Biochem. J. **50**:298–303.
22. **Vagabov, V. M., and Y. A. Shabalin.** 1985. The detection in yeasts of a new fermentative system coupling synthesis of glycoproteins and inorganic condensed phosphates, p. 167–175. *In* G. K. Skryabin (ed.), Problems of the biochemistry and physiology of microorganisms. USSR Academy of Sciences, Pushchino, USSR.

Polyphosphates and Sugar Transport in Yeasts

J. VAN STEVENINCK, J. SCHUDDEMAT, C. C. M. VAN LEEUWEN, AND P. J. A. VAN DEN BROEK

Department of Medical Biochemistry, Sylvius Laboratories, 2300 RA Leiden, The Netherlands

The mechanism of glucose transport in yeasts has remained unclear and controversial, despite many investigations during the past 25 years. Many studies suggested a facilitated diffusion system for glucose in yeasts (6, 9, 12, 14). Other investigations demonstrated proton influx during transport of glucose derivatives (22) and other sugars (5) into yeast cells, suggesting an active sugar-proton symport mechanism. Still other observations support the notion that group translocation is involved in transport of glucose and glucose derivatives in yeast cells. On the basis of circumstantial evidence, a transport-associated phosphorylation has been proposed, with a peripherally localized polyphosphate fraction as phosphate donor. A crucial observation in this context was that specific metal-ion-binding ligands at the outside of the cell membrane were involved in sugar transport. These ligands were tentatively identified as polyphosphate groups (23).

More recently, the presence of polyphosphates at the outside of the plasma membrane of *Kluyveromyces marxianus* has been confirmed by utilizing different experimental approaches, namely, osmotic shock (17), toluidine blue-binding studies (15), 4′,6-diamidino-2-phenylindole fluorescence measurements (16), nuclear magnetic resonance studies (19), and electron microscopy (20). Moreover, pulse-labeling experiments, designed to determine the temporal sequence of appearance of 2-deoxy-*d*-glucose in the intracellular free sugar and sugar-phosphate pools, confirmed that 2-deoxy-*d*-glucose first appears in the cell in the phosphorylated form (8, 10, 13).

Other investigations demonstrated that in the yeast *Saccharomyces cerevisiae* glucose uptake has a high-affinity and a low-affinity component. In mutants lacking both hexokinases and glucokinase, only the low-affinity component was present. These observations suggest the involvement of hexose kinases in the transport-associated phosphorylation of glucose in yeasts (2–4).

To evaluate the contribution of various transport systems, several aspects of glucose transport in yeasts were reinvestigated, as described here.

Results

Transport in early and late stationary phase. During growth of *K. marxianus* on a synthetic medium, several parameters were monitored as shown in Fig. 1. This yeast enters the stationary phase shortly after depletion of the glucose in the medium. Subsequently, the initial transport velocity of 6-deoxy-*d*-glucose increases, reaching a maximum value after about 30 h, whereas the cellular polyphosphate concentration decreases sharply between 25 and 35 h of culture time. The change in 6-deoxy-*d*-glucose transport during this time period was reinvestigated here by studying the transport kinetics of 6-deoxy-*d*-glucose and 2-deoxy-*d*-glucose in early stationary (13 h of culture time) and late stationary (38 h of culture time) yeast cells.

In 13-h cultured *K. marxianus*, 6-deoxy-*d*-glucose transport exhibited the characteristics of facilitated diffusion, with a K_m of about 100 mM and a steady-state intracellular 6-deoxy-*d*-glucose concentration which did not exceed the sugar concentration in the medium. During transport, no shift of the medium pH could be observed (Fig. 2). Uptake of 6-deoxy-*d*-glucose was inhibited by glucose, fructose, and 2-deoxy-*d*-glucose, but not by galactose (Table 1). In 38-h cultured yeast the kinetics of 6-deoxy-*d*-glucose transport were biphasic, indicating the simultaneous operation of two different transport systems (21) with K_m values of 90 and 0.8 mM. Uptake of 6-deoxy-*d*-glucose in these cells was inhibited by glucose, 2-deoxy-*d*-glucose, and galactose, but not by fructose (Table 1). Furthermore, 6-deoxy-*d*-glucose transport in 38-h cultured yeast had the characteristics of a sugar-proton symport system, with alkalinization of the medium during transport (Fig. 2), accumulation of 6-deoxy-*d*-glucose against a concentration gradient, and a high sensitivity to uncouplers like carbonyl cyanide *m*-chlorophenylhydrazone (Fig. 3).

Transport of 2-deoxy-*d*-glucose in 13-h *K. marxianus* cultures occurred by the phosphorylation system. Pulse-labeling experiments demonstrated that 2-deoxy-*d*-glucose first appeared inside the cells in the phosphorylated form, as described previously (10). 2-Deoxy-*d*-glucose uptake via this transport system was inhibited by glucose, fructose, and 6-deoxy-*d*-glucose, but not by galactose (Table 1). Finally, 2-deoxy-*d*-glucose uptake in this yeast was not accompanied by proton influx (Fig. 2). In 38-h cultured yeast cells, 2-deoxy-*d*-glucose transport exhibited characteristics similar to those in 13-h cultured yeast cells, with three important differ-

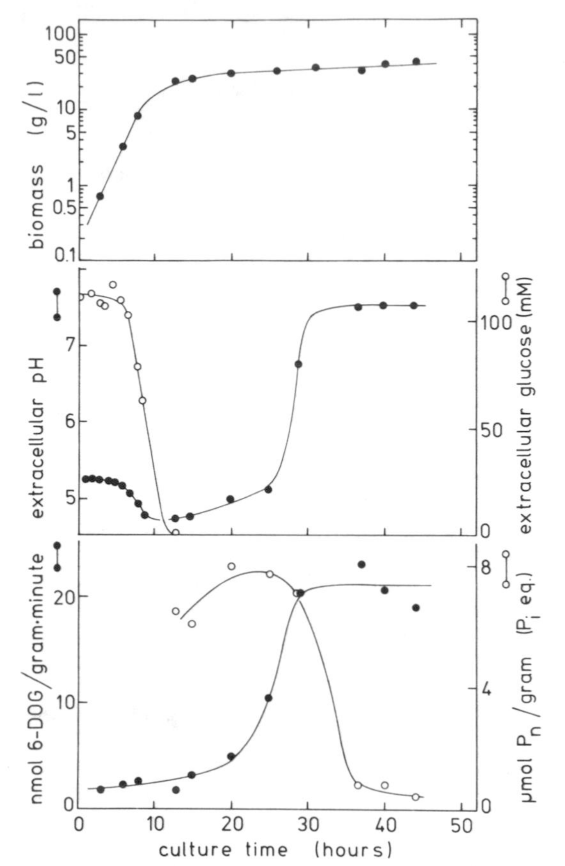

FIG. 1. Characteristics of growth of *K. marxianus* on a glucose-containing medium. Biomass was determined by measuring the weight of cells in 250-ml samples of culture medium. Extracellular glucose and pH were obtained in filtrates of the culture. 6-Deoxy-*d*-glucose (6-DOG) transport was measured aerobically in 0.1 M Tris maleate, pH 4.5, as described before (18). Initial extracellular 6-deoxy-*d*-glucose concentration: 0.1 mM.

ences: (i) 2-deoxy-*d*-glucose transport was now inhibited by galactose (Table 1); (ii) inhibition by fructose was significantly less than in 13-h cultured yeast cells (Table 1); and (iii) 2-deoxy-*d*-glucose transport was now accompanied by a simultaneous influx of protons (Fig. 2).

Involvement of kinases in glucose transport. After *S. cerevisiae* CBS 1172 was cultured for 20 h in a synthetic medium, glucose and 2-deoxy-*d*-glucose were taken up exclusively via a transport-associated phosphorylation system, with no indications of a concomitant sugar-proton symport. Competition studies demonstrated that 6-deoxy-*d*-glucose is taken up via the same transport system, with the characteristics of facilitated diffusion.

As shown by Dela Fuente (7), incubation of *S. cerevisiae* with high concentrations of xylose leads to an irreversible inhibition of hexokinase and glucokinase. In the present experiments incubation of *S. cerevisiae* CBS 1172 with 300 mM xylose during 16 h resulted in an 85% reduction of glucose-phosphorylating activity, whereas the cellular concentrations of polyphosphate, P_i, and ATP were not affected. Concomitantly, transport of 2-deoxy-*d*-glucose into xylose-treated cells was significantly reduced, as a result of an increase of the K_m from 0.5 to 1.6 mM and a decrease of V_{max} from 0.019 to 0.013 μmol/s per g. However, 6-deoxy-*d*-glucose transport was unaffected.

Phosphoryl donor in transport-associated phosphorylation. Aerobic incubation of galactose-grown *K. marxianus* with either 2-deoxy-*d*-glucose or 2-deoxy-*d*-galactose led to massive accumulation of the corresponding sugar phosphates, at the expense of the cellular ATP, polyphosphate, and P_i pools (Fig. 4). With either sugar, there was a preferential decrease of polyphosphates of chain length 15 to 20, with much less decrease of the shorter polyphosphates (Fig. 5). Under these experimental conditions there was, apparently, a fast exchange between phosphate pools, making it impossible to identify the primary phosphoryl donor during sugar phosphorylation.

Under anaerobic conditions, however, the turnover between phosphate pools was strongly

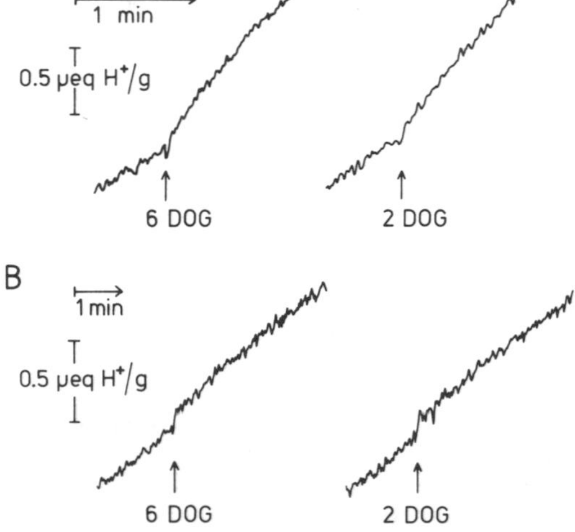

FIG. 2. Influence of sugars on proton fluxes in 13- and 38-h *K. marxianus* cultures. Yeast cells were suspended in 10% (wet w/vol) suspension in 2.5 mM Tris maleate. Anaerobiosis was achieved by flushing with argon. At the moment of sugar addition pH was 5.8. 6-Deoxy-*d*-glucose (6 DOG) and 2-deoxy-*d*-glucose (2 DOG) were added in 5 mM concentration. A, 38-h cultures; B, 13-h cultures. An upward deflection shows a pH increase.

TABLE 1. Deoxyglucose transport in *K. marxianus* cultures[a]

Deoxy-glucose	Culture time (h)	Transport system	K_m (mM)	% Inhibition by: Glucose (5 mM)	Fructose (5 mM)	2-Deoxy-*d*-glucose (5 mM)	6-Deoxy-*d*-glucose 100 mM	6-Deoxy-*d*-glucose 5 mM	Galactose (5 mM)	Medium alkalinization
6-Deoxy-*d*-glucose	13	Facilitated diffusion	100	61	60	67			11	No
	38	Symport	90 and 0.8	97	8	93			97	Yes
2-Deoxy-*d*-glucose	13	Group transfer	0.5	78	60		47		0	No
	38	Group transfer plus symport	1.4	95	32			64	84	Yes

[a] 2-Deoxy-*d*-glucose and 6-deoxy-*d*-glucose transport was assayed at tracer concentrations (18).

diminished or blocked (18). Therefore, further experiments were conducted anaerobically.

As shown previously, 2-deoxy-*d*-galactose enters the yeast cell as the free sugar, followed by intracellular phosphorylation (11). As shown in Fig. 6A, anaerobic uptake during 90 min led to a total 2-deoxy-*d*-galactose uptake of 10 μmol/g, of which 4 μmol/g was phosphorylated. During this incubation, the ATP level dropped from 0.1 to 0.015 μmol/g, whereas the polyphosphate level dropped from 15.0 to 10.9 μmol/g (expressed in P_i equivalents), thus balancing the 2-deoxy-*d*-galactose-phosphate accumulation.

Transport of 2-deoxy-*d*-glucose, taken up via the transport-associated phosphorylation system under the same experimental conditions

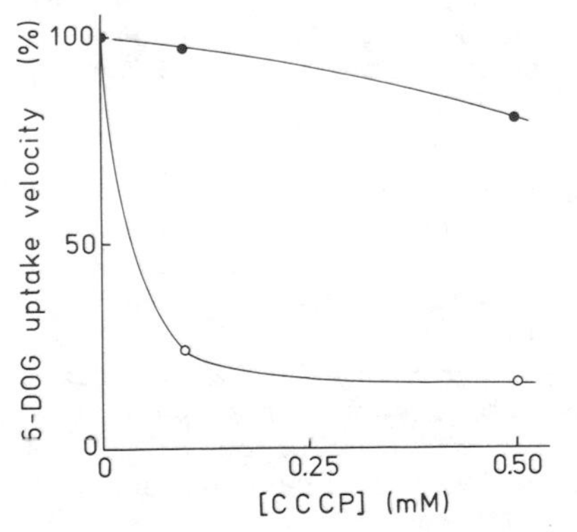

FIG. 3. Influence of carbonyl cyanide *m*-chlorophenylhydrazone (CCCP) on 6-deoxy-*d*-glucose influx. Transport was measured aerobically in 0.1 M Tris maleate, pH 4.5. 6-Deoxy-*d*-glucose (6-DOG) was used at tracer concentration. *K. marxianus* cultures were tested at 13 h (●) and 38 h (○). Influx was determined from the linear uptake during the first 6 to 18 s. Yeast cells were preincubated with carbonyl cyanide *m*-chlorophenylhydrazone for 2 min before 6-deoxy-*d*-glucose was added.

(18), is shown in Fig. 6B. Although the intracellular concentration of free 2-deoxy-*d*-glucose increased slowly during the entire incubation period, the concentration of 2-deoxy-*d*-glucose-phosphate leveled off within 10 min at about 2.3 μmol/g of yeast cells. Concomitantly, the cellular ATP level dropped from 0.1 to 0.01 μmol/g and the polyphosphate concentration decreased from 15.1 to 13.2 μmol/g, balancing the accumulation of 2-deoxy-*d*-glucose-phosphate.

In further experiments, *K. marxianus* cells were preincubated anaerobically with 2-deoxy-*d*-galactose for 30 min, before addition of 2-deoxy-*d*-glucose to the medium. This preincubation with 2-deoxy-*d*-galactose did not noticeably affect phosphorylation of 2-deoxy-*d*-glucose. Again, 2-deoxy-*d*-glucose-phosphate reached an intracellular concentration of about 2.3 μmol/g of yeast cells. In the reverse experiment the results were similar: preincubation with 2-deoxy-*d*-glucose did not affect subsequent phosphorylation of 2-deoxy-*d*-galactose.

Discussion

Two hexose transport systems apparently function in *K. marxianus*, viz., a transport-associated phosphorylation system (recognizing glucose, fructose, 2-deoxy-*d*-glucose, and 6-deoxy-*d*-glucose) and a sugar-proton symport system (transporting glucose, 2-deoxy-*d*-glucose, 6-deoxy-*d*-glucose, and galactose). In the early stationary phase only the first system was present, whereas in the late stationary phase both systems were active. The symport system gradually increased in activity during the culturing period of 10 to 30 h, whereas the cellular polyphosphate concentration decreased sharply in the period from 25 to 35 h (Fig. 1). It is tempting to speculate that there may be a causal relationship between these two phenomena, but the regulations involved are still unclear. These observations clarify some contradictory results in

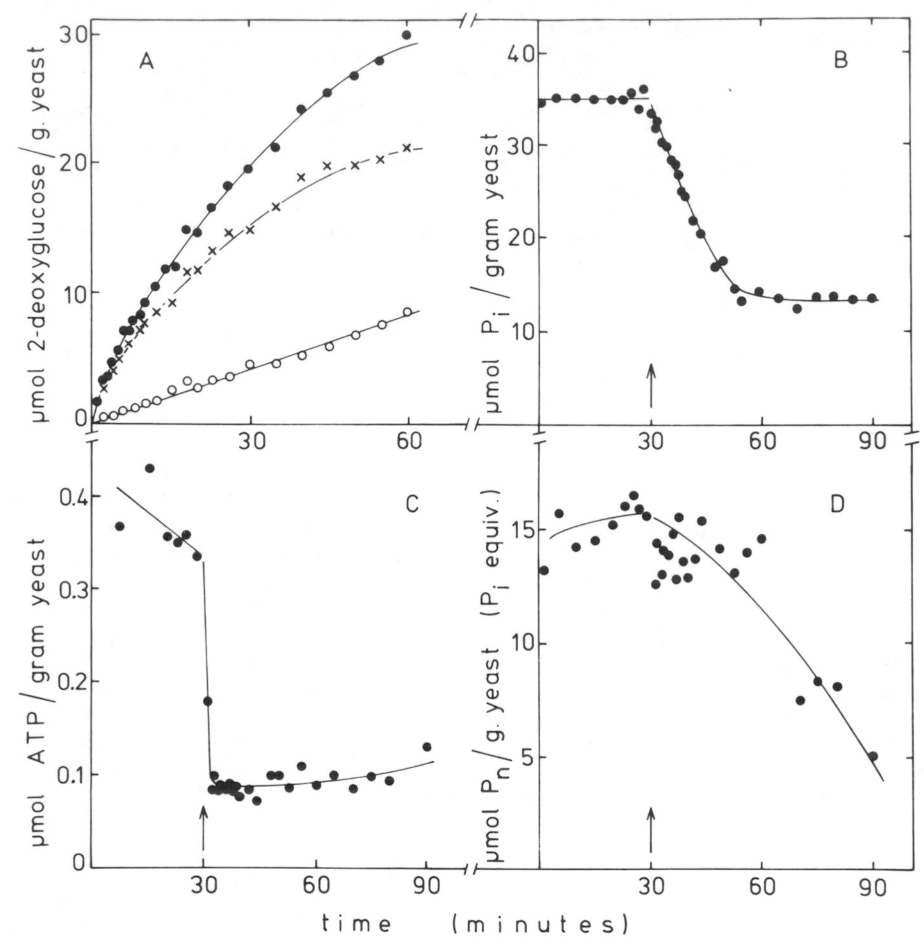

FIG. 4. Uptake of 2-deoxy-*d*-glucose in *K. marxianus* and its influence on the cellular contents of phosphate compounds. The yeast was incubated aerobically in 0.1 M Tris maleate buffer, pH 4.5. Transport was initiated by adding 10 mM 2-deoxy-*d*-glucose at time zero (A) or at 30 min (B, C, D). (A) 2-Deoxy-*d*-glucose uptake curves: ●, total intracellular 2-deoxy-*d*-glucose; ×, 2-deoxy-*d*-glucose-phosphate; ○, free 2-deoxy-*d*-glucose. (B) P_i. (C) ATP. (D) Polyphosphate (expressed in P_i equivalents).

previous studies. As shown before, 2-deoxy-*d*-glucose transport in 20-h cultured *K. marxianus* exhibited the characteristics of transport-associated phosphorylation, but uptake of this sugar was accompanied by a small alkalinization of the medium (22). This apparent proton symport was interpreted in terms of the phosphotransferase mechanism, suggesting that charge compensation during transport was involved. Considering the present results, however, this interpretation was probably incorrect. In early stationary phase the operation of the transport-associated phosphorylation system was not attended with alkalinization of the medium (Fig. 2). It seems likely that the alkalinization observed in the earlier studies should be attributed to the simultaneous activity of a sugar-proton symport system, as can be expected in 20-h cultured *K. marxianus*.

Under anaerobic conditions, transport-associated phosphorylation of 2-deoxy-*d*-glucose occurs with polyphosphate as the ultimate phosphoryl donor, as discussed previously (18). In these studies it was shown that it is very unlikely that this phosphorylation would occur by the hexokinase reaction with simple replenishment of cellular ATP from the total cellular polyphosphate pool via the ATP-polyphosphate phosphotransferase reaction:

$$PP_n + ADP \rightarrow PP_{n-1} + ATP$$

$$ATP + 2\text{-deoxy-}d\text{-glucose}$$
$$\rightarrow 2\text{-deoxy-}d\text{-glucose-phosphate} + ADP$$

This conclusion was based on two considerations. (i) In anaerobic yeasts the cellular ATP concentration is far below its K_m for the hexokinase reaction. Calculations based on the

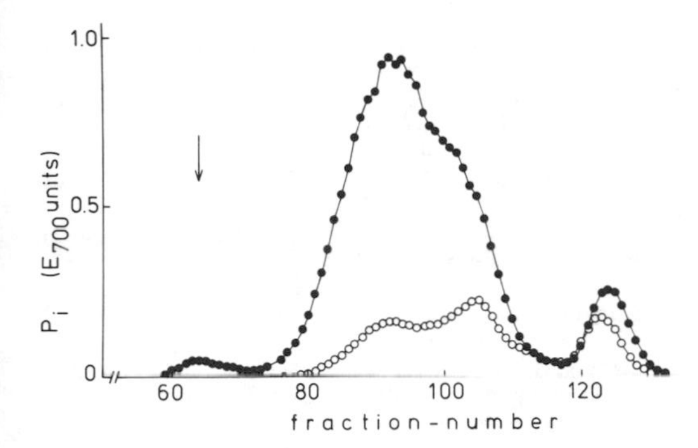

FIG. 5. Gel filtration of cellular polyphosphates. *K. marxianus* was incubated aerobically for 2 h in the presence of 50 mM 2-deoxy-*d*-galactose (○). The control (●) represents the untreated yeast. Separation was carried out essentially as described before (17). The arrow indicates the void volume. Commercial polyphosphate (Sigma Chemical Co.) with mean chain length 5 is found mostly in fraction 105, and chain length 15 is in fraction 90. Fraction volume: 0.55 ml. The amount of phosphate in the fractions is expressed in E_{700} units coming from the P_i determination which followed the 30-min acid hydrolysis of the polyphosphate.

measured ATP and hexokinase levels in the cells revealed that intracellular phosphorylation of 2-deoxy-*d*-glucose would occur with a maximal velocity of 0.4 μmol/g per min, whereas the actual velocity of transport-associated phosphorylation amounts to 6 μmol/g per min. (ii) In anaerobic yeasts transport-associated phosphorylation is limited to about 2.3 μmol of 2-deoxy-*d*-glucose per g of yeast cells. Considering the total cellular polyphosphate pool of about 15 μmol/g of yeast cells, it is difficult to understand why only 2 μmol of polyphosphate per g of yeast cells can be mobilized for ATP synthesis via the ATP-polyphosphate phosphotransferase reaction. Thus, the experimental observations suggest a functional compartmentation of polyphosphate pools. The present results reinforce this conclusion. In galactose-grown yeasts 2-deoxy-*d*-galactose enters the cells as the free sugar, with subsequent intracellular phosphorylation via the galactokinase/ATP reaction (11). As shown in Fig. 6A, 2-deoxy-*d*-galactose phosphorylation under anaerobic conditions continues in time, reaching a value of 4 μmol/g of yeast cells after 90 min. This accumulation of 2-deoxy-*d*-galactose-phosphate is counterbalanced by a similar decrease of cellular polyphosphate.

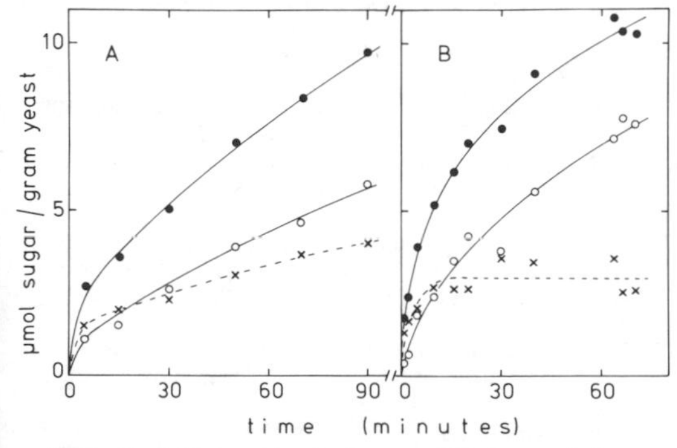

FIG. 6. 2-Deoxy-*d*-galactose and 2-deoxy-*d*-glucose uptake in *K. marxianus* under anaerobic conditions. Initial extracellular concentration: 10 mM. (A) 2-Deoxy-*d*-galactose. (B) 2-Deoxy-*d*-glucose. Symbols: ●, total intracellular sugar; ×, sugar phosphate; ○, free sugar.

As the cellular ATP concentration under these experimental conditions is only 0.1 μmol/g of yeast cells, phosphorylation of 2-deoxy-*d*-galactose probably occurs via the galactokinase reaction, with replenishment of ATP via the ATP-polyphosphate phosphotransferase reaction from the cellular polyphosphate pool. The initial rate of 2-deoxy-*d*-glucose phosphorylation under anaerobic conditions is higher than the initial rate of 2-deoxy-*d*-galactose phosphorylation. Assuming 2-deoxy-*d*-glucose phosphorylation via the hexokinase reaction with the cellular ATP pool as phosphoryl donor, it would be expected that also 2-deoxy-*d*-glucose phosphorylation should continue in the course of time, as a result of replenishment of ATP via the ATP-polyphosphate phosphotransferase reaction. Actually, 2-deoxy-*d*-glucose phosphorylation leveled off after a short time at a final 2-deoxy-*d*-glucose-phosphate concentration of 2.3 μmol/g of yeast cells (Fig. 6B). This observation corroborates the previous conclusion that 2-deoxy-*d*-glucose phosphorylation under anaerobic conditions cannot be explained via the hexokinase reaction, with the cellular ATP pool as phosphoryl donor.

Thus, although both 2-deoxy-*d*-glucose and 2-deoxy-*d*-galactose are phosphorylated at the expense of cellular polyphosphate, the mechanisms involved are basically different. Moreover, it appears that in 2-deoxy-*d*-glucose and 2-deoxy-*d*-galactose phosphorylation different polyphosphate pools are involved.This is indicated by the observation that significantly more 2-deoxy-*d*-galactose than 2-deoxy-*d*-glucose can be phosphorylated under anaerobic conditions. A stronger argument in this context is the finding that anaerobic preincubation with 2-deoxy-*d*-galactose does not affect subsequent phosphorylation of 2-deoxy-*d*-glucose. Apparently, these two deoxy sugars do not deplete each other's phosphoryl donor pools.

Investigations by Bisson and Fraenkel (2–4) demonstrated that in *S. cerevisiae* glucose uptake has a high- and a low-affinity component. In mutants lacking both hexokinases and glucokinase, only the low-affinity component was present, suggesting that the hexose kinases are involved in transport-associated phosphorylation in this yeast. Our experiments with *S. cerevisiae* cells, in which the kinase activity was irreversibly inhibited by xylose incubation, support this hypothesis. 2-Deoxy-*d*-glucose and 6-deoxy-*d*-glucose are transported via the same permease in these cells, as revealed by competition studies. Inhibition of the hexose kinases by xylose did not affect 6-deoxy-*d*-glucose transport (with the characteristics of facilitated diffusion) but caused a significant inhibition of 2-deoxy-*d*-glucose transport, mainly via a pronounced increase of the K_m. This indicates that an active hexose kinase is required for glucose analogs which can be phosphorylated.

These observations are, however, not contradictory with the conclusion that the transport-associated phosphorylation of 2-deoxy-*d*-glucose cannot be explained via the hexokinase reaction, with cellular ATP as final phosphoryl donor. As suggested by Bisson and Fraenkel (2), it is conceivable that the kinases form a functional part of the hexose uptake system. In that case it would be energetically efficient if the actual translocation by the permease system would deliver not only the sugar molecule directly to the kinase, but also the cosubstrate, ATP. The kinase reaction would subsequently release the sugar-phosphate in the cytoplasm, leaving a permease-ADP complex. Experimental evidence indicates that a polyphosphate pool localized outside the plasma membrane is the ultimate phosphoryl donor in transport-associated phosphorylation (23). In the present model this would imply that at the outer membrane face the permease-ADP complex would be converted to a permease-ATP complex, at the expense of the peripherally localized polyphosphate pool, and bind another sugar molecule, as postulated previously (24).

LITERATURE CITED

1. **Addanki, S., J. F. Sotos, and P. D. Rearick.** 1966. Rapid determination of picomole quantities of ATP with a liquid scintillation counter. Anal. Biochem. **14:**261–264.
2. **Bisson, L. F., and D. G. Fraenkel.** 1983. Involvement of kinases in glucose and fructose uptake by *Saccharomyces cerevisiae*. Proc. Natl. Acad. Sci. USA **80:**1730–1734.
3. **Bisson, L. F., and D. G. Fraenkel.** 1983. Transport of 6-deoxy-glucose in *Saccharomyces cerevisiae*. J. Bacteriol. **155:**995–1000.
4. **Bisson, L. F., and D. G. Fraenkel.** 1984. Expression of kinase-dependent glucose uptake in *Saccharomyces cerevisiae*. J. Bacteriol. **159:**1013–1017.
5. **Brocklehurst, R., D. Gardner, and A. A. Eddy.** 1977. The absorption of protons with α-methyl glucoside and α-thioethyl glucoside by the yeast NCYC 240. Biochem. J. **162:**591–599.
6. **Cirillo, V. P.** 1968. Relationship between sugar structure and competition for the sugar transport system in bakers' yeast. J. Bacteriol. **95:**603–611.
7. **Dela Fuente, G.** 1970. Specific inactivation of yeast hexokinase induced by xylose in the presence of a phosphoryl donor substrate. Eur. J. Biochem. **16:**240–243.
8. **Franzusoff, A., and V. P. Cirillo.** 1982. Uptake and phosphorylation of 2-deoxy-D-glucose by wild-type and single-kinase strains of *Saccharomyces cerevisiae*. Biochim. Biophys. Acta **688:**295–304.
9. **Heredia, C. F., A. Sols, and G. Dela Fuente.** 1968. Specificity of the constitutive hexose transport in yeast. Eur. J. Biochem. **5:**324–329.
10. **Jaspers, H. T. A., and J. Van Steveninck.** 1975. Transport-associated phosphorylation of 2-deoxy-D-glucose in *Saccharomyces fragilis*. Biochim. Biophys. Acta **406:**370–385.
11. **Jaspers, H. T. A., and J. Van Steveninck.** 1976. Transport of 2-deoxy-D-galactose in *Saccharomyces fragilis*. Biochim. Biophys. Acta **443:**243–253.
12. **Kotyk, A.** 1967. Properties of the sugar carrier in bakers' yeast. II. Specificity of transport. Folia Microbiol. (Prague) **12:**121–131.
13. **Meredith, S. A., and A. H. Romano.** 1977. Uptake and phosphorylation of 2-deoxy-D-glucose by wild type and respiration deficient bakers' yeast. Biochim. Biophys. Acta **497:**745–759.
14. **Romano, A. H.** 1982. Facilitated diffusion of 6-deoxy-D-glucose in bakers' yeast: evidence against phosphorylation-associated transport of glucose. J. Bacteriol. **152:** 1295–1297.
15. **Tijssen, J. P. F., H. W. Beekes, and J. Van Steveninck.** 1981. Localization of polyphosphates at the outside of the yeast cell plasma membrane. Biochim. Biophys. Acta **649:** 529–532.
16. **Tijssen, J. P. F., H. W. Beekes, and J. Van Steveninck.** 1982. Localization of polyphosphates in *Saccharomyces fragilis*, as revealed by 4′,6-diamidino-2-phenylindole fluorescence. Biochim. Biophys. Acta **721:**394–398.
17. **Tijssen, J. P. F., T. M. A. R. Dubbelman, and J. Van Steveninck.** 1983. Isolation and characterization of polyphosphates from the yeast cell surface. Biochim. Biophys. Acta **760:**143–148.
18. **Tijssen, J. P. F., P. J. A. Van den Broek, and J. Van Steveninck.** 1984. The involvement of high-energy phosphate in 2-deoxy-D-glucose transport in *Kluyveromyces marxianus*. Biochim. Biophys. Acta **778:**87–93.
19. **Tijssen, J. P. F., and J. Van Steveninck.** 1984. Detection of a yeast polyphosphate fraction localized outside the plasma membrane by the method of phosphorus-31 nuclear magnetic resonance. Biochem. Biophys. Res. Commun. **119:**447–451.
20. **Tijssen, J. P. F., J. Van Steveninck, and W. C. De Bruijn.** 1985. Cytochemical staining of a yeast polyphosphate fraction, localized outside the plasma membrane. Protoplasma **125:**124–128.
21. **Van den Broek, P. J. A., and J. Van Steveninck.** 1980. Kinetic analysis of simultaneously occurring proton-sorbose symport and passive sorbose transport in *Saccharomyces fragilis*. Biochim. Biophys. Acta **602:**419–432.
22. **Van den Broek, P. J. A., and J. Van Steveninck.** 1981. Kinetic analysis of 2-deoxy-D-glucose uptake in *Saccharomyces fragilis*. Biochim. Biophys. Acta **649:**305–309.
23. **Van Steveninck, J., and H. L. Booij.** 1964. The role of polyphosphates in the transport mechanism of glucose in yeast cells. J. Gen. Physiol. **48:**43–60.
24. **Van Steveninck, J., and A. Rothstein.** 1965. Sugar transport and metal binding in yeast. J. Gen. Physiol. **49:**235–246.

VIII: PHOSPHATE RESERVES AND ENERGY STORAGE: PYROPHOSPHATES

Introduction

HARLAND G. WOOD

Department of Biochemistry, Case Western Reserve University, Cleveland, Ohio 44106

The biochemical dogma in the past has been that the energy of the anhydride bond of PP_i is not utilized. PP_i is formed in many anabolic reactions, and it has been considered to be removed by hydrolysis, thus making these synthetic reactions energetically favorable. There has been an ever increasing series of developments, at least with some forms of life, which make this dogma untenable and show that PP_i is utilized in metabolism and the high energy of PP_i is not wasted. These developments include the discoveries (i) that PP_i is used in certain reactions in place of ATP, (ii) that PP_i is synthesized during photosynthesis and oxidative phosphorylation, and (iii) that PP_i is used nutritionally as a source of energy. These developments form the subject of this section. The authors of the papers that follow have made major contributions to our understanding of the role of PP_i.

The paper by Reeves summarizes his investigations with *Entamoeba histolytica*. Surprisingly, he has found that this organism, which is a eucaryote, contains enzymes that use PP_i for phosphorylation in place of ATP, just as does a procaryote, *Propionibacterium shermanii*, which has been studied in my laboratory. It has been found that both organisms contain three enzymes which use PP_i in reactions that have a major role in the metabolism of glucose. Reeves' results make it quite clear that PP_i is used by certain organisms as a source of energy.

The discussion then shifts to the role of PP_i in *Rhodospirillum rubrum*. Among the earliest and best-documented examples of energy-linked reactions that involve PP_i are those observed with this photosynthetic bacterium. These studies were initiated in 1966 and Baltscheffsky and others have continued these investigations to the present time. It has been found that the synthesis of PP_i is coupled to light-induced electron transport. It also has been demonstrated that PP_i can serve as an energy source for numerous reactions.

Many have considered that PP_i serves only as a source of energy in microorganisms. However, this is not always the case. Pyruvate phosphate dikinase which is present in bacteria is also present in so-called C_4 plants and has an important role. Black and others have shown that PP_i is utilized via PP_i-phosphofructokinase in plants.

One of the most surprising developments is the discovery that PP_i can be used as a source of nutrient energy when added to the medium for the growth of microorganisms. Growth is minimal in the absence of PP_i. It has been proposed that this growth-promoting action of PP_i may be "a phenomenon representing the simplest ATP-generating system in the biological world." It remains possible, however, that the PP_i is transported from the medium into the cell where it is used directly in reactions such as PP_i phosphofructokinase in anaerobic microorganisms.

This section also includes the results of Roberts et al. concerning the exciting discovery of a cyclic pyrophosphate intermediate in methanobacteria. This compound may serve as a means of energy storage.

It seems likely that the first forms of life may have been anaerobes that used the PP_i which almost certainly was present on the earth's crust, and the forms that evolved later which use oxygen developed other means of generating energy. It still remains possible, however, that some reactions that involve direct phosphorylation with PP_i may be retained by animals. The latter possibility is being actively investigated in a number of laboratories.

Nuclear Magnetic Resonance Studies of Methanogens: What Use Is a Cyclic Pyrophosphate?

M. F. ROBERTS, J. N. S. EVANS, AND C. J. TOLMAN

Department of Chemistry, Massachusetts Institute of Technology, Cambridge, Massachusetts 02139

Methanogens are microorganisms with unique and fascinating biochemistry. They are members of the archaebacteria kingdom (5), a group which is evolutionarily distinct from eucaryotes and eubacteria. The characteristic reaction carried out by methanogens is the reduction of CO_2 to CH_4 in an atmosphere of H_2. The electron transport generated in this process drives ATP synthesis. CO_2 is also the sole carbon source for growth, although some methanogens can use partially reduced C_1 species. *Methanobacterium thermoautotrophicum* is the most commonly studied of the methanogens. It is a thermophile, and CO_2 fixation into cell carbon occurs initially by the formation of acetyl coenzyme A via an asymmetric "activated acetic acid" pathway (4, 7, 8, 15, 17). Neither the Calvin cycle nor the full reductive tricarboxylic acid cycle operates in carbon assimilation in *M. thermoautotrophicum* (2, 6, 8, 18, 20). Instead, there are a number of unusual aspects of carbon fixation in this organism.

A promising technique for examining biochemical pathways is multinuclear high-resolution nuclear magnetic resonance (NMR) spectroscopy. Spectra of different nuclear probes can be obtained in intact cells and in cell extracts. Specific label uptake and distribution and occurrence of unusual metabolites can be monitored. For a wide range of microorganisms, ^{13}C and ^{31}P NMR spectroscopy have been used to study cell metabolism, in particular carbon flow, production of glycolytic intermediates, and development and maintenance of a pH gradient across cell membranes. Our early ^{31}P NMR investigation of *M. thermoautotrophicum* ΔH led us to identify and isolate a previously undetected cyclized PP_i, 2,3-cyclopyrophosphoglycerate (CPP), whose intracellular concentration under growth conditions in excess phosphate is extremely high (11). This compound appears to be unique to methanobacteria and methanobrevibacter (19). Examples of ^{31}P NMR spectra of two methanogens showing the intense CPP multiplet (AB quartet) are shown in Fig. 1. These particular methanogens have an unusual pseudomurein cell wall structure (10, 12) with high demands on carbohydrate metabolism during rapid growth. It can be shown (19) that CPP does not function solely as a high-energy phosphate donor (phosphagen) or phosphate storage compound (14) analogous to polyphosphate. Since it has a simple three-carbon unit, we decided to use ^{13}C NMR spectroscopy to determine (i) how CPP is synthesized, (ii) how labile the newly formed CPP is, and (iii) where CPP-labeled carbons appear upon enzymatic transformation. This spectroscopic approach is easier than more commonly carried out ^{14}C-radiolabeling studies because all soluble ^{13}C-labeled spe-

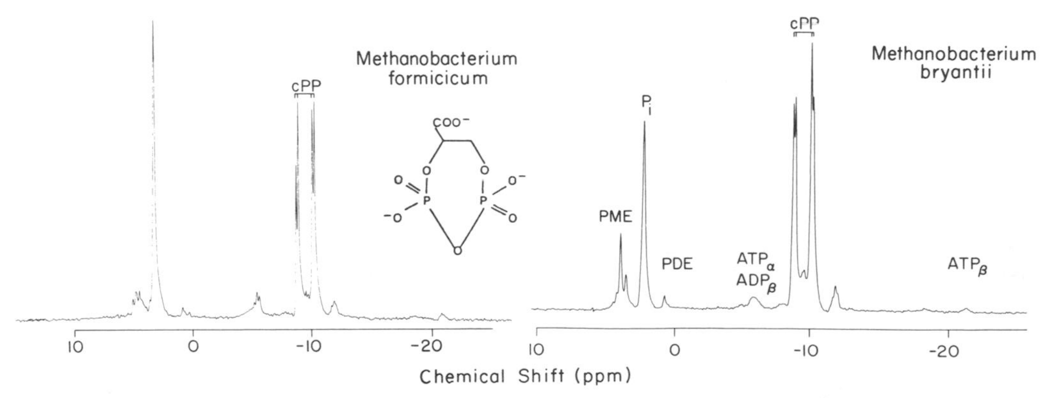

FIG. 1. In vivo ^{31}P NMR spectra (109.3 MHz) of *Methanobacterium formicicum* and *Methanobacterium bryantii*, each with about 1 g of wet-cell paste suspended in 50 mM PIPES [piperazine-*N*,*N*′-bis(2-ethanesulfonic acid)]–10 mM EDTA, pH 7.2. Chemical shifts are referenced with respect to an external capillary of H_3PO_4. PME, Phosphomonoesters; PDE, phosphodiesters.

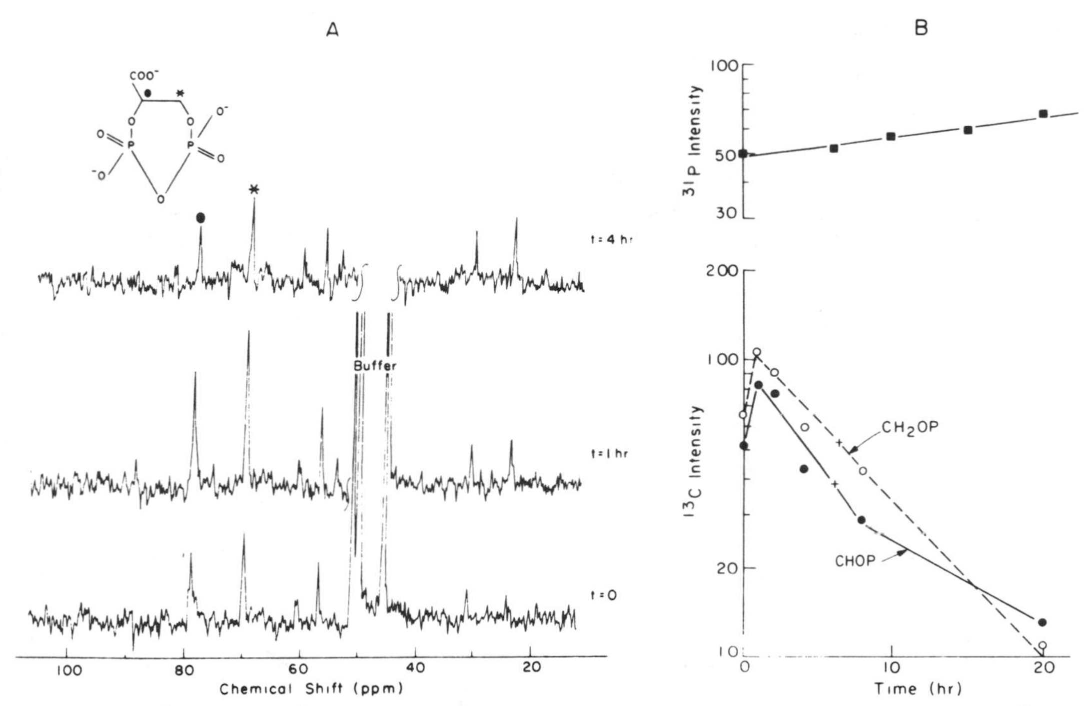

FIG. 2. (A) ^{1}H-decoupled ^{13}C NMR spectra (67.9 MHz) of *M. thermoautotrophicum* cells exposed to a $^{13}CO_2$ pulse (0.5 h) followed by a $^{12}CO_2$ chase for the time indicated. (B) ^{13}C intensity of the C-2 and C-3 of CPP as a function of time after start of the chase. The plus signs denote half-times for CPP in these cells.

cies can be detected without isolation/purification of specific compounds. The ^{13}C chemical shifts and ^{13}C-^{13}C coupling information can uniquely identify compounds without resort to degradation of the molecule. With this technique (both high-resolution solution and solid-state ^{13}C NMR spectroscopy) we were able to show that CO_2, acetate, pyruvate, and phosphoenolpyruvate serve as biosynthetic precursors for CPP and that CPP is connected with carbohydrate metabolism in methanogens.

Experimental Procedures

Cells of *M. thermoautotrophicum* ΔH for pulse-chase and steady-state labeling experiments were cultivated at 62°C as previously described (3, 11). Most pulse-chase experiments were carried out with 1 liter of cells (optical density at 660 nm of 1 for exponentially growing cells and 3.2 for stationary-phase cells) in a 2-liter Applikon fermentor. Cells were extracted with ethanol as outlined previously (3). For studies of solid cell debris using cross-polarization–magic angle sample spinning (CP-MASS) techniques, *M. thermoautotrophicum* cells were grown on ^{13}C-depleted CO_2 (99.96% ^{12}C) to minimize the ^{13}C natural abundance background. The cell debris remaining from the ethanol extracts was dried by lyophilization and used for CP-MASS NMR spectroscopy. Particulate cell wall material was isolated from total cell debris by treatment of the resuspended solid fraction with trypsin and DNase (1).

High-resolution ^{13}C NMR spectra were obtained at 100.6 MHz on a Bruker AM-400 WB instrument (Tufts University Medical School) and at 67.9 MHz on a Bruker 270 instrument (F. Bitter National Magnet Laboratory, Massachusetts Institute of Technology). Lyophilized samples were dissolved in 10 mM potassium phosphate–0.1 mM EDTA (pH 7.2)–50% D_2O. Chemical shifts were referenced to *p*-dioxane.

CP-MASS ^{13}C NMR spectra were obtained at 79.9 MHz on a home-built pulsed spectrometer (F. Bitter National Magnet Laboratory). Spectra were obtained by using cross-polarization with a ^{1}H decoupling field of 100 kHz, a ^{13}C radiofrequency field of 48 kHz, and a contact time of 2.5 ms. Sample (25 to 100 mg) spinning rates varied from 2.7 to 3.5 kHz. Chemical shifts were referenced to external tetramethylsilane.

Results

M. thermoautotrophicum cells exposed to $^{13}CO_2$ for 0.5 h show dramatic labeling of CPP carbons. In fact, CPP is the major small-molecule carbon as well as phosphorus pool in this organism (3). Both C-2 and C-3 of CPP are much more intense than the carboxyl carbon (C-1), making them good markers for CPP turnover experiments. The CPP pool is dynamic, not static. In vivo $^{13}CO_2$-pulse/$^{12}CO_2$-chase experiments (Fig. 2) suggested that ^{13}C-CPP rapidly loses its ^{13}C to an insoluble pool and not to another small molecule, while the steady-state concentration of CPP is maintained (as monitored by ^{31}P NMR spectroscopy). Cells at these high densities have long doubling times (24 h). If, rather than using intact cells, we repeated the experiment at lower cell densities, withdrawing samples from the fermentor, cooling them rapidly, and then extracting soluble material, we obtained better spectral resolution and sensitivity to monitor this decay of the soluble ^{13}C labels (Fig. 3). For this experiment, the cells typically had a doubling time of 12 h and the CPP turnover had a half-life of 1.2 h. CPP carbons were rapidly labeled (δ_c = 70.1 ppm [C-3] and 78.6 ppm [C-2]) during a 15-min $^{13}CO_2$ pulse. The CPP-integrated intensity increased slightly during the first 15 min after the start of the chase as $^{13}CO_2$ in the medium was fixed into CPP. Under these conditions the maximum enrichment was about 12-fold over the natural-abundance CPP background. After 1 h, CPP peaks decreased and no new ^{13}C species appeared. Small amounts of other ^{13}C-labeled species appeared immediately at the start of the chase. Glutamate (δ_c = 27.8 ppm [C-3], 34.3 ppm [C-4], 55.5 ppm [C-2]) did not decrease on the time scale of the chase. Therefore, the glutamate half-life is longer than that for CPP. In contrast, a number of unidentified but well-resolved multiplets (around 20, 23, and 38 ppm) decayed rapidly ($\tau_{1/2}$ = 10 to 15 min).

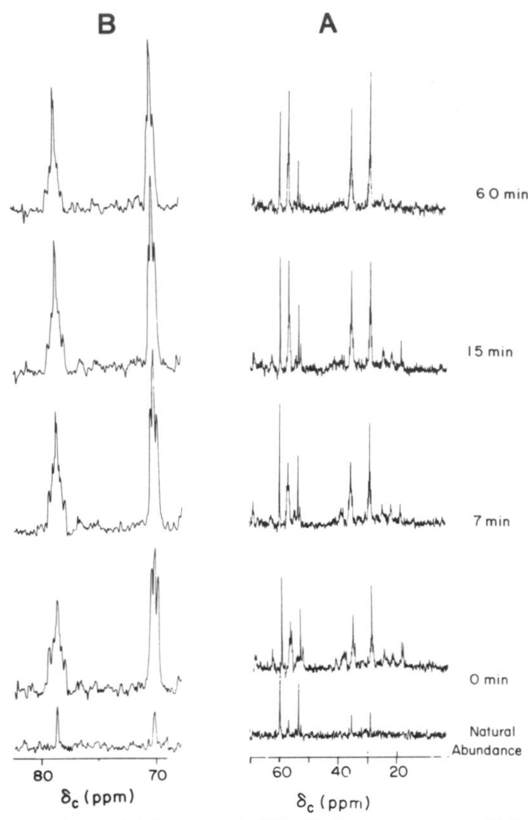

FIG. 3. ^{1}H-decoupled ^{13}C NMR spectra at 67.9 MHz of ethanol extracts of exponentially growing *M. thermoautotrophicum* cells given a 15-min $^{13}CO_2$ pulse followed by a $^{12}CO_2$ chase and harvested at different times after the start of the chase. Two spectral regions are shown: (A) upfield region containing glutamate, alanine, acetyl methyls, etc., and (B) CPP C-2 and C-3 region.

One way of tracking down the fate of ^{13}C from CPP is to use ^{13}C CP-MASS techniques to monitor incorporation of ^{13}C into the particulate matter of the cell. NMR spectra of solids are, with few exceptions, broad featureless lines due to dipolar interactions. Three approaches have been tried to regain resolution in the solidlike state: (i) MASS, (ii) multiple pulse techniques, and (iii) magnetic dilution. All of these methods reduce or eliminate dipolar broadening, and a combination of MASS with dilute spin double resonance (CP-MASS) is commonly used to detect enhanced and narrowed ^{13}C resonances from solids.

Preliminary experiments used ^{13}C CP-MASS solid-state NMR spectroscopy to determine the fate of ^{13}C turned over from the CPP pool after a pulse-chase experiment. The solid material obtained from cells given a 15-min $^{13}CO_2$ pulse followed by a $^{12}CO_2$ chase in bottles and harvested after 30 h showed increased intensity in a region of the spectrum consistent with carbohydrate. However, the natural-abundance background of the solid material limited interpretation of these data. Even though the natural abundance of ^{13}C is only 1.1%, all the carbons in proteins, nucleic acids, cell wall, lipids, etc., contribute to this background spectrum.

Because CO_2 is the sole carbon source for *M. thermoautotrophicum*, we can dramatically suppress the natural abundance ^{13}C background by growing the cells on ^{13}C-depleted $^{12}CO_2/H_2$. Consequently, cells were grown in bottles with ^{13}C-depleted CO_2/H_2 (99.96% ^{12}C). When the optical density of the cell culture reached 1.0, the cultures were subjected to a pulse of $^{13}CO_2$ (99.2%) for 0.5 h prior to switching back to ^{13}C-depleted CO_2. Cells grown in this fashion in pressurized bottles have a doubling time of about 30 h. After 12 and 30 h of the $^{12}CO_2$ chase, the cells were extracted with ethanol, and the soluble as well as the lyophilized cell debris was examined by ^{13}C NMR spectroscopy. The solution-state spectrum shows that CPP labeling by the pulse of $^{13}CO_2$ (Fig. 4B) was substantially above ^{12}C-enriched background (Fig. 4A) for the

12-h time point. In contrast to extracts from cells grown in the fermentor, significantly enriched resonances from 2,3-diphosphoglycerate (67.4 ppm [C-3], 76.5 ppm [C-2]) were observed with the extracts from bottle-grown cells. At 30 h after the start of the chase, 75% of the CPP (and 2,3-diphosphoglycerate) integrated intensity had disappeared from the high-resolution solution spectrum (Fig. 4C). ^{13}C CP-MASS spectra of the corresponding solid phases are shown in Fig. 4D–G. The ^{13}C background in the lyophilized cell debris (Fig. 4D) has been essentially eliminated by growth on ^{13}C-depleted medium. The dramatic incorporation of ^{13}C shown in Fig. 4E reflects carbohydrate, protein, and nucleic acid species whose components were labeled with a 0.5-h $^{13}CO_2$ pulse followed by a 12-h $^{12}CO_2$ chase. Organic solvent extracts of this material did not remove any ^{13}C, implying that label uptake by the lipid fraction is much less on this time scale. After 30 h (Fig. 4F), the largest change over the 12-h spectrum was the increase in intensity at 103 ppm and around 70 ppm. For the resonance at 103 ppm, this can be detected by comparing its peak intensity with that for the carboxyl (~175 ppm) or methyl (~20 ppm) carbons. The intensity ratio of the 103-ppm peak to these others at 30 h has increased to 1.6 times the value observed at 12 h. This indicates a preferential flux of ^{13}C from the solution state to the carbons, giving rise to the resonance at 103 ppm. This region corresponds to anomeric carbons of carbohydrates. The region around 70 ppm is consistent with other carbohydrate ring carbons. Trypsin and DNase treatment of the solid material collected at 30 h removes protein and nucleic acids, but leaves behind the cell wall material (9). The cell wall contains glucosamine, talosamine, lysine, glutamate, and alanine, and resonances consistent with this composition were observed. As verification of the purity of this material, we detected no aromatic carbons in the CP-MASS spectrum of this isolated cell wall preparation. Furthermore, amino acid analyses of acid-hydrolyzed samples showed appreciable levels of lysine, glutamate, and alanine and none of the other amino acids. Therefore, the increased intensity in the solid state which corresponds to the decay of ^{13}C from the soluble

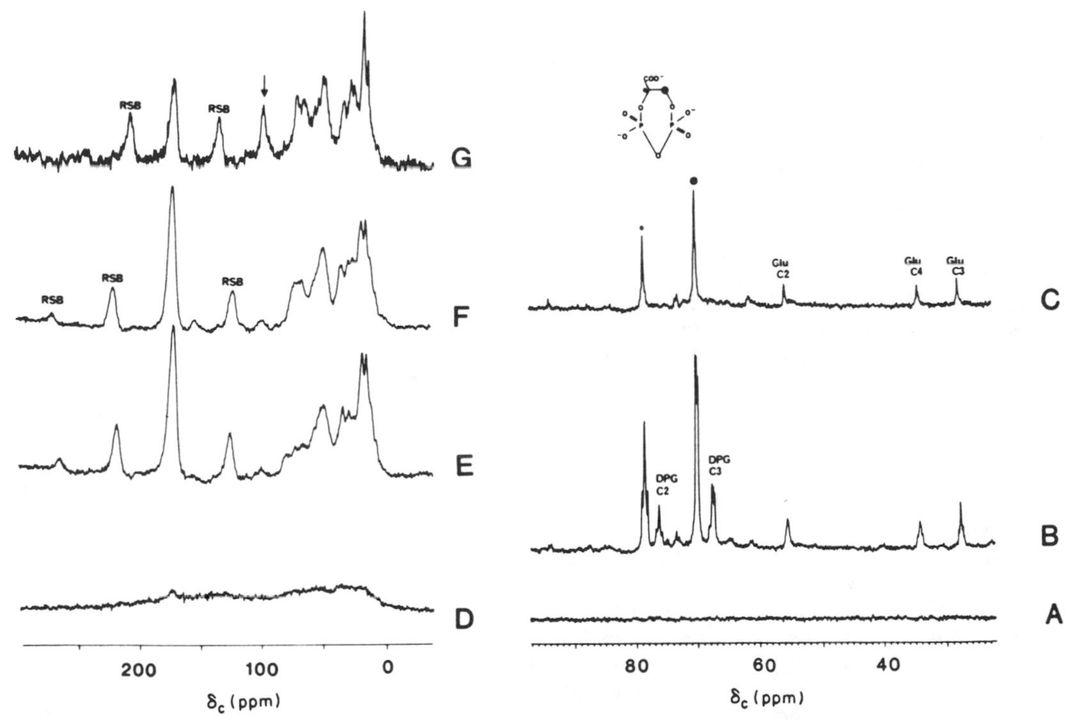

FIG. 4. High-resolution ^{1}H-decoupled ^{13}C NMR (100.6 MHz) spectra of ethanol extracts of cells grown on ^{13}C-depleted CO_2, exposed to a $^{13}CO_2$ pulse for 0.5 h, followed by a $^{12}CO_2$ chase: (A) background of soluble cell material before $^{13}CO_2$ pulse; (B) 12 and (C) 30 h after the start of the chase. ^{13}C CP-MASS NMR spectra (79.9 MHz) of ^{13}C-depleted cell debris from the same cells: (D) before the $^{13}CO_2$ pulse; (E) 12 h and (F) 30 h after the start of the $^{12}CO_2$ chase. In G the cell debris in F was digested with trypsin and DNase to isolate pseudomurein cell wall. The arrow indicates the resonance position of the anomeric carbon of hexoses. Peaks identified as RSB are rotational sidebands and are associated with carbons such as carbonyls that have large chemical shift anisotropy.

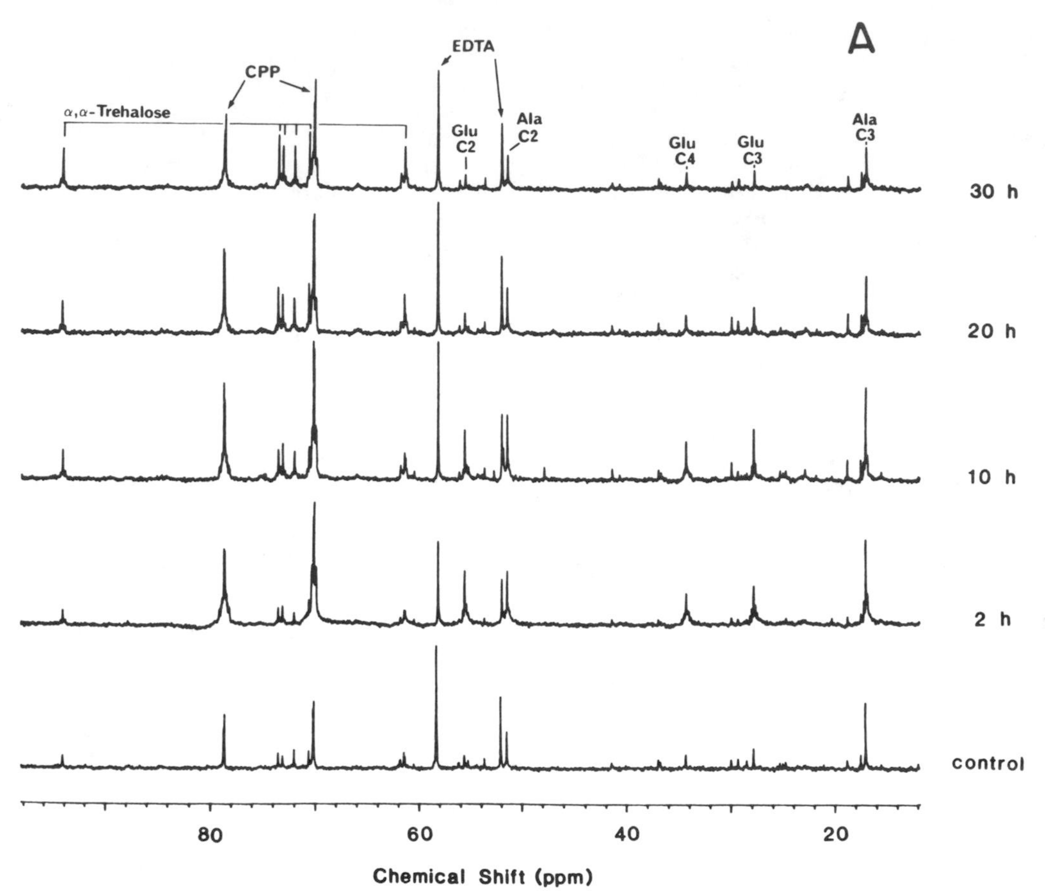

FIG. 5. (A) ^{1}H-decoupled ^{13}C NMR (100.6 MHz) spectra of ethanol extracts of stationary-phase *M. thermoautotrophicum* cells given a 0.5-h $^{13}CO_2$ pulse followed by a $^{12}CO_2$ chase. (B) ^{13}C integrated intensities of C-2 of CPP (●), C-1 of α,α-trehalose (○), alanine C-3 (▲), and glutamate C-3 (△). Natural-abundance intensities have been subtracted from the observed intensities.

phase corresponds to carbohydrate in the pseudomurein cell wall of this organism.

The flux of carbon through CPP depends on cell growth conditions in an interesting fashion. Cells grown to stationary phase and given a 0.5-h $^{13}CO_2$ pulse followed by the $^{12}CO_2$ chase show behavior different from that observed when the same experiment was done with exponentially growing cells (Fig. 5). CPP was labeled but to a much lower level of ^{13}C enrichment (only 2.3-fold above the natural-abundance spectrum which is shown as the control in Fig. 5A). Incorporation of $^{13}CO_2$ was easily detected as ^{13}C-coupled multiplets flanking a ^{13}C singlet. The multiplet structure reflects ^{13}C adjacent to another ^{13}C nucleus. The singlet reflects ^{13}C in the already present natural-abundance level of material (which will be replenished as $^{12}CO_2$ is

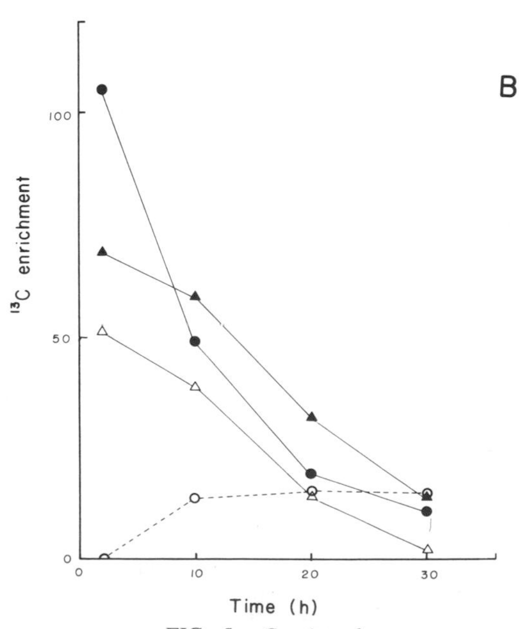

FIG. 5.—*Continued*

incorporated into CPP during the $^{12}CO_2$ chase). Because the level of ^{13}C enrichment was considerably lower than in other experiments, the natural-abundance background ^{13}C intensity must be subtracted from the 2-h through 30-h spectra to monitor the flux of the pulse of $^{13}CO_2$ through these soluble metabolites. The turnover of CPP in these cells has a 9- to 10-h half-life. Additionally, the disaccharide α,α-trehalose (13) became increasingly labeled over the course of the chase. The greatest increase in trehalose intensity corresponded to the initial decrease in CPP carbons. Since the overall CPP concentration as determined by ^{31}P NMR does not vary significantly with growth phase, the difference in CPP turnover must reflect a decreased carbon flux.

If cells are frozen and then slowly thawed, the cell membrane becomes leaky, allowing passage of small molecules into and out of the cell envelope. If enzymes which synthesize CPP are not dependent on the metabolic state of the cell and sufficient cofactors or necessary substrates are added for biosynthesis, then CPP should be labeled when the mixture is incubated with $^{13}CO_2$. Cells permeabilized in this fashion required added buffer components (ATP, coenzyme A, dithiothreitol, $MgCl_2$, P_i, NH_4Cl) to accelerate acetyl coenzyme A biosynthesis as well as methane production in vitro (16). With these permeabilized cells and the addition of the above components, there was very little uptake of $^{13}CO_2$ into CPP after 24 h, certainly much less than that which is seen with intact cells. Synthesis of a number of carbon-containing molecules did still occur with these permeabilized cells. Both glutamate and alanine were significantly labeled. A cell suspension broken in a French pressure cell showed similar behavior.

^{13}C NMR experiments can also be used to establish the biosynthesis of CPP by incubating *M. thermoautotrophicum* with specifically labeled precursors and $^{12}CO_2/H_2$ for methane production. Experiments with ^{13}C-labeled acetate showed that the CH_2O (C-3) of CPP is derived from the acetate methyl group and the CHO (C-2) is derived from the acetate carboxyl carbon (Fig. 6). A similar experiment with [C-1]-pyruvate showed that the CPP carboxyl group (C-1) is specifically derived from the carboxylate of pyruvate (Fig. 7). In vitro experiments with cell suspensions passed through a French pressure cell were also undertaken to see what precursors lead to increased CPP levels. A variety of combinations of acetate, pyruvate, phosphoenolpyruvate, ATP, and ADP were used as carbon and phosphorus donors (3). Limited, but significantly above background, levels of CPP synthesis occurred only with phosphenolpyruvate as the carbon skeleton and ATP or ADP

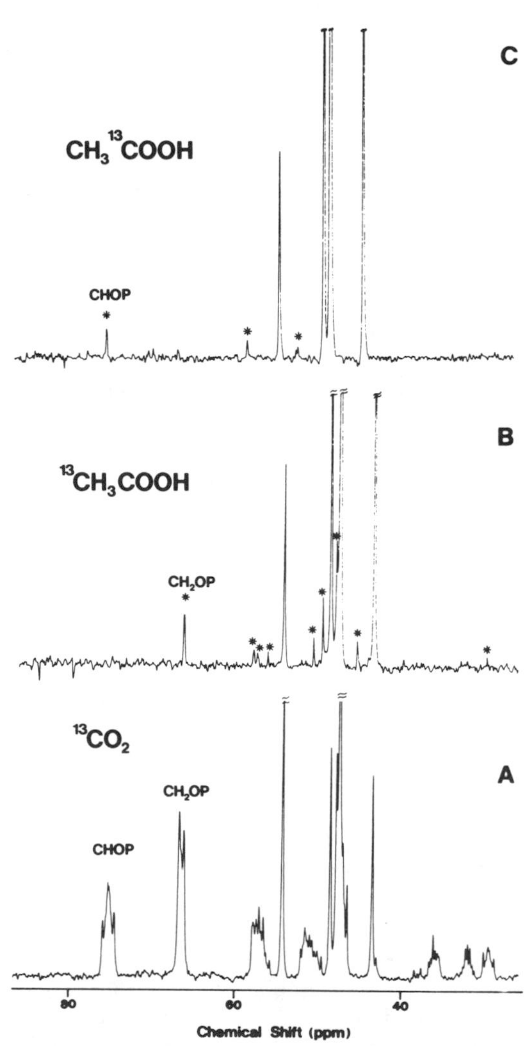

FIG. 6. ^1H-decoupled ^{13}C NMR (100.6 MHz) spectra of extracts of cells exposed to (A) $^{13}CO_2$ (2 h), (B) [2-^{13}C]acetate (20 mM), and (C) [1-^{13}C]acetate (20 mM), all 6 h prior to extraction. Peaks marked * arise from labeled species derived from the specifically labeled acetates; those remaining are buffer peaks.

presumably as the necessary phosphorus donor (an enzyme activity in lysed cells rapidly interconverts these two species).

The structure of CPP, bearing a close resemblance to 2,3-diphosphoglycerate, and its occurrence in cells with a high demand on carbohydrate biosynthesis, together with its chemical transformation into an insoluble pool, are highly suggestive of a role in carbohydrate biosynthesis. If this is so, any [6-^{13}C]glucose incubated with *M. thermoautotrophicum* that is internalized should be processed by intracellular glycolytic enzymes to specifically label the C-3 of CPP (Fig. 8). This does indeed occur. Figure 9 shows

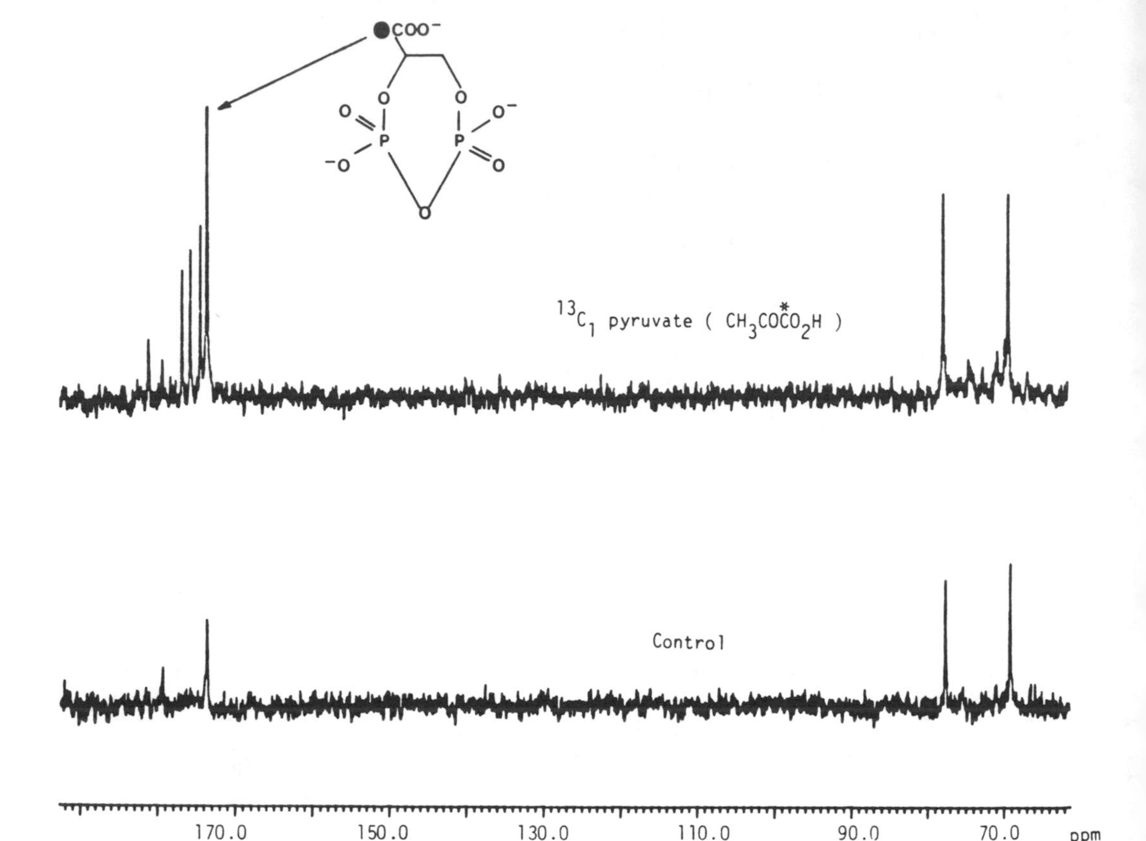

FIG. 7. ^1H-decoupled ^{13}C NMR spectra (100.6 MHz) of extracts from cells incubated for 6 h with $^{12}CO_2$ (control) and $^{12}CO_2$ and [1-^{13}C]pyruvate (20 mM).

the result of incubating cells with [6-^{13}C]glucose and $^{12}CO_2/H_2$ for 6 h. C-3 of CPP was exclusively labeled, establishing that glucose is cleaved by *M. thermoautotrophicum* in the reverse of the gluconeogenic pathway.

Discussion

^{13}C NMR spectroscopy has been shown to be extremely useful in determining the turnover and biosynthesis of a unique metabolite, CPP, in methanobacteria. [^{13}C]CPP label is rapidly metabolized in actively growing cell cultures. Rather than forming a soluble product (which would be expected if it were a phosphagen in an energy cycle), the CPP is converted to particulate material. The use of CP-MASS ^{13}C NMR of ^{13}C-depleted cell debris resulting from a $^{13}CO_2$ pulse-$^{12}CO_2$ chase experiment represents the first application of this technique to a preparation of ^{13}C-depleted cells and allowed us to monitor label transfer from soluble CPP into insoluble cell components. Since the loss of ^{13}C intensity correlates with the appearance of ^{13}C intensity in the carbohydrate resonances of the solid debris (and not in lipidic or proteinaceous material), a relationship between CPP turnover and carbohydrate synthesis was established. Biosynthesis experiments also linked carbohydrate metabolism and CPP metabolism. Glucose, which is not an energy source for *M. thermoautotrophicum*, was metabolized by cells under CO_2/H_2 to generate specifically labeled CPP.

The net result of these studies is that the novel methanogen PP_i CPP is an important metabolite whose turnover is regulated by the metabolic state of the cell. It is not a static reservoir for excess carbon or phosphorus storage. What remains puzzling is the high intracellular concentration of CPP. It is unusual for a biosynthetic intermediate to be present at such high concentrations. Therefore, it is unlikely that CPP is just an aberrant intermediate in carbon assimilation. It is not known how gluconeogenesis is regulated in *M. thermoautotrophicum*. Fuchs has shown that the enzymes which are usually the key regulated steps for carbohydrate metabolism in other bacteria do not appear to be

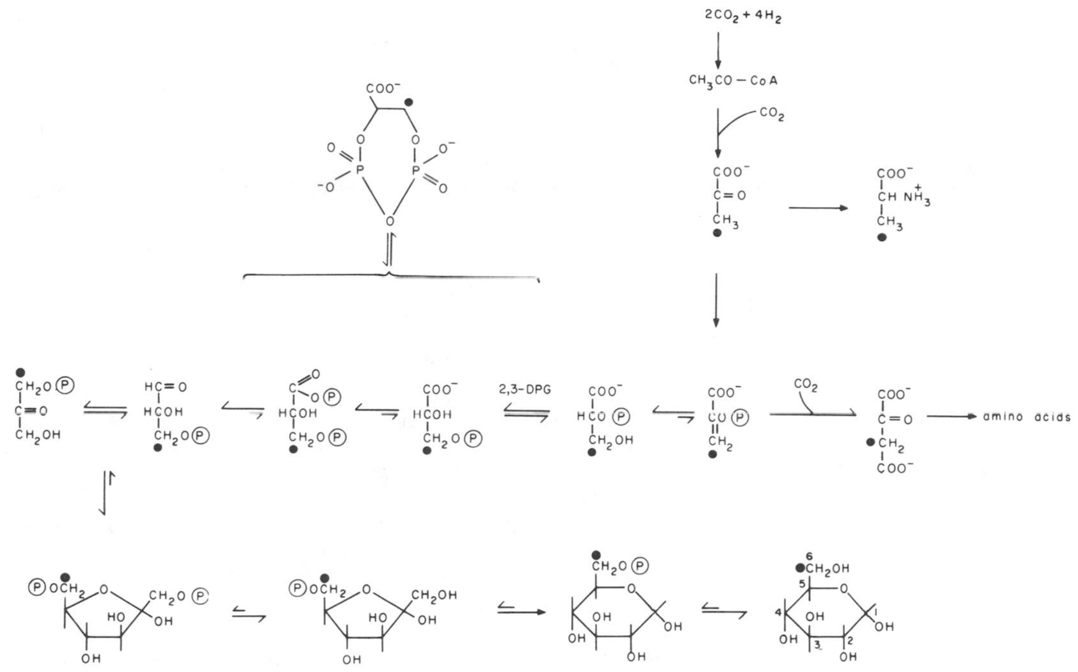

FIG. 8. Gluconeogenesis pathway illustrating the predicted labeling from metabolism of [6-^{13}C]glucose via the reverse of this pathway.

so stringently regulated in this organism. It is possible that CPP may serve as a highly regulated pool for eventual generation of carbohydrates. Perhaps the enzymes which metabolize CPP are the points of control in gluconeogenesis in this methanogen.

LITERATURE CITED

1. **Choi, B. S., J. E. Roberts, J. N. S. Evans, and M. F. Roberts.** 1986. Nitrogen assimilation in *Methanobacterium thermoautotrophicum*: a solution and solid-state ^{15}N NMR study. Biochemistry **25:**2243–2248.
2. **Daniels, L., and J. G. Zeikus.** 1978. One-carbon metabolism in methanogenic bacteria: analysis of short-term

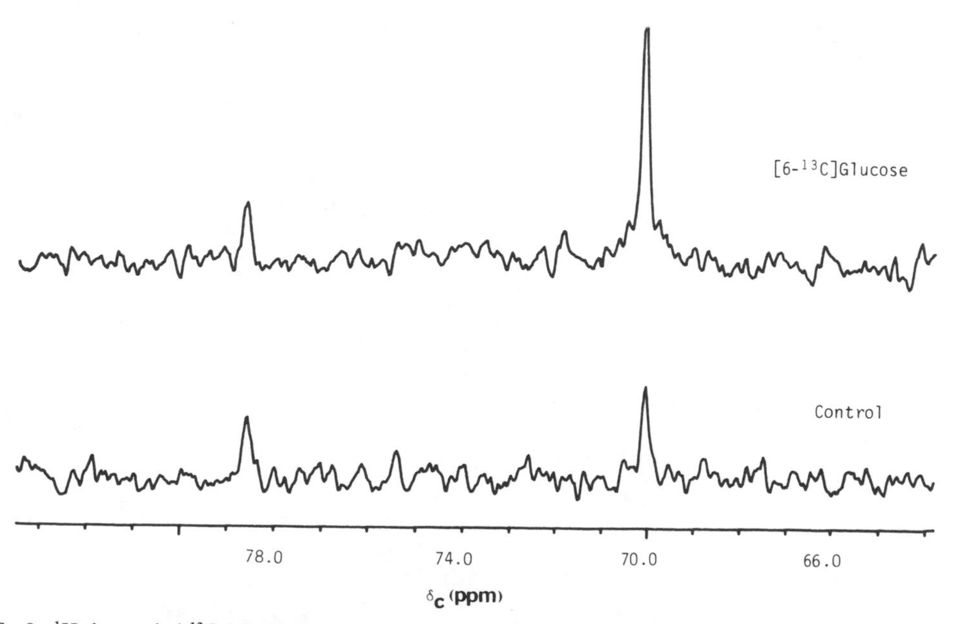

FIG. 9. ^{1}H-decoupled ^{13}C NMR spectra at 100.6 MHz of ethanol extracts of *M. thermoautotrophicum* cells incubated with CO_2/H_2 and CO_2/H_2 plus [6-^{13}C]glucose.

fixation products of $^{14}CO_2$ and $^{14}CH_3OH$ incorporated into whole cells. J. Bacteriol. **136:**75–84.

3. **Evans, J. N. S., C. J. Tolman, S. Kanodia, and M. F. Roberts.** 1985. 2,3-Cyclopyrophosphoglycerate in methanogens: evidence by ^{13}C NMR spectroscopy for a role in carbohydrate metabolism. Biochemistry **24:**5693–5698.
4. **Evans, J. N. S., C. J. Tolman, and M. F. Roberts.** 1986. Indirect observation by ^{13}C NMR spectroscopy of a novel CO_2 fixation pathway in methanogens. Science **231:**488–491.
5. **Fox, G. E., E. Stackebrandt, R. B. Hespell, J. Gibson, J. Maniloff, T. A. Dyer, R. S. Wolfe, W. E. Balch, R. S. Tanner, L. J. Magrum, L. B. Zablen, R. Blakemore, R. Gupta, L. Bonnen, B. J. Lewis, D. A. Stahl, K. R. Luehrsen, K. N. Chen, and C. R. Woese.** 1980. The phylogeny of prokaryotes. Science **209:**457–463.
6. **Fuchs, G., and E. Stupperich.** 1978. Evidence for an incomplete reductive carboxylic acid cycle in *Methanobacterium thermoautotrophicum*. Arch. Microbiol. **118:** 121–125.
7. **Fuchs, G., and E. Stupperich.** 1980. Acetyl CoA, a central intermediate of autotrophic CO_2 fixation in *Methanobacterium thermoautotrophicum*. Arch. Microbiol. **127:**267–272.
8. **Fuchs, G., E. Stupperich, and R. K. Thauer.** 1978. Acetate assimilation and the synthesis of alanine, aspartate, and glutamate in *Methanobacterium thermoautotrophicum*. Arch. Microbiol. **117:**61–66.
9. **Jacob, G. S., J. Schaefer, and G. E. Wilson, Jr.** 1983. Direct measurement of peptidoglycan cross-linking in bacteria by ^{15}N nuclear magnetic resonance. J. Biol. Chem. **258:**10824–10826.
10. **Kandler, O.** 1979. Zellwandstrukturen bei Methan-Bakterien. Naturwissenschaften **66:**95–105.
11. **Kanodia, S., and M. F. Roberts.** 1983. A novel cyclic pyrophosphate from *Methanobacterium thermoautotrophicum*. Proc. Natl. Acad. Sci. USA **80:**5217–5221.
12. **Konig, H., and O. Kandler.** 1979. The amino acid sequence of the peptide moiety and the pseudomurein from *Methanobacterium thermoautotrophicum*. Arch. Microbiol. **121:**271–275.
13. **Pfeffer, P. E., K. M. Valentine, and F. W. Parrish.** 1979. Deuterium-induced differential isotope shift ^{13}C NMR. 1. Resonance reassignments of mono- and disaccharides. J. Am. Chem. Soc. **101:**1265–1274.
14. **Seely, R. J., and D. E. Fahrney.** 1983. A novel diphospho-P,P′-diester from *Methanobacterium thermoautotrophicum*. J. Biol. Chem. **258:**10835–10838.
15. **Stupperich, E., and G. Fuchs.** 1981. Products of CO_2 fixation and ^{14}C labelling pattern of alanine in *Methanobacterium thermoautotrophicum* pulse-labelled with $^{14}CO_2$. Arch. Microbiol. **130:**294–300.
16. **Stupperich, E., and G. Fuchs.** 1984. Autotrophic synthesis of activated acetic acid from two CO_2 in *Methanobacterium thermoautotrophicum*. I. Properties of *in vitro* system. Arch. Microbiol. **139:**8–13.
17. **Stupperich, E., and G. Fuchs.** 1984. Autotrophic synthesis of activated acetic acid from two CO_2 in *Methanobacterium thermoautotrophicum*. II. Evidence for different origins of acetate carbon atoms. Arch. Microbiol. **139:**14–20.
18. **Taylor, C. T., D. P. Kelly, and S. J. Pirt.** 1976. Intermediary metabolism in methanogenic bacteria. p. 173–180. *In* H. G. Schlegal, G. Gottschalk, and N. Pfenning (ed.), Microbial production and utilization of gases (H_2, CH_4, CO). Akademie der Wissenschafter zu Göttingen, E. Goltze, Göttingen.
19. **Tolman, C. J., S. Kanodia, L. Daniels, and M. F. Roberts.** 1986. ^{31}P NMR spectra of methanogens: 2,3-cyclopyrophosphoglycerate is detected only in methanobacteria strains. Biochim. Biophys. Acta **886:**345–352.
20. **Zeikus, J. G., G. Fuchs, W. Kenealy, and R. K. Thauer.** 1977. Oxidoreductases involved in cell carbon synthesis of *Methanobacterium thermoautotrophicum*. J. Bacteriol. **132:**604–613.

Metabolic Energy Supplied by PP_i

RICHARD E. REEVES

Department of Microbiology, Immunology, and Parasitology, Louisiana State University Medical Center, New Orleans, Louisiana 70112

It is now 30 years since Kornberg (9) foresaw the possible discovery of PP_i-utilizing enzyme systems which might conserve the energy of anabolically formed PP_i. A decade has passed since Reeves (18) and Wood (29) separately summarized work reporting several newly discovered PP_i-utilizing processes located in important metabolic pathways, but the concept of the instantaneous hydrolysis of PP_i is still prominent in textbooks. The concept of the ubiquitous PP_i hydrolase has been investigated in microbial metabolism (8) with equivocal results. The idea that the fidelity and irreversibility of anabolic processes depends upon a vanishingly low intracellular concentration of PP_i is supported by a theoretical argument in the case of RNA synthesis (13), but experimental evidence seems inconclusive. The latter idea was more tenable when the only means of identifying PP_i was by a chemical procedure barely capable of detecting 1 part PP_i in the presence of 20 parts P_i, but a specific enzymatic method with 1,000-fold greater sensitivity was described in 1969 (22) and, in modified form, is now commercially available as a test kit.

An excellent review of the role of PP_i in metabolism is that of Kulaev and Vagabov (10). There is now a realization that in many organisms PP_i has a fate more important than simple hydrolysis. The new chapter began in 1962 with the finding by Sui and Wood (28) that an enzyme from propionibacteria catalyzes the reversible reaction linking phosphoenolpyruvate (PEP) with oxalacetate: PEP + P_i + CO_2 = oxalacetate + PP_i (EC 4.1.1.38). Most readers of Wood's extensive work on this PP_i-utilizing enzyme (28–33) may have focused on the formation of a C_4 dibasic acid from a C_3 molecule because the photosynthetic cycle involving the C_4 acids was being vigorously pursued in those years. The next step in the PP_i story occurred in 1968 when three laboratories independently reported the finding of an enzyme now called pyruvate phosphate dikinase, catalyzing the following reaction (5, 7, 15): PEP + AMP + PP_i = pyruvate + ATP + P_i (EC 2.7.9.1). Since one of the laboratories was my own, I was happy to find our results in agreement with those of Wood's group and of M. D. Hatch. This enzyme lay directly in the glycolytic pathway of the experimental organism we were using (15, 23).

The earliest place in glycolysis where large quantities of PP_i could enter metabolism was found to be at the step of phosphorylation of fructose 6-phosphate, the phosphofructokinase reaction (4, 12, 21, 26): fructose 6-phosphate + PP_i = fructose 1,6-bisphosphate + P_i (EC 2.7.1.90). Unlike the reaction in which ATP is the phosphoryl donor, this reaction is physiologically reversible. In the absence of divalent metal ions the PP_i bond is energetically equivalent to the beta or gamma bonds in ATP. In the presence of divalent metal ions their binding with P_i is weaker than that to ADP, with the result that the esterification of a hydroxyl group by ATP is physiologically irreversible while the reaction with PP_i is readily reversible (18, 29, 33). The equilibrium constant of the reaction as written above in the presence of Mg^{2+} is less than 10 at pH 7.

Minor reactions which consume PP_i with conservation of energy are serine pyrophosphokinase (3) and glucose 6-phosphate formation from glucose and PP_i (14). The latter appears to be a phosphotransferase reaction which, in the reverse direction, produces P_i and free glucose in the liver. A true phosphoglucokinase which should be reversible under physiological conditions has not yet been reported.

PP_i metabolism in *Entamoeba histolytica* (19)

Before discussing these reactions, let me mention some unusual aspects of the metabolism of *E. histolytica* which made it possible to glimpse the significance of catabolic reactions involving PP_i. More interesting than what amoebae can do is what they cannot do. Figure 1 shows the glycolytic pathway between glucose and pyruvate. This pathway is much less complex than that of organisms such as *Escherichia coli* and propionic acid bacteria or mammals. We note six paths diverging in or out of the glycolytic pathway which amoebae do not utilize. The absence of many branching pathways made it easier to discern what the PP_i enzymes can accomplish. I cannot say that amoebae are genetically imcompetent to make the enzymes which facilitate these branch pathways, but a diligent search for them has failed to reveal any evidence of their existence in several strains of axenically grown amoebae, whether from stationary-phase cultures or from log-phase cultures.

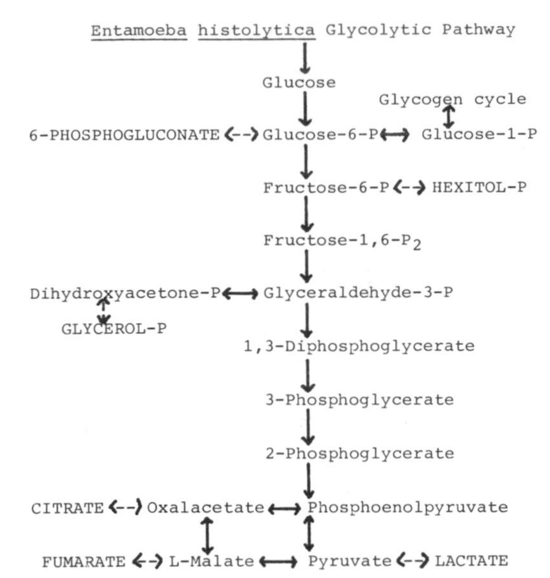

FIG. 1. Glycolytic pathway of *E. histolytica*. Metabolites in capitals appear not to be made by axenically grown cells.

The first step is catalyzed by an ATP-dependent glucokinase, and the second is catalyzed by the usual glucose 6-phosphate isomerase. The third step in the glycolytic pathway gave us years of trouble, but was finally resolved in 1974 by the finding of the PP_i-dependent phosphofructokinase (25, 26). Until this enzyme was found, it was often speculated that the reason amoebae are parasites is that they lack phosphofructokinase.

When a homogenate prepared from fresh amoebae was subjected to mild centrifugation, the resulting crude enzyme solution contained what appeared to be a trivial amount of ATP-dependent phosphofructokinase, which showed exasperatingly ephemeral activity. Because the activity was lost on strong centrifugation, its enzyme seemed to be a big molecule, but since it also vanished on dialysis, a small molecule was involved. After much work we finally tried other nucleoside triphosphates. Some of these gave more activity than we had seen with ATP. Reaction with GTP was much the best, but quantitatively only a small fraction of the GTP seemed to react. Earlier, we had assayed our nucleoside triphosphates for contamination by PP_i, and now it was noted that, while ATP was essentially free from PP_i, GTP was grossly contaminated. When PP_i was substituted for a nucleoside triphosphate in the phosphofructokinase assay, much more activity was found than had been seen with ATP. A column fractionation of a crude homogenate from fresh cells showed a small peak of phosphofructokinase activity elicited with ATP and a larger peak overlapping the smaller, representing activity elicited with PP_i. These results were submitted in a manuscript suggesting that amoebae contain a small amount of ATP-dependent phosphofructokinase activity and a much greater amount of PP_i-dependent activity (26). A reviewer for the journal pointed out that we really had not established the presence of two distinct enzymes since the curves obviously overlapped. In correspondence with the editor my view prevailed, and this was the only occasion when I ever won an argument with a reviewer. Unfortunately, the reviewer was right. Further work eventually showed that what had been regarded as an ATP-dependent enzyme was the functioning of a trace of the glycogen-forming apparatus which had survived the centrifugation step and, by a roundabout process, slowly liberated PP_i from the added ATP (17, 19).

At first glimpse the PP_i-phosphofructokinase had been mistaken for a fructose diphosphatase. It could also have been mistaken for a GTP-dependent phosphofructokinase or for a PP_i-hydrolyase. Its activity supports the total glycolytic flux which, at the hexose level, can reach 5 μmol/min per g of cells. More will be said about this enzyme by later authors in this volume, so I will turn to two other reactions which involve PP_i in amoebic glycolysis.

A feature which we call the "loop" is shown at the bottom of Fig. 1 and in greater detail in Fig. 2. The loop consists of four physiologically reversible enzyme-catalyzed steps. At the top in Fig. 2 is pyruvate phosphate dikinase. Proceeding clockwise, we have malic enzyme, followed by malate dehydrogenase, and finally the Sui and Wood enzyme which they called PEP carboxytransphosphorylase (16, 28). The loop provides amoebae with alternative pathways between PEP and pyruvate. The concentration of metabolites determines how the flux will be divided. One of the loop pathways functions to produce a substrate-level pyridine nucleotide transhydrogenase. A substrate-level pyridine nucleotide transhydrogenase consists of two

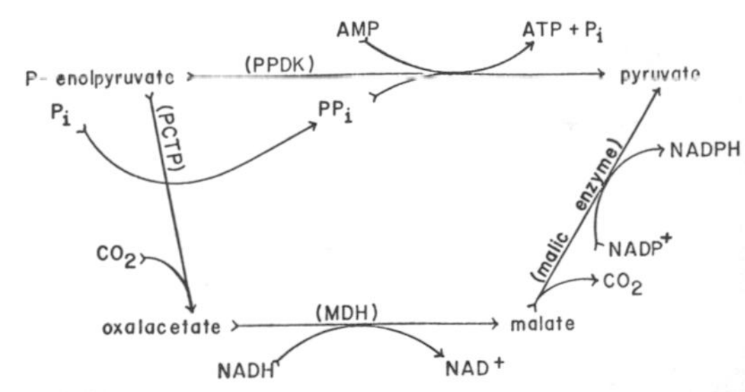

FIG. 2. Loop enzymes of *E. histolytica*. PPDK, Pyruvate phosphate dikinase; malic enzyme, malate dehydrogenate (decarboxylating); MDH, malate dehydrogenase; PCTP, PEP carboxytransphosphorylase.

separate reactions which have a substrate in common. In the loop the common substrate is malate, which is linked to oxalacetate by the NAD-dependent malate dehydrogenase and to pyruvate by the NADP-dependent malic enzyme (malate dehydrogenase, decarboxylating; EC 1.1.1.40). In living cells the glycolytic flux is PEP to pyruvate: the direct pathway via pyruvate phosphate dikinase produces ATP; the longer pathway via oxalacetate and malate produces PP_i and, indirectly, an NADH-to-NADPH transhydrogenation.

E. histolytica are aerotolerant anaerobes. They live in the presence of oxygen but cannot then multiply. They have no known direct link between NADH and oxygen, but a very active link between NADPH and oxygen, producing water in the intact cell (11). The loop provides the organism's defense against molecular oxygen by producing NADPH from NADH. In the presence of oxygen the flux through the loop must favor the oxalacetate-malate pathway producing PP_i, pyruvate, and NADPH. The latter enzymatically reduces a flavin which reacts with oxygen, regenerating the oxidized form of the flavin. The loci most sensitive to damage by oxygen are the iron-sulfur centers contained in the small ferridoxin molecule which is the electron carrier transporting electrons from pyruvate oxidation to NAD. Lacking these iron-sulfur centers the electrons from pyruvate oxidation must go via flavin to oxygen, producing water (or in cell homogenates, producing peroxide). This shuts off the reduction of the acetyl group of acetyl coenzyme A (acetyl-CoA) to ethanol, and dictates that CoA be regenerated from acetyl-CoA by the formation of free acetate (to be considered later). In amoebae glycolytic pathways beyond pyruvate are not directly concerned with PP_i and hence are not included in this brief manuscript. They were discussed in detail in an earlier paper (19).

When studied with the oxygen electrode, added glucose can stimulate oxygen uptake by intact amoebae to the extent of 2 μatom/min per g of cells. This indicates the amount by which flux through oxalacetate and malate by loop enzymes may exceed the direct PEP to pyruvate conversion. In living cells the loop enzymes may never function as a cycle. In aerobically prepared cell homogenates the enzyme called pyruvate synthase (EC 1.2.7.1) is rapidly inactivated by air. This is the enzyme which amoebae employ, instead of pyruvate dehydrogenase which they lack, to convert pyruvate and CoA into acetyl-CoA. With this enzyme inactivated there is no flux of PEP to acetyl-CoA, and the reversible loop enzymes can then be observed to function as a cycle. In assays conducted in vitro, Harlow et al. (6) had found activity which I calculate to be about 0.2 μmol/min per g of cells of an ATP-dependent transhydrogenase activity in the NADH-to-NADPH direction. In experiments similar to the one which follows, the greatest amount of PP_i-dependent transhydrogenation I observed in the NADPH-to-NADH direction was about 0.45 μmol/min per g of cells. These activities, which amount to large fractions of those observed with whole cells, are greater than might have been expected considering the dilution of the four pertinent enzymes in the assay systems. The explanation may be a noncovalent association of the loop enzymes into a functioning complex entity.

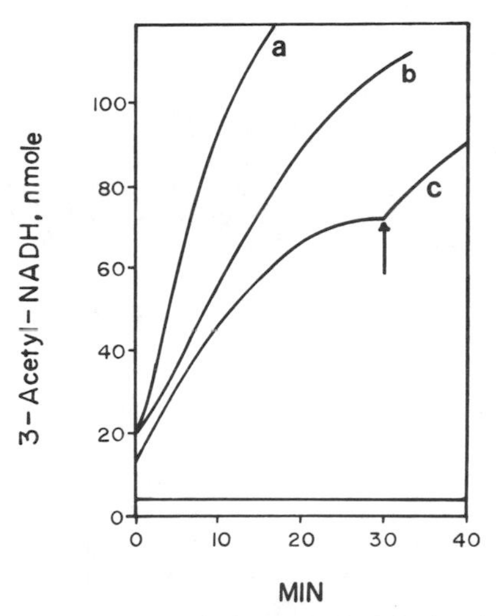

FIG. 3. Evidence of the loop functioning in the direction of transhydrogenation from NADPH to acetyl-NADH. The cuvette of 10-mm optical path was monitored at 375 nm, 20°C, and the molar extinction coefficient for acetyl-NADH was taken to be 5.1 at this wavelength. The complete reaction contained 30 μmol of piperazine-*N,N'*-bis(2-ethanesulfonic acid) buffer (pH 6.9), 1 μmol of $MgCl_2$, 1 μmol of PP_i, 0.2 μmol of acetyl-NAD, 0.1 μmol of NADPH, 10 mmol of isopropanol, 10 μl of saturated $NaHCO_3$ freshly bubbled with CO_2, the enzyme system (supernatant from 12.5 μl of amoebic cells which had been lyophilized, extracted with water, and centrifuged for 120 min at 120,000 × *g*), and water to 0.50 ml. Bottom trace, Minus PP_i. Curves: a, complete plus glucose and yeast hexokinase; b, complete plus 0.4 μmol of AMP; c, complete with more AMP added at the arrow.

Figure 3 shows a rather complicated experiment in which evidence of the loop enzymes functioning in the clockwise direction (transhydrogenation from NADPH to acetyl-NAD to form acetyl-NADH) can be seen. The basic experiment consists of the lowest two curves of Fig. 3; the bottom trace is without added PP_i. The next curve above (c) shows the reaction stimulated by PP_i and the small amount of endogenous AMP in the cell homogenate, causing the reduction of acetyl-NAD by transhydrogenation from NADPH. After about 30 min, this curve leveled off, and the reaction was rekindled by the addition of more AMP (at arrow). Curve b shows the reaction when extra AMP was present from zero time. Curve a shows the reaction when glucose and yeast hexokinase were added to the complete reaction mixture, removing a rate-inhibiting effect caused by the build-up of ATP. There are two aspects of the amoebic cell homogenate and the complete reaction mixture which may make it difficult to conduct this experiment with any other known organism. First, the isopropanol in the reaction mixture causes the amoebic NADP-linked alcohol dehydrogenase (24) to buffer the NADP-NADPH equilibrium to that specified by the acetone-isopropanol equilibrium. Small changes in NADPH concentration which occur as the reaction progresses are compensated for by the acetone/isopropanol equilibrium. Second, the freshly prepared cell homogenate employed had not been dialyzed. Apparently this permitted those loop enzymes with trace metal requirements (Mn, Co) to retain their activity. In keeping with the transhydrogenase experiments of Harlow et al. (6), magnesium ions were added. Not shown in Fig. 3 are other experiments in which ATP instead of AMP was added at the point shown by the arrow. In these instances the apparent amount of acetyl NADH decreased to a new equilibrium located 15 to 30 nmol lower. These experiments and those of Harlow et al. show that the loop enzymes, two of which are linked to PP_i, confer upon amoebae the ability to defend their cell components against damage by oxygen. To accomplish this, the loop need not function as a cycle. It is sufficient that the flux through the loop generates enough NADPH to remove the intruding oxygen, albeit at the expense of some ATP production. The speculative nature of this concept is certainly obvious to all. I will now turn to the problem of acetate formation.

Acetate is one of the end products of carbohydrate metabolism in amoebae. Amoebae have two enzymes which can catalyze formation of free acetate (20, 27): P_i + acetyl-CoA + ADP = acetate + ATP (EC 6.2.1.13) and acetyl-phosphate + P_i = acetate + PP_i (EC 2.7.1.12). Only the first of these reactions is used by amoebae in axenic culture. The presence of the second enzyme in axenically grown amoebae is an enigma. We have not been able to find an alternative way for this organism to make acetyl-phosphate. The enzyme transacetylase catalyzing the reaction acetyl-CoA + P_i = acetyl-phosphate + CoA is absent in axenically grown cells, but it is present at a low level in amoebae grown with certain bacteria. The existence of alternative pathways between acetyl-CoA and acetate, one producing ATP and the other producing PP_i, may help explain the observed enhancement of amoebic growth in the presence of certain bacterial associates. If axenically grown amoebae utilize the PP_i-acetate kinase, it must be for the synthesis of acetyl-phosphate for some unknown purpose. The enzyme called acetyl-CoA synthetase (ADP-forming) (EC 6.2.1.13) differs from the conventional enzyme (EC 6.2.1.1), which employs AMP and PP_i in the direction of ATP formation. Berg (1, 2) has shown that adenyl acetate is an intermediate substrate in the reaction catalyzed by the latter enzyme. AMP was not a substrate for the amoebic enzyme (27), and later work showed that adenyl acetate plus PP_i would not serve to form acetate and ATP in the presence of the enzyme (R. E. Reeves, unpublished data).

Propionibacterium shermanii

Propionic acid bacteria have highly complicated energy-producing pathways that allow them to grow on all monosaccharides, on the corresponding polyols including glycerol, and on pyruvate and lactate. To facilitate such diversity, they possess two separate enzymes for at least three steps of the Embden-Meyerhof pathway: an ATP- and a polyphosphate-glucokinase, an ATP- and a PP_i-phosphofructokinase, and a conventional pyruvate kinase as well as a pyruvate phosphate dikinase (30). In the face of such complexity, it is sometimes difficult to know whether one of the reversible PP_i enzymes is catalyzing the use of PP_i or its formation. In growth on glucose, PP_i is clearly used at the phosphofructokinase step. When growing on pyruvate or lactate, the flux through this step must lie in PP_i formation. The task of charting the flow of intermediates through the web of pathways in this organism is of heroic proportions and has been summarized in a series of review articles (29, 32, 33).

With regard to PP_i utilization by *P. shermanii* I would like to suggest a PP_i-dependent triose kinase. I was unable to find such an enzyme in *E. histolytica*. However, in the absence of a PP_i-dependent glucokinase (32), the existence of a triose kinase plus aldolase does a better job of explaining the diversity of monosaccharides found in nature than does a hexokinase functioning with the gemisch of pentoses and hexoses presumed to be abiotically produced as polymerization products of formaldehyde. The final chapter on energy metabolism of *P. shermanii* cannot yet be written.

LITERATURE CITED

1. **Berg, P.** 1956. Acyl adenylates: an enzymatic mechanism of acetate activation. J. Biol. Chem. **222:**991–1013.

2. **Berg, P.** 1956. Acyl adenylates: the synthesis and properties of adenyl acetate. J. Biol. Chem. **222:**1015–1023.
3. **Cagen, L. M., and H. C. Freedman.** 1972. Enzymatic phosphorylation of serine. J. Biol. Chem. **247:**3382–3392.
4. **Carnal, N. W., and C. C. Black, Jr.** 1979. Pyrophosphate-dependent 6-phosphofructokinase, a new glycolytic enzyme in pineapple leaves. Biochem. Biophys. Res. Commun. **86:**20–26.
5. **Evans, H. J., and H. G. Wood.** 1968. The mechanism of the pyruvate, phosphate dikinase reaction. Proc. Natl. Acad. Sci. USA **61:**1441–1453.
6. **Harlow, D. R., E. C. Weinbach, and L. S. Diamond.** 1976. Nicotinamide nucleotide transhydrogenase in *Entamoeba histolytica*, a protozoan lacking mitochondria. Comp. Biochem. Physiol. **53B:**141–144.
7. **Hatch, M. D., and C. R. Slack.** 1968. A new enzyme for the interconversion of pyruvate and phosphopyruvate and its role in the C_4 dicarboxylic acid pathway of photosynthesis. Biochem. J. **106:**141–146.
8. **Klemme, J.-H.** 1976. Regulation of intracellular pyrophosphatase activity and conservation of the phosphoanhydride-energy of inorganic pyrophosphate in microbial metabolism. Z. Naturforsch. Teil C **31:**544–550.
9. **Kornberg, A.** 1957. Pyrophosphorylases and phosphorylases in biosynthetic reactions. Adv. Enzymol. **18:**191–240.
10. **Kulaev, I. S., and V. M. Vagabov.** 1983. Polyphosphate metabolism in microorganisms. Adv. Microb. Physiol. **24:** 83–171.
11. **Lo, H.-S., and R. E. Reeves.** 1980. Purification and properties of NADPH:flavin oxidoreductase from *Entamoeba histolytica*. Mol. Biochem. Parasitol. **2:**23–30.
12. **O'Brien, W. E., S. Bowien, and H. G. Wood.** 1975. Isolation and characterization of a pyrophosphate-dependent phosphofructokinase from *Propionibacterium shermanii*. J. Biol. Chem. **250:**8690–8695.
13. **Peller, L.** 1975. *In vitro* RNA synthesis should be coupled to pyrophosphate hydrolysis. Biochem. Biophys. Res. Commun. **63:**912–916.
14. **Rafter, G. W.** 1960. Pyrophosphate metabolism in liver mitochondria. J. Biol. Chem. **253:**2475–2477.
15. **Reeves, R. E.** 1968. A new enzyme with the glycolytic function of pyruvate kinase. J. Biol. Chem. **243:**3202–3204.
16. **Reeves, R. E.** 1970. Phosphopyruvate carboxylase from *Entamoeba histolytica*. Biochim. Biophys. Acta **220:**346–349.
17. **Reeves, R. E.** 1974. Glycolytic enzymes in *E. histolytica*. Arch. Invest. Med. **5**(Suppl. 2):411–414.
18. **Reeves, R. E.** 1976. How useful is the energy in inorganic pyrophosphate? Trends Biochem. Sci. **1:**53–55.
19. **Reeves, R. E.** 1984. Metabolism of *Entamoeba histolytica* Schaudinn, 1903. Adv. Parasitol. **23:**105–142.
20. **Reeves, R. E., and J. D. Guthrie.** 1975. Acetate kinase (pyrophosphate). A fourth pyrophosphate-dependent kinase from *Entamoeba histolytica*. Biochem. Biophys. Res. Commun. **66:**1389–1395.
21. **Reeves, R. E., P. Lobelle-Rich, and W. B. Eubank.** 1982. 6-Phosphofructokinase (PP_i) from *Entamoeba histolytica*. Methods Enzymol. **90:**97–102.
22. **Reeves, R. E., and L. K. Malin.** 1969. Enzymic assay method for inorganic pyrophosphate. Anal. Biochem. **28:** 282–287.
23. **Reeves, R. E., R. A. Menzies, and D.-S. Hsu.** 1968. The pyruvate-phosphate dikinase reaction. The fate of phosphate and the equilibrium. J. Biol. Chem. **243:**5486–5491.
24. **Reeves, R. E., F. E. Montalvo, and T. S. Lushbaugh.** 1971. Nicotinamide-adenine dinucleotide phosphate-dependent alcohol dehydrogenase: the enzyme from *Entamoeba histolytica* and some enzyme inhibitors. Int. J. Biochem. **2:**55–64.
25. **Reeves, R. E., R. Serrano, and D. J. South.** 1976. 6-Phosphofructokinase (pyrophosphate). Properties of the enzyme from *Entamoeba histolytica* and its reaction mechanism. J. Biol. Chem. **251:**2958–2962.
26. **Reeves, R. E., D. J. South, H. J. Blytt, and L. G. Warren.** 1974. Pyrophosphate: D-fructose 6-phosphate 1-phosphotransferase. A new enzyme with the glycolytic function of 6-phosphofructokinase. J. Biol. Chem. **249:**7737–7741.
27. **Reeves, R. E., L. G. Warren, B. Susskind, and H.-S. Lo.** 1977. An energy-conserving pyruvate-to-acetate pathway in *Entamoeba histolytica*. Pyruvate synthetase and a new acetate thiokinase. J. Biol. Chem. **252:**726–731.
28. **Sui, P. M. L., and H. G. Wood.** 1962. Phosphoenolpyruvic carboxytransphosphorylase, a CO_2 fixation enzyme from propionic acid bacteria. J. Biol. Chem. **237:**3044–3051.
29. **Wood, H. G.** 1977. Some reactions in which inorganic pyrophosphate replaces ATP and serves as a source of energy. Fed. Proc. **36:**2197–2205.
30. **Wood, H. G., J. J. Davis, and H. Lochmuller.** 1966. The equilibria of reactions catalyzed by carboxytransphosphorylase, carboxykinase and pyruvate carboxylase and the synthesis of phosphoenolpyruvate. J. Biol. Chem. **241:** 5692–5704.
31. **Wood, H. G., J. J. Davis, and J. M. Willard.** 1969. Phosphoenolpyruvate carboxytransphorylase. V. Mechanism of the reaction and the role of cations. Biochemistry **8:**3145–3155.
32. **Wood, H. G., and N. H. Goss.** 1985. Phosphorylation enzymes of propionic acid bacteria and the roles of ATP, inorganic pyrophosphate, and polyphosphates. Proc. Natl. Acad. Sci. USA **81:**312–315.
33. **Wood, H. G., W. E. O'Brien, and G. Michaels.** 1977. Properties of carboxytransphosphorylase; pyruvate phosphate dikinase; pyrophosphate-phosphofructokinase and pyrophosphate-acetate kinase and their roles in the metabolism of inorganic pyrophosphate. Adv. Enzymol. **45:** 85–155.

PP_i in the Energy Conversion System of *Rhodospirillum rubrum*

MARGARETA BALTSCHEFFSKY AND PÅL NYRÉN

Department of Biochemistry, Arrhenius Laboratory, University of Stockholm, S-106 91 Stockholm, Sweden

Three sets of distinct but related data led to our discovery, in an electron transport phosphorylation system, of the first and still the only known alternative to the membrane-bound H^+ ATP synthase, namely, the H^+ PP_i synthase (or "coupling factor" PPase) in chromatophores from *Rhodospirillum rubrum*. The first indication came in the 1960s with our demonstration of an uncoupler-stimulated, membrane-bound PPase activity in isolated, washed chromatophores (M. Baltscheffsky, Abstr. 1st FEBS Meet., London, 1964, p. 67). Soon thereafter, in such chromatophores a photophosphorylation was found, in the absence of added ADP (7), which yielded PP_i as the energy-rich product (1, 2). Subsequently, it was possible to demonstrate that PP_i, when added in the dark, could act as an energy donor, as measured with both reversed electron transport at the cytochrome level (3, 4) and membrane energization shown with the absorbance change of endogenous carotenoids (5). Additional and more direct evidence for the existence and function of a special, alternative factor for energy coupling in the light-induced formation of PP_i was obtained by physical separation when, after sonication, the ATPase became solubilized from the membranes to which the PPase remained bound (8) and also by selective inhibition, with a monospecific antibody, of the synthesis and hydrolysis of ATP (B. C. Johansson, thesis, University of Stockholm, Stockholm, Sweden, 1974).

Early attempts to solubilize and purify the integrally bound coupling factor PPase were made both by others (19) and by us, but it turned out in this respect to be very difficult to handle. However, after long and strenuous years, we now have a pure and rather stable solubilized PPase preparation, which lends itself to functional studies as well as incorporation into liposomes with retained energy transfer reactions (20). The molecular weight is still unknown and the subunit composition is still somewhat ambiguous, because of its very hydrophobic nature and the high content of added detergent in the preparation, but preliminary data indicate that the subunit number might be as low as two with apparent molecular weights of 54,000 and 52,000.

Utilization and Hydrolysis of PP_i

Following our finding that there were clear similarities in the reaction patterns of PP_i and ATP hydrolysis, such as the stimulation by uncoupling agents or low concentrations of certain detergents, it seemed logical to investigate whether PP_i also could, like ATP, serve as an energy donor. We were soon able to show that the addition of PP_i to an Mg^{2+} ion-containing chromatophore suspension in the dark elicited redox changes of at least two endogenous cytochromes. The changes were an oxidation of cytochrome c_2 and a reduction of cytochrome b, which showed that reversed electron transport could be driven by the hydrolysis of PP_i. These experiments not only demonstrated that PP_i could serve as a biological energy donor but also showed, for the first time, reversed electron transport in a procaryotic system (3). Interestingly enough, both the rate and extent of the cytochrome redox changes were greater with PP_i than with ATP as an energy donor. It was also shown by Keister and Yike (9, 10) that the hydrolysis of PP_i could be linked with reversed electron transport from succinate to $NADH^+$ and that the energy-requiring transhydrogenase reaction could be driven with hydrolysis of PP_i.

We also found that the carotenoid band shift, generally assumed to indicate the membrane potential, was generated in the dark as long as PP_i was hydrolyzed (5). Thus, it became obvious that PP_i can indeed serve as an energy donor, in a way comparable to ATP.

The hydrolytic activity was considerably (50 to 100%) stimulated by uncouplers in accordance with the H^+ pumping capability of the enzyme. The PPase activity remained stimulated even at high uncoupler concentrations, in contrast to the ATPase activity, which became inhibited.

Mg^{2+} ions were necessary for the activity, and on the basis of kinetic observations (18; unpublished data), we assume that the true substrate is the Mg-PP_i complex. An excess of free Mg^{2+} was inhibitory, as was an excess of PP_i.

A number of inhibitors of the PPase activity have been found (Table 1), and the known modes of action of these have often given us

TABLE 1. Effect of some substances on membrane-bound and purified PPase[a]

Addition[b]	Concn	Activity of inorganic pyrophosphatase (% of control) dependent on:	
		Membrane-bound enzyme	Purified enzyme
FCCP	1.5 μM	150	100
DCCD (0°C)	100 μM	30	97
NaF	5 mM	83	32
	10 mM	64	11
	20 mM	45	4
Methylenediphosphonate	0.1 mM	84	64
	0.2 mM	74	36
Imidodiphosphate	0.1 mM	48	24
	0.2 mM	34	12
NBF-Cl (0°C)	0.25 mM	33	23
	0.50 mM	18	7
NBF-Cl (30°C)	0.25 mM	82	12
	0.50 mM	61	1
N-Ethylmaleimide (0°C)	1 mM	30	20
	2 mM	13	13
N-Ethylmaleimide (30°C)	1 mM	99	84
	2 mM	85	76
Dio-9 (0°C)	15 μg/ml	80	40
	30 μg/ml	60	30

[a] Particles corresponding to 120 μg of protein and purified enzyme corresponding to 15 μg of protein were assayed. The $MgCl_2$ concentration was 0.75 mM except for the methylenediphosphonate and imidodiphosphate treatments, with which 12.5 mM $MgCl_2$ was used. The PP_i concentration was 0.5 mM. The DCCD, *N*-ethylmaleimide, and NBF-Cl treatments were performed by incubation of particles and enzyme for 10 min at 0 or 30°C. The PPase reaction was initiated by adding PP_i. (From reference 15.)

[b] FCCP, Carbonyl cyanide *p*-trifluoromethoxyhydrazone; DCCD, *N,N'*-dicyclohexylcarbodiimide; NBF-Cl, 4-chloro-7-nitrobenzo-2-oxo-1,3-diazol.

information about the nature of the enzyme. Analogs to PP_i, such as methylenediphosphonate or imidodiphosphate, acted, as could be expected, as competitive inhibitors. *N*-Ethylmaleimide and 4-chloro-7-nitrobenzo-2-oxo-1,3-diazol, on the other hand, were noncompetitive, and by analogy with other inhibitions by these two agents, we may assume that SH groups as well as a tyrosine (or an arginine) are involved in or near the catalytic site. *N,N'*-Dicyclohexylcarbodiimide inhibited the membrane-bound enzyme, but not when it was solubilized. This compound is known to abolish proton pumping activity in other proton translocating enzymes but is rather unspecific in its action.

It can also be seen in Table 1 that many of the catalytic inhibitors were much more efficient with the purified than with the membrane-bound enzyme. A possible explanation for this effect is that, in the membrane-bound enzyme, the catalytic site is somewhat buried below the membrane surface, limiting the accessibility of the inhibitors.

The substrate specificity of the PPase is high, with only PP_i hydrolyzed at any appreciable rate. The only other substrate hydrolyzed was tripolyphosphate, but the rate was only 6% of that of PP_i hydrolysis.

It was first shown by Moyle et al. (12) that the hydrolysis of PP_i is coupled to the uptake of protons from the surrounding medium into the chromatophore membrane. Subsequently, we showed with the purified enzyme that those properties which are likely to be linked to proton pumping ability are lost when the PPase is solubilized but regained when the enzyme is reconstituted into liposomes (20). We also showed, in liposomes containing both the PPase and the F_0F_1 complex from *R. rubrum*, that the proton gradient generated from hydrolysis of PP_i is able to drive the synthesis of ATP (14). Thus, it seems quite clear that the membrane-bound PPase is a proton pump, working in parallel with the ATPase complex.

Synthesis of PP_i

The synthesis of PP_i coupled to photosynthetic electron transport was first described by Baltscheffsky and von Stedingk (1). The synthesis was strictly light dependent and was inhibited by electron transport inhibitors such as antimycin. On the other hand, the ATPase inhibitor oligomycin did not inhibit PP_i synthesis but caused a marked stimulation instead. The rate of PP_i synthesis was considerably lower than that of ATP synthesis, usually only 10 to 15% as great. In a thorough study of PP_i synthesis, Guillory and Fisher (6) found a number of differences between the two phosphorylation systems, among them that the light intensity saturating PP_i synthesis was much lower than that saturating ATP synthesis.

Recently, the measurement of PP_i, even at rather low concentrations, has been greatly facilitated by the introduction of the method of Nyrén and Lundin (16), which uses a coupled assay system in which the PP_i formed is converted to ATP by the enzyme ATP sulfurylase. The ATP is then measured with a luciferin-luciferase assay. Using this method, we have been able to measure both the rate and extent of PP_i synthesis under various conditions. Keister and Yike (9) demonstrated that PP_i could drive the energy-linked transhydrogenase reaction. We have now shown that this reaction can be reversed; i.e., reversal of the transhydrogenation drives PP_i synthesis in the dark (13).

PP_i synthesis appears to have a lower energy requirement than ATP synthesis. We recently found three sets of conditions under which the yield of PP_i was considerably larger than the yield of ATP. One was at extremely low light intensities, less than 10 W/m^2, where the rate of PP_i synthesis was twice that of ATP synthesis (17a). Another was when the driving force was excitation of the photosynthetic electron transport chain by short light flashes. A single 1-ms light flash could support the synthesis of up to 60% more PP_i than ATP (17) (Table 2).

The third set of conditions was following an acid-base jump with which we have very recently been able to induce PP_i synthesis. Leiser and Gromet-Elhanan (11) showed that, in *R. rubrum* chromatophores, ATP synthesis would not occur after an acid-base jump unless the proton gradient thus created was supplemented with a diffusion potential of K^+ ions in the presence of valinomycin. PP_i synthesis, on the other hand, was readily obtained with a pH difference of only about 3 pH units. These results may indicate that an activation step of the ATPase complex, brought about by the membrane potential, is necessary for ATP synthesis, but not required for PP_i synthesis.

TABLE 2. Simultaneous measurement of flash-induced PP_i and ATP synthesis[a]

Conditions	PP_i and/or ATP formed (mol/1,000 μmol of Bchl)
Control	9.6 (100%)
ADP (20 μM)	13.2 (138%)
ADP (20 μM) + NaF (10 mM)	8.0 (83%)
ADP (20 μM) + oligomycin (10 μg/ml)	13.5 (141%)
ADP (20 μM) + NaF (10 mM) + oligomycin (10 μg/ml)	0.0 (0%)

[a] Flash-induced PP_i and ATP syntheses were measured in the same reaction mixture. When NaF was present, only ATP synthesis was measured; when oligomycin was present, only PP_i synthesis was measured. Bacteriochlorophyll (Bchl) concentration was 1.0 μM. (From reference 17.)

TABLE 3. Illustration of the competition between photosynthetic PP_i and ATP formation[a]

Conditions	Amt (μmol) formed per min per μmol of Bchl	
	PP_i	ATP
Dark	0.03	0.04
Light	1.5	10.0
Light + NaF (10 mM)	0.0	12.1
Light + oligomycin (10 μg/ml)	2.1	0.0

[a] PP_i formation and ATP formation were measured in the presence and absence of oligomycin and NaF. Bacteriochlorophyll (Bchl) concentration was 0.1 μM. The illuminating intensity was 625 W/m^2. (From reference 17a.)

A result of the coexistence of two proton-translocating enzymes in the same membrane is that they compete with each other for the available proton motive force. Neither will show optimal activity unless the other one is inhibited since there will be passive proton leaks through inactive enzyme molecules (Table 3).

In recent years it has become evident that membrane-bound PPases are not uncommon in nature. Even if, as we believe, the PPase coupling factor preceded the ATPase in the course of evolution, the PPase probably would not have been retained unless it served a purpose. A possible physiological role would be to conserve energy at a low energy supply, too low to sustain ATP synthesis. Examples of such conditions could be very low light intensities in the case of photosynthetic bacteria or a limited oxygen supply in the soil in the case of plant root mitochondria.

Regardless of its physiological role, the enzyme is a good model system for the study of the electron transport-coupled phosphorylation reaction because of its relative simplicity. It is hoped that the apparent absence of some of the sophisticated regulative properties of the ATPase will make it easier to sort out the core of the matter: the synthesis and hydrolysis of the pyrophosphate moiety.

LITERATURE CITED

1. **Baltscheffsky, H., and L.-V. von Stedingk.** 1966. Bacterial photophosphorylation in the absence of added nucleotide: a second intermediate stage of energy transfer in light-induced formation of ATP. Biochem. Biophys. Res. Commun. **22:**722–728.
2. **Baltscheffsky, H., L.-V. von Stedingk, H.-W. Heldt, and M. Klingenberg.** 1966. Inorganic pyrophosphate: formation in bacterial photophosphorylation. Science **153:**1120.

3. **Baltscheffsky, M.** 1967. Inorganic pyrophosphate and ATP as energy donors in chromatophores from *Rhodospirillum rubrum*. Nature (London) **216:**241–243.
4. **Baltscheffsky, M.** 1967. Inorganic pyrophosphate as an energy donor in photosynthetic and respiratory electron transport phosphorylation systems. Biochem. Biophys. Res. Commun. **28:**270–276.
5. **Baltscheffsky, M.** 1969. Energy conversion-linked changes of carotenoid absorbance in *Rhodospirillum rubrum* chromatophores. Arch. Biochem. Biophys. **130:**646–652.
6. **Guillory, R. J., and R. R. Fisher.** 1972. Studies on the light-dependent synthesis of inorganic pyrophosphate by *Rhodospirillum rubrum* chromatophores. Biochem. J. **129:**471–481.
7. **Horio, T., L.-V. von Stedingk, and H. Baltscheffsky.** 1966. Photophosphorylation in presence and absence of added adenosine diphosphate in chromatophores from *Rhodospirillum rubrum*. Acta Chem. Scand. **20:**1–10.
8. **Johansson, B. C., M. Baltscheffsky, and H. Baltscheffsky.** 1972. Coupling factor capabilities with chromatophore fragments from *Rhodospirillum rubrum*, p. 1203–1209. *In* G. Forti, M. Avron, and A. Melandri (ed.), Proceedings of the Second International Congress on Photosynthesis Research. Junk Publishers, The Hague.
9. **Keister, D. L., and N. J. Yike.** 1966. Studies on an energy-linked pyridine nucleotide transhydrogenase in photosynthetic bacteria. I. Demonstration of the reaction in *Rhodospirillum rubrum*. Biochem. Biophys. Res. Commun. **24:**519–525.
10. **Keister, D. L., and N. J. Yike.** 1967. Energy-linked reactions in photosyntehtic bacteria. I. Succinate-linked ATP-driven NAD^+ reduction by *Rhodospirillum rubrum* chromatophores. Arch. Biochem. Biophys. **121:**415–422.
11. **Leiser, M., and Z. Gromet-Elhanan.** 1974. Demonstration of acid-base phosphorylation in chromatophores in the presence of a K^+ diffusion potential. FEBS Lett. **43:**267–270.
12. **Moyle, J., R. Mitchell, and P. Mitchell.** 1972. Proton translocating pyrophosphatase of *Rhodospirillum rubrum*. FEBS Lett. **23:**233–236.
13. **Nore, B. F., I. Husain, P. Nyrén, and M. Baltscheffsky.** 1986. Synthesis of pyrophosphate coupled to the reverse energy-linked transhydrogenase reaction in *Rhodospirillum rubrum* chromatophores. FEBS Lett. **200:**133–138.
14. **Nyrén, P., and M. Baltscheffsky.** 1983. Inorganic pyrophosphate-driven ATP-synthesis in liposomes containing membrane-bound inorganic pyrophosphatase and F_0-F_1 complex from *Rhodospirillum rubrum*. FEBS Lett. **155:**125–130.
15. **Nyrén, P., K. Hajnal, and M. Baltscheffsky.** 1984. Purification of the membrane-bound proton-translocating inorganic pyrophosphatase from *Rhodospirillum rubrum*. Biochim. Biophys. Acta **766:**630–635.
16. **Nyrén, P., and A. Lundin.** 1985. Enzymatic method for continuous monitoring of inorganic pyrophosphate synthesis. Anal. Biochem. **151:**504–509.
17. **Nyrén, P., B. F. Nore, and M. Baltscheffsky.** 1986. Inorganic pyrophosphate synthesis after a short light flash in chromatophores from *Rhodospirillum rubrum*. Photobiochem. Photobiophys. **11:**189–196.
17a. **Nyrén, P., B. F. Nore, and M. Baltscheffsky.** 1986. Studies on inorganic pyrophosphate formation in *Rhodospirillum rubrum* chromatophores. Biochim. Biophys. Acta **851:** 276–282.
18. **Randahl, H.** 1979. Characterization of the membrane-bound inorganic pyrophosphatase in *Rhodospirillum rubrum*. Eur. J. Biochem. **102:**251–256.
19. **Rao, P. V., and D. L. Keister.** 1978. Energy-linked reactions in photosynthetic bacteria. X. Solubilization of the membrane-bound energy-linked inorganic pyrophosphatase of *Rhodospirillum rubrum*. Biochem. Biophys. Res. Commun. **84:**465–473.
20. **Shakov Yu, A., P. Nyrén, and M. Baltscheffsky.** 1982. Reconstitution of highly purified proton-translocating pyrophosphatase from *Rhodospirillum rubrum*. FEBS Lett. **146:**177–180.

PP_i Metabolism and Its Regulation by Fructose 2,6-Bisphosphate in Plants

CLANTON C. BLACK, JR.,[1] D.-P. XU,[1] S. S. SUNG,[1] L. MUSTARDY,[1] N. PAZ,[1] AND P. P. KORMANIK[2]

Biochemistry Department, University of Georgia, Athens, Georgia 30602,[1] and Institute for Mycorrhizal Research and Development, U.S. Forest Service, Athens, Georgia 30602[2]

A traditional thesis of biochemistry is that PP_i hydrolysis during polymer synthesis provides a favorable thermodynamic pull, perhaps even an irreversible environment for processes such as the biosynthesis of lipids, starch, glycogen, DNA, RNA, sucrose, or other polymers. Pyrophosphatases are thought of as hydrolytic cleavage enzymes, thereby resulting in the loss of the PP_i bond energy. These traditional views must be altered today because of newer discoveries on alternative metabolic roles for PP_i, certainly in protozoans, bacteria, and plants. In this report we concentrate on plant work, but we fully recognize the pioneering work on PP_i metabolism with protozoans and bacteria from the laboratories of R. Reeves and H. Wood, along with work on the energetic roles of PP_i in photosynthetic bacteria.

Two sets of discoveries about plant phosphorus metabolism have resulted in renewed interest in energetics and in sugar phosphate metabolism. First, it was recognized that PP_i was both a phosphate donor and a replacement of nucleotide triphosphates (NTPs) with a plant phosphofructokinase (PFK) (3, 4). During glycolysis with soluble plant extracts, PP_i was shown to drive the conversion of fructose 6-phosphate (F-6-P) through to pyruvate (14). Next, we demonstrated that PP_i was present in plant tissues in amounts comparable to those of ATP (13). In this report, data are presented concerning the PP_i-dependent breakdown of sucrose in plants. Second was the identification of fructose 2,6-bisphosphate (F-2,6-P_2) as a potent regulator, specifically in sugar metabolism, with action at picomole amounts (9, 10). These two discoveries of PP_i as a new energy source in plants and of F-2,6-P_2 as a new regulatory sugar phosphate are the foundations of this presentation.

Enzymes Utilizing PP_i in Biosynthetic Reaction

Today, PP_i is known to be used in at least four ways by plants, as outlined in reactions 1 to 4, which are given in their chronological order of recognition.

$$\text{UDP-Glu} + PP_i \leftrightarrow \text{glucose 1-phosphate} + \text{UTP} \quad (1)$$

$$\text{phosphoenolpyruvate} + \text{AMP} + 2H^+ + PP_i \leftrightarrow \text{pyruvate} + \text{ATP} + P_i \quad (2)$$

$$\text{F-6-P} + PP_i \leftrightarrow \text{F-1,6-}P_2 + P_i \quad (3)$$

$$PP_i + H^+ \rightarrow 2\ P_i + H^+ \quad (4)$$

Reaction 1 is catalyzed by UDP-glucopyrophosphorylase, found three decades ago in plants, but it has traditionally been accepted that the enzyme operates in the direction of UDP-glucose synthesis. One factor in this consideration which had to be evaluated again was the belief that a substrate-level PP_i pool did not exist in plants. Reaction 2 is catalyzed by pyruvate P_i dikinase, which was originally found in leaves with C_4 photosynthesis (1), where it functions in the generation of phosphoenolpyruvate, the CO_2 acceptor of C_4 photosynthesis. Again, the reaction utilizing PP_i is not the accepted direction in plant tissues. Reaction III is catalyzed by a PP_i-dependent F-6-P 1-phosphotransferase (EC 2.7.1.10). This reaction was detected as a PP_i-dependent PFK activity in plants, but it is readily reversible, in sharp contrast to the essentially irreversible plant ATP-PFK (2–4; C. C. Black, L. Mustardy, S. S. Sung, P. P. Kormanik, D.-P. Xu, and N. Paz, Physiol. Plant., in press). Reaction 4 is mediated by a pyrophosphatase-H^+ membrane transporting protein in the plant tonoplast or Golgi (5, 12) and is associated with coupling solute transport to electrogenic proton pumping.

Reactions 1, 2, and 3 all can be considered as part of plant glycolysis and gluconeogenesis from sucrose to pyruvate. Since sucrose is the major translocation form of carbon in plants, we will soon consider sucrose metabolism, especially regarding PP_i.

Detection of PP_i in Plants

Classically, it has been assumed that pyrophosphatase hydrolyzes the PP_i produced in plants. Indeed, it has been calculated that the cellular cytoplasmic concentration of PP_i is very low, e.g., ~10 nM in liver cells when the P_i is ~4 mM (8); plant cells would have similar P_i contents. There were no previous measurements of the PP_i content of plant cells when we reported values for pea, corn, and pineapple tissues (13). Table 1 summarizes some data on measured PP_i contents of several plant tissues. These mea-

TABLE 1. PP_i content of plant cells

Plant tissue	Amt of PP_i (nmol/g of fresh wt)[a]
Pea shoots[b]	16
Pea roots[b,c]	14, 9
Pea cotyledons[b,c]	5, 9
Corn shoots[b]	39
Corn roots[b]	23
Corn scutellums[b]	15
Corn coleoptiles[d]	20
Pineapple leaves[b]	14
Sycamore cell culture[e]	4

[a] By assuming the cytoplasm is 10% of plant cell volume (10, 15), these values range from near 0.04 to 0.04 mM cytoplasmic PP_i.

[b] Data from Smyth and Black (13).

[c] Data from Edwards et al. (7).

[d] Data from Chanson et al. (5).

[e] Data from S. Huber and T. Akazawa (personal communication).

surements all were made possible through the use of the PP_i-dependent PFK to assay low levels of PP_i (13). If it is assumed that the plant cell has about 10% of its volume as cytoplasm, then the values in Table 1 calculate to about 0.04 to 0.4 mM PP_i in the cytoplasm. These PP_i levels are much higher than predicted from pyrophosphatase equilibria (8) and are in the range of K_m values for plant PP_i-utilizing enzymes (reaction 1 to 4).

Roles of Both PP_i- and NTP-Dependent PFK Activities in Plant Cell Cytoplasm

Upon the detection of two types of PFK activities in plants (2–4; Black et al., in press), we asked whether plants had developed alternative sugar metabolism pathways. Indeed, this appears to be the case, even beginning with the importation of sucrose into plant cells, where either invertase or sucrose synthase breaks down the sucrose. Plant PFKs catalyze the following reactions in the indicated cellular compartment:

$$\text{ATP (or NTP)} + \text{F-6-P} \xrightarrow[\text{+ plastid}]{\text{cytoplasm}} \text{F-1,6-P}_2 + \text{NDP} \quad (5)$$

$$\text{PP}_i + \text{F-6-P} \xleftrightarrow{\text{cytoplasm}} \text{F-1,6-P}_2 + \text{P}_i \quad (6)$$

The nucleotide PFK is relatively nonspecific for its triphosphate (reaction 5). Physiologically, the NTP-dependent PFK is irreversible, but the PP_i-dependent PFK is readily reversible. We have presented the following hypothesis for the roles of these two ways to interconvert F-6-P and F-1,6-P_2 in the plant cell cytoplasm: (i) that plants developed a maintenance pathway of sugar metabolism involving an NTP-dependent PFK for glycolysis (reaction 5) and an F-1,6-P_2 1-phosphatase for gluconeogenesis, with both enzymes being irreversible physiologically, and (ii) that plants developed an adaptive pathway of sugar metabolism involving a physiologically reversible PP_i-dependent PFK (Black et al., in press). Thus, two pathways of glycolysis and gluconeogenesis are present in the cytoplasm of plant cells (2; Black et al., in press). We will now concentrate on the PP_i-dependent PFK.

With the PP_i-dependent PFK it was learned in about 1981 that the enzyme was regulated by F-2,6-P_2 (9, 10). In our initial work with the pineapple leaf enzyme we knew that PP_i as the substrate also inhibited the enzyme at concentrations above about 1 mM (3). Figure 1 illustrates the response of the purified leaf PP_i-dependent PFK to increasing levels of PP_i in the presence or absence of F-2,6-P_2 (N. W. Carnal, Ph.D. thesis, University of Georgia, Athens, 1984). F-2,6-P_2 had two effects on the activity, namely, a lowering of the K_m for PP_i and a relief of the inhibition by high levels of PP_i (Fig. 1). The K_ms for the pineapple enzyme are, respectively, 0.03 and 0.016 mM without and with F-2,6-P_2 (Fig. 1); the K_ms for the enzyme from several other plant species also fall in this range. These K_m values are quite compatible with the estimated cytoplasmic levels of PP_i in plant cells, ranging from 0.04 to 0.4 mM (Table 1).

F-2,6-P_2 Regulation

Figure 1 illustrates general regulatory effects of F-2,6-P_2 on the pineapple PP_i-dependent PFK; however, most other plant sources of the enzyme also have a large increase in V_{max} (10) in addition to a lowering of the substrate K_m values when F-2,6-P_2 is present. From a study on

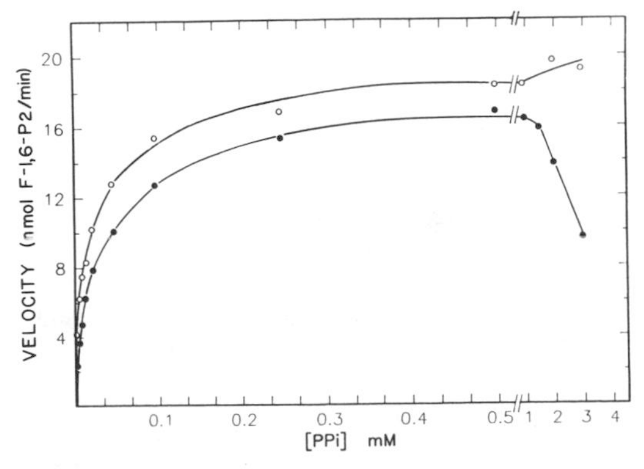

FIG. 1. Reaction velocity for the PP_i-dependent PFK from pineapple leaves as a function of PP_i concentrations in the presence (○) or absence (●) of 0.5 μM F-2,6-P_2. Each assay contained 0.53 μg of purified enzyme. (Data of Carnal, thesis.)

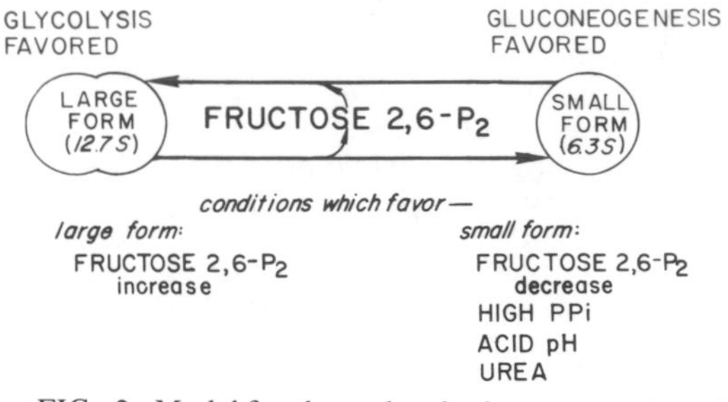

FIG. 2. Model for the molecular interconversion of two forms of PP_i-dependent PFK by F-2,6-P_2. The enzyme has an α and a β subunit and seems to exist primarily as an $\alpha\beta$ dimer (small form) or as an $\alpha_2\beta_2$ tetramer (large form).

development of the pea plant, we learned more about the mechanism of the F-2,6-P_2 regulatory effects on PP_i-dependent PFK (17, 18). In brief, this work demonstrated that PP_i-dependent PFK exists in two molecular forms whose activity depends upon association and dissociation characteristics regulated by F-2,6-P_2. The enzyme exists as two subunits which associate in the presence of F-2,6-P_2. Figure 2 summarizes this work. Thus, glycolytic and gluconeogenic hexose metabolism in plants includes this regulatory mechanism induced by changing F-2,6-P_2 levels. Consistent with this mechanism we, and other workers, have demonstrated that F-2,6-P_2 levels in plants can change rapidly (in periods of minutes), with up to 10-fold changes in concentration (2, 10, 11; Black et al., in press). In the small, dimeric form (Fig. 2) gluconeogenesis is favored about 3:1, but when F-2,6-P_2 concentrations rise, the subunits associate and glycolysis is favored about 3:1 (2, 17, 18).

When F-2,6-P_2 is present, the large, tetrameric form of the enzyme is more active. Indeed, this freely reversible enzyme (see reactions 3 and 6) is activated in both directions by F-2,6-P_2. When the small form converts to the large form (Fig. 2), the reaction toward F-1,6-P_2 is activated 2- to 10-fold; simultaneously, the reaction toward F-6-P is activated 25 to 50%. Hence, F-2,6-P_2 is a powerful regulator of glycolysis and gluconeogenesis at the PP_i-dependent PFK step in plants. In addition, at about 10-fold higher concentrations F-2,6-P_2 partially inhibits F-1,6-P_2 1-phosphatase (9, 10, 15), again regulating hexose phosphate metabolism at the F-6-P interconversion with F-1,6-P_2 site.

PP_i-Dependent Breakdown of Translocated Sucrose

As we were studying alternative pathways of hexose metabolism and their regulation, we realized that sucrose breakdown closely linked into this sugar metabolism. Sucrose is the major translocation form of carbon on which most plant cells live and grow. Therefore, we explored the alternative pathways of sucrose breakdown as outlined in Fig. 3. The breakdown by invertase is well known. In addition, we recently proposed a new sucrose breakdown pathway initiated by sucrose synthase (Fig. 3) because we recognized that a substantial pool of PP_i existed in plant cells as another source of energy (Table 1). Furthermore, the complete sucrose synthase pathway necessitated the involvement of both PFKs (reactions 5 and 6) in a cycle regulated by F-2,6-P_2 (2; Black et al., in press).

Several distinct features are evident from contrasting these two pathways of sucrose breakdown (Fig. 3). First, the sucrose synthase pathway conserves the glycosidic bond energy in sucrose; hence, it requires one-half the ATP equivalent of the invertase pathway to produce two hexose phosphates. Both pathways produce fructose which is phosphorylated by an NTP nonspecific fructokinase. The sucrose synthase pathway has several unique features, including the cycling of PP_i and uridylates. Uridylate cycling is possible because the plant cytoplasmic ATP-dependent PFK (reaction 5) also is nonspecific in regard to NTPs. Therefore, UTP can be cycled to form UDP either by the NTP-dependent PFK or by the fructokinase. The conversion of F-1,6-P_2 to F-6-P then allows a cycle to

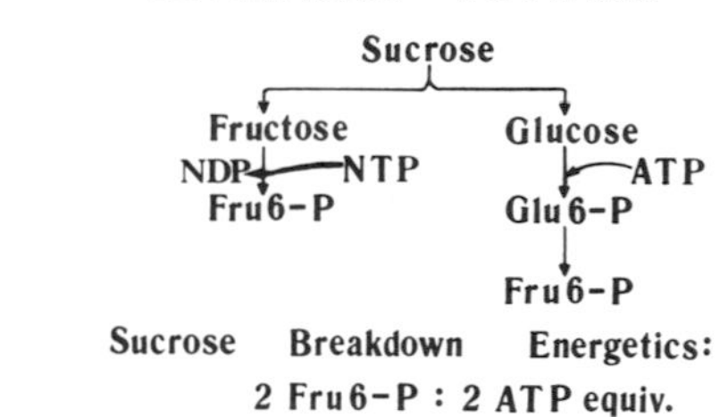

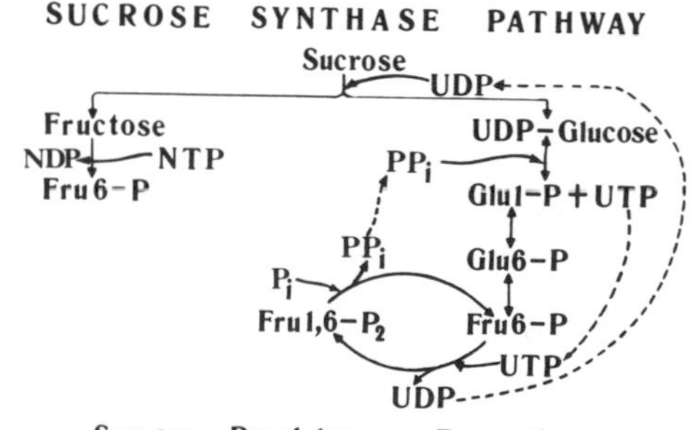

FIG. 3. Alternative routes for sucrose breakdown in plant cells via either the invertase or the sucrose synthase pathway. Note the cycling of PP_i and uridylates in the sucrose synthase pathway.

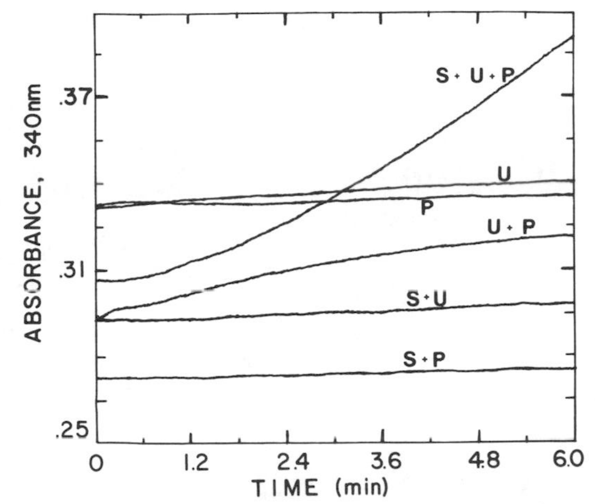

FIG. 4. PP_i- and UDP-dependent breakdown of sucrose by extracts from onion root tips. Spectrophotometer tracings of time courses for NADH formation. In this complete assay mixture, sucrose first is cleaved by sucrose synthase which requires UDP (6). Then UDP-glucose from this reaction is converted to glucose 1-phosphate by UDP-glucopyrophosphorylase which requires PP_i. The three substrates were added in various combinations; omission of any substrate, e.g., sucrose (S), UDP (U), or PP_i (P), resulted in no sustained formation of glucose 1-phosphate. The beginning absorbances are offset to simultaneously record all reaction mixtures. In this assay glucose 1-phosphate is converted to glucose 6-phosphate by excess phosphoglucomutase, and glucose 6-phosphate is assayed by NADH formation in the presence of the *Leuconostoc mesenteroides* glucose 6-phosphate dehydrogenase (S. S. Sung, D.-P. Xu, and C. C. Black, unpublished data).

be completed when catalyzed by PP_i-dependent PFK which forms PP_i and F-6-P. The activity of PP_i-dependent PFK is increased by F-2,6-P_2 (see previous section), which means that this enzyme is a site of regulation by F-2,6-P_2 during sucrose breakdown (2; Black et al., in press).

Previously, most workers postulated that sucrose synthase formed UDP-glucose, but in the absence of PP_i, it was assumed that UDP-glucose was used to synthesize cell walls or other glycosides (16).

To obtain evidence on the sucrose synthase pathway, we developed an assay for sucrose breakdown, using extracts from sucrose-importing tissues such as root and shoot tips. The assay was designed to demonstrate PP_i dependence during the breakdown of sucrose to form glucose 1-phosphate. Figure 4 illustrates this PP_i- and UDP-dependent breakdown of sucrose. Using a soluble extract from onion root tips, we found the breakdown of sucrose to be completely dependent upon the simultaneous presence of sucrose, PP_i, and UDP. Deletion of any of these three components resulted in no sustained formation of glucose 1-phosphate. Furthermore, substitution of UTP or ATP or PP_i was ineffective, and CDP, ADP, IDP, or GDP would not replace UDP. The rates of PP_i-dependent sucrose breakdown were near 50 nmol/min per mg of protein, which is approximately the same as those of the individual enzymes when assayed under optimum assay conditions. Figure 4 is a clear demonstration of the first two steps in the sucrose synthase breakdown pathway in Fig. 3.

In summary, this work shows that plant tissues contain substrate-level amounts of PP_i ranging from perhaps 0.04 to 0.4 mM in the cytoplasm. Three glycolytic enzymes are known that could utilize PP_i as an energy source and phosphate donor. The PP_i-dependent PFK is strongly regulated by F-2,6-P_2. We demonstrate here the PP_i- and UDP-dependent metabolism of sucrose via a new sucrose synthase pathway. Sucrose is the major form of carbon translocated in plants, and as it enters recipient cells, we propose that the efficient sucrose synthase pathway breaks sucrose down to hexose phosphates. During this sucrose breakdown, the required UDP and PP_i are produced via a cycle completed by the interconversion of F-6-P with F-1,6-P_2 in glycolysis.

This research was graciously supported by the National Science Foundation through grant DMB 84-06331 and by U.S./Hungarian Cooperative Research on Photosynthesis program NSF INT-8403748.

We thank H. D. Peck for presenting these data at the conference.

LITERATURE CITED

1. **Andrews, T. J., and M. D. Hatch.** 1969. Properties and mechanism of action of pyruvate, phosphate dikinase from leaves. Biochem. J. **114:**117–125.
2. **Black, C. C., Jr., D. A. Smyth, and M.-X. Wu.** 1985. Pyrophosphate-dependent glycolysis and regulation by fructose 2,6-bisphosphate in plants, p. 361–370. *In* P. W. Ludden and J. E. Burris (ed.), Nitrogen fixation and CO_2 metabolism. Elsevier Science Publishing, Inc., New York.
3. **Carnal, N. W., and C. C. Black, Jr.** 1979. Pyrophosphate-dependent 6-phosphofructokinase, a new glycolytic enzyme in pineapple leaves. Biochem. Biophys. Res. Commun. **86:**20–26.
4. **Carnal, N. W., and C. C. Black, Jr.** 1983. Phosphofructokinase activities in photosynthetic organisms. The occurrence of pyrophosphate-dependent 6-phosphofructokinase in plants. Plant Physiol. **71:**150–155.
5. **Chanson, A., J. Fichmann, D. Spear, and L. Taiz.** 1985. Pyrophosphate-driven proton transport by microsomal membranes of corn coleoptiles. Plant Physiol. **79:**159–164.
6. **Echeverria, E., and T. Humphreys.** 1984. Involvement of sucrose synthase in sucrose catabolism. Photochemistry **23:**2173–2178.
7. **Edwards, J., T. Ap Rees, P. M. Wilson, and S. Morrell.** 1984. Measurement of the inorganic pyrophosphate in tissues of *Pisum sativum* L. Planta **162:**188–191.
8. **Guynn, R. W., D. Velosco, Jr., R. Lawson, and R. L. Veech.** 1974. The concentration and control of cytoplasmic free inorganic pyrophosphate in rat liver *in vitro*. Biochem. J. **140:**369–375.

9. **Hers, H.-G., L. Hue, and E. Van Schaftingen.** 1982. Fructose 2,6-bisphosphate. Trends Biochem. Sci. **7**:329–331.
10. **Huber, S. C.** 1986. Fructose 2,6-bisphosphate as a regulatory metabolite in plants. Annu. Rev. Plant Physiol. **37:** 233–246.
11. **Paz, N., D.-P. Xu, and C. C. Black, Jr.** 1985. Rapid oscillations in fructose 2,6-bisphosphate levels in plant tissues. Plant Physiol. **79:**1133–1136.
12. **Rea, P. A., and R. J. Poole.** 1985. Proton-translocating inorganic pyrophosphatase in red beet (*Beta vulgaris* L.) tonoplast vesicles. Plant Physiol. **77:**46–53.
13. **Smyth, D. A., and C. C. Black, Jr.** 1984. Measurement of the pyrophosphate content of plant tissues. Plant Physiol. **75:**862–864.
14. **Smyth, D. A., M.-X. Wu, and C. C. Black, Jr.** 1984. Pyrophosphate and fructose 2,6-bisphosphate effects on glycolysis in pea seed extracts. Plant Physiol. **76:**316–320.
15. **Stitt, M., G. Meiskes, H.-D. Soling, and H. W. Heldt.** 1982. On a possible role for fructose 2,6-bisphosphate in regulating photosynthetic metabolism in leaves. FEBS Lett. **145:**217–222.
16. **Turner, J. F., and D. H. Turner.** 1980. The regulation of glycolysis and the pentose phosphate pathway, p. 279–316. *In* D. D. Davies (ed.), The biochemistry of plants, vol. 2. Academic Press, Inc., New York.
17. **Wu, M.-X., D. A. Smyth, and C. C. Black, Jr.** 1983. Fructose 2,6-bisphosphate and the regulation of pyrophosphate-dependent phosphofructokinase activity in germinating pea seeds. Plant Physiol. **73:**188–191.
18. **Wu, M.-X., D. A. Smyth, and C. C. Black, Jr.** 1984. Regulation of pea seed pyrophosphate-dependent phosphofructokinase: evidence for interconversion of two molecular forms as a glycolytic regulatory mechanism. Proc. Natl. Acad. Sci. USA **81:**5051–5055.

IX. GLOBAL REGULATORY SYSTEMS IN ENTERIC BACTERIA

Introduction

JONATHAN A. GALLANT

Department of Genetics, University of Washington, Seattle, Washington 98195

Global regulatory systems (1, 2) were first defined rather simply as regulons (9) which comprise a large number of genes. The discovery that phosphate limitation induces the synthesis of alkaline phosphatase (4, 10; A. Torriani, Fed. Proc. **18**:33, 1958) was one of the first examples of repression/derepression to be recognized, and more recent work has shown that phosphate controls a global system consisting of a score of genes (13).

A section on other global control systems in a book devoted primarily to phosphate metabolism is appropriate for three reasons. Firstly, the comparative anatomy of a variety of global control systems can shed light on the organization of each of them. A striking example, set forth in this book, is the occurrence of pairs of homologous protein domains in the genes of several control systems, such as the nitrogen regulation system, the phosphate regulon, and membrane protein genes (see articles in this volume by Merrick et al. and by Shinagawa et al.).

Secondly, the metabolism of phosphate and phosphorylated compounds is such a central matter that it impinges on virtually all global control systems. As Travers (this volume) points out, a variety of nucleotides affect the promoter preferences of RNA polymerase. The levels of these regulatory nucleotides (ATP, ADP, AMP, cyclic AMP, ppGpp and its congeners) are controlled by and in turn feed back upon a variety of global control systems, such as those linked to energy metabolism, carbohydrate transport by the phosphotransferase system, phosphate metabolism, nitrogen metabolism, and the balance of aminoacyl-tRNA formation and demand. An overlapping web of signals thus links the pattern of transcription, via these regulatory nucleotides, to what one might call the leading economic indicators of the cell.

Thirdly, there is what Walker (Paek et al., this volume) calls "cross-talk" between global systems. One kind of cross-talk is exemplified by genes whose functions affect more than one global system, as in the interconnections between the DNA damage repair and heat shock systems that Walker and colleagues recount. Another example is the observation that *phoB* mediates the response of some genes to limitation for nitrogen as well as phosphate (Wanner, this volume). I suspect that more and more such interconnections will come to light.

Not only is there cross-talk, but the domains of different global systems in fact overlap in a number of cases. Some of the genes induced by heat shock are also induced by the stringent control system (3). Some of the genes controlled by phosphate also belong to global systems affected, variously, by nitrogen, carbon, oxidative metabolism, and DNA damage (12), and even by osmolarity (11). I think the overlap of domains would be even more striking if we were to survey a larger number of global control systems, such as those which respond to sulfate, temperature, osmotic stress, oxidation stress, methyl group metabolism, growth rate, and errors in protein synthesis. One can also confidently predict that additional global systems will be discovered that adjust the expression of batteries of genes to still other environmental insults. The National Aeronautics and Space Administration will no doubt soon receive grant applications for investigating the global response to variations in gravitational force, and applications to the Department of Energy for studying the response of bacteria to popular music are long overdue.

Someday, we hope to be able to represent these interconnected and overlapping networks as cleanly as a circuit diagram represents an electronic device. Our biological diagram would consist of boxes representing the activities of genes or gene products and arrows representing positive or negative regulatory effects of box_i (or of environmental stimulus *Z*) on box_j, box_k, etc. In principle, one could then simulate how banging on the network at any given box would affect the magnitude of all the others. Such simulations would tell us what sort of oscillations the system undergoes and what sort of stable states it would reach. A systems analysis approach to global regulation was once only a gleam in the eyes of Maaloe and his colleagues in Copenhagen (7, 8), but I believe we will be in sight of it within a few years.

It is worth noting that system behavior is independent of the physical mechanism of the regulatory relationships. If box_x exerts negative control over box_y, the outcome is the same regardless of whether the control operates on transcription initiation or transcription attenuation, mRNA stability or translation initiation, etc.

Nevertheless, molecular biology will continue to be concerned with specific mechanisms because it is still biology, a descriptive subject. A curious and unanticipated consequence of the DNA revolution is that biology is returning to its roots in what was called natural history in the 18th century. Linnaeus classified organisms according to morphological homologies (6), and today we classify genes according to DNA sequence homologies. DNA taxonomy has revealed certain general rules, such as consensus sequences shared (more or less) by the promoters of all genes subject to a particular global control system. The implications for metazoan differentiation are well appreciated. Just as the *pho* box is a central character in this book, developmental geneticists hope that future books will include such characters as the liver box, the antenna box, or the anterior-compartment-of-any-segment box.

We also need to understand the molecular details of particular control systems if we wish to intervene in the biological world, for medical, agricultural, or industrial purposes. The manipulation of one biological system by another is as old as, for example, endosymbiosis. Fortunately, environmental activists of a certain type were not around when prehistoric eucaryotic and procaryotic cells collaborated in the development of mitochondria, else God would have been subjected to endless litigation.

Incidentally, we should give credit to Japanese microbiology for an early appreciation that microbes can be put to work for human use, exemplified in the production of food additives and antibiotics. The tradition of applied microbiological chemistry can be summarized in the following way. If there is some thermodynamically possible chemical reaction we wish done, human chemists can probably devise some exceedingly expensive way to do it with platinum catalysts and high temperatures, but somewhere there are soil bacteria which toss off the reaction with ease. In other words, evolution has enabled microorganisms to become much better chemists than we are.

How did this come about? The unicellular life style is very ancient, and microbial generation times can be very short, so there has been a vast number of generations during which the shaping force of natural selection has adapted microorganisms to various ecological niches. The selective pressures in turn reflect the physics and chemistry of each particular niche. Koch has long been interested in the evolutionary implications of the ecosystem inhabited in nature by the molecular biologist's favorite, *Escherichia coli*. He concludes this section with an essay on this ecosystem and a model of *E. coli* population growth in relation to the system's flow characteristics.

Many of this book's readers unconsciously assume that the normal environment of *E. coli* is glucose minimal medium, perhaps supplemented with the β-galactosidase indicator X-Gal. (An advertising flyer from a biotechnology firm once observed that colonies of *E. coli* are naturally blue.) It is salutary to be reminded of the natural world and of the ecology to which *E. coli* has become adapted in that world.

Ecological modeling of this sort would seem scarcely believable if it dealt with some of the more dramatic habitats occupied by other microbial species. These include saturated ammonium sulfate, 10 mM sulfuric acid, 10% urea at pH 11, snow and ocean water at temperatures only a few degrees above the freezing point, and hot springs at temperatures not far below the boiling point (5). The astonishing adaptive radiation of microorganisms into environments which verge on science fiction poses interesting questions for evolutionary biology. In every environment, microbial growth and metabolism will be found to be fine-tuned by global regulatory systems which we are only beginning to discover and analyze at the molecular level.

LITERATURE CITED

1. **Gottesman, S.** 1984. Bacterial regulation: global regulatory networks. Annu. Rev. Genet. **18:**415–441.
2. **Gottesman, S., and F. C. Neidhardt.** 1983. Global control systems, p. 163–184. *In* J. Beckwith, J. Davies, and J. Gallant (ed.), Gene function in prokaryotes. Cold Spring Harbor Laboratory, Cold Spring Harbor, N.Y.
3. **Grossman, A. D., W. E. Taylor, Z. F. Burton, R. R. Burgess, and C. A. Gross.** 1985. Stringent response in *Escherichia coli* induces expression of heat shock proteins. J. Mol. Biol. **186:**357–365.
4. **Horiuchi, T., S. Horiuchi, and D. Mizuno.** 1959. A possible negative feedback phenomenon controlling formation of alkaline phosphatase in *E. coli*. Nature (London) **183:** 1529–1530.
5. **Kushner, D. J.** 1978. Microbial life in extreme environments. Academic Press, Inc., New York.
6. **Linnaeus, C.** 1735. Systema naturae, sive rezna tria

naturae systematice proposita per classes, ordines, genera, et species (Luzduni Batavorum).
7. **Maaloe, O.** 1979. Regulation of the protein-synthesizing machinery: ribosomes, tRNA, factors, and so on, p. 487–542. *In* R. F. Goldberger (ed.), Biological regulation and development, vol 1. Plenum Publishing Corp., New York.
8. **Maaloe, O., and N.-O. Kjeldgaard.** 1966. Control of macromolecular synthesis. Benjamin, New York.
9. **Maas, W. K., and A. J. Clark.** 1964. Studies on the mechanism of repression of arginine biosynthesis in *Escherichia coli*. II. Dominance of repressibility in diploids. J. Mol. Biol. **8:**365–370.
10. **Torriani, A.** 1960. Influence of inorganic phosphate in the formation of phosphatases by *Escherichia coli*. Biochim. Biophys. Acta **38:**460–470.
11. **Villarejo, M., J. L. David, and S. Granett.** 1983. Osmoregulation of alkaline phosphatase synthesis in *Escherichia coli* K-12. J. Bacteriol. **156:**975–978.
12. **Wanner, B.** 1983. Overlapping and separate controls on the phosphate regulon in *Escherichia coli* K12. J. Mol. Biol. **166:**283–308.
13. **Wanner, B., and R. McSharry.** 1982. Phosphate-controlled gene expression in *Escherichia coli* K12 using *MudI*-directed *lacZ* fusions. J. Mol. Biol. **158:**347–363.

Global Regulation of Carbon Metabolism in Bacteria: Networks of Interacting Systems

MILTON H. SAIER, JR., MAMORU YAMADA, AND A. MICHAEL CHIN

Department of Biology, University of California, San Diego, La Jolla, California 92093

The phosphoenolpyruvate:sugar phosphotransferase system (PTS) found in a wide variety of bacterial species has been intensively characterized. In most strict and facultatively anaerobic bacteria in which it is found, it is a complex, multicomponent system which initiates the metabolism of many sugars. In most bacteria it functions in sugar recognition, transport, and phosphorylation. It is integrated into the cellular metabolic system controlling the utilization of several carbon sources whose metabolism is not initiated by the PTS.

In this review we discuss the regulatory ramifications of the PTS, emphasizing the most recent advances from our laboratory indicating that the system regulates the utilization of many gluconeogenic substrates by a novel mechanism. The presence of previously undetected proteins of the system and the possible implications of their presence are also discussed. The reader is referred to a recent monograph (6) as well as a review (5) for more detailed accounts of all but the most recent advances concerning these subjects.

Distribution and Specificity of the PTS in Diverse Bacterial Genera

Different sugars are utilized via the PTS in several different gram-negative and gram-positive bacterial genera (13). Many photosynthetic, nitrogen-fixing, and heterotrophic bacteria can utilize only one sugar, fructose, via a phosphoenolpyruvate-dependent mechanism. *Rhodopseudomonas* is a representative bacterial genus containing this system. Although the proteins of these fructose-specific systems are poorly characterized, the systems appear to differ from those found in other bacteria in that only a single, high-molecular-weight protein (molecular weight, 200,000), in addition to the integral membrane constituent, has been identified (10). Proteins resembling enzyme I and HPr seem to be lacking, and no enzymatic cross-reactivity between these systems has been demonstrated (10). We have recently cloned the genes encoding the fructose PTS proteins from *R. capsulata* and hope to sequence these genes in the near future.

All other bacteria studied to date contain a recognizable enzyme I dimer with a 60,000 monomeric molecular weight and an HPr (heat stable phosphocarrier protein) of about 9,000 molecular weight. Gram-negative bacterial genera which possess enzyme I- and HPr-containing PTSs include *Megasphaera*, which can utilize only two sugars, fructose and glucose; *Escherichia, Salmonella*, and *Klebsiella*, which utilize many sugars via the PTS; and *Spirochaeta*, which utilizes only mannitol via a phosphoenolpyruvate-dependent mechanism (4). In *Spirochaeta aurantia*, synthesis of all four known constituents of the mannitol catabolic system (enzyme I, HPr, and enzyme II^{mtl} of the PTS, as well as mannitol phosphate dehydrogenase) is inducible about 200-fold by growth in the presence of mannitol (11).

Except in the spirochetes, enzyme I and HPr are general, energy-coupling proteins of the PTS, utilized for the uptake of several sugars. These two proteins are synthesized in nearly invariant amounts, regardless of growth conditions, being induced maximally two- to threefold by growth in the presence of a PTS sugar. Sugar-specific enzymes II and enzyme II-III pairs serve as the sugar-specific recognition components of the PTS. The syntheses of these proteins are specifically induced by growth in the presence of their sugar substrates. *Escherichia coli* utilizes at least 10 sugars via the PTS; *Arthrobacter*, 2; *Bacillus*, 6; *Staphylococcus*, 8; and *Streptococcus*, 10. In each of these cases, growth of the respective bacteria in the presence of any one of these carbohydrates usually enhances synthesis of the relevant enzyme II, or enzyme II-III pair, 10- to 100-fold. It is also worthy of note that some sugars not utilized for growth by most bacteria can be utilized via the PTS in certain bacteria. For example, xylitol and ribitol are PTS sugars in *Lactobacillus* species. Finally, species of *Mycoplasma* can utilize both glucose and fructose via a PTS which differs from conventional bacterial PTSs in that enzyme I is a multisubunit complex (molecular weight, 220,000), and separate HPr-like proteins function in the exclusive phosphorylation of only one of the two substrates, glucose or fructose. The regulatory functions of PTS proteins are not known in most of the bacteria in which the PTS has been detected.

TABLE 1. Functional comparison of the PTS and the lactose permease in *E. coli*

Function	System	
	PTS	*lac* permease
Sugar		
Transport	+	+
Phosphorylation	+	−
Chemoreception	+	−
Regulation of:		
Other permeases	+	−
Catabolic enzymes	+	−
Adenylate cyclase	+	−
Gene transcription	+	−
Gluconeogenesis	+	−
Regulated by:	ATP-dependent protein phosphorylation	PTS

Functional Complexity of the PTS

The structural complexity of the PTS is equaled only by its functional complexity. To illustrate this point, it is valuable to compare it with the well-characterized lactose permease of *E. coli*. As revealed in Table 1, the PTS not only phosphorylates its sugar substrates and transports them across the bacterial cell membrane, but it also serves as a chemoreceptor-effector system, directing (via a transmission system of unknown nature) the motile behavior of the organism. The chemotactic signal elicited by the PTS causes the bacteria to respond positively to concentration gradients of all PTS sugar substrates, thereby allowing them to seek optimal extracellular concentrations of these compounds. By contrast, the lactose permease of *E. coli*, which functions by a proton symport mechanism, serves the sole function of transmembrane permeation (Table 1). Lactose is not a chemoattractant in *E. coli*.

In addition to its primary functions of sugar chemoreception, transport, and phosphorylation, the PTS in *E. coli* also functions as a complex regulatory system, controlling the activities of several transmembrane non-PTS permeases such as the maltose, melibiose, and lactose permeases, at least one cytoplasmic catabolic enzyme (glycerol kinase), and the cyclic AMP biosynthetic enzyme adenylate cyclase (7). The lactose (*lac*) permease lacks all of these functions and is itself regulated by the PTS. The PTS in gram-positive bacteria is regulated by a protein kinase-mediated mechanism specific to the PTS (see next section). These considerations suggest that the PTS must play a central and exceptionally important role in the coordination of carbon metabolism in bacteria capable of anaerobic carbohydrate utilization.

Influence of Exogenous Nitrogen on the Regulation of Cyclic AMP Metabolism by the PTS

Early studies by the Paris group showed that β-galactosidase synthesis is influenced by the exogenous source of nitrogen available for growth. Specifically, β-galactosidase activity was much less sensitive to what we now know to be PTS-mediated repression in the presence of a "good" nitrogen source such as ammonium sulfate than in a medium containing a poor nitrogen source such as glutamate or glycyl glutamate (3). Further, recent work has shown that the regulation of adenylate cyclase, giving rise to growth arrest and initiation of sporulation in *Saccharomyces cerevisiae*, is due to a nitrogen deficiency (15). It is not known whether the effect of the nitrogen source on β-galactosidase induction reported by Contesse et al. (3) was mediated by cyclic AMP.

We have reinvestigated this age-old problem with some interesting and definitive results (M. H. Saier and B. U. Feucht, unpublished data). The observations are as follows. (i) Four different *E. coli* strains of different genetic backgrounds showed similar behavior with respect to β-galactosidase synthesis: repression by glucose was stronger in the presence of glutamate or glycyl glutamate than in the presence of ammonium sulfate. The difference was two- to threefold. (ii) In all cases, repression was largely overcome by addition of exogenous cyclic AMP (5 mM). (iii) The effects were similar in strains lacking and containing cyclic AMP phosphodiesterase (due to *cpd* mutations). Therefore, regulation by nitrogen could not be explained by an effect influencing the rate of cyclic AMP degradation. (iv) Intracellular cyclic AMP levels correlated with β-galactosidase induction: glycyl glutamate-grown cells contained about one-third the amount of cyclic AMP contained by ammonium sulfate-grown cells. (v) When total production of cyclic AMP was measured as a function of time, a surprising result was obtained: net production was substantially greater in the glycyl glutamate-grown cells than in the ammonium sulfate-grown cells.

These results suggest that the nitrogen source exerts its action primarily by regulating the level or activity of the cyclic AMP transporter (9) rather than adenylate cyclase (8). The molecular basis of this unexpected observation is not currently understood, but the results suggest that the nitrogen source regulates cyclic AMP levels by a mechanism which is independent of the PTS.

Autoregulation of PTS Protein Synthesis and Function

The proteins of the PTS not only regulate the synthesis and activities of non-PTS carbohydrate permeases and catabolic enzymes, but they also control their own syntheses and activities. Several regulatory mechanisms have been demonstrated (5). For example, competition for phospho-HPr and competition for exogenous sugar binding to an enzyme II provide two mechanisms for controlling sugar uptake. Inhibition of PTS function by intracellular metabolites and inhibition by chemiosmotic energy provide two additional control mechanisms.

Work by J. Reizer and J. Deutscher in our laboratory first demonstrated that in gram-positive bacteria the PTS is autoregulated by an HPr kinase-phosphatase mechanism. The small phosphocarrier protein of the PTS, HPr, is phosphorylated on a seryl residue, and this phosphorylation event strongly inhibits the transport of PTS substrates (for a review, see reference 7). Because mutants defective specifically in the regulatory kinase and phosphatase have not been characterized in sufficient detail, the physiological consequences of this unique mechanism of bacterial regulation have not been fully defined. Such studies are currently in progress in our laboratory.

The probable involvement of PTS proteins in autoregulation of gene transcription has also been discussed (6, 7). Just one example will be cited here. Lactose is transported across the cytoplasmic membrane of *Staphylococcus aureus* by the lactose PTS. The uptake and phosphorylation of lactose via this system requires the functional integrity of enzyme I, HPr, enzyme II^{lac}, and enzyme III^{lac}. The product, cytoplasmic lactose 6-phosphate, is then cleaved to glucose and galactose 6-phosphate by a phospho-β-galactosidase. The latter compound is the true inducer of the regulon which codes for the lactose-specific constituents of the system, enzyme II^{lac}, enzyme III^{lac}, and the phospho-β-galactosidase. This phosphate ester can induce the operon when added to a cell suspension at high concentration, even when the PTS is nonfunctional.

Mutants defective for lactose utilization were isolated from two *S. aureus* parental strains (14). One was the lactose-inducible wild-type strain, and the other was a lactose-constitutive strain derived from the inducible parent. The two strains each yielded three identical classes of lactose-negative mutants defective in enzyme I, enzyme II^{lac}, or phospho-β-galactosidase. When the inducible parent was studied, a fourth class of mutants lacked all three of the proteins of the lactose regulon, enzyme II^{lac}, enzyme III^{lac}, and phospho-β-galactosidase. Extensive attempts to isolate mutants lacking only enzyme III^{lac} from the inducible parent gave negative results. In contrast, the constitutive parental strain did not yield mutants lacking the three proteins of the lactose regulon, but yielded a major class lacking only enzyme III^{lac}. The difference in behavior of the inducible and constitutive strains can be explained if it is assumed that enzyme III^{lac} plays a role in regulating transcription of the lactose regulon in *S. aureus*. Thus, the pleiotropic class of mutants in the inducible background results from a lack of functional enzyme III^{lac} because induction is presumed to be dependent on the presence of this protein.

We have recently initiated experiments aimed at establishing the involvement of enzyme III^{lac} in the transcriptional regulation of the staphylococcal lactose operon. To this end we have cloned the *lac* operon from *S. aureus* in *E. coli*. The achievement of this goal was possible because the enzymes of the staphylococcal PTS function with the energy-coupling enzymes of *E. coli*. As a consequence, transfer of the cloned *S. aureus lac* genes to an *E. coli* strain deleted for the chromosomal *lac* genes allowed utilization of lactose. We hope to use this cloned system in an in vitro transcription-translation system to establish the involvement of enzyme III^{lac} in the transcriptional regulation of the operon which encodes this protein.

Detection of Novel Energy-Coupling Proteins of the PTS in *E. coli* and *Salmonella typhimurium*

Our recent experiments have provided evidence for two novel energy-coupling proteins of the PTS, termed FPr (fructose-inducible HPr-like protein) and enzyme I* (enzyme I-like protein). The former protein has been purified to near homogeneity as a complex with the fructose-specific enzyme III (III^{fru}; S. Sutrina and M. H. Saier, Jr., unpublished data). FPr is present in crude extracts both as the free protein and as the FPr-III^{fru} complex, and it can function with enzyme I for the phosphorylation of all sugar substrates of the PTS (7, 12).

A novel regulatory mutation, designated *fruR51*::Tn*10*, which renders expression of the fructose regulon constitutive and appears to destroy the fructose repressor, renders FPr synthesis constitutive (1). In a *ptsH* mutant (which lacks HPr), such *fruR* mutants can grow on all PTS sugars since FPr can substitute for HPr. The *fruR* mutants exhibit regulatory characteristics to be described in the next section.

When deletion mutants of *S. typhimurium* which lack all or part of the structural genes encoding HPr and enzyme I (Δ*ptsHI* mutants) were plated on EMB fructose medium, strains capable of fructose utilization appeared. These strains grew well on fructose, grew poorly on glucose and mannose, and gained enzyme I activity. The mutations giving rise to this novel enzyme I (termed enzyme I*) defined a gene, *ptsJ* (2), located about 0.1 min from the *pts*

TABLE 2. Effect of *fruR51*::Tn*10* mutation on the utilization and fermentation of a variety of carbon sources in several different genetic backgrounds[a]

Carbon source	LT-2 strains:		*ptsH15* strains:		Δ*cysK ptsHI41* strains:		*crp*-771* strains:	
	LJ709	LJ712	SB1475	LJ711	LJ703	LJ704	PP914	LJ713
fruR51::Tn*10*	−	+	−	+	−	+	−	+
Fructose	+	+	+	+	−	−	+	+
Mannitol	+	+	−	+	−	−	−	+
Glycerol	+	+	−	+	−	−	+	+
Galactose	+	+	+	+	+	+	+	+
Phosphenolpyruvate	+	+	+	+	+	+	+	+
Acetate	+	−	+	−	+	−	+	−
Pyruvate	+	−	+	−	+	−	+	−
Alanine	+	−	+	−	+	−	+	−
Citrate	+	−	+	−	+	−	+	−
Isocitrate	+	−	+	−	−	−	+	−
Fumarate	+	−	+	−	−	−	+	−
Malate	+	−	+	−	−	−	+	−
Oxolacetate	+	−	+	−	+	−	+	−

[a] LT-2, Wild-type *S. typhimurium; ptsH15,* a mutation which specifically eliminates HPr activity; Δ*cysK ptsHI41,* a deletion mutation which eliminates the *cysK* gene as well as the *pts* operon; *crp*-771,* a mutation which renders the cyclic AMP receptor protein independent of cyclic AMP.

operon, between *pts* and *cysA*. The protein exhibited properties which were nearly indistinguishable from those of wild-type enzyme I. These properties included molecular size, association to the dimeric state, chemical reagent and heat sensitivities, and elution profiles from ion-exchange columns. Surprisingly, *Salmonella* strains carrying the *ptsJ* mutation were resistant to growth stasis by 2-deoxyglucose. This nonmetabolizable glucose analog prevented growth of isogenic strains which carried the wild-type *ptsJ* allele. This result and others suggested that the *ptsJ* gene (and consequently enzyme I*) exhibits regulatory characteristics, allowing the inhibitory effect of 2-deoxyglucose to be overcome when the *ptsJ* gene product is altered.

Regulation of Gluconeogenesis by the PTS

The *fruR* mutants described in the previous section were found to be unable to utilize a variety of organic acids, including acetate, pyruvate, alanine, and all Krebs cycle intermediates tested (Table 2) (1). This growth inhibition resulted from the absence of *fruR* function regardless of the genetic background. Thus, the presence or absence of enzyme I, HPr, or an altered cyclic AMP receptor protein, which could promote transcription of catabolite-sensitive operons without cyclic AMP, did not allow *fruR* mutants to utilize these organic acids (Table 2). *fruR* mutants were found to be constitutive for the enzymes of the fructose regulon, but markedly deficient for two gluconeogenic enzymes, phosphoenolpyruvate synthase and phosphoenolpyruvate carboxykinase. The activities of these enzymes were partially restored by mutations which resulted in reduced activities of the fructose-specific PTS enzymes, while the *fruR51*::Tn*10* mutation was retained. This result suggested that the enzymes of the fructose PTS functioned in the regulation of gluconeogenesis.

We have also found that *E. coli* strains which overproduce the enzymes of the glucitol (*gut*) operon (which includes the glucitol enzymes II and III) exhibit the same growth phenotypes as do the *fruR* mutants and are similarly deficient in phosphoenolpyruvate carboxykinase. A single mechanism may be involved. Preliminary evidence suggests that, while the FPr-enzyme III^{fru} complex may function in regulation of gluconeogenesis in *fruR* mutants, enzyme III^{gut} may be the important constituent for regulation following overproduction of the glucitol operon gene products. While the results clearly point to the direct involvement of PTS proteins in the regulation of gluconeogenesis, identification of the proteins and the exact mechanism involved remains to be elucidated.

Concluding Remarks

Evidence for the integration of the phosphotransferase systems of *E. coli* and *S. typhimurium* into the carbon and energy metabolic system of the bacterium is as follows. (i) Because the end product of glycolysis, phosphoenolpyruvate, is the energy source for sugar uptake and phosphorylation, the glycolytic scheme is a cyclic rather than a linear process. (ii) Because the PTS catalyzes the first irreversible reaction of glycolysis, it is subject to allosteric regulation. (iii) Because the PTS catalyzes the most efficient mechanism for the uptake and phosphorylation of sugars, it allosterically regulates other sugar permease systems. (iv) Because the metabolism of any sugar, initiated by the PTS, ensures the availability of all sugars required for biosynthetic purposes, the PTS regulates gluconeogenesis.

Bacteria can metabolize sources of carbon via several pathways. These pathways include glycolysis and the Krebs cycle for the generation of energy and biosynthetic intermediates from carbohydrates, as well as the glyoxalate shunt plus the enzymes of gluconeogenesis for the generation of carbohydrates from a variety of metabolic intermediates. The first catalytic element of glycolysis is the PTS which detects, transports, and phosphorylates exogenous carbohydrates (PTS sugars). By virtue of its central role in detecting carbohydrates and initiating glycolysis, the system regulates a variety of transport systems and catabolic enzymes which initiate metabolism of other carbohydrates (non-PTS sugars). The system also regulates the intracellular concentration of cyclic AMP which is under both nitrogen and carbon control. As a consequence of its role in the regulation of cytoplasmic inducer and cyclic AMP levels, the PTS controls transcription of a large class of genes involved with energy metabolism and carbon flux.

The PTS appears to autoregulate its own functions by influencing rates of gene transcription and enzyme catalysis. Evidence suggests that some PTS proteins function directly in the control of gene transcription, while in gram-positive bacteria a protein kinase-phosphatase system controls the activity of the PTS by phosphorylation of a seryl residue in the phosphocarrier protein of the PTS, HPr.

Recent work in our laboratory has shown that the PTS also regulates the utilization of gluconeogenic substrates, including amino acids and Krebs cycle intermediates. The mechanism appears to be distinct from those established previously for the utilization of non-PTS sugars and may involve regulation of two gluconeogenic enzymes, phosphoenolpyruvate carboxykinase and phosphoenolpyruvate synthase. The recent detection of new proteins of the PTS with regulatory functions leads to the suggestion that the system may be structurally and functionally more complex than was previously supposed.

LITERATURE CITED

1. **Chin, A. M., B. U. Feucht, and M. H. Saier, Jr.** 1986. Evidence for the regulation of gluconeogenesis by the fructose phosphotransferase system in *Salmonella typhimurium*. J. Bacteriol. **169:**897–899.
2. **Chin, A. M., S. Sutrina, D. A. Feldheim, and M. H. Saier, Jr.** 1986. Genetic expression of enzyme I* activity of the phosphoenolpyruvate:sugar phosphotransferase system in *ptsHI* deletion strains of *Salmonella typhimurium*. J. Bacteriol. **169:**894–896.
3. **Contesse, G., M. Crepin, F. Gros, A. Ullmann, and J. Monod.** 1970. On the mechanism of catabolite repression, p. 401–415. *In* J. R. Beckwith and D. Zipser (ed.), The lactose operon. Cold Spring Harbor Laboratory, Cold Spring Harbor, N.Y.
4. **Dills, S. S., A. Apperson, M. R. Schmidt, and M. H. Saier, Jr.** 1980. Carbohydrate transport in bacteria. Microbiol. Rev. **44:**385–418.
5. **Postma, P. W., and J. W. Lengeler.** 1985. Phosphoenolpyruvate:carbohydrate phosphotransferase system of bacteria. Microbiol. Rev. **49:**232–269.
6. **Saier, M. H., Jr.** 1979. The role of the cell surface in regulating the internal environment, p. 167–227. *In* J. R. Sokatch and L. N. Ornston (ed.), The bacteria, vol. VII. Academic Press, Inc., New York.
7. **Saier, M. H., Jr.** 1985. Mechanisms and regulation of carbohydrate transport in bacteria. Academic Press, Inc., New York.
8. **Saier, M. H., Jr., and B. U. Feucht.** 1975. Coordinate regulation of adenylate cyclase and carbohydrate permeases by the phosphoenolpyruvate:sugar phosphotransferase system in *Salmonella typhimurium*. J. Biol. Chem. **250:**7078–7080.
9. **Saier, M. H., Jr., B. U. Feucht, and M. T. McCaman.** 1975. Regulation of intracellular adenosine cyclic 3′:5′-monophosphate levels in *Escherichia coli* and *Salmonella typhimurium*. J. Biol. Chem. **250:**7593–7601.
10. **Saier, M. H., Jr., B. U. Feucht, and S. Roseman.** 1971. Phosphoenolpyruvate-dependent fructose phosphorylation in photosynthetic bacteria. J. Biol. Chem. **246:**7819–7821.
11. **Saier, M. H., Jr., M. J. Newman, and A. W. Rephaeli.** 1977. Properties of a phosphoenolpyruvate:mannitol phosphotranserase system in *Spirochaeta aurantia*. J. Biol. Chem. **252:**8890–8898.
12. **Saier, M. H., Jr., R. D. Simoni, and S. Roseman.** 1970. The physiological behavior of enzyme I and heat-stable protein mutants of a bacterial phosphotransferase system. J. Biol. Chem. **245:**5870–5873.
13. **Saier, M. H., Jr., and M. Yamada.** 1986. Evolutionary considerations concerning the bacterial phosphorenol-pyruvate: sugar phosphotransferase system, p. 196–214. *In* J. Reizer and A. Peterkopsky (ed.), Sugar transport and metabolism in gram-positive bacteria. Ellis Horwood Publisher, New York.
14. **Simoni, R. D., and S. Roseman.** 1973. Sugar transport. VII. Lactose transport in *Staphylococcus aureus*. J. Biol. Chem. **248:**966–976.
15. **Tripp, M. L., R. Piñon, J. Meisenhelder, and T. Hunter.** 1986. Identification of phosphoproteins correlated with proliferation and cell cycle arrest in *Saccharomyces cerevisiae*: positive and negative regulation by cAMP-dependent protein kinase. Proc. Natl. Acad. Sci. USA **83:**5973–5977.

Regulation of Nitrogen Assimilation in Enteric Bacteria

M. J. MERRICK, S. AUSTIN, M. BUCK, R. DIXON, M. DRUMMOND, A. HOLTEL, AND S. MACFARLANE

AFRC Unit of Nitrogen Fixation, University of Sussex, Brighton BN1 9RQ, United Kingdom

All microorganisms can use a variety of nitrogen compounds as sources of cellular nitrogen. Although the preferred source is usually ammonia, the possible sources range from dinitrogen gas through nitrate and nitrite to complex organic compounds such as amino acids. In virtually all cells glutamate and glutamine serve as nitrogen donors for biosynthetic reactions, and in enteric bacteria the assimilation of nitrogen into glutamate and glutamine is controlled at the transcriptional level by a complex regulatory system known as the nitrogen regulation (*ntr*) system. The existence of such a regulatory mechanism was first indicated in the early 1970s by Magasanik and co-workers, who studied regulation of the enzyme histidase in *Klebsiella aerogenes* (36). However, it was not until some years later that the primary *ntr* regulatory genes were identified and a clear picture of the mechanism of nitrogen control began to emerge (28). Parallel studies have been carried out in four enteric bacteria, *Escherichia coli, Salmonella typhimurium, K. aerogenes*, and *Klebsiella pneumoniae*, and the overall picture in all four organisms is very similar. More recently, homologous *ntr* systems have been identified to various degrees in *Azotobacter* (39), *Rhizobium* (2), *Pseudomonas* (14), and *Rhodopseudomonas* (24) species, and it therefore seems likely that this particular regulatory system may be quite widespread in procaryotes.

Nitrogen Assimilation

Enteric bacteria have two primary routes of ammonia assimilation. Ammonia can be utilized for glutamate synthesis by means of glutamate dehydrogenase.

$$\text{2-ketoglutarate} + \text{NADPH} + NH_3 \rightleftharpoons \text{glutamate} + \text{NADP}^+$$

This enzyme has a relatively high K_m (of the order of 1 mM) for ammonia and is consequently relatively ineffective in nitrogen assimilation under conditions of nitrogen limitation. The formation of glutamate dehydrogenase is repressed in N-starved cells of *K. aerogenes* and *K. pneumoniae* but not in comparable cultures of *E. coli* or *S. typhimurium*; the molecular basis of this regulation is presently unknown.

In all microorganisms capable of growing with ammonia as the N source, an alternative pathway is found comprising the enzymes glutamine synthetase (GS) and glutamate synthase (GOGAT).

$$\text{glutamate} + NH_3 + \text{ATP} \xrightarrow{\text{GS}} \text{glutamine} + \text{ADP} + P_i$$

$$\text{glutamine} + \text{2-ketoglutarate} + \text{NADPH} \xrightarrow{\text{GOGAT}} \text{2 glutamate} + \text{2 NADP}^+$$

These reactions allow assimilation of ammonia present in the media at concentrations of less than 0.1 mM, and in some organisms, e.g., *Bacillus subtilis*, they constitute the only pathway for utilization of ammonia.

In enteric bacteria and many other organisms, the enzymatic activity of GS is regulated by reversible adenylylation of a specific tyrosyl on each of the enzyme's 12 subunits. The biosynthetic activity of GS is inversely proportional to the number of adenylylated subunits and is controlled by a complex cascade. Adenylylation is catalyzed by an adenylyltransferase encoded by *glnE*. Adenylylation by adenylyltransferase is in turn stimulated by a small regulatory protein, P_{II}, encoded by *glnB*. P_{II} can itself be uridylylated, and the uridylylated P_{II} ($P_{II\text{-}UMP}$) enhances deadenylylation. Finally, the interconversion of P_{II} and $P_{II\text{-}UMP}$ is catalyzed by uridylyl transferase and a uridylyl-removing enzyme, probably both encoded by *glnD*. Not only the activity, but also the intercellular concentration, of GS is regulated by the availability of nitrogen, and the products of *glnB* and *glnD* appear to have an independent role, which will be discussed later, in this transcriptional regulation, which is mediated by the *ntr* system.

Nitrogen Regulation

A wide range of enzymes in enteric bacteria are known to be subject to nitrogen control, and as expected, these are enzymes which allow the cell to obtain ammonia or glutamate from other nitrogen sources. They include nitrogenase, histidase, urease, and a number of amino acid permeases. Of these, only the regulation of the nitrogenase (*nif*) system (12) and that of GS itself (28) are characterized in detail at the molecular level, although some information is also available for histidase (*hut*) (33) and the arginine and histidine transport systems (*argT* and *dhuA*) (17). The following summary of the nitrogen regulation (*ntr*) system draws on studies on all these systems.

Nitrogen Regulation (*ntr*) Genes

Three primary regulatory genes have been identified in enteric bacteria. These genes are designated *ntrA*, *ntrB*, and *ntrC*, although in *E. coli* and *K. aerogenes* they are referred to as *glnF, glnL*, and *glnG*, respectively. The products of these genes can effect both positive and negative control at a variety of nitrogen-regulated promoters.

The *ntrB* and *ntrC* genes are part of a complex operon, *glnA ntrBC*, and *ntrBC* can be expressed either by transcription initiating upstream of *glnA* and reading through *glnA* or by transcription initiating at a promoter between *glnA* and *ntrBC* (1, 23, 35). Under conditions of nitrogen limitation, *ntrBC* expression is primarily due to transcription reading through *glnA*, but in nitrogen excess, when *glnA* expression is reduced approximately 10-fold, *ntrBC* is primarily expressed from the *ntrB* promoter. The majority of transcripts reading through *glnA* terminate at a position coincident with a Rho-independent terminator in the *glnA-ntrB* intergenic region (26, 40). In *E. coli* this region also contains a copy of a DNA element known as the repetitive extragenic palindrome sequence.

Many copies of the repetitive extragenic palindrome are present in the chromosomes of *E. coli* and *S. typhimurium*, and they are invariably located extragenically. The sequence is believed to have a function in stabilizing mRNA, possibly from exonucleolytic attack (20), but the significance of its presence between *glnA* and *ntrB* in *E. coli* but not in *S. typhimurium* or *K. pneumoniae* is not understood. Data on expression of *ntrBC* vary. In *E. coli* these genes are apparently expressed at a higher level in N excess than in N deficiency, while in *K. pneumoniae* the complex regulatory pattern appears to result in a relatively constitutive level of expression (1, 35).

The *ntrA* gene is unlinked to *ntrBC*, and the precise transcriptional organization of *ntrA* is presently unclear. Sequencing data from *K. pneumoniae* indicate potential open reading frames both upstream and downstream of *ntrA*, with no distinct terminator sequences between *ntrA* and these adjacent open reading frames. Hence, *ntrA* may be part of an operon in which the identity and function of the other genes are presently unknown (30). Studies in both *E. coli* and *K. pneumoniae* indicate that *ntrA* expression is not regulated by the N status of the cell, and the gene appears to be expressed at a low constitutive level (10, 32).

Negative Control

Genetic analysis of the *glnA ntrBC* operon first indicated that the *ntrC* product (NtrC) could function as a negative regulator of *glnA* transcription (29, 35). Subsequent analysis has identified a DNA sequence in both the *glnA* and *ntrB* promoter regions to which purified NtrC binds to prevent transcription initiation (17, 19, 40). These sequences are present in *E. coli, S. typhimurium*, and *K. pneumoniae* and conform to a consensus with dyad symmetry which is typical of binding sites for regulatory proteins (13). The promoters at which NtrC binds conform to the consensus −10, −35 type of sequences characteristic of promoters in enteric (and many other) bacteria, and in each case the binding site for NtrC is very close to the −10 region such that binding of NtrC will exclude RNA polymerase binding and prevent transcription initiation.

The negative control mediated by NtrC can be modulated by the *ntrB* product (NtrB). In the absence of NtrB the *ntrB* promoter is repressed by NtrC under conditions of both excess and limiting N (1). However, in the presence of NtrB maximal repression of *ntrBCp* is observed only in limiting N; in excess N NtrB apparently acts to relieve NtrC-mediated repression (1). Hence, it appears that NtrB can modulate the repressive effects of NtrC in response to the N status of the cell (1, 29).

Positive Control

Promoters subject to positive control by the *ntr* system are transcriptionally active in conditions of limiting N and are not expressed in excess N, i.e., high ammonia. These promoters are quite unlike the consensus bacterial promoters which are characterized by recognition sequences for RNA polymerase at positions −10 and −35. Promoters that can be activated by *ntr* conform to a consensus sequence TGGCAC N_5 TTGCA, which is found at positions −26 to −11 with respect to the transcript start, and within this sequence the GG at −24/25 and the GC at −12/13 are invariant in enteric bacteria (13). The best characterized of these promoters are the downstream promoter for *glnA* (13, 38) (the tandem upstream promoter being negatively controlled by NtrC) and the *nif* promoters of *K. pneumoniae* (4). Activation of transcription at these promoters requires the *ntrA* product (NtrA) and the product of either *ntrC* or *nifA* (25, 31). Hence, mutations in either *ntrA* or *ntrC* can result in an Ntr^- phenotype and failure to grow on a number of poor N sources, including arginine, histidine, and dinitrogen. The *K. pneumoniae nifA* product (NifA) is a *nif*-specific positive activator, but it can substitute for NtrC at the downstream *glnA* promoter and the *nifLA* promoter, which are normally activated by NtrC (15, 31).

The atypical consensus sequence found in *ntr*-activatable promoters suggested that these promoters may be recognized by a modified form of RNA polymerase in which the normal sigma subunit (the product of *rpoD*) is replaced by another sigma factor which confers *ntr* specificity (11, 32). In vitro transcription experiments with purified NtrA have now confirmed that this protein does indeed act as an alternative sigma factor to RpoD, and it has been proposed that *ntrA* should be renamed *rpoN* (21, 22). NtrA probably plays a role in recognition of the −11, −26 consensus sequence. By in vitro mutagenesis of *nif* promoters, the invariant GG and GC residues in the consensus have been shown to be essential for promoter activity (6, 34). Mutation of any one of these residues reduces promoter activity to less than 5% of that observed in the wild type. The spacing of the GG and GC residues by precisely 10 base pairs ensures that the invariant residues lie on the same face of the DNA helix. This separation is critical, as removal of just one of the nonconserved residues has been shown (in the *nifH* promoter of *K. pneumoniae*) to eliminate promoter activity (5). The role of the activator protein (NtrC or NifA) in transcription initiation is presently unclear. This protein may also act by binding to specific DNA sequences in the promoter region, or it may act by protein-protein interaction with RNA polymerase.

Most *nif* promoters contain a second conserved sequence positioned some 100 base pairs upstream of the −11, −26 consensus. This sequence, which conforms to a dyad symmetrical structure often found in binding sites for regulatory proteins, is essential for NifA-specific activation and is probably a NifA binding site. NifA-mediated activation of the *nifH* promoter is still detectable when the upstream activator sequence is placed 2,000 base pairs away from the −11, −26 region, so the sequence is effective over very considerable distances (7). The precise way in which NifA-specific activation occurs is, however, still unclear, and to date the upstream activator sequences have been found only in *nif* promoters.

The regulation, in response to N status, of promoters subject to positive control by *ntr* is mediated by NtrB. In a cell lacking NtrB and synthesizing NtrC constitutively, such promoters, e.g., *nifLA*, are constitutively expressed regardless of the N status of the medium (31). Hence, NtrB is required to prevent transcriptional activation by NtrC in excess N.

Response to the N Status

The response of both negatively and positively controlled promoters to N status is apparently mediated by the same protein, NtrB. In limiting N, NtrB appears to have little effect, but in excess N, NtrB both prevents activation (e.g., at *nifLAp*) by NtrC and relieves repression (e.g., at *ntrBCp*) by NtrC (1, 30). The way in which this sensing is achieved is not clear, but genetic data suggest that the products of *glnB* (P_{II} protein) and *glnD* (uridylyl transferase) play a role (8). These proteins are known to respond to N status to regulate the adenylylation of GS, but they may also affect *ntr* regulation of transcription. Some mutations in *glnB* apparently lock the P_{II} protein in a form in which it cannot be uridylylated, and consequently GS is permanently adenylylated. Such mutants are glutamine auxotrophs, but their auxotrophy is largely a consequence of the fact that they show a significantly reduced level of *glnA* transcription. At the same time, they show significantly increased transcription from *ntrBC*, and hence they behave as if in N excess regardless of the actual N status of the medium (A. Holtel, unpublished data). Such mutants are also Ntr^-; i.e., they fail to grow on histidine, arginine, dinitrogen, etc., as the sole N source. Mutations in *glnD* which impair synthesis of uridylyl transferase have a similar Gln^- Ntr^- phenotype, presumably because they also fail to convert P_{II} to $P_{II\text{-}UMP}$ (18). These mutant phenotypes are consistent with a model in which the nonuridylylated form of P_{II}, which is present in the cells when nitrogen is in excess, causes NtrB to inactivate NtrC (8). This would impair both the activator and repressor functions of NtrC. A number of authors have suggested that NtrB and NtrC may interact directly, forming a protein-protein complex, and such a proposition is supported by the description of *ntrC* suppressor mutations which map in *ntrB* (41).

There is no evidence that NtrA plays any role in the response of nitrogen-regulated systems to changes in the availability of ammonia. The protein appears to be synthesized at a low constitutive level, and its ability to promote transcription initiation is solely dependent on the status of the NtrB and NtrC proteins (31).

Sequences of the *ntr* Genes

All three primary *ntr* genes have been sequenced in *K. pneumoniae* (9, 16, 26, 30), and these sequences provide some interesting information about possible structure/function relationships in the respective protein products.

The *ntrA* product is a very acidic polypeptide with a molecular weight of 75,000 on sodium dodecyl sulfate-polyacrylamide gel electrophoresis (32). The DNA sequence of *ntrA* confirms the acidity of the protein but predicts a molecular weight of only 54,000. This discrepancy may be due to the unusual charge properties of the protein. The primary amino acid sequences of a number of sigma factors have been determined from DNA sequences, and in many cases, in-

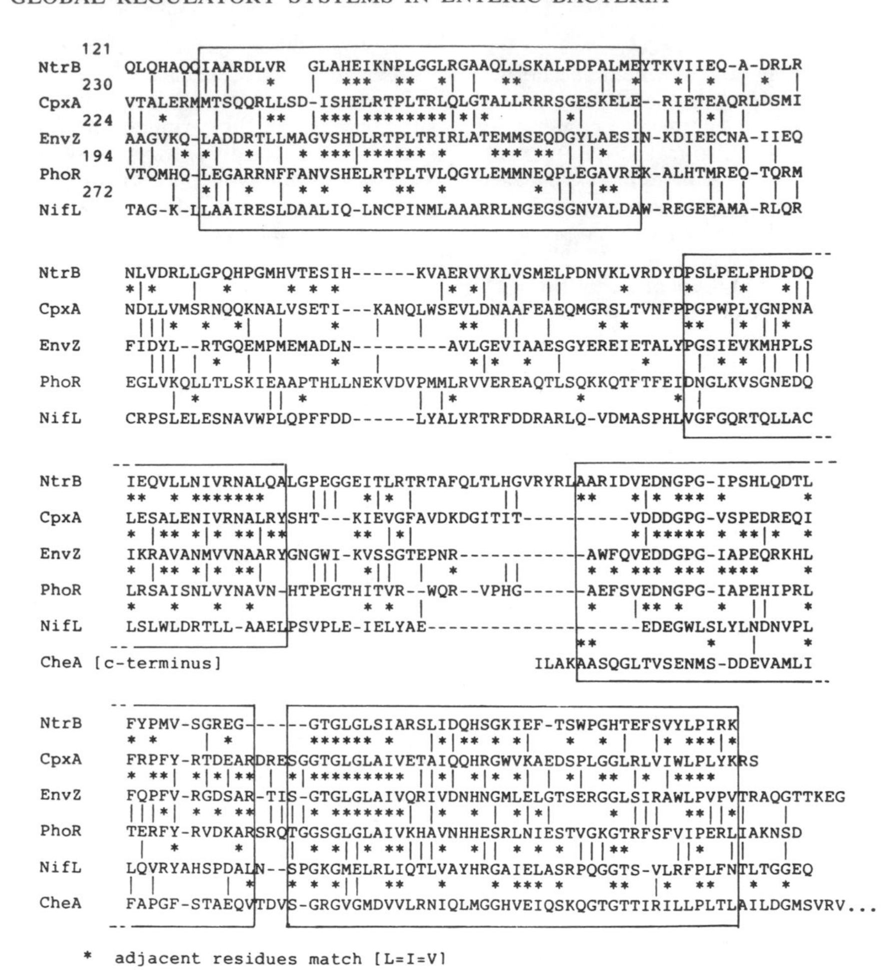

FIG. 1. Alignment of the C-terminal region of NtrB with homologous regions of other procaryotic regulatory proteins. Asterisks denote identities in adjacent sequences (residues I, V and L are considered identical); vertical bars denote identities in nonadjacent sequences. Boxes delineate regions of maximum sequence conservation.

cluding *E. coli rpoD* and *htpR* (*rpoH*) and *B. subtilis rpoD, spoIIG*, and *spoIIAC*, these show a high degree of homology. NtrA shows no significant homology at the amino acid level to any of these proteins (30), and hence the mode of action of NtrA may be quite distinct from that of other sigma factors.

The *ntrB* product has a molecular weight of 36,000 on sodium dodecyl sulfate-polyacrylamide gel electrophoresis which agrees well with that of 38,000 predicted from its DNA sequence. Comparison of the amino acid sequence of NtrB with sequences of other proteins derived from DNA data banks has identified a number of homologous regulatory proteins. The C-terminal 225 amino acids of NtrB show significant homology with the comparable regions of EnvZ, PhoR, CpxA, CheA, and (to a lesser degree) NifL (Fig. 1). This C-terminal region is separated from the N terminus by an interdomain linker of approximately 40 amino acids. A potential DNA-binding (helix-turn-helix) motif is present within the N-terminal domain (26), but its role, if any, is presently unknown.

The current model for the role of NtrB suggests that it may interact with both NtrC and P_{II}. Previous genetic analysis suggested that most *ntrB* mutations showing loss of negative function were at the N terminus (27). However, recent in vitro mutagenesis of *ntrB* has demonstrated that mutations with similar phenotypes are not localized to one part of the gene, suggesting that different regions of the protein are not necessarily functionally distinct (S. MacFarlane, unpublished data).

The *ntrC* product has a molecular weight of 53,000 on sodium dodecyl sulfate-polyacrylamide gel electrophoresis and 52,000 by DNA sequence and behaves as a dimer when purified (17, 19, 37). Analysis of the *ntrC* DNA sequence

FIG. 2. Alignment of the N-terminal domain of NtrC with homologous regions of other procaryotic regulatory regions. Symbols are as defined in the Fig. 1 legend.

has identified four potential domains (16). The N-terminal domain (of approximately 120 amino acids) shows significant homology to a number of proteins, including OmpR, PhoB, Dye, CheB, CheY, Spo0A and Spo0F, and the functional independence of this domain is demonstrated by the fact that CheY and Spo0F comprise only this region (Fig. 2). This domain is followed by a short linker sequence of approximately 20 amino acids and then two further potential domains of 240 and 65 amino acids, respectively. These last two regions are homologous to comparable regions in NifA. The C-terminal domains of both NtrC and NifA contain potential, but not identical, DNA-binding motifs. In NtrC this may play a role in negative control and recognition of the consensus sequence in the *ntrBC* and *glnA* promoters. Mutational analysis of *ntrC* is consistent with the C terminus of NtrC being responsible for negative control (27). In NifA the DNA-binding motif may recognize the consensus upstream activator sequence in *nif* promoters. The conserved central regions of NtrC and NifA may be concerned with positive control and recognition of either NtrA or RNA polymerase core enzyme, or both.

The sequence homology to other regulatory proteins detected in the C terminus of NtrB and the N terminus of NtrC is particularly interesting. In many cases these homologous regions occur in pairs of proteins which are believed, on the basis of genetic analysis, to interact functionally. This is true of EnvZ-OmpR, PhoR-PhoB, and CheA-CheB,CheY. The precise nature of this interaction is not known, but such apparent conservation of protein structure would be quite consistent with a protein-protein interaction such as has been proposed for NtrB and NtrC. The interactions of these regulatory proteins are clearly complex and are likely to take some time to unravel. However, the occurrence of what is potentially a similar relationship between pairs of regulatory proteins in a number of different systems is likely to hasten our understanding of the role of these conserved protein structures.

Another Regulatory Gene

In *K. aerogenes* a gene termed *nac* (nitrogen assimilation control), closely linked to *his*, has been shown to be essential in coupling regulation of histidase (*hut*) and glutamate dehydrogenase (*gdh*) to the Ntr system (3). Nac^- mutants fail to activate *hutUH* in response to N limitation and fail to show the characteristic repression of *gdh* under conditions where *glnA* is derepressed. Expression of *nac* is itself subject to *ntr* control (A. Macaluso and R. Bender, personal communication), and therefore it may couple the Ntr system to other nitrogen-regulated genes in a manner comparable to the coupling of *nif* to Ntr by regulation at *nifLAp*. The identification of *nac* adds a further level of complexity to the already complex Ntr regulatory circuits and emphasizes the need for studies of other systems subject to Ntr control.

Conclusions

Our understanding of the mechanism of nitrogen regulation in enteric bacteria has advanced significantly in the past 10 years, but there are still many questions to be answered. We still do not know how the N status of the cell is transmitted to the Ntr proteins, and much remains to be elucidated in the unique positive control system mediated by NtrA and NtrB/C. The function of NtrA is of particular interest, as it is clearly somewhat different from other sigma factors such as RpoD, HtpR, and SpoIIG. We do not yet understand why this control mechanism is so distinct and what selective advantage such a mechanism has over other well-characterized forms of gene regulation. The recent identification of *ntr* systems in a number of other bacterial genera has demonstrated that this somewhat atypical regulatory system is likely to be widespread in procaryotes and is probably not restricted to control of genes whose expression is linked to the nitrogen status of the cell. The foundation being laid in analysis of the *ntr* system in enteric bacteria may therefore promote a better understanding of gene regulation in many quite diverse procaryotic systems.

We thank H. Shinagawa for providing sequence data on *phoR* prior to publication.

ADDENDUM IN PROOF

NtrB has recently been shown to mediate the covalent modification (by phosphorylation) of NtrC (A. J. Ninja and B. Magasanik, Proc. Natl. Acad. Sci. USA **83**:5909–5913, 1986). Phosphorylated NtrC is the form of the protein which is active in promoting transcription from *ntr*-activatable promoters.

LITERATURE CITED

1. **Alvarez-Morales, A., R. Dixon, and M. Merrick.** 1984. Positive and negative control of the *glnAntrBC* regulon in *Klebsiella pneumoniae*. EMBO J. **3**:501–507.
2. **Ausubel, F., W. J. Buikema, C. D. Earl, J. A. Klingensmith, T. Nixon, and W. W. Seto.** 1985. Organisation and regulation of *Rhizobium meliloti* and *Parasponia bradyrhizobium* nitrogen fixation genes, p. 165–171. *In* H. Evans, P. J. Bottomley, and W. E. Newton (ed.), Nitrogen fixation research progress. Martinus Nijhoff, Boston.
3. **Bender, R. A., P. A. Snyder, R. Bueno, M. Quinto, and B. Magasanik.** 1983. Nitrogen regulation system of *Klebsiella aerogenes:* the *nac* gene. J. Bacteriol. **156**:444–446.
4. **Beynon, J., M. Cannon, V. Buchanan-Wollaston, and F. Cannon.** 1983. The *nif* promoters of *Klebsiella pneumoniae* have a characteristic primary structure. Cell **34**:665–671.
5. **Buck, M.** 1986. Deletion analysis of the *Klebsiella pneumoniae* nitrogenase promoter: importance of spacing between conserved sequences around positions −12 and −24 for activation by the *nifA* and *ntrC* (*glnG*) products. J. Bacteriol. **166**:545–551.
6. **Buck, M., H. Khan, and R. Dixon.** 1985. Site-directed mutagenesis of the *Klebsiella pneumoniae nifL* and *nifH* promoters and *in vivo* analysis of promoter activity. Nucleic Acids Res. **13**:7621–7638.
7. **Buck, M., S. Miller, M. Drummond, and R. Dixon.** 1986. Upstream activator sequences are present in the promoters of nitrogen fixation genes. Nature (London) **320**:374–378.
8. **Bueno, R., G. Pahel, and B. Magasanik.** 1985. Role of *glnB* and *glnD* gene products in regulation of the *glnALG* operon of *Escherichia coli*. J. Bacteriol. **164**:816–822.
9. **Buikema, W. J., W. W. Szeto, P. V. Lemley, W. H. Orme-Johnson, and F. M. Ausubel.** 1985. Nitrogen fixation specific regulatory genes of *Klebsiella pneumoniae* and *Rhizobium meliloti* share homology with the general nitrogen regulatory gene *ntrC* of *K. pneumoniae*. Nucleic Acids Res. **12**:4539–4555.
10. **Castano, I., and F. Bastarrachea.** 1984. *glnF-lacZ* fusions in *Escherichia coli:* studies on *glnF* expression and its chromosomal orientation. Mol. Gen. Genet. **195**:228–233.
11. **de Bruijn, F. J., and F. M. Ausubel.** 1983. The cloning and characterisation of the *glnF* (*ntrA*) gene of *Klebsiella pneumoniae*: role of *glnF* (*ntrA*) in the regulation of nitrogen fixation (*nif*) and other nitrogen assimilation genes. Mol. Gen. Genet. **192**:342–353.
12. **Dixon, R.** 1984. The genetic complexity of nitrogen fixation. J. Gen. Microbiol. **130**:2745–2755.
13. **Dixon, R.** 1984. Tandem promoters determine regulation of the *Klebsiella pneumoniae* glutamine synthetase (*glnA*) gene. Nucleic Acids Res. **12**:7811–7830.
14. **Dixon, R.** 1986. The xylABC promoter from the *Pseudomonas putida* TOL plasmid is activated by nitrogen regulatory genes in *Escherichia coli*. Mol. Gen. Genet. **203:** 129–136.
15. **Drummond, M., J. Clements, M. Merrick, and R. Dixon.** 1983. Positive control and autogenous regulation of the *nifLA* promoter in *Klebsiella pneumoniae*. Nature (London) **301**:302–307.
16. **Drummond, M., P. Whitty, and J. Wootton.** 1986. Sequence and domain relationships of *ntrC* and *nifA* from *Klebsiella pneumoniae*: homologies to other regulatory proteins. EMBO J. **5**:441–447.
17. **Ferro-Luzzi Ames, G., and K. Nikaido.** 1985. Nitrogen regulation in *Salmonella typhimurium*. Identification of an *ntrC* protein-binding site and definition of a consensus binding sequence. EMBO J. **4**:539–547.
18. **Foor, F., R. J. Cedergren, S. L. Streicher, S. G. Rhee, and B. Magasanik.** 1978. Glutamine synthetase of *Klebsiella aerogenes*: properties of *glnD* mutants lacking uridylyltransferase. J. Bacteriol. **134**:562–568.
19. **Hawkes, T., M. Merrick, and R. Dixon.** 1985. Interaction of purified NtrC protein with nitrogen regulated promoters from *Klebsiella pneumoniae*. Mol. Gen. Genet. **201**:492–498.
20. **Higgins, C., and N. H. Smith.** 1986. Messenger RNA processing, degradation and the control of gene expression, p. 179–198. *In* I. R. Booth and C. F. Higgins (ed.), Regulation of Gene Expression. Cambridge University Press, Cambridge.
21. **Hirschman, J., P.-K. Wong, K. Sei, J. Keener, and S. Kustu.** 1985. Products of nitrogen regulatory genes *ntrA* and *ntrC* of enteric bacteria activate *glnA* transcription *in vitro*: evidence that the *ntrA* product is a σ factor. Proc. Natl. Acad. Sci. USA **82**:7525–7529.
22. **Hunt, T. P., and B. Magasanik.** 1985. Transcription of *glnA* by purified *Escherichia coli* components: core RNA polymerase and the products of *glnF, glnG* and *glnL*. Proc. Natl. Acad. Sci. USA **82**:8453–8457.
23. **Krajewska-Grynkiewicz, K., and S. Kustu.** 1984. Evidence that nitrogen regulatory gene *ntrC* of *Salmonella typhimurium* is transcribed from the *glnA* promoter as well as from a separate *ntr* promoter. Mol. Gen. Genet. **193**:135–142.
24. **Kranz, R. G., and R. Haselkorn.** 1985. Characterisation of *nif* regulatory genes in *Rhodopseudomonas capsulata* using *lac* gene fusions. Gene **40**:203–215.
25. **Kustu, S., D. Burton, E. Garcia, L. McCarter, and N. McFarland.** 1979. Nitrogen control in *Salmonella*: regulation by the *glnR* and *glnF* gene products. Proc. Natl. Acad. Sci. USA **76**:4576–4580.
26. **MacFarlane, S. A., and M. Merrick.** 1985. The nucleotide sequence of the nitrogen regulation gene *ntrB* and the *glnA-ntrBC* intergenic region of *Klebsiella pneumoniae*. Nucleic Acids Res. **13**:7591–7606.
27. **MacNeil, T., G. P. Roberts, D. MacNeil, and B. Tyler.** 1982. The products of *glnL* and *glnG* are bifunctional regulatory proteins. Mol. Gen. Genet. **188**:325–333.
28. **Magasanik, B.** 1982. Genetic control of nitrogen assimilation in bacteria. Annu. Rev. Genet. **16**:135–168.
29. **McFarland, N., L. McCarter, S. Artz, and S. Kustu.** 1981. Nitrogen regulatory locus "glnR" of enteric bacteria is composed of cistrons *ntrB* and *ntrC*: identification of their protein products. Proc. Natl. Acad. Sci. USA **78**:2135–2139.
30. **Merrick, M., and J. R. Gibbins.** 1985. The nucleotide sequence of the nitrogen-regulation gene *ntrA* of *Klebsiella pneumoniae* and comparison with conserved features in bacterial RNA polymerase sigma factors. Nucleic Acids Res. **13**:7607–7620.
31. **Merrick, M. J.** 1983. Nitrogen control of the *nif* regulon in *Klebsiella pneumoniae*: involvement of the *ntrA* gene and analogies between *ntrC* and *nifA*. EMBO J. **2**:39–44.
32. **Merrick, M. J., and W. D. P. Stewart.** 1985. Studies on the regulation and function of the *Klebsiella pneumoniae ntrA* gene. Gene **35**:297–303.
33. **Nieuwkoop, A. J., S. A. Boylan, and R. A. Bender.** 1984. Regulation of *hutUH* operon expression by the catabolite gene activator protein-cyclic AMP complex in *Klebsiella aerogenes*. J. Bacteriol. **159**:934–939.
34. **Ow, D. W., Y. Xiong, Q. Gu, and S. C. Shen.** 1985. Mutational analysis of the *Klebsiella pneumoniae* nitrogenase promoter: sequences essential for positive control

by *nifA* and *ntrC* (*glnG*) products. J. Bacteriol. **161:**868–874.
35. **Pahel, G., D. Rothstein, and B. Magasanik.** 1982. Complex *glnA-glnL-glnG* operon of *Escherichia coli*. J. Bacteriol. **150:**202–213.
36. **Prival, M. J., and B. Magasanik.** 1971. Resistance to catabolite repression of histidase and proline oxidase during nitrogen-limited growth of *Klebsiella aerogenes*. J. Biol. Chem. **246:**6288–6296.
37. **Reitzer, L. J., and B. Magasanik.** 1983. Isolation of the nitrogen assimilation regulator NR_1, the product of the *glnG* gene of *Escherichia coli*. Proc. Natl. Acad. Sci. USA **80:**5554–5558.
38. **Reitzer, L. J., and B. Magasanik.** 1985. Expression of *glnA* in *Escherichia coli* is regulated at tandem promoters. Proc. Natl. Acad. Sci. USA **82:**1979–1983.
39. **Toukdarian, A., and C. Kennedy.** 1986. Regulation of nitrogen metabolism in *Azotobacter vinelandii*: isolation of *ntr* and *glnA* genes and construction of *ntr* mutants. EMBO J. **5:**399–407.
40. **Ueno-Nishio, S., S. Mango, L. J. Reitzer, and B. Magasanik.** 1984. Identification and regulation of the *glnL* operator-promoter of the complex *glnALG* operon of *Escherichia coli*. J. Bacteriol. **160:**379–384.
41. **Wei, G. R., and S. Kustu.** 1981. Glutamine auxotrophs with mutations in a nitrogen regulatory gene *ntrC*, that is near *glnA*. Mol. Gen. Genet. **183:**392–399.

Regulation of Stable RNA Transcription Initiation in Enterobacteria

ANDREW A. TRAVERS

MRC Laboratory of Molecular Biology, Cambridge, CB2 2QH, England

In *Escherichia coli* promoter utilization is globally controlled, the activity of particular classes of promoter, such as the stable RNA promoters, being regulated in concert. The stable RNA promoters direct the expression of two essential components of the translation apparatus, rRNA and tRNA, and in general promoter activity is tightly coupled to the cellular requirement for protein synthesis. Two modes of such regulation have been described. The archetype is the stringent response. When bacterial cells are starved for an amino acid, the rate of transcription of rRNA and tRNA, together with that of the mRNA species for ribosomal proteins, falls abruptly by about 10-fold. By contrast, the rate of total mRNA synthesis is only modestly reduced after starvation, and that of certain species, e.g., *his* mRNA and *trp* mRNA, is actually increased. There is now substantial evidence that a major effector of this selective regulation is the nucleotide ppGpp, which accumulates to millimolar levels after amino acid starvation (3). By contrast to stringent control, in which large and abrupt changes in the rates of transcription of stable RNA genes are observed, in growth rate control the expression of these genes varies over a wide range such that in general this expression is positively correlated with growth rate (1, 5, 12, 16). Again, however, the control is selective, the transcription of other genes such as *lac* showing a negative correlation with growth rate while that of ribosomal protein mRNA species is independent of growth rate (12, 21).

A further essential characteristic of stable RNA genes is a high maximum rate of expression, which ensures that the production of the corresponding RNA species is maintained at the high levels necessary for maximal growth. This requirement is necessarily reflected in high rates of transcription such that at the highest growth rates more than half of the actively transcribing RNA polymerase molecules are producing nascent transcripts of stable RNA genes. Under these conditions the corresponding promoters are thus among the most active in the bacterial cell.

Promoter Structure of Stable RNA Genes

A plethora of mechanisms have been proposed to explain the selective control of stable RNA expression during both stringent and growth-rate regulation. The available evidence now indicates that modulation of transcription initiation is the principal control, although contributory roles for other mechanisms have not been excluded (reviewed in reference 10).

Regulation of transcription initiation implies that the different promoters under stringent and growth-rate control must share some conserved structural features. Two such features have been identified, a guanine-plus-cytosine-rich discriminator close to the transcription start point within the region of DNA melted by RNA polymerase (17) and an "entry" region, the −35 box, which deviates from the canonical sequence of the most highly conserved base pairs. In addition, some of the most active promoters have a suboptimal 16-base-pair separation between the −35 and −10 boxes. All these features suggest that in these genes this "core promoter" region is, unlike the promoters of many bacteriophages, a weak binding site for RNA polymerase. Yet, paradoxically, these same promoters encode the capacity for high rates of initiation.

To investigate this apparent paradox and to delimit the sequences necessary for the regulation of stable RNA transcription, we chose to study the promoter of a particular *E. coli* stable RNA gene, *tyrT*, encoding a major $tRNA^{Tyr}$ species (Fig. 1). Deletion analysis showed that the high in vivo activity of the *tyrT* promoter is not an intrinsic property of the primary RNA polymerase binding site itself, but instead requires sequences located within a region spanning 40 to at least 98 base pairs upstream of the transcription start point (Fig. 2). This upstream element is believed to positively activate transcription at the normal start point. Deletions lacking this element but retaining the primary polymerase binding site have only a modest transcriptional activity in vivo. This activity is comparable to that of the fully induced *lac* promoter and is thus substantially less than that of strong phage promoters such as λ p_L. This result thus confirms the intrinsic weakness of the *tyrT* polymerase binding site.

Although upstream sequences are required for the optimal activity of the *tyrT* promoters, their deletion does not remove the regulatory response to amino acid starvation in vivo. Stringent reg-

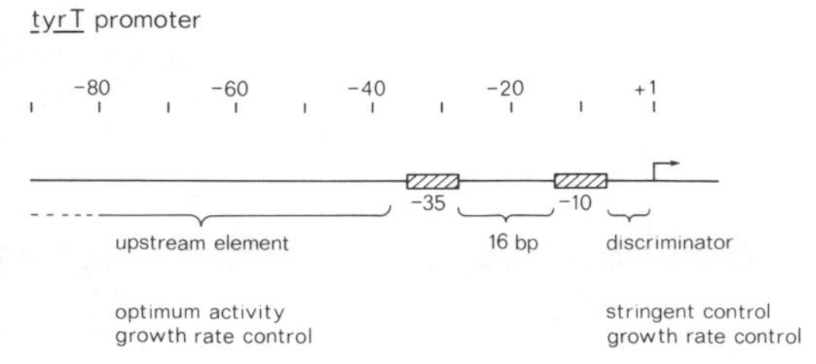

FIG. 1. Structure and function of the *E. coli tyrT* promoter. +1, Transcription start point.

ulation is thus a property of the core promoter. This regulation is abolished by a 4-base-pair mutation in the discriminator region which changes the sequence from GCGCCCCGCT to GCGTTAAGCT (where the underline indicates the transcription start point). Transcription from the mutant promoter, like that from the *bla* gene, is unresponsive to amino acid starvation (Fig. 3). This result shows, first, that the discriminator region is necessary for stringent regulation, although it may not be sufficient, and, second, that since the mutation lies outside the transcribed region this regulation does in fact occur in vivo at the level of transcription initiation.

By contrast to stringent regulation, the sequences in the *tyrT* promoter that determine growth-rate control are located in at least two regions (20). A major element that determines this mode of regulation is again the discriminator. The same mutation that abolishes stringent control also alters the growth-rate dependence of *tyrT* expression such that the correlation with growth rate is changed from positive to negative (Fig. 4). The simplest interpretation of this result is that the mechanisms of growth-rate control and stringent control share a common target molecule which interacts with the discriminator region of the promoter. A second region of the *tyrT* promoter which influences the growth-rate dependence is the upstream element. Deletion of sequences in this element results in *tyrT* transcription having a higher positive correlation with growth rate. In other words, the weaker the promoter, the lower its relative activity is at low growth rates.

The *tyrT* promoter is, of course, a single example of a stable RNA promoter and may well possess regulatory characteristics that are peculiar to itself. However, limited studies on other stable RNA promoters suggest that certain major features that determine the transcriptional regulation of the *tyrT* promoter are common to these other promoters. Thus, sequences upstream of the primary polymerase binding site are also necessary for the optimal utilization of both the *E. coli rrnB* P_1 promoter (6) and the *Salmonella typhimurium hisR* promoter (2). Further, growth rate control of transcription from *rrnB* P_1 depends, at least in part, on sequences between −26 and −48, a region that overlaps sequences determining this mode of regulation of the *tyrT* promoter. However, as yet, no further in vivo analysis of the sequence determinants for stringent control has been reported.

RNA Polymerase Interactions at Stable RNA Promoters

What are the ultimate target molecules for stringent and growth-rate control? A major intracellular signal for stringent control is the nucleotide ppGpp, whose synthesis is triggered

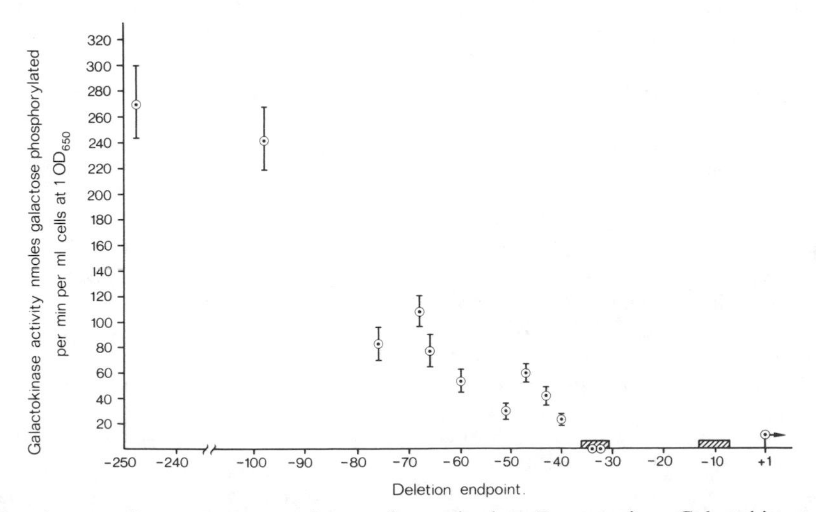

FIG. 2. Requirement for an upstream element for optimal *tyrT* expression. Galactokinase activities of different *tyrTp-galK* fusions are shown in relation to the amount of *tyrT* sequence present upstream of the transcription start point. OD_{650}, Optical density at 650 nm. (Taken from reference 8.)

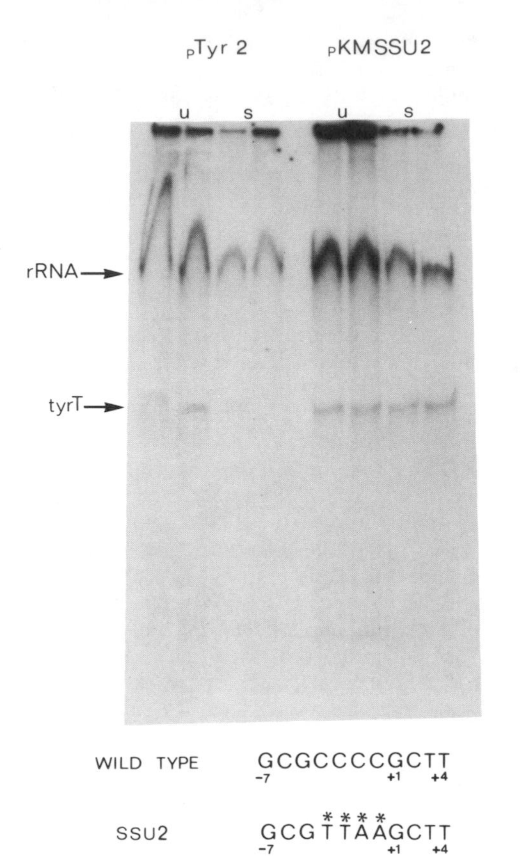

FIG. 3. Stringent control of transcription from the wild-type *tyrT* and a discriminator mutant (SSU2). Transcript levels were measured in normal medium (u) and after starvation for amino acid (s) as described previously.

by the binding of uncharged tRNA to ribosomes. In a highly purified in vitro transcription system, ppGpp selectively inhibits RNA chain initiation at stable RNA promoters while not reducing the activity of other promoters, such as *lac*, that are not stringently regulated in vivo (7). This selective inhibition is abolished by mutations in the discriminator regions of both the *tyrT* and *tufB* promoters (11, 18). The in vivo phenotype of such mutations can thus be reproduced in vitro in a simple system containing RNA polymerase and DNA as the sole macromolecular components. Since ppGpp binds to RNA polymerase (4), this result strongly suggests that the ultimate target of the stringent control mechanism is RNA polymerase and that ppGpp exerts a selective effect on transcription by directly altering the interaction of RNA polymerase with promoters.

Since the growth-rate regulation of *tyrT* transcription is also altered by a discriminator mutation, it follows, on this argument, that RNA polymerase is also a target for growth-rate control. However, we know that the response of stable RNA promoters to growth rate is modulated by sequences upstream of the primary RNA polymerase binding site. In the *tyrT* promoter and other stable RNA promoters, this upstream region contains sequences homologous to a normal RNA polymerase binding site. The precise position and orientation of these sequences is variable, but in the *tyrT* promoter one such site is found in the opposite orientation to the primary site with putative −10 and −35 sequences, TAATTT and TTGAGA, located, respectively, at 67 and 44 base pairs upstream of the transcription start point. Two lines of evidence suggest that this upstream region can interact directly with RNA polymerase. First, upstream sequences are protected from DNase I digestion by this enzyme (19) and, second, a double mutation at positions −44 and −45, which changes the sequence TTGAGA to CCGAGA, is a strong down mutation both in vivo and in vitro (20). In addition, a 3-base-pair deletion at position −70 in the *hisR* promoter

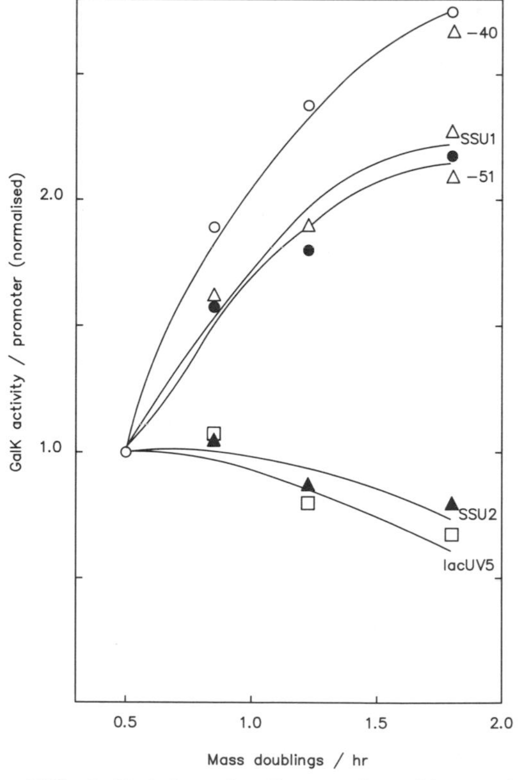

FIG. 4. Variation of *tyrT* expression with growth rate. Galactokinase activities from various *tyrTp-gal* fusions and a *lacUV5-galK* fusion were determined in different media, and the data were normalized to the lowest growth as described previously (20).

also reduces transcription in vivo and in vitro (2). These observations thus suggest that RNA polymerase itself may act as a positive activator of stable RNA transcription by binding to an upstream promoter element and increasing the rate of initiation by binding cooperatively through protein-protein contacts to a second polymerase molecule bound at the primary promoter site. On this model the effect of upstream sequences on growth-rate control can again be explained by an alteration of the interactions of RNA polymerase with its binding sites. Clearly, however, such a mechanism does not rule out the possibility that other regulatory factors, as yet uncharacterized, might interact with the upstream element.

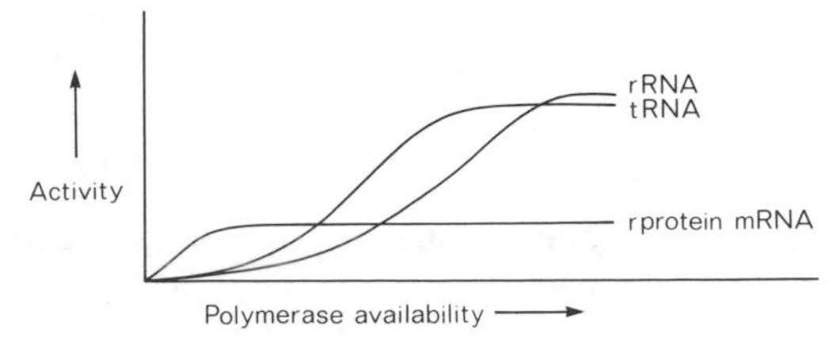

FIG. 5. Variation of rate of expression of different transcription units with RNA polymerase availability. Individual rRNA, tRNA, and ribosomal protein operons would be expected to differ in their response to variations in polymerase availability.

Model for Transcriptional Regulation In Vivo

I have argued that RNA polymerase is one, possibly the major, target molecule for the stringent and growth-rate control mechanisms regulating stable RNA transcription initiation. Although the structural heterogeneity of stable RNA promoters suggests that their regulation may be achieved by mechanisms that differ in detail from promoter to promoter, it seems probable that they all depend on interactions that are highly sensitive to the availability of RNA polymerase. Availability is defined as a quantity which is a function of both the absolute concentration of the free enzyme and its binding affinity. This model thus implies that in the bacterial cell the transcription of stable RNA genes is limited by RNA polymerase under most growth conditions, with the possible exception of those favoring the highest growth rates. This view of transcriptional regulation is clearly distinct from the ribosome feedback model proposed by Nomura et al. (13), which requires that RNA polymerase be in functional excess under all growth conditions.

When applied to stable RNA transcription, the model requires that the polymerase availability at stable RNA promoters decreases both on amino acid starvation and at reduced growth rates (Fig. 5). This is consistent with the observation that the activity of weaker stable RNA promoters shows a greater positive correlation with growth rate. However, the *lac* and *his* promoters show the opposite pattern of transcriptional regulation. The transcription of both increases on amino acid starvation, while mutations which increase the activity of these promoters decrease the negative correlation of activity with growth rate (21). We know also that in vitro ppGpp stimulates transcription from the *his* promoter in a manner which is consistent with a direct influence on RNA polymerase (14, 15). To reconcile the reciprocal patterns of transcriptional regulation of stable RNA promoters on the one hand and those including the *his* promoter on the other, the simplest postulate is that in vivo RNA polymerase exists as an equilibrium mixture of two forms, one form which has a high affinity for stable RNA promoters and a low affinity for *his*-type promoters and a second form with the converse promoter preference. A predominance of the first form at high growth rates and of the second at low growth rates and during amino acid starvation is sufficient to explain the global regulation of transcription from both classes of promoter. Furthermore, by changing the proportions of the two forms of RNA polymerase, the rate of transcription initiation at polymerase-limited promoters would be varied over a wide range (Fig. 6).

If the availability of a particular form of RNA polymerase limited the expression of stable RNA genes and hence limited the growth of the cell, any significant increase in the number of binding sites for this form and not for any other form in the bacterial cell should result in a reduction in growth. This is precisely what is observed. When cells containing multiple copies of a plasmid bearing the wild-type *tyrT* promoter are grown under conditions of nutritional limita-

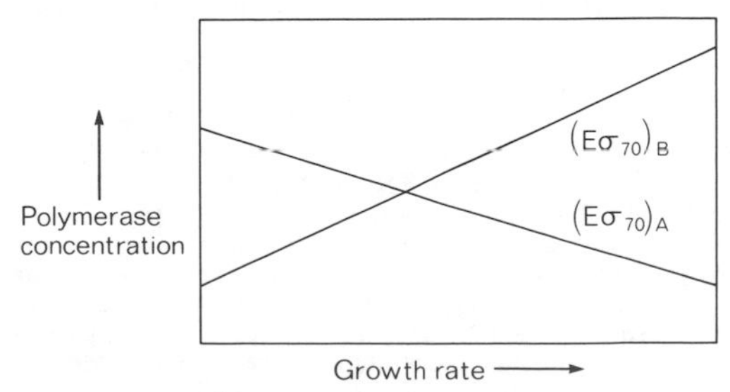

FIG. 6. Model for global transcriptional regulation as a function of growth rate. Changes in the concentration of two forms of RNA polymerase holoenzyme, $(E\sigma_{70})_A$ with a preference for *his*-type promoters and $(E\sigma_{70})_B$ with a preference for stable RNA promoters, are shown.

tion, a substantial reduction in growth rate occurs relative to untransformed cells (20). This growth inhibition can be relieved both by a mutation in the upstream region which reduces promoter activity and by the same mutation in the discriminator region that alters stringent and growth-rate control of *tyrT* transcription and concomitantly increases promoter activity. The growth inhibition is thus not correlated with promoter activity but is related to promoter sequences known to interact with RNA polymerase. This result is completely consistent with the view that the population of RNA polymerase molecules in vivo is functionally partitioned between those molecules which interact with the wild-type *tyrT* promoter and those that interact with the discriminator mutant. Sequestration of the former by the plasmid-borne promoters results in growth inhibition; sequestration of the latter does not. Since the regulation of the *tyrT* discriminator mutant parallels that of *lac* with regard to both stringent control and growth rate, it seems probable that these two promoters preferentially interact with the same form of RNA polymerase.

What is the biological logic of this type of regulatory system? At high growth rates transcription of the stable RNA genes utilizes a major proportion of the transcriptional capacity of the cell. Direct regulation of RNA polymerase provides a mechanism both for tuning the initiation rate of stable RNA promoters and for switching the available RNA polymerase molecules for one class of promoter to another. Such direct regulation also implies that RNA polymerase must sense the physiological state of the cell by interaction with molecules, such as ppGpp, whose concentration is related to variations in the functional efficiency of essential metabolic processes.

This review summarizes work carried out at Cambridge over the past 5 years. I particularly acknowledge Paul Debenham, who initiated the study of *tyrT* transcription, and also Angus Lamond, Hilary Mace, and John Weeks, who were all involved in the later experiments.

LITERATURE CITED

1. **Berman, M. L., and J. Beckwith.** 1979. Fusions of the *lac* operon to the transfer RNA gene *tyrT* of *Escherichia coli*. J. Mol. Biol. **130:**285–301.
2. **Bossi, L., and D. M. Smith.** 1984. Conformational change in the DNA associated with an unusual promoter mutation in a tRNA operon of *Salmonella*. Cell **39:**643–652.
3. **Cashel, M., and J. Gallant.** 1969. Two compounds implicated in the function of the RC gene of *Escherichia coli*. Nature (London) **221:**838–841.
4. **Cashel, M., E. Hamel, P. Shapshak, and M. Bouquet.** 1976. Interaction of ppGpp structural analogues with RNA polymerase, p. 279–290. *In* N. O. Kjeldgaard and O. Maaløe (ed.), Control of ribosome synthesis. Munksgaard, Copenhagen.
5. **Duester, G., R. M. Elford, and W. M. Holmes.** 1982. Fusion of the *Escherichia coli* $tRNA_1^{Leu}$ promoter to the *galK* gene: analysis of sequences necessary for the growth rate dependent regulation. Cell **30:**855–864.
6. **Gourse, R. L., H. A. de Boer, and M. Nomura.** 1986. DNA determinants of rRNA synthesis in E. coli: growth rate dependent regulation, feedback inhibition, upstream activation, antitermination. Cell **44:**197–205.
7. **Kajitani, M., and A. Ishihama.** 1984. Promoter selectivity of *Escherichia coli* RNA polymerase. Differential stringent control of the multiple promoters from ribosomal RNA and protein operons. J. Biol. Chem. **259:**1951–1957.
8. **Lamond, A. I., and A. A. Travers.** 1983. Requirement for an upstream element for optimal transcription of a bacterial tRNA gene. Nature (London) **305:**248–250.
9. **Lamond, A. I., and A. A. Travers.** 1985. Genetically separable functional elements mediate the optimal expression and stringent regulation of a bacterial tRNA gene. Cell **40:** 319–326.
10. **Lamond, A. I., and A. A. Travers.** 1985. Stringent control of bacterial transcription. Cell **41:**6–8.
11. **Mizushima-Sugano, J., and Y. Kaziro.** 1985. Regulation of the expression of the *tufB* operon: DNA sequences directly involved in the stringent control. EMBO J. **4:**1053–1058.
12. **Miura, A., J. H. Krueger, S. Itoh, H. A. de Boer, and M. Nomura.** 1981. Growth rate dependent regulation of ribosome synthesis in *E. coli*: expression of the *lacZ* and *galK* genes fused to ribosomal promoters. Cell **25:**773–782.
13. **Nomura, M., R. Gourse, and G. Baughman.** 1984. Regulation of the synthesis of ribosomes and ribosomal components. Annu. Rev. Biochem. **53:**75–117.
14. **Primakoff, P., and S. W. Artz.** 1979. Positive control of *lac* operon expression *in vitro* by guanosine 5′-diphosphate 3′-diphosphate. Proc. Natl. Acad. Sci. USA **76:** 1726–1730.
15. **Stephens, J. C., S. W. Artz, and B. N. Ames.** 1975. Guanosine 5′-diphosphate 3′-diphosphate (ppGpp): positive effector for histidine operon transcription and general signal for amino-acid deficiency. Proc. Natl. Acad. Sci. USA **72:**4389–4393.
16. **Sarmientos, P., and M. Cashel.** 1983. Carbon starvation and growth rate-dependent regulation of the *Escherichia coli* ribosomal RNA promoters: differential control of dual promoters. Proc. Natl. Acad. Sci. USA **80:**7010–7013.
17. **Travers, A. A.** 1980. Promoter sequence for the stringent control of bacterial ribonucleic acid synthesis. J. Bacteriol. **141:**973–976.
18. **Travers, A. A.** 1980. A $tRNA^{Tyr}$ promoter with an altered *in vitro* response to ppGpp. J. Mol. Biol. **141:**91–97.
19. **Travers, A. A., A. I. Lamond, H. A. F. Mace, and M. L. Berman.** 1983. RNA polymerase interactions with the upstream region of the E. coli tyrT promoter. Cell **35:**265–273.
20. **Travers, A. A., A. I. Lamond, and J. R. Weeks.** 1986. Alteration of the growth-rate-dependent regulation of *Escherichia coli tyrT* expression by promoter mutations. J. Mol. Biol. **189:**251–255.
21. **Wannar, B. L., R. Kodaira, and F. C. Neidhardt.** 1977. Physiological regulation of a decontrolled *lac* operon. J. Bacteriol. **130:**212–222.

Global Analysis of Phosphorylated Metabolites to Search for Alarmones

BARRY R. BOCHNER†

Department of Fermentation Research and Process Development, Genentech, Inc., South San Francisco, California 94080

Proposed Functions for Alarmones in Microbial Cells

Cells of multicellular, higher organisms live in a fairly constant, regulated environment. These cells have evolved to perform specific and highly specialized functions while being relieved of the metabolic burden of dealing with highly variable environments. In so doing, cells of higher organisms have become fragile and highly interdependent. In contrast, microbial cells are typically single-celled, independent life forms which live in environments that are changing constantly. Just by their own growth, these cells change their environment as they assimilate nutrients, expell metabolic by-products, and generate heat. Extracellular events cause even more sudden and drastic changes. Thus, microbial cells must have evolved efficient and effective means for regulating their growth and coordinating their metabolic processes to deal with a constantly changing environment.

The cell is composed of numerous metabolic processes involved in catabolism, biosynthesis, macromolecular assembly, and many specialized physiological functions. The sum total of all of these processes is referred to as global metabolism. The global metabolism of a well-studied microorganism such as *Escherichia coli* is usually thought of and depicted as an extremely complex network of lines and arrows. Although this model is too complex to work with, it is useful in illustrating the problems that *E. coli* must handle in achieving proper coordination of this complex network in the face of constant change.

Known regulatory mechanisms can account for smooth and coordinated control of most functionally "coupled" metabolic pathways. To illustrate this, let us consider an *E. coli* cell that is growing in metabolic balance with all of its pathways functioning at coordinated, appropriate rates. The cell suddenly exhausts all of the histidine from its environment, and protein synthesis must slow as a result of an amino acid insufficiency. To get back into metabolic balance, the appropriate cellular response is to increase the histidine supply (release feedback inhibition of biosynthesis, derepress synthesis of histidine biosynthetic enzymes, increase high-affinity histidine transport, increase proteolysis of histidine-containing peptides) while slowing down the rest of cellular metabolism. The coupled pathways involving biosynthesis of the other 19 amino acids can sense the problem quickly and respond appropriately. When histidine is in short supply, the ribosomes must work more slowly, causing the pools of the other 19 amino acids to swell and turn down their own rates of synthesis by feedback inhibition, attenuation, and repression.

However, all other metabolic pathways that are not functionally coupled to protein synthesis also need to be altered to the histidine shortage so that they can respond appropriately. In 1975, Stephens et al. (27) coined the term alarmone to generically describe a set of signal molecules that would serve this perceived need of alerting the global metabolism of the cell to the onset of a metabolic imbalance or stress. A particular alarmone would signal the onset of a particular stress, and every metabolic system could evolve to respond to or to ignore an alarmone, as was physiologically appropriate. The postulated set of alarmones could then overcome the problems of coordinating this complex, "weakly coupled" metabolic system.

A prototypical alarmone is ppGpp (guanosine 5′-diphosphate 3′-diphosphate). In our example of histidine insufficiency, this unusual nucleotide is synthesized from the GTP normally used to form peptide bonds, when the ribosome is stalled, waiting for histidyl-tRNA. As postulated, it turns up the expression of the histidine biosynthetic operon (27) and turns down many other cellular processes, most notably stable RNA synthesis (10). The other previously discovered example of what Ames would term an alarmone is cyclic AMP. In *E. coli* adenyl cyclase activity is activated when enzyme III-glucose of the phosphoenolpyruvate:sugar phosphotransferase system is idled as a result of insufficient carbohydrate supply (22). The pleiotropic effects of cyclic AMP on gene expression in *E. coli* are well documented (28).

† Present address: Biolog, Inc., 3447 Investment Blvd., Hayward, CA 94545.

Alarmones may have functions and utility which go beyond keeping the cell in metabolic balance. For example, alarmones could be involved in triggering developmental processes, such as sporulation, or secondary metabolic processes, such as antibiotic biosynthesis. Dormant viruses and other mobile genetic elements could also sense the levels of cellular alarmones to help them decide when to activate their own genetic programs. In signaling the beginning, the end, and perhaps the severity of a stress, the alarmone may also interact with integrative control processes which help the cell decide whether to wait out a stress with short-term solutions or activate a genetic program aimed at long-term adaptation.

Searching for More Alarmones by Global Analysis of Phosphate Metabolism

Because it seemed unlikely that ppGpp and cyclic AMP were the only alarmones in *E. coli*, I devised a systematic procedure to search for more. Initially, I restricted my search to include only those cellular metabolites that are acid soluble and that contain phosphate. Although there is no reason to presume that alarmones must be phosphorylated small molecules, the focus was on this area of metabolism for three reasons: (i) the two prototypical alarmones are phosphorylated metabolites, (ii) phosphorylated metabolites can be labeled with high-specific-activity $^{32}P_i$ and detected at very low levels by autoradiography, and (iii) the techniques of Randerath and Randerath (23) provided an initial thin-layer chromatographic technology for resolving and measuring pools of phosphorylated metabolites.

After several years of work, the technology for globally surveying the pools of phosphorylated metabolites has been thoroughly developed. The improved thin-layer chromatographic methods (3) were also aided by the development of efficient techniques for precipitation of P_i (2) and detection of chromatographic standards (6). The end result is a method that allows one to resolve and measure the pools of all the principal phosphorylated metabolites and observe changes in these pools over time as cells are exposed to a particular stress. The technique is designed for true global analysis in that the chromatographic conditions and subsequent chemical and enzymatic analyses ensure detection of all acid-soluble phosphorylated metabolites, so long as their intracellular level exceeds approximately 5 μM.

As an essential starting point in searching for "unusual" metabolites that might be alarmones, I identified and cataloged levels of phosphorylated metabolites in cells undergoing balanced growth. This work has been completed for all cellular nucleotides (3) and is quite far along for the nonnucleotide metabolites as well (B. Bochner, unpublished data). By sampling and analyzing pool changes in a time-course experiment, we can look for the appearance of new spots, potential alarmones, whose synthesis correlates with the onset of a particular physiological stress. This global, dynamic analysis of cellular phosphate metabolism has other uses also, which are described briefly at the end of this article.

Discovery of Two New Putative Alarmones

The application of these techniques has led to the discovery of two unusual nucleotides whose synthesis correlates with the onset of a specific stress. The proposed biological function of these nucleotides remains to be proved or disproved. They will be referred to as putative alarmones.

ZTP (5-amino 4-imidazole carboxamide ribonucleoside 5′-triphosphate) is synthesized specifically in bacterial cells deficient in C_1 or folic acid metabolism (4). It accumulates rapidly to levels of nearly 400 μM. In vitro and presumably in vivo it is synthesized by a backwards reaction of the enzyme phosphoribosylpyrophosphate synthetase (25). ZTP has now been found in both procaryotes and eucaryotes. It has been found in C_1-folate-deficient gram-negative (*E. coli, Salmonella typhimurium* [4]) and gram-positive (*Bacillus subtilis* [B. Bochner, unpublished data]) bacteria, fungi (24), and human cells (26, 31).

AppppA (diadenosine 5′,5‴-P^1,P^4-tetraphosphate) and a family of adenylylated nucleotides are synthesized specifically in bacterial cells during oxidation stress (5, 19). Pools rise rapidly to levels that can exceed 300 μM when cells are exposed to a wide variety of oxidizing chemicals, including quinones, peroxides, and heavy metals. In vitro and presumably in vivo the adenylylated nucleotides are synthesized by a backwards reaction of aminoacyl tRNA synthetases (30). Several laboratories have now confirmed the accumulation of AppppA during oxidation stress in eucaryotic cells, including fungi (14), *Drosophila* cells (9), and human cells (1).

Our group has also found AppppA synthesis in heat-shocked bacteria (20), and there is strong evidence of overlap between the effects of and the cellular response to heat shock and oxidation stress (11, 20). Increased levels of adenylylated nucleotides during heat shock have also been found in *Saccharomyces* sp. (13), *Xenopus* (15), and *Drosophila* (9) cells. The very recent results of VanBogelen et al. (29), in which nucleotide pool changes and stress protein induction were followed simultaneously, were interpreted as being inconsistent with a simple role for AppppA in triggering the heat-shock response, but

TABLE 1. Conditions which unbalance metabolism and which might trigger alarmone synthesis

Stress	Metabolites that accumulate
Insufficient synthesis of macromolecules	
Protein	ppGpp, pppGpp
DNA	All deoxyribonucleotides, all mono- and diphosphoribonucleotides
RNA	All ribonucleoside triphosphates
Membrane	Not examined
Wall	Not examined
Insufficient supply of nutrients	
Carbohydrate	Cyclic AMP, ppGpp, pppGpp
C_1-folate	ZTP, ZMP, sZMP (4)
Ammonia	ppGpp pppGpp, NaAD[a], XMP
Phosphate	Not examined
Sulfate	Not examined
Electron acceptors	Not examined
K^+, Mg^{2+}, Fe^{2+}, etc.	Not examined
Environmental stress	
Oxidation	AppppA and other adenylylated nucleotides (5)
Heat shock	AppppA and other adenylated nucleotides (20)
Cold shock	NADP, all mono- and diphosphoribonucleotides
High osmolarity	NADP, GDPMan[b], ADPMtl[c], spot S3 (3)
Low osmolarity	Not examined
High pH	Not examined
Low pH	Not examined

[a] Nicotinic acid-adenosine dinucleotide (i.e., deamidated NAD).
[b] GDP-mannose.
[c] ADP-mannitol.

consistent with a role in triggering the oxidation-stress response.

The existence of enzymes that specifically hydrolyze dinucleoside polyphosphates suggests that the levels of these nucleotides are regulated and further supports the idea that they have a biological function. These hydrolases have now been found in bacterial, fungal, plant, and animal cells (reviewed in reference 12). The enzyme from *E. coli* has been purified and characterized (16), and its coding sequence has been cloned (21). The strategy of using a cloned gene, overproducing AppppA hydrolase, to manipulate pool levels in vivo (21) should permit testing for a role of AppppA in regulation of gene expression during oxidation stress. Microinjection of AppppA (15) is another promising approach being used to study effects of AppppA in eucaryotic cells. Present results, however, are inconclusive.

Perspectives on the Occurrence of Unusual Phosphorylated Metabolites

Table 1 summarizes the current status of our survey of the occurrence of unusual phosphorylated metabolites in *E. coli* or *S. typhimurium* under stress conditions that we thought might lead to synthesis of an alarmone. Although our survey is not complete, we have examined many of the stresses during which a defined regulon of proteins is known to be induced, suggestive of alarmone involvement.

An example of our search for a DNA alarmone (i.e., the inducer of the SOS regulon) is shown in Fig. 1. In the example shown, streptonigrin was used to induce the SOS systems, but similar results have been obtained with bleomycin and mitomycin C. No substantial change was observed in the nucleotide pools of these cells until approximately 15 min after exposure to the SOS inducing agent; then there was a dramatic rise in the pools of all deoxyribonucleotides (Fig. 1C). Mono- and diphosphoribonucleoside levels also increased. No unusual nucleotides, such as the dinucleotides described by Irbe et al. (17), were detected.

We have also examined pool levels during high-osmolarity shock and during ammonia starvation. In the former instance some increase was seen in the pools of NADP and several nucleoside-diphosphate sugars. In the latter instance, ppGpp, nicotinic acid-adenosine dinucleotide (the deamidated precursor of NAD), and XMP (the deaminated precursor of GMP) accumulated.

Other Uses for Global Analysis of Phosphate Metabolism

Although this series of analytical methodologies was developed in the hope of discovering new alarmones, there are many other interesting applications of global analysis of phosphate metabolism. These applications have been listed in a previous publication (3), but I will illustrate three with examples.

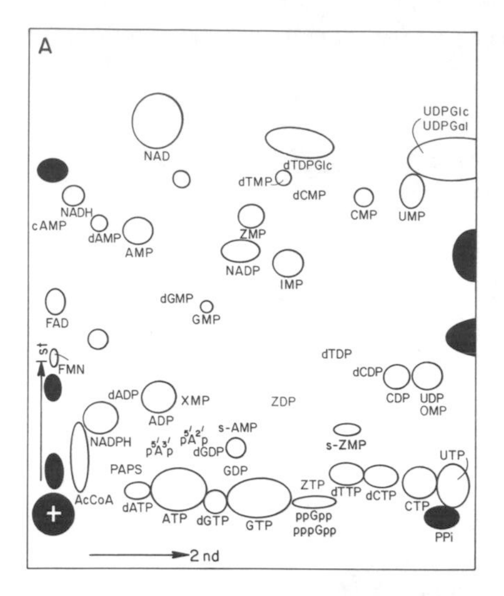

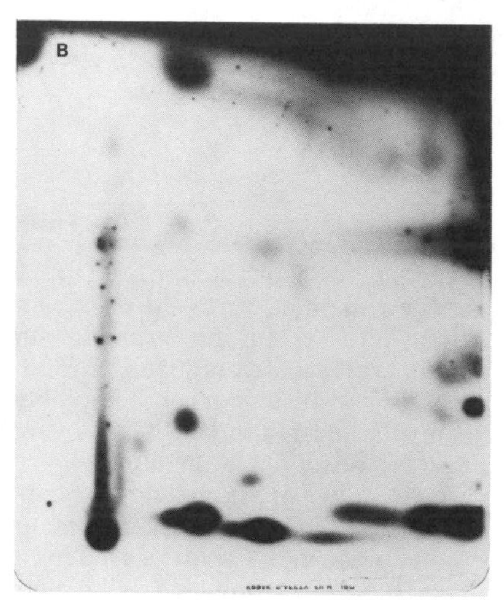

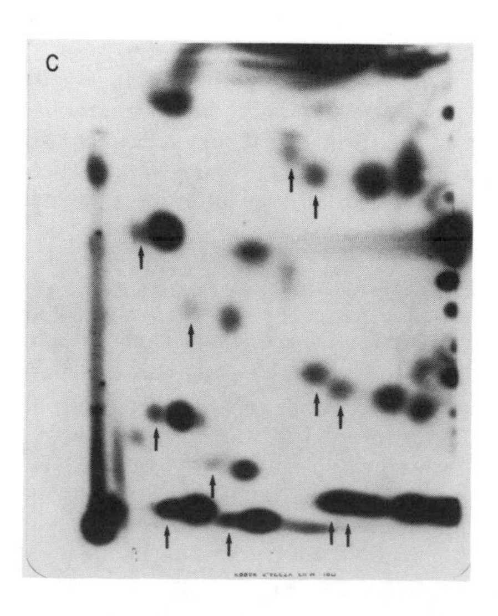

FIG. 1. Autoradiograms exposed from two-dimensional thin-layer chromatograms developed with solvent Ta followed by solvent Sb, as described previously (see Fig. 3A in reference 3). (A) Diagrammatic representation showing the identities of the metabolites resolved. The black spots at the periphery do not absorb to charcoal and therefore are presumed to be nonnucleotides. (B and C) Autoradiograms showing $^{32}P_i$-labeled pools of *S. typhimurium* LT7 before (B) and 50 min after (C) the addition of 10 μg of streptonigrin per ml. The spots indicated with arrows are deoxyribonucleotides that accumulated after streptonigrin was added. Their identities are indicated in A.

(i) Surveying the catalytic activity of a protein against a globally representative set of phosphorylated metabolites. The *E. coli* DnaK protein is an important regulatory protein involved in stress response and DNA replication. The enzyme was shown to have weak activity as an ATPase. However, incubating a $^{32}P_i$-labeled *E. coli* extract with purified DnaK protein and analyzing the spots that disappeared from the autoradiogram showed that the protein is actually a nonspecific 5′-nucleotidase with a preference for tri- and monophosphates over diphosphates and for purines over pyrimidines (7). By a similar analysis, the unactivated, oncogenic *ras* protein was shown to be quite specific for GTP and dGTP (B. Bochner, unpublished data).

(ii) Following the in vivo metabolism of a chemical that is phosphorylated. Purine is an analog of adenine. The ribonucleoside 5′-triphosphate of purine inhibits *E. coli* RNA polymerase, but is not a substrate that is incorporated into RNA (8). In the second dimension of our chromatographic system, the behavior of a nucleo-base or a nucleoside is predictive of the behavior of the corresponding phosphorylated nucleotide (3). This allows us to predict that, if purine is metabolized to a nucleoside mono-, di-, and triphosphate, we should see the appearance of a series of spots running between the guanine and the thymine series on the autoradiogram. This expected result was in fact seen, as shown in Fig. 2B. After a 50-min exposure to purine, it was clear that the cells had phosphorylated it to a ribonucleoside triphosphate and a deoxyribonucleoside triphosphate. In addition to the phosphorylated purine nucleotides, several other new "shadow" spots appeared. This may indicate that purine can partially replace adenine in adenine-containing metabolites (e.g., NAD, NADPH).

(iii) Examining the specificity in vivo of an antimetabolite. Guanazole is thought to act as a one-electron acceptor and specifically inhibit the B2 subunit of the enzyme ribonucleoside diphosphate reductase (18). This reductase is the key step in biosynthesis of deoxyribonucleotides. By

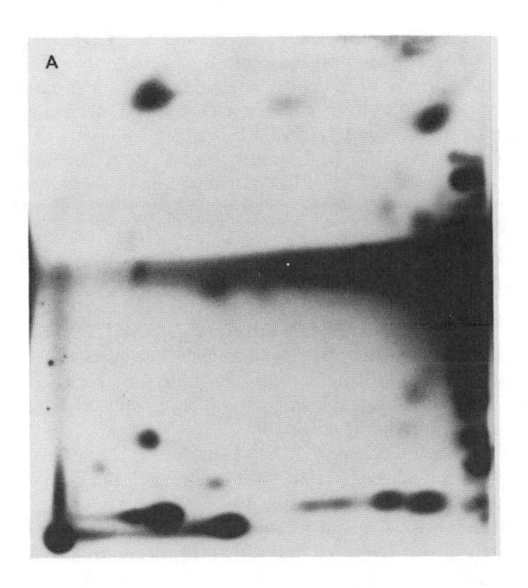

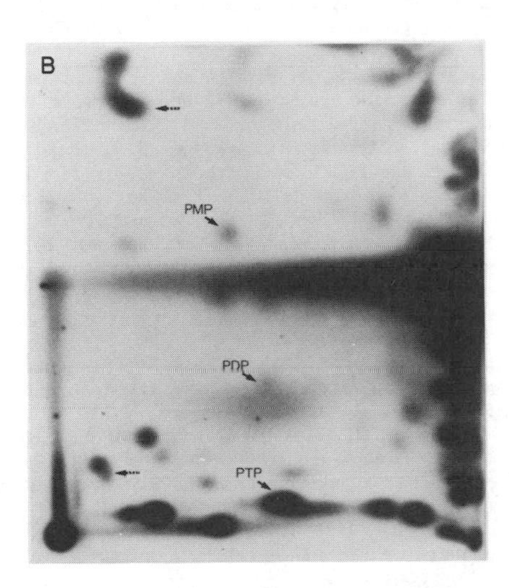

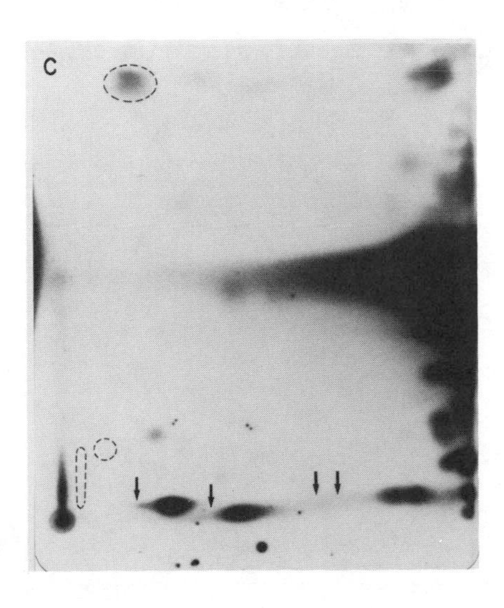

FIG. 2. Autoradiograms showing $^{32}P_i$-labeled pools of *S. typhimurium* LT2 before any additions (A), 50 min after the addition of 400 μg of purine per ml (B), or 5 min after the addition of 20 mg of guanazole per ml (C). In B the spots indicated with solid arrows are believed to be purine ribonucleoside mono-, di-, and triphosphates. The spots indicated with dashed arrows are "shadow" spots of NAD (top) and NADPH (bottom), presumably formed by incorporation of purine in place of adenine. In C the missing spots indicated with arrows are the four deoxyribonucleoside triphosphates whose synthesis is inhibited by guanazole. The dashed circles and ovals highlight the decreased pools of (top to bottom) NAD, NADPH, and acetyl coenzyme A. Refer to Fig. 1 and its legend for additional details.

looking at pool changes in *S. typhimurium* after a 5-min exposure to guanazole, we were able to confirm the predicted inhibition of deoxyribonucleotide synthesis by the disappearance of the deoxynucleoside triphosphate spots (Fig. 2C). However, guanazole also appeared to have secondary effects on cellular metabolism since the pools of NAD, NADPH, and acetyl coenzyme A also decreased. Presumably, its oxidizing capability also interfered with cellular respiration.

Conclusions

A technique has been developed which permits a global and comprehensive analysis of cellular pools of acid-soluble, phosphate-containing metabolites. An analysis can be performed of kinetically sampled cells to visualize pool changes throughout the time course of a stress. Thus far, the technique has been utilized principally to search for new alarmones in gram-negative bacteria. Two putative alarmones, ZTP and AppppA, have been discovered. However, our survey results also suggest that the number of previously undiscovered phosphorylated metabolites in these bacteria is likely to be rather small. The same global analysis techniques are also useful for other metabolic and enzymatic studies, and several of these alternative uses have been demonstrated.

This work was performed in the laboratory of Bruce Ames at the University of California, Berkeley. I thank Bruce Ames for his years of support and encouragement. I also thank Genentech, Inc., for continuing support.

LITERATURE CITED

1. **Baker, J. C., and M. K. Jacobson.** 1986. Alteration of adenyl dinucleotide metabolism by environmental stress. Proc. Natl. Acad. Sci. USA **83:**2350–2352.
2. **Bochner, B. R., and B. N. Ames.** 1982. Selective precipitation of orthophosphate from mixtures containing labile phosphorylated metabolites. Anal. Biochem. **122:**100–107.
3. **Bochner, B. R., and B. N. Ames.** 1982. Complete analysis of cellular nucleotides by two-dimensional thin layer chromatography. J. Biol. Chem. **257:**9759–9769.

4. **Bochner, B. R., and B. N. Ames.** 1982. ZTP (5-amino 4-imidazole carboxamide riboside 5′-triphosphate): a proposed alarmone for 10-formyl-tetrahydrofolate deficiency. Cell **29**:929–937.
5. **Bochner, B. R., P. C. Lee, S. W. Wilson, C. W. Cutler, and B. N. Ames.** 1984. AppppA and related adenylylated nucleotides are synthesized as a consequence of oxidation stress. Cell **37**:225–232.
6. **Bochner, B. R., D. M. Maron, and B. N. Ames.** 1981. Detection of phosphate esters on chromatograms: an improved reagent. Anal. Biochem. **117**:81–83.
7. **Bochner, B. R., M. Zylicz, and C. Georgopoulos.** 1986. The *Escherichia coli* DnaK protein possesses a 5′-nucleotidase activity that is inhibited by Ap_4A. J. Bacteriol. **168**:931–935.
8. **Bohr, V.** 1978. Effects of purine riboside on nucleic acid synthesis in ascites cells. Biochim. Biophys. Acta **519**: 125–137.
9. **Brevet, A., P. Plateau, M. Best-Belpomme, and S. Blanquet.** 1985. Variation of Ap_4A and other dinucleoside polyphosphates in stressed *Drosophila* cells. J. Biol. Chem. **260**:15566–15570.
10. **Cashel, M., and K. E. Rudd.** 1986. The stringent response, p. 1410–1438. *In* F. C. Neidhardt, J. L. Ingraham, K. B. Low, B. Magasanik, M. Schaechter, and H. E. Umbarger (ed.), *Escherichia coli* and *Salmonella typhimurium*: cellular and molecular biology. American Society for Microbiology, Washington, D.C.
11. **Christman, M. F., R. W. Morgan, F. S. Jacobson, and B. N. Ames.** 1985. Positive control of a regulon for defenses against oxidative stress and some heat-shock proteins in *Salmonella typhimurium*. Cell **41**:753–762.
12. **Costas, M. J., J. C. Cameselle, and A. Sillero.** 1986. Mitochondrial location of rat liver dinucleoside triphosphatase. J. Biol. Chem. **261**:2064–2067.
13. **Denisenko, O. N.** 1984. Synthesis of diadenosine-5′,5‴-p^1,p^3-triphosphate in yeast at heat shock. FEBS Lett. **178**: 149–152.
14. **Garrison, P. N., S. A. Mathis, and L. D. Barnes.** 1986. In vivo levels of diadenosine tetraphosphate and adenosine tetraphospho-guanosine in *Physarum polycephalum* during the cell cycle and oxidative stress. Mol. Cell. Biol. **6**: 1179–1186.
15. **Guedon, G., D. Sovia, J. P. Ebel, N. Befort, and P. Remy.** 1985. Effect of diadenosine tetraphosphate microinjection on heat shock protein synthesis in *Xenopus laevis* oocytes. EMBO J. **4**:3743–3749.
16. **Guranowski, A., H. Jakubowski, and E. Holler.** 1983. Catabolism of diadenosine 5′,5‴-P^1,P^4-tetraphosphate in procaryotes. J. Biol. Chem. **258**:14784–14789.
17. **Irbe, R. M., L. M. E. Morin, and M. Oishi.** 1981. Prophage (φ80) induction in *Escherichia coli* K-12 by specific deoxyoligonucleotides. Proc. Natl. Acad. Sci. USA **78**: 138–142.
18. **Larsen, I. K., B.-M. Sjoberg, and L. Thelander.** 1982. Characterization of the active site of ribonucleotide reductase of *Escherichia coli*, bacteriophage T4 and mammalian cells by inhibition studies with hydroxyurea analogues. Eur. J. Biochem. **125**:75–81.
19. **Lee, P. C., B. R. Bochner, and B. N. Ames.** 1983. Diadenosine 5′,5‴-P^1,P^4-tetraphosphate and related adenylylated nucleotides in *Salmonella typhimurium*. J. Biol. Chem. **258**:6827–6834.
20. **Lee, P. C., B. R. Bochner, and B. N. Ames.** 1983. AppppA, heat-shock stress, and cell oxidation. Proc. Natl. Acad. Sci. USA **80**:7496–7500.
21. **Mechulam, Y., M. Fromant, P. Mellot, P. Plateau, S. Blanchin-Roland, G. Fayat, and S. Blanquet.** 1985. Molecular cloning of the *Escherichia coli* gene for diadenosine 5′,5‴-P^1,P^4-tetraphosphate pyrophosphohydrolase. J. Bacteriol. **164**:63–69.
22. **Postma, P. W., and J. W. Lengeler.** 1985. Phosphoenolpyruvate:carbohydrate phosphotransferase system of bacteria. Microbiol. Rev. **49**:232–269.
23. **Randerath, K., and E. Randerath.** 1967. Thin-layer separation methods for nucleic acid derivatives. Methods Enzymol. **12A**:323–347.
24. **Sabina, R. L., P. Dalke, A. R. Hanks, J. M. Magill, and C. W. Magill.** 1981. Changes in nucleotide pools during conidial germination in *Neurospora crassa*. Can. J. Biochem. **59**:899–905.
25. **Sabina, R. L., E. W. Holmes, and M. A. Becker.** 1984. The enzymatic synthesis of 5-amino-4-imidazolecarboxamide riboside triphosphate (ZTP). Science **223**:1193–1195.
26. **Sabina, R. L., K. H. Kernstine, R. L. Boyd, E. W. Holmes, and J. L. Swain.** 1982. Metabolism of 5-amino-4-imidazolecarboxamide riboside in cardiac and skeletal muscle. J. Biol. Chem. **257**:10178–10183.
27. **Stephens, J. C., S. W. Artz, and B. N. Ames.** 1975. Guanosine 5′-diphosphate 3′-diphosphate (ppGpp): positive effector for histidine operon transcription and general signal for amino-acid deficiency. Proc. Natl. Acad. Sci. USA **72**:4389–4393.
28. **Ullmann, A., and A. Danchin.** 1980. Role of cyclic AMP in regulatory mechanisms in bacteria. Trends Biochem. Sci. **5**:95–96.
29. **VanBogelen, R. A., P. M. Kelley, and F. C. Neidhardt.** 1986. Differential induction of heat-shock, SOS, and oxidation-stress regulons and accumulation of nucleotides in *Escherichia coli*. J. Bacteriol. **169**:26–32.
30. **Zamecnik, P.** 1983. Diadenosine 5′,5‴-P^1,P^4-tetraphosphate (Ap_4A): its role in cellular metabolism. Anal. Biochem. **134**:1–10.
31. **Zimmerman, T. P., and R. D. Deeprose.** 1978. Metabolism of 5-amino-1β-D-ribofuranosyl-imidazole-4-carboxamide and related five-membered heterocycles to 5′-triphosphates in human blood and L5178Y cells. Biochem. Pharmacol. **27**:709–716.

Global Responses of *Escherichia coli* to DNA Damage and Stress

KYUNG-HEE PAEK, CHRIS C. DYKSTRA, DIANE E. SHEVELL, JOHN R. BATTISTA, LORRAINE MARSH, AND GRAHAM C. WALKER

Biology Department, Massachusetts Institute of Technology, Cambridge, Massachusetts 02139

Analyses of the responses of *Escherichia coli* to DNA damage and stress have resulted in a number of insights into cellular strategies for global regulation. Furthermore, these studies have revealed several novel molecular mechanisms for implementing this regulation. A recurring theme in these studies has been the identification of proteins that have both mechanistic and regulatory functions. In this paper we briefly summarize certain interesting features of global regulation that have been discerned in studies of the SOS, adaptive, and heat shock responses of *E. coli*.

SOS Response

Exposure of *E. coli* to agents that damage DNA or interfere with DNA replication results in the induction of a diverse set of physiological responses, termed the SOS responses, which include (i) an increased capacity to reactivate UV-irradiated bacteriophage (Weigle reactivation) (24, 31), (ii) a capacity to mutate UV-irradiated bacteriophage (Weigle mutagenesis) (24, 31), (iii) the induction of functions that allow bacteria to be mutated by UV light and a variety of agents, (iv) filamentous growth, (v) an increased capacity to repair double-stranded breaks, (vi) an alleviation of restriction, and (vii) a capacity to carry out long patch DNA excision repair (15, 28–30). These physiological responses are due to the induction of more than 17 genes which have often been referred to as *din* (damage-inducible) genes (8).

The expression of the genes in the SOS regulatory circuit is controlled by a complex circuitry involving the RecA and LexA proteins (15, 28–30). The LexA protein apparently serves as the direct repressor of every SOS gene that has been identified to date (Fig. 1). Exposure of cells to DNA-damaging agents (e.g., UV light or mitomycin C), or to treatments which interfere with DNA replication (e.g., shifting certain mutants that are temperature sensitive for DNA replication to the restrictive temperature), generates an inducing signal which activates RecA molecules. When activated RecA (the black form in Fig. 1) interacts with a LexA monomer, an alanine-glycine peptide bond in the LexA molecule is cleaved. As the LexA molecules in a cell are inactivated by this proteolytic cleavage, the various SOS genes are expressed at increased levels, and the SOS responses mediated by the products of these genes are observed. The repressors of bacteriophages such as lambda have homology to LexA (25) and are similarly cleaved at an alanine-glycine peptide bond when they interact with activated RecA protein, leading to prophage induction. As DNA repair helps the cells recover from the DNA-damaging treatment, the inducing signal disappears so that RecA molecules cease to be activated (13). LexA molecules then accumulate in the cells and repress the SOS genes.

RecA

The RecA protein is an example of a protein with dual regulatory and mechanistic functions. It is needed for homologous recombination and for UV and chemical mutagenesis (discussed below), and it is the key positively acting regulatory element governing the expression of the genes in the SOS network (28, 29). RecA is the molecule that senses the intracellular signals generated in response to an SOS-inducing treatment. In vitro studies have shown that RecA becomes activated for SOS induction when it finds single-stranded DNA and a nucleoside triphosphate. A variety of genetic and physiological experiments are consistent with the in vivo inducing signal being single-stranded DNA or double-stranded DNA containing a gap which is generated by processing of damaged DNA or by interference with replication fork movement (28). It is interesting that the biochemical properties of RecA apparently responsible for sensing the inducing signal are related to these properties required for its role in homologous recombination.

Our view of the molecular mechanism by which RecA mediates the proteolytic cleavage of LexA has been modified by Little's discovery (14) that incubation of highly purified LexA and lambda repressor under alkaline conditions but in the absence of RecA resulted in the cleavage of the same alanine-glycine peptide bond normaly cleaved at neutral pH only in the presence of RecA. Rather than acting as the direct catalytic agent in the cleavage of this alanine-glycine

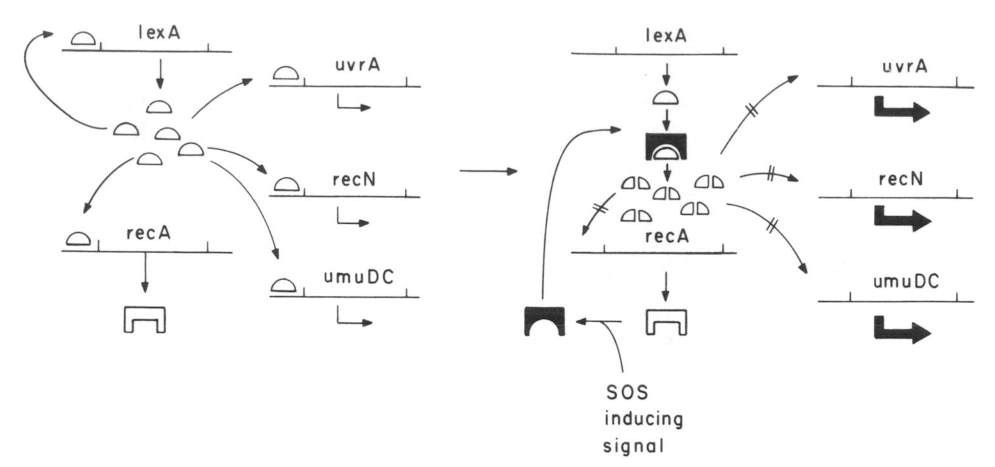

FIG. 1. Model of the SOS regulatory system (29). LexA is the repressor of at least 17 genes on the *E. coli* chromosome. The generation of inducing signal as a consequence of DNA damage leads to an activation of RecA. The interaction of activated RecA with LexA results in the cleavage of LexA. As the LexA pools decrease, the SOS genes are expressed at higher levels.

peptide bond, it seems that RecA may interact with LexA in such a way as to facilitate some previously latent capability of LexA for autodigestion.

A Protein Required for UV Mutagenesis Shares Homology with LexA

One of the SOS responses is an induced capability to be mutated by UV light and a variety of chemicals; the system that processes damaged DNA in such a way that mutations result is often referred to as error-prone repair or SOS processing (2, 24, 28, 29). Alleles of *recA* and *lexA* that prevent induction of the SOS system render *E. coli* nonmutable by UV light and various chemicals. The products of the *umuD* and *umuC* genes apparently play key mechanistic roles in error-prone repair; *umuD* and *umuC* mutants are nonmutable by UV light and various chemicals (9). These two genes are arranged in an operon, *umuDC*, which is repressed by LexA, and code for proteins of approximately 16,000 and 45,000 daltons, respectively (4). A homologous and functionally equivalent operon, *mucAB*, is located on the mutagenesis-enhancing plasmid pKM101 (23). DNA sequencing of the two operons has revealed that they have undergone significant evolutionary divergence, with the deduced amino acid sequences of UmuD and MucA having 41% exact homology and those of UmuC and MucB having 55% exact homology (22).

We have recently made the intriguing observation that UmuD and MucA share homology with the carboxyl-terminal domain of LexA (22). The homology includes the alanine-glycine peptide cleavage site of LexA (alanine-glycine in MucA and cysteine-glycine in UmuD) as well as most of the residues in the carboxyl-terminal domain of LexA that had previously been shown by Sauer et al. (25) to be conserved in the repressors of the bacteriophages lambda, 434, and P22. A series of experiments have suggested that this homology has functional significance and furthermore that UmuD (and MucA) interacts with activated RecA. Our most striking observations were that (i) modest overproduction of wild-type MucA and MucB inhibits SOS induction of a wild-type strain by UV light and mitomycin C, possibly by competing with LexA for the pool of activated RecA, and (ii) equivalent overproduction of wild-type MucB and an altered MucA, which has the alanine-glycine changed to alanine-glutamate, fails to exhibit this inhibition (16a). We do not yet know whether the 16,000-dalton UmuD or MucA proteins are cleaved upon interaction with RecA. However, these observations take on special significance in the light of genetic experiments which indicate that RecA has a second role in error-prone repair besides causing the induction of *umuDC* by the cleavage of LexA (28). Error-prone repair may represent an example of a system in which elements of protein architecture that are crucial to transcriptional regulation may also prove crucial at a mechanistic level in the induced process.

Adaptive Response

Exposure of *E. coli* to low levels of methylating agents induces the expression of a set of repair processes, collectively termed the adaptive response, that repair DNA damage induced by methylating and ethylating agents. The regulatory network consists of at least four genes, *ada, alkA, alkB*, and *aidB*, that are positively regulated by the *ada* gene product (11, 28, 29). The product of the *alkA* gene is a broad-spectrum glycosylase that removes 3-methyladenine, 3-methylguanine, 7-methylguanine, O^2-methylcytosine, and O^2-methylthymine (17). The product of the *ada* gene repairs O^6-methylguanine, O^4-methylthymine, and phosphotriester lesions (3, 18). Aspects of the biochemistry and regulation of the adaptive response have recently been reviewed (11, 28, 29).

Ada: a Protein with Dual Regulatory and Mechanistic Activities

The *ada* gene product is a remarkable 39,000-dalton protein that has two independent DNA repair activities and a regulatory activity that acts positively to increase transcription of the genes in the adaptive response network (7, 10). O^6-Methylguanine and O^4-methylthymine lesions are repaired by transfer of the methyl group to a cysteine residue located near the carboxyl terminus of the Ada protein; the transfer of the methyl group to the cysteine is apparently irreversible (3, 12, 26). A 19,000-dalton carboxyl-terminal fragment of Ada, which apparently arises from proteolytic cleavage following cell disruption, carries the methyltransferase activity for repairing O^6-methylguanine and O^4-methylthymine residues in DNA (3). Methyl phosphotriester lesions are repaired by transfer of the methyl group to a cysteine residue located near the amino terminus of Ada; once again, this transfer is apparently irreversible (18). Teo et al. (26) recently demonstrated that the transfer of a methyl group from a methyl phosphotriester to Cys-69 in Ada activates Ada to be a positive effector of transcription. Interestingly, we have found that a series of *ada* derivatives truncated at their 3′ end are able to cause extremely high constitutive levels of β-galactosidase production in a strain carrying an *ada′-lacZ*$^+$ operon fusion in its chromosome (10; D. E. Shevell, P. K. LeMotte, and G. C. Walker, unpublished data). It is not yet clear whether truncation of the carboxy terminus of Ada mimics a conformational change normally caused by the methylation of Ada or whether there might possibly be some type of proteolytic processing of the methylated Ada in vivo.

Heat Shock Response

Shifting *E. coli* from 30 to 42°C results in increased synthesis of a set of more than 17 proteins which are termed the heat shock proteins. The regulation of the heat shock response has recently been reviewed (21). Many of the genes induced during the heat shock response have been identified, and these include *groEL, groES, dnaK, dnaJ, lon*, and *lysU*. The induction of these genes requires the product of the *rpoH* (*htpR*) gene (20, 32). The product of this gene, sigma-32, functions as an alternative sigma subunit of RNA polymerase and directs the transcription of the heat shock genes (5). After a shift from 30°C to 42°C, the synthesis of the heat shock proteins increases, reaches a maximum at approximately 7 to 10 min, and then rapidly decreases. *dnaK* mutants fail to exhibit this rapid shutoff of the synthesis of heat shock proteins, implying that the DnaK protein negatively modulates the heat shock response (27).

DnaK

The DnaK protein appears to represent yet another example of a protein that plays both mechanistic and regulatory roles. DnaK is a 70,000-dalton protein that (i) is homologous to HSP70 proteins of *Drosophila* and humans (1, 6), (ii) is needed for bacteriophage lambda replication both in vivo and in vitro (33, 34), (iii) has a nonspecific 5′-nucleotidyl phosphatase activity (C. Georgopoulos and B. Bochner, personal communication), and (iv) autophosphorylates itself (33). Although temperature-sensitive lethal *dnaK* mutants had been isolated, a result which suggested that *dnaK* is an essential gene, we have been able to delete *dnaK*; the *dnaK* null mutant is temperature sensitive for growth, implying that DnaK is needed only at high temperature (21a). The mechanism(s) by which *dnaK* modulates the heat shock response is not completely understood, but at least one component of its action may be related to a proteolytic degradation of the sigma-32 protein. Both C. Gross (personal communication) and C. Georgopoulos and K. Tilly (personal communication) have observed that the half-life of sigma-32 is longer in *dnaK* mutants. This result is consistent with other observations, in both our laboratory (C. C. Dykstra and G. C. Walker, unpublished data) and that of C. Gross (personal communication), implicating DnaK in aspects of generalized cellular proteolysis in *E. coli*.

Proteolysis in Regulation and Cellular Physiology

A recent line of experimentation in our laboratory, still imperfectly understood, has focused our attention on the possible importance of proteases in aspects of regulation and cellular physiology. We previously reported (16) that a *lexA*(Def) strain carrying a multicopy *umuD*$^+$ *umuC*$^+$ plasmid grew only at 43°C; shifting the cells to 30°C resulted in a block of DNA replication and cell death. Interestingly, this cold sensitivity could be suppressed by *lon* and *rpoH* mutations. We recently extended these results and found that the cold sensitivity of a *lexA51*(Def) strain carrying a multicopy *umuD*$^+$ *umuC*$^+$ plasmid could be suppressed by *dnaK, groEL*, and *groES* mutations as well as by a series of new mutations termed *css* (cold-sensitivity suppressor) that were generated by Tn*5* insertions (Dykstra and Walker, unpublished data). To date, the mutations we have identified that suppress cold sensitivity seem to cause a defect in proteolysis as judged by the fact that the mutants share certain of the following phenotypes: (i) a deficiency in the turnover of puromycyl fragments, (ii) a stabilization of the LacZ(X90) amber fragment, and (iii) an apparent stabilization of the LexA51 protein inferred from studies of the expression of a *umuC'-lacZ'* fusion (Dykstra and Walker, unpublished data). Although these studies are in a relatively early phase, they suggest the existence of a genetically complex system that controls protein turnover in *E. coli*.

There are a few clues that proteolytic degradation may be important in regulating the action of certain induced proteins. For example, overproduction of the SOS-regulated proteins SulA and UmuD/C is lethal, causing filamentation and a blockage of DNA replication, respectively. Thus, the syntheses of these proteins represent examples of situations in which *E. coli* chooses to transiently synthesize potentially toxic proteins to help it survive an emergency situation. Since the accumulation or persistence of such proteins would be deleterious, it is perhaps not too surprising to find that the synthesis of the SulA and UmuDC proteins is inducible and that the proteins themselves have relatively short half-lives. In the case of SulA, its degradation has been shown to involve the product of the *lon* gene (19). It is interesting that the Lon protein, a product of a heat shock gene, seems to be involved in regulating the action of SulA, a product of an SOS gene, so that there is crosstalk between these two regulatory networks.

The facts that the *rpoH* gene product (sigma-32) has a short half-life and that its half-life is longer in *dnaK* mutants suggest that proteolytic turnover of regulatory elements may be a mechanism used by *E. coli* in the regulation of gene expression.

This work was supported by Public Health Service grants CA2161 and GM28988 from the National Institutes of Health and by grant NP-461C from the American Cancer Society. C.C.D. and J.R.B. were supported by postdoctoral fellowships awarded by the American Cancer Society and the Massachusetts Division of the American Cancer Society, respectively. D.E.S. and L.M. were supported by the National Institutes of Health training grant T32ES07020.

LITERATURE CITED

1. **Bardwell, J. C. A., and E. A. Craig.** 1984. Major heat-shock gene of *Drosophila* and *Escherichia coli* heat-inducible *dnaK* gene are homologous. Proc. Natl. Acad. Sci. USA **81:**848–852.
2. **Defais, M., P. Faquet, M. Radman, and M. Errera.** 1971. Ultraviolet reactivation and ultraviolet mutagenesis of lambda in different genetic systems. Nature (London) **304:** 466–468.
3. **Demple, B., B. Sedgwick, P. Robins, N. Totty, M. D. Waterfield, and T. Lindahl.** 1985. Active site and complete sequence of the suicidal methyltransferase that counters alkylation mutagenesis. Proc. Natl. Acad. Sci. USA **82:** 2688–2692.
4. **Elledge, S. J., and G. C. Walker.** 1983. Proteins required for ultraviolet light and chemical mutagenesis: identification of the products of the *umuC* locus of *Escherichia coli*. J. Mol. Biol. **164:**175–192.
5. **Grossman, A. D., J. W. Erickson, and C. A. Gross.** 1984. The *htpR* gene product of *E. coli* is a sigma factor for heat-shock promoters. Cell **38:**383–390.
6. **Hunt, C., and R. I. Moriomoto.** 1985. Conserved features of eukaryotic *hsp70* gene revealed by comparison with the nucleotide sequence of human *hsp70*. Proc. Natl. Acad. Sci. USA **82:**6455–6459.
7. **Jeggo, P.** 1979. Isolation and characterization of *Escherichia coli* K-12 mutants unable to induce the adaptive response to simple alkylating agents. J. Bacteriol. **139:**783–791.
8. **Kato, T., and Y. Shinoura.** 1977. Isolation and characterization of mutants of *Escherichia coli* deficient in induction of mutations by ultraviolet light. Mol. Gen. Genet. **156:**121–131.
9. **Kenyon, C. J., and G. C. Walker.** 1980. DNA-damaging agents stimulate gene expression at specific loci in *Escherichia coli*. Proc. Natl. Acad. Sci. USA **77:**2819–2823.
10. **LeMotte, P. K., and G. C. Walker.** 1985. Induction and autoregulation of *ada*, a positively acting element regulating the adaptive response to alkylating agents. J. Bacteriol. **161:**888–895.
11. **Lindahl, T.** 1982. DNA repair enzymes. Annu. Rev. Biochem. **51:**61–87.
12. **Lindahl, T., B. Demple, and P. Robins.** 1982. Suicide inactivation of the *E. coli* O^6-methyl-guanine-DNA methyltransferase. EMBO J. **1:**1359.
13. **Little, J. W.** 1983. The SOS regulatory system: control of its state by the level of *recA* protease. J. Mol. Biol. **167:** 791–808.
14. **Little, J. W.** 1984. Autodigestion of LexA and phage lambda repressors. Proc. Natl. Acad. Sci. USA **81:**1375–1379.
15. **Little, J. W., and D. W. Mount.** 1982. The SOS regulatory system of *Escherichia coli*. Cell **29:**11–22.
16. **Marsh, L., and G. C. Walker.** 1985. Cold sensitivity induced by overproduction of UmuDC in *Escherichia coli*. J. Bacteriol. **162:**155–161.

16a. **Marsh, L., and G. C. Walker.** 1987. New phenotypes associated with *mucAB*: alteration of a MucA sequence homologous to a LexA cleavage site. J. Bacteriol. **169:**1818–1823.

17. **McCarthy, T. V., P. Karran, and T. Lindahl.** 1984. Inducible repair of O-alkylated DNA pyrimidines in *Escherichia coli*. EMBO J. **3:**545–550.
18. **McCarthy, T. V., and T. Lindahl.** 1985. Methyl phosphotriesters in alkylated DNA are repaired by the Ada regulatory protein of *Escherichia coli*. Nucleic Acids Res. **13:** 2683–2698.
19. **Mizusawa, S., and S. Gottesman.** 1983. Protein degradation in *Escherichia coli*: the *lon* gene controls the stability of SulA protein. Proc. Natl. Acad. Sci. USA **80:**358–362.
20. **Neidhardt, F. C., and R. A. vanBogelen.** 1981. Positive regulatory gene for temperature-controlled proteins in *Escherichia coli*. Biochem. Biophys. Res. Commun. **100:** 893–900.
21. **Neidhardt, F. C., R. A. vanBogelen, and V. Vaughn.** 1984. The genetics and regulation of heat-shock proteins. Annu. Rev. Genet. **18:**295–329.

21a. **Paek, K.-H., and G. C. Walker.** 1987. *Escherichia coli dnaK* null mutants are inviable at high temperature. J. Bacteriol. **169:**283–290.

22. **Perry, K. L., S. J. Elledge, B. M. Mitchell, L. Marsh, and G. C. Walker.** 1985. *umuDC* and *mucAB* operons whose

products are required for UV light- and chemical-induced mutagenesis: UmuD, MucA, and LexA proteins share homology. Proc. Natl. Acad. Sci. USA **82**:4331–4335.

23. **Perry, K. L., and G. C. Walker.** 1982. Identification of plasmid (pKM101)-encoded proteins involved in mutagenesis and UV resistance. Nature (London) **300**:278–281.
24. **Radman, M.** 1974. Phenomenology of an inducible mutagenic DNA repair pathway in *Escherichia coli*: SOS repair hypothesis, p. 128–142. *In* L. Prakash, F. Sherman, M. Miller, C. Lawrence, and H. W. Tabor (ed.), Molecular and environmental aspects of mutagenesis. Charles C Thomas, Publisher, Springfield, Ill.
25. **Sauer, R. T., R. R. Yocum, R. F. Doolittle, M. Lewis, and C. O. Pabo.** 1982. Homology among DNA-binding proteins suggests use of a conserved super-secondary structure. Nature (London) **298**:447–451.
26. **Teo, I., B. Sedgwick, M. W. Kilpatrick, T. McCarthy, and T. Lindahl.** 1986. The intracellular signal for induction of resistance to alkylating agents in *Escherichia coli*. Cell **45**: 315–324.
27. **Tilly, K., N. McKittrick, M. Zylicz, and C. Georgopoulos.** 1983. The *dnaK* protein modulates the heat-shock response of *Escherichia coli*. Cell **34**:641–646.
28. **Walker, G. C.** 1984. Mutagenesis and inducible responses to deoxyribonucleic acid damage in *Escherichia coli*. Microbiol. Rev. **48**:60–93.
29. **Walker, G. C.** 1985. Inducible DNA repair systems. Annu. Rev. Biochem. **54**:425–457.
30. **Weigle, J.** 1953. Induction of mutation in a bacterial virus. Proc. Natl. Acad. Sci. USA **39**:628–636.
31. **Witkin, E. M.** 1976. Ultraviolet mutagenesis and inducible DNA repair in *Escherichia coli*. Bacteriol. Rev. **40**:869–907.
32. **Yamamori, T., and T. Yura.** 1982. Genetic control of heat-shock protein synthesis and its bearing on growth and thermal resistance in *Escherichia coli* K-12. Proc. Natl. Acad. Sci. USA **79**:860–864.
33. **Zylicz, M., and C. Georgopoulos.** 1984. Purification and properties of the *Escherichia coli dnaK* replication protein. J. Biol. Chem. **259**:8820–8825.
34. **Zylicz, M., J. H. LeBowitz, J. McMacken, and C. Georgopoulos.** 1983. The *dnaK* protein of *Escherichia coli* possesses an ATPase and autophosphorylating activity and is essential in an *in vitro* DNA replication system. Proc. Natl. Acad. Sci. USA **80**:6431–6435.

Why *Escherichia coli* Should Be Renamed *Escherichia ilei*

ARTHUR L. KOCH

Department of Biology, Indiana University, Bloomington, Indiana 47405

Many aspects of the molecular genetics of *Escherichia coli* suggest that this species has evolved to cope with an additional environment quite distinct from the colon, feces, soil, and water environments where it is usually studied. It has elegant regulatory systems to alter its ability to metabolize the lactose that occurs in only very small amounts in any of the above locations. Consequently, the argument was advanced (18; A. L. Koch, *in* P. Calow, ed., *Evolutionary Physiological Ecology*, in press) that this coliform, and possibly all coliforms, was adapted to life in the small intestine, where, for example, there are opportunities to capture and consume the lactose before the mammalian host does. Because the small intestine feeds the large intestine, it can be argued that the coliforms that enter the colon grow to a larger population, but inherit the genotype optimized for growth in the ileum.

An additional aspect, appropriate to this symposium, is the pH optimum of alkaline phosphatase. Since the pancreatic secretion is alkaline and intestinal secretions are slightly alkaline, an enzyme designed to function in the milieu of the small intestine should have an alkaline pH optimum appropriate to that environment (like the alkaline phosphatase of the brush border of the intestinal epithelium). This pH environment is distinct from that of the colon, where the pH becomes slightly acidic as a result of progressive fermentation of carbon compounds and exchange of Na^+ and K^+ for H^+ by the brush border cells (1, 6, 23).

Bacterial Physiology and the Mammalian Gut

The proposition that the small intestine is an important habitat for coliforms is supported by the known importance of adhesion to the small intestine in the pathology of *E. coli* in infants and piglets, but this proposition may be difficult to accept because the number of organisms in the upper bowels is small compared with bacterial populations lower in the intestinal tract (4, 6–9, 27). Estimates of the enterobacteria in the ileum are variable (10^2 to 10^7 CFU/ml), whereas there are 100 to 1,000 times as many in the colon contents and feces. Simon and Gorbach (27) noted that in the distal ileum the gram-negative bacteria begin to outnumber the gram-positive bacteria and that the redox potential, E_h, becomes sufficiently negative to support the growth of anaerobes. Thus, the big expansion of the coliform population may well occur just upstream of the region of the expansion of the anaerobic population in the ascending colon.

The maximal specific growth rate of many laboratory strains of *E. coli* corresponds to a doubling time of about 20 min under optimal conditions. With a strain that had not been subjected to mutagenesis, I have reported doubling times in rich medium of 15 to 16 min (17) at 37°C. But even this rapid growth is not fast enough to allow the organism to maintain itself in the lumen of the small intestine (2, 3, 11, 25, 26). Peristalsis moves the contents at an average speed of 2 cm/min. This velocity is calculated from the length of the small bowel in humans and the time for barium meal to pass from the stomach to the ileo-colonic valve. Segmental motility cannot provide an adequate backflow that could predominate over the propulsive events.

After a meal is taken, a series of peristaltic waves quickly moves at about 1 cm/s to empty the ileum and fill the cecum and ascending colon with chyme. This decreases by a large factor the number of organisms remaining in the fluid phase of the small intestine.

The total flow is indeed great, although it varies a good deal from point to point. Typically, daily flows into the adult small intestine are 1 liter of ingested fluid, 1 liter of saliva, 2 liters of gastric juice, 2 liters of bile, and 3 liters of intestinal secretion (3, 11, 25, 26). There is, consequently, a total daily efflux of 9 liters from the ileum. Of this, 8.5 liters is reabsorbed into the bloodstream. Thus, the output of the small bowel is 0.5 liter per day. This leaves 0.4 liter to be absorbed in the colon (mainly in the ascending portion) and only 0.1 liter of water to be excreted with the feces.

The total volume of feces excreted in a day, of course, depends on the size of the animal, but even when size is not a factor, excretion depends on habits and diet. For example, Englishmen on average excrete 300 g per day, while Nigerians excrete 2,000 g per day. However, the transit time may be comparable in these two cases. Herbivores, omnivores, and carnivores are significantly different continuous culture devices from the point of view of their coliform inhabitants.

Serotypes and Electrophoretic Types of *E. coli*

Consider the body of data on the serotypes and electrophoretic types of *E. coli* accumulated over the past 40 years (12). Clearly, there are valid reports of resident strains that remain in an individual human for several years while transient coliforms come and go, and there is clear evidence in numerous other studies (Koch, in press) for the repeated temporary appearance of the same strains episodically. The appearance, disappearance, and reappearance of such clones have been observed in the careful studies from Linton's laboratory with weanling pigs (22) and one human (13). These reports plus the observation that humans, at least in affluent, hygienic societies (5), may be devoid of *E. coli* a significant portion of the time suggest that there are protected sites where coliforms persist for long periods of time. Occasionally, inoculation from such a colony results in growth in the lumen and a wave of organism growth passing through and out of the tract.

Chronic Starvation Phases

From the beginning of the transverse colon the concentration and distribution of organisms vary little on a CFU per gram of dry weight basis. Consequently, the passage of coliforms through the rest of the colon and through the external ecosystem occurs largely in a state of chronic starvation. Evidently, during most of the time the bacteria spend outside the mammalian host they are also chronically starved. This chronic phase of slow growth starts approximately at the transverse colon. This is the coliform's fate for the indefinite period until a few rare organisms once again grow luxuriantly while being passed from the duodenum to the transverse colon of a new host.

Even though a particular phase of a species' existence is not represented in nature often, either because it is short or does not occur with every duplication of the organism, every necessary phase for the species is as essential as every other. Coliforms, at the very least, must find new hosts at about the rate that old hosts die. Of course, the evidence is clear that they do find new hosts much more frequently than that, but still many fission cell cycles occur for every host-finding, colonization event. Of the billions of emigrants, many are lost in the soil and water habitat or destroyed in the stomach. Slow growth in the environment outside the mammalian intestinal tract may be important for coliforms, such as *E. coli*, even though they grow more poorly than many autochthonous soil organisms in these environments. Probably more important to the species is its short-term survival in extreme environments such as the surface of the mother's breasts and other anatomical features. This may be the major phase of the species existence when UV damage occurs and may be the phase for which photoreactivation has evolved to protect the cell. But I suggest that persisting in the upper reaches of the small intestine, largely in a nongrowing or slowly growing state, is one of the most critical phases of the coliform's existence.

Passage into the Intestine

It should be noted that the passage through the stomach process is considerably more efficient in babies, individuals in poor health, especially those with diarrhea, and individuals without enough gastric acidity, whether a result of overuse of antacids or otherwise. Of course, the chance for ingestion of coliforms is also much higher in babies, animals, and many third-world humans (4). On the other hand, the gastrointestinal tract of happy and healthy adult Europeans and North Americans might be a suitable but not fully occupied habitat for coliforms (5).

Expansion Era

After the gastric acidity has been overneutralized by the alkaline bile, the few immigrants which survive passage through the stomach and the few organisms shed from protected sites grow rapidly, competing mainly with absorptive processes of the host. In the several hours in the jejunum and ileum many-fold growth probably occurs. Growth continues in the cecum and lower ascending colon, where the competition is mainly with the anaerobes. The era of rapid growth ends when the anaerobes have sequestered the bulk of remaining resources and have created an unsuitable environment for rapid growth of the coliforms because of the high concentration of fermentation products.

Plug Flow Growth with Mixing

As a model illustrating an important aspect of the gut ecosystem, consider a plug flow system in which some mixing occurs (a tubular reactor in the chemical engineer's terminology [24]). Food enters at one end of the tube and passes at a constant rate, and feces are continuously excreted. We start with a sterile gut and inoculate the input at an instant of time. We will first consider the input of organisms as only occurring together with the passage of food. Below we consider the case in which the organisms had succeeded in colonizing the surface of the small bowel, from time to time to be extruded and inoculate the gut from their protected environments.

If there is no mixing, then the incoming cells will grow during the time that it takes the colon contents to traverse the favorable part of the

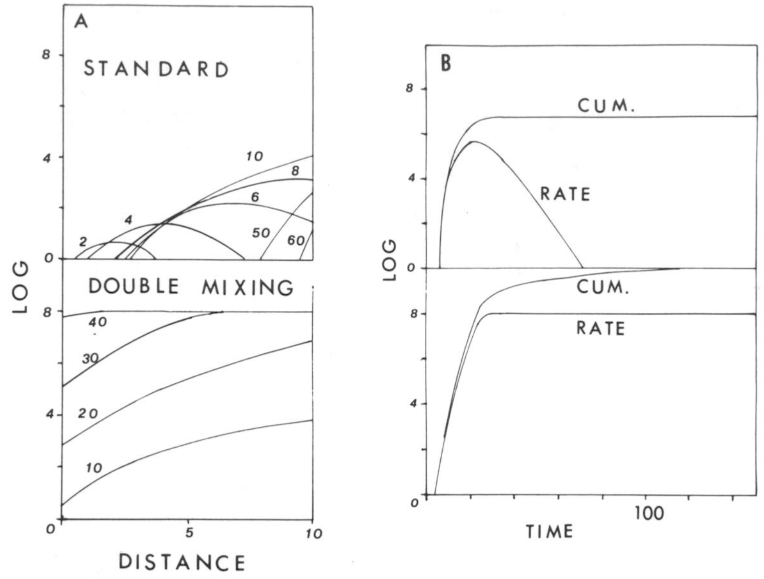

FIG. 1. Distribution of organisms in the hypothetical gut. The distribution of organisms along a plug flow reactor in which some mixing takes place is shown in A. The time course for the rate of excretion and cumulative excretion is shown in B. The upper half shows the standard situation designated by asterisks in Table 1. In this case the system eventually becomes sterile. The bottom half shows a case in which mixing and growth lead to a stable population of organisms. In this case the growth capabilities are sufficient to consume the limiting nutrient so that the maximum number of organisms is 10^8. A steady-state growth occurs only in the first part of the system, and starvation occurs in the remainder. Thus, the organisms are subject to a "feast and famine" existence. The program in BASIC that was used in the simulation will be supplied to those wishing it.

tract. In this case, an amplified yield of organisms will be produced and will appear as a pulse of bacteria in the fecal contents. If some mixing occurs, then the yield is larger, but eventually the system will become "washed out," as does a chemostat when the flow is too fast. The top part of Fig. 1A shows a simulation of the concentration along such a growth tube at various times after the inoculation event. The rate of organism excretion and the cumulative number of organisms excreted are shown on the top part of Fig. 1B. These curves are for a case in which the inoculation produces a wave of organisms and eventually the system becomes sterile. Taking this case of minimal mixing as the standard situation, some variations are shown in Table 1.

In the standard situation (designated with an asterisk in Table 1) one bacterium results in $10^{6.8}$ or 6.3 million total descendants leaving our imaginary gut. The parameters were chosen in the hope that they might represent a typical situation. The "log of yield" is 4.8 or a yield of 63,000 bacteria when there is only plug flow and no mixing. The only parameter directly within the control of the microorganism is its growth rate. In the simulation a quantity related to growth rate is designated by *F*. This is the fold increase of cells during the time required for fluid to pass one segment of the intestine, usually taken to be 10 segments long. Given the other standard parameters of the system, *F* would have to increase from the standard value of 3 to 5 for the enteric population to maintain itself. In such a case, the population density would be limited by nutrient availability. The value of *F* is also dependent on the properties of the host. If flow were slower, then the growth factor for a unit of movement along the system would be greater. Therefore, intestinal stasis can be very important.

The mixing properties of the mammalian part of the system can be critical, as can be seen from Table 1. For example, note that a four-fold lengthening of *NI*, the number of segments in the hypothetical gut, from 5 segments to 20 segments (the difference between carnivores and herbivores) causes a 10^{10}-fold increase in total output. This is $10^{6.2}$ times greater than if there were no mixing. While the number of organisms pushed upstream by mixing as determined by the *R* parameter is important, the distance that they are pushed as determined by the *NJ* parameter is even more important. The parameter *R* is the fraction of organisms removed from the main cohort by mixing as the intestinal contents move a segment's distance; the parameter *NJ* is the number of segments on each side of the central

TABLE 1. Intestine as a plug flow fermentor with mixing[a]

Variable	Log of yield[b]
Variation in growth rate	
$F = 2.5$	5.2
$F = 3.0$	6.8*
$F = 3.5$	8.6
Variation in mixing degree	
$R = 0.0$	4.8
$R = 0.1$	5.4
$R = 0.2$	6.8*
$R = 0.3$	11.0
Length of growth system	
$NI = 5$	3.4
$NI = 10$	6.8*
$NI = 20$	13.6
Length of mixing	
$NJ = 0$	4.8
$NJ = 1$	6.8*
$NJ = 2$; $R = 0.10$	∞
$R = 0.05$	∞
$R = 0.03$	∞
$R = 0.02$	5.9
$R = 0.02$; max = 4	5.5
$R = 0.02$; max = 3	4.6
$R = 0.10$; max = 8	∞

[a] Standard situation (indicated in the table by an asterisk): length (*NI*), 10 segments; mixing length (*NJ*), 1 segment; movement (*NK*), 1 segment per time unit; mixing amount (*R*), 0.2; growth factor (*F*), threefold growth while traversing segment; max, log maximum growth in cells per segment.

[b] Logarithm of cumulative yield after the system becomes sterile.

cohort to which there is an even redistribution. Doubling the length of the segmental peristalsis (back and forth mixing) from $NI = 1$ to $NJ = 2$ leads to a permanent population, even if the total number of organisms moved proximally remains the same.

A major reason for carrying out this stimulation was to see whether stable populations with an increasing number of organisms along the tract could arise in a plug flow system with mixing, but with no continuous inoculation. However, without continuous inoculation with food or from protected surface sites, this state is very difficult, probably impossible, to achieve. The last line of Table 1 corresponds to the lower parts of Fig. 1; it is for a case in which the rate of excretion becomes constant, limited by the level of some nutrient. This is a "feast and famine" situation (14, 15). At the steady state all growth takes place at the mouth of the system, and little (or none in our idealized gut) takes place thereafter. Obviously, the greater selective pressure would occur in the narrow region where most growth takes place. In the actual intestine, the appropriate model for residual growth in the latter part of the colon would be that of a chemostat where small amounts of nutrient are formed by degradation. The selective pressure to evolve in distal parts of the system depends on the degree to which growth is possible on minor resources or on resources made available by other organisms or by digestive enzymes of the host. However, the important point is that the growth at the initial part of the system does not correspond to growth as in a chemostat culture because, in the steady state, the bacterial population is nearly maximal near the input to the system and zero elsewhere.

The second and third lines from the bottom of Table 1 represent a special case: that of growth limited to a low maximal concentration. In these cases the population is spatially distributed along the growth system and is gradually washed out because distal growth does not produce enough organisms so that mechanically mixing sufficiently reinoculates the upstream regions. In no case was it possible to achieve a stable distribution without continued inoculation.

In summary of this section, the implication of plug growth with mixing for bacterial growth in the intestines is that, from an inoculation event, a wave of organisms grows and passes through the system, resulting in an amplified output of organisms. The gain of the system depends very sensitively on the amount of mixing, the specific growth rate, the length of the system, and the value of the length mixing parameter. Above critical values of these parameters, stable growth is achieved. However, it is different from what is observed in the gut in that the bulk of the growth occurs very near to the mouth of the system and is limited by the amount of readily available nutrient easily accessible to the organisms.

Importance of Starting Upstream

As before, imagine that the flow rate is constant throughout a system consisting of a pipe of constant diameter. Imagine further that no nutrients are created or lost through the walls of the tube and that nutrients enter only at one end and leave at the other. However, in this case, inoculation is imagined to occur episodically from sites within the system. The simulation result is that position of the inoculation site along the tract, i.e., the location where organisms are maintained in protected sites, is extremely important (Fig. 2). In our hypothetical gut, a single organism introduced at the distal end shows no growth within the system, while one introduced at the proximal end yields, over a period of time, $10^{13.6}$ organisms. This applies if there is no limitation on the concentration of the bacteria per unit volume. In terms of the normal biomass of the feces and the number of coliforms excreted during a wave that may last days, this is not an outrageous number.

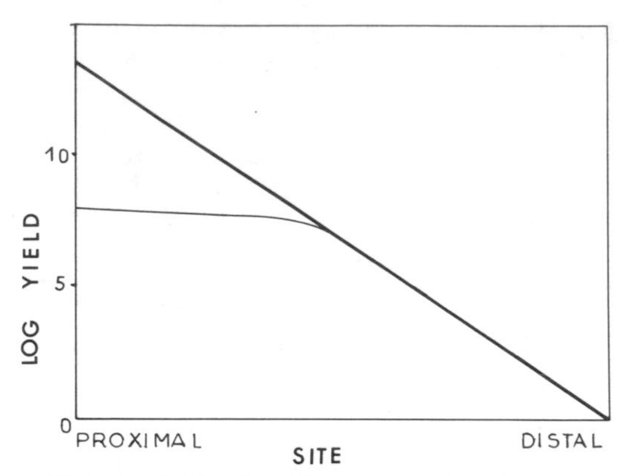

FIG. 2. Yield of coliforms as a function of the length of the hypothetical gut. The heavy line shows the cumulative yield from the growth system resulting from the inoculation with a single cell. The thin line shows the yield if the logarithm of the maximum concentration is 8.

Note that the relationship of yield versus position is linear on this semilogarithmic plot. This relationship arises since time is proportional to distance (as a result of the assumption of constant diameter and flow rate). Because mixing takes place and a few organisms are lost prematurely at the distal end and a few are destroyed at the proximal end, this linearity is not exact, but the deviations are negligible. Of course, the wave of organisms passing through the system is delayed and spread out in time, in proportion to the distance of the inoculation site from the distal end. The total yield of organisms resulting from a single bacterium when the maximum number per unit volume is restricted to 10^8 is also shown in Fig. 2. Even with this nutritional restriction, it is still the case that a wave passes down the system and the organisms become "washed out," but now the advantage for a clone having the most proximal site is less. However, this is true only in those situations in which it is not likely that two different protected clones inoculate the bulk phase nearly simultaneously in such a way as to lead to competition for resources. In that case growth from a cell from a downstream clone could be almost completely blocked by an upstream competitor.

Discussion

The basic model of the gut ecosystem is grossly oversimplified as a result of the assumptions of constant diameter and flow rate of the plug flow reactor. However, we have incorporated certain essential features, i.e., that inoculation of coliforms into the gastrointestinal tract occurs only sporadically either through influx into the bowel or through detachment of organisms from protected sites on the walls of the intestines. It has been assumed that the organisms quickly shift to their maximal growth rate (14–17, 20) and maintain that rate until the readily utilized substrates are consumed. It is not realistic, of course, to assume that growth ceases abruptly as the nutrient concentration falls; rather, growth rate probably decreases more gradually according to some kinetic law (16, 21). Also, in this analysis the complex human gut ecosystem with its 450 known bacterial species is reduced to but one or a few types of coliforms. Another simplifying assumption used here is that no organism dies.

Nonetheless, the simulations presented above will, I hope, fix ideas concerning the precarious existence of a coliform clone in its natural setting. To maintain itself in an individual mammalian host, an organism is fortunate if it finds a protected unoccupied site, and the farther up the tract the better. It can then persist, but in humans and weanling pigs any particular clone is only infrequently and temporarily a major contributor to the fecal coliform population; most of the time some other serotype dominates. The number of different clones embedded in protected sites within any animal appears to be large. The evidence for this is that the number of natural markers of serotypes is large in a single mammalian organism growing in a clean environment. The main goal of the simulation presented here is to point out the relevance of this kind of growth process and how the gastrointestinal tract behaves as an amplifier. The observed facts about the gastrointestinal tract in toto support the concept of episodic inoculation from some internal points and stand as evidence against some form of continuous culture maintaining the coliform population.

Elsewhere I have cataloged the modes of growth that could be responsible for the maintenance of polymorphism, such as the variety of serotypes or electrophoretic types in an ecosystem (19). Repeated inoculation from protected sites is only one of the nine possibilities, but from the consideration presented here it is the most likely for this case.

Work in my laboratory was supported by Public Health Service grant GM 34222 from the National Institute of General Medical Sciences. Most of the writing and simulations were carried out while I was a visiting investigator in the laboratory of Richard D'Ari at the Institut Jacques Monod, Centre National de la Recherche Scientifique, Université Paris VII.

LITERATURE CITED

1. **Binder, H. J., and H. Murer.** 1986. Potassium/proton exchange in the brush border or rat ileum. J. Membr. Biol. **91:**78–84.
2. **Creamer, B.** 1974. The small intestine. Heinemann, London.
3. **Davenport, H. U.** 1966. Physiology of the digestive tract, 2nd ed. Year Book, Chicago.

4. **Drasar, B. S., and M. J. Hill.** 1974. Human intestinal flora. Academic Press, London.
5. **Finegold, S. M., and V. L. Sutter.** 1978. Fecal flora in different populations with special reference to diet. Am. J. Clin. Med. **31:**S116–S122.
6. **Foster, E. S., J. P. Hayslett, and H. J. Binder.** 1984. Mechanism of active potassium absorption and secretion in the rat colon. Am. J. Physiol. **246:**G611–G617.
7. **Freter, R., H. Brickner, M. Botney, D. Cleven, and A. Aranki.** 1983. Mechanisms that control bacterial populations in continuous-flow culture models of mouse large intestinal flora. Infect. Immun. **39:**676–685.
8. **Freter, R., H. Brickner, J. Fekete, M. M. Vickerman, and K. E. Carey.** 1983. Survival and implantation of *Escherichia coli* in the intestinal tract. Infect. Immun. **39:**686–703.
9. **Freter, R., E. Stauffer, D. Cleven, L. V. Holdeman, and W. E. C. Moore.** 1983. Continuous-flow cultures as in vitro models of the ecology of large intestinal flora. Infect. Immun. **39:**666–675.
10. **Freter, R.** 1983. Mechanisms that control the microflora in the large intestines, p. 33–54. *In* D. J. Hentges (ed.), Human intestinal microflora in health and disease. Academic Press, Inc., New York.
11. **Granger, D. N., J. A. Barrowman, and R. P. Kvietys.** 1985. Clinical gastrointestinal physiology. The W. B. Sanders Co., Philadelphia.
12. **Hartl, D. L., and D. E. Dykhuizen.** 1984. The population genetics of *Escherichia coli*. Annu. Rev. Genet. **18:**31–68.
13. **Hartley, C. L., H. M. Clements, and K. B. Linton.** 1977. *Escherichia coli* in the fecal flora of man. J. Appl. Bacteriol. **43:**261–269.
14. **Koch, A. L.** 1971. The adaptive responses of *Escherichia coli* to a feast and famine existence. Adv. Microb. Physiol. **6:**147–217.
15. **Koch, A. L.** 1976. How bacteria face depression, recession, and derepression. Perspect. Biol. Med. **20:** 44–63.
16. **Koch, A. L.** 1979. Microbial growth in low concentrations of nutrients, p. 261–279. *In* M. Shilo (ed.), Strategies of microbial life in extreme environments. Dahlem Konferenzen–1978. Verlag-Chimie, Berlin.
17. **Koch, A. L.** 1980. Inefficiency of ribosomes functioning in *Escherichia coli* growing at moderate rates. J. Gen. Microbiol. **116:**165–171.
18. **Koch, A. L.** 1981. Evolution of antibiotic resistance gene function. Microbiol. Rev. **45:**355–378.
19. **Koch, A. L.** 1985. The macroeconomics of bacterial growth, p. 1–43. *In* M. M. Fletcher and G. D. Floodgate (ed.), Bacteria in their natural environment. Special publication of the Society for General Microbiology no. 16. Academic Press, Inc., Orlando, Fla.
20. **Koch, A. L., and M. Schaechter.** 1985. The world and ways of *E. coli*, p. 1–25. *In* A. L. Demain and N. A. Solomon (ed.), Biology of industrial microorganisms, vol. 1. Addison-Wesley, Reading, Mass.
21. **Koch, A. L., and C. H. Wang.** 1982. How close to the theoretical diffusion limit do bacterial uptake systems function? Arch. Microbiol. **131:**36–42.
22. **Linton, A. H., B. Handley, and A. D. Osborn.** 1978. Fluctuations in *Escherichia coli* O-serotypes in pigs throughout life in the presence and absence of antibiotic treatment. J. Appl. Bacteriol. **44:**285–298.
23. **Murer, H., U. Hopfer, and R. Kinne.** 1976. Sodium/proton antiport in the brush border membrane vesicles isolated from rat small intestine and kidney. Biochem. J. **154:**541–604.
24. **Pirt, S. J.** 1975. Principles of microbe and cell cultivation. Blackwell Scientific Publications, Oxford.
25. **Sanford, P. A.** 1982. Digestive system physiology. University Park Press, Baltimore.
26. **Sernka, T., and E. Jacobson.** 1983. Gastrointestinal physiology—the essentials, 2nd ed. The Williams Co., Baltimore.
27. **Simon, G. L., and S. L. Gorbach.** 1981. Intestinal flora in health and disease, p. 1361–1380. *In* L. R. Johnson (ed. in chief), Physiology of the gastrointestinal tract, vol. 2. Raven Press, New York.

HISTORICAL PERSPECTIVE

Gene-Protein Relationships in *Escherichia coli* Alkaline Phosphatase: Competition and Luck in Scientific Research†

FRANK G. ROTHMAN

Division of Biology and Medicine, Brown University, Providence, Rhode Island 02912

The relationships between genes and proteins were for the most part elucidated in the 25-year period 1941 through 1966. Although the view that a gene controlled an enzyme was clear in the earlier work of Garrod (22) on inborn errors of human metabolism and of Beadle and Ephrussi (3) on eye pigment synthesis in *Drosophila*, the one-gene-one-enzyme concept crystallized out of the work of Beadle and Tatum (1, 2, 4) on *Neurospora* species. Twenty-five years later, the genetic code was known, as were the biochemical mechanisms of suppressor mutations and genetic complementation.

In this article I review and reflect on the gene-protein studies on *Escherichia coli* alkaline phosphatase which were started in Cyrus Levinthal's laboratory at the Massachusetts Institute of Technology in 1957. I also illustrate the wonderfully open nature of the molecular biology community at the time, which deserves to be cultivated today in spite of the phenomenal growth of our discipline.

Becoming a Molecular Biologist

> ". . . .they [the characters in this book] come out larger than life, perhaps, and as different one from another as Caterpillar and Mad Hatter. Watson's childlike vision makes them seem like the creatures of a Wonderland, all at a strange contentious noisy tea party which made room for him because for people like him, at this particular kind of party, there is always room."
>
> P. B. Medawar, in a review of J. D. Watson's *The Double Helix (New York Review of Books*, 28 March 1963)

As an organic chemist I had been largely unaware of the new field of molecular biology until 1956. Then I happened to hear of the fascinating phenomenon of induced enzyme synthesis (13), when a microbiologist colleague at Walter Reed Army Medical Center asked me how to synthesize β-thiomethyl-D-galactoside. My interest blossomed after I took up a postdoctoral position in the Chemistry Department at the University of Wisconsin. I looked up Harlyn Halvorson, who extended the hospitality of his laboratory and arranged for me to visit Sol Spiegelman, with whom I spent a few days and evenings doing experiments on β-galactosidase synthesis in shocked protoplasts. I knew little about the field, but being excited about it was enough to be welcomed. Halvorson also took me to Evanston to a 1-day meeting of Midwest molecular biologists. The informality was stupendous. There was no program. At the beginning of the day, the convener stood at the board and asked, "Who wants to talk today?" Everyone who wanted to did. At the meeting I renewed my acquaintance with Alan Garen, whom I had known at Harvard, and heard Cyrus Levinthal lecture about phage replication.

Shortly thereafter, I was fortunate to meet Leo Szilard, the nuclear physicist who had switched to biology and was working on induced enzyme synthesis. He gave me much encouragement and advice, including that I should take the phage course at Cold Spring Harbor and then go to work with Max Delbruck or Cy Levinthal. Somewhat different advice was offered by Joshua Lederberg. Rather than doing postdoctoral work in molecular genetics, he thought it would make more sense for me to remain a chemist, but work on the important chemical problems confronting molecular biologists. For postdoctoral mentors he suggested Arthur Kornberg or Paul Zamecnik.

I had the good fortune to be admitted to the phage course at the last minute because of a

† This article is adapted from remarks presented at the symposium banquet.

cancellation. The course was taught by George Streisinger and Salvador Luria. The combination of intensity and informality was amazing. The laboratory was open 24 h a day, and we went in on demand of the experiments. Streisinger lectured in his bathing suit. Following the lead of Max Delbruck, who had founded the course, the idea of a physical scientist turning to genetics was accepted without discussion (9).

From Cold Spring Harbor I traveled to Boston for an interview with Zamecnik. He was cordial, but he had no space available for that year. From Zamecnik's office I called Levinthal and was lucky that he was in and free to see me before my return flight. He had just moved to the Massachusetts Institute of Technology, and I found him poring over blueprints for his new laboratory. The interview was successful.

I can think of no better way to capture the flavor of how I became a molecular biologist than with the quote from Medawar reproduced above. Szilard, Spiegelman, Lederberg, Halvorson, Levinthal, Garen, Streisinger, Luria, and the other molecular biologists I had met were indeed having a particular kind of party, and room had been made for me.

Gene-Protein Relations in Alkaline Phosphatase

What was the state of knowledge concerning gene-protein relations in 1958? Genetic analysis had been extended from fungi to *Escherichia coli* and coliphage. Benzer's studies on the bacteriophage T4 *r*II system approached the ultimate in fine-structure analysis (5). He showed that a gene (cistron) consisted of hundreds of mutational sites in a strictly linear order (6). On the protein chemistry side, techniques for sequencing had been worked out (31), and Ingram had shown that hemoglobin S differed from hemoglobin A in only one amino acid residue (23). It was a small step from the results of Benzer and Ingram to the hypothesis that the gene and its polypeptide product were colinear.

Very little was known about the genetic code. Methods for DNA sequencing, which together with protein sequencing would have made elucidation of the genetic code a rather routine problem, were nonexistent (and did not become available for over 15 years). Although some stimulating theoretical papers had been published (reviewed in reference 11), experimental approaches to the genetic code were hard to come by. One approach was to examine the amino acid changes caused by mutations. If this approach was pursued by isolating mutants in a gene in which fine-structure mapping was possible, the colinearity hypothesis could also be tested. By looking at amino acid changes caused by forward and reverse mutations at the same site, families of code words related by single base changes could become known. Finally, if mutagens could be discovered which affected only one of the four DNA bases, one might actually be able to discover the coding dictionary.

An obvious system to use for studies of this kind was T4 *r*II, since hundreds of mutants were already at hand and had been mapped. However, before coming to the Massachusetts Institute of Technology, Alan Garen had worked in Benzer's laboratory and had, in Benzer's words, "established the tradition of not finding the elusive *r*II protein" (7). About this time Torriani came to the United States with news of her discovery of *E. coli* alkaline phosphatase (37; this volume). Levinthal and Garen were impressed with the likelihood that this enzyme could readily be purified, since (i) its rate of synthesis was dramatically high under growth conditions of derepression (i.e., when phosphate was the limiting growth factor); (ii) it was quite stable to heating; and (iii) it could be assayed by a simple colorimetric reaction (the hydrolysis of nitrophenyl phosphate). They therefore decided to try to use *E. coli* alkaline phosphatase to study gene-protein relations, focusing initially on the colinearity hypothesis.

Garen and Levinthal (16, 25) readily worked out procedures for scoring mutant colonies lacking alkaline phosphatase (*pho* mutants) by spraying colonies with nitrophenyl phosphate and for selecting wild-type recombinants in crosses between two *pho* mutants by plating on glycerol phosphate as the only phosphate source. Using these techniques, a *pho* locus was mapped on the *E. coli* map, and a fine-structure map was readily constructed (16). All of the 13 *pho* mutants mapped within 0.3 recombination unit, i.e., in a region comprising 0.01% of the total map of the *E. coli* chromosome.

One anomalous genetic result did emerge. To cross two *pho* mutants, each mutant had to be transferred from the original Hfr strain to an F^- strain. Nine of the 22 mutations which showed no alkaline phosphatase activity in the original Hfr background showed activity when transferred to the F^- strain (25). I shall return to these mutants later.

The protein chemistry also got off to a good start. Under conditions of phosphate deprivation, about 6% of the total protein synthesized was found to be alkaline phosphatase, making purification quite simple (19). Heating to 80°C, a temperature at which the enzyme was stable, provided a convenient step (16). The simplicity of the purification of alkaline phosphatase caused chagrin among the biochemists on the floor below, who thought they would have a good laugh on this pair of physicists/biophysicists turned geneticists, who thought that purifying enzymes was a piece of cake.

With the wild-type enzyme pure, I developed a method for fingerprinting it. From the number of peptides in a tryptic digest, and the amino acid composition and molecular weight of the enzyme, we were able to deduce that the enzyme was composed of two identical polypeptide subunits, each containing about 400 amino acids (30). This information, together with the fact that the size of the *pho* gene corresponded to 0.3 recombination unit, and an estimate of 5,000 nucleotide pairs per recombination unit (24), allowed the first calculation of the coding ratio using experimental data (16, 25). Considering the assumptions in the calculation, the value of 4 that we obtained was surprisingly close to the correct value of 3.

We next looked for altered alkaline phosphatase in extracts of *pho* mutants by treating with antiserum against the purified wild-type phosphatase. The results gave us a rude surprise. Of the first 22 *pho* mutants isolated which totally lacked enzyme activity, only one made cross-reacting material (25). (By the spring of 1961, we had screened 100 non-"leaky" mutants, and only 6 of them produced an amount of cross-reacting material approximating that produced by the wild-type strain [26]). Therefore, we developed a supplementary strategy, namely, to screen revertants of the *pho* mutants for alkaline phosphatase with altered electrophoretic mobility and to fingerprint those altered enzymes. The revertants which made altered enzymes were also backcrossed to the wild-type strain to determine the genetic distance between forward and reverse mutations (16, 25). This strategy provided us with more altered phosphatases to fingerprint, but the extra steps required were time-consuming, and working with two-step mutants complicated the analysis.

During the years 1957 through 1961, approaches similar to ours were being carried out with three other gene-protein systems. Charles Yanofsky, a biochemical geneticist well versed in all aspects of the work, had chosen the tryptophan synthetase (*trpA*) cistron of *E. coli*. Sydney Brenner, a molecular biologist, chose the major subunit of the head protein of phage T4. George Streisinger, a phage geneticist, was collaborating with William Dreyer, a protein chemist, in examining the lysozyme of phage T4.

At the first session of the 1961 Cold Spring Harbor Symposium on Cellular Regulatory Mechanisms, Yanofsky (41) and Streisinger (34) lectured, and I (29) presented the results of the Massachusetts Institute of Technology group (20) as an invited discussant. Results directly relevant to colinearity were meager, as noted by Monod (27) in his concluding lecture:

> "A few years ago, following the beautiful work of Benzer which demonstrated the linear structure of the genetic material at the ultrafine level, and the work of Ingram on sickle cell hemoglobin, it seemed that the basic assumption of all coding hypotheses, namely, colinearity, would soon be proved. The only proof that has been obtained so far is that optimism is essential for the development of science; colinearity still remains to be formally demonstrated. However, the reports of Yanofsky, of Streisinger, and of Rothman at the conference, and what is known of the work of other laboratories, notably Brenner's, again encourages optimism; one feels confident that the final demonstration will soon be at hand."

Streisinger reported the isolation and mapping of mutants which affected the lysozyme of phage T4, but no fingerprint comparisons. Yanofsky and I both reported fingerprint differences between mutant and wild-type proteins. Yanofsky had found that two mutants which mapped at the same site, but were phenotypically distinguishable, had identical fingerprints which differed from the wild type by having an extra peptide. I reported that two closely linked mutations resulted in the absence of the same tryptic peptide; however, I had not identified any altered peptides due to these mutations. No one had defined the amino acid difference between a mutant and wild type.

Of great interest to me was Yanofsky's finding that, of the first 100 UV-induced mutants examined, 32 made cross-reacting material. At this point we had only a few mutants which formed cross-reacting material, and I was working mostly on the chemistry of revertants. Yanofsky's mutants had all been mapped, and it seemed likely that the large number of altered proteins in hand would allow him and his colleagues to establish colinearity along a substantial portion of the gene much sooner than we could.

This indeed turned out to be the case, and Yanofsky's now classic work on colinearity moved steadily to completion by 1964 (40). Also in 1964, Brenner's group showed the colinearity of the gene for the T4 head protein and the polypeptide coded for it in a novel way, namely, by correlating the length of the chain-terminated fragment produced by nonsense mutants with the genetic distance of the mutation from one end of the map (32). This method could not even have been proposed in 1958, since nonsense mutants had not been identified.

Two papers which appeared in 1961 marked the turning point in the effort to elucidate the genetic code. Elegant genetic experiments by Crick et al. (12), using insertion and deletion mutants in the phage T4 *r*II system, established the triplet nature of the code and provided information concerning frameshift mutations, degeneracy, nonsense, and punctuation. On the chemical side, Nirenberg and Matthaei (28) reported that addition of polyuridylic acid to a cell-free protein-synthesizing system stimulated incorporation of phenylalanine into protein, thereby establishing the codon UUU for phenylalanine and opening the way for identification of other codons. By the end of 5 more years, the genetic code was known essentially in its entirety (15).

Work on gene-protein relations in alkaline phosphatase after 1961 turned away from testing the colinearity hypothesis and focused on supressor mutations and intragenic complementation. As noted above, the early genetic studies had turned up an anomalous result, namely, a class of *pho* mutations in the original Hfr strain which showed a Pho^+ phenotype when transferred to the F^- strain. These mutants turned out to be among the first nonsense mutants identified (21), and the Pho phenotypes were explained by the presence of nonsense suppressor genes in the F^- strain, but not in the Hfr strain which Garen and Levinthal happened to choose. Garen followed up this discovery (after moving to the University of Pennsylvania in 1961 and later to Yale). The investigations of his group in the 1960s made crucial contributions to the identification of the UAG (amber) (38) and UAA (ochre) (39) nonsense codons, and to the mechanism of action of nonsense suppressors (17).

Intragenic (or interallelic) complementation refers to situations in which two mutated forms of the same gene complement each other. A complementation test between two mutants defective in the same enzyme involves crossing the two mutants and observing whether there is enzymatic activity in the heterozygote. Brenner (8) and Fincham (14) proposed a mechanism for this phenomenon: genes in which intragenic complementation occurs code for proteins which are dimers (or multimers) of a polypeptide. In a heterozygote, some of the protein molecules coded for by the gene in question will contain one (or more) polypeptide chain coded for by each parental allele. If these heterodimer (or multimer) molecules are enzymatically active (unlike the homodimers or multimers), complementation is observed. The earlier discovery that *E. coli* alkaline phosphatase was a dimer of two identical subunits (30) made phosphatase an attractive system in which to test this mechanism. In vitro studies by Schlesinger and Levinthal (33), and in vivo ones by Garen and Garen (18), proved that the enzymatically active molecules formed in a complementation test were the heterodimers containing one chain coded for by each parental phosphatase gene, as predicted by Brenner and Fincham.

Reflections on Luck and Competition

Insofar as the experiments of the Yanofsky, Levinthal and Garen, Brenner, and Streisinger groups were directed toward testing the colinearity hypothesis, the winners of the competition were clearly Yanofsky and Brenner. In comparing the progress of Yanofsky and our own group, using very similar approaches, I have pointed out the striking difference in the fraction of *trp* and *pho* mutants which made cross-reacting material. The paucity of such *pho* mutants certainly made our progress slower. One cause of this difference turned out to be entirely fortuitous and was not known until a number of years after the work was started, when nonsense mutants and their suppressors were discovered. As noted above, our parental strain K10 carried no nonsense suppressors, and we therefore picked up nonsense mutants (which do not form cross-reacting material) among our *pho* mutants. Yanofsky's wild-type strain Ymel carried two nonsense (amber) suppressor genes, *supE* and *supF*. He therefore did not pick up any of this major class of nonsense mutants, and his mutant collection was enriched for missense mutants (which usually form cross-reacting material).

If the work on the four systems is viewed in terms of all aspects of gene-protein relations, then all four groups of investigators must be considered winners. Each made important contributions, some of which could not have been made readily with the other systems. Yanofsky's group not only showed colinearity, but also discovered and explained supression of missense mutations (10, 41). The latter work provided the first example of a suppressor mutation which acted by changing the specificity of the protein translation machinery. Brenner and his co-workers (32) proved chemically that nonsense mutations cause chain termination and, as previously noted, used the fragments made by nonsense mutants to demonstrate colinearity. Their experiments could only have been done readily in a system like the phage head protein where the fragments did not have to be purified. In addition, his group (36) made important contributions to identifying nonsense codons and understanding nonsense suppression. Streisinger and co-workers (35) used the T4 lysozyme to show the effects of frameshift mutations on the amino acid sequence and provided the first assignment of the particular codon used in vivo for amino acids coded for by more than one codon.

T4 lysozyme was the only one of the four systems in which proflavin mutagenesis led to viable frameshift mutants. The contributions of the alkaline phosphatase group to the understanding of intragenic complementation and nonsense suppression have already been described. The fact that alkaline phosphatase is a dimer of identical subunits was essential to testing the mechanism proposed for intragenic complementation.

Two lessons may be drawn from this case history. (i) Luck does enter into scientific research (which we all knew anyway). As noted above, the unknown genotype with respect to nonsense suppressors of the original strains significantly affected the progress and direction of research on gene-protein relationships in tryptophan synthetase and alkaline phosphatase. (ii) Competition does not necessarily lead to any losers. In the cases discussed above, an important scientific question was investigated by several research groups, using different biological systems. The diversity in the properties of the systems resulted in each research group making unique contributions.

LITERATURE CITED

1. **Beadle, G. W.** 1945. Biochemical genetics. Chem. Rev. **37:** 15–96.
2. **Beadle, G. W.** 1966. Biochemical genetics: some recollections, p. 23–32. *In* J. Cairns, G. S. Stent, and J. D. Watson (ed.), Phage and the origins of molecular biology. Cold Spring Harbor Laboratory, Cold Spring Harbor, N.Y.
3. **Beadle, G. W., and B. Ephrussi.** 1936. The differentiation of eye pigments in *Drosophila* as studied by transplantation. Genetics **21:**225–247.
4. **Beadle, G. W., and E. L. Tatum.** 1941. Genetic control of biochemical reactions in *Neurospora*. Proc. Natl. Acad. Sci. USA **27:**499–506.
5. **Benzer, S.** 1957. The elementary units of heredity, p. 70–93. *In* W. D. McElroy and B. Glass (ed.), The chemical basis of heredity. Johns Hopkins Press, Baltimore.
6. **Benzer, S.** 1959. On the topology of the genetic fine structure. Proc. Natl. Proc. Sci. USA **45:**1607–1620.
7. **Benzer, S.** 1966. Adventures in the rII region, p. 157–165. *In* J. Cairns, G. S. Stent, and J. D. Watson (ed.), Phage and the origins of molecular biology. Cold Spring Harbor Laboratory, Cold Spring Harbor, N.Y.
8. **Brenner, S.** 1959. The mechanism of gene action, p. 304–316. *In* G. E. W. Wolstenholme and C. M. O'Connor (ed.), Ciba Foundation symposium on biochemistry of human genetics. Churchill, London.
9. **Cairns, J., G. S. Stent, and J. D. Watson (ed.).** 1966. Phage and the origins of molecular biology. Cold Spring Harbor Laboratory, Cold Spring Harbor, N.Y.
10. **Carbon, J., P. Berg, and C. Yanofsky.** 1966. Missense suppression due to a genetically altered tRNA. Cold Spring Harbor Symp. Quant. Biol. **31:**487–496.
11. **Crick, F. H. C.** 1966. The genetic code—yesterday, today, and tomorrow. Cold Spring Harbor Symp. Quant. Biol. **31:**3–9.
12. **Crick, F. H. C., L. Barnett, S. Brenner, and R. J. Watts-Tobin.** 1961. General nature of the genetic code for proteins. Nature (London) **192:**1227–1232.
13. **Davies, R., and E. F. Gale (ed.).** 1953. Adaptation in microorganisms. Third Symposium of the Society of General Microbiology. Cambridge University Press, Cambridge, England.
14. **Fincham, J. R. S.** 1960. Genetically controlled differences in enzyme activity. Adv. Enzymol. **22:**1–43.
15. **Frisch, L. (ed.)** 1966. The genetic code. Cold Spring Harbor Symp. Quant. Biol. **31:**1–762.
16. **Garen, A.** 1960. Genetic control of the specificity of the bacterial enzyme, alkaline phosphatase. Symp. Soc. Gen. Microbiol. **10:**239–247.
17. **Garen, A.** 1968. Sense and nonsense in the genetic code. Science **160:**149–159.
18. **Garen, A., and S. Garen.** 1963. Complementation *in vivo* between structural mutants of alkaline phosphatase from *E. coli*. J. Mol. Biol. **7:**13–22.
19. **Garen, A., and C. Levinthal.** 1960. A fine-structure genetic and chemical study of the enzyme alkaline phosphatase of *E. coli*. I. Purification and characterization of alkaline phosphatase. Biochim. Biophys. Acta **38:**470–483.
20. **Garen, A., C. Levinthal, and F. Rothman.** 1961. Alterations in alkaline phosphatase induced by mutations. J. Chim. Phys. **58:**1068–1071.
21. **Garen, A., and O. Siddiqi.** 1962. Suppression of mutations in the alkaline phosphatase structural cistron of *E. coli*. Proc. Natl. Acad. Sci. USA **48:**1121–1127.
22. **Garrod, A. E.** 1909, 1923. Inborn errors of metabolism. Oxford University Press, London.
23. **Ingram, V. M.** 1957. Gene mutations in human hemoglobin: the chemical difference between normal and sickle cell hemoglobin. Nature (London) **180:**326–328.
24. **Jacob, F., and E. L. Wollman.** 1958. Genetic and physical determinations of chromosomal segments in *Escherichia coli*. Symp. Soc. Exp. Biol. **12:**75–92.
25. **Levinthal, C.** 1959. Genetic and chemical studies with alkaline phosphatase of *E. coli*. Brookhaven Symp. Biol. **12:**76–83.
26. **Levinthal, C., A. Garen, and F. Rothman.** 1963. Relationship of gene structure to protein structure: studies on alkaline phosphatase of *E. coli*, p. 196–203. *In* Proceedings of the Fifth International Congress of Biochemistry I. Pergamon Press, Oxford.
27. **Monod, J., and F. Jacob.** 1961. General conclusions: teleonomic mechanisms in cellular metabolism, growth, and differentiation. Cold Spring Harbor Symp. Quant. Biol. **26:**389–401.
28. **Nirenberg, M., and J. Matthaei.** 1961. The dependence of cell-free protein synthesis in *E. coli* upon naturally occurring or synthetic polynucleotides. Proc. Natl. Acad. Sci. USA **47:**1588–1602.
29. **Rothman, F.** 1961. Discussion. Cold Spring Harbor Symp. Quant. Biol. **26:**23–24.
30. **Rothman, F., and R. Byrne.** 1963. Fingerprint analysis of alkaline phosphatase of *Escherichia coli* K12. J. Mol. Biol. **6:**330–340.
31. **Sanger, F.** 1956. The structure of insulin, p. 434–459. *In* D. E. Green (ed.), Currents in biochemical research. Interscience Publishers, New York.
32. **Sarabhai, A. S., A. O. W. Stretton, S. Brenner, and A. Bolle.** 1964. Colinearity of the gene with the polypetide chain. Nature (London) **201:**13–17.
33. **Schlesinger, M. J., and C. Levinthal.** 1963. Hybrid protein formation of *E. coli* alkaline phosphatase leading to *in vitro* complementation. J. Mol. Biol. **7:**1–12.
34. **Streisinger, G., F. Mukai, W. J. Dreyer, B. Miller, and S. Horiuchi.** 1961. Mutations affecting the lysozyme of phage T4. Cold Spring Harbor Symp. Quant. Biol. **26:**25–30.
35. **Streisinger, G., Y. Okada, J. Emrich, J. Newton, A. Tsugita, E. Terzaghi, and M. Inouye.** 1966. Frameshift mutations and the genetic code. Cold Spring Harbor Symp. Quant. Biol. **31:**77–84.
36. **Stretton, A. O. W., S. Kaplan, and S. Brenner.** 1966. Nonsense codons. Cold Spring Harbor Symp. Quant. Biol. **31:**173–179.
37. **Torriani, A.** 1960. Influence of inorganic phosphate in the formation of phosphatases by *Escherichia coli*. Biochim. Biophys. Acta **38:**460–469.
38. **Weigert, M. G., and A. Garen.** 1965. Base composition of

nonsense codons in *E. coli*. Nature (London) **206:**992–994.
39. **Weigert, M. G., E. Lanka, and A. Garen.** 1967. Base composition of the nonsense codons in *Escherichia coli*. J. Mol. Biol. **23:**391–400.
40. **Yanofsky, C., B. C. Carlton, J. R. Guest, D. R. Helinski, and U. Henning.** 1964. On the colinearity of gene structure and protein structure. Proc. Natl. Acad. Sci. USA **51:**266–272.
41. **Yanofsky, C., D. R. Helinski, and B. Maling.** 1961. The effects of mutation on the composition and properties of the A protein of *Escherichia coli* tryptophan synthetase. Cold Spring Harbor Symp. Quant. Biol. **26:**11–23.

Author Index

Subject Index